Magnesium Technology 2016

TMS2016

145th Annual Meeting & Exhibition

FEBRUARY 14-18 DOWNTOWN NASHVILLE, TENNESSEE MUSIC CITY CENTER

New proceedings volumes from the TMS2016 Annual Meeting:

- 7th International Symposium on High-Temperature Metallurgical Processing
- CFD Modeling and Simulation in Materials Processing 2016
- Characterization of Minerals, Metals, and Materials 2016
- Energy Technology 2016: Carbon Dioxide Management and Other Technologies
- EPD Congress 2016
- Light Metals 2016
- Magnesium Technology 2016
- Rare Metal Technology 2016
- REWAS 2016
- Shape Casting: 6th International Symposium
- TMS 2016 Supplemental Proceedings

Magnesium Technology 2016

Proceedings of a symposium sponsored by

Magnesium Committee of the Light Metals Division of

The Minerals, Metals & Materials Society (TMS)

held during

TMS2016

145th Annual Meeting & Exhibition

FEBRUARY 14-18 DOWNTOWN NASHVILLE, TENNESSEE MUSIC CITY CENTER

Edited by:

Alok Singh, Kiran Solanki, Michele V. Manuel, and Neale R. Neelameggham

Editors
Alok Singh
Kiran Solanki
Michele V. Manuel
Neale R. Neelameggham

ISBN 978-3-319-48620-8
ISBN 978-3-319-48114-2 (eBook)
DOI 10.1007/978-3-319-48114-2

Chemistry and Materials Science: Professional

Published by Springer International Publishers, Switzerland, 2016
Reprint of the original edition published by John Wiley & Sons, Inc., 2016, 978-1-119-22580-5

Printed on acid-free paper

This Springer imprint is published by Springer Nature
The registered company is Springer International Publishing AG
The registered company address is: Gewerbestrasse 11, 6330 Cham, Switzerland

TABLE OF CONTENTS
Magnesium Technology 2016

Magnesium Technology 2016

Keynote Session

Solidification and Casting

Keynote Session Part II and Primary Production and Recycling

Alloy Development, Diffusion and Joining

Magnesium-Rare Earth Alloys

LPSO Alloys and Composites

Twinning and Plasticity

Texture and Formability

Corrosion

Poster Session

Magnesium-based Biodegradable Implants

Materials and Processing / Surface Modification and Corrosion

Corrosion / Market and Clinic

Strip Casting of Light Metals

Strip Casting Process

Strip Casting: Properties

Poster Session

PREFACE

It is our great pleasure to bring you this volume of *Magnesium Technology 2016*, which is the proceedings of the Magnesium Technology Symposium held at the 145th TMS Annual Meeting & Exhibition in Nashville, Tennessee, February 14–18, 2016. With contributions from 16 countries, representing the latest trends in the field of magnesium research, this volume can be regarded as a central repository of the finest research carried out in magnesium technology from around the world. With the tradition of presenting the most recent and highest quality work, all presenters have submitted their work to this edited proceedings volume or other peer reviewed TMS journals. All papers included in this volume were peer reviewed by the best possible experts in the concerned fields of magnesium research. The reviewers' contribution continues to be important to the success of this symposium. These contributions were presented in 9 sessions including a plenary session. Extended abstracts of the six keynote lectures given in the plenary session are also included here.

Going by the current trends, the contributions are classified into primary production and recycling, solidification and casting, alloy development, joining (welding) and diffusion, magnesium–rare-earth alloys, long period stacking ordered (LPSO) alloys and composites, twinning and plasticity, texture and formability, and corrosion. Challenges to improve plasticity and formability continue to excite researchers; most contributions deal with twinning, dislocation slip and texture. A fairly large number of contributions deal with the improvement of properties by the addition of rare-earth elements, investigating different aspects. Alloys forming the LPSO phase are also a part of this effort. Efforts are also seen in the areas of joining and corrosion, which are crucial for the application of magnesium alloys. Primary production, recycling and solidification remain as important as ever.

Over the years, a vast amount of scientific insights have been gained into magnesium and its alloys from various perspectives, as witnessed in the volumes of *Magnesium Technology*. How to put all this expertise into practical applications and commercialization is the main focus of the plenary session this year. The talk by Martyn Alderman (Magnesium Elektron), who has years of experience in industry and R&D, examines "A Perspective: Potential Growth in the Global Magnesium Industry–Where Is Our Research Leading Us?" Karl U. Kainer (MagIC – Helmholtz Zentrum Geesthacht), a well-known leader in magnesium research, looks at "Challenges for Implementation of Magnesium into More Applications." An extensive and critical application of magnesium alloys has come up in high speed trains in China. Given the large network of these trains in China, to be potentially followed in other countries, this is a large scale application. This will be discussed by Eric Nyberg (Pacific Northwest National Laboratory). South Korea has been in the forefront of innovations for applications of magnesium alloys. Nack J. Kim (Pohang University of Science and Technology) will tell us about this in his talk entitled "Korea's R&D Activities Towards the Applications of Wrought Mg Alloys." Many of the intermetallic phases used in the strengthening of magnesium alloys have very fascinating structures, such as the quasicrystalline phase and the LPSO phase found in Mg-Zn-RE (rare-earth) alloys. LPSO phase magnesium alloys have attracted much attention in the past several years. Eiji Abe (University of Tokyo) will talk about "Fascinating LPSO-structured Mg Alloys." Issues of plasticity and other aspects can be skirted by making the matrix amorphous, instead of crystalline, as described by Kevin Laws (University of New South Wales) in his talk "The Developments in the High Magnesium Content Bulk Metallic Glasses and Future Possibilities."

In parallel, two specialized symposia are being held, one on Magnesium-based Biodegradable Implants, being co-sponsored by (TMS) Functional Materials Division, Light Metals Division, Structural Materials Division, Biomaterials Committee and our own Magnesium Committee, and another on Strip Casting of Light Metals (second in the series, following last year's success), co-sponsored by Aluminum Committee and Magnesium Committee. Organized by Wim Sillekens (European Space Agency), Martyn Alderman (Magnesium Elektron), Patrick K. Bowen (Michigan Technological University), Jaroslaw Drelich (Michigan Technological University), and Petra Maier (University of Applied Sciences, Stralsund), 16 presentations will be made in the biodegradable implants symposium. The written accounts of eleven of them appear in this volume, while the other five are scheduled for publication in the April 2016 Issue of *JOM*. There are also three manuscripts from the Strip Casting of Light Metals Symposium, which is organized by Kai F. Karhausen (Hydro Aluminium Rolled Products GmbH), Dietmar Letzig (MagIC – Helmholtz Zentrum Geesthacht), Jan Bohlen (MagIC – Helmholtz Zentrum Geesthacht), and Murat Dundar (Assan Aluminium).

We would like to gratefully acknowledge and thank all authors for contributing to the symposium and this volume with their high-quality work, reviewers who put in their best efforts to ensure the quality of papers, plenary session speakers for giving their insights into current research and pointing to the future directions, session chairs, judges for posters, and everyone else who have volunteered to make this symposium a great success and in bringing out this high quality proceedings.

2016 Magnesium Technology Conference Organizers

Alok Singh, Chair
Kiran N. Solanki, Vice Chair
Michele V. Manuel, Past Chair
Neale R. Neelameghham, Advisor

ABOUT THE LEAD EDITOR

ALOK SINGH
MAGNESIUM TECHNOLOGY 2016 LEAD EDITOR

Alok Singh is a Chief Researcher in the Structural Materials Unit of National Institute for Materials Science in Tsukuba, Japan. He studied metallurgical engineering at undergraduate, masters and doctoral levels. He worked on kinetics of phase transformation in steels for his master's thesis at the Indian Institute of Technology at Kanpur. His Ph.D. work, at the Indian Institute of Science, was on the study of quasicrystalline and related intermetallic phases in aluminum alloys by transmission electron microscopy (TEM). He tackled the complex structures and reciprocal space of quasicrystals and its indexing problems.

Subsequently, he worked several years studying advanced materials by TEM at the Indira Gandhi Center for Atomic Research, and visited National Research Institute for Metals in Japan. He studied the effect of interfaces on melting and solidification of embedded nanoparticles. He has extensively studied the structure of interfaces between simple crystals and quasicrystals by TEM.

In 2002 he moved to his present working place National Institute for Materials Science and started working on magnesium alloys with special emphasis on Mg-Zn-RE alloys containing stable quasicrystal phase. His work has demonstrated very high strength with ductility in these alloys. These high mechanical properties have been analyzed with respect to microstructural characteristics. He has employed TEM to study dislocations, grain boundaries and twins, and interactions among these, to understand deformation behavior of magnesium alloys. Recently, he is applying advanced TEM techniques of scanning transmission electron microscopy (STEM) to study severely plastically deformed (SPD, a current trend in materials to achieve nano-scale microstructures) magnesium alloys, which is a challenge for the conventional TEM because of the strong contrast from the high amount of mechanical strain. He has over 90 refereed publications, over 30 contributions to proceedings, and several patents on magnesium alloys.

As a TMS member, he is a regular attendee of TMS annual meetings and has been involved with the Magnesium Committee for many years. He has been *JOM* representative and Vice Chair of the Magnesium Committee. He received the TMS Magnesium Fundamental Research Award for year 2009 along with his coworkers.

MAGNESIUM TECHNOLOGY 2016 EDITORS

Kiran Solanki is an Assistant Professor of Mechanical Engineering in the SEMTE at Arizona State University. Prior to coming to ASU, he was an Associate Director for the Center for Advanced Vehicular Systems at Mississippi State University. Dr. Solanki received his Ph.D. from Mississippi State University in December 2008. Dr. Solanki's research interest is at the interface of solid mechanics and material science, with a focus on characterizing and developing microstructure-based structure–property relationships across multiple length and time scales. To date, he has coauthored more than 50 journal articles, four book chapters, and more than 35 conference proceedings with faculty and students at ASU and MSU. In addition, his paper published in *Engineering Fracture Mechanics* was recognized as one of the most highly cited papers from year 2002 to 2005. For his efforts to promote the education of engineering students in the area of fatigue technology, he was awarded the SAE Henry O. Fuch Award by the SAE Fatigue Design & Evaluation Committee. In 2011, Dr. Solanki received the TMS Light Metals Magnesium Best Fundamental Research Paper Award for his work on predicting deformation and failure behavior in magnesium alloys using a multiscale modeling approach. Recently, he received the 2013 TMS Light Metals Division Young Leader Professional Development Award, the 2013 Air Force Office of Scientific Research Young Investigator Research Award, and the 2013 ASME "Orr Award" for Early Career Excellence in Fatigue, Fracture, and Creep.

Michele V. Manuel is an Associate Professor in the Department of Materials Science and Engineering at the University of Florida. She received her Ph.D in Materials Science and Engineering at Northwestern University in 2007 and a B.S. in Materials Science and Engineering at the University of Florida in 2002. Prior to arriving at the University of Florida, she performed her postdoctoral work at the General Motors Technical Center in Warren, Michigan. She is the recipient of the NSF CAREER, NASA Early Career Faculty, ASM Bradley Stoughton Award for Young Teachers, AVS Recognition for Excellence in Leadership, TMS Early Career Faculty, TMS Young Leader Professional Development, and TMS/JIM International Scholar awards. Her research lies in the basic understanding of the relationship between processing, structure, properties, and performance. She uses a systems-based materials design approach that couples experimental research with theory and mechanistic modeling for the accelerated development of materials. Her current research is focused on the use of systems-level design methods to advance the development of new materials through microstructure optimization. In the TMS Magnesium Committee, she has previously served in the roles of Chair, Vice-Chair, Secretary, *JOM* Advisor, *JOM* Advisor in Training, and currently serves in the position as the Past Chair. Additionally, she serves as the Vice-Chair on the Content Development and Dissemination Committee, as well as a member of the Education, Young Leaders, Integrated Computational Materials Engineering (ICME), Biomaterials, and the Women in Science Committees.

Neale R. Neelameggham is 'The Guru' at IND LLC, involved in technology marketing and international consulting in the field of light metals and associated chemicals (boron, magnesium, titanium, lithium, and alkali metals), rare earth elements, battery and energy technologies, etc. He was a visiting expert at Beihang University of Aeronautics and Astronautics, Beijing, China. He was a plenary speaker at the Light Metal Symposium in South Africa – on low carbon dioxide emission processes for magnesium.

He has over 38 years of expertise in magnesium production and was involved in process development of its startup company NL Magnesium through to the present US Magnesium LLC, UT from where he retired in 2011. He is developing thiometallurgical processes – a new concept of using sulfur as the reductant and/or fuel. He has published a heat transfer model for global anthropogenic warming based on thermal emissions independent of energy conversion source.

Dr. Neelameggham holds 16 patents and patent applications, and has published several technical papers. He has served in the Magnesium Committee of LMD since its inception in 2000, chaired it in 2005, and in 2007 he was made a permanent co-organizer for the Magnesium Symposium. He has been a member of the Reactive Metals Committee, Recycling Committee, Titanium Committee, and Programming Committee Representative of LMD and the LMD Council.

Dr. Neelameggham was the inaugural chair, when in 2008, LMD and EPD created the Energy Committee, and has been a co-organizer of the Energy Technology symposium through the present. He received the LMD Distinguished Service Award in 2010. While he was the chair of Hydro- and Electrometallurgy Committee he initated Rare Metal Technology symposium in 2014. He is co-organizer for the 2016 symposia on Magnesium Technology, Energy Technology, Rare Metal Technology and light metals section of REWAS 2016.

MAGNESIUM-BASED BIODEGRADABLE IMPLANTS 2016 EDITORS

Wim H. Sillekens is a project manager in the Strategic & Emerging Technologies Team at the research and technology center of the European Space Agency (ESA–ESTEC), where he is currently acting as the coordinator of the European Community research project ExoMet. He obtained his Ph.D. from Eindhoven University of Technology, Netherlands, on a subject relating to metal-forming technology. Since that time he has been engaged in aluminum and magnesium research, amongst others on (hydro-mechanical) forming, recycling/refining, (hydrostatic) extrusion, forging, magnesium-based biodegradable implants, and as of late on light-metal-matrix nanocomposites. His professional career includes positions as a post-doc researcher at his alma mater and as a research scientist/project leader at the Netherlands Organization for Applied Scientific Research (TNO). International working experience covers a placement as a research fellow at MEL (now AIST) in Tsukuba, Japan, and – more recently – shorter stays as a visiting scientist at GKSS (now HZG) in Geesthacht, Germany, and at Pacific Northwest National Laboratory (PNNL) in Richland WA, USA. He has (co)-authored a dissertation, book chapters, journal papers, patents, conference papers, (keynote/invited/contributed) oral presentations, and so on (about 150 entries to date). Other professional activities include an involvement in association activities (amongst others, as the lead organizer of TMS Magnesium Technology 2011), international conference committees, and as a peer reviewer of paper manuscripts for scientific journals and conference proceedings as well as of research proposals. Research interests are in physical and mechanical metallurgy in general and in light-metals technology in particular.

Martyn Alderman is Divisional Director of Technology for the Magnesium Elektron Group Worldwide. He obtained a Masters Degree in Material Science from the University of Cambridge (United Kingdom) in 1979 and for the next twenty years he worked in aluminum rolling mills producing and developing Al-Cu-Zr, Al-Mn, Al Mg-Mn, Al-Zn-Mg, and Al-Li-Cu alloys for superplastic forming. He also has significant operational experience in direct chill casting and in the extrusion of high strength aluminum alloys.

In 2003 he joined Magnesium Elektron during the acquisition of their large Madison, Illinois rolling operation in the USA. For the next few years he travelled between the USA and Europe encouraging the wider use of magnesium sheet in transport applications, in particular by use of superplastic forming, and later in managing the post acquisition integration of the Revere North American Graphic Arts business. He chaired the 2010 and 2015 International Magnesium Association Conferences in Hong Kong and Vancouver, WA, USA, and in 2011–2012 was involved in editing and producing a 270 page *Handbook on Designing with Magnesium Alloys*.

More recently he has been involved in a U.K. Technology Strategy Board Project aimed at the Design of a Magnesium Intensive Vehicle in conjunction with the Morgan Motor Company; as an advisor to the International Magnesium Association Life Cycle Study on automotive and aerospace applications; and in a National Aerospace Technology Exploitation Programme focused on improving "buy to fly" ratios in magnesium aircraft seat components.

As a member of TMS, SAE and AMS, he is a regular attendee at worldwide magnesium conferences keeping abreast of global technology development with the aim of maintaining Magnesium Elektron's position as a world leader in magnesium alloys and their end use including biomedical applications; during 2016 he will again be chairing the International Magnesium Association Conference in Rome.

Patrick K. Bowen studied Materials Science and Engineering at Michigan Technological University, receiving his B.S. in 2011, and his Ph.D. in 2015. His research has spanned a wide array of topics—from archaeometric studies on ceramic artifacts, to magnesium biocorrosion kinetics and mechanism, to the application of zinc in bioabsorbable materials applications. He has published nearly twenty articles in peer-reviewed journals since 2011 with more than 230 combined citations, and he has been an active member of TMS since 2010 and a member of the Biomaterials Committee since 2013. He will begin employment as a Senior Research Metallurgist with Deringer-Ney, Inc. (Bloomfield, Connecticut) beginning in January 2016.

Jaroslaw W. Drelich received his B.S. degree in chemistry and M.S. degree in chemical technology from the Technical University of Gdansk (TUG), Poland, in 1983, and earned his Ph.D. degree in metallurgical engineering from the University of Utah in 1993. He came to Michigan Technological University (Michigan Tech) in 1997 and currently is a professor of materials science and engineering. His main research interests are in applied surface chemistry and interfacial engineering for ore dressing and materials processing, materials characterization, formulation, modification and testing of biomaterials and antimicrobial materials. In the last few years, Dr. Drelich's research focused on bioabsorbable implant materials including magnesium, and he is co-inventor of a new class of biodegradable zinc-based materials for vascular stent applications. Aside from teaching several courses on characterization and processing of materials at Michigan Tech, Dr. Drelich has edited six books, published more than 170 technical papers (cited nearly 4,000 times according to Google Scholar), holds nine patents and has more than 50 conference presentations, including several keynote addresses, to his credit.

Dr. Drelich is the Editor-in-Chief for the *Surface Innovations* journal. He is the active member of The Minerals, Metals & Materials Society (TMS), Society for Mining, Metallurgy and Exploration (SME), and American Chemical Society (ACS), and has served on a number of different committees;

currently, he serves as Past-Chair for the TMS Energy Committee. Dr. Drelich is also an advocate of interdisciplinary capstone senior design projects, and promotes interdisciplinary projects between engineering and business to encourage the students to think about both technical and business viability and interconnections. In recognition of his entrepreneurial activities, Dr. Drelich has been awarded with the 2012 Food Safety Innovation Award from Great Lakes Entrepreneur's Quest in Michigan.

Petra Maier received her doctoral degree from Loughborough University, U.K. in 2002 in Materials Science, especially in the field of grain boundary segregation in steel. After completing her Ph.D. she worked at the University of Applied Sciences Wildau in Germany as a postdoctoral fellow under supervision of Prof. Asta Richter with a focus on microhardness measurements by nanoindentation. From 2004 to 2006 she worked as a research associate at the Helmholtz Zentrum Geesthacht in Germany in the Institute of Materials Research led by Prof. Karl Ulrich Kainer, responsible for the operational division "Magnesium Innovation Centre". In the group of Dr. Norbert Hort, she worked in the area of Magnesium recycling and high temperature alloy development. From 2006 to 2008 Petra Maier was a research associate at the Technical University Berlin in Germany in the Institute of Material Sciences and Technologies, Department of Materials Engineering of Prof. Claudia Fleck. There, her research specialities included corrosion fatigue on magnesium. Since 2008 Petra Maier has been a Professor of Materials and Production Engineering in the School of Mechanical Engineering at the University of Applied Sciences Stralsund in Germany. She enjoys working in the field of magnesium-based biodegradable implants. Her research interests are focused on corrosion under stress and crack propagation influenced by the microstructure. The current research relates to investigating innovative magnesium–rare-earth alloys in the form of wires, with a possible application as pins, rods, and clips for the temporary fixation of bone fracture or for forming stents.

STRIP CASTING OF LIGHT METALS 2016 EDITORS

Kai F. Karhausen is head of the research department Rolling Technology at the R&D Centre of Hydro Aluminium Rolled Products GmbH in Bonn, Germany. He graduated in 1994 with a Ph.D. in Materials Science from the University of Technology RWTH Aachen. He joined the Hydro Aluminium R&D Centre in Karmøy, Norway and conducted research in the field of aluminum extrusion before moving to the R&D Centre in Bonn to work on the rolling technology of aluminum. He is the author of a large number of papers with a focus on metal forming technology, numerical simulation, and integration of material models into process simulation tools. He served as chair of the Aluminum Processing Committee of TMS for five years.

Dietmar Letzig is head of the Wrought Magnesium Alloys Department at the Magnesium Innovation Centre of the Helmholtz Zentrum Geesthacht in Germany. He is the author of many papers in peer-reviewed scientific journals and proceedings of international conferences. His area of expertise is microstructure and mechanical properties of wrought magnesium alloys processed by severe production processes. He is leading fundamental as well as application-oriented research projects in metal forming processes and an expert in the processing of magnesium sheet including twin roll casting and the optimisation of the magnesium sheet performance.

Jan Bohlen is a research scientist with the Magnesium Innovation Centre (MagIC) at the Helmholtz-Zentrum Geesthacht in Germany. He graduated in Material Physics in 1996 and obtained his Ph.D from the University of Göttingen (Germany) in 2000. His work in the wrought magnesium alloys group at MagIC is focused on the development of new wrought magnesium alloys and understanding the effect of processing on microstructural development.

Murat Dündar holds the position of Director of Technology in Assan Aluminum. He joined Assan Aluminum in May 1999 as a research specialist. He has primarily focused on performance improvement of aluminum foil and sheet products produced out of Twin Roll Casting Technology (TRC), managing research and development projects on developing new alloys compatible with TRC technology, characterization of as-cast structures and related casting defects, interface between liquid metal and caster shell surface, solidification in TRC and related microstructures, tailoring microstructural features starting from casting and in further downstream operations, improvement in productivity of casting process and finally casting of high Mg-bearing and 6000 series alloys with TRC.

He holds a B.Sc. degree in Metallurgical Engineering from Middle East Technical University, Turkey, an M.Sc. degree in Materials Science from State University of New York at Buffalo, USA and a Ph.D. degree in Materials Engineering, from New Mexico Institute of Mining and Technology, USA.

MAGNESIUM TECHNOLOGY 2016
Session Chairs

Keynote Session
Alok Singh,
National Institute for Materials Science

Kiran Solanki,
Arizona State University

Solidification and Casting
Norbert Hort,
MagIC – Helmholtz Zentrum Geesthacht

Tracy Berman,
University of Michigan

Keynote Session Part II and Primary Production and Recycling
Neale R. Neelameggham,
IND LLC

Dmytro Orlov,
Lund University

Kiran Solanki,
Arizona State University

Alloy Development, Diffusion, and Joining
Sean Agnew,
University of Virginia

Miroslav Sahul,
Slovak University of Technology Bratislava

Magnesium-Rare Earth Alloys
Mark Easton,
RMIT University

Francesco D'Elia,
MagIC – Helmholtz Zentrum Geesthacht

LPSO Alloys and Composites
Manoj Gupta,
National University of Singapore

Hyunkyu Lim,
Korea Institute of Technology KITECH

Twinning and Plasticity
Tyrone Jones,
US Army Research Laboratory

Peifeng Li,
Nanyang Technological University

Texture and Formability
Jan Bohlen,
MagIC – Helmholtz Zentrum Geesthacht

Nitin Chandola,
University of Florida

Corrosion
Michele Viola Manuel,
University of Florida

MAGNESIUM-BASED BIODEGRADABLE IMPLANTS
Session Chairs

Materials and Processing / Surface Modification and Corrosion
Petra Maier,
Fachhochschule Stralsund

Jarek Drelich,
Michigan Technological University

Corrosion / Market and Clinic
Pat Bowen,
Michigan Technological University

Martyn Alderman,
Magnesium Elektron

STRIP CASTING OF LIGHT METALS
Session Chairs

Strip Casting Process
Kai F. Karhausen,
Hydro Aluminium Rolled Products

Jan Bohlen,
MagIC – Helmholtz Zentrum Geesthacht

Strip Casting: Properties
Murat Dundar,
Assan Aluminium

Dietmar Letzig
MagIC – Helmholtz Zentrum Geesthacht

MAGNESIUM TECHNOLOGY 2016
Reviewer Pool

ADLAKHA, Ilaksh	Arizona State University
ALDERMAN, Martyn	Magnesium Elektron
BAI, Jing	Southeast University
BARTHELAT, François	McGill University
BEALS, Randy	Cosma
BERMAN, Tracy	University of Michigan
BHATIA, Mehul	Arizona State University
BOHLEN, Jan	MagIC – Helmholtz Zentrum Geesthacht
BOWEN, Patrick	Michigan Technological University
DECKER, Ray	Nanomag
DIERINGA, Hajo	MagIC – Helmholtz Zentrum Geesthacht
DOHERTY, Kevin	US Army Research Laboratory
ESEN, Ziya	Çankaya University
GHADERI, Alireza	Deakin University
GUPTA, Manoj	National University of Singapore
HASELHUHN, Amberlee	Michigan Technological University
HERING, Britta	Leibnitz Universität Hannover
HORT, Norbert	MagIC – Helmholtz Zentrum Geesthacht
JAIN, Vipin	National Physical Laboratory, New Delhi
JORDAN, Brian	University of Alabama
JOSHI, Vineet	Pacific Northwest National Laboratory
KADA, Sitarama	Deakin University
KARKAYA, Ishak	Middle East Technical University
KRUZIC, Jamie	Oregon State University
LAWS, Kevin	University of New South Wales
LETZIG, Dietmar	MagIC – Helmholtz Zentrum Geesthacht
LI, Bin	University of Nevada
LYON, Paul	Magnesium Elektron
MAIER, Petra	Fachhochschule Stralsund – University of Applied Sciences
MANSOOR, Bilal	Texas A&M University at Qatar
MANUEL, Michele	University of Florida
MENDIS, Chamini	MagIC – Helmholtz Zentrum Geesthacht
MISRA, Devesh	The University of Texas at El Paso
NEELAMEGGHAM, Neale R.	IND LLC
PRENTICE, Leon	CSIRO
PULUGURTHA, Syamala R.	Medtronic Inc.
RAU, Helen	Michigan Technological University

RAVINDRA, Nuggahalli	New Jersey Institute of Technology
ROHATGI, Aashish	Pacific Northwestern University
SAAL, James	QuestTek Innovations LLC
SCHAFFER, Jeremy	Fort Wayne Metals
SCHARNAGL, Nico	MagIC – Helmholtz Zentrum Geesthacht
SEITZ, Jan-Marten	Syntellix AG
SEVER, James	Alpha/Omega Engineering
SILLEKENS, Wim	European Space Agency
SINGH, Alok	National Institute for Materials Science
SINGH, Raman	Monash University
SOLANKI, Kiran	Arizona State University
SRINIVASAN, A.	National Institute for Interdisciplinary Science and Technology
SZAKÁCS, Gábor	MagIC – Helmholtz Zentrum Geesthacht
TOLNAI, Domonkos	MagIC – Helmholtz Zentrum Geesthacht
WANG, Bin	University of California, San Diego
WILKS, Tim	Magnesium Elektron
YANG, Tian	Kunming University
ZANDER, Daniela	RWTH Aachen University
ZHENG, Yufeng	Peking University

Magnesium Technology 2016

SYMPOSIUM:

Magnesium Technology 2016

Magnesium Technology 2016

Keynote Session

Magnesium Technology 2016
Edited by: Alok Singh, Kiran Solanki, Michele V. Manuel, and Neale R. Neelameggham
TMS (The Minerals, Metals & Materials Society), 2016

CHALLENGES FOR IMPLEMENTATION OF MAGNESIUM INTO MORE APPLICATIONS

K.U. Kainer
Institute of Materials Research, Magnesium Innovation Center,
Helmholtz-Zentrum Geesthacht, Max-Planck-Straße 1,
D-21502 Geesthacht, Germany

Abstract

As a result of the demands made on the transportation and the communication industries to introduce lighter materials, it is necessary to completely utilize the potentials of many different light structural materials. The different classes of light metal materials have to compete with each other as well as with polymers and steels. 200 years after its discovery magnesium is still one of the most promising materials for light weight constructions. Especially in the automotive industries the reduction of the weight of vehicles is a must. Car manufacturers have to decrease CO_2 emissions in accordance to the legislation, especially that of the EU. Additionally, with the increasing fuel prices consumers expect a decrease in fuel consumption either by a decrease of vehicle weight or by increasing the engine efficiency. Mg alloys can contribute to both, especially in reduction in vehicle weight. Moreover Mg alloys can be fully recycled, which fulfils the requirements from the EU directive for the end of life of vehicles and also helps to lower the CO_2 emissions associated with the processing of Mg.

Since the end of the last millennium, Mg alloys have made inroads into the applications in automotive industries as well as in consumer, computer and communication (3C) applications. Their favourable property profile – high specific strength, good machinability, recyclability – promote increased usage. Still, the full potential of Mg as a structural material has not been achieved. There a number of reasons can be specified:

1. Market issues: Price, protective duties, dependence on the Chinese Mg market, availability etc.
2. Reservation against Mg due to misinterpretation of properties: "magnesium corrodes", "magnesium burns" etc.
3. Inadequate properties i.e. strength, ductility, corrosion behaviour, crash worthiness etc.
4. Limited number of suitable Mg alloys may restrict wider application of Mg alloys.
5. Life cycle assessment issues.
6. Despite magnesium alloys being in service for almost one hundred years, there is still a lack of knowledge potential of Mg alloys compared with other metallic materials such as steel or aluminium: Design, processing, potential etc.

Mg and its alloys face some specific obstacles in relation to its mechanical and physical properties that need to be addressed before wider application of Mg would be envisaged. Examples of properties of Mg that restrict the application of Mg include the flammability of the metal [1] with strong exothermic reaction, the poor formability especially at the room temperature [2] and the corrosion properties. Additionally, in classical forming processes such as machining, forming or surface protection require specific knowledge on how magnesium may be handled, e.g. handling and storage of Mg scrap from machining.

Flammability is not only an issue during casting of Mg alloys. It requires specific attention during any subsequent processing. Special measures should be considered during machining of the alloys and processing parameters must be adjusted with respect to the alloys used. There needs to be education on handling Mg during processing and subsequent use, e.g. automotive or aerospace require knowledge about handling Mg fire during service.

The specific nature of wrought Mg with its distinct microstructures and strong textures [3] requires adjustment of forming procedures during manufacturing. Engineering knowledge on handling of the specific needs of Mg needs to be developed. It is noteworthy that scientific and engineering knowledge on handling Mg is concurrently developed at present so that there is potential for direct interaction between scientific knowledge developed and the technical needs of application. Example of such concurrent development in knowledge is the formability of Mg alloys in form of forming limits, the applicability in form of forming faults as well as the resulting properties of the formed parts which have to meet the application requirements.

While all these issues can be overcome with the theoretical and technical knowledge on deformation of Mg, there is a broad distribution of knowledge on understanding the differences between Mg alloys and other metallic materials. It may be hypothesised that - being a metal - Mg and its alloys are prone to be misunderstood just as a metal, an issue which is not faced by other classes of materials such as reinforced polymers, plastics in general, wood, or textiles.

While the fundamental understanding on materials sciences are applicable for almost any metals and alloys, materials sciences and engineering classes often deal only with a limited number of materials. In terms of metallic materials steels are studied in detail as they form the basis of many structural applications. However, the exposure of students to other metals and alloys depend on the preference and education of the lecturer. Additionally text books are also written based on the interest and preferences of the author and therefore Mg often is not a topic in lectures any more than in text books which deal with metals and alloys. As a consequence engineers often do not know about Mg alloys and its advantages or has misleading information (Mg burns! This is from the science classes at secondary school, showing propagation of wrong or misrepresented information). The missing knowledge or wrong statements on Mg alloys contribute significantly to the prevention of their use.

When Mg alloys are compared with other materials often e.g. their mechanical properties are directly compared. However, this is like comparing an apple with an orange; both are round but have vastly different properties even though they are both fruit. A similar concept should also be applied for alloys e.g. for strength:

specific strength. In this case ultimate tensile strength (UTS) or yield tensile strength (YTS) are divided by the density of the alloy and the earth acceleration. The outcome would be a value with the dimension km. For UTS and YTS this would mean the length where the material fails or shows first plastic deformation under its own load. This concept can also be applied to Young's modulus but would lead to more or less the same value for all metals and alloys. Applying this to engineering approaches would immediately show that Mg alloys would have advantageous properties in comparison with Al alloys and even steels.

Price is often also a determining factor in considering if a material is used or not. When comparing Al to Mg it could be shown that approximately the same amount of energy is necessary for the primary production, machining and recycling for both. However, Mg alloys perform better during casting and especially during use e.g. in automotive applications. Fuel savings (= less CO_2 emissions) or an extended range are especially important for the consumer more than for the car manufacturers. Mg is still produced in much smaller quantities compared with other common alloys and therefore often more expensive. This in combination with non-existent knowledge on the properties and processing of Mg alloys contributes significantly to hindering the use of Mg alloys.

Since the interest in magnesium alloys for automotive application was reawakened in the middle of the 1990s, there is a rapid increase in the number of R&D activities on Mg alloys. Driven from the demands on Mg alloys for light structural application, research programs for the improvement in the property profile of Mg alloys and the development of production technology were launched. The research topics include fundamental and applied research with the aim to extend the potential use of magnesium components in different application sectors.

In the last 25 years the research on magnesium alloys and their technology escalated. Simple search shows the effect: Increase from about 408 papers per annum in 1990 to 2654 in 2014 [4]. In special topics such as magnesium biomaterials the number of papers has increased from about 2 in 1990 to more than 124 in 2014 [5]. The outcome from this research boom was a sustainable development magnesium technology through the development of new alloys and the development of advanced processes. The property profile of modern magnesium alloys shows a high potential for the increased use of magnesium in different application sections. To achieve this goal the dissemination of information on novelties, potential, property profiles and limitations of magnesium alloys is a necessary key factor. Conferences within the magnesium communities on magnesium alloys and their applications do not satisfy this demand. The proceedings from the biggest magnesium conference in 2015 for example contain more than 377 papers [6]. More than 53% of papers dealt with magnesium wrought materials which have a market share of less than 5%. Papers on cast alloys and casting processes only contributed to 14% of the papers and the important topic of corrosion and corrosion protection less than 10% of the total contributions. Design issues and consideration for applications are rarely discussed.

The biggest challenge for the Mg industry is the improvement of the visibility of Mg showing the high potential of Mg alloy as a potential light weight material. In commercially oriented conferences on lightweight materials Mg plays a minor role. Mg alloys are not even named as a potential lightweight material while aluminium, fibre reinforced polymers and even high performance steel are considered to be lightweight materials. In this sense Mg shows a shadowy existence. The Key factor for the improved visibility of Mg alloys is the education of designers, engineers and technicians on issues, challenges and potential of Mg alloys. Educational seminars, webinars, creating of data books and design guidelines together with revision of existing resource books [7] and textbooks and including complete and new information on Mg alloys and Mg alloy processing will provide an important contribution to extend the knowledge of Mg and its alloys and this will help with the visibility of Mg alloys leading and increased potential to introduce Mg into a wider range of applications.

Magnesium Technology 2016
Edited by: Alok Singh, Kiran Solanki, Michele V. Manuel, and Neale R. Neelameggham
TMS (The Minerals, Metals & Materials Society), 2016

DEVELOPMENT OF MAGNESIUM ALLOYS FOR HIGH SPEED TRAINS IN CHINA

Eric Nyberg[1], Jian Peng[2], Neale R. Neelameggham[3]
[1]Pacific Northwest National Laboratory, Richland, WA, U.S.A.,
[2]Chongqing University, National Engineering Research Center for Magnesium Alloys, Chongqing, China,
[3]Ind LLC, Salt Lake City, UT, U.S.A.

Extended Abstract

In 1939, Dow Chemical had advertisements discussing the use of magnesium for trains, following their successful use of Dowmetal truck bodies in 1935. Despite the fact that the concept of using magnesium (Mg) alloys in trains has existed for at least 75 years, their remains significant challenges in alloy development and forming technologies before commercial application will find mass application.

In Dec. 2011, China Daily reported that a test bullet-train reached a world record-breaking speed of 500 km/hr (311 mph). Prior to this announcement, the fastest passenger train was the Beijing Shanghai High Speed Railway, operating at a top speed of 300 kilometers per hour. The new test train includes six passenger carriages, and the front end is tapered to a fine point, "similar to (the) sharp edge of an ancient Chinese sword". The power of this train is 22,800 kilowatts and is constructed of lightweight plastic, magnesium alloy and reinforced with carbon fiber. Earlier versions of China Railway's High-Speed (CRH) trains run at 9,600 kilowatts, with demonstrated speed of about 300 kmph (190 mph) [2].

A 2012 study indicated that use of ZK60A and AZ31B in seat back material in high speed trains will have a reduction in weight of 33% compared to the aluminum alloy seat back material [A7003-T5 or 5052] [3]. Although this may be an overly conservative estimate, because using Mg alloys may require engineering design with added strengthening features, it is still likely that more than 20% weight savings can be achieved compared to aluminum alloys. Similarly, when an aluminum alloy with Young's modulus of 70GPa was used to replace steel with a modulus of 210GPa, the train body and its components were redesigned to bear the vertical, lateral and torsional load. It should be expected that similar redesigns will be required when important structural parts of a train are substituted with Mg alloy having a Young's modulus not more than 45 GPa. Meanwhile,the main products used on high speed train are hollow, thin-wall extrusion profiles and sheet with thickness of approximately 4mm [4]. Additional efforts are still required to develop new alloys and advance manufacturing techniques to meet the demands of strength, straightness, flatness, corrosion resistance, and joining.

Gaofeng Quan, et al. described the use of Mg alloys will provide reduced vibration and noise in high speed trains - mainly by the superior damping capacity of magnesium alloys [5]. Such new alloys of high damping performance with associated high strength were developed in 2010 by Pan and Wang [6]. Such developments were made while continuing to push the limits for developing larger and larger magnesium alloy extrusion profiles. An additional advantage for using Mg alloys in high speed trains is its prominent ability to provide electromagnetic shielding [7], particular for the magnetic levitation trains, with the Mg alloy shielding the high intensity electromagnetic radiation produced when the trains start.

One of the Chinese patents shows the use of a magnesium alloy composite application as an air vent grill used in the high-speed trains and other vehicles. This composite is unique in using a magnesium structure with ribs covered on both the top and bottom with a rubber coating. This composite provides a high strength, impact resistant, light weight, sound, strong magnesium alloy composite structural panel [8]. Further Mg alloy applications in the demonstration train used forged ZK60 alloy for the frame, AZ31 sheet in the flooring, and extruded AZ31B profile for arm rests.[9]

Still today, the large-scale application of Mg alloy on trains is still in development. Some magnesium companies continue efforts to suplly samples to train manufacturing plants for assembly operational tests. In May 2015 Baosteel, China announced its entry into magnesium alloy market by a joint venture with BAIC Motor and Amgain Shandong Magnesium Co. with the main goal of providing components for transportation areas in automobile and rail transit [10]. Recently Amgain Shandong Magnesium Co. had provided about 50 large Mg alloy extrusion profiles for rail transit applications to Nanche Group and China CNR Corporation. Applications include stiffener or stringers with a weight of 7.5 kg, grills with a weight of 2.62 kg, and a baggage holder etc. The baggage holders had a fixed length of 7.5 meters, and could be produced up to 25 meters as needed [11]. The seat arm, reading lamp holder, seat frame, etc. were manufactured by Shandong Yinguang Yuyuan Light Metal Precise Forming Co., Ltd. Using semi-solid forming technology to produce a dense and refined microstructure with improved properties compared to similar die casting products [12]. The largest use of Mg alloys will likely come in the form of wrought alloy applications such as the structural stringers, seat frames and mounting profiles.

References:

[1] DowMetal –The Metal Travels Light, The Michigan Technic, vol 57 -58 May 1939.
[2] Tuan Nguyen, China Daily, Topic: Innovation, December 28, 2011.
[3] L. Song, Y. H. Zhao, J. Z. Liu, J. H. Mu, X. C. Guo, "Study on the ZK60A and AZ31B Magnesium-Alloy in the Application on High-Speed Train Seats", Advanced Materials Research, Vols. 690-693, pp. 53-57, May. 2013.
[4] Li Rui-chun. Application and Development of Magnesium Used for High-speed Train. Railway Locomotive & Car, 2011, 31(6): pp. 59-62.
[5] Gaofeng Quan, Ruichum Li, Siu Gu and Zhaoming Liu, 'Magnesium alloys- new materials for high-speed train with reduced vibration and noise", The 1st International Workshop on High-speed and Intercity Railways (IWHIR 2011), Yi-Qing Ni and Xiao-Wei Ye (Eds.),Shenzhen, July 2011 .p 349-356.

[6] J.F.Wang, S.Gao, P.F.Song, X.F.Huang, Z.Z.Shi, F.S.Pan. J. Alloys Compd. 509 (2011) 8567-8572.
[7] Xianhua Chen, Lizi Liu, Juan Liu, Fusheng Pan. Microstructure, electromagnetic shielding effectiveness and mechanical properties of Mg - Zn - Y - Zr alloys. Materials & Design, 2015, 65, pp 360-369.
[8] Chinese Patent, "Magnesium alloy composite structural slab for high-speed trains and rail transit vehicles," CN 201971017 U.
[9] May 2015 Press release by Baosteel, http://www.baosteel.com/group_en/contents/2863/78669.html Accessed 9/14/2015.
[10] http://www.rautomead.co.uk/158_Copper+Magnesium+Rod.html, Accessed 9/14/2015.
[11] Tianchang Huo. The application of magnesium alloy profiles in rail transit. The 17th annual meeting of magnesium industry session, 2014.
[12]Qianjin Wang. Application progress of high speed rail components with semi-solid casting. The 17th annual meeting of magnesium industry session, 2014.

Magnesium Technology 2016
Edited by: Alok Singh, Kiran Solanki, Michele V. Manuel, and Neale R. Neelameggham
TMS (The Minerals, Metals & Materials Society), 2016

KOREA'S R&D ACTIVITIES TOWARDS THE APPLICATION OF WROUGHT Mg ALLOYS

Nack J. Kim
Pohang University of Science and Technology (POSTECH); Pohang, Korea

Keywords: Magnesium alloys, alloy development, twin-roll casting, extrusion, applications.

Extended Abstract

Lightweighting of automobiles is possible through the application of materials such as advanced high strength steels and low density alloys including Al and Mg alloys. Among these, Mg alloys having a density about one-fourth of steel and two-thirds of Al can offer greater weight reduction over other materials, provided that Mg alloys have comparable mechanical properties to those of steels and Al alloys. Due to their advantage of being the lightest structural alloys, the use of large amounts of Mg alloys in automobiles has been frequently predicted, but their actual application, particularly of wrought products, in automobiles is quite limited. Such limited application of wrought Mg alloys has been mainly due to their several shortcomings such as high cost, poor mechanical properties including formability, poor corrosion resistance, etc. The widespread application of wrought Mg alloys in automobiles depends on how these challenges are addressed. At present, several countries are actively pursuing the R&D to overcome these challenges and the present talk discusses the Korea's R&D activities towards the applications of wrought Mg alloys.

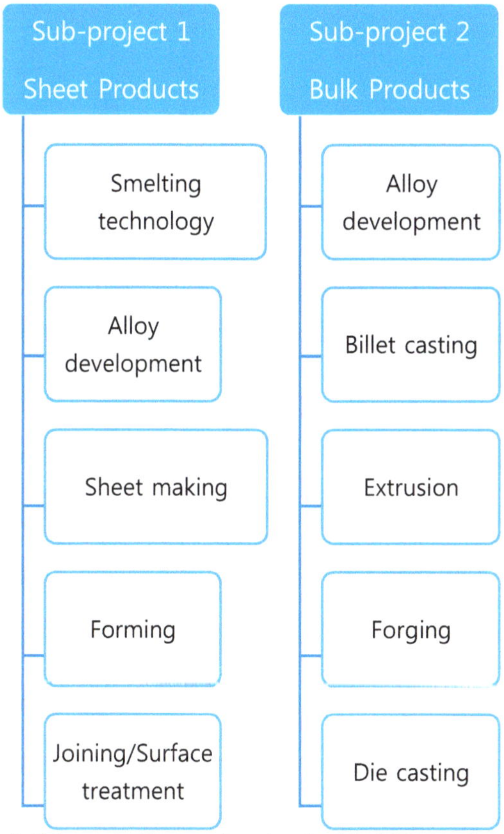

Fig. 1. Schematic diagram showing the Mg-WPM project.

The major R&D project on Mg alloys in Korea is the so-called "Mg materials R&D project for the super-light vehicles", a part of the World Premier Materials (WPM) program. Led by POSCO, it started in 2010 (until 2019 for a duration of 9 years) and 28 organizations are currently involved in the project, including 3 automotive makers, 14 automotive component manufacturers, 7 universities, and 4 research institutes. The project consists of 2 sub-projects; sheet products and bulk products as schematically shown in Fig. 1.

The sub-project on sheet products is aimed at developing low cost, high strength and high formability Mg alloy sheets and deals with various aspects of Mg alloy sheet productions and applications such as the development of efficient smelting technology, new alloy developments, hot/warm coil rolling process, forming, joining and surface treatments. The key part of this sub-project is the utilization of twin-roll casting (TRC) process, which allows the continuous fabrication of thin sheet products and therefore improves productivity and reduces the cost associated with sheet production. TRC process also gives several metallurgical benefits to the alloys such as refinement of microstructural constituent phases, allowing the development of novel alloys utilizing alloying elements that have limited solid solubilities in Mg. Utilizing TRC, several conventional alloys such as AZ, ZK, and AM series alloys have been fabricated and new alloy systems have also been developed with improved tensile properties and formability, Some of these alloys have found application in automobiles and are currently being used in several automobiles including the Porsche 911, Renault-Samsung SM7, and Renault's concept car, Eolab. POSCO's Mg plant boasts a 2,000 mm wide twin-roll caster and a few 600 mm wide twin-roll caster, along with tandem rolling mill.

The second sub-project on bulk products is aimed at developing high strength Mg alloys and related processes with improved productivity. Similar to the sub-project on sheet products, it deals with new alloy development and process design for extrusions, forgings and castings. Regarding the extrusions, the main emphases have been placed on developing high strength extrusion alloys that can be extruded at a fast extrusion speed as well as ultra-high strength extrusion alloys. One of the developed alloys can be extruded at an extrusion speed of 24 m/min without forming any surface defects. Another alloy system shows tensile strength of ~ 450 MPa and elongation of ~ 13% and more importantly does not show tension-compression anisotropy. These alloys have been successfully forged into various automotive parts such as arm, link, and bumper.

Although not supported by the WPM program, there are several other programs on Mg alloys in Korea. One of them is the development of non-flammable Mg alloys with both excellent mechanical properties and ignition/corrosion resistance, led by the scientists at the Korea Institute of Materials Science. They showed that the combined addition of Ca and Y leads to significant improvements in non-flammability and tensile properties, compared to conventional Ca containing Mg alloys. These alloys also have better corrosion resistance than any other commercial alloys or Ca containing alloys

This talk will summarize the current R&D efforts on Mg alloys in Korea and will provide an insight on future R&D directions for commercialization of Mg alloys.

Acknowledgements

This work was supported by the WPM Program funded by the Ministry of Industry, Trade and Energy, Korea. The author would like to thank all the participants of the program.

Magnesium Technology 2016
Edited by: Alok Singh, Kiran Solanki, Michele V. Manuel, and Neale R. Neelameggham
TMS (The Minerals, Metals & Materials Society), 2016

FASCINATING LPSO-STRUCTURED Mg ALLOYS

Eiji Abe
Department of Materials Science & Engineering,
University of Tokyo, 7-3-1, Hongo, Bunkyo-ku, Tokyo, 113-8656, Japan

Keywords: Mg alloys, Structure, Phase stability, First-principles calculations, Electron Microscopy

Recent successful alloy-design showed that Mg alloys with addition of a small amount of Zn and Y (or rare-earth elements) reveal excellent mechanical properties including remarkably improved strength with a reasonable ductility [1,2]. One of the prominent microstructural features, which are believed to contribute for these excellent properties, is formation of a novel type of long-period structures [3-6]. The structures are fundamentally long-period stacking derivatives of a hexagonal close-packed structure (hcp-Mg), and the resultant stacking polytypes accompany a unique chemical order that occurs to synchronize with the corresponding stacking order; i.e., the synchronized long-period stacking/order (LPSO) structure (Fig.1) [5].

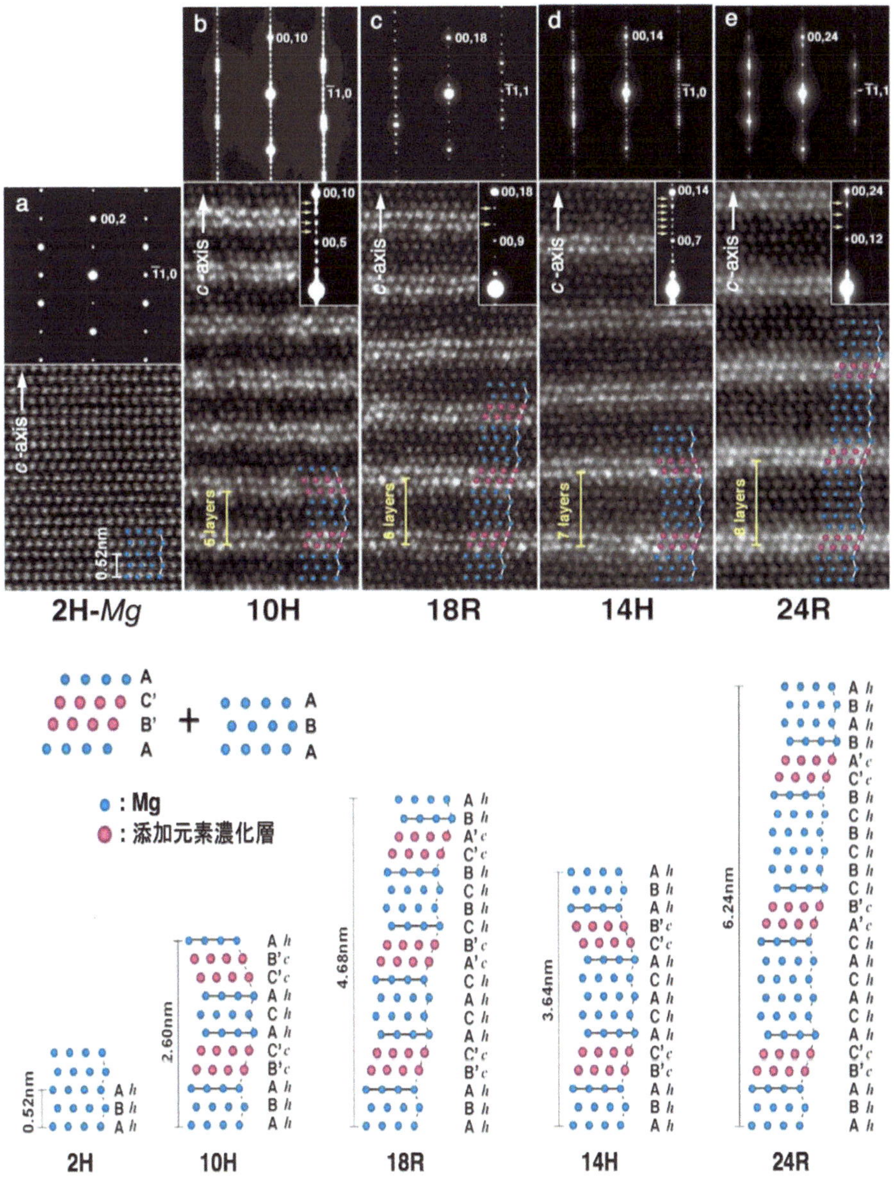

Fig.1 Z-contrast STEM images and model structures of a series of synchro-LPSO crysatals; (a) hcp-Mg, (b) 10H, (c) 18R, (d) 14H, and (e) 24R (adapted from ref. 5).

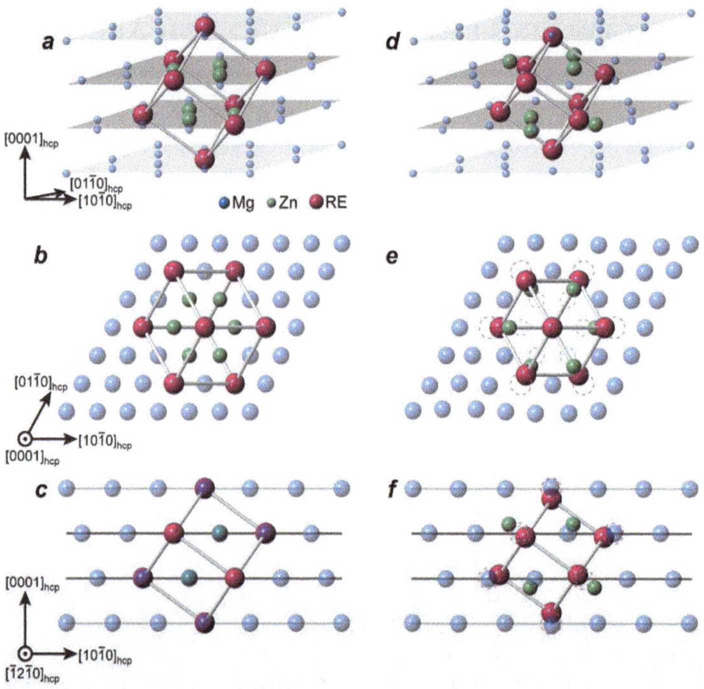

Fig.2 Local Zn-Y configurations before (left) and after (right) the energetic relaxations (adapted from ref. 6).

LPSO-structured ternary Mg-TM-RE (TM: transition metals, RE: rare-earth elements and Y) alloys reveal remarkable performance even at elevated temperatures, providing an extended opportunity for various applications. To accelerate further developments of the alloys, understanding their thermodynamic stability becomes important. With this aim, we have attempted to evaluate phase stability as well as to construct model structures of the complex LPSO crystals, based on the-state-of-the-art electron microscopy and first-principles calculations. It has become apparent that the ordered characteristics of the LPSO structures are well represented by the TM_6RE_8 clusters with a $L1_2$-type configuration. Interestingly, significant displacements occur for the Zn_6Y_8 clusters after energetic relaxations (Fig.2) [6], causing generations of internal 'voids' at the cluster centers, which are sufficiently capable for incorporating an extra atom as interstitials. Introduction of such interstitials (*I*) into the close-packed LPSO structures turns out to increase remarkably the thermodynamic stability for several LPSO-Mg alloys [7]. These theoretical predictions have been experimentally verified with the aid of a novel ultrahigh-sensitive electron microscopy imaging [8] and convergent-beam electron diffraction (CBED) analysis, leading to a conclusion that Mg atoms indeed exist at the relaxation-induced *I*-sites in the several LPSO-Mg alloys. We believe that the present results provide an interesting interstitial-aided stability mechanism. We will also describe several LPSO-related structures identified in ternary Mg-TM-RE alloys, including unique zones similar to those frequently observed in Al-TM alloys (i.e., GP zones).

References

[1] Y. Kawamura et al, Mater. Trans. **42** (2001) 1172.
[2] Y. Kawamura, M. Yamasaki, Mater. Trans. **48** (2007) 2986.
[3] E. Abe et al, Acta Mater. **50** (2002) 3845.
[4] Y. M. Zhu, A. J. Morton, J. F. Nie, Acta Mater. **58** (2010) 2936.
[5] E. Abe et al, Philos. Mag. Lett. **91** (2011) 690.
[6] D. Egusa, E. Abe, Acta Mater. **60** (2012) 166.
[7] J. E. Saal et al. Acta Mater. **68** (2014) 325.
[8] R. Ishikawa et al., Nature Mater. **10** (2011) 278

Magnesium Technology 2016
Edited by: Alok Singh, Kiran Solanki, Michele V. Manuel, and Neale R. Neelameggham
TMS (The Minerals, Metals & Materials Society), 2016

DEVELOPMENTS IN HIGH MAGNESIUM-CONTENT BULK METALLIC GLASSES AND FUTURE POSSIBILITIES

Kevin J. Laws, Karl F. Shamlaye, Jörg F Löffler, Michael Ferry
[1]School of Materials Science and Engineering, UNSW, Sydney, NSW 2052, Australia
[2] Laboratory of Metal Physics and Technology, Department of Materials, ETH Zurich, 8093 Zurich, Switzerland

Keywords: Bulk Metallic Glass, Amorphous Alloys, Magnesium Alloys

Abstract

Amorphous alloys or metallic glasses have been earmarked as *"the most significant development in materials science since the discovery of plastics over 50 years ago"* (Mike Ashby, 2011) and are gaining significant attraction as *'Next Generation'* materials [1].
When compared to crystalline alloys, amorphous alloys possess superior strengths (approximately three-times that of their crystalline counter-parts), which approach the theoretical strength maximum [1], the highest elastic limits of all metallic materials (at least twice that of regular metals), improved corrosion resistance, and amorphous alloys based on palladium exhibit the highest damage-tolerance of all materials known [1,2]. Unlike crystalline metal alloys, metallic glasses exhibit a glass transition temperature, above which the alloy exists in a 'super-cooled liquid' state, where viscosity or flow stress is reduced by several orders of magnitude. Here, superplastic forming techniques (similar to plastics and ceramic glasses) such as forging, extrusion, drawing and blow moulding can be performed [3].

Magnesium alloys exhibit the lowest density of all engineering materials and continue to play an increasingly significant role in aerospace, automotive, consumer electronics and biomedical applications as a result of their high specific strength, stiffness and biocompatibility. Since the discovery of the $Mg_{70}Zn_{30}$ metallic glass in 1977 [4], numerous Mg-based amorphous alloys or 'bulk' metallic glasses have been developed. Due to ideal thermodynamic, kinetic and electronic characteristics, some Mg-based alloy compositions can exhibit exceptional 'glass-forming ability' when cast from a melt. To date, majority of Mg-based glass-forming compositions are based on Mg–TM–RE or Mg–TM–Ca ternary systems (where TM = transition metals Cu, Ni, Zn, Ag and RE = rare-earth metals Y, Gd, La, Nd, Ce, etc.) [4-8] with critical casting sizes of up to 27 mm [8]. However, Mg-based metallic glasses have traditionally been rich in these solute elements making them relatively heavy (dense) compared to crystalline Mg alloys.

Another key issue plaguing the use of Mg-based glasses - at least in small-scale structural applications is their inherent low fracture toughness or brittle nature. This shortfall in mechanical performance is inevitably due to a low activation energy of shear banding - which is directly associated with the chemistry of these glasses, and the rapid structural relaxation effects they exhibit [8]. Due to their low glass transition temperatures (T_g) severe embrittlement can occur even at room temperature over relatively short timescales, strongly compromising their potential usefulness and commercial viability [8]. However, there are some examples that do point to a more positive design direction for these high-potential materials.

Considerable ductility of amorphous ribbons has been reported by Gu *et al.* in Mg–Zn–Ca alloys with Mg concentrations from 68 to 85 at% and by Guo *et al.* in in the Mg–Cu–Ca, Mg–Cu–Y and Mg–Ni–Y systems for alloys with a Mg content >85 at.% [8]. Modest 'bulk' glass ductility has also been observed in Mg-rich Mg–Ni–Gd-based [6] and Mg–Ni–Ca [9] alloy systems, pointing to higher Mg-contents (which simultaneously lowers their density) and lessening the degree of electronic charge transfer between elements (i.e. the degree of covalency in bonds) as a first step to improving the bulk ductility of these materials [8].

The ductility of Mg-based metallic glasses can also be significantly improved by the incorporation of ductile crystalline phases [10]. However, a significant volume fraction of crystallites in such a composite microstructure often results in the deterioration of strength and corrosion resistance, further, composite microstructures are often difficult to produce, requiring specific cooling rates or specialised processing techniques [10].

Here, we present relative comparisons to crystalline Mg-alloys, recent results, developments and design strategies in overcoming the former shortfalls of Mg-based bulk metallic glasses and future possibilities for a range of new, light-weight, 'ductile' Mg-rich bulk metallic glasses.

References

[1] M.F. Ashby, A.L. Greer *"Metallic glasses as structural materials"* ***Scripta Materialia*** 54 (2006) 321-326
[2] M.D. Demetriou *et al.*, *"A damage-tolerant glass"* ***Nature Materials*** 10 (2011) 123-128
[3] J. Schroers *"Processing of Bulk Metallic Glass"* ***Advanced Materials*** 21 (2010) 1-32
[4] A. Calka, M. Madhava, D.E. Polk, B.C. Giessen, H. Matyja, J. Vander Sande, *"A Transition-Metal-Free Amorphous Alloy Mg70Zn30"*, ***Scripta Metallurgica,*** 2 (1977) 65-70
[5] K.J. Laws, K.F. Shamlaye, J.D Cao, J.P. Scicluna, M. Ferry *"Locating new Mg-based BMGs free of rare-earth elements"* ***Journal of Alloys Compounds*** 542 (2012) 105-110
[6] E.S. Park, H.J. Chang, D.H. Kim *"Mg-rich Mg–Ni–Gd ternary bulk metallic glasses with high compressive specific strength and ductility"* ***Journal of Materials Research*** 22 (2007) 334-338
[7] Q. Zheng, J. Xu, E. Ma *"High glass-forming ability correlated with fragility of Mg–Cu(Ag) –Gd alloy"* ***Journal of Applied Physics*** 102 (2007) 113519
[8] K.J. Laws, D. Granata, J.F. Löffler, *"Alloy design strategies for sustained ductility in Mg-based amorphous alloys – Tackling structural relaxation"* ***Acta Materialia*** (2015) DOI:*10.1016/j.actamat.2015.08.077*

[9] K.J. Laws, J.D. Cao, C. Reddy, K.F. Shamlaye, B. Gun, M. Ferry *"Ultra magnesium-rich, low-density Mg–Ni–Ca bulk metallic glasses"* ***Scripta Materialia*** 88 (2014) 37-40
[10] M. Ferry, K.J. Laws, C. White, D. Miskovic, K.F. Shamlaye, W. Xu, O. Biletska; *"Recent developments in ductile BMG composites"* ***MRS Communications*** 3 (2013) 1-12.

Magnesium Technology 2016

Solidification and Casting

Magnesium Technology 2016
Edited by: Alok Singh, Kiran Solanki, Michele V. Manuel, and Neale R. Neelameggham
TMS (The Minerals, Metals & Materials Society), 2016

IN SITU SYNCHROTRON RADIATION DIFFRACTION OF THE SOLIDIFICATION OF Mg-Dy(-Zr) ALLOYS

Domonkos Tolnai*[1], Peter Staron[1], Andreas Staeck[1], Helmut Eckerlebe[1], Norbert Schell[1], Martin Müller[1], Joachim Gröbner[2], Norbert Hort[1]
[1]Institute of Materials Research, Helmholtz-Zentrum Geesthacht, Max-Planck Str. 1, Geesthacht, D21502, Germany
[2]Institute of Metallurgy, Clausthal University of Technology, Robert-Koch-Str. 42, D-38678 Clausthal-Zellerfeld, Germany

*Corresponding author. Tel: +49 4152 871974
e-mail address: domonkos.tolnai@hzg.de

Keywords: Mg alloys, solidification, in situ, synchrotron diffraction

Abstract

Mg-Dy alloys are attractive for biomaterial applications. Their mechanical property profile is close to that of cortical bone, they are non-toxic, osseoconductive and degradable. Their macroscopic characteristics depend on their microstructure, which can be tailored through the alloy composition and the solidification parameters. *In situ* synchrotron radiation diffraction is a tool to unequivocally follow the phase formation and grain growth during cooling, thus determining the solidification sequence. In the present study Mg alloys containing Dy and Zr were investigated to characterize the solidification phenomenon during cooling from 660°C to 200°C. Samples, contained in steel crucibles, were melted in a modified induction furnace for *in situ* synchrotron radiation measurements at the HZG beamline P07B (HEMS) at PETRA III, DESY, with the temperature controlled by type K thermocouples during the measurements. The results give an experimental validation of the thermodynamic calculations and input for refining the existing thermodynamic models. This contributes to a better understanding of the microstructure evolution thus to control desirable macroscopic characteristics.

Introduction

Mg alloys with Rare Earth (RE) additions are attractive for biomaterial applications as degradable bone implants [1]. Their mechanical property profile is close to that of cortical bone, they are non-toxic, osseoconductive and they degrade with time, sparing the implant removal surgery [2]. Besides being biocompatible Dy has a high solubility (25.4 wt.%) in Mg, which allows the wide range tailoring of the macroscopic mechanical properties. The UTS of Mg5Dy is 180 MPa while of Mg25Dy is 402 MPa [3]. Mg has three intermetallic compounds with Dy, $Mg_{24}Dy_5$, Mg_2Dy and MgDy as shown in the Mg-Dy binary phase diagram in Figure 1.

In the case of multiphase materials the mechanical property profile depends strongly on the chemical composition, volume fraction and spatial distribution of the intermetallic phases [4]. These alloys are produced mainly in the form of castings and, as such, these microstructural parameters are determined during the solidification [5,6] and the following thermo-mechanical processing [7]. Therefore, the understanding of the sequence of formation and evolution of the meta-stable and stable phases during solidification is a prerequisite to achieve control of the microstructure of these alloys.

Performing *in situ* diffraction experiments while cooling the molten sample to the fully solidified state, allows identifying the microstructural phases, to determine their solidification sequence [8,9,10] and to use the results for experimental validation of the existing thermodynamic databases.

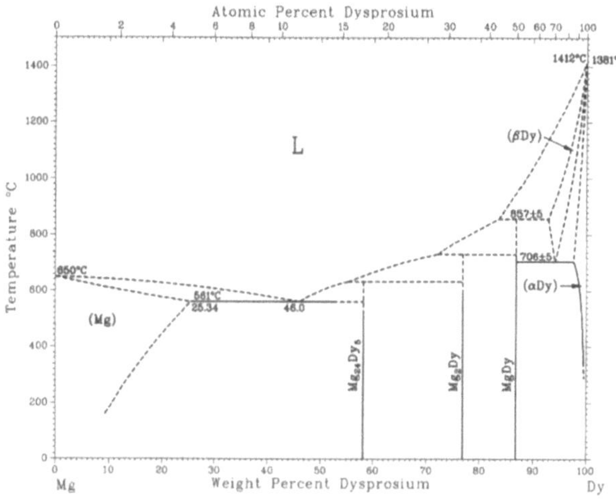

Figure 1. Mg-Dy binary phase diagram [11].

The aim of this investigation is to perform *in situ* synchrotron diffraction during solidification of a Mg20Dy alloy unmodified and modified with Zr, and to compare the experimental results with the existing thermodynamic data on this system.

Experimental methods

<u>Alloy production</u>
The alloys Mg20Dy(-0.4Zr) were chosen because of the particular interest in biomaterial applications. The alloy Mg20Dy was prepared by permanent mould indirect chill casting as described in [12]. The materials were melted in an electric resistance furnace under protective atmosphere of Ar with 2% SF_6. Dy was added as a pure element. To modify the alloy with Zr, it was remelted and casted into steel moulds.

<u>Microstructural characterization</u>
The microstructures of the alloys were investigated with a Carl Zeiss Gemini Ultra 55 Scanning Electron Microscope (SEM) operated at 15kV and equipped with an EDAX energy dispersive X-ray spectrometer (EDXS).

<u>*In situ* synchrotron diffraction</u>
The samples for the synchrotron diffraction measurements were cut into small cubes and encapsulated in steel crucibles. The

experiments were carried out at the HZG beamline, HEMS P07B beamline at the PETRA III at the Deutsches Elektronen-Synchrotron (DESY). The measurements were performed in the chamber of a furnace, heated by an induction coil. There are two windows on the sides covered by Kapton foils, to prevent interference with the X-ray beam, while the chamber itself is floated with Ar. The measurement setup is shon in Figure 2. During the tests the samples were held in a sealed stainless steel crucible in order to hold the molten metal and to isolate it from the surrounding atmosphere. The temperature was regulated by a type K thermocouple welded on the steel crucible containing the sample. To melt the samples the system was heated up to 660 °C, held for 5 min and then cooled to 200 °C at a cooling rate of 5K/min, 10 K/min and uncontrolled cooling with disabled coil, respectively. The uncontrolled cooling was significantly faster then the controlled cooling. The *in situ* diffraction experiment was done in transmission geometry using a beam cross section of 1 x 1 mm^2. In order to penetrate the sample, high-energy X-rays were used with a photon energy of 87 keV, corresponding to a wavelength of $\lambda = 0.014$ nm.

Figure 2. Furnace and sample holder for the *in situ* solidification tests.

During the experiment the Debye-Scherrer diffraction rings were recorded every 25 s with an acquisition time of 5 s by a Perkin Elmer 1622 Flatpanel (2048 x 2048) pixels with a pixel size of 200 μm^2. Conventional diffraction patterns (line profiles) were achieved by an azimuthal integration of the Debye-Scherrer rings. The line profiles were quantified by the Rietveld method with the Maud software. The thermodynamic simulation of the solidification was carried out with the software Pandat 8.1.

Results and Discussion

Microstructural characterization

The secondary electron micrographs of Mg20Dy are shown in Figure 3. and Figure 4. The microstructure consists of α-Mg dendritic structures, segregated areas rich in Dy and $Mg_{24}Dy_5$ intermetallic particles.

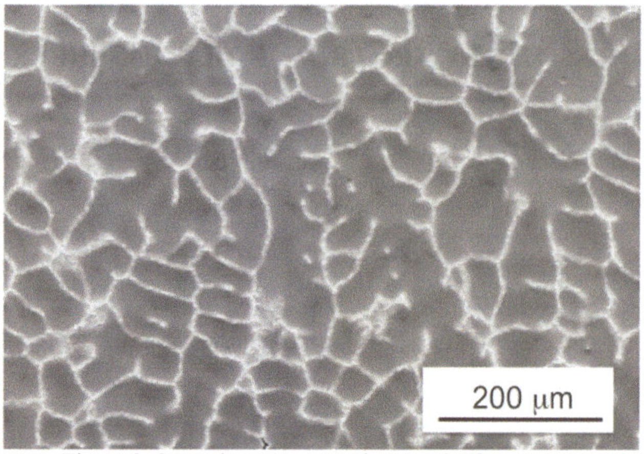

Figure 3. Secondary electron micrograph of Mg20Dy

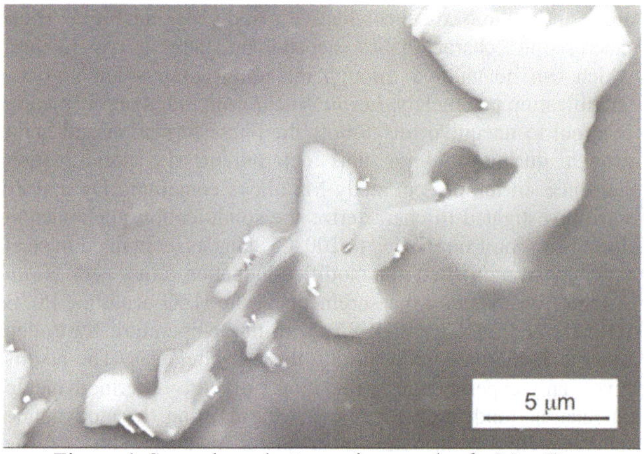

Figure 4. Secondary electron micrograph of a $Mg_{24}Dy_5$ intermetallic particle.

In situ synchrotron diffraction

The diffraction pattern of the unmodified and the Zr modified Mg20Dy are shown in Figure 5.

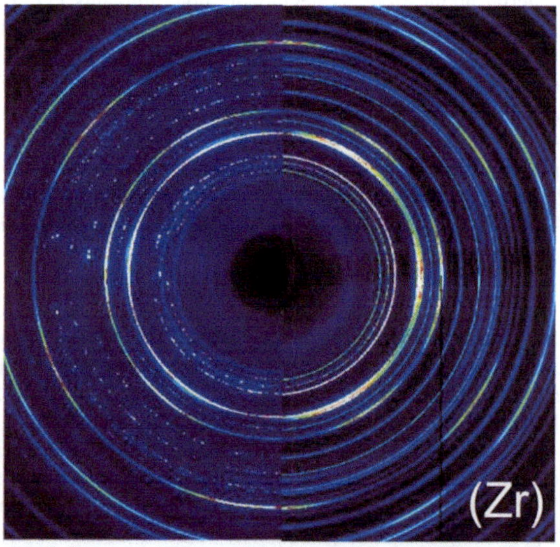

Figure 5. Detector images of the unmodified Mg20Dy (left) and the Mg20Dy0.4Zr (right).

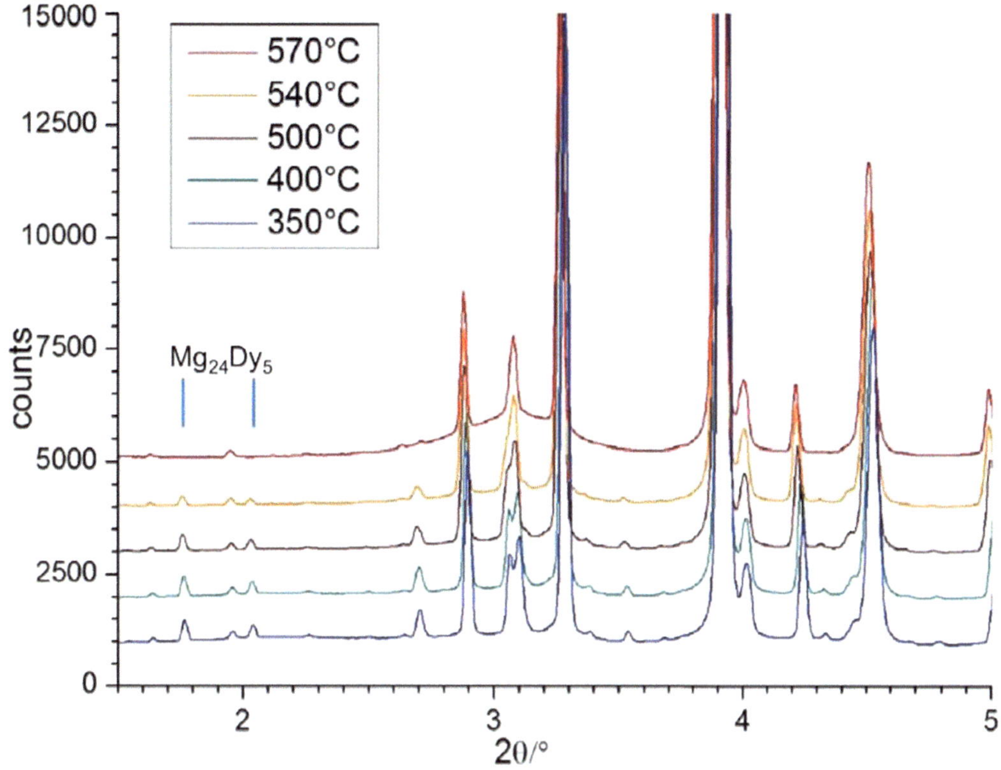

Figure 6. Line profiles, at different temperatures during the solidification of Mg20Dy, showing the formation of the $Mg_{24}Dy_5$ intermetallic compound.

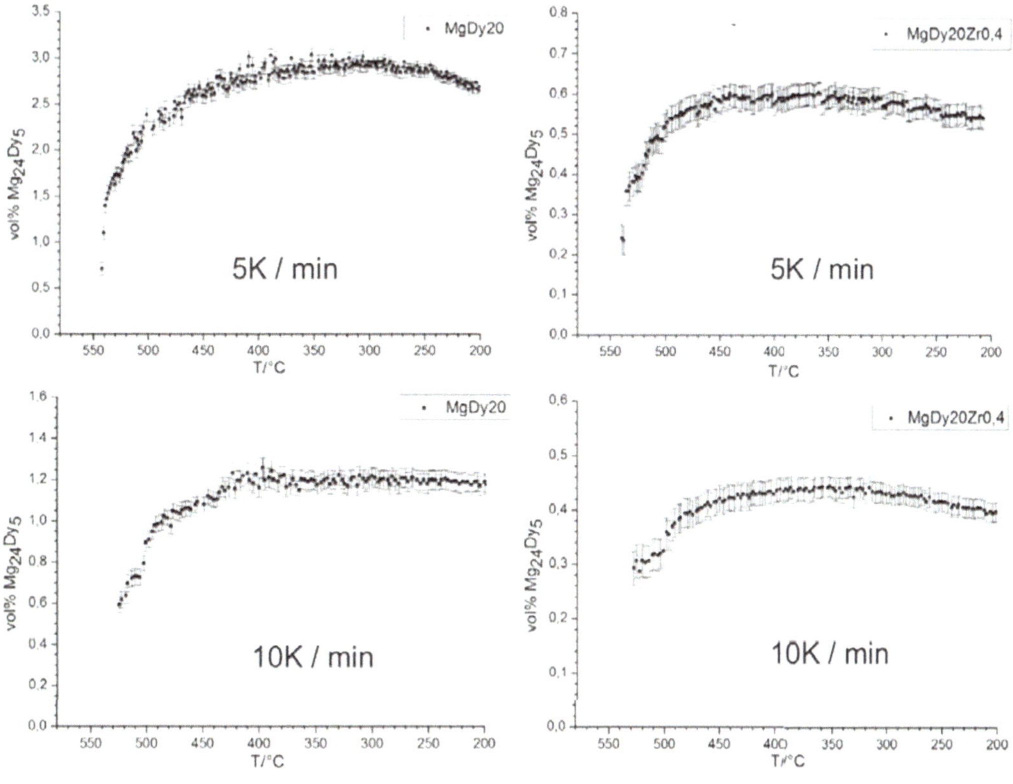

Figure 7. Change in the volume fraction of Mg24Dy5 in the alloys during solidification.

The Mg20Dy show individual spots along the Debye-Scherrer rings indicating a coarse grain structure from the casting. The continuous rings correspond to the reflections of the stainless steel crucible. The addition of Zr as a grain refiner has a pronounced effect on the diffraction patterns, full Debye-Scherrer rings can be observed for the Mg phase, indicating finer grains in the microstructure. The XRD line diagrams resulting from the azimuthal integration of the 2D diffraction patterns are shown in Figure 6. Based on the peak positions the solidification sequence of the phases can be determined. The solidification starts with the nucleation and growth of the α-Mg dendrites at a measured temperature of 595 °C for both the Mg20Dy and the Mg20Dy0.4Zr as well. As the cooling advances the intermetallic compound $Mg_{24}Dy_5$ can be observed at 540 °C and 537 °C in the case of Mg20Dy and Mg20Dy0.4Zr, respectively.

The change of the volume fraction of the intermetallic phase during solidification, obtained by the Rietveld analysis of the diffraction line profiles is shown in Figure 7.

The volume fraction starts to grow below 540 °C, where the solidification of $Mg_{24}Dy_5$ starts. As the solidification finishes the volume fraction reaches a stationary value, which remains nearly constant during further cooling. The final volume fraction values for the different cooling rates are shown in Table 1.

Table 1. Final volume fraction of $Mg_{24}Dy_5$, after solidification and cooling.

Alloy	5 K/min	10 K/min	Uncontrolled cooling
Mg20Dy	0.51 ± 0.03 %	0.39 ± 0.02 %	0.89 ± 0.06 %
Mg20Dy0.4Zr	2.65 ± 0.06 %	1.14 ± 0.05 %	0.93 ± 0.05 %

The volume fraction of the intermetallic compound increases with the decrease of the cooling rate. The exception from this trend is the uncontrolled cooling of Mg20Dy, which has a very high volume fraction of $Mg_{24}Dy_5$ compared to the controlled cooling rates.

Thermodynamic modeling

The results of the thermodynamic modeling of the two alloys are shown in Table 2.

Table 2. Onset temperatures of solidification of the phases calculated by the lever rule and for non-equilibrium conditions.

Phase	Lever rule	Scheil
α-Mg (Mg20Dy)	609 °C	609 °C
$Mg_{24}Dy_5$ (Mg20Dy)	-	559 °C
α-Mg (Mg20Dy0.4Zr)	611 °C	611 °C
$Mg_{24}Dy_5$ (Mg20Dy0.4Zr)	450 °C	559 °C

The determination of the solidification sequence by *in situ* synchrotron diffraction has limitations. The cooling rate must be low enough for the acquisition and the desired temperature-time resolution. The crystallographic orientation of the solidifying grains limits the detection as well. As the nucleated small grains float, their changing orientation and their small size hinders their detection. It is likely that they can be detected properly only after the system reaches the dendritic coherency point, the temperature where a continuous dendrite skeleton is built up. Despite these limitations, the onset temperatures, obtained experimentally by *in situ* diffraction and that predicted theoretically from the thermodynamic calculations, for the solidification of different phases correlate reasonably well.

Conclusions

In situ solidification experiments have been performed by synchrotron diffraction to determine the solidification sequence of Mg20Dy and Mg20Dy0.4Zr with different cooling rates.

- The solidification starts with the nucleation of a-Mg dendrites at a measured temperature of 595 °C in the case of both alloys.
- The dendritic solidification is followed by the eutectic solidification of a-Mg/$Mg_{24}Dy_5$ at 540 °C and 537 °C for the Mg20Dy and the Mg20Dy0.4Zr, respectively.
- The addition of Zr increased the final volume fraction of the intermetallic compound.
- The decrease in cooling rate resulted in the increase of the volume fraction of the intermetallic compound except in the case of uncontrolled cooling of Mg20Dy.
- The thermodynamic modeling of the cooling resulted in temperatures which correlate reasonably well with those obtained experimentally in the case of the binary Mg20Dy, while it can be seen that the addition of Zr to the modeled system resulted in significantly lower onset temperature for the intermetallic $Mg_{24}Dy_5$, than experimentally observed.

Acknowledgements

The authors acknowledge the Deutsches Elektronen Synchrotron for the provision of beamtime and the facilities of Petra III.

[1] H. Hermawan, D. Dubé, D. Mantovani, „Developments in metallic biodegradable stents" Acta Biomaterialia 6 (2010) 1693-1697.

[2] F. Feyerabend et al., „Evaluation of short-term effects of rare earth and other elements used in magnesium alloys on primary cells and cell lines" Acta Biomaterialia, 6 (2010) 1834-1842.

[3] LL. Rokhlin, Magnesium Alloys Containing Rare Earth Metals. Moscow: Taylor & Francis, 2003.

[4] G. Requena et al., „The effect of the connectivity of rigid phases on strength of Al-Si alloys" Advanced Engineering Materials 13 (2011) 674-684.

[5] D. Tolnai et al.," In situ synchrotron tomographic investigation of the solidification of an AlMg4.7Si8 alloy" Acta Materialia 60 (2012) 2568-2577.

[6] S. Terzi et al., „In situ study of nucleation and growth of the irregular α-Al/β-Al_5FeSi eutectic by 3-D synchrotron X-ray microtomography" Acta Materialia 58 (2010) 5370–5380.

[7] D. Tolnai et al., "Sub-micrometre holotomographic characterisation of the effects of solution heat treatment on an AlMg7.3Si3.5 alloy" Materials Science and Engineering A 550 (2012) 214-221.

[8] S.S. Babu et al., „Time-resolved X-ray diffraction investigation of primary weld solidification in Fe-C-Al-Mn steel welds" Acta Materialia 50 (2002) 4763-4781.

[9] O. Shuleshova et al., "In situ observations of solidification in Ti–Al–Ta alloys by synchrotron radiation" Intermetallics 19 (2011) 688.

[10] D. Tolnai et al., "In situ synchrotron diffraction of the solidification of Mg4Y3Nd" Materials Letters 102-103 (2013) 62-64.
[11] M Avedesian, H. Baker, Magnesium and Magneisum Alloys. ASM Specialty Handbook, ASM International, 1999
[12] F.R. Elsayed et al., "Magnesium permanent mold Castings optimization" Materials Science Forum 690 (2011) 65-68.

As Solidified Microstructure Investigation of Mg15Y and MgxYyGd (x+y=15 wt.%) Ternary Alloys

G. Szakács[1]*, C.L. Mendis[1], D. Tolnai[1], M. Vlcek[2], F. Lukác[2], I. Stulíková[2], B. Smola[2], M. Wolff[1], R. Schmid-Fetzer[3], N. Schell[1], K.U. Kainer[1] and N. Hort[1]

[1]Institute of Materials Research, Helmholtz-Zentrum Geesthacht, Max-Planck-Straße 1, D-21502 Geesthacht, Germany
[2] Faculty of Mathematics and Physics, Charles University in Prague, Ke Karlovu 3, 12116 Prague 2, Czech Republic
[3]Institute of Metallurgy, Clausthal University of Technology, Robert-Koch-Str. 42, D-38678 Clausthal-Zellerfeld, Germany
*Corresponding author. Tel: +49 4152 871963, Fax: +49 4152 871909,
e-mail address: gabor.szakacs@hzg.de

Keywords: Mg-RE alloys, phase formation, *in situ* synchrotron radiation diffraction, solidification, microstructures

Abstract

MgxYyGd (x+y=15 wt.%) alloys were produced via permanent mould casting to investigate the microstructure evolution during solidification of the ternary system. The microstructure of the as-solidified samples was characterized with scanning electron microscopy (SEM) and transmission electron microscopy (TEM). *In situ* synchrotron radiation diffraction experiments were performed during the solidification of the alloys at the P07 (HEMS) Beamline of PETRA III at DESY. The phase evolution observed during controlled cooling at 20 and 100 K/min and the resultant microstructures were compared with the as-cast conditions. The experimental results were correlated with the calculations from the Pandat thermodynamic software. In the case of the ternary alloys the equilibrium phase diagram suggests the formation of the $Mg_{24}Y_5$ phase at elevated temperatures followed by the formation of the Mg_5Gd phase at eutectic temperatures. However, the experiment shows only the formation of $Mg_{24}Y_5$ phase at eutectic temperatures even with a cooling rate (CR) of 100 K/min.

Introduction

Due to their light weight and high specific strength, Mg-RE (Rare earth) alloys are important for structural applications **[1]**, and recently have become attractive in applications as a degradable biometal **[24]**. Mg-Y-RE based alloys, such as WE43 and WE54 form an important class of cast magnesium alloys used in elevated temperature applications with superior mechanical properties at both room and elevated temperatures **[5]**.

Investigations based on the binary Mg-Y alloys resulted in the development of a number of commercial alloys with high strength **[1]**. The binary Mg-Y system is reported to contain: MgY (cP2-CsCl type, T_{max}= 935 °C), Mg_2Y (hP12-$MgZn_2$ type, T_{max}= 780 °C) and $Mg_{24}Y_5$ (CI58-αMn, T_{max}= 605 °C) **[6]**. The effect of the addition of other RE elements on the phase evolution in Mg-Y-RE systems have been investigated with thermodynamic modelling methods **10]**, based on the available experimental results from the literature. In case of bio-magnesium the biocompatibility of the RE additions plays an important role. Amongst the RE elements Gd shows a good biocompatibility **[11]**, and it is used currently in medicine as a contrast agent in MRI (magnetic resonance imaging) **[2]**. Furthermore, Gd enhances mechanical and corrosion properties of Mg-Y alloys. Gd has a maximum solubility of 23.49 wt.% at 548 °C **[12]** and offers both solid solution and precipitation strengthening. The Mg_5Gd (space group cF440, and lattice parameter a = 2.234 nm) **[13]** phase forms at the grain boundaries during casting of Mg-Gd alloys **[7]**. The isothermal section of ternary Mg-Y-Gd system at 500 °C was investigated by Giovannini *et al.* and it was reported that Y and Gd show mutual solubility in the binary intermetallic particles, and that no new ternary phases form. Therefore, a range of intermetallics with various amounts of Mg, Y and Gd solubility are observed **[13]**. Based on the available experimental results, the thermodynamic description of the Mg-Y-Gd system was investigated by Guo *et al.* **[14]** using the CALPHAD technique. Yongchun *et al.* reported that in the Mg rich corner of the Mg-Y-Gd system the alloys contain a combination of $Mg_{24}Y_5$ and Mg_5Gd phases depending on the concentration of the alloy. All literature on Mg-Y-Gd system considers the equilibrium states and provide necessary information for thermodynamic databases. Most Mg alloys are produced in the form of castings, thus their microstructure and consequently their macroscopic mechanical properties are determined during solidification and the following thermo-mechanical treatment, which can be modeled based on the thermodynamic databases **[15]**. The solidification paths may be calculated using thermodynamic data bases using both equilibrium and non-equilibrium solidification paths accurately **[7]** However, experimental results are necessary to clarify or modify these databases and the defined solidification paths. An understanding of the sequence of the formation and the evolution of intermetallic phases during solidification in complex systems is a prerequisite to control the microstructure of such alloys. Formerly only *ex situ* methods have been used, i.e. interrupting the solidification at different temperatures (different stages of microstructure development) by quenching the sample, to perform the characterization **[16]**, to determine the evolution of the microstructure during solidification. The recent development of the X-ray acquisition systems at synchrotron radiation sources provides a unique tool to characterize the phase formation and evolution during the solidification *in situ*. Recording the diffraction patterns while cooling the alloy system allows identification of the internal phases, to determine the solidification sequence **[17-23]** and to use these results for the validation of the existing thermodynamic databases and contribute to their refinement. Furthermore, this method allows us to study the microstructure evaluation with faster cooling rates.

The present study reports results from the microstructure investigation of as cast MgxYyGd alloys (x+y= 15 wt.%). In order to determine the solidification sequences of the alloys, *in situ* synchrotron radiation diffraction studies were conducted during solidification with cooling rates (CR) of 20 and 100 K/min. The experimental results were compared with the prediction from the thermodynamic data available for this system.

Experimental methods

The alloys were cast at HZG-MagIC using permanent mould casting and elementally pure materials Mg (99.99 %), Gd (99.95 %) and Y (99.95 %). After melting the pure Mg the melt was kept at 720 °C for 15 minutes and stirred continuously after the addition of the alloying elements. During the casting protective atmosphere was used containing (2 wt.% SF_6 in Ar). The melt was poured into a steel mould (d = 12 mm, h = 200 mm) preheated to 350 °C. The compositions of the alloys used in this study were analysed with X-ray fluoresce spectroscopy.

Specimens for microstructural evaluation were ground with SiC paper followed by polishing with 1 µm diamond suspension and OP-S suspension. The microstructures of the alloys in the as-cast condition were examined with a Mira I Tescan scanning electron microscope (SEM) equipped with a Bruker energy dispersive X-ray spectrometer (EDXS) operating at 20 kV in the backscattered electron (BSE) mode. The microstructures of the as-solidified specimens were examined with light optical microscopy (LOM) using a Leica DMI500 microscope, and SEM was performed with a VEGA3 Tescan microscope equipped with an EDAX EDXS operating at 15 kV in BSE mode.

For TEM investigation of the as-cast alloys, slices with a thickness of 0.5 mm, were cut with a slow speed saw. Samples with a diameter of 3 mm were cut with a spark erosion machine. The samples were ground to a thickness of~150 µm before further thinning with electrolytic polishing. Samples were electro polished using a twin-jet e TENUPOL 5 electro polisher in 20 % $HClO_4$ in ethanol at 20V at -36 °C to -40 °C and thoroughly cleaned in ethanol.

For solidification experiments, small samples were cut from the as-cast material in order to fit perfectly in the graphite crucibles designed for the synchrotron diffraction experiments (d = 3 mm, h = 5 mm). The measurements were conducted at the Petra III P07 High Energy Materials Science (HEMS) Beamline of Helmholtz Zentrum Geesthacht (HZG) at the Deutsches Elektronen-Synchroton (DESY). The measurements were performed in the chamber of a DIL 805 A/D dilatometer in Ar flow. The dilatometer has been modified for *in situ* synchrotron radiation measurements, there are two windows on the sides covered by Kapton foil, which is transparent for the X-ray beam, and the induction coil open in the middle so the beam passes through only the sample and the crucible **[24]**. During the solidification experiements the graphite crucible was closed with stainless steel caps to hold the molten metal and isolate it from the surrounding atmosphere. Type S thermocouples were welded onto the cap. The samples were heated to 750 °C and held for 5 min, to ensure melt homogeneity then cooled to 200 °C with controlled cooling rates of 20 and 100 K/min. The *in situ* diffraction experiments were conducted in transmission geometry using a beam with a cross section of 1×1 mm^2. The beam was set to the photon energy of 100 keV, corresponding to a wavelength of λ = 0.0124 nm. During the experiments the 2D diffraction patterns were recorded with an acquisition time of 3 or 0.1 s, which corresponds to a 1 K temperature resolution with a CR of 20 and 100 K/min. respectively, using a Perkin Elemer XRD 1621 flat panel detector with a pixel size of 200x200 μm^2 **[25]**. The 2048 × 2048 pixel diffraction patterns were integrated to azimuthal line profiles by using the fit2d software **[26]**.

The information on the intermetallic phases was obtained from the Pearson's Crystal Structure Database **[27]**, the d-spacings and 2θ angles for the phases were calculated using CaRIn Crystallography 3.1^{TM} software.

The transformation temperatures during solidification were correlated with Differential Thermal Analysis (DTA) using a Mettler DTA 851 at a cooling rate of 20 K/min, and the experiments were performed under an Ar atmosphere.

The experimentally recorded temperatures and determined phases were compared with the solidification simulations (non-equilibrium Scheil-model) performed with Pandat 2014^{TM} thermodynamic software with PanMg2014 database.

Results and discussion

The actual chemical compositions of the alloys determined by X-ray fluorescence are presented in Table 1, below.

Table 1: Chemical compositions of the investigated alloys.

Composition	Y [wt.%]	Gd [wt.%]	Y [at.%]	Gd [at.%]
Mg15Y	14,10	-	4,30	-
Mg10Y5Gd	10,30	5,40	3,20	0,95
Mg7.5Y7.5Gd	6,90	7,20	2,10	1,25
Mg5Y10Gd	5,70	10,90	1,80	1,95

<u>As-cast microstructure</u>

The SEM-BSE micrographs of the microstructures of the alloys in as-cast condition are shown in Figure 1. The microstructures show similar volume fractions of intermetallic phases and segregation of solute elements near the dendritic boundaries.

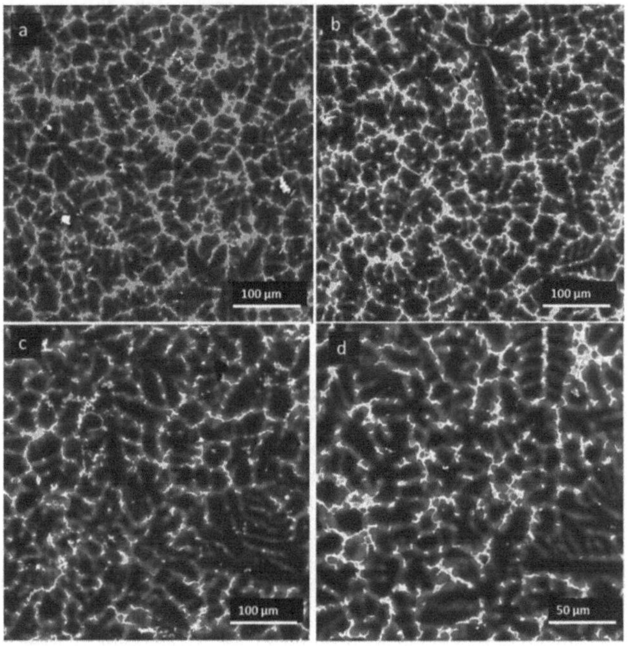

Figure 1: SEM-BSE micrographs of as-cast: (a) Mg15Y, (b) Mg10Y5Gd, (c) Mg7.5Y7.5Gd and (d) Mg5Y10Gd alloys.

The alloys consist of intermetallic phases with high Y and Gd content and are likely to be $Mg_{24}(Y,Gd)_5$. In the ternary alloys, despite of the high solubility of the Gd, the ratio of the segregated areas are comparable. In order to determine the chemical composition of the intermetallic phases, EDXS point analysis was obtained from each of these samples. The results of the point analysis are shown in Table 2. The examination of eutectic structure does not show heterogeneities in the distribution of Gd and Y suggesting that there is only one intermetallic phase found in the microstructure.

Table 2: EDX analysis of Gd and Y (in wt.% and at.%) from the eutectic structure of the alloys.

Composition	Concentration in the eutectic structure			
	Y [wt.%]	Gd [wt.%]	Y [at.%]	Gd [at.%]
Mg15Y	28,05	-	9,63	-
Mg10Y5Gd	24,12	9,88	8,90	2,06
Mg7.5Y7.5Gd	19,01	20,28	7,53	4,54
Mg5Y10Gd	14,96	23,08	5,87	5,12

The X-ray line profiles, from the synchrotron radiation investigation prior to melting of the samples, show the alloys in the as-cast condition (not included here). These show the presence of only $Mg_{24}Y_5$ phase in the microstructure regardless of Gd concentration.

In order to determine the structures of the intermetallic phases, in the cast alloys, TEM investigations were performed. The TEM results reveal that regardless of the chemical composition of the ternary alloys, the eutectic structure contains only one intermetallic phase. This phase can be indexed according to the $Mg_{24}Y_5$ phase with some solubility of Gd in the intermetallic phase as observed by SEM-EDXS point analysis

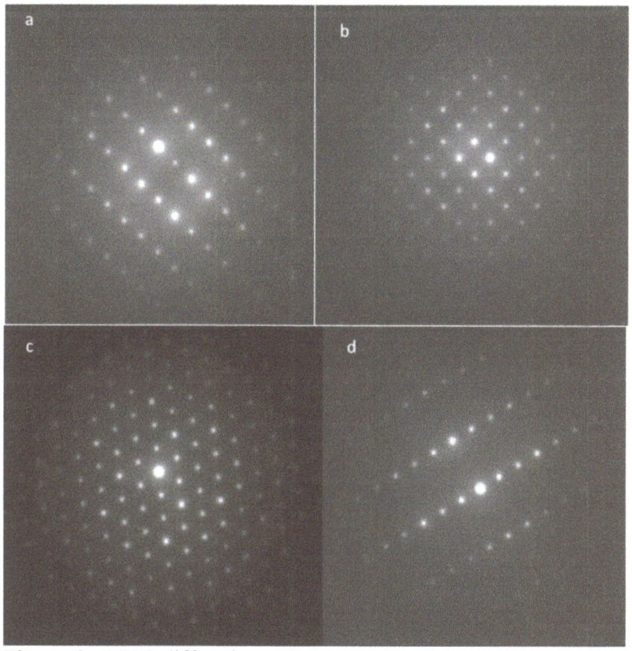

Figure 2: TEM diffraction patterns of the $Mg_{24}Y_5$ intermetallic phase in Mg15Y <113> a, <001> b and Mg5Y10Gd <111> c and <112> d.

In situ synchrotron radiation diffraction

The azimuthally integrated line profiles, of the 2D diffraction patterns acquired during the solidification experiments, are shown in Figures 3 and 4. Figure 3 shows the binary $Mg_{15}Y$ alloy while Figure 4 shows the $Mg_5Y_{10}Gd$ alloy. In terms of the type of intermetallic phases observed the alloys are similar, with the exception of Y_2O_3 and F_3Y phases. When the samples are molten, a diffuse background originating from the melt and diffraction peaks of the graphite crucible can be observed (red lines). At the liquidus temperature α-Mg starts to form from the liquid. With the decrease in temperature, the Mg peaks become more pronounced and Mg is the only phase that forms from the melt before the eutectic temperature (blue lines). After the eutectic temperature formation of only one intermetallic phase is observed. This phase is the $Mg_{24}Y_5$. At this temperature the system is fully solidified and there is no further phase transformation observed in the alloys (black lines).

As these solidified phases (e.g. α-Mg) float in the melt without fixed orientation their detection is difficult resulting in an uncertain temperature for the existence of the phase detected. Most likely the α-Mg can be detected at the first time as the system reaches the dendritic coherency point, the temperature where a continuous dendritic skeleton is built up. It can result in 5 °C error. Despite these limitations, the onset temperatures, obtained experimentally with *in situ* diffraction and this method provide a unique tool to understand the microstructural evolution. Furthermore the presence of extra phases (F_3Y, Y_3O_2) was proven experimentally. Although the thermodynamic calculations predict the formation of Mg_5Gd phase at eutectic temperature in all ternary alloys, it could not be observed experimentally.

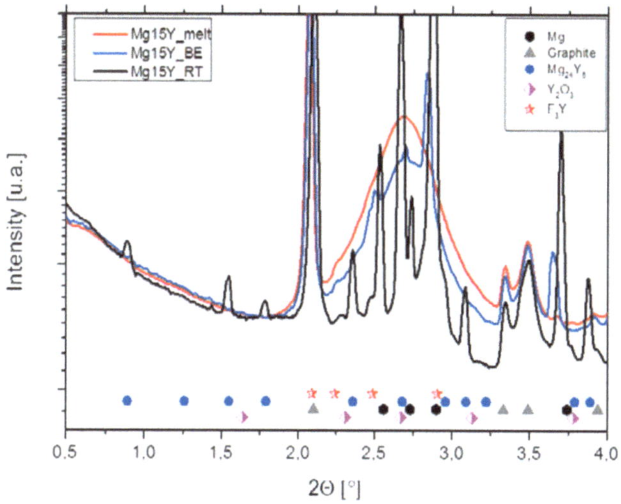

Figure 3: XRD line profiles obtained by the integration of the diffraction pattern recorded during solidification of $Mg_{15}Y$ at 750°C (melt), 5 °C above eutectic temperature (BE) and at room temperature (RT).

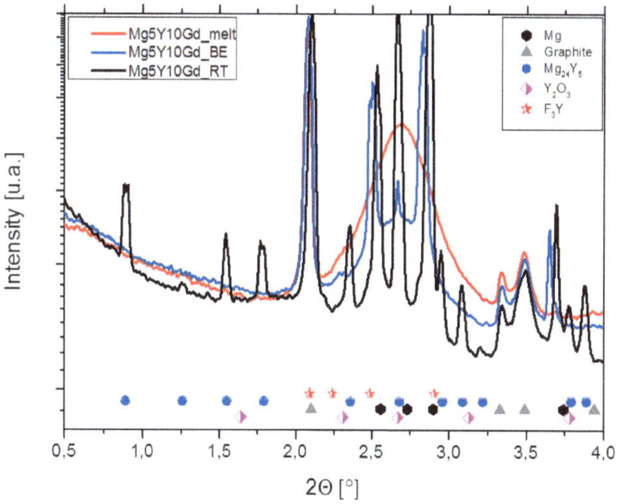

Figure 4: XRD line diagram obtained by the integration of the diffraction pattern recorded during solidification of $Mg_5Y_{10}Gd$ at 750°C, 5°C above eutectic temperature (BE) and room temperature (RT).

As solidified microstructures

The microstructure of the as-solidified samples contains particles with higher concentration of Y, O and F (EDX point analysis). These particles can be distinguished from other particles in the microstructure as shown for the Mg7.5Y7.5Gd alloy in Figure 5. The EDX point analysis results of the particles in the microstructure are shown in Table 3. The morphology of the intermetallic particles in the as-solidified microstructure is similar regardless of the cooling rate. The amount and distribution of intermetallic particles vary with the chemical composition. The only source of F is from the SF_6 gas and Y forming YF_3. The formation of Y_2O_3 phase during the solidification of the Y containing alloys has been shown previously due to the reaction with O during casting **[28]**. These phases can be observed both on SEM micrographs and X-ray diffraction line profiles.

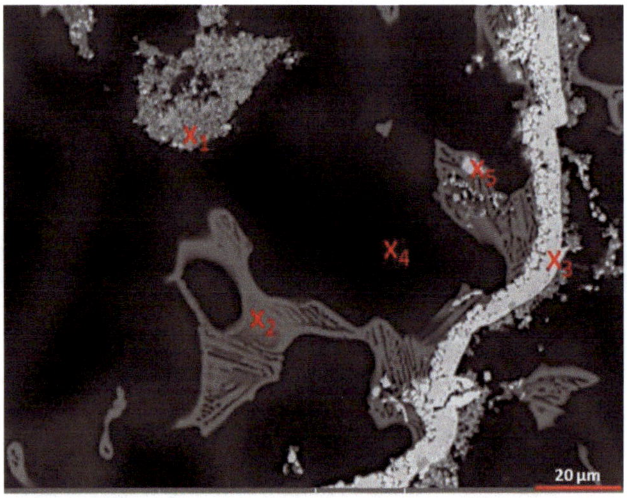

Figure 5: SEM-BSE micrograph of the Mg7.5Y7.5Gd.

Table3: EDX point analysis results of the marked points of the Figure 3 SEM micrograph.

Point	Concentration [wt.%]					Concentration [at.%[				
	Mg	Y	Gd	F	O	Mg	Y	Gd	F	O
1	40,5	52,9	0,4	-	6,2	62,8	22,4	0,1	-	14,7
2	69,9	29,7	0,5	-	-	89,5	10,4	0,2	-	-
3	2,0	80,3	0,2	6,5	11,0	4,1	44,8	0,1	17,0	34,0
4	92,4	7,3	0,2	-	-	97,8	2,1	0,1	-	-
5	14,2	65,6	0,3	10,0	9,9	23,6	29,9	0,1	21,4	25,1

The experimental results from both the DTA and the synchrotron radiation experiments were compared with the thermodynamic predictions from the Pandat thermodynamic databases. The results are presented in Table 4a (liquidus temperatures) and Table 4b (eutectic temperatures).

In the case of the ternary alloys, the thermodynamic calculations predict the formation of $Mg_{24}Y_5$ phase 3-4 °C above the eutectic temperature and Mg_5Gd phase formation at the eutectic temperature. The presented experimental results don't support these predictions. Although it is possible that the volume fraction of the predicted Mg_5Gd phase is so low that neither of the techniques (DTA, Synchrotron radiation, SEM, TEM) is able to detect the presence/formation of the phase but in this case the predicted volume fractions are much higher than the actual ones.

Table 4a and b: The experimentally measured and theoretically calculated (a) liquidus temperatures and (b) eutectic temperatures of the alloys investigated.

Composition	Liquidus temperature [°C]			
	Experimental [CR]			Pandat
	diff. 100	diff. 20	DTA 20	
Mg15Y	617	605	613	614
Mg10Y5Gd	611	622	614	615
Mg7.5Y7.5Gd	614	618	624	620
Mg5Y10Gd	611	605	616	615
Composition	Eutectic temperature [°C]			
	Experimental [CR]			Pandat
	diff. 100	diff. 20	DTA 20	
Mg15Y	564	553	562	574
Mg10Y5Gd	550	561	556	545
Mg7.5Y7.5Gd	547	549	553	545
Mg5Y10Gd	546	543	551	545

Summary of findings

The microstructure characterisation of as-cast condition MgxYyGd (x+y= 15 wt.%) alloys were performed with SEM and TEM. Solidification experiments with synchrotron diffraction and DTA were utilized to clarify the solidification paths of the alloys investigated. The results from the solidification experiments were correlated with thermodynamic calculations. From the results the following conclusions can be drawn.

- During cooling from 750 °C to 200 °C the solidification sequence could be determined experimentally.
- The change of the cooling rate from 20 to 100 K/min. does not change the evolution of the phase formation in the examined alloys.
- α-Mg solidifies first in each of the alloys at the start of the solidification, The temperatures predicted with Pandat 2014^{TM} software show a similar trend in the formation temperatures determined experimentally, albeit, not the same temperatures.
- After the nucleation of the α-Mg, only one intermetallic phase formation was observed and this occurs at the eutectic temperature. This intermetalic phase is the $Mg_{24}Y_5$.
- The presence of $Mg_{24}Y_5$ phase in the microstructure was confirmed with TEM from the as-cast condition in each alloy case. Since the EDXS measurements show Gd enrichment in the eutectic structure in the ternary alloys, thus it indicates that the Gd can be exchanged for Y in the $Mg_{24}Y_5$ phase.
- The thermodynamic calculations predict the formation of Mg_5Gd phase in each of the ternary alloys, however this is not observed in the analyses of the as-cast or as-solidified microstructures.
- The solidification studies show the presence of Y_2O_3 and YF_3 particles during the experiment. These particles probably form during the casting and remaining in the microstructure.

Acknowledgement

The authors acknowledge the Deutsches Elektronen-Synchrotron for the provision of synchrotron radiation facilities in the framework of the proposal I-20130434. G. Sz. acknowledges the European Union FP7-PEOPLE-ITN-2011 program under REA

Grant Agreement No 289163: MagnIM project for the financial support.

References

[1] L.L. Rokhlin, Magnesium Alloys Containing Rare Earth Metals, Structure and Properties, Taylor & Francis, (2003).
[2] N. Hort, Y. Huang, D. Fechner, M. Störmer, C. Blawert, F. Witte, C. Vogt, H. Drücker, R. Willumeit, K.U. Kainer, F. Feyerabend; Acta Biomaterialia 6 (2010) 1714-1725.
[3] F. Witte, N. Hort, C. Vogt, S. Cohen, R. Willumeit, K.U. Kainer, F. Feyerabend; Current Opinion in Solid State and Materials Science 12 (2008) 63-72.
[4] F. Witte, Acta Biomaterialia 6(2010) 1680-1692.
[5] G. Barucca, R. Ferragut, F. Fiori, D. Lussana, P. Mengucci, F. Moria, G. Riotino, Acta Materialia 59 (2011) 4151-4158.
[6] M. Giovannini, A. Saccone, R. Marazza, R. Ferro, Metallurgical and Materials Transactions A 26 (1995) 5-10.
[7] J. Gröbner, R. Shmid-Fetzer, Scripta Materialia 63 (2010) 674-679.
[8] Z. Du, C. Guo, C. Li, W. Zhang, Rare Metals 25 (2006) 492-500.
[9]C. Guo, Z. Du, C. Li, International Journal of Materials Research 99 (2008) 650-668.
[10] C. Guo, Z. Du, C. Li, Calphad: Computer Coupling of Phase Diagrams and Thermochemistry 32 (2008) 177-18.
[11] F. Feyerabend, J. Fischer, J. Holtz, F. Witte, R. Willumeit, H. Drücker, C. Vogt, N. Hort, Acta Biomaterialia 6 (2010) 1834-1842.
[12] T.B. Massalski, Binary Alloy Phase Diagrams, ASM International Materials Park (1990).
[13] M. Giovannini, A. Saccone, H. Flandorfer, P. Rogl, R. Ferro, Zeitschrift fuer Metallkunde/Materials Research and Advanced Techniques 88 (1997) 372-378.
[14] C. Guo, Z. Du, C. Li, Computer Coupling of phase Diagrams and Thermochemistry 31 (2007) 75-88.
[15] W.J. Boettinger, J.A. Warren, C. Beckermann, A. Karma, Annual Review of Materials Research 32 (2002) 164-194.
[16] Y.L. Liu, S.B. Kang; Journal of Materials Science 32 (1997) 1443-1447.
[17] S.S. Babu, J.W. Elmer, J.M. Vitek, S.A. David, Acta Materialia 50 (2002) 4763-4781.
[18] O. Shuleshova, D.H. Moritz, A. Voss, W. Löser, Intermetallics 19 (2011) 688-692
[19] D.Tolnai, G. Szakács, G. Requena, A. Stark, N. Schell, K.U. Kainer, N. Hort, Materials Science Forum 765 (2013) 286-290.
[20] D. Tolnai, C.L. Mendis, A. Stark, G. Szakács, B. Wiese, K.U. Kainer, N. Hort, Magnesium Technology 2013, 253-257.
[21] D. Tolnai, C.L. Mendis, A. Stark, G. Szakács, B. Wiese, K.U. Kainer, N. Hort, Materials Letters 102-103 (2013) 62-64.
[22] D. Tolnai, P. Townsend, G. Requena, L. Salvo, J. Lendvai, H.P. Degischer, Acta Materialia Volume 60 (2012) 2568-2577.
[23] G. Szakacs, C.L. Mendis, D. Tolnai, A. Stark, N. Schell, H. Ovri, M. Wolff, K.U. Kainer, J. Gröbner, R. Schmid-Fetzer, N. Hort, Magnesium Technology 2015, 79-84
[24] P. Staron, T. Fischer, T. Lippmann, A. Stark, S. Daneshpour, D. Schnubel, E.Uhlmann, R. Gerstenberger, B. Camin, W. Reimers, E. Eidenberger, H. Clemens, N. Huber, A. Schreyer, Advanced Engineering Materials 13 (2011) 658-663.
[25] http://www.perkinelmer.com/pdfs/downloads/009299B_01%20PRD.pdf.
[26] http://www.esrf.eu/computing/scientific/FIT2D/index.html.
[27] Pearson's Crystal Structure Database for Inorganic Compounds (on CD-ROM), Release 2014, Ed.: Pierre Villars and Karin Cenzual.
[28] G. Szakacs, B. Wiese, C.L. Mendis, D. Tolnai, A. Stark, N. Schell, M. Nair, K.U. Kainer, N. Hort, Magnesium Technology 2014, 213-218.

Magnesium Technology 2016
Edited by: Alok Singh, Kiran Solanki, Michele V. Manuel, and Neale R. Neelameggham
TMS (The Minerals, Metals & Materials Society), 2016

DEVELOPMENT OF THE NEW HIGH SHEAR TECHNOLOGY FOR CONTINUOUS PROCESSING OF Mg-ALLOYS FOR INGOT CASTING

Jayesh B Patel[1], Peter Lloyd[1], Guosheng Peng[1] and Zhongyun Fan[1]

[1]BCAST, Brunel University London, Kingston Lane, Uxbridge, UB8 3PH, United Kingdom

Keywords: Melt shearing, Grain refinement, Molten metal pump, Ingot casting

Abstract

Melt quality is crucial for both continuous and shape casting of Magnesium alloys. The recently developed melt conditioning technology where intensive shearing is applied to the melt either by twin-screw or rotor/stator type processing, in order to achieve uniform temperature and chemical composition, together with dispersion and distribution of the harmful inclusions such as oxides that already exist in the melt. Melt conditioning by application of intensive melt shearing of Mg-alloys has shown that MgO films/skins that already exist in the melt can be dispersed into individual particles which can act as potent nucleation sites for the α-Mg phase. Due to this enhanced nucleation during solidification, significant grain refinement is achieved in Mg-alloys including Mg-Al alloys without any specific chemical inoculation. Currently, the melt conditioning technology is limited to batch processing only, hence the need to further develop it into a continuous high shear processing technology which can be used in conjunction with various continuous casting processes such as ingot casting. Here, we report on the development of the new continuous type high shear device with grain refined microstructures of ingots cast after continuous shearing. The preservation of the grain refinement in the melt conditioned ingots after re-melting and casting again is also reported and discussed.

Introduction

Alloys of light metals such as Magnesium are widely used in various industrial sectors mainly automotive for the purpose of weight saving [1]. Melt quality is crucial for both continuous and shape casting of these light alloys. Gas, oxides and other inclusions in the melt usually deteriorate the quality of the casting products. Liquid metal treatment prior to solidification processing is crucial for ensuring the high quality of the cast products regardless of which casting process is used. The existing methods for liquid metal treatment mainly include melt filtering, mechanical stirring by an impeller, electromagnetic stirring, and rotary degassing. However, such processes are time consuming, energy intensive and high cost. Melt conditioning by intensive melt shearing in a twin-screw machine has been studied and used for treating both liquid and semi-solid light alloys [2-5]. Improved microstructure and mechanical properties of both wrought and cast light alloys were achieved by the application of this technology. Based on the similar principle, a new technology [6] but with simpler equipment was developed for treating liquid metals using a high shear unit. The new technology uses a simple rotor-stator unit to provide intensive melt shearing, which effectively disperses the harmful inclusions into fine particles to enhance nucleation during the subsequent solidification processing. Here we report on the development of the high shear technology for continuous processing and its application to ingot casting.

The High Shear Technology

Intensive melt shearing for melt treatment can be achieved in a rotor-stator unit [6]. A schematic of the rotor/stator high shear unit is shown in Fig. 1. This unit comprises of a set of rotor and stator attached to an electrical motor with a speed control. During its operation, the motor passes the power to the rotor by the shaft and drives the rotor to rotate and shear the liquid metal in the gap between the rotor and the stator and also in the openings of the stator. The rotation speed can be in the order of 5000-15000 rpm providing a shear rate of up to $10^5 s^{-1}$. The new high shear device provides macro-flow in a volume of melt for distributive mixing and, intensive shearing near the tip of the device for dispersive mixing. The main advantages of the high shear device include; significantly enhanced kinetics for chemical reactions or phase transformations, uniform dispersion, distribution and size reduction of solid particles & gas bubbles, improved homogenisation of chemical composition and temperature fields and also forced wetting of usually difficult-to-wet solid particles in the liquid metal. Therefore, the high shear device can be used for; physical grain refinement by dispersing naturally occurring oxides, for degassing of melts, for the preparation of metal matrix composites and also for preparation of semi-solid slurries. Hence, the newly developed high shear technology can be applied to benefit various industrial casting processes such as ingot casting, sand casting, low & high pressure die casting, twin-roll casting and direct chill casting [7-9].

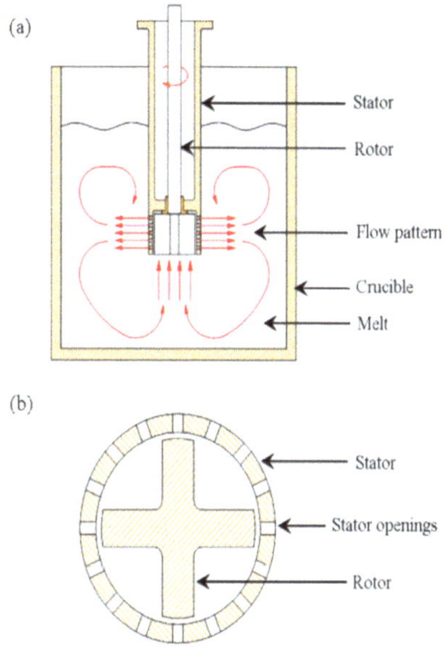

Fig. 1. The new rotor-stator high shear device for conditioning of liquid melts by application of intensive melt shearing, (a) inside a crucible of melt illustrating the macroscopic flow pattern and (b) bottom view of the high shear device.

Grain Refinement by Intensive Melt Shearing

In order to demonstrate the effect of intensive melt shearing on the grain refinement of Mg-alloys, AZ91D alloy was sheared at two different temperatures, 650°C and 605°C, and then cast in a TP-1 standard mould [10]. The TP-1 mould was prepared under standard conditions for grain size observation with a constant cooling rate of ~3.5K/s. Reference samples without shearing were also cast at the same temperatures, respectively. The resulting microstructures are shown in Fig. 2. It is clearly observed, that with intensive melt shearing, the microstructure becomes more uniform and the grain size is significantly reduced at both casting temperatures. When measured, the as-cast grain size was reduced from 550μm to 170μm when cast at 650°C, and from 180μm to 90μm when cast at 605°C. It is interesting to note that the grain size of Mg-alloys has a strong dependence on melt superheat and that such superheat effect is suppressed significantly by intensive melt shearing.

The New Multi-Stage High Shear Pump

Currently the rotor/stator type high shear device is only applicable to batch processing, since it can only apply intensive shearing to the volume of melt inside a particular crucible. Pumping of molten metal is sometimes necessary in order to quickly transfer melt from a melting furnace to a holding furnace with minimal exposure to air and to avoid unnecessary turbulence due to pick-up of inclusions and also to minimize oxidation. The rotor/stator type high shear device can be utilized as a pump for transfer of melt to a holding furnace or directly to a casting facility such as the 'shot sleeve' in case of high pressure die-casting process or directly to the hot-top in case of the direct chill casting process.

A schematic illustration of the multi-stage high shear pump is shown in Fig. 3. The high shear device not only pumps the molten metal but also applies intensive melt shearing at the same time in order to condition the melt. In the multi-stage high shear unit, when the rotor is running at high speed, a volume of melt is sucked upwards into the first stator chamber, then forced out through the stator holes, then sucked back into the second stator chamber and so on, until it is discharged completely in a fully sheared state out of the high shear pump. The intensively sheared melt can be pumped out of the furnace to feed an insulated laundering system or simply collected and poured for casting. The discharge rate or flow rate can be controlled by the speed of the rotor. There will be a minimum speed required for the sucking and pumping, after which, increasing the speed would directly increase the flow rate. In comparison with the single stage high shear device, Fig. 1, the multi-stage high shear pump ensures that only intensively sheared melt is discharged. In the case of a single stage high shear mixer placed in a large crucible and volume of melt, there is a chance of some melt not been fully sheared. This multi-stage high shear pump can also be allowed to discharge back into the furnace in case of large quantities of melt to ensure that all the melt in the crucible goes through the shearing stages at some point or multiple times. The discharge pipe can be submerged below the surface of the melt so as to avoid unnecessary turbulence on the melt surface to minimise the creation of new oxides films.

A 10kg melt of AZ91D alloy was sheared continuously at 660°C using the new multi stage high shear pump where the melt was let to discharge back into the furnace. TP-1 samples were taken at various intervals during continuous running which were cast at the same temperature of 660°C. The resulting microstructures from the TP-1 samples are shown in Fig. 4. Compared with the microstructure of the sample taken before starting the pump, the sample taken after 5 minutes of pumping already showed some degree of grain refinement. Grain refinement was significantly improved with pumping time, until after 20 minutes of continuous shearing, the entire volume of melt was fully grain refined.

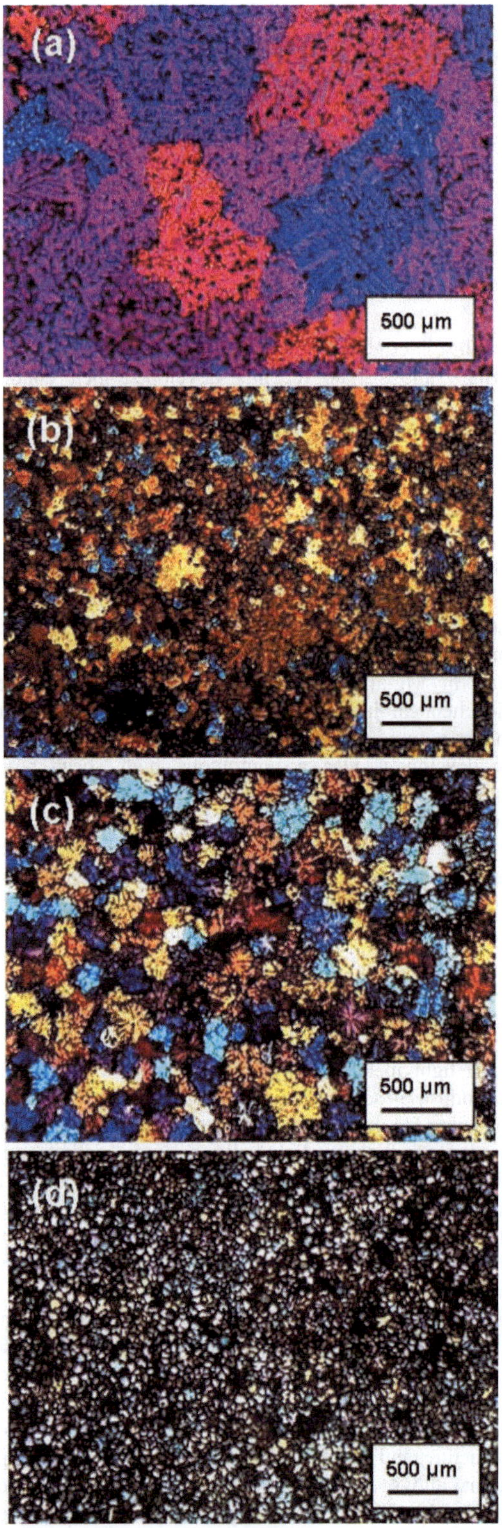

Fig. 2. Microstructures of AZ91D magnesium alloy cast (a) at 650°C without shearing, (b) at 650°C after shearing, (c) at 605°C without shearing and, (d) at 605°C after shearing.

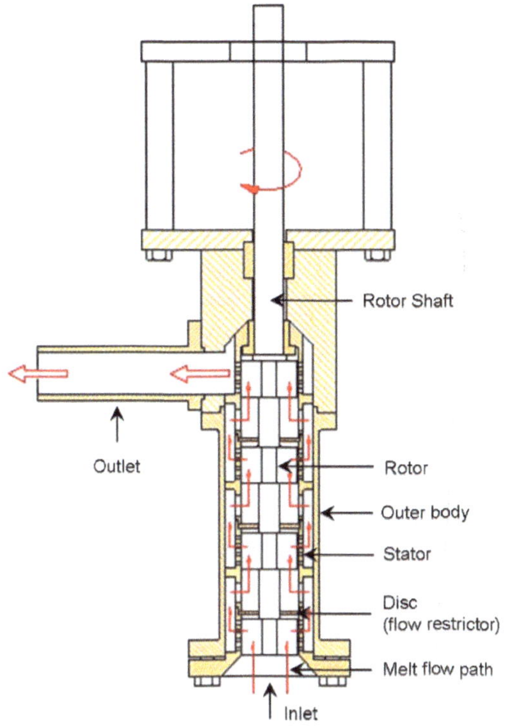

Fig. 3. Schematic illustration of the multi-stage high shear pump for supply of fully intensively sheared melt.

Ingot Casting and Re-melting

It has been shown that the grain structure of an as-cast sample after intense shearing is highly refined. A 10kg melt of AZ31 alloy was fully sheared and then was cast in ingot form at 650°using steel moulds of 300mm length by 100mm width by 60mm depth. The ingots were allowed to solidify and cool down naturally to room temperature. One of the as-cast ingots was then sectioned to examine the grain structure. The macro-etched section of the ingot is shown in Fig. 5. The solidified ingot exhibited a fine and uniform grain structure throughout the thickness from top to bottom and edge to edge.

An investigation was carried out to see if the grain refinement was preserved upon re-melting and re-casting the fully sheared ingot. The AZ31 sheared ingot was re-melted carefully avoiding any unnecessary oxidation during re-heating to ~660°C and thereafter a TP-1 sample was cast at 650°C. The remaining melt was again cast into an ingot. This same ingot cooled down naturally to room temperature and was then remelted again with the same procedure and condition for taking a TP-1 sample. In total, the same ingot was re-melted and re-cast 3 times, respectively. Reference TP-1 samples from non-sheared AZ31 alloy ingot were also cast with the same procedure and at same re-melting and re-casting temperatures. The microstructures from all the TP-1 samples taken after shearing and re-melting, including the reference samples, are shown in Fig. 6. The microstructure after shearing is highly refined compared to the corresponding reference sample. More importantly, the microstructure appears to be still refined after re-melting and re-casting, not only once but up to 3 times. In the case of non-sheared samples, the grain size was measured to be 410μm (±33), 402μm (±49) after 1st re-melt, 369μm (±36) after 2nd re-melt and 299μm (±20) after 3rd re-melt.

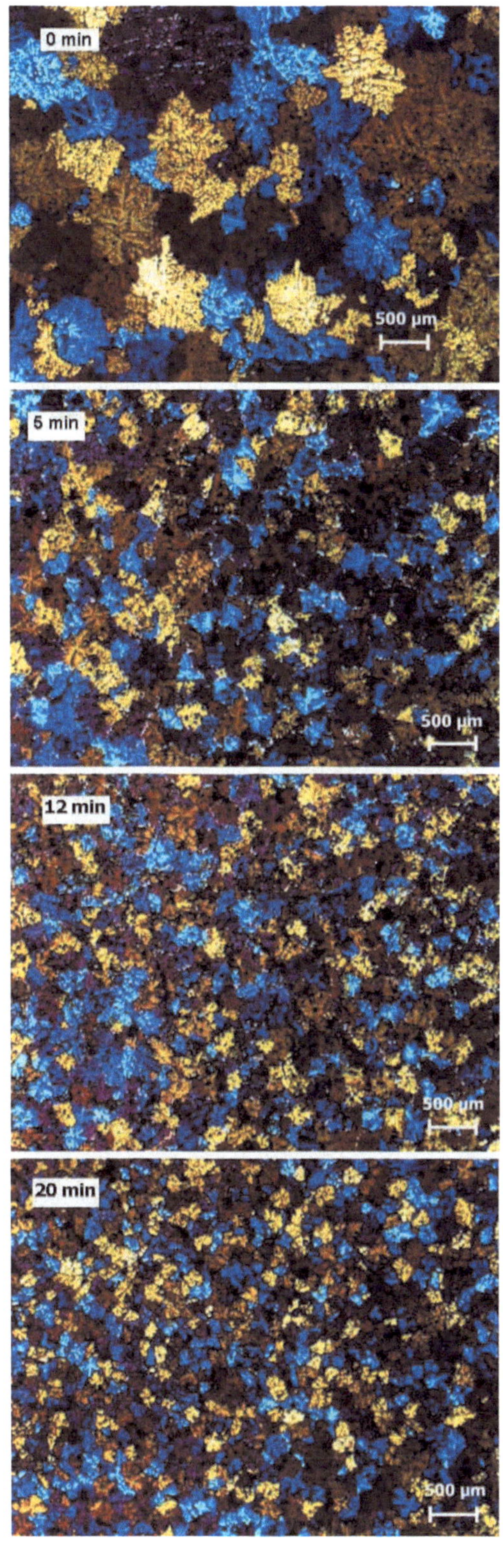

Fig. 4. A 10kg melt of AZ91D Magnesium alloy was conditioned using the high shear pump at 660°C, and TP-1 samples were cast after; 0mins, 5mins, 12mins and 20mins.

Upon re-melting the non-sheared ingot, the grain size gradually decreased from 410μm to 299μm. In the case of sheared samples, the grain size was measured to be 189μm (±9) after shearing,

225μm (±9) after 1st re-melt, 225μm (±6) after 2nd re-melt and 244μm (±7) after 3rd re-melt. Upon re-melting of the sheared ingot, the grain size was more stable from 189μm to 225μm until the 2nd re-melt, but slightly increased after the 3rd re-melt to 244μm. So, for the sheared ingot, it means that, once the oxide particles/films are dispersed by intensive shearing, they can stay dispersed even upon re-melting of the ingot, without the need for applying intensive shearing again. This in-turn means that the new multistage high shear technology can be used for ingot casting in order to supply various conventional casting processes. This also benefits the high shear technology, in a way that there is no need to integrate the technology with the final casting processing machines, but just have it integrated at the melting or holding furnace. So therefore, the new multi-stage high shear pump can be utilized for conditioning of melts directly inside a furnace.

Fig. 5. AZ31 Mg-alloy ingot cast after application of intensive shearing, showing fine and uniform grain structure upon final solidification.

Since the intensive shearing can disperse oxide films into individual particles, this can also benefit recycling of scrap ingots as well. It would be still possible to achieve grain refinement even if the fully sheared ingot is mixed with non-sheared ingot, as long as there are enough dispersed oxide particles to facilitate grain refinement. In order to investigate this, an experiment was performed to show that, grain refinement is still achieved by mixing sheared melt with non-sheared melt of AZ91D alloy. TP-1 samples were taken at 650°C after mixing the two melts in different quantities. The samples were examined for their microstructure and grain size measurements were taken. The plot of grain size corresponding to the mixture of sheared melt and non-sheared melt is given in Fig. 7. It is observed that, the sheared melt controls the grain refinement for up to a 50% mixture of non-sheared melt, after which the grain size starts to become coarser. Hence, this could make in-house scrap recycling more efficient, by the need to apply intensive shearing only to the scrap batch of melt instead of having to process a batch of fresh primary melt.

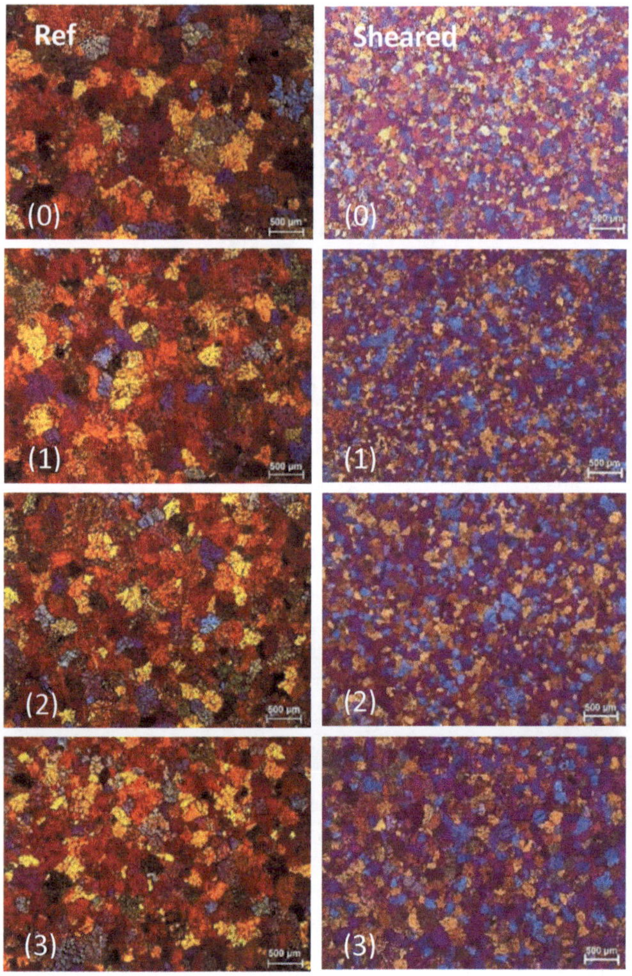

Fig. 6. Microstructures of reference (non-sheared) compared with sheared AZ31 Mg-alloy ingots: (0) as-cast, (1) 1st re-melt, (2) 2nd re-melt, and (3) 3rd re-melt, respectively.

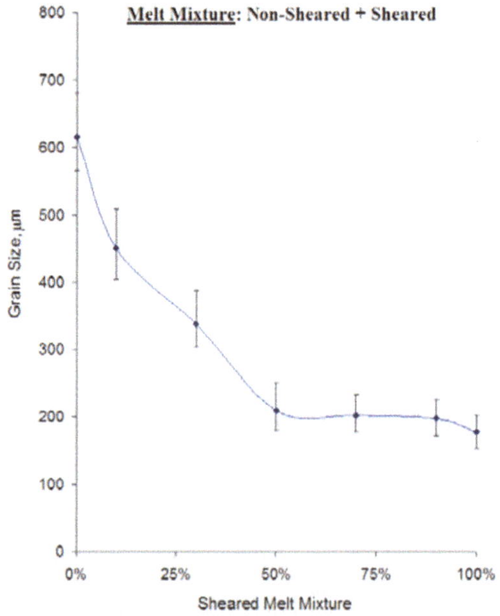

Fig. 7. The plot of grain size corresponding to the mixture of fully sheared melt with non-sheared melt of AZ91D alloy.

Discussion

Liquid melts of Magnesium alloys inevitably contain oxide in forms of clusters or films with a non-uniform distribution. Such oxides may act as substrates for nucleation but are not effective for grain refinement due to their poor wettability and low number density. However, such oxide clusters and films can be dispersed effectively by intensive melt shearing. It has been confirmed that the well dispersed oxide particles can be as effective nucleation sites [5, 11]. The high shear device not only provides intensive melt shearing in the gap between the rotor and the stator and in the openings of the stator, which can disperse oxide clusters or films to individual particles, but also forced convection in the melt, which can distribute particles uniformly in the entire volume of the melt. Intensive melt shearing is also believed to be helpful for improving wettability between the liquid alloy and the dispersed oxide particles. It is therefore concluded that the intensive melt shearing can enhance nucleation and make significant contribution to the formation of fine and uniform microstructures. It has been well demonstrated that intensive melt shearing prior to solidification processing is capable of significant grain refinement of Mg alloys, and this grain refining effect has been attributed to the enhanced heterogeneous nucleation by the dispersed oxide particles. The macroscopic flow pattern that is generated by the rapidly escaping liquid from the openings on the stator provides a distributive mixing action which homogenizes the temperature and chemical composition of the melt. As a consequence, the melt temperature and chemical composition will be uniform as well. In addition, this macroscopic melt flow will be stronger near the tip of the stator and weaker near the surface of the melt. This will prevent the occurrence of gas entrapment from the melt surface. The anticipated increase in particle number density by intensive melt shearing can be as high as 3 orders of magnitude [12]. It is well known that increased nucleation events will lead to grain refinement. With regards to enhanced heterogeneous nucleation by the dispersion of potent oxide particles, the application of intensive melt shearing can facilitate recycling of scrap Mg-alloys. Mg-alloy scrap is usually discarded due to their high content of oxides and other inclusions. Since there are no external chemical additions required to achieve grain refinement by intensive melt shearing, it should be noted that almost any kind of melt can be conditioned in order to produce better quality recycled Mg-alloy castings. In addition, the intensive melt shearing would disperse and make a huge number of potent oxide particles, of which only a small fraction of them will get an opportunity to participate during solidification. Hence, after intensive shearing of a melt which contains an increased amount of oxide clusters and films than usual, there will be surplus amount of well dispersed potent oxide particles left over in the melt upon solidification. It is believed, that these surplus potent oxide particles can get another chance to participate in the nucleation event, when the already solidified casting or ingot is re-melted and cast again. Grain refinement can still be achieved just by re-melting the once sheared ingot without applying intensive melt shearing again. For the same reason, an intensively sheared melt can also be mixed with primary melt that has not been intensively sheared and still achieve grain refinement upon solidification. The multi-stage high shearing technology ensures that a specific quantity of melt is fully sheared, which in turn means the oxides have been fully dispersed, therefore only well dispersed particles exist in the melt. These particles will ofcourse have a size range, of which only a small fraction will participate in nucleation depending on the undercooling generated during the casting process. It is also known that the time needed for particles to come together and form agglomerates or long films is very long due to the existence of substantially low natural force between each particle. Hence, the dispersion of these oxides can be preserved upon re-melting and casting.

Summary

Liquid metal engineering by application of intensive melt shearing can be achieved by the newly developed Rotor/Stator high shear technology to produce high quality cast products. This new Rotor/Stator type high shear device provides distributive and dispersive mixing and exhibits; significantly enhanced kinetics for phase transformations, uniform dispersion, distribution and size reduction of solid particles, improved homogenisation of chemical composition and temperature fields, forced wetting of usually difficult-to-wet solid particles in the liquid metal and most importantly, provides physical grain refinement by dispersing naturally occurring oxide particles. In addition, the new Rotor/Stator high shear concept can also be combined into a multi-stage high shear pump, which is highly applicable for continuous processing and has shown that it can benefit ingot casting in general and facilitate efficient recycling of in-house Mg-alloy scrap. More importantly, the continuously processed sheared ingots can be re-melted and benefit conventional casting processes.

Acknowledgements

Financial support from the EPSRC-UK under the grant for the EPSRC Centre LiME and contributions by Dr. Mingxu Xia and Dr. Yubo Zuo, is gratefully acknowledged.

References

[1] J. Polmear; Light Alloys, Metallurgy of the light metals, 3rd edition, (1995).

[2] Z. Fan, M. Xia, H. Zhang, G. Liu, J.B. Patel, Z. Bian, I. Bayandorian, Y. Wang, H.T. Li, G.M. Scamans ; Int. J. Cast Met. Res., 22, (2009), 103-107.

[3] J.B. Patel, Y.Q. Liu, G. Shao, Z. Fan; Mat. Sci. Eng. A, 476, (2008), 341-349.

[4] S. Tzamtzis, N.S. Barekar, J.B. Patel, N. Hari Babu, Z. Fan, Composites A, 40 (2009), 144–151.

[5] Z. Fan, Y. Wang, M. Xia, S. Arumuganathar; Acta Mater., 57, (2009), 4891-4901.

[6] Z. Fan, Y. B. Zuo, B. Jiang; Apparatus and method for liquid metals treatment, Application No.1015498.7, (2010), UK Patent.

[7] J.B. Patel, A.K. Prasada Rao, B. Jiang, Y.B. Zuo, Z. Fan; 9th Int. Conf. Magnesium Alloys and their Applications, Vancouver, BC, Canada, (2012), 731-736.

[8] J.B. Patel, H.T. Li, M. Xia, S. Jones, S. Kumar, K. O'Reilly, Z. Fan, Mater. Sci. Forum, 794, (2014), 149-154.

[9] H.T. Li, J.B. Patel, H.R. Kotadia, Z. Fan, Mat. Sci. Forum, 828-829, (2015), 43-47.

[10] Standard test procedure for aluminium alloy grain refiners (TP-1), The Aluminum Association, Washington, DC, (1990).

[11] Y. Wang , Z. Fan , X. Zhou, G.E. Thompson, Philosophical Magazine Letters, 91.8, (2011), 516-529.

[12] H. Men, B. Jiang, Z. Fan, Acta Mater., 58, (2010), 6526.

Magnesium Technology 2016
Edited by: Alok Singh, Kiran Solanki, Michele V. Manuel, and Neale R. Neelameggham
TMS (The Minerals, Metals & Materials Society), 2016

DENDRITE MORPHOLOGY AND GROWTH ORIENTATION OF MAGNESIUM ALLOYS: SIMULATION BY PHASE-FIELD AND 3-D CHARACTERIZATION BY SYNCHROTRON X-RAY TOMOGRAPHY

Manhong Yang[1], Shoumei Xiong[1, 2], Zhipeng Guo[1, *]
[1]School of Materials Science and Engineering, Tsinghua University; Beijing, 100084, China
[2]State Key Laboratory of Automobile Safety and Energy, Tsinghua University

Keywords: Magnesium Alloys, Synchrotron X-ray Tomography, Dendritic Growth, Three-dimensional

Abstract

Upon solidification, magnesium alloys with hexagonal close-packed (HCP) structure form dendrite morphologies which are supposed to grow along $\langle 11\bar{2}0\rangle$ directions. However, our previous studies showed that the Mg-Sn, Mg-Gd and Mg-Y dendrites have eighteen primary branches and six of them grow with a crystallographic orientation of $\langle 11\bar{2}0\rangle$ and twelve others lie in the non-basal plane growing along $\langle 11\bar{2}3\rangle$. In this study, based on the analysis of the 3-D dendrite morphology of α-Mg, a predictive anisotropy growth function was proposed and developed to describe the growth of an α-Mg dendrite in the Mg-Zn alloy. The 3-D phase-field simulation was then performed and the 3-D morphology of α-Mg dendrite was characterized by synchrotron X-ray tomography in combination with the determination of the growth orientation of the dendrites *via* electron backscatter diffraction. Most interestingly, from the reconstructed dendrites of Mg–45 wt.% Zn, we found that α-Mg (Zn) dendrites have twelve branches and grow along $\langle 11\bar{2}2\rangle$ orientation in non-basal plane, which indicated great similarity to the simulation result by phase-field.

Introduction

Light metals such as magnesium alloys have an increasing interest in the transportation and electronics industries due to the superior properties such as high strength and good ductility [1]. The performance of magnesium alloy castings is extensively influenced by the microstructure during solidification. An understanding of the microstructure in these alloys is vital, especially the dendrite morphology. When nucleation and growth happen in large undercooled melts, dendritic microstructure may come into being. However, the solidification process in the solid-liquid interface is rather complex, and the evolution of dendritic structures is still not fully understood [2], especially for the magnesium alloy with the hexagonal closed-packed (*hcp*) structure.

In recent years, numerical modelling and simulation has been rapidly developed as a powerful tool to predict the time-dependent microstructure evolution during solidification of magnesium alloys [3]. Böttger *et al* [4] integrated a hexagonal anisotropy function into the phase-field model (PF), and simulated the three-dimensional dendrite growth of magnesium, which had six prism orientations in the {0001} basal plane and two basal orientations in the ⟨0001⟩ directions. Eiken [5] investigated the growth texture evolution of magnesium by PF method. They found that grains grew faster in $\langle 11\bar{2}0\rangle$ orientation and retarded in ⟨0001⟩ orientation. The grain arrangement is plate-like within the basal plane.

In addition to numerical simulation, experimental methods were also applied to study the dendrite morphology of magnesium alloys. Pettersen et al. [6] studied the AZ91 magnesium alloy under directional solidification, and suggested that the stem grew along $\langle 11\bar{2}0\rangle$ with six secondary arms or $\langle 22\bar{4}5\rangle$ with three secondary arms around, which depends on the ratio of temperature gradient and growth velocity. Wang et al. [7] found that the three-dimensional dendrite morphology of Mg–40 wt.% Zn alloy grew with six secondary arms around the stem which was similar to the $\langle 11\bar{2}0\rangle$-type stems in reference [6]. However, based on our previous studies [8], the Mg-Sn, Mg-Gd and Mg-Y dendrites had eighteen primary branches with six growing along a crystallographic orientation of $\langle 11\bar{2}0\rangle$ and twelve others lying in the non-basal plane along $\langle 11\bar{2}3\rangle$. It is worth stressing that $\langle 11\bar{2}3\rangle$ direction is quite close to $\langle 22\bar{4}5\rangle$ as reported by Pettersen and co-workers [6]. Furthermore, the different additional element has an effect on the angle between the $\langle 11\bar{2}0\rangle$ and $\langle 11\bar{2}3\rangle$.

For binary Al-Zn alloys, a phenomenon termed dendrite orientation transition (DOT) could take place [9-11]. An increase in the zinc content in Al-Zn alloys continuously changes the dendrite growth direction from ⟨100⟩ to ⟨110⟩ in {001} planes. Zinc, an *hcp* element, exhibits the anisotropy of about 30 % between the c-axis and the basal plane [12], which can perturb the weak anisotropy of aluminum. However, it is unclear whether the DOT phenomenon occurs in magnesium-based alloys.

This current study is thus aimed at understanding better the dendrite morphologies and growth orientation of magnesium-based alloy. Combining the literature on Al-Zn alloys [9, 10] and our previous studies [8], a predictive anisotropy growth function was developed. The 3-D phase field model was performed to simulate the dendrite growth and then validated by synchrotron X-ray tomography experimental results of Mg–45 wt.% Zn.

Modelling and Experimental Methods

Phase Field Model

The isothermal phase field model developed by Echebarria et al. [13] was adopted here, and the governing equations are:

$$\tau\frac{\partial\phi}{\partial t}=\nabla\cdot\left(W(\vec{n})^2\nabla\phi\right)+\frac{\partial}{\partial x}\left(\left|\nabla\phi\right|^2 W(\vec{n})\frac{\partial W(\vec{n})}{\partial\phi_x}\right)$$
$$+\frac{\partial}{\partial y}\left(\left|\nabla\phi\right|^2 W(\vec{n})\frac{\partial W(\vec{n})}{\partial\phi_y}\right)+\frac{\partial}{\partial z}\left(\left|\nabla\phi\right|^2 W(\vec{n})\frac{\partial W(\vec{n})}{\partial\phi_z}\right)$$
$$+\phi\left(1-\phi^2\right)-\lambda\left(1-\phi^2\right)^2\left(\theta+kU\right)$$
(1)

$$\left(\frac{1+k}{2}-\frac{1-k}{2}\phi\right)\frac{\partial U}{\partial t}=\nabla\cdot\left(D\frac{1-\phi}{2}\nabla U-\vec{j}_{at}\right)+\frac{1}{2}\left[1+(1-k)U\right]\frac{\partial \phi}{\partial t}$$
(2)

where, τ is the relaxation time, ϕ is the phase field, $W(\vec{n})$ is the anisotropic width of the diffuse interface with $\vec{n}$ as the unit normal vector pointing out into the liquid, k is the partition coefficient for the solute, and D is the solute and thermal diffusivities, respectively. To efficiently solve Eqs. (1) and (2), a recently developed Para-AMR algorithm with the abilities of adaptive mesh refinement and parallel computing was employed. More details relating the governing equation and the algorithm can be found in references [14].

Experimental procedures

Mg–45 wt.% Zn alloy was selected and prepared as follows: magnesium (99.95 wt.%) and zinc (99.99 wt.%) were first melted, mixed and poured into a permanent mould to obtain the as-cast ingots; then, cylindrical specimen of 10 mm in diameter and 30 mm in length was machined from the former ingots; thirdly, the specimen was sealed in a quartz tube with argon as protective atmosphere, remelted and quenched between the liquidus and eutectic temperatures; finally, cylindrical specimens of 1.0 mm in diameter and 10 mm in length were machined from the specimen for X-ray tomography experiments [8].

Tomography experiments were carried out at the BL13W1 beam line of Shanghai Synchrotron Radiation Facility (SSRF). An X-ray energy of 28 KeV is used to penetrate the specimens, and 1080 projection are taken between 0° and 180° with an exposure time of 1.2 s. The distance between the specimen and the camera is 20 cm. The final spatial resolution is 0.65 μm.

A software package namely PITRE is used to convert the projected images to 8 bit slice images. The dendrites are segmented from subvolume of about 400 × 300 × 460 pixels, i.e. 0.26 × 0.2 × 0.3 mm after 3-D median filter, thresholding and segmentation [8].

Results and Discussion

A Predictive Growth Model of the α-Mg (Zn) Dendrite

According to the research by Haxhimali and co-workers [9], a continuous transition from ⟨100⟩ to ⟨110⟩ in {001} planes occurs for α-Al dendrites as an increase in the zinc content in binary Al-Zn alloy. They attributed the DOT to a modification of the week anisotropy of solid-liquid interfacial energy γ_{sl} of aluminum by the additional element of zinc, which is an *hcp* element, and exhibits the anisotropy of about 30 % between the c-axis and the basal plane. Moreover, Dantzig et al. [11] also found that the interfacial energy anisotropy plays an important role on the selection of dendrite orientation. Magnesium also has a weak anisotropy of γ_{sl} (about 1.2 %), that can be possibly perturbed by the addition of an alloying element with a strong anisotropy like Zn.

Based on our previous studies, the prominent characteristic of dendritic microstructure in Mg-based alloys, i.e. Mg-Sn, Mg-Gd and Mg-Y is the eighteen-branch structure. However, there is a big difference between the Mg-Sn and Mg-Gd/Y dendrite morphology. For the Mg-Sn alloy, the branches have the same length along $\langle 11\bar{2}0\rangle$ and $\langle 11\bar{2}3\rangle$. For the Mg-Gd/Y alloy, the branches along $\langle 11\bar{2}3\rangle$ are much longer than that along $\langle 11\bar{2}0\rangle$, which means the $\langle 11\bar{2}3\rangle$-type branch grows faster than the others to some extent. From the point of interface energy, the alloying elements have different intensity of anisotropy which may induce different effects on the growth of α-Mg dendrites.

Anyway, irrespective of the six-branch [5] (Fig. 1a) or the eighteen-branch structure [8] (Fig. 1b), the understanding of 3-D dendrites of magnesium is limited to the six-fold symmetric structure owing to the *hcp* crystal structure. Therefore, with the vital role of the interfacial energy anisotropy on the dendrite orientation and the assumption that the anisotropy have little effect on the six-fold symmetric structure in mind, we believe that the 3-D morphology of α-Mg dendrites could also be twelve branches along $\langle 11\bar{2}x\rangle$ orientation without the six branches on the basal plane, compared to the eighteen-branch morphology as shown in Fig. 1c.

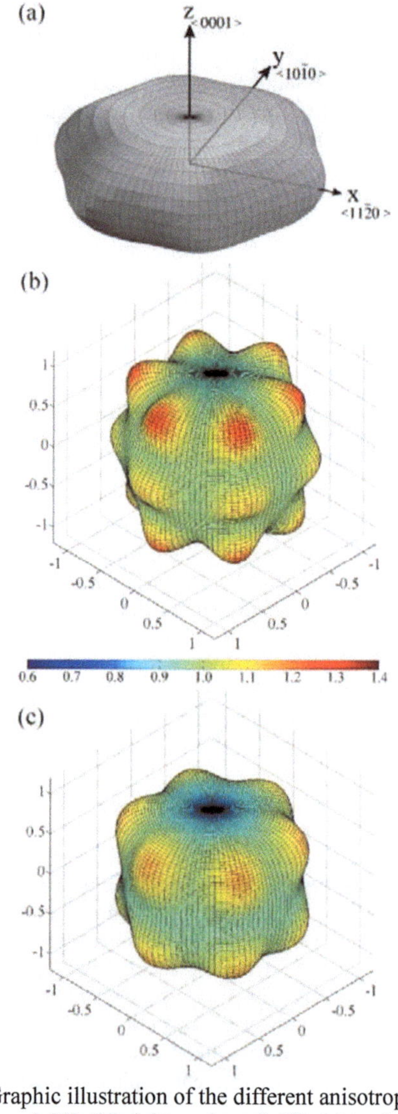

Figure 1. Graphic illustration of the different anisotropy function: (a) six-branch [5]; (b) eighteen-branch [8]; (c) twelve-branch.

An anisotropy function was developed by combing and modifying the Y_{43} and Y_{20} terms of the spherical harmonics to perform the twelve-branch dendrite morphology:

$$\gamma(\theta,\varphi)=\gamma_0\left[1+\varepsilon_1\left(3n_z^2-1\right)^2+\varepsilon_2\left(n_x^3-3n_xn_y{}^2\right)^2\left(n_z\right)^2\right] \quad (3)$$

where ε_1 and ε_2 are the anisotropic parameters to describe the anisotropic strength along different directions.

Simulation of the α-Mg (Zn) Dendrite in 3-D

The simulation was performed using a hierarchical mesh with 5 levels of grids. The computing domain was 819.2 × 819.2 × 819.2, i.e. equivalent to 1024 × 1024 × 1024 cells if a uniform grid of size dx = 0.8 was employed. A uniform temperature θ = – 0.10 was set across the domain, and other key parameters used for the simulation included k = 0.15, λ = 30, ε_1 = –0.035 and ε_2 = 0.15. Simulation was initiated by planting a solid seed with radius of R_0 = 3.0 in the centre of the domain.

Fig. 2 shows the simulated twelve-branch dendrite morphology of magnesium alloy. To achieve a better illustration, dendrite morphology at different perspectives was shown in detail in Fig. 2b-d. As shown, the simulated dendrite exhibited twelve primary branches with six branches lying upper or lower with respect to the basal plane, respectively. Fig. 2e shows the 2-D projection of the microstructure retrieved from the centre of the simulated dendrite in the view of Fig. 2d. As seen, for the primary branch, three secondary arms can exist, i.e. three $\langle 11\bar{2}x\rangle$ directions. Moreover, the simulated dendrite morphology strikes great symmetry according to the modelling condition.

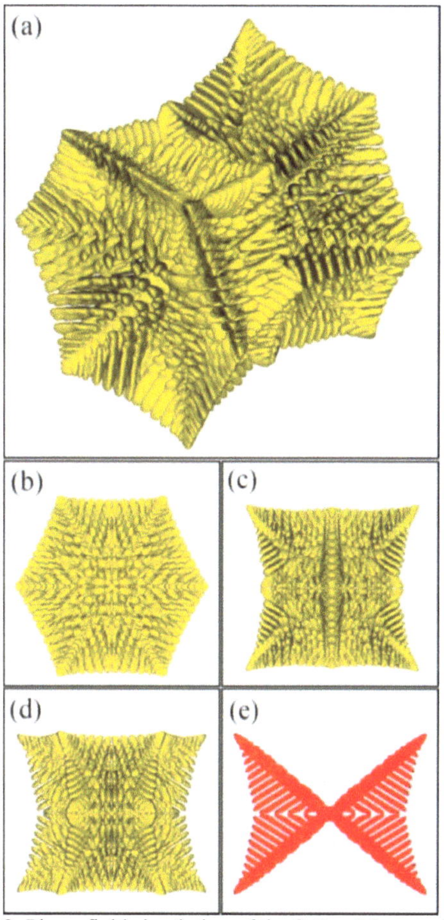

Figure 2. Phase field simulation of the 3-D α-Mg morphology with respect to different view directions and the structure pattern by cutting the dendrite along certain plane: (a) $\langle 111\rangle$; (b) $\langle 0001\rangle$; (c) $\langle 11\bar{2}0\rangle$; (d) $\langle 10\bar{1}0\rangle$; (e) 2-D project on $(10\bar{1}0)$.

The 3-D reconstructed α-Mg (Zn) Dendrite

The reconstructed 3-D α-Mg (Zn) dendrite is shown in Fig. 3a, and a typical dendrite size is about 200 – 300 μm. To achieve a better illustration of the dendrite morphology, the 3-D dendrite was further investigated by applying different planes to cut the dendrite to reveal the internal 2-D structures. As shown in Fig. 3b, three planes were used and the angles between S_1, S_2 and S_3 were 60°. For the 2-D projections on S_3, as shown in Fig. 3e, the dendrite had four branches, which lay at upper and lower sides. Similar patterns can also be observed on planes S_1 and S_2. In this respect, it can be concluded that the α-Mg (Zn) dendrite comprised of twelve primary branches which tilted away from the basal plane in a growth direction of $\langle 11\bar{2}x\rangle$ or $\langle 10\bar{1}x\rangle$.

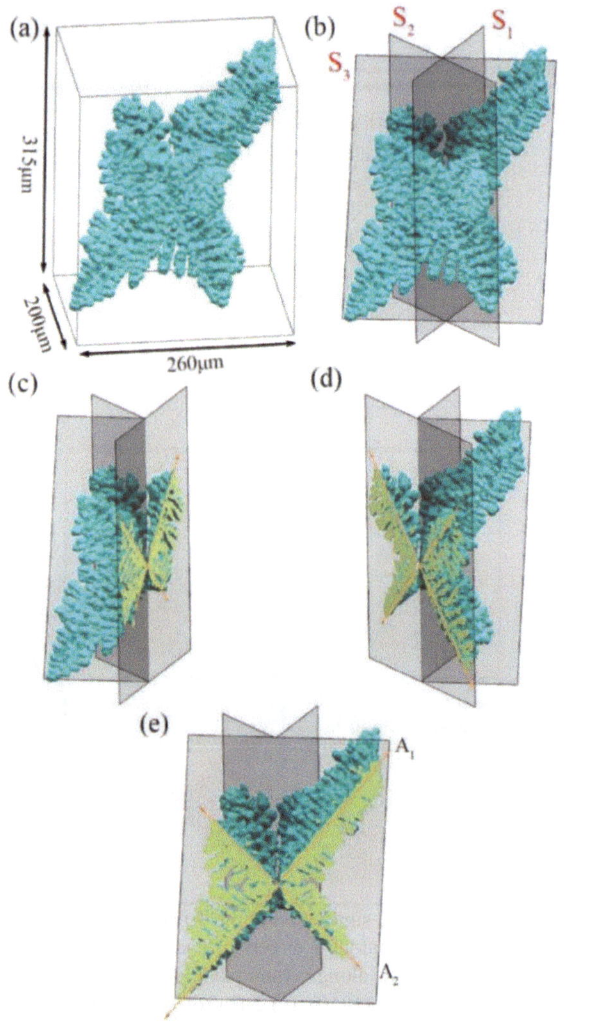

Figure 3. The 3-D morphology of α-Mg (Zn) dendrite: (a) the dendrite morphology, (b) configuration of the three plane i.e. S_1, S_2 and S_3 used to cut the dendrite, and (c-e) the according 2-D projections of the dendrite patterns on the above planes.

EBSD experiment was performed to determine the crystallographic orientation of the preferred growth direction. As shown in Fig. 4, the crystallographic orientation was $\langle 11\bar{2}2\rangle$. It can be concluded that the growth direction of the primary branches can be characterized as $\langle 11\bar{2}2\rangle$.

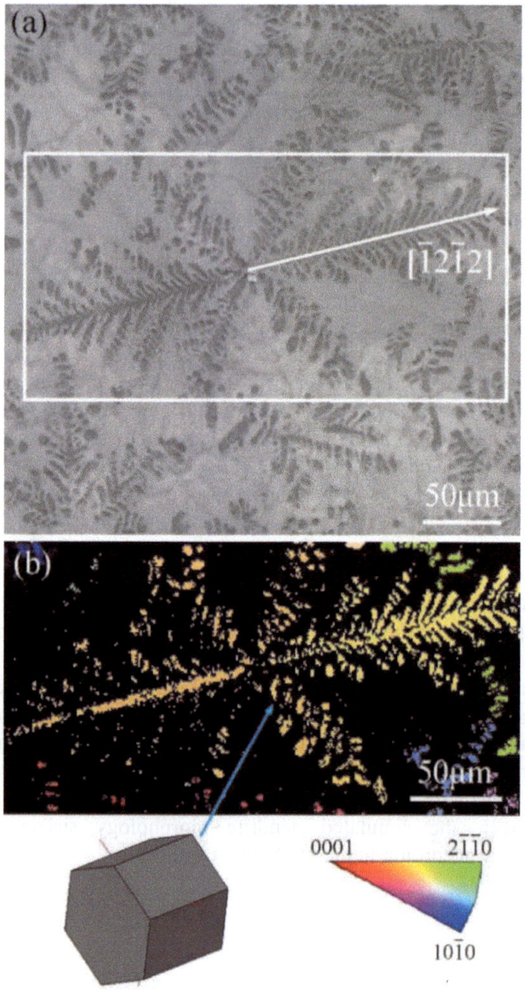

Figure 4. EBSD analysis on the dendrite preferred growth directions of Mg–45 wt.% Zn alloy: (a) dendrite morphology and (b) the according $(13\bar{4}4)$ orientation map.

Considering the simulated and the experimental dendrite morphology of Mg-Zn alloy shown in Figure 2 and 3, both of them exhibited twelve primary branches and stuck great similarity to each other. Remarkably, by cutting the dendrite on the $(10\bar{1}0)$plane, the patterns obtained in the experiment exhibits four branches (Figure 3e) and can be successfully validated by the current PF simulation (Figure 2e). It is worth stressing that the twelve-branch structure is different from our previous studies [8] and the results in literature [5], and needs further investigation.

Summary and Future Work

A twelve-branch morphology of α-Mg (Zn) dendrite was proposed. Based on the newly developed anisotropy function, the 3-D phase field simulation was then performed and the results were verified with those from synchrotron X-ray tomography experiment of Mg–45 wt.% Zn alloy.

Further investigation on the effect of alloying element on the dendrite morphology of Mg-based alloy is required to determine whether the DOT takes place in magnesium alloy and how it affects the morphology.

Acknowledgements

This research is funded by The National Natural Science Foundation of China (51205229 and 51275269), and the Tsinghua University Initiative Scientific Research Program (20121087918). The authors would also like to thank the Shanghai Synchrotron Radiation Facility for the provision of beam time and the National Laboratory for Information Science and Technology in Tsinghua University for access to supercomputing facilities.

References

[1] M. K. Kulekci, "Magnesium and its alloys applications in automotive industry." *Int J Adv Manuf Technol*, 39 (9-10) (2008), 851-865.

[2] M. Y.Wang et al., "Dendritic morphology of α-Mg during the solidification of Mg-based alloys: 3D experimental characterization by X-ray synchrotron tomography and phase-field simulations." *Scr Mater*, 65 (10) (2011), 855-858.

[3] J. A. Dantzig, M. Rappaz, *Solidification* (EPFL press, 2009).

[4] B. Böttger, J. Eiken, and I. Steinbach, "Phase field simulation of equiaxed solidification in technical alloys." *Acta Mater*, 54 (10) (2006), 2697-2704.

[5] J. Eiken, "Dendritic growth texture evolution in Mg-based alloys investigated by phase-field simulation." *Int J Cast Metal Res*, 1 (4) (2009), 86-89.

[6] K. Pettersen, O. Lohne, and N. Ryum, "Dendritic solidification of magnesium alloy AZ91." *Metall Trans A*, 21 (1) (1990), 221-230.

[7] M. Y. Wang et al., "Growth orientations and morphologies of α-Mg dendrites in Mg–Zn alloys." *Scr Mater*, 67 (2012), 629-632.

[8] M. Yang, S. M. Xiong, and Z. Guo, "Characterisation of the 3-D dendrite morphology of magne sium alloys using synchrotron X-ray tomography and 3-D phase-field modelling." *Acta Mater*, 92 (0) (2015), 8-17.

[9] T. Haxhimali et al., "Orientation selection in dendritic evolution." *Nature Mater*, 5 (8) (2006), 660-664.

[10] J. Friedli et al., "Dendritic Growth Morphologies in Al-Zn Alloys—Part I: X-ray Tomographic Microscopy." *Metall Mater Trans A*, 44 (12) (2013), 5522-5531.

[11] J. A. Dantzig et al., "Dendritic Growth Morphologies in Al-Zn Alloys—Part II: Phase-Field Computations." *Metall Mater Trans A*, 44 (12) (2013), 5532-5543.

[12] A. Semoroz, M. Rappaz, S. Henry, "Application of the phase-field method to the solidification of hot-dipped galvanized coatings." *Metall Mater Trans A*, 31 (2) (2000), 487-495.

[13] B. Echebarria et al., "Quantitative phase-field model of alloy solidification." *Phys Rev E*, 70 (2004), 61604.

[14] Z. Guo, S. M. Xiong. "On solving the 3-D phase field equations by employing a parallel-adaptive mesh refinement (Para-AMR) algorithm." *Comput Phys Commun*, 190 (0) (2015), 89-97.

Magnesium Technology 2016
Edited by: Alok Singh, Kiran Solanki, Michele V. Manuel, and Neale R. Neelameggham
TMS (The Minerals, Metals & Materials Society), 2016

INFLUENCE OF HOT ISOSTATIC PROCESSING ON THE MICROSTRUCTURE AND TENSILE BEHAVIOR OF HPDC AM50

Erin Deda[1], John Allison[1]
[1] Materials Science and Engineering, 2300 Hayward St, University of Michigan, Ann Arbor, MI 48109, USA

Keywords: AM50, High Pressure Die Casting, Microstructure, Mechanical Properties, Hot Isostatic Pressing

Abstract

The mechanical properties of high pressure die cast (HPDC) magnesium can be highly variable and dependent on the manufacturing process history and on location within a casting. This investigation provides experimental input to modeling activities for the development of an ICME capability, to assess and quantify the impact of section thickness on the microstructure and tensile behavior of HPDC Mg AM50 alloy. AM50 2.5 mm thick plates were super vacuum die cast and characterized via quantitative metallography and uniaxial tensile testing. Plates were processed by hot isostatic pressing (HIP) to identify the effect of different microstructural features, specifically β-$Mg_{17}Al_{12}$ and shrinkage porosity, on the mechanical behavior. This allows investigation specifically on the effects of casting defects, such as oxide bifilms on the microstructure and tensile behavior of these plates.

Introduction

Increasing fuel economy requirements have generated interest in the use of magnesium alloys for weight saving applications [1]. Magnesium alloys high strength to density ratio and good castability makes them desirable for increased use in automobiles. However, prior work has shown the high variability associated with the mechanical properties, specifically ductility, of high pressure die castings currently limits usage and potential for weight savings [2, 3].

One common alloy for high pressure die castings for automotive use is AM50 (magnesium-aluminum-manganese). The microstructures of AM50 are reasonably well understood, however, quantifying the location-dependent and casting process-history dependent microstructure remains an open issue [3]. This is particularly the case for those microstructural features (e.g. porosity) that are thought to affect the mechanical behavior, specifically in this study, the ductility. Porosity is generally believed to have a deleterious effect on tensile ductility [3]–[7], however, there is no consistent approach for quantifying or predicting this effect. Campbell has discussed the deleterious effect of porosity on aluminum castings [8]. Oxide bifilms have been shown in aluminum alloy A206 to have a strongly negative effect on ductility and fatigue life due to the lack of cohesion between layers of the film, acting as a crack in the material [9].

An improved understanding of the origins and extremes of ductility in these alloys would allow the design of improved parts as well as a better understanding of the variability associated with die castings. In order to better understand the extremes, in this study, we have processed the as-cast plates with hot isostatic pressing (HIP). Hot isostatic pressing allows isolation of the intrinsic behaviors of the plate castings by removing the β-$Mg_{17}Al_{12}$ and shrinkage porosity from the plate microstructure. In order to completely understand the effects of HIP, we fully characterize the microstructure of the as cast and HIP plates. The influence of HIP is relatively unstudied in Mg AM series alloys. The studies that have been conducted have been on high pressure die cast AZ91 [6, 7]. In the current study we focus on the influence of HIP on microstructures and tensile properties in the super vacuum die cast AM series Mg alloys.

Experimental

Magnesium-aluminum-manganese AM50 plates were super vacuum die cast (SVDC) by Ford Research and Innovation Center at MagTech Corporation. Sulfur hexafluoride was used as the cover gas to minimize oxidation. Plates were cast in two thicknesses, 2.5 and 5 mm. This paper focuses on the 2.5mm plate results. The nominal composition of the plates was 4.6 wt.% Al, 0.4 wt.% Mn; balance Mg as measured by optical emission spectroscopy.

In order to remove the effects of shrinkage porosity and β-$Mg_{17}Al_{12}$, plates were processed with HIP. HIP was completed at Bodycote, at a temperature of 410°C under a pressure of 100 MPa for a duration of 2 hours, at which point they were cooled to room temperature. This was completed in a 17" diameter chamber in which the plates were stacked horizontally to prevent distortion.

Metallographic specimens were prepared by mounting through thickness samples of each condition. These samples were prepared by grinding with SiC following standard metallographic procedures. Polishing was performed with a 3 μm and 1 μm diamond paste, with Buehler MetaDi fluid for lubrication. Specimens were etched for 3 seconds in chilled acetic-nitric solution (15 mL Acetic acid, 5 mL nitric acid, 60 mL ethanol, 20 mL water). Microstructure characterization was conducted in a TESCAN Mira FEG SEM using backscatter imaging (BSE). BSE was used to quantify the β-$Mg_{17}Al_{12}$ eutectic phase in as polished samples. Quantification was completed using Image J. EBSD was used to characterize the grain size in etched samples.

Operating conditions for the EBSD were an accelerating voltage of 30 keV, working distance of 20, and step size of 0.5μm. Data sets were cleaned using the TSL OIM data analysis software using a procedure that consisted of 1) neighbor orientation correlation, 2) one iteration of grain dilation, and 3) confidence index (CI) correlation. Points with a CI less than 0.1 were dropped. These points appear black on the EBSD maps, indicating porosity, grain boundaries, β-$Mg_{17}Al_{12}$, or areas of poor surface quality.

Tensile dogbone specimens which had a gauge length of 25.4 mm and a cross section of 6.4mm x 2.5mm were machined from the as-cast and HIP plates according to ASTM Standards E8/B557

[10, 11]. Room temperature tensile tests were performed on a hydraulic MTS load frame, with an initial strain rate of $1x10^{-3}s^{-1}$, and a constant displacement rate of 2 mm/min until sample failure. An extensometer with a 25.4mm gauge length was attached to the gauge section to measure tensile elongation. Tensile testing was completed on at least 18 samples per condition.

Results and Discussion

Microstructure

The primary microstructural features in AM50 SVDC plates that were characterized are the pore area fraction, the area fraction and distribution of eutectic phases, the grain size distribution, and presence of oxide films. The microstructure of the as-cast plate can be separated into two regions, the skin, or the surface of the casting, and the core. The skin is composed of fine α-Mg grains surrounded by a divorced β-$Mg_{17}Al_{12}$ eutectic. The core is composed of fine α-Mg in-mold grains interspersed with large externally solidified crystals (ESCs), with β-$Mg_{17}Al_{12}$ eutectic along the edges. The fine α-Mg grains are those that form during solidification in the mold, and ESCs form prior to injection into the die, when the liquid metal is poured into the shot sleeve. Throughout the casting, there is a uniform distribution of AlMn particles that appear as bright white particles in BSE images, as in **Error! Reference source not found.**. The skin and core structure for the 2.5mm as cast plates has a skin thickness of approximately 300 μm. This was characterized by the first occurrence ESCs, and β-$Mg_{17}Al_{12}$ shape transition, which is detailed in the following sections.

Porosity

Porosity content was evaluated through area fraction as observed in SEM BSE. Porosity in the as cast plates had an area fraction between 0.5 and 1.5%. This observed porosity can be characterized as predominantly shrinkage porosity, with minimal porosity due to entrapped gas. This was confirmed by the observation that the area fraction of porosity was significantly reduced in the HIP plates. After HIP, there was less than 0.3% area fraction porosity in the 2.5mm AM50 plates.

β-$Mg_{17}Al_{12}$ characterization

The β-$Mg_{17}Al_{12}$ eutectic phase changes shape and distribution through the thickness of the plates in the as cast condition. The average area fraction of the β-$Mg_{17}Al_{12}$ phase was determined to be 1.4% ± 0.3% in the 2.5mm as cast condition. One method of predicting the amount of eutectic phases formed during non-equilibrium solidification is through alloy partitioning, as described by the Scheil-Gulliver solidification model. This has been used previously in MgAl alloys to calculate the amount of eutectic and intermetallic phases [14] The amount of β-$Mg_{17}Al_{12}$ can be calculated using the Thermo-Calc Scheil solidification module. Using the Thermo-Calc TCMG3 database, for non-equilibrium AM50, a β-$Mg_{17}Al_{12}$ phase fraction of 5.1% is predicted under Scheil solification conditions. The as-cast condition has a significantly lower (approximately 25%) β-$Mg_{17}Al_{12}$ phase fraction compared to that predicted by the Scheil solidification model. This is presumably due to the non-equilibrium solidification that occurs at the much higher solidification rates that exist in super vacuum die casting.

It can be observed from Figure 1 (a) that the skin region consists largely of β-$Mg_{17}Al_{12}$ particles. In the center of the plates however, the β-$Mg_{17}Al_{12}$ phase transitions from the isolated particles to a divorced eutectic. This transition begins approximately 300 μm from the surface of the casting. In contrast, the core has a combination of particle and divorced eutectic β-$Mg_{17}Al_{12}$ distributed along the grain boundaries, which can be seen more clearly in Figure 2 (a).

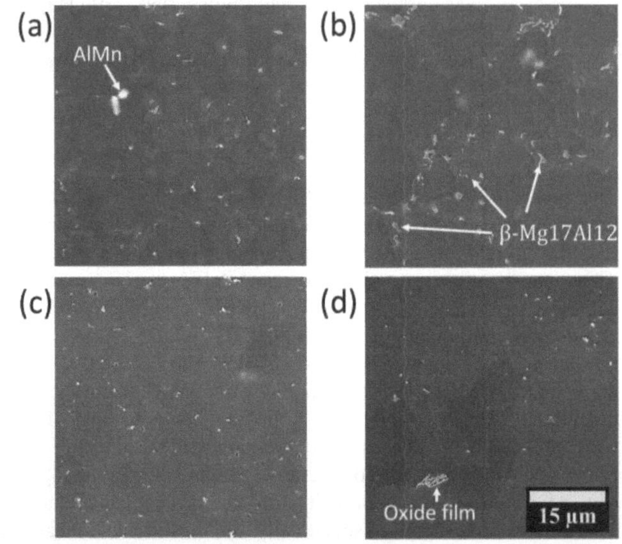

Figure 1: Detailed SE images of the microstructure of etched, as cast AM50 2.5mm thick (a) as-cast edge, (b) as-cast center, (c) HIP edge, and (d) HIP center

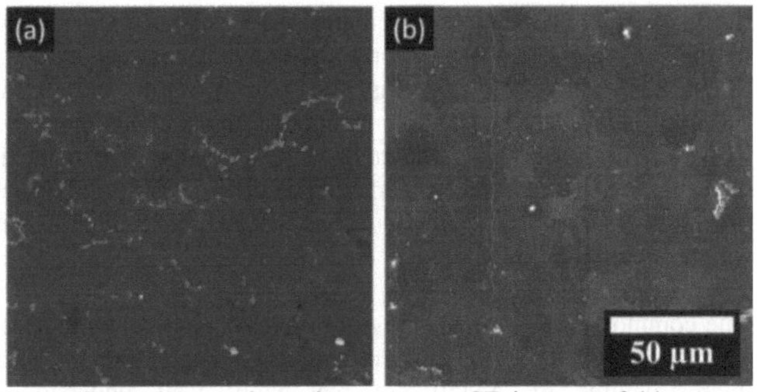

Figure 2: Representative microstructure SE images of (a) as-cast and (b) HIP AM50 from the center of the plates.

This can then be compared to the HIP microstructures. As observed in Figure 1 (c) and (d), no β-$Mg_{17}Al_{12}$ was observed in the as-HIP condition. The β-$Mg_{17}Al_{12}$ is fully dissolved during the HIP process. This is consistent with published phase diagrams that show that β-$Mg_{17}Al_{12}$ is not an equilibrium phase at 410°C [15]. The representative view in Figure 2 (b) shows a larger field of view, confirming the dissolution of the eutectic phase.

Grain size characterization

Grain size was characterized using EBSD OIM. The results are summarized in Table 1. The α-Mg in-mold grain size was consistently between 8-10 μm for both skin and core of all conditions of the AM50 plates. This fine skin grain size can be observed in Figure 3 (a) and (c) for both the as cast and HIP 2.5mm conditions respectively. No grain growth was observed in the HIP condition. ESCs were defined to be grains larger than 30 μm, based on image analysis work by Forsmark et al [16]. ESC sizes ranged from 30 μm up to 150 μm in grain diameter. The average ESC size consistently averaged 45 μm in diameter. ESCs can be seen in Figure 3 (b) and (d), where the fine in-mold grains surround the ESCs. The area fraction of ESCs did vary significantly depending on the observed area. The distribution of area fraction of grains is fitted with a lognormal distribution in Figure 4. The distributions for the core are broader than the skin for both the as-cast and HIP conditions.

After HIP, the average grain diameter was unchanged compared to the as cast condition. The grain size distribution remains the same after HIP for both edge and center. The EBSD results were much more distinct ("clean") in the as-HIP condition due to the dissolution of β-$Mg_{17}Al_{12}$.. This is observed in Figure 3 (c) and (d), where there is significantly less non-indexing locations compared to the EBSD images shown for the as-cast conditions in Figure 3 (a) and (b).

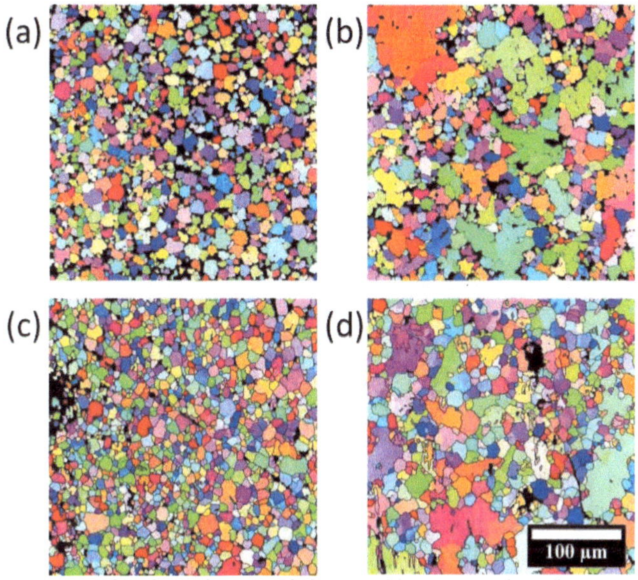

Figure 3: EBSD OIM for AM50 2.5mm (a) as cast edge, (b) as cast center, (c) HIP edge, and (d) HIP center.

Condition	In mold Edge (μm)	In mold center (μm)	ESC center (μm)	ESC fraction (%)
AM50 2.5mm AC	8.2	9.2	44	30
AM50 2.5mm HIP	8.4	9.9	45	22

Table 1: Average in mold and ESC grain diameter (in μm) for each condition and ESC area fraction in percent.

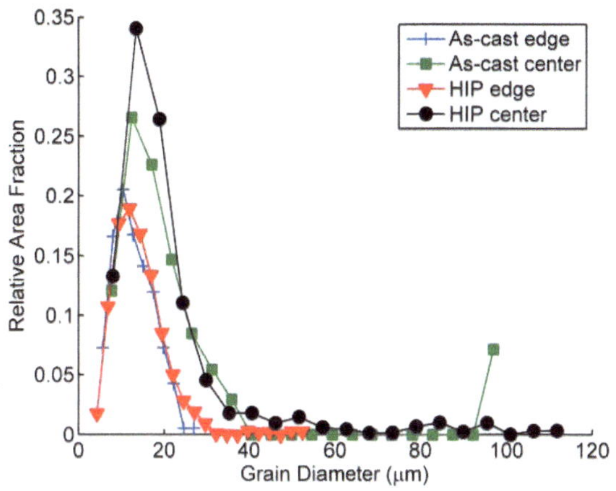

Figure 4: Lognormal grain size distribution for as-cast and HIP conditions, edge and center.

Tensile Behavior

Tensile properties were measured for each condition. Representative tensile curves are shown in **Error! Reference source not found.**. The tensile properties are summarized in Table 2. Each value in this table is the average of at least 18 samples. The as cast 2.5mm plates have a yield strength of 110 MPa, and the HIP 2.5mm plates a yield of 119 MPa. This increase in yield strength in the HIP 2.5mm plate relative to the as cast 2.5mm plate is attributed to an increase in solid solution strengthening due to Al after the dissolution of the of β-$Mg_{17}Al_{12}$ phase during HIP.

There is also significant variation in the average elongations of the different conditions. The average elongation to failure was 13.3% for the as cast 2.5mm plates, and 18.1% for the HIP 2.5mm. This represents an average elongation increase of 36% for the HIP plates. Commensurate with the increase in average elongation, a 10% increase in the fracture strength was also observed. The increase in elongation was attributed to the substantial reduction in the amount of shrinkage porosity that was observed in the HIP condition.

Although the average elongation increased with HIP the standard deviation also increased slightly. HIP tended to increase the upper bound however the lower bound was only moderately affected, as demonstrated in the box plots for elongation to failure for each condition in Figure 6. When reading the box plot, the box represents the 25% quartile and 75% quartile, lower and upper bounds of the box respectively. The average is shown as a circle, and the median given as the central line. The whiskers of the plot represent 1.5 times the interquartile range, and the plus symbols are any outlying data points beyond 1.5 times the interquartile range. This allows for a graphical interpretation of the distribution of elongations to failure for each condition with a statistical description.

This increased variability can likely be attributed to the ductility dependence on the presence, location and size of oxide films in tensile samples. The presence of large oxide films is characteristic of the samples with low ductility. The upper bound of ductility by contrast is limited only by the intrinsic ductility of the material in the absence of casting defects including oxides and porosity. The difference between the high and low end of the ductility can be clearly observed on the fracture surface, where the upper bound ductility samples commonly fail in a shear mode, as observed in Figure 7 (a), with dimpled features. A sample with a ductility of 11.5% from the HIP condition by contrast also includes a large oxide film as seen in Figure 7 (b) and (c). The dotted circle in Figure 7 (c) indicates the extent of the oxide immediately adjacent to the surface of the tensile bar, likely leading to the low observed ductility.

Condition	YS (MPa)	Fracture Strength (MPa)	Avg Elong. (%)	n-value
AM50 2.5mm AC	110 ± 2	233 ± 12	13.3± 2.4	.34
AM50 2.5mm HIP	119 ± 2	256 ± 12	18.1± 3.1	.33

Table 2: Tensile Properties of AM50 as cast and HIP conditions

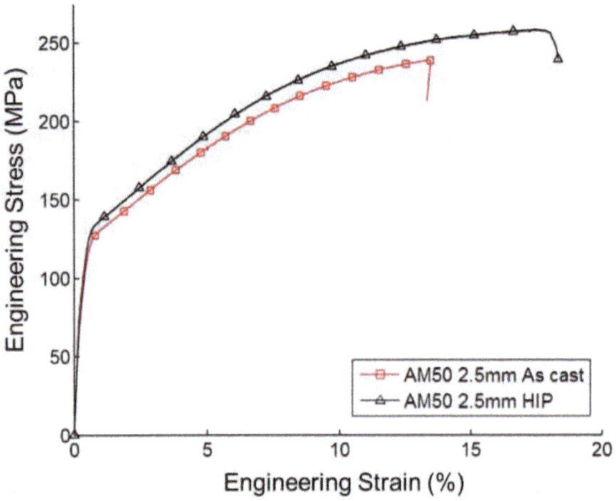

Figure 5: Representative tensile curves for AM50 2.5mm thick plates for the as cast and HIP conditions.

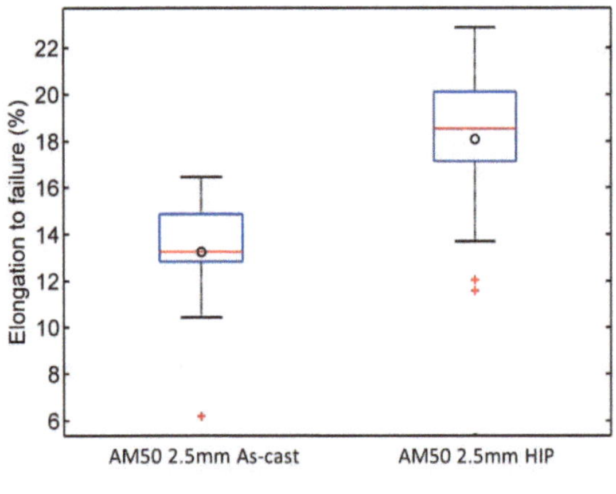

Figure 6: Box plot of elongation to failure for the as-cast (L) and HIP (R) conditions. The average is represented by the central circle, median central line, and the box represents the interquartile range (IQR). Whiskers extend to 1.5 times the IQR, and outliers are marked with a plus sign (+) to denote points that lay beyond the whiskers.

Summary

By processing SVDC AM50 plates with hot isostatic pressing (HIP), the effects of porosity and β-$Mg_{17}Al_{12}$ can be reduced or eliminated. This allows investigation of the effects of casting defects, such as oxide bifilms on the microstructure and tensile behavior of these plates. It also allows determination of the upper bound properties that are intrinsic to the alloy. It has been demonstrated that HIP completely eliminates shrinkage porosity and β-$Mg_{17}Al_{12}$ without causing grain growth. Elimination of porosity and of β-$Mg_{17}Al_{12}$ phase by HIP leads to an increase in the average yield strength, fracture strength and elongation at failure. However HIP also increases the variability in elongation. This is due to an increase in the upper value (from elimination of pores) while the lower bound is unchanged due to the presence of oxides. By isolating the effects of microstructure on the tensile behavior, we can better understand the sources of property variability and better control it for optimizing casting designs. Further characterization of the frequency and size distribution of oxide films in these alloys is required to understand the lower bound of ductility in SVDC castings.

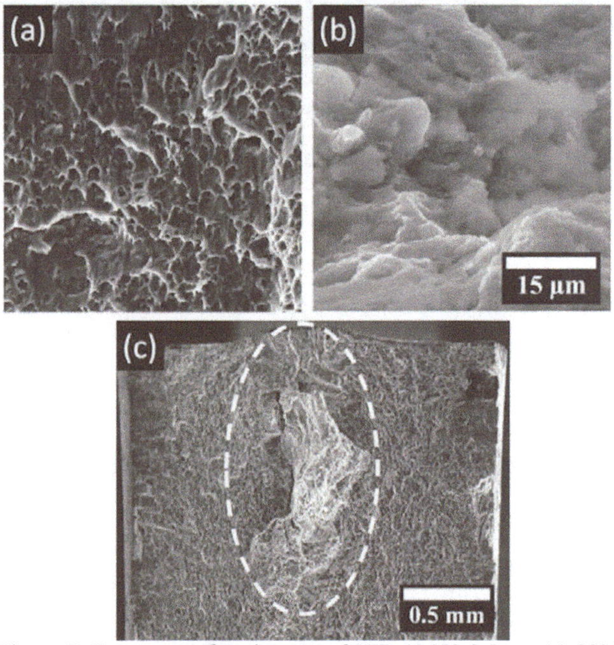

Figure 7: Fracture surface images of HIP AM50 2.5mm (a) 22% elongation to failure, as compared with (b) 11.5% elongation at failure, where (c) is a low magnification of a large oxide immediately subsurface.

Acknowledgements

This material is based upon work funded by the Department of Energy Vehicle Technologies Office under the Automotive Lightweighting Materials Program managed by William Joost. The authors express their appreciation to M. Li, J. Forsmark, J. Zindel, and L. Godlewski at Ford Motor Company Research and Innovation Center for providing the SVDC alloys and helpful discussions and to X. Sun and E.I. Barker at Pacific Northwest

National Labs for their support and insightful discussions. We also express our gratitude to T.D. Berman at University of Michigan for her contributions to this work.

References

[1] H. E. Friedrich and B. L. Mordike, Eds., *Magnesium Technology: Metallurgy, Design Data, Applications*. Berlin: Springer, 2006.

[2] R. Alain, T. Lawson, P. Katool, G. Wang, J. Jekl, R. Berkmortel, and L. Miller, "Robustness of Large Thin Wall Magnesium Die Castings for Crash Applications," *SAE Int.*, no. 724, 2004–01–0131, 2004.

[3] J. H. Forsmark, J. W. Zindel, L. Godlewski, J. Zheng, J. E. Allison, and M. Li, "Using quality mapping to predict spatial variation in local properties and component performance in Mg alloy thin-walled high-pressure die castings: an ICME approach and case study," *Integr. Mater. Manuf. Innov.*, vol. 4, March, 2015.

[4] J. P. Weiler and J. T. Wood, "Modeling the tensile failure of cast magnesium alloys," *J. Alloys Compd.*, vol. 537, pp. 133–140, Oct. 2012.

[5] K. S. Choi, D. Li, X. Sun, M. Li, and J. E. Allison, "Effects of Pore Distributions on Ductility of Thin-Walled High Pressure Die-Cast Magnesium," *SAE Int.*, no. 2013–01–0644, Apr. 2013.

[6] G. Chadha, J. E. Allison, and J. W. Jones, "The Role of Microstructure on Ductility of Die-Cast AM50 and AM60 Magnesium Alloys," *Metall. Mater. Trans. A*, vol. 38, no. 2, pp. 286–297, Feb. 2007.

[7] C. Do Lee, "Tensile properties of high-pressure die-cast AM60 and AZ91 magnesium alloys on microporosity variation," *J. Mater. Sci.*, vol. 42, no. 24, pp. 10032–10039, Sep. 2007.

[8] J. Campbell, *Castings*, 3rd ed. Oxford: Butterworth-Heineman, 1991.

[9] J. T. Staley, M. Tiryakioğlu, and J. Campbell, "The effect of hot isostatic pressing (HIP) on the fatigue life of A206-T71 aluminum castings," *Mater. Sci. Eng. A*, vol. 465, no. 1–2, pp. 136–145, Sep. 2007.

[10] T. Schubert, P. Kumar, and B. Kieback, "IMPROVING MECHANICAL PROPERTIES OF MAGNESIUM PRESSURE DIE CASTINGS," in *Mg2012: 9th International Conference on Magnesium alloys and their Applications*, 2012, pp. 1223–1228.

[11] M. Ostermeier, M. Brummer, and E. Werner, "Hot isostatic pressing of magnesium castings," *Int. Foundry Res.*, vol. 61, no. 3, pp. 2–7, 2009.

[12] *Standard Test Methods for Tension Testing of Metallic Materials*, ASTM E8/E8M - 09, ASTM International, West Conshohocken, PA.

[13] *Standard Test Methods for Tension Testing Wrought and Cast Aluminum- and Magnesium-Alloy Products*, ASTM B557M - 10, ASTM International, West Conshohocken, PA.

[14] K. V. Yang, C. H. Cáceres, and M. A. Easton, "Strengthening Micromechanisms in Cold-Chamber High-Pressure Die-Cast Mg-Al Alloys," *Metall. Mater. Trans. A*, vol. 45, no. 9, pp. 4117–4128, May 2014.

[15] Y. Zhong, M. Yang, and Z.-K. Liu, "Contribution of first-principles energetics to Al–Mg thermodynamic modeling," *Calphad*, vol. 29, no. 4, pp. 303–311, Dec. 2005.

[16] J. H. Forsmark, Z. Dowling, K. Gibson, C. Mueller, L. Godlewski, J. Zindel, and J. M. Boileau, "An Investigation of the Effects of Cast Skin on the Mechanical Properties of an AM60 Die-Cast Magnesium Alloy," *SAE Int. J. Mater. Manuf.*, vol. 8, no. 3, 2015.

Magnesium Technology 2016
Edited by: Alok Singh, Kiran Solanki, Michele V. Manuel, and Neale R. Neelameggham
TMS (The Minerals, Metals & Materials Society), 2016

Microsegregation in High Pressure Die Cast AM70

T.D. Berman[1], E. Deda[1], J. Miao[1],M. Li[2], J.E. Allison[1]

[1]Materials Science and Engineering, University of Michigan, 2300 Hayward St; Ann Arbor, MI, 48109, USA
[2]Ford Motor Company; Dearborn, MI, USA
3

Keywords: Die Casting, Macrosegregation, Microsegregation, Microstructures

Abstract

High pressure die casting (HPDC) is the predominant manufacturing method for magnesium alloy automotive components. Cooling rates during HPDC may reach 300 °C/s; however, the phase transformation kinetics in this regime are poorly understood. In this study, a description of microsegregation in HPDC AM70 is obtained by electron probe micro-analysis (EPMA) mapping. It has been found that the solute profiles are in good agreement with a Scheil solidification model at mid-thickness, but diverge from Scheil solidification at the edges of the plate. The skin is Al-rich compared to the bulk of the plate; this macrosegregation persists through solution treatment. As a consequence, less Al is available in the bulk for strengthening.

Introduction

High pressure die casting (HPDC) is the predominant manufacturing method for magnesium alloy automotive components, however, the phase transformation kinetics that occur in this process, are poorly understood. Prior work has shown that casting cooling rate has significant effects on as-cast microstructure of these advanced alloys [1]. Under such extreme conditions, the solidification kinetics, phase transformations and the redistribution of alloying elements cannot be predicted using equilibrium thermodynamics nor the typical modifications to predict alloy partitioning, such as those represented by a Scheil model [2]. The redistribution of alloying elements during non-equilibrium solidification leads to micro-scale segregation across the dendrite/cell and this micro-scale segregation is strongly dependent on the cooling rate. Although microsegregation has been the subject of a limited number of investigations in Mg alloys [3–8] there is no known microsegregation study for Mg alloys under HPDC conditions in the open literature. Previous studies of microsegregation in Mg alloys have been obtained in directionally solidification castings or in samples cast under moderate cooling rate conditions. Cooling rates for automotive HPDC component are in the range of 10 to 1000 °C/s [9]. At these very high cooling rates, i.e. when the solidification front velocity is on the order of the solute diffusion rate in the melt, the solute atoms are caught by the advancing interface even though there is a thermodynamic driving force for them to escape the interface. This phenomenon is known as "solute trapping". The challenge in predicting the microsegregation and phase transformation kinetics in HPDC is how to treat the solute trapping at the interface. An improved understanding of microsegregation in HPDC magnesium alloys will provide a foundation for future alloy and process development.

Experimental

Ford Research and Innovation Center at MagTech Corporation supplied Super vacuum die cast (SVDC) AM70 plates in two thicknesses, 2.5 and 5 mm. Through thickness samples of each plate were mounted and ground with SiC following standard metallographic procedures. Polishing performed with 6 μm, 3 μm, and finally 1 μm diamond paste while using Buehler MetaDi fluid as lubrication. Prior to SEM or EBSD examination, samples were etched for 3 to 5 s in a 5 °C solution of 60 mL ethanol, 20 mL water, 15 mL glacial acetic acid, and 5 mL of nitric acid. Samples were not etched before electron probe microanalysis (EPMA).

Standards of at least 99.9% pure Al, Mg, and Mn were used to calibrate the EPMA. EPMA was performed using a Cameca SX-100 electron microprobe equipped with wavelength-dispersive spectrometers. The accelerating voltage and beam current were 15 kV and 10 nA, respectively. The spectrometer crystal and background levels used for each element are given in Table I. The Kα X-ray line was used for all elements.

EPMA measurements were collected using either a 20 x 20 grid pattern (following the method of Gunger [10]) or a line scan. Grid scans were centered at mid-thickness or 300 μm from the edge of the sample, and are referred to as center and near edge, respectively. Line scans were parallel to the surface of the castings at a distance of approximately 10 μm. A step size of 10 μm was

Table I: EPMA parameters. Pure metal standards were used for each element.

Element	Crystal	Background
Mg	TAP	-1600/1100
Al	LTAP	-900/1100
Mn	LLIF	-1200/1600

used, therefore line scans covered a linear region 4000 μm in length and grid scans covered an area of 190 μm x 190 μm. Only points with a total concentration of 99 to 100.5 wt% were retained for analysis. Multiple scans were collected on many of the specimen, resulting in 300 to 900 quality composition measurements per condition. Fraction solid was assigned to each composition measurement using the Weighted Interval Rank and Sort (WIRS) technique developed by Ganesan et al. [11] as applied to AM50 by Mirković et al. [6]

General operating conditions for EBSD were an accelerating voltage of 30 keV and a working distance of 20 mm. The beam conditions results in a spot size of ~50 nm and were chosen in order to give the best quality of patterns at a rate of 30 to 40 patterns per second. The cleaning procedure was performed in the TSL OIM data analysis software and consisted of (1) neighbor orientation correlation, (2) one iteration of grain dilation, and (3) grain Confidence Index (CI) standardization. After cleaning, points with a CI $<$ 0.1 were dropped from the dataset. Dropped points (indicative of regions of β-phase, grain boundaries, porosity, or regions of poor surface quality) appear black in the EBSD maps.

Results and Discussion

Microstructure

The microstructure consists of α-Mg grains surrounded by a divorced β-$Mg_{17}Al_{12}$ eutectic and β-particles (Fig. 1). The edge of the plate has the lowest volume fraction of β-phase and more refined particles. There is some porosity, which is concentrated in defect bands ~400 μm from the edge of the sample. The pore volume is less than 2%.

Another constituent of the microstructure is larger dendritic crystals, referred to as externally solidified crystals (ESCs) or externally solidified grains (Figure 2). ESCs form when the liquid metal is poured into the shot sleeve, prior to injection into the die. The ESCs may be up to several hundred micrometers in diameter. The majority of the ESCs are segregated in the center of the castings, though they are occasionally found near the casting surface as well, which is consistent with observations by Laukli et al. [12].

The grain size distribution at the edge and center locations is similar in the 5.0 mm thick plates (Figure 3). The average size of the in-mold grains is approximately 10 μm in the 5.0 mm plates and in the center region of the thinner plate. The average grain size is finer at the edge of the thinner plate, at 7 μm in diameter. A lack of ESCs at the edge of the thinner plate, in addition to a higher cooling rate, leads to a narrower, lognormal distribution. In the other three conditions, in which ESCs were observed, a bimodal lognormal distribution is more appropriate.

Microsegregation in the as-Cast Plates

Figure 4 allows for the comparison of microsegregation in the 5mm thick AM70 plates, as determined by EPMA, with Scheil solidification modeling results generated in Thermo-Calc. Microsegregation behavior at the center location is in agreement with the Scheil model up to a high solid fraction (around 0.8). As the measurement moves closer to the sample edge, the agreement between measurement and prediction using the Scheil model is significantly reduced.

More solute is trapped during solidification at the edge of the casting. As a result the HPDC plates have an Al rich skin. The mean Al content of all composition measurements for the 5.0 mm AM70 plates is 4.8, 6.5, and 6.8 wt% for the center, near edge, and edge regions, respectively. As the center region makes up a majority of the volume of the plate, EPMA indicates that the AM70 plate has less Al than the alloy designation implies.

A comparison between the microsegregation in the two plate thicknesses is shown in Figure 5. The microsegregation profiles show little thickness dependence. Location within the casting has a larger influence on microsegregation than thickness.

One concern when studying HPDC alloys with EPMA is the relative size of the x-ray interaction volume (~4.5 μm) and the grain size (~10 μm). As the beam size increases with respect to the grain size, there is an increased probability of sampling the β-phase particles and solute-rich regions along the grain boundaries, thereby raising the measured Al concentration. Future work will address this issue using a forward model which simulates the EPMA results from a material with a given solidification profile based on the grain size distribution, beam size, and β-phase fraction. Nevertheless,

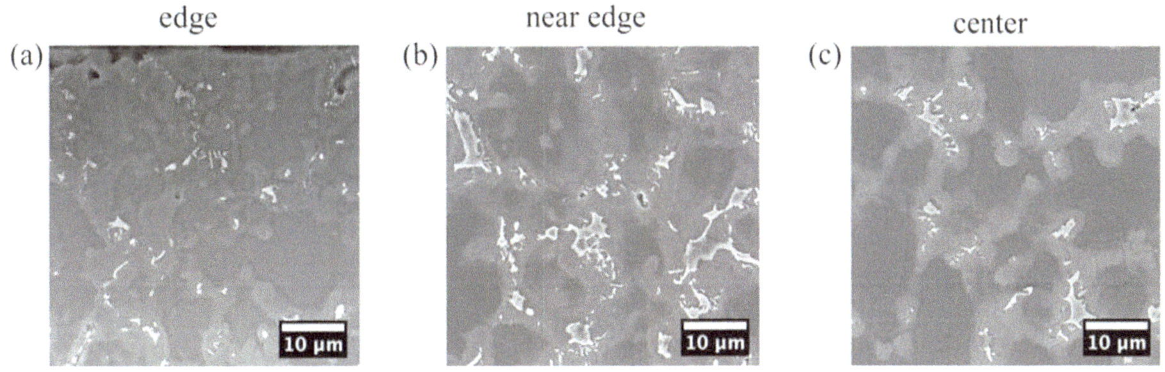

Figure 1: Secondary electron images of the microstructures (a) within 50 μm of the edge of the plate, (b) 200 to 400 μm from the edge ("near edge"), and (c) at mid-thickness in the plate.

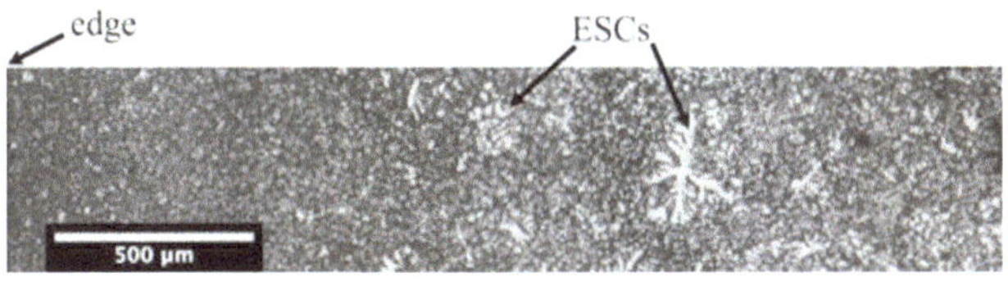

Figure 2: Distribution of ESCs in the 5.0 mm thick plates as observed by optical microscopy. ESCs segregate to the center (right side).

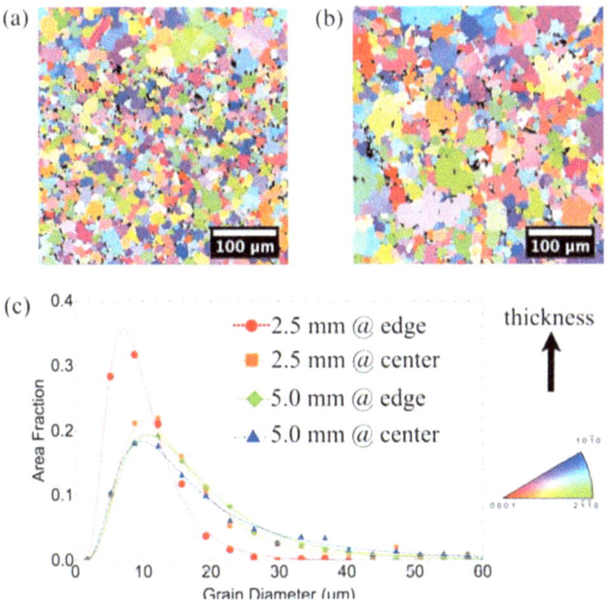

Figure 3: EBSD Inverse pole figure maps at the (a) edge and (b) center of the 5.0 mm thick plates. (c) The grain size distributions at the edge and center for both plate thicknesses.

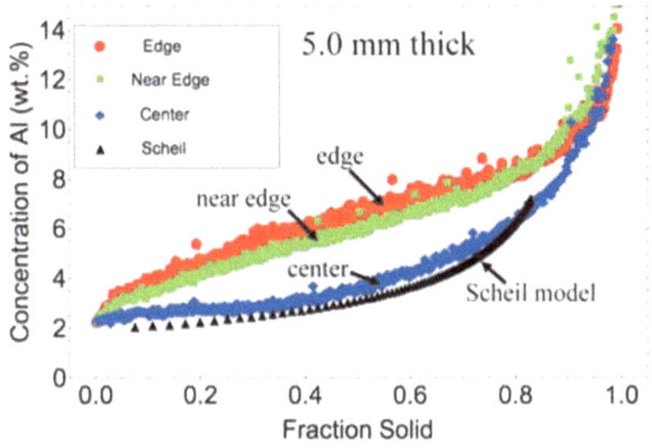

Figure 4: Summary of microsegregation in 5 mm thick AM70 plate.

examination of the microstructure and microsegregation in the two plate thicknesses validates the EPMA microsegregation profiles. Since the grain size distribution at the edge and center of the 5.0 mm plate are similar, we can conclude that the higher Al concentration measured at the edge is real and not a consequence of the beam interaction volume. In addition, despite the smaller grain size at the edge of the 2.5 mm thick plates compared to the 5.0 mm thick plates, the microsegregation profiles are comparable. At the edge of the plates, where solidification is most rapid, more Al is trapped in the α-Mg phase, resulting in a lower β-phase volume fraction as observed in Figure 1.

Microsegregation following Annealing

To begin to study changes in microsegregation profiles during heat treatment, two solution treatments (20 h and 44 h at 420 °C) were performed on the 5 mm thick AM70 plates. Examination in SEM showed complete dissolution of the β-phase after 20 h, though Al-Mn intermetallics remained (Figure 6). Following

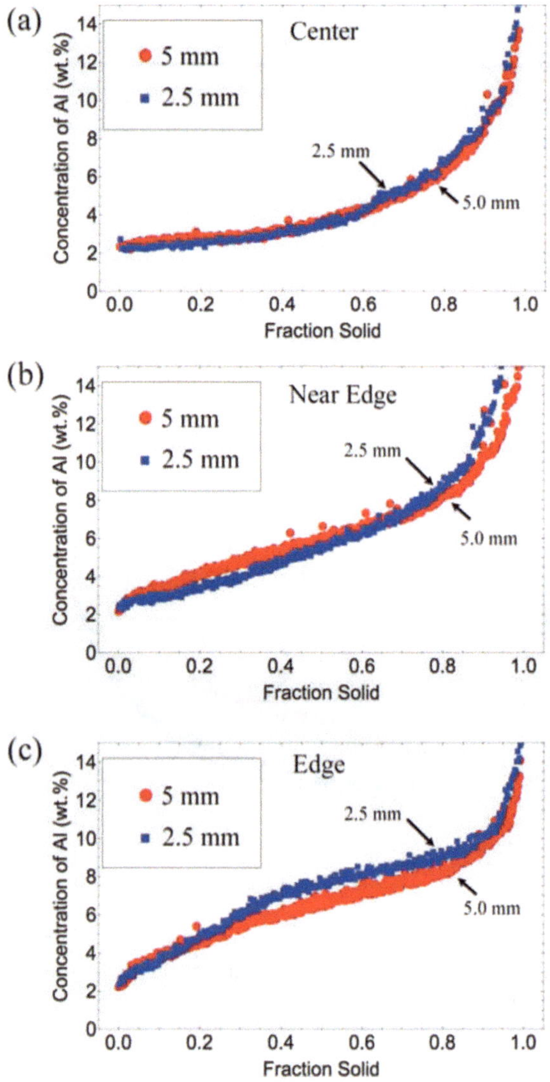

Figure 5: Comparison of microsegragtion at the (a) center, (b) near edge, and (c) edge for the two plate thicknesses.

both solution treatments, microsegregation is almost completely eliminated; however, macrosegregation (differences through the plate thickness) remain (Figure 7). As shown in Table II, the average Al content in the center region increased from 4.8 wt% in the as-cast condition to 5.7 wt% following both solution treatments. Elimination of macrosegregation would require heat treatment durations that are impractical in an industry setting. As a consequence, HPDC parts will have less Al available in the bulk for solid solution strengthening or for aging treatments in heat treatable components produced by Super Vacuum Die Casting.

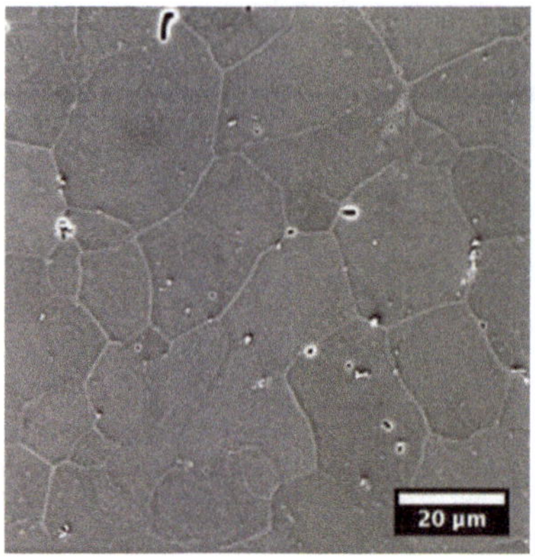

Figure 6: Secondary electron images of the microstructure at the center of the 5.0 mm thick plates following solution treatment at 420 °C for 20h.

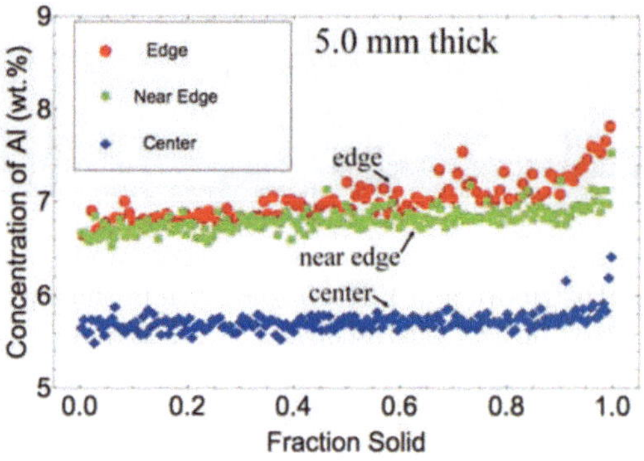

Figure 7: Summary of microsegregation in 5 mm thick AM70 plate following solution treatment at 420 °C. For clarity only the results from the 44h treatment are shown, differences between the two durations were negligible.

Summary

It has been demonstrated that microstructure and microsegregation are casting location dependent in HPDC AM70. At the edge of the plate, where the cooling rate is the highest, more Al is trapped in the α-Mg matrix, leading to a lower β-phase volume fraction. Towards the center of the plate, i.e., as the cooling rate decreases, microsegregation increases. In addition, the microstructure tends to become coarser, though this is partially due to segregation of ESCs to the center of the casting. At mid-thickness the microsegregation can be adequately described by the Scheil model, however in order to describe microsegregation at the edge of the casting, and macro segregation through the plate thickness, solute trapping will need to be taken into account.

Table II: Macrosegregation in 5.0 mm thick AM70 plates in the as-cast state and following annealing at 420 °C for 44 h. Aluminum concentration is given in wt%.

	as-cast	solution treated
edge	7.1	7.2
near edge	6.5	6.9
center	4.8	5.7

Acknowledgements

This material is based upon research supported by the Department of Energy under contract DE-EE0006434. The authors express their gratitude to S.K. Makineni and B. Qu for their contributions to this work.

References

1. N. Saddock, A. Suzuki, J. TerBush, E. Heinninger, J. Zindel, J. Allison, T. Pollock, and J. Jones: in A. Luo, N. Neelameggham, and R. Beals, editors, *Magnesium Technology*, The Minerals, Metals & Materials Society, Warrendale, PA, vol. 4, 2006 pp. 77–82.

2. D. Porter and K. Easterling: *Phase Transformations in Metals and Alloys*, CRC Press, Boca Raton, Fl, 3 edn., 2009.

3. X. Zheng, A. A. Luo, C. Zhang, J. Dong, and R. A. Waldo: *Metallurgical and Materials Transactions A*, 2012, vol. 43, pp. 3239–3248.

4. M. I. Khan, A. O. Mostafa, M. Aljarrah, E. Essadiqi, and M. Medraj: *Journal of Materials*, 2014, vol. 2014, Article ID 657647, pp. 1–18.

5. C. Zhang, D. Ma, K.-S. Wu, H.-B. Cao, G.-P. Cao, S. Kou, Y. Chang, and X.-Y. Yan: *Intermetallics*, 2007, vol. 15, pp. 1395–1400.

6. D. Mirković and R. Schmid-Fetzer: *Metallurgical and Materials Transactions A*, 2009, vol. 40, pp. 958–973.

7. J. R. TerBush, N. D. Saddock, J. W. Jones, and T. Pollock: *Metallurgical and Materials Transactions A*, 2010, vol. 41, pp. 2435–2442.

8. M. Paliwal, D. H. Kang, E. Essadiqi, and I. H. Jung: *Metallurgical and Materials Transactions A*, 2014, vol. 45, pp. 3596–3608.

9. H. Gjestland and H. Westengen: *Advanced Engineering Materials*, 2007, vol. 9, pp. 769–776.

10. M. N. Gungor: *Metallurgical Transactions A*, 1989, vol. 20, pp. 2529–2533.

11. M. Ganesan, D. Dye, and P. D. Lee: *Metallurgical and Materials Transactions A*, 2005, vol. 36, pp. 2191–2204.

12. H. I. Laukli, C. M. Gourlay, and A. K. Dahle: *Metallurgical and Materials Transactions A*, 2005, vol. 36, pp. 805–818.

Magnesium Technology 2016
Edited by: Alok Singh, Kiran Solanki, Michele V. Manuel, and Neale R. Neelameggham
TMS (The Minerals, Metals & Materials Society), 2016

PREDICTING SOLIDIFICATION PROPERTIES OF MAGNESIUM BY MOLECULAR DYNAMICS SIMULATIONS

Ebrahim Asadi[1] and Mohsen Asle Zaeem[2]

[1] Department of Mechanical Engineering, The University of Memphis, Memphis, TN 38152, USA

[2] Department of Materials Science and Engineering, Missouri University of Science and Technology, Rolla, MO 65409, USA

Keywords: Solidification; Magnesium; Capillary Fluctuation Method; Molecular Dynamics.

Abstract

In this work, we studied the solid-liquid coexistence properties in Mg by using the modified-embedded atom method (MEAM) in molecular dynamics (MD) simulations. The solid-liquid coexisting approach was used, and the melting point of 937.9 K, latent heat of 10.2 kJ/mol, and liquid density of 0.037 atom/Å^3 were predicted, which are in good agreement with the experimental data. The capillary fluctuation method (CFM) was used to determine the solid-liquid interface free energy and anisotropy parameters. Eight slip and twinning planes (basal, two prismatic, two pyramidal, and three twinning planes) were used as the solid-liquid interface planes. The average solid-liquid interface free energy of 122.2 mJ/m^2 was predicted.

Introduction

Magnesium is a light weight metal and is one of the most abundant metals in Earth. Mg alloys, because of their desirable mechanical properties, are suitable candidates for many applications in different industries such as automobile and aerospace industries [1]. The mechanical properties of cast Mg alloys in many ways depend on the solidification properties and the subsequent microstructures [2]. Therefore, understanding the solidification of Mg is of extreme importance to optimize the manufacturing process of Mg alloys and ensure enhanced mechanical properties for different applications. The current advancement in high performance computing has made nanoscale computational modeling an excellent technique to study nano-structural evolution in materials, and molecular dynamics (MD) method is a well-established computational tool for this purpose [3].

To the date, the only MD effort dealing with the solidification simulations of Mg has been the work of Sun et al. [4] which was based on the capillary fluctuation method (CFM). CFM is based on equilibrating a coexisting solid-liquid structure at the exact melting point of the material, determining the location of the solid-liquid interface, finding the Fourier modes of the solid-liquid interface, and using them to determine the solid-liquid properties [3, 5-7]. They used embedded-atom method (EAM) [8] as the interatomic potential in their MD simulations; EAM is a semi-empirical many-body potential for the atomistic simulations of metallic systems. Until now and to the knowledge of the authors, the work of Sun and coworkers is the only MD work to determine solidification properties of Mg. The other works related to the MD simulations of HCP-liquid coexisting systems include the study of the melting [9-11], interface properties [12] and kinetic mobility [13, 14].

In this work, MD simulations of solidification of Mg are performed using a different class of interatomic potentials, modified embedded-atom method (MEAM) potentials [15]. MEAM adds the directionality of bonding in covalent materials to the EAM formalism. MEAM potentials are widely used in the computational materials science and engineering community to simulate unary, binary, ternary, and multi-component metallic systems [16]. Since only EAM-MD simulations have been used to determine the solidification properties of Mg, using MEAM-MD simulations in this study provides another perspective and more confidence in using MD simulations for this purpose. In fact, we recently used MEAM to study the solid-liquid coexistence of BCC and FCC metals [17, 18] and we showed that the calculated solidification properties are in a very good agreement with the experimental data. We study the solidification properties of Mg more comprehensively by including all of the deformation planes in HCP crystals (eight different slip and twining planes) as the HCP-liquid interface planes; this was missing in the previous studies. In this work, we first calculate the melting point and properties of Mg using the solid-liquid coexistence approach which is crucial for the rest of the simulations. Then, we use CFM with HCP-symmetry adopted spherical harmonics to determine the solid-liquid interface stiffness, solid-liquid interface free energy, and surface anisotropy for Mg.

Interatomic Potential

We use the second nearest-neighbor (2NN) formalism of MEAM [19, 20] as the interatomic potential for Mg in this work. The description of the formalism of 2NN-MEAM is well-documented in the related published works and the interested reader are directed to Ref. [21]. MEAM for single element system has fourteen parameters in its formalism which need be determined for the selected material. We use the parameters which were presented previously by Kim et al. [22]; they showed that using these parameters result in reasonable calculations of the low and high temperature properties of Mg, which are comparable to the experimental and *ab initio* MD data. The properties considered for the development of these MEAM parameters for Mg were: cohesive energy, lattice parameters, elastic constants, structural energy differences, vacancy formation and migration energies, divacancy formation energy, self-interstitial energy, stacking fault energies, surface energies, specific heat, thermal expansion, melting point, latent heat, and expansion in melting. In our simulations, the large-scale atomic/molecular massively parallel simulator (LAMMPS) MD code [23] is used. Periodic boundary conditions and a time step size of 0.002 ps are used for all the MD simulations.

MD Simulations and Results

In order to accurately calculate the melting point for Mg corresponding to the utilized interatomic potential, we employ the solid-liquid coexisting approach [24]. In this approach the melting point is calculated as a temperature at which a two-phase coexisting solid-liquid structure exists. In the other words, the number of solid and liquid atoms remain constant as the whole structure equilibrates at that given temperature. We use eight different planes as the solid-liquid interface in this work which are schematically shown at Fig. 1. The labeling shown in Fig. 1 will be used throughout this work to refer to different crystallographic configurations.

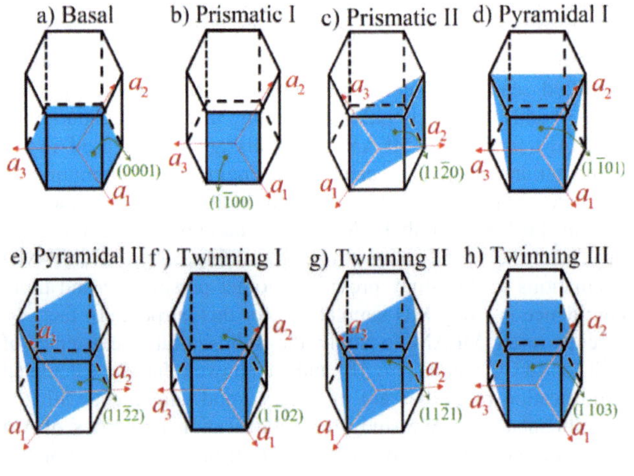

Figure 1. The schematic of the eight different HCP-liquid interface planes.

Here, we explain the method to accurately calculate the melting point of Mg using MD simulations. First, we calculate the lattice parameters of solid and liquid versus temperature variation using a standard NVT (canonical) ensemble. These lattice parameters will be used in the next simulations to equilibrate relaxed solid and liquid Mg at different temperatures. Second, a simulation cell consisting of $12\times15\times120$ lattice cells is initially equilibrated at the initial guess of the melting point T_M^G at the the corresponding lattice parameters obtained from the previous step. The longer direction (<0001>) is normal to the solid-liquid interface and it is called normal direction hereafter. The central half of the simulation cell is held fixed in a NVT ensemble, while the other half is melted at a high temperature (□ $1.6T_M^G$) using Nòse-Hoover thermostating in NVT ensemble. Then the melted half of the simulation cell is equilibrated at T_M^G using an NPT ensemble, while the length of the cell in the normal direction is allowed to relax and the central half is held fixed. The required simulation time for this step is less than 0.5 ns. Finally, the entire simulation cell is allowed to relax at the normal direction for 1 ns using an NPT ensemble at T_M^G to reduce the normal pressure to atmospheric pressure. Fig. 2 shows the average energy per atom versus different T_M^G for the last simulation step. The simulation cell solidifies entirely for temperatures of 930 K and 940 K after 0.1 ns. The same pattern exists for all the temperatures except for the exact melting point. The best fit lines for 960 K and 940 K are also depicted by solid black lines in Fig. 2, where the negative slope changes to a positive slope, respectively, indicating that the accurate melting temperature should be between these two temperatures (the slope of the best fit line is theoretically zero for the melting point). It is difficult to target the accurate melting temperature using this simulation though. Therefore, the solid-liquid coexisting structure at 935 K is further equilibrated in an isenthalpic ensemble (NPH) while allowing the normal direction to relax and the accurate melting point is calculated as the average temperature at a relatively high simulation time (~10 ns). The calculated melting point is used as T_M^G and the whole simulation is repeated until the melting temperature is converged.

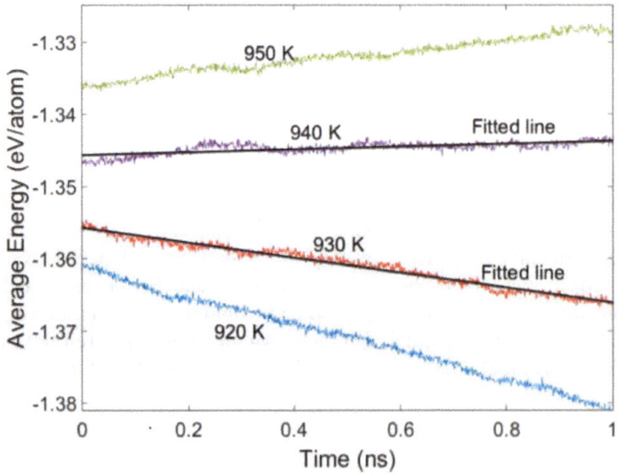

Figure 2. Plot of the average energy versus simulation time for an initially coexisting solid-liquid structure of Mg equilibrated at different temperatures. The solid-liquid interface plane is the basal plane.

A snapshot of solid-liquid coexisting structure where basal plane is the solid-liquid interface is shown at Fig. 3. The explained procedure was repeated for all the eight orientations shown at Fig 1 and the melting point is calculated as the average value of the all calculations (937.9 K). It is worth mentioning that the variation of the calculated melting point for different orientations was less than 1% which shows that the calculated melting point is orientation independent. Also, we conducted a convergence study regarding the size of the simulation box and confirmed that the calculated melting point is also orientation independent. Finally, we calculate the latent heat as 10.2 kJ/mol, liquid density as 0.037 atom/Å^3, and expansion in melting as 1.84 Å^3/atom for Mg which are in reasonable agreement with their experimental counterparts; i.e. 10.1 kJ/mol, 0.037 atom/Å^3, and 1.80 Å^3/atom, respectively.

Figure 3. A snapshot of solid-liquid coexisting structure where basal plane is the solid-liquid interface.

We utilize CFM [3] to determine solid-liquid interface free energy and its anisotropy for Mg. The first step in using CFM is to construct two-phase solid-liquid coexisting slabs where the solid-liquid interface plane can be one of the eight planes denoted in

Fig. 1. The details of using CFM requires a spate article and we will only review this method in here. First, we construct a solid-liquid coexisting slab of Mg representing directional solidification. For instance, a snapshot of the constructed solid-liquid coexisting slab for basal plane is shown at Fig. 4.

Figure 4. A snapshot of a solid-liquid coexisting slab used in CFM to calculate the interface stiffness of Mg. The solid red line shows the location of solid-liquid interface.

The second step in using CFM, is to identify the solid-liquid interface location by defining an order parameter to identify atoms as solid or liquid. The third step is to determine the Fourier amplitude of the solid-liquid interface function and use it to calculate solid-liquid interface stiffness for each orientation. Fourth, HCP-adopted spherical harmonics are used to explain the solid-liquid interface free energy and surface anisotropy as

$$\gamma(\theta,\phi) = \gamma_0[1 + \varepsilon_{20}\, y_{20}(\theta,\phi) + \varepsilon_{40}\, y_{40}(\theta,\phi) + \varepsilon_{60}\, y_{60}(\theta,\phi) + \varepsilon_{66}\, y_{66}(\theta,\phi)], \quad (1)$$

where

$$y_{20} = \frac{1}{4}\sqrt{\frac{5}{\pi}}[3\cos(\theta)^2 - 1],$$

$$y_{40} = \frac{3}{16}\sqrt{\frac{1}{\pi}}[35\cos(\theta)^4 - 30\cos(\theta)^2 + 3],$$

$$y_{60} = \frac{1}{32}\sqrt{\frac{13}{\pi}}[231\cos(\theta)^6 - 315\cos(\theta)^4 + 105\cos(\theta)^2 - 5],$$

$$y_{66} = \frac{1}{64}\sqrt{\frac{6006}{\pi}}\sin(\theta)^6\cos(6\phi) \quad (2)$$

and ε_{20}, ε_{40}, ε_{60}, and ε_{66} are anisotropy parameters, γ_0 is the average interface free energy. Finally, the solid-liquid interface free energy and surface anisotropy are determined by comparing the interface stiffness calculated using CFM and HCP-symmetry adopted spherical harmonics.

Using the explained method, we calculated the solid-liquid interface free energy of 122.2 mJ/m^2 which is in the order of the calculations of Sun et al. [4] as 89.9 mJ/m^2. Also, we calculated the surface anisotropy parameters for Mg as: $\varepsilon_{20} = -3.45\%$, $\varepsilon_{40} = -0.47\%$, $\varepsilon_{20} = -0.06\%$, $\varepsilon_{20} = -0.46\%$. Also, the preferred dendrite growth direction may be obtained by calculating the trace of stiffness tensor from the solid-liquid interface free energy and anisotropy parameters [4]. Using these calculations, we predict the basal plane and $[11\bar{2}0]$ direction as the primary preferred growth direction (primary dendrite arms direction). Also, the the least preferred growth direction is $[0001]$. These findings are in agreement with the EAM-MD calculations of Sun at al. [4]. In addition, we predict $[33\bar{6}2]$ as the second preferred dendrite growth direction for Mg.

Conclusion

Solidification properties of Mg were determined by MEAM-MD simulations. In the concept of CFM, MEAM interatomic potentials were used in the MD simulations, and the pyramidal and twinning (as well as basal and prismatic) orientations were considered for the first time to determine the solid-liquid interface properties of Mg. The average interface free energy was calculated to be $\gamma_0 = 122.2$ mJ/m^2, and the solid-liquid interface anisotropic parameters were calculated to be $\varepsilon_{20} = -3.45$, $\varepsilon_{40} = -0.47$, $\varepsilon_{66} = -0.46$, and a negligible ε_{60}. Furthermore, the calculations of the HCP-liquid interface free energy and anisotropy predicts the primary dendrite growth direction of $[11\bar{2}0]$ and a secondary dendrite growth direction of $[33\bar{6}2]$.

Acknowledgment

The authors are grateful for computer time allocation provided by the Extreme Science and Engineering Discovery Environment (XSEDE).

References

[1] E.F. Emley. Principles of magnesium technology, (1966).

[2] K. Pettersen, O. Lohne, N. Ryum. Dendritic solidification of magnesium alloy AZ91, Metallurgical Transactions A 21 (1990) 221-230.

[3] J. Hoyt, M. Asta, A. Karma. Method for computing the anisotropy of the solid-liquid interfacial free energy, Physical review letters 86 (2001) 5530.

[4] D. Sun, M. Mendelev, C. Becker, K. Kudin, T. Haxhimali, M. Asta, J. Hoyt, A. Karma, D. Srolovitz. Crystal-melt interfacial free energies in hcp metals: A molecular dynamics study of Mg, Physical Review B 73 (2006) 024116.

[5] J. Hoyt, M. Asta. Atomistic computation of liquid diffusivity, solid-liquid interfacial free energy, and kinetic coefficient in Au and Ag, Physical Review B 65 (2002) 214106.

[6] J. Hoyt, M. Asta, A. Karma. Atomistic and continuum modeling of dendritic solidification, Materials Science and Engineering: R: Reports 41 (2003) 121-163.
[7] D. Sun, M. Asta, J. Hoyt. Crystal-melt interfacial free energies and mobilities in fcc and bcc Fe, Physical Review B 69 (2004) 174103.
[8] M.S. Daw, M.I. Baskes. Embedded-atom method: Derivation and application to impurities, surfaces, and other defects in metals, Physical Review B 29 (1984) 6443.
[9] P. Bavli, J. Adler. Parallel codes for simulating elastic constants and melting in Ar and Mg, Physics Procedia 4 (2010) 3-7.
[10] P. Bavli, E. Polturak, J. Adler. Molecular dynamics study of melting of the hcp metal Mg, Physical Review B 84 (2011) 235442.
[11] P. Bavli, M. Sc. Melting in HCP Lattices. Technion-Israel Institute of Technology, Faculty of Physics, 2009.
[12] Y. Bai. MOLECULAR DYNAMICS SIMULATION STUDY OF SOLID-LIQUID INTERFACE PROPERTIES OF HCP MAGNESIUM, (2012).
[13] Y. Gao, Y. Yang, D. Sun, M. Asta, J. Hoyt. Molecular dynamics simulations of the crystal–melt interface mobility in HCP Mg and BCC Fe, Journal of Crystal Growth 312 (2010) 3238-3242.
[14] Z. Xia, D. Sun, M. Asta, J. Hoyt. Molecular dynamics calculations of the crystal-melt interfacial mobility for hexagonal close-packed Mg, Physical Review B 75 (2007) 012103.
[15] M. Baskes. Modified embedded-atom potentials for cubic materials and impurities, Physical Review B 46 (1992) 2727.
[16] B. Jelinek, S. Groh, M. Horstemeyer, J. Houze, S. Kim, G. Wagner, A. Moitra, M. Baskes. Modified embedded atom method potential for Al, Si, Mg, Cu, and Fe alloys, Physical Review B 85 (2012) 245102.
[17] E. Asadi, M.A. Zaeem, S. Nouranian, M.I. Baskes. Two-phase solid–liquid coexistence of Ni, Cu, and Al by molecular dynamics simulations using the modified embedded-atom method, Acta Materialia 86 (2015) 169-181.
[18] E. Asadi, M.A. Zaeem, S. Nouranian, M.I. Baskes. Quantitative modeling of the equilibration of two-phase solid-liquid Fe by atomistic simulations on diffusive time scales, Physical Review B 91 (2015) 024105.
[19] B.-J. Lee, M. Baskes. Second nearest-neighbor modified embedded-atom-method potential, Physical Review B 62 (2000) 8564.
[20] B.-J. Lee, M. Baskes, H. Kim, Y.K. Cho. Second nearest-neighbor modified embedded atom method potentials for bcc transition metals, Physical Review B 64 (2001) 184102.
[21] S. Nouranian, M.A. Tschopp, S.R. Gwaltney, M.I. Baskes, M.F. Horstemeyer. An Interatomic Potential for Saturated Hydrocarbons Based on the Modified Embedded-Atom Method, Physical Chemistry Chemical Physics 16 (2014) 6233-6249.
[22] Y.-M. Kim, N.J. Kim, B.-J. Lee. Atomistic modeling of pure Mg and Mg–Al systems, Calphad 33 (2009) 650-657.
[23] S. Plimpton. Fast parallel algorithms for short-range molecular dynamics, Journal of computational physics 117 (1995) 1-19.
[24] E. Asadi, M.A. Zaeem, M.I. Baskes. Phase-Field Crystal Model for Fe Connected to MEAM Molecular Dynamics Simulations, JOM 66 (2014) 429-436.

Magnesium Technology 2016

Keynote Session Part II and Primary Production and Recycling

Magnesium Technology 2016
Edited by: Alok Singh, Kiran Solanki, Michele V. Manuel, and Neale R. Neelameggham
TMS (The Minerals, Metals & Materials Society), 2016

A PERSPECTIVE: POTENTIAL GROWTH IN THE GLOBAL MAGNESIUM INDUSTRY – WHERE IS OUR RESEARCH LEADING US?

Martyn Alderman

Magnesium Elektron, Magnesium Technology Centre, Rake Lane, Swinton, Manchester, UK

Keywords: Primary Market, Aerospace, Automotive, Biomedical, Additive Manufacturing

Abstract

Magnesium enjoyed considerable growth in the last decade of the twentieth century due to an increase in automotive die casting applications. In spite of considerable research activity into sheet applications, the use of magnesium sheet is largely limited to non-structural applications, primary batteries and small electronic devices (cameras- cell phones- lap-tops & tablets). The major volume use for magnesium remains as an alloying element for aluminium, and as a chemical reducing or modifying agent in the production of titanium, steel and SG iron. Powders are currently used in organic chemical reactions and as thermal decoy devices rather than for any structural purpose.

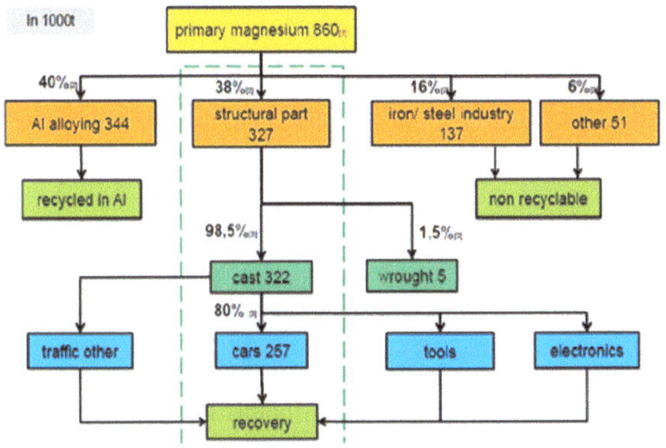

Figure 1. Magnesium usage globally in thousands of metric tonnes – Source DLR from 2013 IMA data

A number of magnesium primary projects in Australia failed to meet the basic scale economics in the 1990's and are closed or still on the drawing board. Electrolytic smelters in Canada and Norway have closed over the past 15 years only to resurface as projects in China. The economics of the Dead Sea Magnesium smelter have been brought into question by natural resource taxation. Some new electrolytic capacity is about to come on stream in China in addition to Pidgeon Process movements that have followed the economies of coal and ferrosilicon production from Province to Province, but the primary industry economics remain weak outside of the USA.

Today the Chinese are focused on "deep processing" to promote higher value added magnesium end use within their own economy. Will the addition of capacity in China and newly emerging Pidgeon capacity in Turkey[1] and Iran[2] bring low cost magnesium to the market place to compete with aluminium in the automotive and aerospace industries, or are there more fundamental issues beyond the "cost" of pure magnesium that continue to restrict applications?

The focus of this paper is to discuss whether our research can drive further magnesium growth, or is it an issue of macroeconomics and environmental pressures that will determine the breadth of our future applications.

Discussion

The most significant structural field of application of magnesium alloys remains in the field of automotive die castings, where the mass saving and consequent fuel efficiency provide a life cycle benefit in terms of reduced carbon dioxide emissions, although this life cycle benefit[3] is very dependent upon the source of primary energy. For this reason the US Government has channeled new ARPA-E funding[4] into alternative processes for the generation of Primary Magnesium outside of China, but it remains to be seen whether these will be commercialized in the timescale required to have a significant impact on the future of low carbon vehicles.

The CVM primary smelter based on horizontal retort Pidgeon process in Malaysia has been beset with technical difficulties and has failed to reach its scheduled 60,000tpa capacity[2]. Nevada Clean Magnesium & ScanMag together have proposed a new primary reduction plant in both Nevada and Glomfjord in Norway[5] that will use a modified Bolzano process. But in spite of being "shovel ready" the SilMag electrolytic plant in Norway that has been on the table for 10 years now has struggled to find adequate financial backing[6], in the face of a Chinese dominated primary price that has continued to fall in the past two years. The large Qinghai plant[7] scheduled to bring another 100,000t (and potentially a further 350,000 t) is likely either to drive other Pidgeon capacity out of business or threaten these other low carbon projects on the table in the West.

Although the lightweighting benefits of magnesium in large castings continues to be exploited in the rotorcraft industry to enhance performance, payload and time on station, the use of magnesium in commercial fixed wing structures declined in the 1970's with increasing engine power and concern about corrosion and flammability being the detractors for the use of the lightest structural metal. The past seven years of work to change the opinion of the Federal Aviation Authority about the acceptability of magnesium inside the aircraft cabin has now resulted in the "magnesium alloys must not be used" statement being removed from SAE AS8049C[8].

The life cycle analysis of increased magnesium use in the cabin suggests that, in aviation fuel saving alone, a payback will be achieved in less than a month in the air, but with lower oil prices currently influencing the payback in the lightweighting calculations, the industry will need to continue to address the challenges of alternative materials such as carbon fibre composites and aluminium lithium alloys if it is to make headway into the structure of the aircraft. Higher performing alloys may well be necessary to compete against the structural aluminium alloys and it is here that an ICME approach has the potential to offer shorter development timescales if the fundamental thermodynamic databases are available to support it.

There are however other applications for magnesium alloys beyond aerospace and automotive; we continue to see a significant number of papers at TMS in the field of Magnesium Biotechnology. The medical sector is however a very tightly regulated environment and the barriers to entry remain high and volumes extremely low. There are certainly good applications in cardiac stents, bone plates and wires for those at the forefront of the technology but volumes in this market sector will never demand a new smelter.

The Technical and Economic Press talks incessantly about the game changing nature of 3D printing or Additive Layer Manufacturing in our world, and with the emerging availability of Magnesium Alloys Powders the potential for lightweight innovative structures incorporating magnesium alloys is beginning to be explored. The key markets here are likely to be specialist aerospace (with the emphasis on space) applications and on the biomedical sector for customized implants.

The Magnesium Industry continues to need innovation in alloy design to remain competitive with alternative materials, and understanding deformation mechanisms in alloy systems with limited room temperature ductility and methods of reducing anisotropy and asymmetry remain important in widening structural applications whether it be for automotive or aerospace.

Ultimately if the Technology presented in the volume is to be useful it must be affordable in its target market, so it makes sense to continue development work in the field of twin roll casting and room temperature formable alloys for the automotive industry, but as the aerospace industry has demonstrated in the past, a lowering in fuel prices impacts behavior. In today's economy it appears that only when CAFÉ or EU CO_2 emission standards demand it will the issues of incorporating mixed materials into vehicles be addressed to allow magnesium sheet use beyond internal closure type components. There has been talk of electric vehicles needing increased range driving lightweighting, but the question remains whether battery technology will outpace the need for lightweighting in the way that jet engines drove magnesium out of the airframe in the 1960's and 1970's.

The magnesium industry caught a cold in 2008 after the Beijing Olympics when uncertainty over supply drove prices beyond the expectations of the automotive industry, which was then hit by a Global Economic crisis. Seven years later the automotive industry is again considering magnesium for programmes that are 5 years out, the technology challenges remain achieving Class A surfaces, and integration technologies off the back of a small installed base of suppliers. There also remains the question whether extrusion productivity and asymmetry of properties can be addressed adequately to make magnesium extrusions competitive with aluminium for transport applications in both the automotive and rail sectors.

Good work has been carried out in Japan on the development of extruded alloys for large rail section components[9] which are being encouraged by the Japanese Government but the installed infrastructure to produce these large sections is very limited outside of Russia and China. Magnesium market consumption still remains at only 1.7% of Global Aluminium Consumption making it difficult for wrought capacity expansions to be justified unless underwritten by a major supply contract.

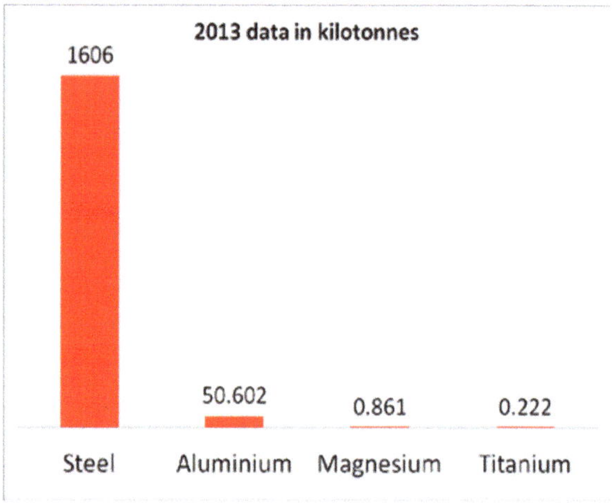

Figure 2. Global metal consumption in kilotonnes for major structural materials

The question remains whether the results of our past twenty years of research activity and that in the pipeline will yield a volume transformation that the industry has been looking for or will growth continue at marginally above GDP rates?

References

1. New Magnesium Metal Production Plant In Turkey – Magnesium Monthly Review -March-April 2015 MMR 10 No. 3
2. Iran Has Ribbon Cutting On New Magnesium Metal Production Plant – Magnesium Monthly Review – Dec 2014-Jan 2015 No. 12
3. Life Cycle Assessment of Magnesium Components in Vehicle Construction – International Magnesium Association 2013 www.intlmag.org
4. http://www.arpa-e.energy.gov/ sites/default/files/documents/files/METALS%26REMOTE_Project_Descriptions_091913.pdf
5. Nevada Clean Magnesium Signs Joint Venture Agreement With ScanMag AS – The Newswire - June 22, 2015 www.thenewswire.ca
6. Hydro and Serenity sign agreement for possible restart of magnesium production at Herøya - Norsk Hydro http://www.hydro.com/en/Press-room/News/Archive/2013/Hydro-and-Serenity-sign-agreement-for-possible-restart-of-magnesium-production-at-Heroya/[03/10/2013]
7. 2013 June Presentation of the Qinghai Salt Lake Project at the 70th IMA Conference in Xi'an - http://magontec.com/magontec/index.php?navID=55
8. Magnesium in Aircraft Seats –SAE Update - International Aircraft Materials Fire Test Working Group https://www.fire.tc.faa.gov/pdf/materials/Feb15Meeting/Lyon-0215-ModifiedAircraftSeats.pdf
9. Japanese Developing Magnesium for High Speed Trains – Magnesium Monthly Review –August 2014 – MMR10 No7

Magnesium Technology 2016
Edited by: Alok Singh, Kiran Solanki, Michele V. Manuel, and Neale R. Neelameggham
TMS (The Minerals, Metals & Materials Society), 2016

Study on Mechanism of Magnesia Production by Reversion Reaction Process in Vacuum

Yang TIAN[1, 2, 3,4], Bao-qiang Xu[1, 2,3,4], Cheng-bo YANG [1, 2,3,4], Bin YANG[1, 2,3,41], Da-chun LIU[1, 2,3,4], Tao QU[1, 2,3,4],Yong-nian DAI[1, 2,3,4]

(1 State Key Laboratory of Complex Nonferrous Metal Resources Clear Utilization, Kunming University of Science and Technology,Kunming 650093, China;
2 National Engineering Laboratory for Vacuum Metallurgy, Kunming University of Science and Technology, Kunming 650093, China;
3 Key Laboratory of Non-Ferrous Metals Vacuum Metallurgy of Yunnan Province, Kunming 650093, China
4 College of Metallurgy and Energy engineering, Kunming University of Science and Technology, Kunming 650093, China)

The mechanism of magnesia production was investigated experimentally in reversion reaction process in vacuum. The results showed that the reversion reactions were exothermic reaction, the gas-phase reversion will commence as soon as a saturated gas mixture was cooled down. MgO and Mg_2C_3 were detected in the condensing tower by Mg vapors and CO during reversion reaction process below 1373K. The reversion reaction indicated that Mg and CO was not only occurred in MgO, but also yielded got Mg_2C_3. The main reaction was 6Mg(g)+4CO= Mg_2C_3+4MgO+C. A high purity metal magnesium product was obtained in the the vacuum distillation process. The calculated peak value of reversion reaction ratio was less than 9%. Contrast the condensing product and purity magnesium distillation product, the condensing product had a filament crystallization and loose structure. But the purity magnesium product had a lump structure and good crystalline feature.

Keywords: Magnesia; Carbothermic reduction; Reversion reaction ratio; Mg_2C_3;

[1]**Foundation item**: Project supported by National Natural Science Foundation of China(NO.51304095); Science and Technology Planning Project of Yunnan Province, China (NO.2013FZ029); The Program for Innovative Research Team in University of Ministry of Science and Technology of China[No.2014RA4018]. Fund of Yunnan Collaborative Innovation Center of Complex Nonferrous Metal Resources Clear Utilization(NO. 2014XTZS005).
Corresponding author: Bin YANG; Tel: +86-871-65161583; E-mail:kgyb2005@126.com.

1 Introduction

In recent years, the annual output of magnesium has increased remarkably in China, and magnesium alloys are being applied to motorcycles, automobiles, electric devices and so on[1-2].Magnesium is produced by two principal processes: electrolysis of molten magnesium chloride and thermal reduction of magnesia[3] In recent years, production of magnesium in China by the Pidgeon process has dominated world production. In this process, ferrosilicon reduces magnesia from calcined dolomite under vacuum. The process suffers from high energy usage and low productivity[4]. The carbothermic reduction of magnesia to produce magnesium offers the potential of a lower energy and higher productivity route for metal production compared to existing industrial routes[5].The main difficulties were considered to be due to the high temperature(about 2000℃)that must be maintained to reduce magnesia with carbon at atmospheric pressure and to the necessity to quench the gases leaving the reacting chamber to avoid the back-reaction between magnesium vapors and carbon monoxide[6].Recent years, carbothermic reduction magnesia under vacuum conditions had been studied at KMUST[7-10], the initial reaction temperature decreased with reducing the total pressure. The fringe crystal of magnesium is obtained by the confirmation experiments in vacuum furnace made for magnesia reduction. Rencently, the great majority of work is concentrating on carbothermic reduction process, particularly focusing on analysing reverse reaction. In this paper, reverse reaction under vacuum conditions has been studied. The purpose of the study is to establish the product of the reverse reaction, and to calculate the reverse reaction ratio by experiment investigated.

2 Experimental details

2.1Raw material

Analytical-grade CaF_2, magnesia and carbon were used as the raw materials in our experiment.

2.2Schematic diagram of vacuum furnace

The experiments were carried out in a self-made vacuum furnace shown in Figure.1.

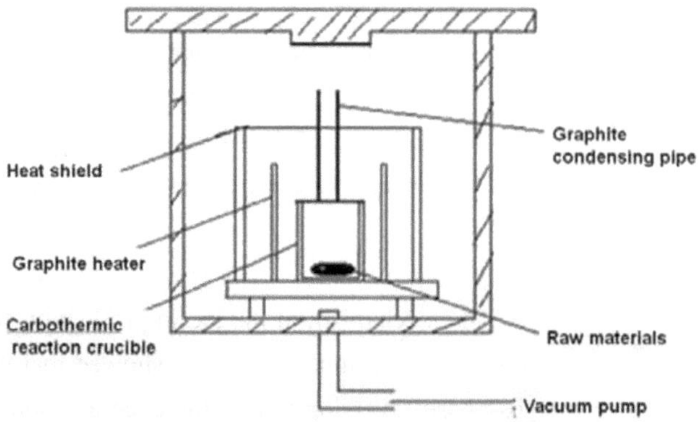

Fig.1 Schematic diagram of vacuum furnace

2.3Experimental procedure

Magnesia(20g), carbon(12g) as the starting materials were mixed in a molar ratio($MgO/C=1/2$) with addition of (5%)CaF_2 and were made into pieces of Φ30mm×10mm under 8MPa. Mixture pieces were put in graphite crucibles. The heating rate of the system temperature was about $15K\times min^{-1}$. The system temperature needs to be kept at 1723K for 2-5h. At high reduction temperature, Mg vapors and CO were produced and moved into the upper condensing towers. The reversion reaction occurred at low temperature in condensing tower.

2.4Analysis method

The crystalline phase of the products was identified by X-ray diffraction (XRD) instrument (D/max-3B) using Cu K_α radiation with a scanning rate of 2 (°)/min which was made by Rigaku Corporation of Japan. The structural features and chemical composition were characterized by scanning electron microscopy and EDAX pattern, respectively, which were made by Philips of Holland.

The reversion reaction ratio(γ) is defined as

$$\gamma = \frac{W_1}{W_0} \times 100\% \qquad (1)$$

Where W_0 represents the total mass of magnesium and magnesia in the condensing product, and W_1 the mass of magnesia in distillation residual.

3 Results and discussion

3.1Thermodynamic analysis of reversion reaction in vacuum

A number of possible reverse reactions could take place during cooling of an Mg/CO/inert gas mixture:

gas phase reverse reaction:

$Mg_{(g)}+CO_{(g)}=MgO_{(s)}+C_{(s)}$ (2)

$5Mg_{(g)}+3CO_{(g)}=Mg_2C_{3(s)}+3MgO_{(s)}$ (3)

liquid-phase reverse reaction:

$Mg_{(l)}+CO_{(g)}=MgO_{(s)}+C_{(s)}$ (4)

$5Mg_{(l)}+3CO_{(g)}=Mg_2C_{3(s)}+3MgO_{(s)}$ (5)

solid-phase reverse reaction:

$Mg_{(s)}+CO_{(g)}=MgO_{(s)}+C_{(s)}$ (6)

$5Mg_{(s)}+3CO_{(g)}=Mg_2C_{3(s)}+3MgO_{(s)}$ (7)

The Gibbs free energy of reaction (2),(4),(6) at different temperatures and pressure(60Pa) were evaluated and the results reported in Fig.2 Thermodynamically, these reverse reactions were exothermic reaction, the gas-phase reaction will commence as soon as a saturated gas mixture was cooled.The magnesium condensation and liquid-phase reverse reaction were favored below 1373K, and magnesium solidification and liquid-phase reverse reactions were favored below 973K. It can be seen that these reversion reactions will be favored over the metal-forming reactions during cooling, meaning that the relative kinetics of each reaction will be critical to any process designed to produce metal.

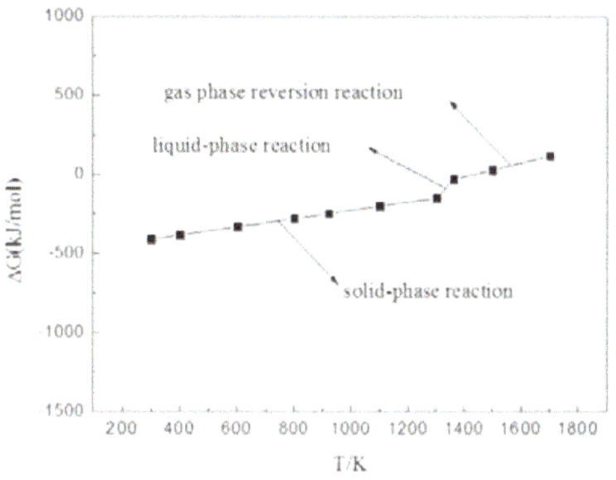

Fig. 2 Free energy changes related to the temperature of reaction(2),(4),(6) under different pressure

The Gibbs free energy of reaction (3),(5),(7) at different temperatures and pressures were calculated, as shown in Fig.3. Thermodynamically, these reverse reactions were also exothermic reaction, the gas-phase reaction will commence as soon as a saturated gas mixture is cooled. The magnesium condensation and liquid-phase reverse reaction were favored below 1373K, and magnesium solidification and liquid-phase reverse reaction were favored below 973K. The conclusion consisted with the reaction (2),(4),(6) absolutely. It can be seen that all of the reverse reactions will be favored over the metal-forming reactions during cooling, meaning that the product of reverse reaction is not only to obtain MgO, but also to get Mg_2C_3.

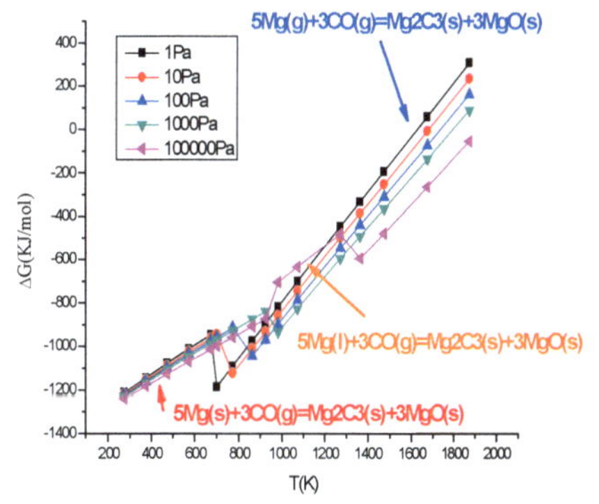

Fig. 3 Free energy changes related to the temperature of reaction(3),(5),(7) at 60Pa

The main reverse reaction was repetitive addition as: $6Mg(g)+4CO=Mg_2C_3+4MgO+C$. The study found that in the reverse reaction process 4mol MgO obtained per mol Mg_2C_3 produced. Fig.4 was XRD patterns of the condensing product in carbothermic reduction process, as shown in this figure, magnesium appeared in condensing tower. MgO and Mg_2C_3 were detected in the condensing tower too, it accorded with the thermodynamic calculation, the main condensing product in carbothermic reduction process included Mg, MgO and Mg_2C_3.

3.2Analysis of reverse reaction ratio

It can be deduced that reverse reaction occured at low temperature($\leq$1373K) during 30~100Pa in condensing process of magnesium vapor through the thermodynamic calculation and experiments. The reverse reaction could not be avoided in the condensing process of carbothermic reduction, therefore, the reaction level of reverse reaction was the key to industrial application of carbothermic reduction. In order to obtain the high purity and good crystallization magnesium and know the level of reverse reaction, it needed to calculate the reverse reaction ratio by separated the metal magnesium and MgO, Mg_2C_3.

The method of vacuum distillation was used to separate the metal magnesium and MgO, Mg_2C_3.In the distillation process, the condensing product was used as raw material and heated from room temperature to 1373K. After distillation for 1h or 2h, the temperature of distillation product cooled to room temperature and took out. After the vacuum distillation process, MgO, Mg_2C_3 and carbon were stayed loaded into the reaction crucible, high purity metal magnesium was produced in the condensing tower, which called distillation product.

Fig.4 shows the XRD patterns of the condensing product in carbothermic reduction, it can be seen that metal magnesium was obtained. Meanwhile, magnesia was also found. The SEM and EDS images of condensing product which were obtained from condensing tower are shown in Fig.5, the condensing product had a linear connecting structure. As shown in these EDS images, the content of magnesium in condensing product was 92.03 wt%, and oxygen was 7.97 wt%.

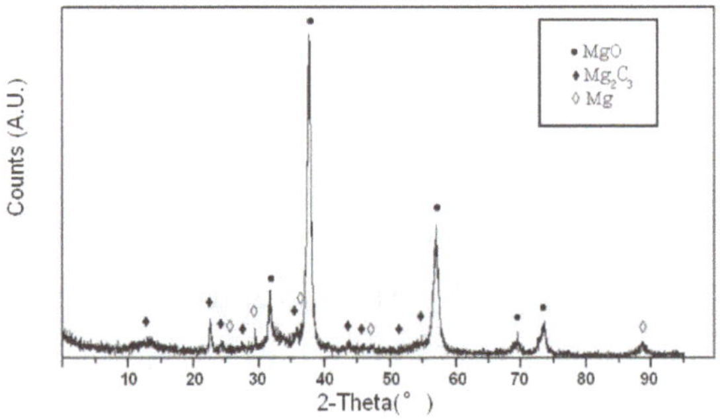

Fig. 4 XRD patterns of the condensing product in carbothermic reduction process

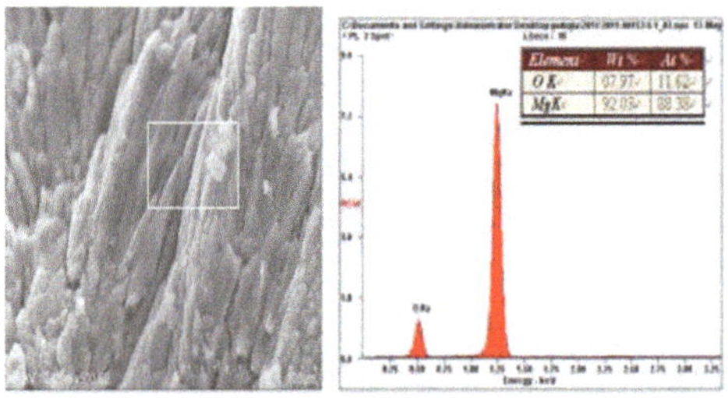

Fig.5 SEM and EDS images of condensing product

As can be seen from the XRD patterns of vacuum distillation product shown in Fig.6, it demonstrated that all peaks were sharp and well-defined, which indicated the high purity of the metal magnesium. The SEM and EDS images of distillation product which were obtained from condensing tower were shown in Fig.7, the distillation product had a lamellar structure. As shown in these EDS images, the content of magnesium was up to 98.91 wt%, and oxygen decreased to1.09 wt%.

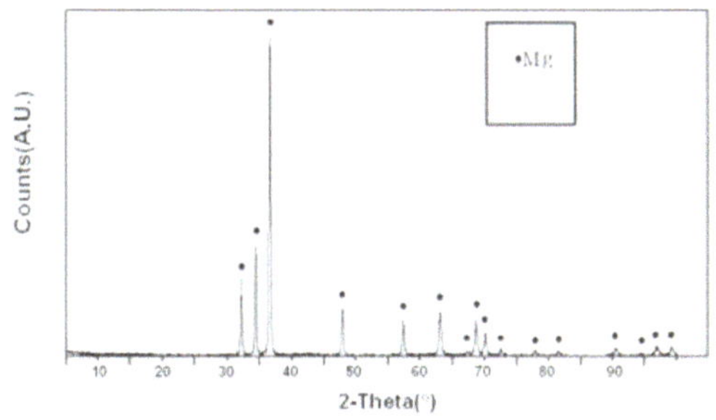

Fig.6 XRD patterns of distillation product in vacuum distillation proces

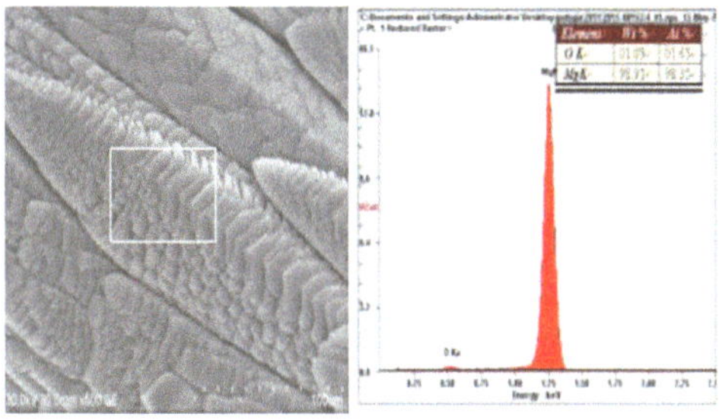

Fig.7 SEM and EDS images of distillation product

The reverse reaction ratio was calculated using the formula(1), as shown in Fig.8, it is clear that the reverse reaction ratio changed slowly with the incremental molar ratio, the reverse reaction ratio increased with increasing reaction time. However, the peak value was less than 9%, it can be deduced that most magnesium vapors transformed into metal magnesium, the condensing condition of the vacuum equipment was appropriate for the carbothermic reduction.

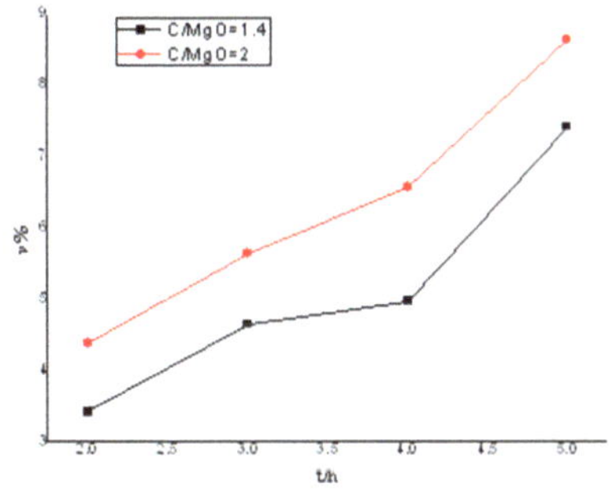

Fig.8 Relationships between reversion ratio and reaction time in different molar ratio

3.3Analysis of reversion reaction influence on condensate and crystallization of magnesium

As is known[11], the magnesium vapors should be cooled and condensed in such manner as to minimize the chemical interaction between the magnesium and the carbon monoxide which inevitably occured during the condensation. It resulted in influencing on condensate and crystallization of magnesium.

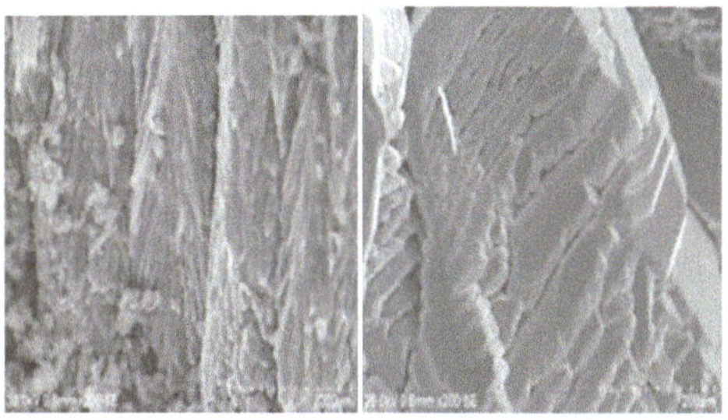

(a) (b)

Fig.9 SEM images of condensing product(a) and distillation product of purity magnesium(b)

Fig.9 shows the SEM images of condensing product and high purity distillation product. It can be seen from Fig.9(a) that the condensing product had a filament crystallization, and chinky structure. It means the carbon which obtained by reverse reaction block the atomic collision and coacervation, then, the nucleation and growth of magnesium crystallization was not well. As shown in Fig.9(b), through the vacuum distillation process, the product had a lump structure and good crystalline feature. According to the analysis result, the MgO, Mg_2C_3 and C which produced by the reverse reaction blocked the crystallization and growth of magnesium atoms, therefore, the structure of condensing product was poor, it was not beneficial for obtaining metal magnesium.

4 Conclusion

1) The reverse reactions were exothermic reaction, the gas-phase reversion will commence as soon as a saturated gas mixture was cooled. The reverse reactions were favored below 1373K and 30~100Pa during carbothermic reduction. The products of reverse reaction were MgO and Mg_2C_3.
2) The distillation product of vacuum distillation process produced a high purity metal magnesium product, through the calculation of reverse reaction, the peak value of γ was less than 9%.
3) Contrast the condensing product and distillation product, the distillation product had a lump structure and good crystalline feature. According to the analysis result, the MgO, Mg_2C_3 and C which produced by the reverse reaction blocked the crystallization and growth of magnesium atoms, therefore, the structure of condensing product was poor, it was not beneficial for obtaining metal magnesium.

Acknowledgments

Project supported by National Natural Science Foundation of China(NO.51304095); Science and Technology Planning Project of Yunnan Province, China (NO.2013FZ029); The Program for Innovative Research Team in University of Ministry of Science and Technology of China(No.2014RA4018). Fund of Yunnan Collaborative Innovation Center of Complex Nonferrous Metal Resources Clear Utilization(NO. 2014XTZS005).

Reference

[1] WU Lu, PAN Fu-sheng, YANG Ming-bo, WU Ju-ying, LIU Ting-ting.As-cast microstructure and Sr-containing phases of AZ31 magnesium alloys with high Sr contents[J].***Transaction of Nonferrous Metals Society of China***,2011,21:784-789.

[2] PAN Fu-sheng, ZHANG Jing, WANG Jing-feng, YANG Ming-bo, HAN En-hou, CHEN Rong-shi. Key R&D activities for development of new types of wrought magnesium alloys in China[J].***Transaction of Nonferrous Metals Society of China***,2010,20:1249-1258.

[3] LI Rong-ti, PAN Wei, Masamichi Sano.Kinetics and mechanism of carbothermic reduction of magnesia[J].***Metallurgical and materials transactions***,2003,8:433-437.

[4] Ramakrishnan,S.;Koltun,P.Resource.Conservation. ***Recycle***.2004,vol.42,pp.49-64.

[5] Geoffrey Brooks, Simon Trang, Peter Witt, etc. The Carbothermic Route to Magnesium[J].***JOM***, 2006,58,5:51-55.

[6] Winand R.Production of Magnesium by Vacuum Carbothermic Reduction of Calcined Dolomite[J]. ***Transactions of the Institution of Mining and Metallurgy Section*** (Section C),1990:105-111.

[7] Yu Qing-chun, Yang Bin, Ma Wen-hui, Li Zhi-hua, Dai Yong-nian.Study of carbothermic reduction of magnesia in vacuum[J]. ***Chinese Journal of Vacuum Science and Technology***, 2009, 5(29): 68-71.

[8] Yu Qing-chun, Yang Bin, Xu Bao-qiang, Liu Da-chun, Li Zhi-hua, Dai Yong-nian. Study on the thermodynamic and experimental carbothermicreduction magnesia[C]. ***Proceedings of the 9th Vacuum Metallurgy and Surface Engineering Conference***,2009:428-433.

[9] Tian Yang, Liu Hong-xiang, Yang Bin, Qu Tao, Dai Yong-nian.Process of magnesium production by magnesia carbothermic reduction in vacuum[J]. ***Chinese Journal of Vacuum Science and Technology***, 2012,32(4):43-48.

[10] Li Zhi-hua, DAI Yong-nian, XUE Huai-sheng.Thermodynamical analysis and experimental test of magnesia vacuum carbothermic reduction[J].***Nonferrous metals***,2005,1(57):56-59.

[11] Roy C.Kirk, ***U.S.patent 2,257,910***(7 October 1941)

Magnesium Technology 2016
Edited by: Alok Singh, Kiran Solanki, Michele V. Manuel, and Neale R. Neelameggham
TMS (The Minerals, Metals & Materials Society), 2016

Thermodynamic Description of Reactions between Mg and CaO

Rainer Schmid-Fetzer[1], Artem Kozlov[1], Björn Wiese[2], Chamini L. Mendis[2], Domonkos Tolnai[2] , Karl U. Kainer[2] , Norbert Hort[2]
[1]Institute of Metallurgy, Clausthal University of Technology, Robert-Koch-Str. 42,
D-38678 Clausthal-Zellerfeld, Germany
schmid-fetzer@tu-clausthal.de
[2]Magnesium Innovation Centre, Helmholtz-Zentrum Geesthacht,
Max Planck Straße 1, D-21502 Geesthacht, Germany

Keywords: Magnesium alloys, CaO, *in situ* XRD, thermodynamic evaluation

Abstract

CaO is considered as possible replacement for cover gases such as SF_6 during melting and casting of Mg alloys. Such CaO additions to molten Mg increase the ignition resistance by forming a protective oxide layer. The actual reactions between liquid Mg and CaO are not well understood. An approach based on chemical reaction equations cannot capture the "CaO dissolution" process. This work presents the development of a consistent thermodynamic description of the ternary Mg-Ca-O alloy system. To that end a revision of the thermodynamic data of key oxides, CaO and MgO, has been performed based on original experimental work so far not considered in thermodynamic databases or tabulations. The formation of a liquid Mg-Ca-[O] alloy during the reaction is predicted from the thermodynamic calculations at melting temperatures; solidification simulations are also performed. These predictions from thermodynamic simulations are validated by experimental data using *in situ* synchrotron radiation diffraction.

Introduction

CaO is used as a cheaper replacement for pure Ca as an alloying addition in Mg alloys. The main motivation for the use of CaO is to reduce the use of cover gases such as SF_6 during melting and casting of Mg alloys. A new group of alloys named as ECO-Mg (Environment COnscious magnesium) with Ca additions to achieve this has been developed [1]. The CaO dissociation (or reduction) in Mg and Mg alloys is reported in literature, but no or insufficient explanation is given [2-4]. According to the Ellingham diagram CaO is more stable than MgO at temperatures of interest [5, 6], therefore, it is often expected that CaO remains unreacted during melting of Mg. In contrast, it is observed that CaO dissolves during the melting procedure [2, 3] and has a protective effect on Mg melts [7-9].

Kondoh et al. [4] proposed a calculation to explain the disassociation of CaO in Mg-Al alloys. They used tabulated Gibbs energy data for the reaction between pure Mg (plus pure Al) and CaO. As reaction products MgO and Ca or the Laves phases, Mg_2Ca and Al_2Ca, where considered. Their key finding is that pure Mg cannot react with CaO because of a positive Gibbs energy of reaction. Experimental results for the stability of Mg + CaO were also mentioned but not presented [4]. They only presented results for the reaction of a Mg-Al-Zn alloy, AZ61B, mixed in different ratios with CaO. In this case CaO was reduced, essentially because Al_2Ca may form.

Gourishankar et al. [10] measured the Gibbs energy and enthalpy of formation, $\Delta H°$, for CaO and MgO. Their derived values for $\Delta H°$(298 K) of CaO and MgO are -602 and -635 kJ/mol, respectively. This differs significantly from the JANAF-tables [6] by +33 kJ/mol for CaO and -34 kJ/mol for MgO. This is important because, in this case, MgO and CaO swap the positions in the Ellingham diagram. These data published in 1993 [10], however, have never been acknowledged in any recent tabulated thermodynamic data compilation [11-13] or the leading electronic databases for pure substances, SGTE [14] or FactPS [15] or Calphad assessments of oxide systems [16].

In addition to this problem with the oxide data it will be shown that the approach based on chemical reaction equations cannot capture the "CaO dissolution" process. That is because multicomponent solution phases, especially liquid, will govern the reaction. The purpose of this work is to demonstrate that the Mg + CaO reaction can only be understood by development of a consistent thermodynamic description of the ternary Mg-Ca-O alloy system and phase diagram. These calculations are verified by dedicated experimental investigation of the Mg + CaO reactions and the subsequent alloy solidification.

Experimental procedure

Sample preparation
The sample Mg20CaO was prepared from chips of pure Mg (99.9 %) and CaO powder (Calcium oxide ≥96 % powder from CARL ROTH). The nominal composition of the sample investigated was Mg + 20 wt.% CaO. The CaO was dried for 6 h at 350°C and mixed with Mg chips for 3 min to achieve a uniform mixture. The mixture was pressed with a hydraulic press to cylinders (diameter of 4 mm and a height of 3 mm) with an applied pressure was 100 MPa and held for 1 min under pressure.

In situ synchrotron radiation diffraction measurement
The *in situ* studies were conducted at P07 hutch of the High Energy Materials Science (HEMS) beam line at PETRA III [17, 18] at DESY (Deutsches Elektronen-Synchotron) in Hamburg Germany. The synchrotron radiation diffraction study was conducted with a beam energy of 100 keV, (λ = 0.0124 nm) in transmission geometry with a beam cross section of 1x1 mm².

The sample was heated in an Ar atmosphere to 750°C at 10 K/min, held at 750°C for 5 min to ensure melt homogeneity, and then cooled at 10 K/min to 200°C (fully solidified state) before air-cooling to room temperature. The molten samples were held in position in an inverted graphite crucible with a steel lid. The temperature-time-curve was controlled with an S type thermocouple welded to the lid of the crucible. The crucible with sample was rotated by 90° during acquisition to improve statistics. The two dimensional (2D) diffraction patterns were recorded every 12 s at an acquisition time of 1 s with a PerkinElmer XRD 1622 Flatpanel at a sample-to-detector-distance of 1162.7 mm (calibrated with LaB_6 reference). The temperature resolution of the measurement was 2 K. X-ray line profiles were obtained by azimuthal integration of the 2D diffraction patterns through 360° using FIT2D V12.077 software. For phase identification simulated line profiles were generated with CaRIne 3.1 Crystallography™ with crystal structure data from the Pearson's Crystallography Database [19].

Measurement of O content of Mg chips

For this measurement the chips were pressed with a hydraulic press to cylinders with a diameter of 4.0 mm, under an applied pressure of 500 MPa and held for 1 min under pressure. The Mg and O content was measured with a TESCAN VEGA III, equipped with an EDAX energy-dispersive X-ray spectrometer (EDXS) operating at 20 kV. Three separate measurements were made with EDXS for 10 min.

Metallographic investigation

The specimens for SEM were prepared by grinding with SiC paper to 2500 grit at a speed of 150 min^{-1}. Samples were then polished in a 3 µm diamond suspension followed with a mixture of 1 µm diamond suspension and OPS™ (≈ 0.1 µm) anhydrous suspension.

Microstructures were analyzed with SEM (TESCAN VERGA III) and the local compositions were determined with EDXS compositional maps. The compositional maps are recalculated with the software Iridium Ultra in wt.%.

Experimental Results

In situ synchrotron radiation diffraction

The X-ray line profiles for selected temperatures, integrated from 2D diffraction patterns recorded during solidification experiments, are shown in Figure 1 and the pertinent temperatures at which phase transformations are detected are listed in Table I and the line profiles (LP) at different temperatures are illustrated in Fig. 1.

Table I: Temperatures where various phase transformation were observed during solidification of Mg20CaO.

Reaction	Experimental data ± 5 [°C]
Appearance of Mg start	570
Appearance of Mg_2Ca start	511
Appearance of Mg end	509
Appearance of Mg_2Ca end	509

The X-ray line profiles show that peaks due to $Ca[OH]_2$, CaO, Mg and graphite crucible are present in the beginning of the experiment. During heating the main peak of CaO becomes more visible when heated above 291°C, due to the thermal expansion of Mg. This is followed by the stabilisation of the intensity of CaO peaks up to 396°C and a second increase in the peak intensity was observed at 502°C to a maximum of intensity. The intensity of CaO decreased as the sample was heated above 568°C. During heating the peaks of MgO are detected initially at 368°C. The peaks of $Ca[OH]_2$ starts to disappear at 408°C and are undetectable at 500°C. The MgO peaks did not increase further beyond 618°C. In the range between 620 and 637°C the Mg peaks disappear. With the disappearance of solid Mg the diffuse signal of the melt was detected. From a temperature of 664°C the CaO peaks stabilize and show no significant changes in intensity.

During cooling the first peaks of Mg are detected at 570°C and no intensity changes were detected below 509°C. The Laves phase Mg_2Ca was detected at 511°C with no further intensity changes below 509°C.

As the samples were rotated during the measurement results are not limited to single view of the sample but provide better statistics on the amount of various phases.

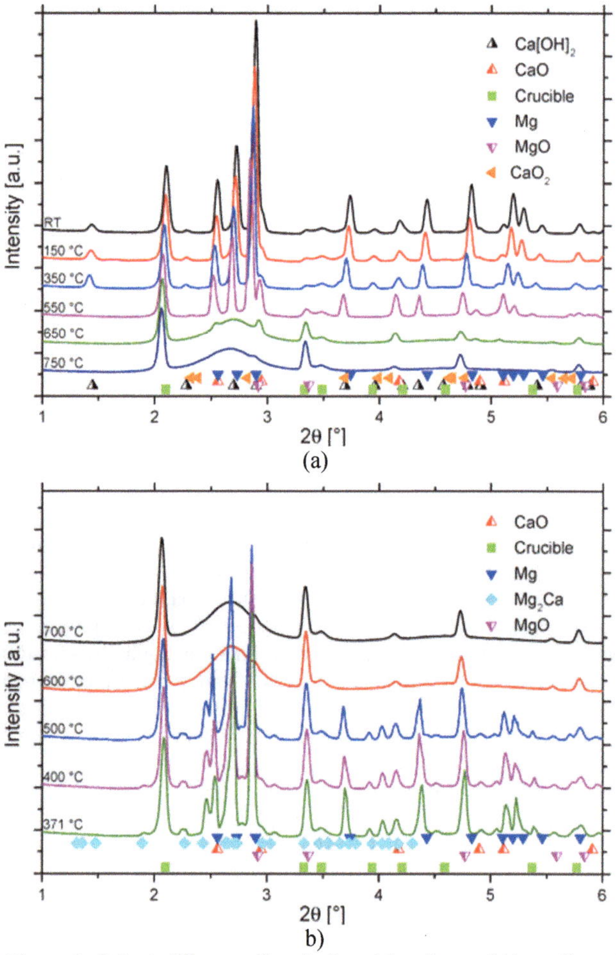

Figure 1: Selected line profiles during a) heating and b) cooling of the Mg20CaO sample with the identity of detected phases.

Microstructures of the solidified samples

The microstructure of the sample used for *in situ* investigations was analyzed with SEM. The EDXS maps show the distribution of the Mg, Ca and O in the as-solidified sample in Fig. 2. From the microstructure the solidification sequence can be understood. The solidification begins with Mg grains and ends finally with the eutectic solidification of an Mg- and Ca-rich lamellar structure. The O agglomerates in the eutectic as clusters of oxide particles with a higher concentration of Mg compared to the intermetallic phase The composition of the as-solidified sample was measured with EDXS to be 6.8 wt.% Ca, 8.7 wt.% O and balance Mg. This composition deviates from the nominal content of 14.3 wt.% Ca, 5.7 wt.% O and balance Mg.

The composition of the pure Mg chips, measured by EDXS, was 97.73±0.23 wt.% Mg and 2.27±0.23 wt.% O.

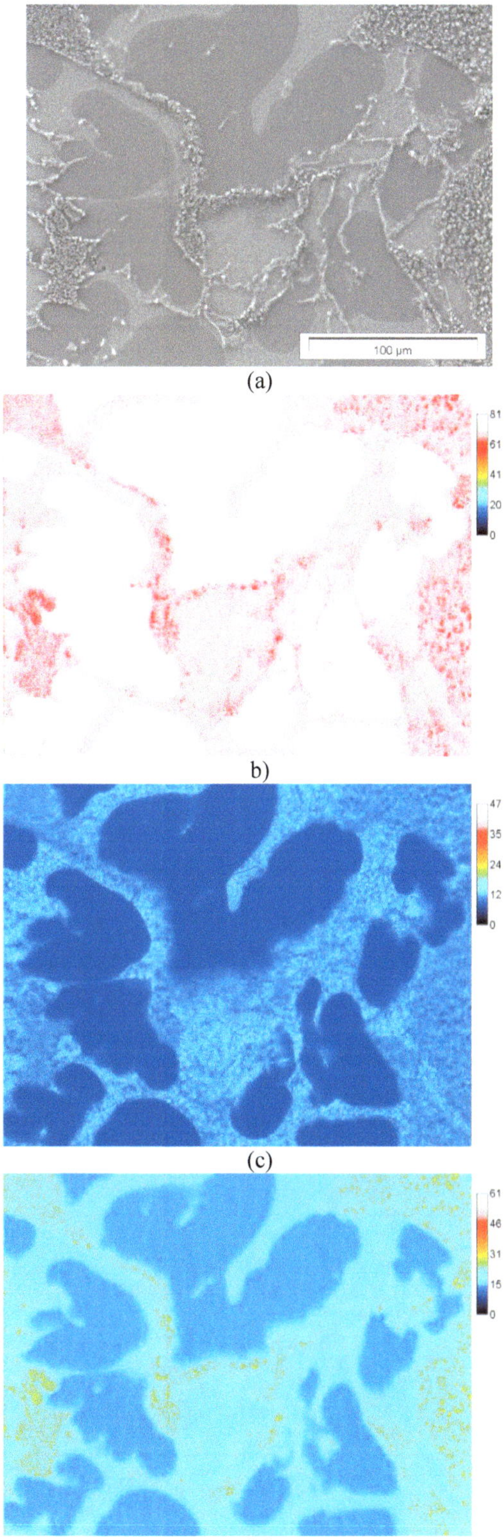

Figure 2: The SE microstructure (a) and EDXS maps for (b) Mg, (c) Ca and (d) O of the Mg20CaO sample, composition in wt.%.

Thermodynamic description of reactions in the ternary Mg-Ca-O system

Two thermodynamic databases of the ternary Mg-Ca-O system were developed in this study. Both are based on the thermodynamic functions on the well established Mg-Ca alloy system and the gas phase from the PanMg2014 database [20]. Two different sets of Gibbs energy functions for the solid oxides, MgO, CaO, and CaO_2, were incorporated. In the first database the currently established data from the leading electronic databases for pure substances, SGTE [14] or FactPS [15] with the numerical values given in [16] were used. In the second database a refined version of the Gibbs energy functions for the solid oxides, based on experimental data from [10], was used and more details of this development will be presented elsewhere [21]. All calculations were done using the Pandat software package [22].

The ternary phase diagrams at 700, 600, and 500°C calculated from the first database are shown in Fig. 3, those calculated from the latter, which are accepted as correct ones in this work, in Fig. 4. The oxide CaO_2 is not stable at these temperatures. The single-phase regions of liquid are shown as thick blue lines for easy reading, the quantitative oxygen solubility in Mg-Ca alloys is negligible on this scale.

The basic idea of clearly understanding any reaction of Mg with CaO by reading these phase diagrams is explained for the nominal composition of sample Mg20CaO, indicated by the solid red dot. In Fig. 3a, at 700°C, this dot is located in the three-phase equilibrium L + MgO + CaO, thus, CaO will be partly reduced. The calculated liquid composition is 87.5Mg12.5Ca (wt.%). The equilibrium phase fractions, also obtained by applying the lever rule to this tie-triangle, are presented in the following "reaction equation", assuming a total mass of 100 g:

80 g Mg + 20 g CaO

⇨ 38.1 g L(87.5Mg12.5Ca) + 45.4 g MgO + 16.5 g CaO (Fig. 3a)

At lower temperatures, 600 and 500°C, we obtain from Figs. 3b-c:

80 g Mg + 20 g CaO

⇨ 38.5 g L(88.0Mg12.0Ca) + 46.6 g MgO + 14.9 g CaO (Fig. 3b)

⇨ 80g Mg + 20g CaO (Fig. 3c)

That is, after completion of solidification, Mg and CaO should not react if the currently established oxide data were correct.

In contrast, the second thermodynamic database developed and accepted as correct one in this work gives the following results that can be read from the equilibrium phase regions at the solid red dot in Fig. 4:

80 g Mg + 20 g CaO

⇨ 85.6 g L(83.3Mg16.7Ca) + 14.4 g MgO (Fig. 4a)

At lower temperatures, 600 and 500°C we obtain from Figs. 4b-c:

80 g Mg + 20 g CaO

⇨ 85.6 g L(83.3Mg16.7Ca) + 14.4 g MgO (Fig. 4b)

⇨ 54.7 g (Mg)(99.4Mg0.6Ca) + 14.4 g MgO + 30.9 g Mg_2Ca (from Fig. 4c)

Thus, CaO will be completely reduced at all temperatures, forming MgO and Mg-rich liquid or, after solidification, Mg_2Ca and (Mg) solid solution.

It is obvious from this simple example that any approach based on classical chemical reaction equations cannot capture the "CaO dissolution" process. The composition of the liquid phase forming would be unknown without the phase diagrams and it will change with temperature. Moreover, for different Mg/CaO initial ratios the results are easily read from the ternary phase diagrams because the solid red dot just shifts along the mass balance line connecting Mg and CaO. It will, thus, fall in the corresponding phase regions in Figs. 3 or 4, clearly indicating the assembly of equilibrium products. Without these phase diagrams one would have no clue how to write down classical

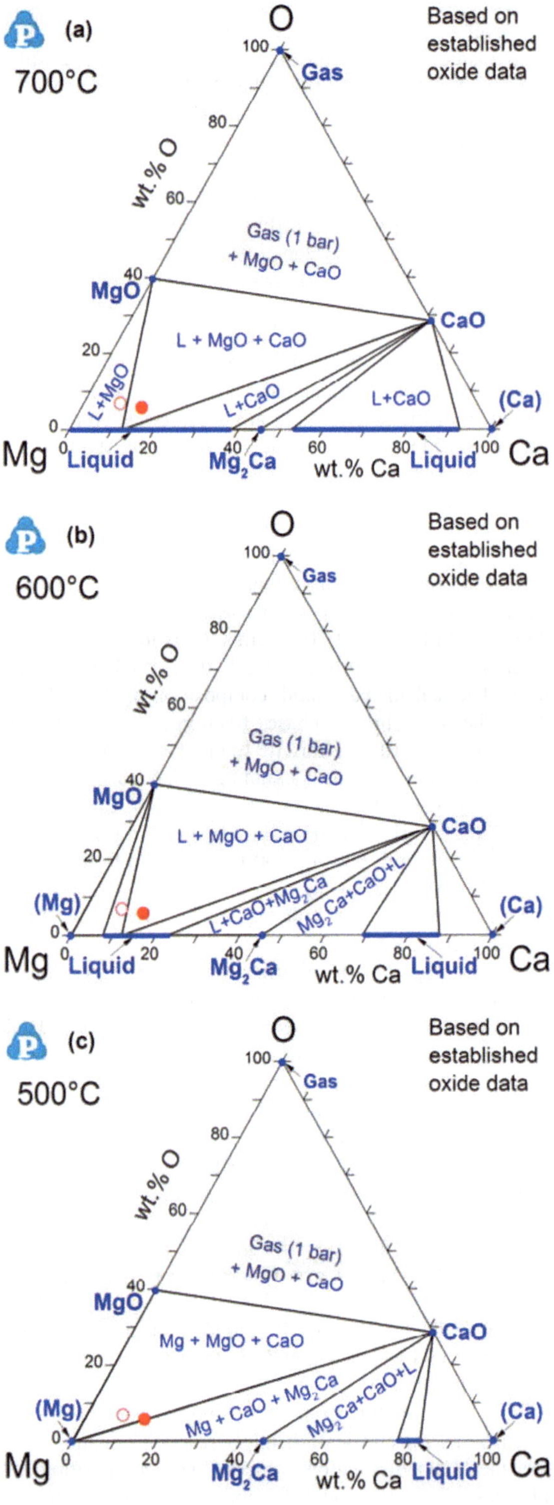

Figure 3: Calculated isothermal sections of the Mg-Ca-O phase diagram from the first thermodynamic database constructed using oxide data [16] established in current literature; (a) 700°C, (b) 600°C, (c) 500°C; the solid red dot indicates the nominal sample composition Mg20CaO and the open circle its composition measured by EDXS.

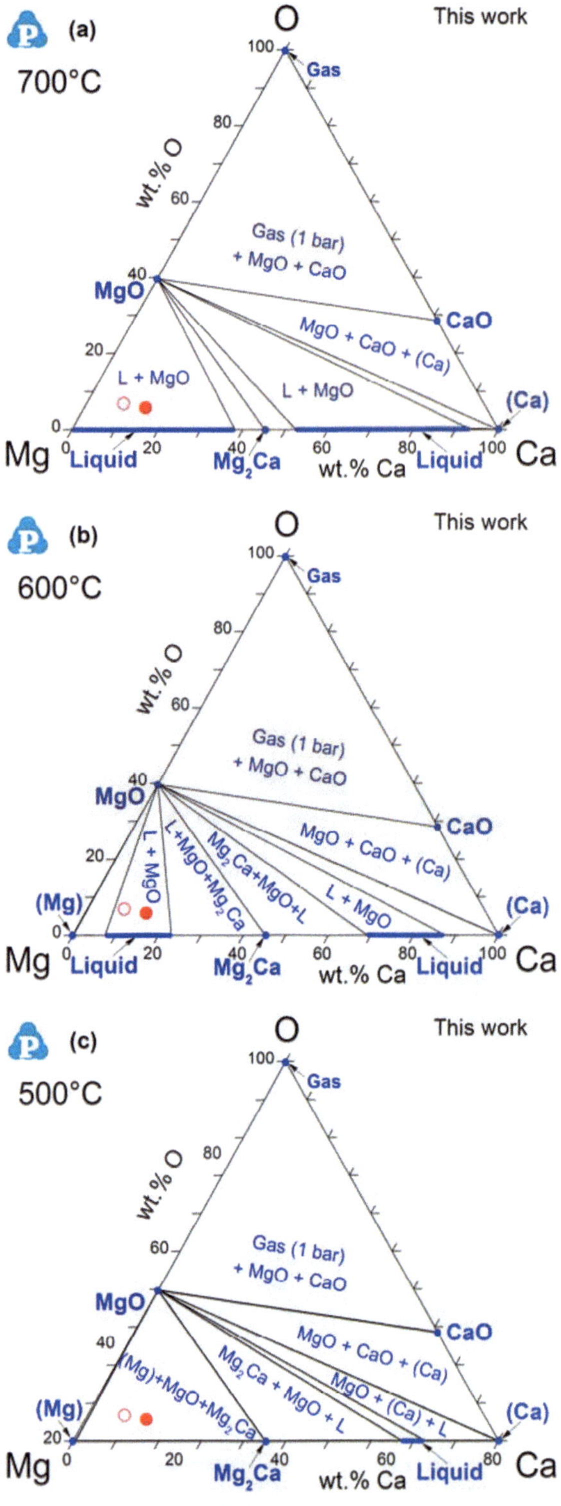

Figure 4: Calculated isothermal sections of the Mg-Ca-O phase diagram from the second thermodynamic database, developed using oxide data from[10] in this work [21]; (a) 700°C, (b) 600°C, (c) 500°C; the solid and open red circles are the same as in Fig. 3.

chemical reaction equations, which also need to be different for any temperature and Mg/CaO initial ratio, as seen from the six example "equations" above.

It is further emphasized that consideration of binary phase diagrams only, as tried by Kondoh et al. [4], is not viable. The equilibrium phase assembly is governed by the tie line distribution in the ternary phase diagrams as shown in Figs. 3 and 4. The two thermodynamic databases differ essentially in the result that a tie line between Mg-rich metallic phase and CaO does not exist at any temperature in Fig. 4, in contrast to Fig. 3.

Discussion of reactions and solidification of sample Mg20CaO

The experimentally observed results in this work can be discussed best by comparison with calculated equilibrium phase fractions of sample Mg20CaO as shown in Fig. 5.

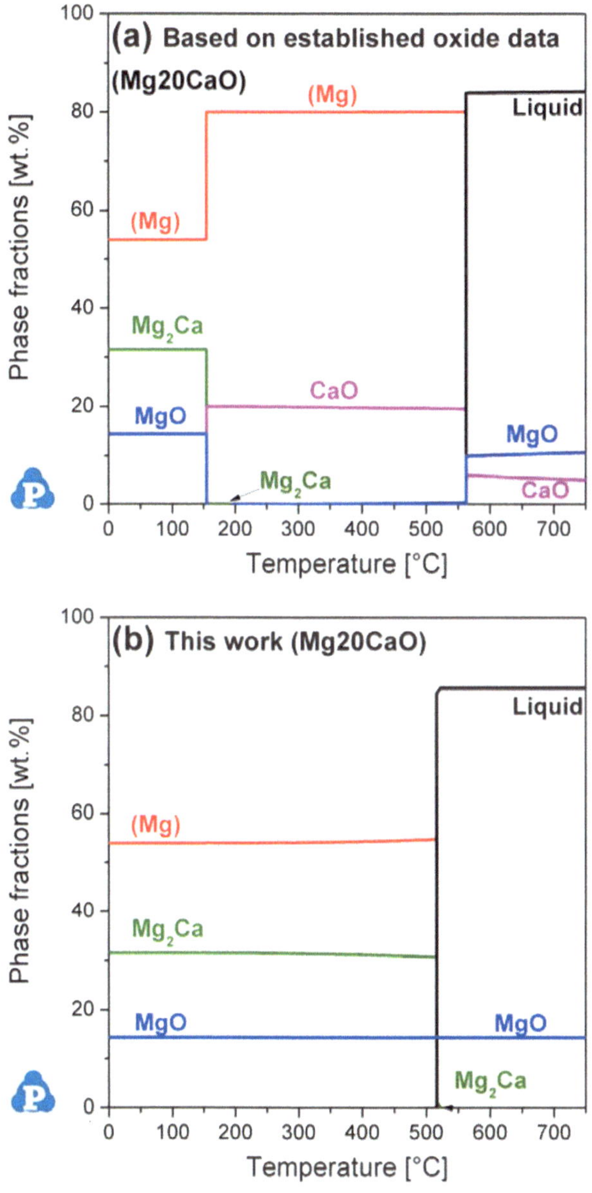

Figure 5: Calculated equilibrium phase fractions for the nominal composition of sample Mg20CaO, Mg-14.3Ca-5.7O (wt.%); (a) Using the first database, see Fig. 3, (b) Using the second database, accepted in this work, see Fig. 4.

Using the first thermodynamic database, Fig. 5a shows that for example at 500°C the sample should consist of 80% (Mg) and 20% CaO. That is consistent with Fig. 3c and the lever rule applied to the solid red dot, the nominal sample composition. That, however, is in stark contrast to the experimental result demonstrating that after melting and equilibration the initially strong XRD signals of CaO essentially disappear and dominant signals of MgO appear. Moreover, the formation of Mg_2Ca is predicted below 200°C only, in a solid state reaction, by using the first database. That is again in conflict with the experimental finding that Mg_2Ca is formed from the liquid phase during solidification, starting at 511°C. This disapproves the validity of the first database.

In contrast, the second database correctly predicts in Fig. 5b that CaO is completely reduced and MgO forms as equilibrium phase at all temperatures. It also correctly predicts the formation of significant amount of Mg_2Ca (30%) from the liquid phase in a eutectic reaction L = (Mg) + Mg_2Ca (+MgO) at 516.5°C. This invariant four-phase reaction is degenerate to the binary Mg-Ca because of the negligible content of oxygen in liquid, thus, the amount of 14.4% MgO remains unchanged in that eutectic. As a detail in Fig. 5b, it is see that a small amount of 0.9% Mg_2Ca should crystallize as primary phase below 522°C, which is not observed in the microstructure, Fig. 2.

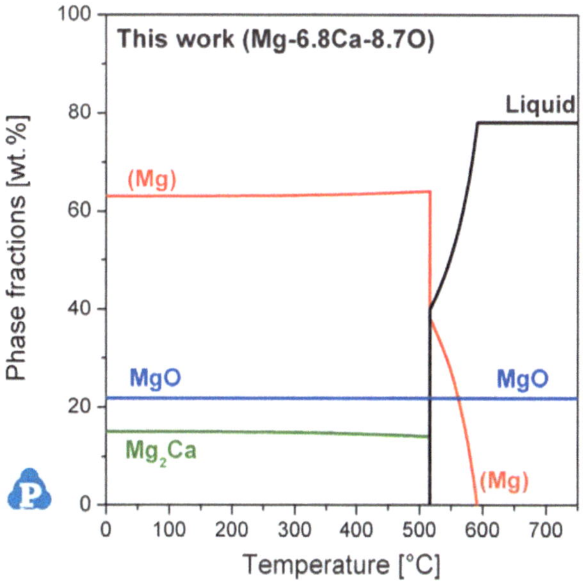

Figure 6: Calculated equilibrium phase fractions for the composition of sample Mg20CaO measured by EDXS, Mg-6.8Ca-8.7O (wt.%) using the second thermodynamic database, accepted in this work, see Fig. 4.

This leads us to the calculated phase fractions of the shifted sample composition Mg-6.8Ca-8.7O (wt.%), measured by EDXS and marked by the open symbol in Fig. 4, using the second thermodynamic database. At this sample composition primary crystallization of (Mg) is predicted to start at 591°C from the liquid with composition L(91.3Mg8.7Ca). That is consistent with the experimental finding of primary (Mg) by *in situ* XRD, and microstructure observation, Fig. 2. The temperature for start of (Mg) in Table I, 570°C, is somewhat below the predicted value. However, the accuracy of the EDXS measurement has also to be taken into account. Thermodynamic calculations reveal that for an assumed liquid composition L(88.9Mg11.1Ca) an exact match of the primary (Mg) precipitation at 570°C is obtained. That corresponds to a shifted sample composition of Mg-6.8Ca-15.3O or Mg-8.3Ca-10.0O (wt.%), or any point along the tie line L(88.9Mg11.1Ca) + MgO, they all show primary (Mg) at 570°C.

It is seen from Fig. 4a that any initial Mg/CaO mixture with overall composition inside the L + MgO region will react to these two phases by melting at 700°C. The composition of the Mg-Ca liquid that forms is obtained by plotting the tie line from MgO to L that cuts through the initial composition point. Inevitably, the Mg-content of L will be higher than that of the initial mixture.

That construction also reveals that any initial mixture located on that same tie line forms the same liquid composition, just the MgO/L phase amount ratio changes. The amount of MgO formed during melting and equilibration will remain constant during the subsequent solidification of that essentially binary Mg-Ca liquid.

For the alloy compositions with reduced Ca-content compared to the nominal one, such as the EDXS value assumed in Fig. 6, the calculation with the second database also correctly predicts the formation of Mg_2Ca in a eutectic microstructure observed clearly in the fine lamellar structure of Fig. 2a and the corresponding composition values of Figs 2b-c. Specifically, this is the degenerate eutectic reaction L = (Mg) + Mg_2Ca (+MgO) at 516.5°C. The experimentally observed temperature from *in situ* XRD measurements in Table I, 511°C, is somewhat lower and may be due to some non-equilibrium undercooling but is still in agreement within experimental accuracy. It is also confirmed that Mg_2Ca forms in an invariant reaction, given the small spread of only 2 K between start and end of Mg_2Ca signal appearance. This final eutectic solidification step is also confirmed by the disappearance of the diffuse signal from the molten metal.

Conclusion

Dedicated thermodynamic calculations and experimental work reveal that:

- The previously claimed stability of pure Mg with CaO [4] and thermodynamic data of CaO and MgO currently considered established [14, 15] are disapproved.
- In a Mg + 20 wt.% CaO sample the CaO is reduced during melting. The microstructure of the solidified sample resembles that of a binary Mg-Ca liquid with finely dispersed MgO particles.
- The Mg + CaO reactions cannot be understood by classical chemical reaction equations but only through the ternary Mg-Ca-O phase diagrams.
- The second thermodynamic database developed in this work correctly predicts these reactions during melting and subsequent solidification. It is implemented in the updated version of the PanMg database [20], PanMg2015.
- This approach enlightens the CaO "dissolution" in Mg melts and the protective effect of CaO on Mg melts.

Acknowledgment

The authors acknowledge the Deutsches Elektronen-Synchrotron for the provision of access to the synchrotron radiation facilities in the framework of the proposal I-20130330.

References

1. Shae K. Kim, “Design and Development of High-Performance Eco-Mg Alloys”, in *Magnesium Alloys - Design, Processing and Properties*, ed. F. Czerwinski (InTech, Rijeka, Croatia, 2011) 431-468.

2. S.-H. Ha et al., "Behavior of CaO and Calcium in Pure Magnesium,". *Rare Metals*, 25 (2006), 150-154.

3. S.-H. Ha, J.-K. Lee, S. K. Kim, "Effect of CaO on Oxidation Resistance and Microstructure of Pure Mg," *Mater. Trans.*, 49 (2008), 1081-1083.

4. K. Kondoh et al., "Thermo-dynamic analysis on solid-state reduction of CaO particles dispersed in Mg-Al alloy," *Mater. Chem. Phys.*, 129 (2011), 631-640.

5. I. Barin, O. Knacke, and O. Kubaschewski, *Thermochemical properties of inorganic substances, Volume 1* (Springer-Verlag Berlin, Heidelberg, 1977) 861.

6. M. W. Chase et al., *JANAF Thermochemical Tables Third Edition, Volume 14* (Journal of Physical and Chemical Reference Data, 1985) 1856.

7. B.-S. You, W.-W. Park, and I.-S. Chung, "The effect of Calcium additions on the oxidation behavior in Magnesium Alloys," *Scripta Mater.*, 42 (2000), 1089-1094.

8. J.-K. Lee, S. K. Kim, "Effect of CaO composition on oxidation and burning behaviors of AM50 Mg alloy," *Trans. Nonferrous Met. Soc. China*, 21 (2011), 23-27.

9. J.-K. Lee, S. K. Kim, "Effect of CaO Addition on the Ignition Resistance of Mg-Al Alloys," *Trans. Nonferrous Met. Soc. China*, 52 (2011), 1483-1488.

10. K. V. Gourishankar, M. K. Ranjbar, and G. R. St. Pierre, "Revision of the enthalpies and gibbs energies of formation of calcium oxide and magnesium oxide," *J. Phase Equilib.*, 14 (5) (1993), 601-611.

11. O. Kubaschewski, C.B. Alcock, and P.J. Spencer. *Metallurgical thermochemistry: 6th Edition* (Pergamon Press, Oxford) 376.

12. M. W. Chase, *NIST-JANAF Thermochemical Tables 4th Edition, Monograph 9 (Part I and Part II)* (Journal of Physical and Chemical Reference Data, 1998) 1951.

13. M. Binnewies, E. Milke, *Thermochemical Data of Elements and Compounds, 2nd ed.*, (Wiley-VCH, Weinheim, 2002) 928.

14. SGSUB substances database, version 13.1, 2013, Scientific Group Thermodata Europe (SGTE), Saint Martin d'Hères, France.

15. Factsage substances database, FactPS-2013, (www.crct.polymtl.ca/fact/documentation)

16. W. Huang, M. Hillert, and X. Wang, "Thermodynamic assessment of the CaO–MgO–SiO2 system," *Metall. Mater. Trans.* 26 (A) (1995), 2293-2310.

17. K. Balewski et al., PETRA III: A New High Brilliance Synchrotron Radiation Source At DESY, Proceedings of EPAC 2004 (2004), 2302-2304.

18. Deutsche Elektronen-Synchrotron DESY, http://photon-science.desy.de (2014)

19. P. Villars and K. Cenzual, *Pearson's Crystal Data - Crystal Structure Database for Inorganic Compounds*, (on CD-ROM), Release 2013/14, ASM International (2014)

20. PanMg database, Computherm LLC, Madison, WI, (www.computherm.com)

21. A. Kozlov and R. Schmid-Fetzer, to be published.

22. W. Cao et al., "PANDAT Software with PanEngine, PanOptimizer and PanPrecipitation for Materials Property Simulation of Multi-Component Systems," *Calphad*, 33 (2009), 328-342.

Magnesium Technology 2016
Edited by: Alok Singh, Kiran Solanki, Michele V. Manuel, and Neale R. Neelameggham
TMS (The Minerals, Metals & Materials Society), 2016

ATOMIC-LEVEL MECHANISMS OF MAGNESIUM OXIDATION

Sandra Gardonio[1], Mattia Fanetti[1], Matjaz Valant[1], Dmytro Orlov[2]
[1]Materials Research Laboratory, University of Nova Gorica, Nova Gorica, SI-5000, Slovenia
[2]Division of Materials Engineering, Department of Mechanical Engineering, LTH, Lund University, Lund, 223 63, Sweden

Keywords: Magnesium, Oxidation, X-ray photoelectron spectroscopy, Synchrotron radiation

Abstract

Magnesium has been recently becoming an increasingly popular material for various applications. However, excessive chemical reactivity, and oxidation rate in particular, is a major obstruction on the way of Mg to become widely adopted. A significant problem causing the lack of Mg reactivity control is insufficient understanding of mechanisms involved in the oxidation of magnesium surface.

Herewith we present the investigation of atomic-level mechanisms of oxidation initiation and propagation in pure Mg. Namely, X-ray photoelectron spectroscopy at synchrotron Elettra was used as a surface sensitive direct method to determine the valence of Mg and O and the valence band states at the early stage of oxide formation over a principal, most densely packed, crystallographic plane (0001) in pure Mg. The mechanisms of oxygen adsorption on magnesium free surface followed by oxidation (i.e. initiation and kinetics of MgO formation) are clarified.

Introduction

Magnesium (Mg) is an extremely interesting material, which recently also becomes increasingly popular for various applications [1-3]. However, the great expectations often fall short when final Mg products are to be exposed to aggressive environments. The excessive chemical reactivity is still a major obstruction on the way of Mg to become widely adopted.

A significant problem causing the lack of Mg reactivity control is insufficient understanding of mechanisms involved in the oxidation of crystalline magnesium surface. The process of oxidation may also significantly depend on the crystallographic orientation of the material since magnesium has very anisotropic hexagonal close-packed (HCP) lattice, as illustrated schematically in Fig.1. The anisotropic behavior of Mg corrosion was reported in Ref. [4]. Furthermore, some evidence of anisotropic corrosion in a Mg-Zn-Zr alloy has been found in [5]. Finally, authors of a very recent review paper on Mg corrosion suggest that the valence of Mg during the corrosion process is in fact one of the controversial aspects of the corrosion models proposed for Mg [6]. In order to be well understood, these phenomena as well as the models require comprehensive investigation of underlying fundamental mechanisms. Such understanding will allow developing a clear roadmap for the control of corrosion behavior in pure Mg and Mg alloys.

One of the main obstacles in revealing atomic-level information about the oxide formation on Mg surface is a very small volume of the material directly involved in the process. This interacting volume covers a very few top-most atomic layers, which is at the scale of just a few nano-meters. Such a volume is normally too small to produce sufficient signal in conventional laboratory-scale techniques for material characterization like Scanning Electron Microscopy, Energy Dispersive X-ray analysis, X-ray Diffraction or Transmission Electron Microscopy. Some other, more surface-sensitive, techniques allowing direct *in-situ* analysis of the process are necessary for revealing basic mechanisms and kinetics of Mg oxidation.

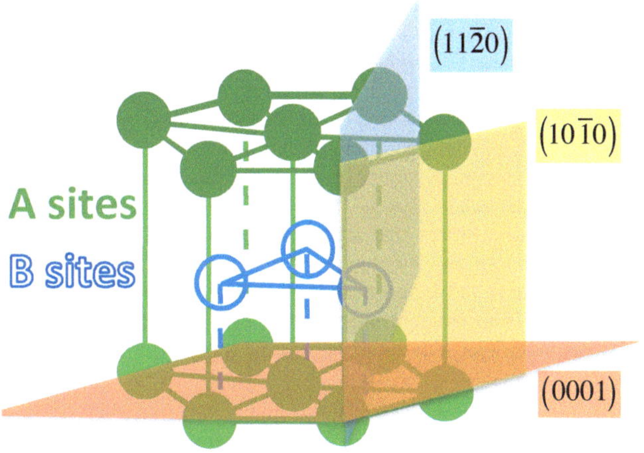

Figure 1. A schematic of Mg unit cell along with an indication of principal planes and the configuration of atoms on them.

Another major problem in the study of oxide formation on reactive surfaces is the speed of reaction. Initial stages of oxidation occur almost instantly upon exposure of freshly prepared Mg surface to the environment. Therefore, it is extremely hard to study in detail the initial stages of magnesium oxidation. To avoid this problem, clean atomically flat surface preparation must be carried out in ultra-high vacuum (UHV) conditions prior to *in-situ* exposure to a well-defined gas atmosphere and the analysis. The fulfillment of these conditions will allow accurate step-by-step monitoring of the initiation and kinetics of oxidation process in Mg.

The main purpose of the present report is to introduce to the magnesium community a synchrotron radiation based experimental technique having all the aforementioned capacities as well as to communicate our first highly promising results on the initiation and kinetics of Mg oxidation. This technique is X-ray Photoelectron Spectroscopy (XPS) [7], which allows (i) to reduce the thickness of material volume under scrutiny down to approximately 3 nm and (ii) to follow *in-situ* the oxidation state of detected elements.

Information obtained in the XPS experiments includes elemental analysis, chemical composition, and chemical states. The principle of this technique is based on resolving at high resolution the binding energy (BE) of core level electrons produced by photoemission under X-ray irradiation. The BEs of core levels are characteristic of an element, which makes possible the qualitative analysis of elements in material surface. In addition, BEs evolve as functions of the oxidation state and the chemical environment of a probed element.

As regards the surface sensitivity of XPS, X-ray irradiation penetrates deep into the material, but the chemical information is carried only by emitted photoelectrons that have a very small inelastic mean free path. The latter strongly depends on a material and on the kinetic energy of photoemitted electrons, and varies between 5 to 30 Å for experimental conditions typical in XPS analysis. The detection limit of X-ray photoemission spectroscopy is usually considered to be 1 at. %.

Experimental material and methods

XPS experiments were carried out at the VUV-Photoemission beamline of Elettra synchrotron (Trieste, Italy). This beamline is designed primarily for characterization of surfaces and solid-state materials by means of high-resolution photoemission experiments. The light source is an undulator with a photon energy range from 17 to 900 eV. The end station of VUV-Photoemission beamline is composed of two interconnected UHV chambers separated by a valve: preparation chamber and measurement chamber. The preparation chamber is a UHV set-up dedicated to the *in-situ* preparation of surfaces that are to be characterized by means of photoemission in the measurement chamber. A fast entry load lock available in the experimental set-up was used for transferring the samples from ambient pressure to UHV without affecting the vacuum conditions. The base pressure in our preparation chamber was 10^{-10} mbar.

Material for the experiment was a commercially supplied Mg single crystal having at least 99.99 % purity. The preparation of (0001) surface of Mg single crystal were done in the preparation chamber by repeated cycles of Ar^{+} sputtering at E~600eV for 20 minutes, followed by annealing at up to 200°C for approximately 15 minutes. The crystalline quality and the cleanliness of Mg surface were monitored by Low Energy Electron Diffraction (LEED) and by XPS.

The XPS experiments were done in the measurement chamber at a base pressure of 10^{-11} mbar. The initial stages and progress of oxidation on Mg surface were studied by measuring the Mg 2*p* and the oxygen (O) 1*s* core levels, and the valence band state (angle integrated mode). The measurements were carried out on the clean samples and after small doses (typically at pressures in the range of $1x10^{-9}$ mbar) of pure oxygen were delivered. Research-grade 6.0 oxygen was inlet in the measurement chamber via a variable leak valve.

SCIENTA R4000 spectrometer with more than 30° acceptance angle, mounted at 45° to the beam direction was used in the experiments as the electron analyzer. Mg 2*p* core levels were measured using a photon energy of 100 eV to enhance the sensitivity to the surface. During the measurements, the energy resolution of the analyzer was better than 10 meV.

Results and discussion

Photoemission spectra of the Mg 2p core levels were measured before and after different doses of oxygen up to 12L. Most important stages of the oxidation process are presented in Fig.2.

In the case of Mg 2*p* core level on the initially clean Mg(0001) surface, the spin-orbit (SO) splitting corresponds to 240 meV. The line shape of Mg 2*p* core levels of clean Mg surface reveals at least two SO doublets: surface (S) and bulk (B) components that differ in energy for approximately 120 meV.

Such energy difference between the surface and the bulk components is known in the literature as Surface Core Level Shift (SCLS). For crystalline surfaces prepared *in-situ*, the presence of B and S components is a fingerprint of the good quality of prepared surface. The presence of surface spectral components in clean surfaces is ascribed to the fact that when a crystal is truncated, the surface atoms have lower coordination number than the bulk due to the smaller number of nearest neighbors [8, 9]. The reduction of atomic coordination number affects the charge distribution in the surroundings of surface atoms and causes lattice relaxation of the first top-most atomic layers of the crystal. The related electrostatic potential shifts the core level towards higher (or lower) binding energies due to charge redistribution, which causes the SCLS effect.

In the case of Mg surfaces, the calculated value of SCLS for the 2*p* core levels is of the order of a few hundred meV, and is dependent on the crystallographic orientation of surface [10, 11]. Thanks to the high energy resolution available during our measurements, the B and S components are better, and thus more accurately, resolved than in previous XPS studies that reported a SCLS of 140 meV [10]. These make our results not only unique, but also essential for further development of realistic analytical and *ab-initio* models describing the phenomena. We do not exclude that more than one surface component is present – only accurate fitting analysis of the measured spectra, complemented with *ab-initio* calculations of the SCLS, will give better understanding of the surface relaxation that occurs on Mg surfaces with different orientations. It is known from the literature that the first interlayer spacing of Mg(0001) expands [8], while Mg $(10\bar{1}0)$ shows an oscillatory multilayer relaxation [9].

After dosing 2 L of O_2, the Mg 2*p* core level spectra change their line shape with respect to the clean surface: there is a decrease of the S component with respect to the B and an increase of the spectral weight at high BE. At 6L, the S component is still visible, but a broad oxide component appears at approximately 1.3 eV higher BE than the metal lines. Even a qualitative analysis at this stage reveals that the broad structure consists of at least two components R_1 and R_2 at about 50.5 eV and 50.9 eV BE, respectively (see Fig. 2). By increasing the oxygen exposure to 10L, the O-related components increase and the intensity of the two features is similar. After dosing to 12L, the R_2 component becomes more intense than the R_1.

Similar two oxygen-related components were observed previously during the oxidation of Mg $\left(1\bar{1}00\right)$ surface [12]. The one at low BE (R_1) was identified as the MgO species while the other at higher BE (R_2) was attributed to oxygen chemisorbed at the surface of the oxide layer in formation.

These results correspond quite well and further elaborate earlier experimental and theoretical evidences [13-17]. Namely, the oxidation process begins with the dissociation of an O_2 molecule, which directly goes below the top Mg layer, followed by the formation of a subsurface two-dimensional 2D-oxide layer. When the oxide reaches a certain thickness, it transforms into the cubic rock-salt type structure of MgO.

So far, the attention has been paid to formation of stoichiometric magnesium oxide MgO, whereas much less attention has been given to the dissociative adsorption of O_2 molecule and the migration of O atom into the bulk. There are several key features that make a difference between oxide nucleation on more familiar late transition metal and free-electron like magnesium. We have studied the initial stages of different low index Mg surfaces, not that in other HCP metals, but it is known from literature, e.g. Ref. [19] and references therein, that usually the oxide formation on transition metals have been described as a four-step process which is different from the case of Mg surfaces. At the first stage of oxidation in transition metals, O_2 molecules dissociates on the metal surface to form chemisorbed oxygen. These oxygen atoms remain at the surface, they repeal each other and form surface superstructures. Additional oxygen sticks onto the chemisorbed oxygen layer until a critical coverage above which oxygen atoms penetrate into the subsurface layers but remain near the surface and do not dissolve into the bulk of the metal.

The initial dissociative chemisorption of O_2 over magnesium results in *subsurface* adsorption rather than surface adsorption, while in transition metals the result is surface adsorption. While oxygen ingress is a rate-limiting step in late transition metals, it appears to be spontaneous in magnesium. Oxygen, apparently repulsive over late transition metals, seems to be attractive over magnesium [18].

These results reveal details of the initial stages of Mg oxidation for only one principal crystallographic plane (0001), which has the closest atomic packing (the highest atom density). The mechanisms of oxidation initiation and propagation on the other two principal crystallographic planes, $\left(10\bar{1}0\right)$ and $\left(11\bar{2}0\right)$, can be significantly different because of completely different density and topology of atomic packing. The difference between the two less densely packed planes would be the presence ($\{11\bar{2}0\}$ case) or the absence ($\{10\bar{1}0\}$ case) of sub-surface B-sites atoms under the center of surface A-sites atoms of an elemental unit cell, see Fig.1. Therefore, they should also be investigated in detail in dedicated studies that are reserved for our future reports.

The split of $2p$ core levels and their evolution during the oxidation process revealed at such a high energy resolution in our study also allow to be considered as the evidence of a variation in the valence of Mg at different stages of surface oxidation.

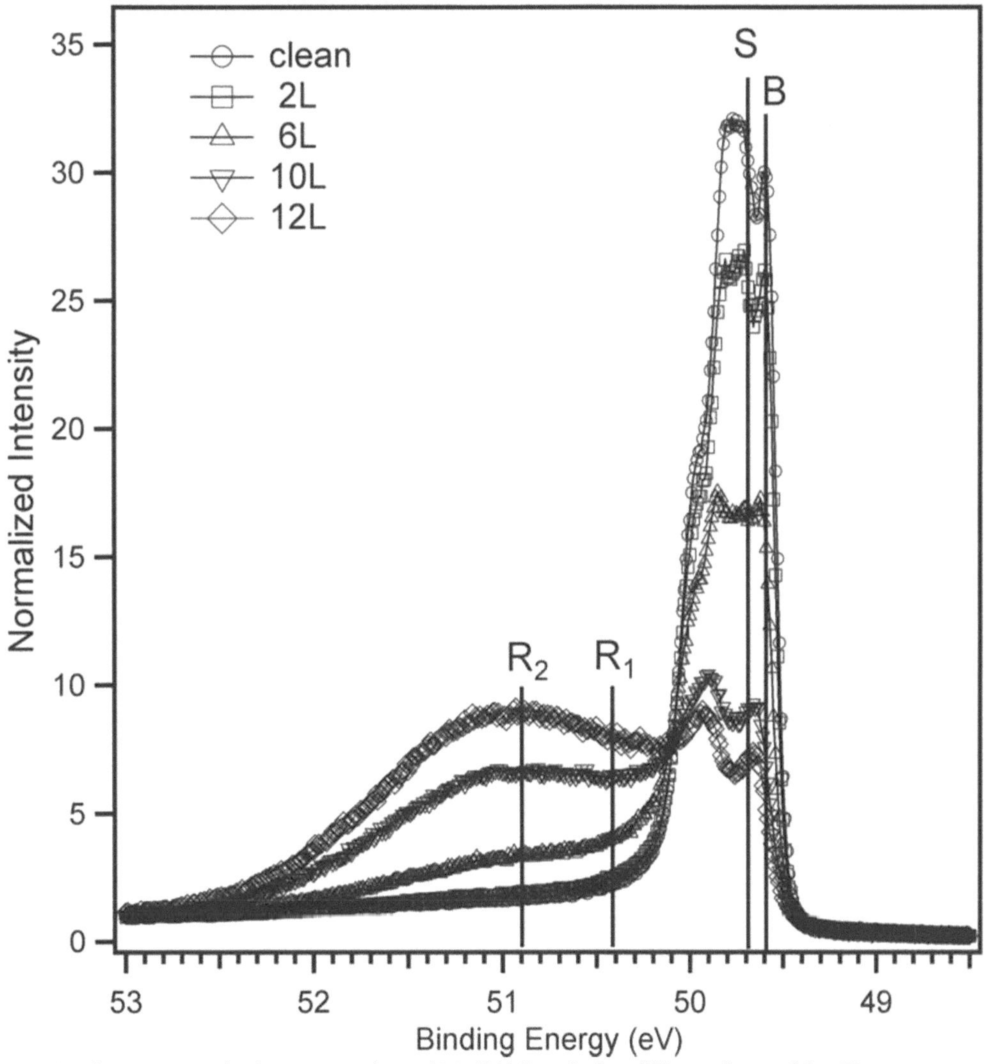

Figure 2. Mg 2p core level spectra acquired on *in-situ* cleaned Mg(0001) and after different doses of O_2. The spectra were measured with 100eV incident photon energy.

Conclusions

The X-ray Photoemission Spectroscopy experiments introduced here can make significant contributions to the fundamental atomic-level understanding of oxidation mechanisms in Mg. The main distinctive feature of our 'surface science' approach is that it deals with crystal surfaces that are well defined both compositionally and crystallographically. The surface sensitivity of XPS, using synchrotron radiation, provided direct information about the electronic and chemical structures of the top-most few atomic layers of Mg(0001) surface. The high-resolution energy measurements reported here allow very accurate determination of the SCLSs of Mg(0001) surface and the monitoring of evolution in different BE components during oxidation. This does not only present important and reliable means for understanding the structure of Mg surface, but also offers new opportunities for revealing the fundamental interactions involved in the process of magnesium oxidation.

Oxygen molecules (O_2) were added intentionally in accurately-controlled sub-monolayer amounts. By comparing the Mg 2*p* core level spectra before and after the O_2 exposure, we found the presence of more than one oxygen-related components. The detailed fitting analysis of the Mg 2*p* core level spectra as a function of O_2 dose along with a comparison with *ab-initio* calculations should clarify further details of the oxidation mechanisms in Mg, and in particular the valence of Mg at different stages of surface oxidation.

References

[1] Pollock TM. Science 2010;328:986.
[2] Agnew SR, Nie JF. Scripta Materialia 2010;63:671.
[3] Witte F. Acta Biomaterialia 2010;6:1680.
[4] McCall CR, Hill MA, Lillard RS. Corrosion Engineering, Science and Technology 2005;40:337.
[5] Orlov D, Ralston KD, Birbilis N, Estrin Y. Acta Mater. 2011;59:6176.
[6] Atrens A, Song G-L, Liu M, Shi Z, Cao F, Dargusch MS. Advanced Engineering Materials 2015;17:400.
[7] Hüfner S. Photoelectron Spectroscopy Principles and Applications: Springer-Verlag Berlin Heidelberg, 2003.
[8] Sprunger PT, Pohl K, Davis HL, Plummer EW. Surface Science 1993;297:L48.
[9] Cho J-H, Ismail, Zhang Z, Plummer EW. Physical Review B 1999;59:1677.
[10] Kammerer R, Barth J, Gerken F, Kunz C, Flodstrøm SA, Johansson LI. Physical Review B 1982;26:3491.
[11] Cho J-H, Kim KS, Lee S-H, Kang M-H, Zhang Z. Physical Review B 2000;61:9975.
[12] Thiry PA, Ghijsen J, Sporken R, Pireaux JJ, Johnson RL, Caudano R. Physical Review B 1989;39:3620.
[13] Hellman A. Physical Review B 2005;72:201403.
[14] Schröder E, Fasel R, Kiejna A. Physical Review B 2004;69:115431.
[15] Schröder E, Fasel R, Kiejna A. Physical Review B 2004;69:193405
[16] Carley AF, Davies PR, Jones RV, Harikumar KR, Roberts MW. Chemical Communications 2002:2020.
[17] Driver SM, Lüdecke J, Jackson GJ, Woodruff DP. J. Electron Spectrosc. Rel. Phenom. 1999;98–99:235.
[18] Cheng S-T, Todorova M, Freysoldt C, Neugebauer J. Physical Review Letters 2014;113:136102.
[19] Reuter K. Nanometer and sub-nanometer thin oxide films at surfaces of late transition metals. In: Heiz U, Hakkinen H, Landman U, editors. Nanocatalysis: Principles, Methods, Case Studies. 2005.

Magnesium Technology 2016

Alloy Development, Diffusion and Joining

Magnesium Technology 2016
Edited by: Alok Singh, Kiran Solanki, Michele V. Manuel, and Neale R. Neelameggham
TMS (The Minerals, Metals & Materials Society), 2016

DEVELOPMENT OF Mg-Al-Sn-Si ALLOYS USING A CALPHAD APPROACH

Andrew Klarner[1], Weihua Sun[1], Janet Meier[1], Alan Luo[1, 2]
[1]Department of Materials Science & Engineering, The Ohio State University, Columbus, Ohio, 43210 , USA
[2]Department of Integrated Systems Engineering, The Ohio State University, Columbus, Ohio 43210, USA

Keywords: Mg-Al-Sn-Si Alloy, Alloy design, CALPHAD, Aging

Abstract

Following the previous development of Mg-Al-Sn alloys, the influence of Si on the microstructure and properties of Mg-Al-Sn alloys is investigated. Thermodynamic modeling of the Mg-Al-Sn-Si system based on the CALPHAD (CALculation of Phase Diagrams) approach was used to predict the microstructure and design the heat treatment schedules. Mg-Al-Sn-Si alloys were prepared by gravity casting using a cylindrical steel mold. The microstructure and mechanical properties of the as-cast and heat-treated samples were characterized and compared with those of Mg-Al-Sn alloys and against the predictions from thermodynamic calculations.

Introduction

The low density of magnesium has led to the development of many structural magnesium alloys for use in lightweighting applications. Currently, Mg-Al-based alloys are the most widely used magnesium alloys because of their excellent castability and mechanical properties, largely due to the addition of Al [1]. However, Mg-Al-based alloys do not show significant response to age-hardening as compared to many age-hardening aluminum alloys. This has been mostly attributed to the orientation and coarseness of the $Mg_{17}Al_{12}$ precipitates [2]. The binary $Mg_{17}Al_{12}$ phase forms on the grain boundaries and has a low softening temperature of about 120°C which does not prevent growth of the grains during aging. Starting with this binary alloy, many alloying additions have been researched to find further improvements to this alloy family. Recently the influence of Sn on Mg-Al alloys was studied by Luo et al. [3] and it was found that the addition of Sn increased the yield strength because of the formation of a third binary phase Mg_2Sn. The Mg_2Sn phase has an anti-fluorite-type (CaF_2) AB2 structure and has high stability with a melting point of 770°C [4]. Mg-Sn-based alloys also are favorable for age hardening as the solubility of Sn in magnesium at the eutectic temperature of 561°C is 14.85 wt.% but drops sharply to 0.4 wt% at 200°C. However Mg-Sn alloys have a sluggish aging response which is unfavorable for commercial applications. The addition of Si has attracted much attention in the development of magnesium alloys due to its improvement on the strength and its low cost [5-8]. To further improve the mechanical properties and age hardening response of previous developed Mg-7Al-2Sn (wt.%) alloy, Si was introduced into this system to investigate its influence on the microstructure and mechanical properties on the alloys in this paper.

In recent decades, CALPHAD approach based on computational thermodynamics has been frequently used in alloy design to accelerate the alloy development [9]. In the present paper, preliminary research on the development of Mg-Al-Sn-Si alloy with the aid of CALPHAD modeling is presented.

Experimental Procedure

Table 1 shows the nominal compositions of the alloys prepared in this work using commercial Mg-7Al-2Sn, pure Sn, pure Mg, and an Al-50Si master alloy. The materials were melted in a graphite crucible under a protective gas atmosphere of CO_2 + 0.5% SF_6 in an induction furnace. The melt was stirred several times to ensure homogeneity of the alloy. After being held at 750°C for 15 min the melt was cast into a steel permanent mold, which was preheated to 200°C, to produce cylindrical ingots of 19mm diameter and 90mm length.

Table 1: Nominal composition (wt.%) of the alloys prepared.

Alloy Designation	Al	Sn	Si	Mn	Mg
AT72	7.15	2.00	0.00	0.43	bal.
ATS0.25	7.34	1.99	0.25	0.42	bal.
ATS1.5	7.39	1.96	1.32	0.37	bal.

The as-cast alloys were sectioned and encapsulated in glass tubes, followed by solution treatment in a furnace at 420°C for 10 hours. After solution treatment the cylindrical ingots were sectioned into 3mm-thick discs using electrical discharge machining (EDM, FANUC). These discs were then aged in an oil bath at 200°C for varying lengths of time (1 - 600hours). The hardness tests were performed with the Rockwell 15T scale (1/16 inch steel ball, 15kg load) and the hardness value was obtained from an average of 10 measurements.

The microscopy samples were mounted and polished using 0.05μm colloidal silica as the final step. The microstructures of the as-cast and solution treated samples were characterized using an optical microscope (OM, Olympus GX71) and scanning electron microscopes (SEM, FEI Quanta 200 and Philips XL-30).

Results and Discussion

Figure 1 shows the solidification path of the alloys calculated using Pandat software and its Mg thermodynamic database PanMg [10]. The calculation is based on the classical Scheil-Gulliver model [11], which assumes complete mixing in liquid and no diffusion in solid. As shown in Figure 1, the solidification sequence of Mg-7Al-2Sn is as follows: The primary (Mg) will form at 609°C, followed by the formation of eutectic Mg_2Sn at 441°C. As the temperature goes down to 430°C, the solidification ends at the invariant reaction L $\leftrightarrow$ Mg_2Sn + (Mg) + $Mg_{17}Al_{12}$ and $Mg_{17}Al_{12}$ phase will form. The addition of Si will lead to the formation of a new binary phase Mg_2Si. Since the thermal stability of the Mg_2Si phase is higher than that of the Mg_2Sn phase, the Mg_2Si will form before Mg_2Sn. As the Si content increases, the primary phase in the solidification will become Mg_2Si instead of (Mg).

Figures 2 and 3 present the concentration distributions of Al and Sn in the primary (Mg) phase, which indicates the segregation of Al and Sn in (Mg). According to the calculation for the center of the grain, where the solidification occurs firstly, the concentration of (Al) and (Sn) are around 2.2 wt.% and 0.75 wt.% respectively. However, as the (Mg) grain grows, the alloying elements in the grains will become inhomogeneous. The Al and Sn concentrations near the grain boundary will increase to be above 10 wt.% and 2.5 wt.% respectively, indicating significant segregation of alloying elements in these alloys.

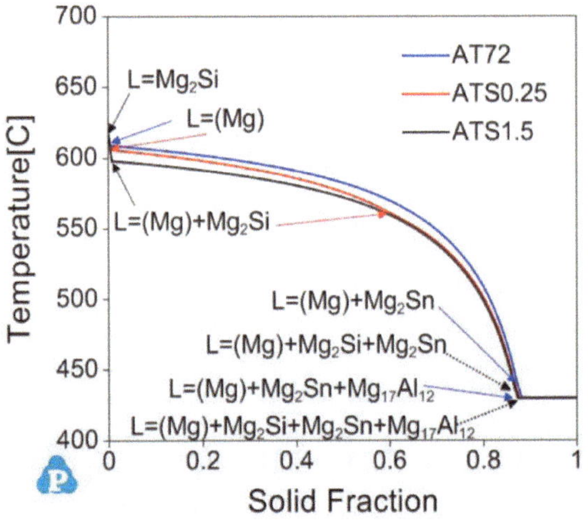

Figure 1: Predicted solidification path of the three alloys using the Scheil-Gulliver model. The calculation is based on the Mg-Al-Sn-Si system using the PanMg thermodynamic database.

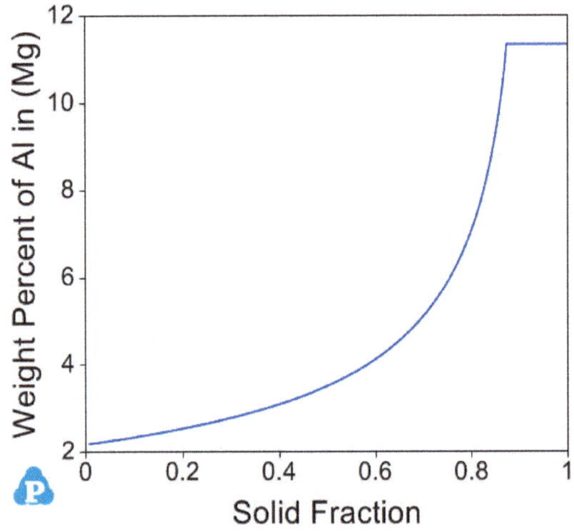

Figure 2: Predicted concentration profile of Al in primary (Mg) for the ATS1.5 alloy based on the Scheil-Gulliver model.

Figures 4 and 5 show the as-cast microstructure of both AT72 and ATS1.5 alloys. These alloys contain small amount of Mn, which improves the corrosion resistance of the alloy. The Mn-rich intermetallics will form as the primary phase during the solidification process and will have a relatively large particle size. Both $Mg_{17}Al_{12}$ and Mg_2Sn phases are found in interdendritic regions. The $Mg_{17}Al_{12}$ phase has a lamellar morphology while Mg_2Sn forms as very small particles. According to the contrast in the back scatter electron (BSE) image, which is sensitive to the atomic weight, it could be clearly seen that the "segregation cloud" of Al and Sn is very obvious near the (Mg) boundary area. As shown in Figure 5, the addition of Si will lead to the formation of Mg_2Si. The microstructure observed in the experiment agrees with the predictions from the CALPHAD modeling.

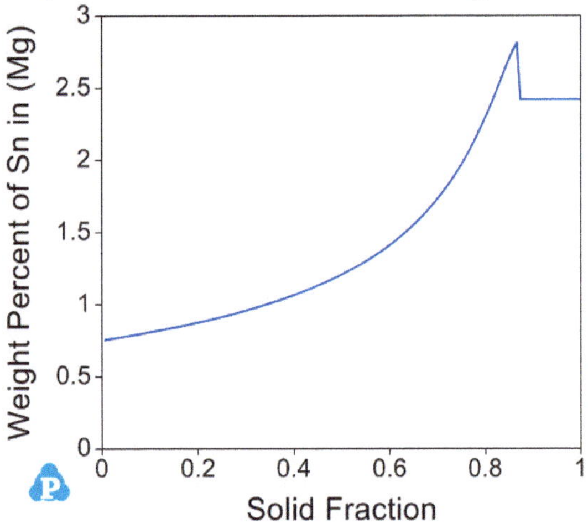

Figure 3: Predicted concentration profile of Sn in primary (Mg) for ATS1.5 alloy based on the Scheil-Gulliver model.

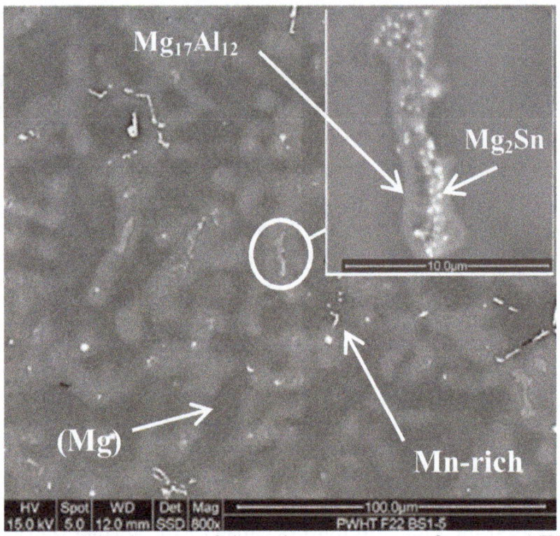

Figure 4: BSE image of the microstructure of as-cast AT72.

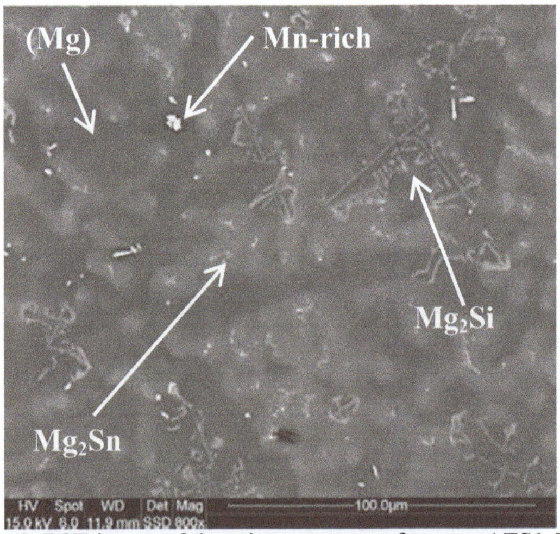

Figure 5: BSE image of the microstructure of as-cast ATS1.5.

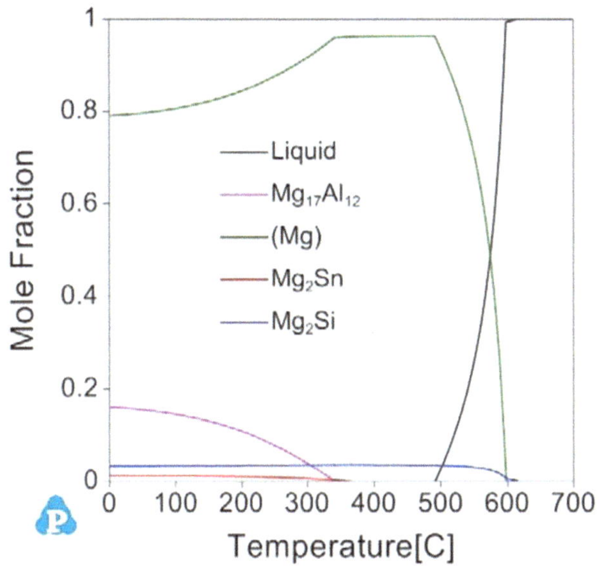

Figure 6: Equilibrium phase fractions versus temperature for ATS1.5.

Figure 6 shows the calculated equilibrium phase fraction vs. temperature, which can be used to select solution treatment and aging temperatures. According to the prediction, the solution treatment temperature should be above 340°C and 370°C to dissolve $Mg_{17}Al_{12}$ and Mg_2Sn respectively, and should not exceed 490°C to avoid the formation of liquid phase. The thermal stability of Mg_2Sn is higher than that of $Mg_{17}Al_{12}$ and also the diffusion of Sn is slower than that of Al, which leads to the slow kinetics process of dissolving Mg_2Sn compared with that of $Mg_{17}Al_{12}$. So the solution treatment temperature used should not be lower or just above the minimum temperature predicted but instead at a higher temperature.

In the present experiment, the solution treatment temperature of 420°C was selected. At this moderate temperature, both the $Mg_{17}Al_{12}$ and Mg_2Sn phases could be dissolved into the (Mg) matrix. In addition, the grain growth of the primary (Mg) can be controlled by using a moderate temperature rather than one closer to the temperature at which a liquid phase forms. The thermal stability of the Mg_2Si phase is very high and there is almost no solubility of Si in the (Mg) matrix, so Mg_2Si will not dissolve. Figures 7 and 8 show the microstructures of AT72 and ATS1.5 alloys after solution treatment at 420°C for 10 hours. Compared with the microstructure of the as-cast alloys, it is obvious that the "segregation cloud" at the (Mg) grain boundary area disappears. EDS analysis shows that the concentration of Al and Sn in the (Mg) is rather uniform, which means the diffusion of Al and Sn in (Mg) matrix during the solution treatment at 420°C could lead to a homogeneous distribution of the alloying elements. The $Mg_{17}Al_{12}$ phase at the grain boundary has been completely dissolved into the (Mg) matrix and the small Mg_2Sn particles have been almost completely dissolved. As expected, the experimental results indicate that it is easier to dissolve the $Mg_{17}Al_{12}$ phase over the Mg_2Sn phase. Both the Mn-rich and Mg_2Si phases do not dissolve due to their comparatively high stability.

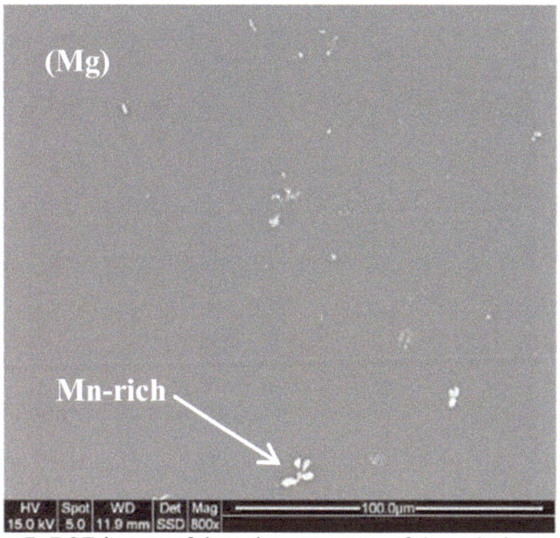

Figure 7: BSE image of the microstructure of the solution treated AT72.

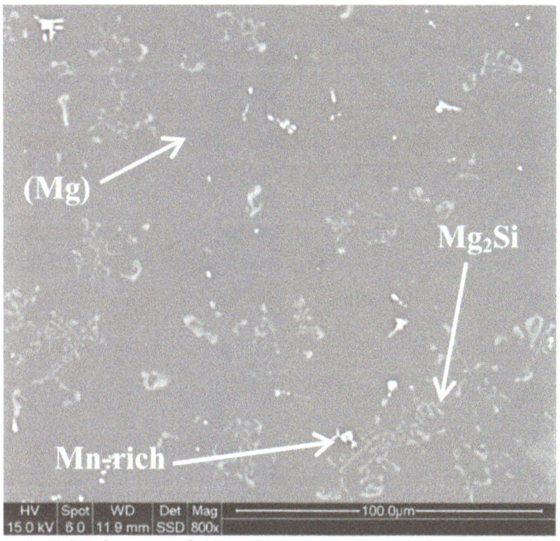

Figure 8: BSE image of the microstructure of the solution treated ATS1.5.

Table 2: Rockwell 15T Hardness of as-cast and solution treated AT72 and ATS alloys.

	Hardness (15T)		
Alloy	As Cast	Solution Treated	Peaked Aged
AT72	64.75 ± 1.5	61.60 ± 1.3	72.55 ± 1.5
ATS0.25	66.25 ± 0.6	62.07 ± 1.0	73.07 ± 1.0
ATS1.5	70.05 ± 0.8	67.93 ± 1.4	76.95 ± 0.6

A common aging temperature found in the aging of magnesium alloys (200°C) was chosen for this study. According to the calculation, the maximum solubility of Al and Sn at 200°C in the (Mg) matrix are about 3.1 wt.% and 0.3 wt.% respectively. Almost all the Sn and a large fraction of Al are predicted to precipitate as Mg_2Sn and $Mg_{17}Al_{12}$ phases at 200°C. **Error! Reference source not found.** shows the hardness values of the ATS alloys in the as-cast conditions. As expected, the hardness increases with increasing amount of Si because of the increase fraction percent of the hard Mg_2Si phase.

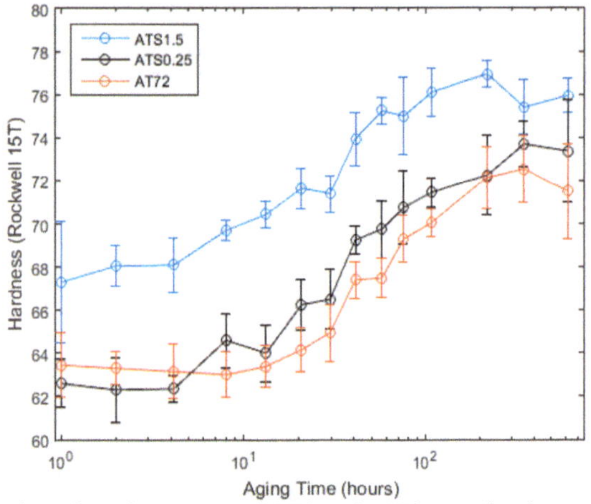

Figure9: Aging curves of ATS and AT72 samples that were solution treated at 420°C for 10 hours and aged at 200°C.

Figure9 shows the age-hardening curves for the ATS and AT72 alloys aged at 200°C after being solution treated for 10 hours at 420°C. As seen in the figure, the response to age hardening is obvious due to the precipitation of the $Mg_{17}Al_{12}$ and Mg_2Sn phases. The peak hardness is seen to increase with increasing amount of Si. ATS1.5 is found to have a peak hardness of 77, which is a 13.3% increase from the solution treated hardness while ATS0.25 has a hardness of 73.7 which is a 17.7% increase. AT72 has a 17.8% increase to a peak hardness 72.6. Again the ATS1.5 alloy has the highest hardness value, as shown in Table 2.

Conclusion

The preliminary results on the development of Mg-Al-Sn-Si alloys supported by the CALPHAD modeling are presented. Microstructural analysis of the alloys in the as-cast, solution treated, and aging condition are presented and the experimental results are in agreement with the predictions. Hardness tests show that the Si addition has a positive influence on the mechanical properties. More research is being carried out to further investigate the influence of Si on the microstructure and mechanical properties.

Acknowledgements

The authors acknowledge the collaboration with General Motors and Meridian Lightweight Technologies, especially the many helpful discussions with Drs. Jon Carter and Anil Sachdev of GM R&D Center. This material is based upon work supported by the United States Department of Energy under Award Number DEEE005753 to GM and a sub-award to The Ohio State University.

References

1. B.L. Mordike and T. Ebert, "Magnesium, Properties — applications — potential", *Materials Science and Engineering,* A302 (2001), 37–45.

2. S. Celotto, "TEM study of continuous precipitation in Mg–9 wt%Al–1 wt%Zn alloy," *Acta Mater.*, 48 (2001), 1775–1787.

3. Alan. A. Luo, Penghuai Fu, Liming Peng, Xiaoyu Kang, Zhenzhen Li, and Tianyu Zhu, "Solidification Microstructure and Mechanical Properties of Cast Magnesium-Aluminum-Tin Alloys," *Metall. Mater. Trans. A* (2011), *43*, 360–368.

4. Hongmei Liu, Yungui Chen, Yongbai Tang, Shanghai Wei, and Gao Niu, "The microstructure, tensile properties, and creep behavior of as-cast Mg–(1–10)%Sn alloys," *Journal of Alloys and Compounds,* 440 (2007), 122–126.

5.B. Bronfin, M. Katsir, and E. Aghion, "Preparation and solidification features of AS21 magnesium alloy," *Materials Science and Engineering*, A302 (2001), 46-50.

6.Dae H. Kang, Sung S. Park and Nack J. Kim, " Development of creep resistant die cast Mg-Sn-Al-Si alloy," *Materials Science and Engineering*, A 413-414 (2005), 555-560.

7.Li Xin-lin, Chen Yan-bin, Wang Xiang, and Ma Guo-rui, " Efffect of cooling rates on as-cast microstructures of Mg-9Al-xSi (x=1,3) alloys" *Trans Nonferrous Met. Soc. China*, 20(2010), s393-s396.

8.Haoran Geng, Peng Liu, Huawei Jiang, and Bo Teng, "Effects of Si on Structure and Properties of Mg-Sn Alloys," *Applied Mechanics and Materials*, Vol 320 (2013), 303-307.

9. A. A. Luo, "Material design and development: From classical thermodynamics to CALPHAD and ICME approaches", CALPHAD, 50(2015), 6-22.

10. Pandat software package and PanMagnesium thermodynamic database (Madison, WI: CompuTherm LLC)

11. E. Scheil, "Bemerkungen zur schichtkristallbildung", Z. Metallkd, 34 (1942) 70–72.

Magnesium Technology 2016
Edited by: Alok Singh, Kiran Solanki, Michele V. Manuel, and Neale R. Neelameggham
TMS (The Minerals, Metals & Materials Society), 2016

Lattice Ordering and Microstructure of Ultra-high Strength Mg-Ca-Zn Alloys

Alok Singh[1], Althaf B. Dudekula[1], Naoko Ikeo[2], Hidetoshi Somekawa[1], Toshiji Mukai[2]

[1]Structural Materials Unit, National Institute for Materials Science, Sengen 1-2-1, Tsukuba 305-0047, Japan
[2]Department of Mechanical Engineering, Kobe University, Rokkodai, Kobe, Japan

Keywords: Mg-Ca-Zn alloys, extrusion, transmission electron microscopy (TEM)

Abstract

Magnesium alloying with calcium is very promising for mechanical strength. One of the promising alloys systems is ternary Mg-Zn-Ca. Recent results by the first principles calculation suggest that the combination of calcium and zinc possibly enhances the cohesive energy of grain boundary. In this work, Mg-Zn-Ca alloys with low alloying amounts of 0.1-0.3at% Ca and 0.2-0.6at% Zn were studied. Very fine grain size of smaller than 0.5 micron was achieved after extrusion, with finer substructure of the order of 100 nm. Microstructure showed sharp grain boundaries A possible ordering of atoms in the matrix was observed. These ultra-fine and complex grain structures have been studied with the aid of advanced transmission electron microscopy techniques.

Introduction

Magnesium alloys offer the advantage of being the lightest structural metals. It is very attractive for application in transportation for weight reduction for fuel saving and consequent environmental conservation. It is also an advantage that magnesium is abundantly available on earth. In addition, magnesium is bio-compatible, being an essential element of our body. By the advantage of being light and bio-compatible, it offers a good implant material in our body [1, 2]. Furthermore, it can dissolve in human body at a rather quick rate, saving the need for another surgery for its removal. To control the strength and other properties such as the degradation rate in the body, magnesium can be alloyed with other bio-compatible elements such as aluminum, zinc and calcium. A common alloy system that has emerged in recent investigations is Mg-Zn-Ca [3]. This alloy system contains several binary phases such as Mg_2Ca and $MgZn_2$, and a ternary phase $Mg_6Zn_3Ca_2$ [4, 5, 6]. Often, the alloys developed in this system are viewed as modification of the Mg-Zn alloys by trace additions of calcium, or vice-versa [7, 8]. In most of these investigations, the ternary $Mg_6Zn_3Ca_2$ phase has been utilized to pin grain boundaries for obtaining fine grain size by hot working such as by extrusion [9, 10, 11, 12]. However, presence of intermetallic phases can complicate the phase equilibrium process (during hot working, heat treatments and degradation). Therefore other strategies for controlling microstructures also need to be tried. Furthermore, high amounts of solute content can interfere with bio-compatibility by excessive gas formation [13]. Therefore very low alloying element alloys need to be developed.

Segregation of the solute elements at grain boundaries has been predicted by atomistic simulations [14]. This segregation can lower the mobility of grain boundaries to achieve fine grain size during hot working, and increase the cohesivity at the boundaries to impart more strength and ductility. With this concept, Mg-Zn-Ca alloys with trace amounts of zinc and calcium have been designed and investigated in the present study. Ultra-fine grain size has been achieved by process of extrusion.

Experimental Procedure

Four alloys of composition (in at%) Mg-0.1Ca, Mg-0.3Ca, Mg-0.1Ca-0.2Zn and Mg-0.3Ca-0.6Zn were prepared by melting in an electric furnace and cast in a steel mold. The cast alloys were then machined into billets and extruded at final temperature of about 250°C into rods of 4 mm diameter.

For transmission electron microscopy (TEM) observations, the extruded samples were sliced parallel to the extrusion direction (longitudinally) with a low speed diamond saw, followed by mechanical thinning to about 80 μm and finally by ion milling. The samples were observed on JEOL 2000FX and 2100F microscopes operated at 200kV, and Tecnai G2 F30 microscope operated at 300kV.

(a)

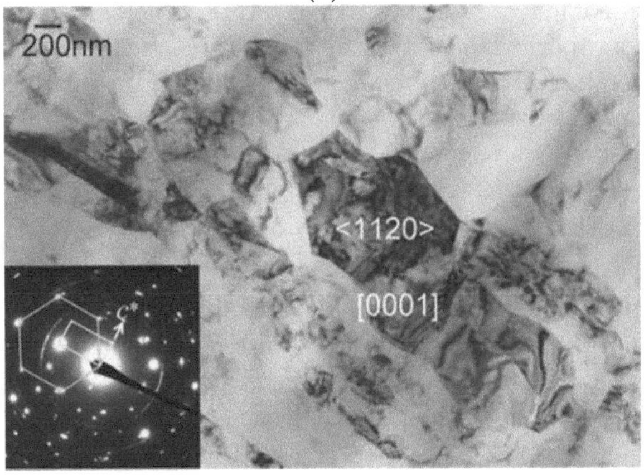

(b)

Figure 1: Bright field micrographs of extruded Mg-0.1Ca alloy showing (a) grain structure and (b) a special boundary

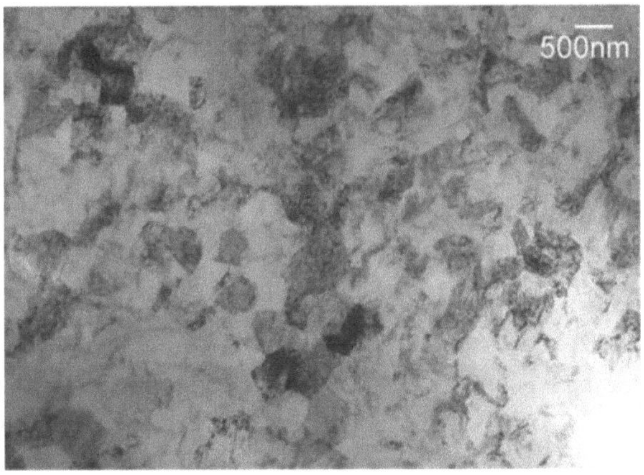

(a)

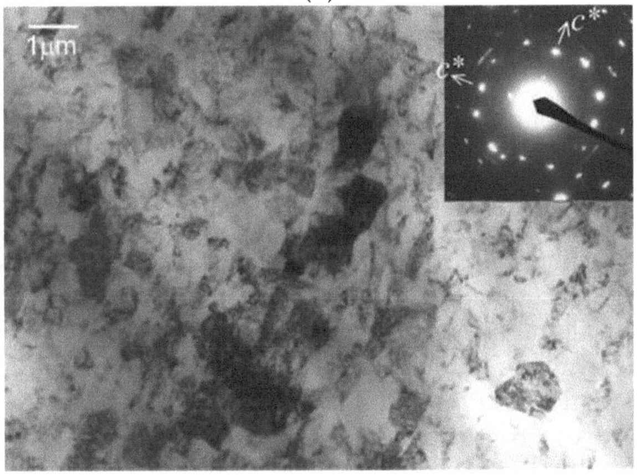

(b)

Figure 2: Bright field micrographs of extruded Mg-0.3Ca alloy showing (a) grain structure and (b) grains and unrecrystallized regions with twinning.

Results

Microstructure of the alloys after extrusion as studied by TEM is described in the following. All the alloys were extruded with similar extrusion conditions. Depending on the composition, the degree of recrystallization and grain size varied.

Binary Mg-Ca alloys

Fig. 1(a) shows general microstructure of extruded Mg-0.1Ca alloy. Some grains of about a micron in size can be observed, while large regions are unrecrystallized. Highly deformed unrecrystallized region is observed at upper right. Fig. 1(b) shows that the sharp boundaries observed are special boundaries. In this case (as shown by the composite diffraction pattern inset) a grain oriented along zone axis $\langle 11\bar{2}0\rangle$ makes boundaries with another grain oriented along [0001] zone axis. The c axis of the former grain is along a $\langle 1\bar{1}00\rangle$ direction of the latter grain. This does not describe a known twin, and therefore is called a special boundary here.

Fig. 2 shows the grain structure in Mg-0.3Ca alloy. Much more recrystallization is observed, compared to Mg-0.1Ca alloy. In Fig. 2(b), strained recrystallized regions are observed along with grains of about a micron size. Inset diffraction pattern suggests presence of a $\{10\bar{1}2\}$ type twin, because two c^* axes approximately perpendicular to each other are observed.

A high resolution micrograph along [0001] zone axis is shown in Fig. 3. Uneven contrast and Moire fringes clearly suggest presence of mechanical strain. Fast Fourier Transform (FFT) from a region marked by a square near the center of the micrograph is shown in an inset (top). Two of the $\{10\bar{1}0\}$ spots

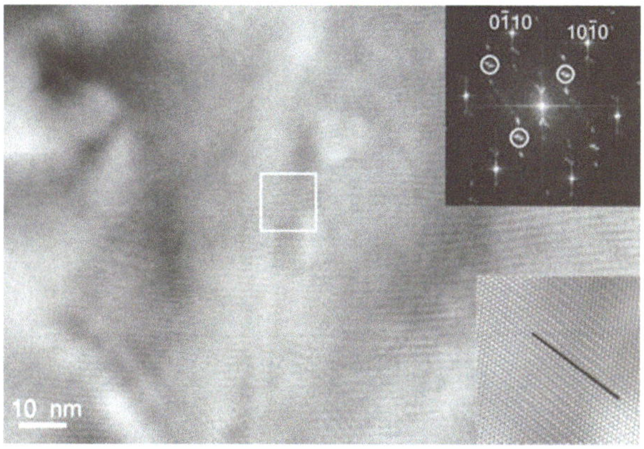

Figure 3: High resolution micrograph along [0001] zone axis from extruded Mg-0.3Ca alloy. A Fast Fourier Transformation (FFT) from the region marked by a square is inset (top), in which extra spots are marked by circles. An Inverse FFT of this area is also inset (bottom)

are indexed. Some extra spots occur at $\frac{1}{2}(10\bar{1}0)$ positions (marked with circles), indicating a doubling of the lattice spacing of one set of prismatic planes. An inverse of FFT is shown in the inset at bottom. It clearly shows fairly well ordered $(10\bar{1}0)$ planes with double the interplanar spacing. This could arise from strain, or indeed by ordering of calcium atoms in magnesium lattice.

Ternary Mg-Ca-Zn alloys

Fig. 4(a) shows a scanning transmission electron microscope (STEM) annular dark field (ADF) micrograph taken from extruded Mg-0.1Ca-0.2Zn alloy with a relatively large camera length of 300mm in Tecnai microscope. Deformation strain, including strained regions and subgrains, is observed due to diffraction contrast. Another STEM micrograph at a smaller camera length of 80cm (making high angle annular dark field (HAADF) contrast) is shown in Fig. 4(b), eliminating diffraction contrast. In this micrograph, grain boundaries are observed clearly. The grain size is about 0.5 μm, and the grain boundaries are straight and sharp. Since there is no strain contrast in this image, the grain boundary contrast is most likely to be from composition, i.e., elemental contrast. Therefore, a line scan by EDS with 5 points was performed across a grain boundary, as marked in this micrograph. The compositions are plotted in Fig. 4(c). A maxima in the concentrations of Ca and Zn occur at the grain boundary. This confirms

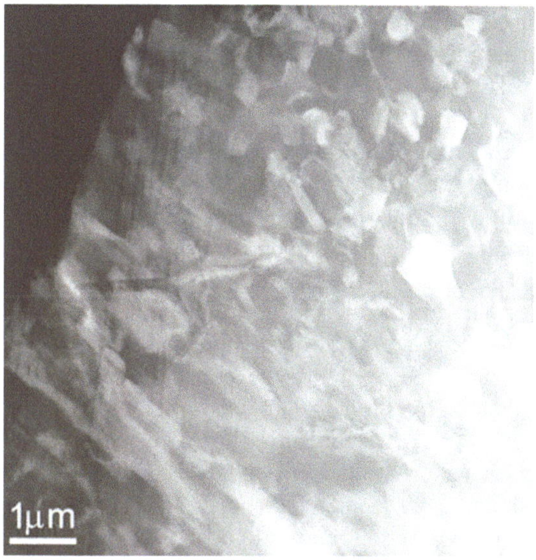

(a)

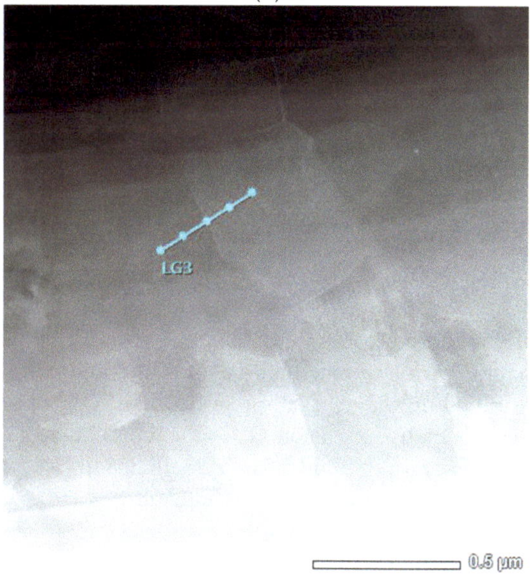

(b)

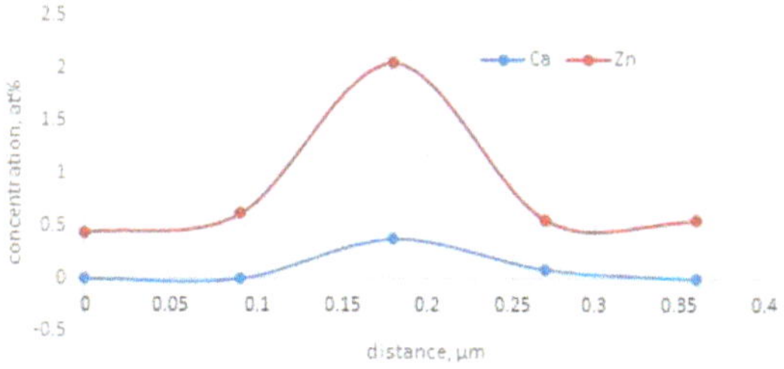

(c)

Figure 4: STEM micrographs from extruded Mg-0.1Ca-0.2Zn alloy showing (a) strain contrast of deformation structures in the matrix, (b) grain boundaries and (c) EDS composition profile across a grain boundary marked in (b).

segregation of Ca and Zn at the boundaries. It also appears that there is very little solubility of Ca in the matrix α-Mg itself.

Fig. 5(a) shows STEM-ADF image (camera length 300 cm) of extruded Mg-0.3Ca-0.6Zn alloy. It shows sharp grain boundaries which are bright in contrast, as well as strain and subgrain boundaries inside the grains. Grain size are less than 500 nm, while the subgrain structure is about 100 nm in scale. Another figure, Fig. 6, examines microstructure in highly strained regions. In this micrograph, there are some dark regions surrounded by lighter contrast regions. The dark regions were confirmed to be in oriented along $\langle 11\bar{2}0\rangle$ zone axis. A composite diffraction pattern is inset, which shows that while the dark regions are oriented along $\langle 11\bar{2}0\rangle$ zone axis, the surrounding regions are along $\langle 11\bar{2}3\rangle$ zone axis. This shows that the two regions are related by $\{10\bar{1}2\}$ type of twinning. Fig. 5(b) is an STEM image showing only grain boundaries in HAADF contrast. Bright contrast at the boundaries indicate solute segregation. Composition profile measured by EDS across this grain boundary is shown in Fig. 5(c). It confirms that a high concentration of Ca and Zn exists at the grain boundary.

Discussion

The effect of the amount of solute element calcium, as well as a combination of zinc and calcium in trace amounts, is studied on the degree of recrystallization and grain refinement by extrusion. It is known that higher amount of solute can aid in recrystallization. This effect is confirmed from the present results.

Mg-Ca-Zn ternary alloys with similar compositions as those studied here have been reported by Hofstetter et al. [11, 12]. Their alloys ZX00 and ZX10 have compositions close to Mg-0.1Ca-0.2Zn and Mg-0.2Ca-0.4Zn in at%. These alloys have been called high strength low alloy (HSLA), following steel terminology. ZX00 has been stated to be precipitate-free, while precipitation has been shown in ZX10. No regular precipitates are observed in the alloys in the present study, either in Mg-0.1Ca-0.2Zn (at%), which corresponds to ZX00, or in Mg-0.3Ca-0.6Zn, because of the processing conditions. This is confirmed by STEM-HAADF images.

Although a large amount of strain exists in both the ternary alloys after extrusion, grain boundaries can be clearly observed by STEM-HAADF imaging. These grain boundaries are confirmed to be sharp and planar. Observation in a large areas, compared to the

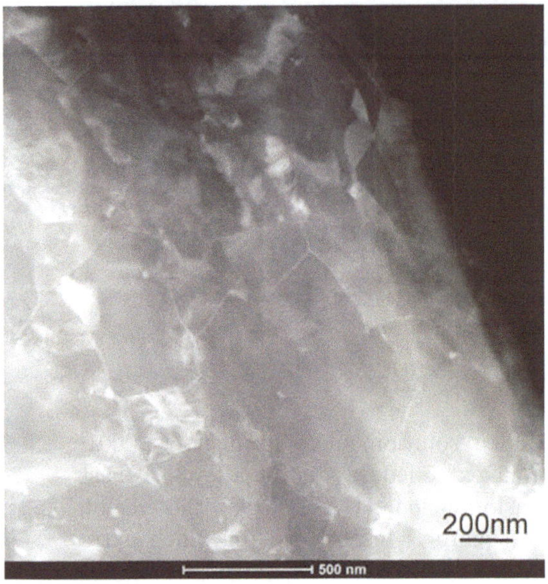

(a)

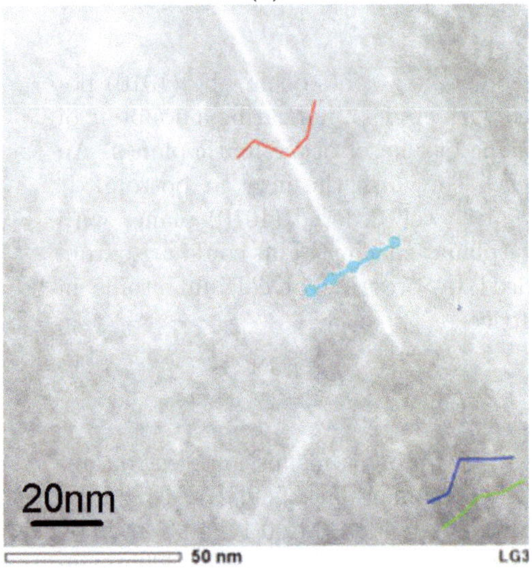

(b)

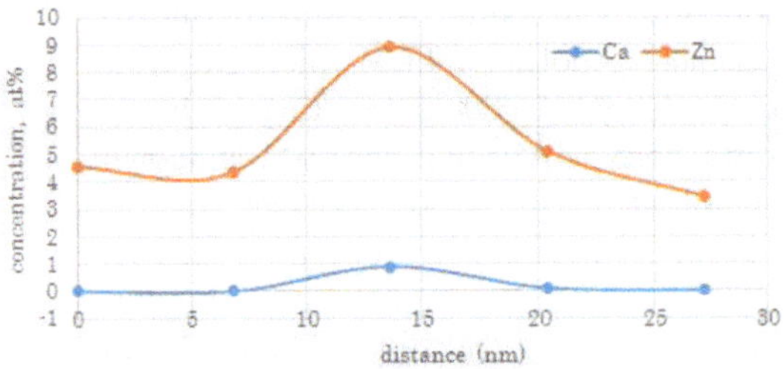

(c)

Figure 5: (a) Grain boundaries and subgrains observed in Mg-0.3Ca-0.6Zn alloy observed in STEM mode, (b) a grain boundary and (c) composition measured across this boundary by EDS.

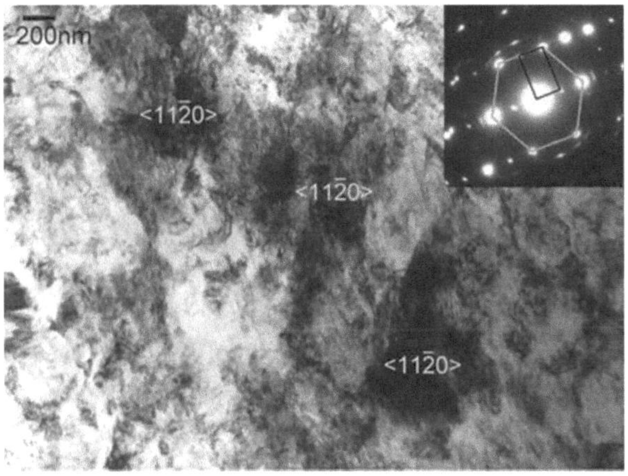

Figure 6: Bright field micrographs of extruded Mg-0.3Ca-0.6Zn alloy showing strain concentrated regions (diffraction inset).

size of the grains, shows a complete recrystallization of the ternary alloys after extrusion. Subgrains of much finer scale exist within these grains, which are expected to enhance the strength of the alloys further than just by grain size.

Fine grain size of about half a micron and solute segregation on the grain boundaries shows that solute is also effective in pinning grain boundaries, preventing grain growth after recrystallization. Thus it is possible to obtain ultra-fine grain size and high strength with low amount of alloying without precipitation. It is also possible to obtain a hierarchical microstructures with ultra-fine grain and subgrain structures.

The planar nature of the grain boundaries and well formed grain boundary junctions, such as observed in Fig. 5, indicates near equilibrium. It cannot be ruled out that solute segregation results in lower grain boundary energies.

Furthermore, we would like to comment that it is possible to study such fine and hierarchical microstructures at a very fine scale by advanced TEM imaging techniques using annular detectors. The conventional TEM techniques, which use diffraction contrast, work well with perfect crystals or discrete defects, but are not suitable for strained materials. Strong diffraction contrast prevents observation of underlying microstructural features. The present work demonstrates how ADF techniques can look past the strain contrast to study microstructure on a very fine scale, as detailed elsewhere[15].

Conclusions

Extruded magnesium alloys by micro alloying with calcium and zinc elements have been designed, aimed at bio-implant applications. The microstructure of the extruded alloys was studied by transmission electron microscopy techniques, which includes advanced STEM techniques such as HAADF, ADF and nano EDS. The following conclusions were drawn:

1. Binary Mg-0.1at%Ca alloy showed few recrystallized grains after extrusion at 250°C in largely unrecrystallized regions of high mechanical strain. Sharp boundaries of special orientation relationships existed. Size of recrystallized grains was about 1.0 μm.

2. Binary Mg-0.3at%Ca alloy showed more extensive recrystallization after extrusion at 250°C. Existence of twins was indicated by electron diffraction.

3. Extruded ternary alloy Mg-0.1at%Ca-0.2at%Zn showed grains of about 0.5 μm in size with subgrains of the order of 100 nm. High amount of strain accumulated inside the grains was observed. Grain boundaries were planar and sharp. Segregation of calcium and zinc was confirmed at the grain boundaries.

4. Similarly, Mg-0.3at%Ca-0.6at%Zn alloy showed grain size of less than 0.5 μm with subgrain structure of 100 nm, where grain boundaries were well defined, showing segregation of calcium and zinc.

5. Thus ultra-fine grained magnesium alloys can be obtained without aid of precipitation in very low alloying in Mg-Ca-Zn system.

Acknowledgment

This work was supported by JSPS KAKENHI Grant No. 25246012 in part.

References

[1] F. Witte, The history of biodegradable magnesium implants: A review, Acta Biomaterialia, 6 (2010), 1680-1692.

[2] H. Hornberger, S. Virtanen, A.R. Boccaccini, Biomedical coatings on magnesium alloys A review, Acta Biomaterialia, 8 (2012), 2442-2455.

[3] E. Zhang, L. Yang, Microstructure, mechanical properties and bio-corrosion properties of MgZnMnCa alloy for biomedical application, Materials Science and Engineering, 497A (2008), 111-118.

[4] G. Levi, S. Avraham, A. Zilberov, M. Bamberger, Solidification, solution treatment and age hardening of a Mg1.6 wt.% Ca3.2 wt.% Zn alloy, Acta Mater. 54 (2006) 523-530.

[5] K. Kubok, L. Litynska-Dobrzynska, J. Wojewoda-Budka, A. Gral, A. Debski, Investigation of Structures in As-Cast Alloys from the Mg-Zn-Ca System, Archives of Metallurgy and Materials 58 (2013), 329-333.

[6] K. Kubok, L. Litynska-Dobrzyska, A. Wierzbica-Miernik, J. Wojewoda-Budka, "Age Hardening of Mg-3Zn-xCa (X = 0, 0.5, 1.0) wt.% Alloys", Materials Science Forum 765 (2013) 481-485.

[7] K. Oh-ishi, R. Watanabe, C.L. Mendis, K. Hono, Age-hardening response of Mg0.3 at.%Ca alloys with different Zn contents, Mater. Sci. Engg. A 526 (2009) 177184

[8] Brian Langelier, Shahrzad Esmaeili "The Effect of Zn Additions on Precipitation Hardening of Mg-Ca Alloys" Magnesium Technology 2010 (eds. S. Agnew, N. Neelameghham, E. Nyberg), The Minerals, Metals & Materials Society Society (TMS), 2010, pp. 505-5010.

[9] A.C. Hanzi, F.H. Dalla Torre, A.S. Sologubenko, P. Gunde, R. Schmid-Fetzer, M. Kuehlein, J.F. Lffler, and P.J. Uggowitzer, Philos. Mag. Lett. 89 (2009) 377.

[10] A.C. Hanzi, A. Sologubenko, P. Gunde, M. Schinhammer, and P.J. Uggowitzer, Philos. Mag. Lett. 92 (2012) 417.

[11] J. Hofstetter, M. Becker, E. Martinelli, A. M. Weinberg, B. Mingler, H. Kilian, S. Pogatscher, P. J. Uggowitzer, J. F. Lffler, High-Strength Low-Alloy (HSLA) MgZnCa Alloys with Excellent Biodegradation Performance, The Journal of The Minerals, Metals & Materials Society (TMS) 66 (2014) 566-572.

[12] J. Hofstetter, S. Ruedi, I. Baumgartner, H. Kilian, B. Mingler, E. Povoden-Karadeniz, S. Pogatscher, P. J. Uggowitzer, J. F. Lffler, Processing and microstructure-property relations of High-Strength Low-Alloy (HSLA) MgZnCa Alloys, Acta Mater. 98 (2015) 423-432.

[13] T. Kraus, S.F. Fischerauer, A.C. Hnazi, P.J. Uggowitzer, J.F. Loffler, A.M. Weinberg, Magnesium alloys for temporary implants in osteosynthesis: In vivo studies of their degradation and interaction with bone, Acta Biomaterialia 8 (2012) 1230-1238.

[14] T. Hase, T. Ohtagaki, M. Yamaguchi, N. Ikeo, T. Mukai, Effect of aluminum and zinc solute addition on enhancing impact fracture toughness in Mg-Ca, (submitted)

[15] A.B. Dudekula, J.M. Rosalie, H. Somekawa, T. Miyawaki, A. Singh, K. Tsuchiya, Microstructure study of a severely plastically deformed Mg-Zn-Y alloy by application of STEM low angle annular dark field diffraction (LAADF) contrast imaging, (submitted)

Magnesium Technology 2016
Edited by: Alok Singh, Kiran Solanki, Michele V. Manuel, and Neale R. Neelameggham
TMS (The Minerals, Metals & Materials Society), 2016

PRE-STRAINING EFFECT ON PRECIPITATION BEHAVIOUR OF AZ31B

Panthea Sepehrband[1], Matthew Lee[1], Aaron Burns[1]
[1]Department of Mechanical Engineering, Santa Clara University; 500 El Camino Real; Santa Clara, CA 95053, USA

Keywords: Mg-Al-Zn, Precipitation hardening, Non-isothermal, Pre-straining

Abstract

Thermomechanical processing techniques consisting of a deformation process followed by an annealing treatment is a common approach to manipulate mechanical properties. In this research a combination of pre-straining and non-isothermal aging process is combined to study precipitation-hardening behavior of an AZ31B magnesium alloy. Effect of heating rate during non-isothermal aging and pre-straining on the kinetics of precipitation and evolution of hardness is investigated.

Introduction

Magnesium (Mg) alloys are known for their low density, high specific strength, and high specific stiffness; meaning that they have a high strength-to-weight ratio when properly manufactured [1, 2]. Considering these mechanical characteristics, magnesium alloys have a great potential for many uses in the automotive, aerospace, and electronics industries. However, they cannot always be used in the same structural applications as steel or aluminum alloys because of their comparatively less ductility. Most Mg-alloys have limited ductility due to their native hexagonal close-pack (HCP) structure. Traditional Mg-alloys are effective when great strength is not necessary, and they are often used to create intricate castings where more thickness or stiffness is needed. While the Mg-alloys are good for casting, they are more difficult to work with when attempting stamping or press forming manufacturing techniques due to their lack of ductility [3, 4]. Ductility of Mg has been shown to be strongly grain size dependent [5]. Thermomechanical processing techniques consisting of a deformation process followed by heat-treatment is a common approach to manipulate grain structure characteristics in various alloys. Precise control over the progress of precipitation enables controlling post deformation recrystallization and, as a result, modifying texture and grain structure evolutions [6, 7].

Studies have shown that the Mg-Al-Zn family of Mg-alloys has a great potential for alloy development and hardening compared to other families of Mg-alloys. From the Mg-Al binary alloy phase diagram, aluminum has a maximum solid solubility of 12.9 wt% in magnesium [8]. Addition of zinc to Mg-Al reduces solid solubility of aluminum in magnesium and makes it precipitation hardenable [9]. Precipitates are the equilibrium phase, β, with the stoichiometric composition of $Mg_{17}Al_{12}$. These phases in the as-received material have been observed to be relatively coarse and not evenly distributed in the matrix [10]. Combination of mechanical and thermal treatment can be used to control the side and distribution of precipitates.

Precipitate evolution in Mg-Al-Zn alloys has been investigated in a number of studies [10-14]. During aging, the β-$Mg_{17}Al_{12}$ phase precipitate through either homogeneous nucleation (continuous precipitation) or heterogeneous nucleation (discontinuous precipitation). Discontinuous precipitation results in formation of second phase particles at high-angle grain boundaries. Continuous precipitation forms second phase particles in the remaining regions of the matrix. Most of the age hardening in Mg-Al-based alloys is attributed to continuous precipitation [11].

Among the few AZ commercial alloys, AZ31 have a good combination of strength, ductility and cost [3]. However, the volume fraction of the precipitates (i.e. β-$Mg_{17}Al_{12}$) that may form in the alloy is comparatively small [10]. The β-$Mg_{17}Al_{12}$ phase resists movement of the planes during plastic flow, adding strength to the alloy, if they are properly orientated in the matrix [3, 4, 13]. Regular isothermal aging process usually promote discontinues precipitation and formation of precipitate at the grain boundary due to the lower interfacial energy in the boundary region.

Non-isothermal aging have been shown to be a strong tool for promoting homogeneous nucleation of precipitates, and improving precipitation behavior of aluminum alloys [6, 7]. Although many studies have been devoted on isothermal heat treatment of the Mg-Al-Zn alloys, to the best knowledge of the authors, there has not been any systematic research on the non-isothermal heat treatment of this family of alloys. This is despite the significant importance of the issue from industrial and scientific point of view. This article, which is part of larger ongoing research, is the first step toward understanding and characterizing this phenomenon, in addition to investigating effect of pre-straining on the precipitation hardening behavior of a commercially available AZ31B magnesium alloy.

Experimental Methodology

A commercially available, 2mm thick AZ31B Mg-alloy in fully annealed condition, provided by MetalMart Inc., has been used for this investigation. Chemical composition of the as-received sheet is provided in Table I. The as-received samples have been thermally or thermo-mechanically processed to achieved different levels of precipitation.

Table I. Composition of as-received AZ31B.

Alloy	Weight %							
	Mg	Al	Zn	Mn	Si	Cu	Ca	Fe & Ni
AZ31B	bal.	2.5-3.5	0.6-1.4	0.2	0.1	0.05	0.04	0.005

In all cases, the solution heat-treatment consisted of holding samples at 400°C for 15hr followed by a quench into a water bath at room temperature. Pre-straining of samples occurred immediately after quenching. Pre-straining was imposed in uniaxial tension using an Instron Model 1123. Samples have been pre-strained to 13% elongation in length at a strain rate of 2mm/s. As-quenched and pre-strained samples were non-isothermally aged from room temperature to various end temperature at heating rates of 0.5 °C/min, 1°C/min, 5°C/min, and 10 °C/min. The heat treatment process used for sample preparation is schematically presented in Figure 1.

The heat-treated samples were cold mounted in an acrylic resin, ground and polished to 1μm finish. Microhardness measurements are conducted using a Leco hardness tester and 500 gf load applied for 10 seconds. Each reported hardness value is the average of seven readings.

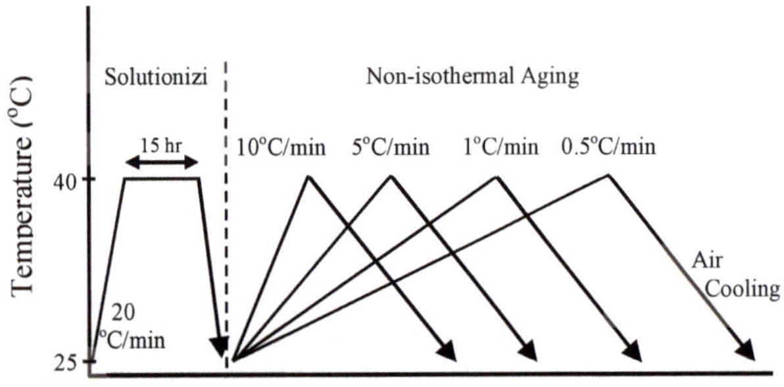

(a)

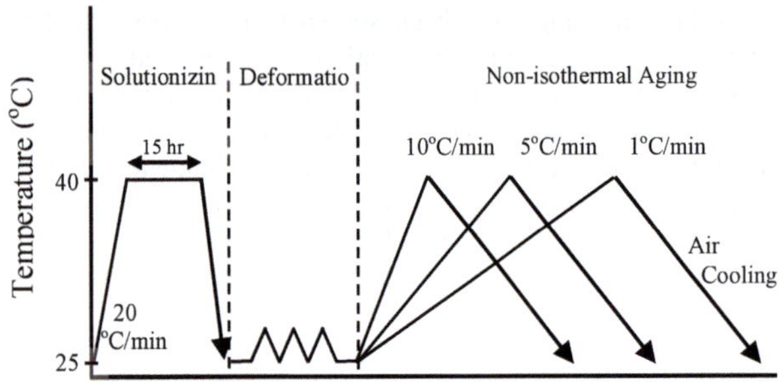

(b)

Figure 1. Schematic presentation of the non-isothermal annealing routes for a) non-deformed samples and b) pre-strained samples.

Results and Discussion

Variation of hardness with temperature during non-isothermal aging of the non-deformed samples are shown in Figure 2. The results provide information on the precipitation hardening behavior and the effect of heating rate on precipitation behavior of the alloy. For all heating rates, the hardness of the samples increases by temperature till it reaches its maximum value; aging beyond this point results in a sharp decrease of hardness followed by a plateau. The increase in hardness during aging of non-deformed samples is purely related to precipitate formation.

Before starting the non-isothermal aging process, samples have been solutionized. Through solutionizing, all of the existing precipitates have been dissolved in the magnesium matrix. By quenching the sample from solutionizing temperature, this microstructural state is preserved producing a metastable supersaturated solid solution. Then, during non-isothermal aging the aluminum atoms diffuse and form β-$Mg_{17}Al_{12}$ precipitates. These precipitates hinder motion of dislocations and alloy is strengthened. As number density of precipitates increases, material gets stronger till it reaches the maximum level of hardness (Hardness Peak). Aging above this state results in a decrease of strength due to the coarsening of the precipitates.

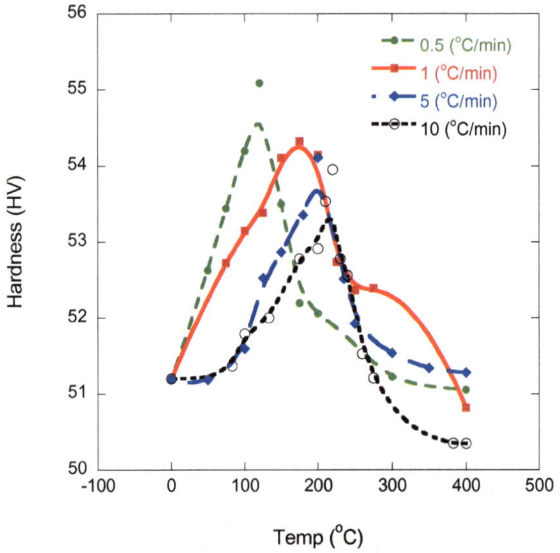

Figure 2. Vickers hardness of non-deformed AZ31B non-isothermally aged at 0.5, 1.0, 5.0, and 10.0 °C/min heating rates.

Figure 2 shows a clear trend between the heating rate and corresponding peak hardness values and temperatures at which the peak hardness was reached. To better represent this trend the peak of hardness, temperature of the peak and the aging time required to reach peak of hardness for each heating rate are presented in Figure 3. It is seen that the peak hardness is inversely related to the heating rate: as the heating rate increased, the peak hardness decreased. Conversely, as the heating rate increased, the temperature at which the peak hardness occurred also increased. Increase in hardness by reducing heating rate, may be related to favoring homogeneous nucleation of precipitates versus discontinuous heterogeneous precipitation on grain boundaries.

The ratio of homogenous and heterogeneous nucleation is given by Equation 1.

$$\frac{N_{het}}{N_{hom}} = \frac{C_1}{C_0} exp\left(\frac{\Delta G^*{}_{hom} - \Delta G^*{}_{het}}{kT}\right) \quad (1)$$

where N_{hom} is the rate of homogenous nucleation and N_{het} is the rate of heterogeneous nucleation. Parameter C_0 is the concentration of homogenous nucleation sites per unit volume, C_1 is the concentration of heterogeneous nucleation sites per unit volume, ΔG^*_{hom} is the driving force for homogenous nucleation, ΔG^*_{het} is the driving force for heterogeneous nucleation, k is Boltzmann's constant, and T is temperature in Kelvin. The ratio of nucleation sites, i.e. C_1/C_0, is a fixed value for a given microstructure and does not vary with temperature. The value of the exponential term $exp\left(\frac{\Delta G^*{}_{hom} - \Delta G^*{}_{het}}{kT}\right)$ decreases when the aging temperature is lower due to the inverse relationship between the heterogeneous/homogenous driving force and ΔT, Equation 2:

$$\Delta G^* \propto \frac{1}{\Delta T^2} \quad (2)$$

where $\Delta T = T_{solutionized} - T_{aging}$. $T_{solutionized}$ is the solutionizing temperature, and T_{aging} is the target aging temperature. Lowering T_{aging} increases the value of ΔT, making ΔG^* smaller. Therefore, a lower aging temperature reduces the value of $\Delta G^*_{hom} - \Delta G^*_{het}$, meaning that the dominant term in Equation 1 becomes C_1/C_0. The magnitude of C_0 is always much larger than the magnitude of C_1. Therefore, reducing aging temperature favors homogenous nucleation. Heterogeneous nucleation still occurs at low aging temperatures, but homogenous nucleation of precipitates will be the dominant mechanism.

Precipitation is a diffusion-controlled phenomenon. Therefore its extent is dictated by time and temperature. When non-isothermal aging is conducted at low heating rate, alloy is kept for a long time at low temperatures, therefore the main portion of precipitation nucleation occurs at lower temperatures. As a result, homogeneous nucleation is favored compared to ageing at higher heating rate. Homogeneous nucleation leads to formation of evenly dispersed, fine precipitates. Therefore, a higher peak of hardness is achieved for samples aged at lower heating rates compared to the samples aged at higher heating rates.

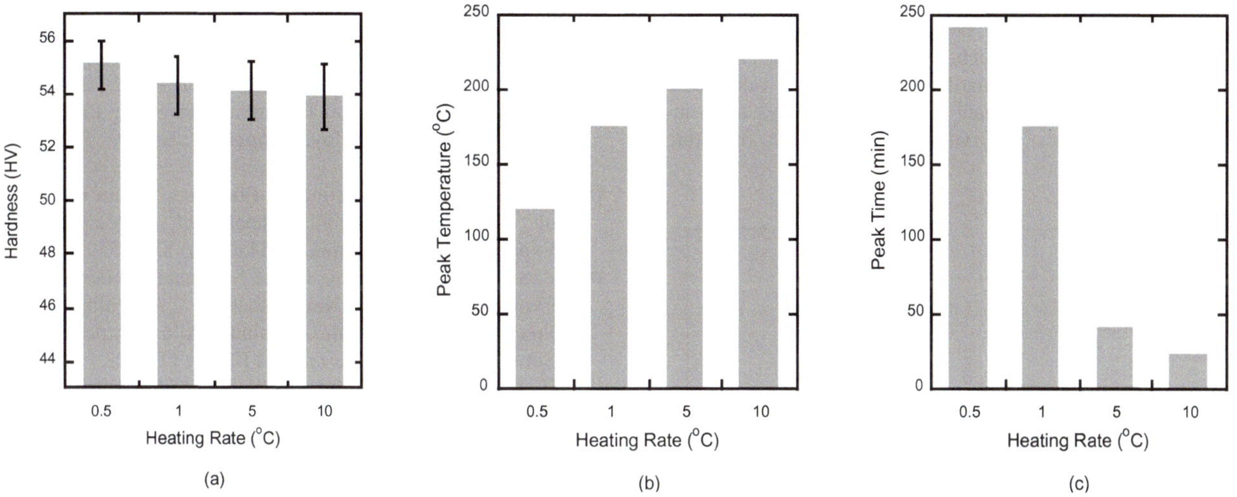

Figure 3. Peak hardness conditions for non-deformed samples non-isothermally aged at different heating rates: a) Hardness value achieved at the peak, b) Temperature at which peak of hardness achieved, and c) Aging time taken to reach the hardness peak.

To study effect of deformation on precipitation, sets of samples have been pre-strained prior to non-isothermal aging process. Evolution of hardness as a function of aging time for these samples is presented in Figure 4. Similar to non-deformed sample, for all heating rates, hardness initially increases; reach a maximum level and then decreases. However, in contrast to non-deformed samples, there is no clear trend between the hardness results for the three heating rates studied. For all cases, hardness peak is achieved at the same temperature (i.e. 150°C). Maximum amount of hardness achieved by heating at 10 °C/min; magnitude of hardness for the other two heating rates (1 and 5 °C/min) is almost the same.

To have a better perspective on the effect of deformation on the precipitation behavior, values of hardness achieved at the peak as well as the aging time to reach to the peak of hardness are plotted in Figure 5. Peak temperature is not included in the image, as it is the same for all heating rates. It is interesting to note that, in contrast to non-deformed samples; the highest level of hardness for pre-strained sample is achieved at the highest heating rate (i.e. 10 °C/min). It may be explained considering the fact that hardness evolution in the pre-strained sample represents the overlapping effects of precipitation hardening and annealing softening. As shown in Figure 5(b), it takes 150min to reach to the peak hardness, while it only takes 15min to the same level of temperature for the 10°C/min aging condition. Holding samples at elevated temperature for longer time lead to higher extent of annealing softening. Therefore, lower values of hardness are achieved for the lower heating rates.

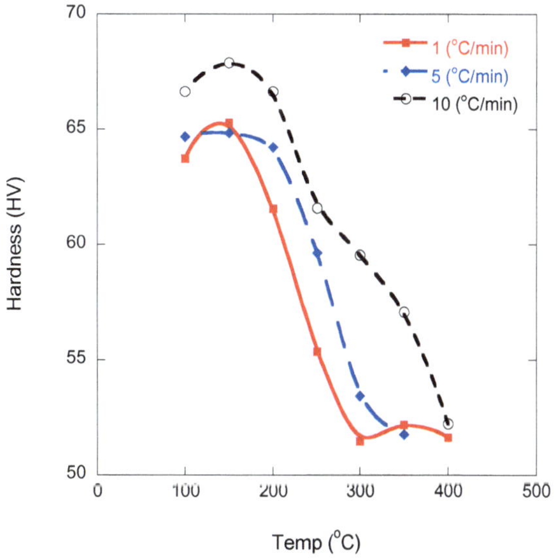

Figure 4. Vickers hardness of non-deformed AZ31B non-isothermally aged at 0.5, 1.0, 5.0, and 10.0 °C/min heating rates.

Comparing hardness evolution of the deformed and non-deformed samples two main points emerge: a) deformed samples show a higher level of hardness compared to non-deformed samples, b) peak of hardness appears at lower aging times and temperatures for deformed samples. During deformation of AZ31 alloy, the material will strain hardened through introduction of new dislocations, in addition to formation of twin boundaries. Dislocations and twin boundaries affect precipitation behavior in two ways: i) acting as heterogeneous nucleation sites and ii) providing easy diffusion paths for diffusing atoms (i.e. aluminum) that make the precipitates. Combination of more nucleation sites and faster diffusion leads to peak formation at lower time and temperature. The higher hardness values for deformed samples may be originated from strain hardening effect, enhanced precipitation on dislocations, or both. However, strain hardening should be the dominant mechanism, considering the low contribution of precipitation hardening for the non-deformed samples. Identifying the dominant mechanism requires more in-depth microstructural analysis, which is the subject of future research.

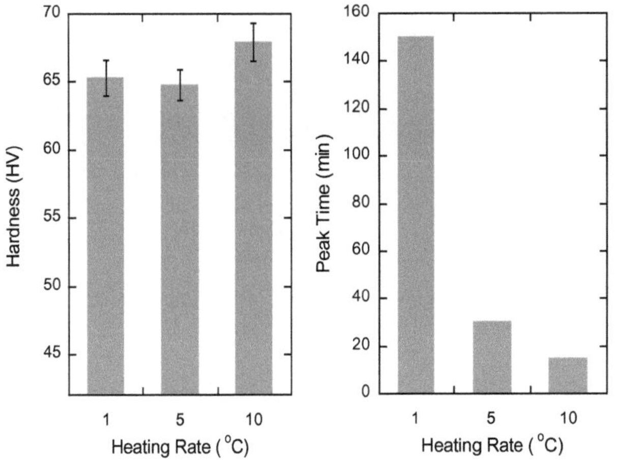

Figure 5. Peak hardness conditions for non-deformed samples non-isothermally aged at different heating rates: a) Hardness value achieved at the peak, b) Aging time taken to reach the hardness peak.

Summary and conclusions

Effect of pre-deformation and heating rate on precipitation hardening behavior of a commercially available AZ31 during non-isothermal aging is investigated. The results provide evidence for promotion of homogeneous nucleation when low heating rates are adopted during annealing of non-deformed samples. For the deformed samples, the highest value of hardness has been achieved when samples are non-isothermally annealed at high heating rates. The high level of hardness in deformed samples is related to the effect of deformation in providing larger number of precipitate nucleation sites, enhancing diffusional mechanism, and strain hardening.

Acknowledgment

This work was supported by Santa Clara University (SCU) FY1415 Kuehler Grant, and 2013-2014 School of Engineering Internal Grant.

References

[1] Edward F. Emley, *Principles of Magnesium Technology* (London, Pergamon Press, 1966).

[2] Roberts C. Sheldon, *Magnesium and Its Alloys* (New York , NY: John Wiley and Sons, 1960).

[3] M. Manuel, L.G. Hector, R. Verma, W. Tong, "Microstructural effects of AZ31 magnesium alloy on its tensile deformation and failure behaviors," *Materials Science and Engineering*, A418 (2006) 341-356.

[4] L. Jing-Yuan, J.F. Wang, and X.K. Wang. "Effects of Annealing Treatment on Texture and Stamping Properties of AZ31 Magnesium Alloy." *Materials Science Forum* 686 (2011) 101-106.

[5] R. Gunther, R. Ch. Harting, R. Bormann, "Grain Refinement of AZ31 by (SiC)P: Theoretical Calculation and Experiment," *Acta Materialia* 54 (2006) 5591-5597.

[6] P. Sepehrband, X. Wang, H. Jin, S. Esmaeili, Microstructural Evolution During Non-isothermal Annealing of a Precipitation-Hardenable Aluminum Alloy: Experiment and Simulation," *Acta Materialia* 94 (2015) 111-123.

[7] B. Poorganji, P. Sepehrband, S. Esmaeili, "Effect of Cold Work and Non-isothermal Annealing on the Recrystallization Behavior and Texture Evolution of a Precipitation-Hardenable Aluminum Alloy," *Scripta Materialia*, 63 (2010), 1157-1160.

[8] Murray, J. L., in *Phase Diagrams of Binary Magnesium Alloys*, ed. A. A. Nayeb-Hashemi and J. B. Clarke. ASM International, Metals Park, OH, 1988, 17-34.

[9] Charlie R. Brooks, *Heat Treatment, Structure and Properties of Nonferrous Alloys* (Metals Park, OH: ASM, 1984).

[10] L. Shang, S. Yue, E. Essadiqi, A. Javaid, R. Verma, "Effect of Pre-aging Treatment on Hot Compression Behavior of AZ31 Magnesium Alloy," Magnesium Technology (2008) 263-66.

[11] S. Celotto, "TEM study of continuous precipitation in Mg–9 wt%Al–1 wt%Zn alloy," *Acta Materialia*, 48 (2000), 1775-1787.

[12] W. Diqing, J. Wang, G. Wang, L. Lin, Z. Feng, G. Yang. "Precipitation and Responding Damping Behavior of Heat-treated AZ31B Magnesium Alloy," *Acta Metallurgica Sinica* 22 (2009) 1-6.

[13] F.Y. Hung, C.C. Shih, L.H. Chen, T.S. Lui, "Microstructures and High Temperature Mechanical Properties of Friction Stirred AZ31-Mg Alloy," *Journal of Alloys and Compounds* 428 (2007) 106–114.

[14] Z.Z. Yong, F.R. Cao, R.G. Guan. "Effect of Heat Treatment on the Microstructures and Mechanical Properties in AZ31B Magnesium Alloy Processed by Continuous Rheo-Extrusion," *Advanced Materials Research* 189-193 (2011) 4221-226.

Magnesium Technology 2016
Edited by: Alok Singh, Kiran Solanki, Michele V. Manuel, and Neale R. Neelameggham
TMS (The Minerals, Metals & Materials Society), 2016

The Effect of Ageing on the Compressive Deformation of Mg-Sn-Zn-Na Alloy

Ehsan Bahrami Motlagh*, Alireza Ghaderi, Sitarama Raju Kada, Peter A. Lynch, and Matthew R. Barnett

Institute for Frontier Materials, Deakin University, 75 Pigdons Road, Waurn Ponds, VIC. 3216, Australia.
*Correspondence e-mail: ebahrami@deakin.edu.au

Keywords: precipitates, twinning, synchrotron, X-Ray diffraction

Abstract

The influence of ageing on deformation twinning in an extruded Mg-6Sn-3Zn-0.04Na alloy is investigated. In-situ compression tests have been carried out using high resolution synchrotron X-Ray Diffraction (XRD) to measure the influence of precipitates on twining activity. Synchrotron experiments revealed the increase in the critical resolved shear stress of twinning with ageing. The compressive yield strength (along the extrusion direction) of the aged sample increased by ~150% over the non-aged specimen. To obtain statistical insight into the twinning activity, the microstructure of the non-aged and aged samples (200°C, 24 hours) deformed up to ~1% plastic strain was studied using optical microscopy. A higher number of thinner twins were observed in the microstructure of the aged sample compared to the non- aged sample.

Introduction

The recently developed Mg-6Sn-3Zn-0.2Na [1] alloy offers a significant improvement in creep resistance which extends the application range of magnesium alloys to higher temperatures [2]. However, magnesium alloys generally show poor formability due to their yield asymmetry [3] which has been attributed to the activation of deformation twinning [4]. Introducing precipitations into the matrix may alter the yield asymmetry by hardening the so-called easy deformation modes [3]. For example, basal plate shaped precipitates in AZ91 reduce the yield asymmetry while rod-like precipitates in Mg-5Zn (Z5) alloy exacerbate asymmetry [3]. This is due to the fact that in the former, precipitates increase the critical resolved shear stress of tensile twinning more than prismatic slip while in the later, precipitates have the opposite effect. Therefore, it appears that precipitate morphology is an important parameter controlling the yield asymmetry. Theoretical predictions have revealed the constructive effect of spherical precipitates on the yield asymmetry in magnesium alloys [3, 4]. However, there is limited experimental evidence reporting on the impact of spherical (or equiaxed) precipitates in terms of twin activation and consequently yield asymmetry. This study aims to explore the influence of equiaxed precipitates on the hardening of $(10\bar{1}2)$ twinning in an extruded Mg-6Sn-3Zn-0.2Na alloy based on in-situ synchrotron XRD compression test results.

Methodology

The magnesium alloy (nominal composition of 6 Wt.% Sn, 3 Wt.%Zn and 0.2 Wt.% Na) was received in the as cast condition. Cylindrical samples (diameter 29.5 mm, height 39 mm) were solution treated in two steps. The samples were heated to 335°C and held for two hours to dissolve the potential low melting point Mg-Zn based intermetallics. This was followed by a second heat treatment at 530°C for 20 hours. This heat treatment was followed by a final water quenching step. The solution treated billets were extruded at 475°C with a ram speed of 0.1 mm/s and an extrusion ratio of ~30.

Extruded samples were then aged at 200°C for different times. The age hardening response of the samples during isothermal ageing at 200°C is shown in Fig. 1 (a). As can be seen, the hardness remains relatively unchanged for the first hour of ageing then gradually increases to a final Vickers hardness of 70±5 with 24hrs of ageing.

To measure the macroscopic stress-strain curve, a non-aged and 24-hours aged samples (diameter 8 mm height 12 mm) were machined. Samples were prepared with the extrusion direction parallel to the compression axis- to enable easy activation of deformation twinning [5-7]. Compression testing was conducted using an Instron machine with a 30kN load cell. Fig. 1 (b) shows the true stress-true strain curves of the non-aged and aged samples. The 0.2% offset compressive yield strength values are 64 MPa and 164 MPa for the non-aged and aged samples, respectively. An increase of 150% in the yield stress is observed. This observed strengthening is on par with the most striking strength increment in magnesium alloys which has been observed in AZ91 alloy[5]. The dramatic difference in the two stress-strain curves is attributed to the precipitate hardening of the deformation twinning mode.

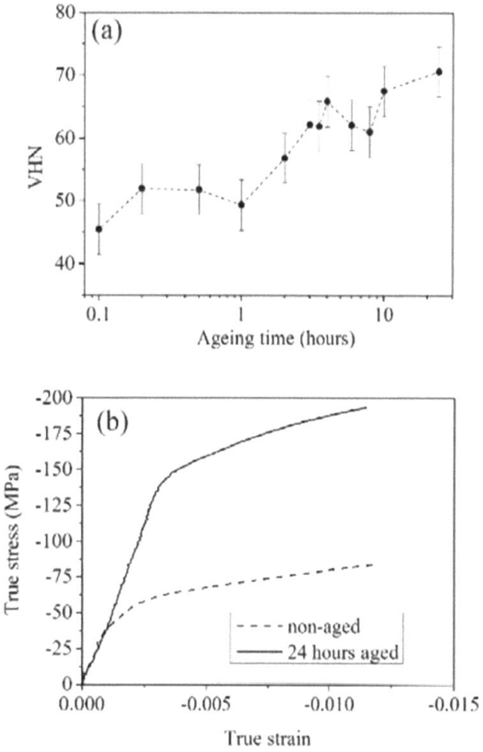

Fig. 1a). Age hardening response of extruded material at 200°C, b). The true stress- true strain curve for the non-aged and 24 hour aged samples.

Application of the Australian synchrotron Powder diffraction beamline was used to conduct the in-situ compression tests to quantify the critical resolved shear stress of deformation twinning in the non-aged and 24hr aged samples. For this purpose, cylindrical samples with 2.5 mm diameter and 3.9 mm height were prepared with the compression direction in line with the extrusion direction. The compression experiments were performed using a 5kN Deben Microtest tensile/ compression stage (Deben UK Ltd) with the loading applied in displacement control mode.

Monochromatic X-rays with a beam energy of 20 keV were employed for the transmission diffraction experiments. The beamline is equipped with a microstrip detector to enable rapid data collection. The entire diffraction pattern can be acquired in a single exposure. To monitor the plastic deformation process in-situ, the diffraction measurements rely on precise determination of the individual Bragg peak position ($2\theta^{hkil}$) and integrated intensity with imposed load. The measured Bragg peak position is then related to the inter-planar spacing (d^{hkil}) via Bragg' law,

$$n\lambda = 2d^{hkil}sin\theta^{hkil} \quad \text{eqn. 1}$$

Where n is the reflection order and λ is the incident X-ray wavelength. Then from the measured lattice spacing the elastic lattice strain is given as,

$$\varepsilon^{hkil} = \frac{d^{hkil}-d_o^{hkil}}{d_o^{hkil}} \quad \text{eqn. 2}$$

Where d_o^{hkil} is the strain free lattice spacing. As implemented in Kada et al [8] the strain free condition is determined from the first set of diffraction measurements at nominal imposed load.

To complement the in-situ diffraction studies, a series of optical microscopy measurements were performed on post mortem macroscopic compression samples which have been deformed up to ~1% plastic strain. This provided additional information regarding the twin size, morphology and distribution.

Results & Discussion

The typical microstructure of the Mg alloy obtained from the thermomechanical and wrought processing steps is shown in Fig. 2a. The microstructure consists of equiaxed grains with a mean linear intercept grain size of 37 µm. The initial crystallographic texture of the extruded samples were determined from a series of electron backscattered diffraction (EBSD) maps. Fig. 2b shows the characteristic extrusion texture in the form of an Extrusion Direction (ED) inverse pole figure. It is clear from the figure that the majority of grains have their c-axes perpendicular to the extrusion direction which favours deformation twinning based on compression along the extrusion axis [6].

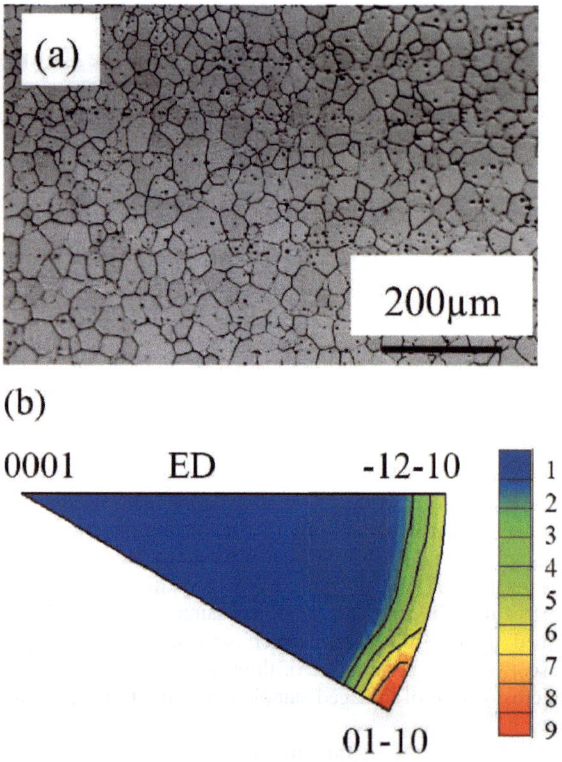

Fig. 2 a). Optical microstructure of the extruded alloy. b). Extrusion direction inverse pole figure; contour lines refer to 2, 4, 6, and 8 times random and maximum intensity is ~ 8.5.

Fig. 3a shows the transmission XRD profiles of the $(10\bar{1}0)$ and (0002) reflections obtained during in-situ compression testing along the extrusion direction of the non-aged alloy. In accordance with the imposed loading direction, the line profiles move to higher 2θ angles- signaling a decrease in the measured lattice plane spacing. As a result, a systematic increase in the peak position 2θ is realised with applied load (Fig. 3b). Owing to the strong $(10\bar{1}0)$ fiber texture (Fig. 2b) the intensity of the $(10\bar{1}0)$ peak is very strong at nominal applied load while the peak corresponding to (0002) grain orientation is absent. Due to the reorientation of basal poles by ~90° the (0002) peak appears upon yielding (Fig. 3a). Upon further loading, the intensity of the $(10\bar{1}0)$ peak decreases while the intensity of the (0002) peak increased. This was attributed to the nucleation and growth of deformation twinning during compressive deformation [8].

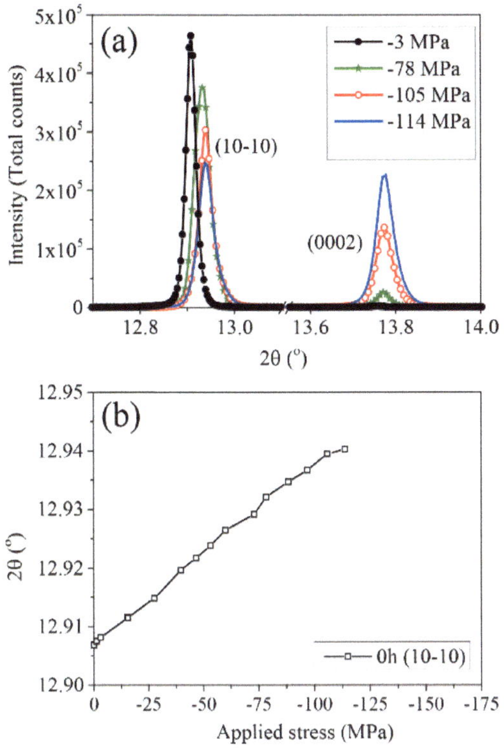

Fig. 3 a). In-situ synchrotron transmission XRD profiles for the $(10\bar{1}0)$ parent and (0002) twin reflections. b). The change in the peak position of $(10\bar{1}0)$ parent reflection as a function of applied load in the non-aged sample.

For analysis of the measured diffraction data, peak profile fitting of the $(10\bar{1}0)$ and (0002) reflections was performed using TOPAS line profile analysis software [9]. Line profiles were fitted using a Pseudo-Voigt (PV) function to extract the peak positions ($2\theta^{hkil}$) and integrated intensity as a function of the applied stress. By using the methodology described above, the evolution of elastic lattice strain for the $(10\bar{1}0)$ and (0002) reflections was calculated according to eqn. 2. The results are plotted as a function of applied stress for the two samples in Fig. 4a. It is reassuring to note that the lattice strain evolution of the $(10\bar{1}0)$ reflections in the early stages of loading followed the predicted elastic line of ~ 45GPa for both data sets. Meanwhile, at higher levels of applied load the (0002) reflections appear at an applied stress of ~ -73 MPa and ~ -113 MPa for the non-aged and aged samples, respectively. This delay in the twin onset is attributed to precipitate hardening. It can be seen from Fig. 4a that twinning, as evidenced by the observed increase in intensity of the (0002) peak, occur with relaxed lattice strains of ~ -0.00120 and ~-0.00207 in the non-aged and aged alloys respectively. Upon further loading, the strain in the newly formed twins increases steadily but it is still lower than the strains measured in the original parent grains.

Nucleation and growth of tensile twinning is quantified by plotting the changes in the measured peak intensities as a function of applied load. The change in the intensity of the parent $(10\bar{1}0)$ reflections is shown as the percentage change in the intensity with respect to the peak intensity at nominal applied load. Alternatively, the change in the integrated intensity of the (0002) twin reflections is shown for the non-aged and aged alloy in Fig. 4b. It is clear, above macroscopic yielding the intensity of the $(10\bar{1}0)$ reflections decreases with a simultaneous increase in the (0002) reflections in both conditions. The first sign of an increase in the (0002) peak intensity in the non-aged sample is at an applied load of ~ -73 MPa which continues with additional load. However, the change in the intensities were retarded by approximately ~40 MPa in the aged alloy- the intensity of (0002) starts to increase at ~ -111 MPa. These intensity variation with applied stress are attributed to the nearly 90° crystallographic reorientation of basal planes due to tensile twinning [10].

From measurement of the change in integrated intensities of the $(10\bar{1}0)$ and (0002) profiles, (Fig. 4 (b)) the critical resolved shear stress (CRSS) for twinning can be ascertained. The CRSS for the twin growth was calculated by multiplying the applied stress at the onset of twinning with the Schmid factor for twinning [8, 11, 12]. The Schmid factor for twinning in the $(10\bar{1}0)$ parent grain orientation is 0.49. Therefore, the CRSS for twinning in the non-aged and 24 hour aged sample is estimated to be 36 ±5 MPa and 58 ±5 MPa, respectively. As a result, following 24 hours of ageing the critical resolved shear stress of twinning increases by 20 ±5 MPa.

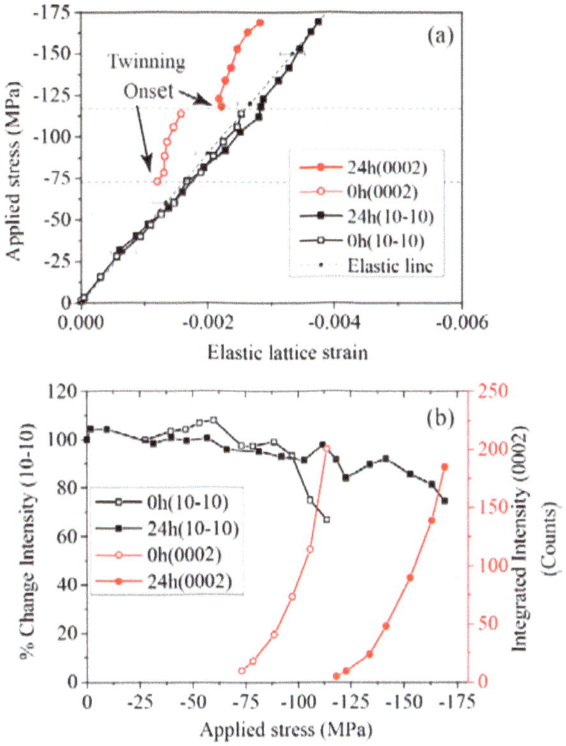

Fig. 4. Summary of the $(10\bar{1}0)$ and (0002) diffraction data for the aged and non-aged compression tests. a). Evolution of elastic lattice strain with applied load. b). The observed change in intensity as a function of applied load.

Typical microstructure of the samples deformed to ~1% macroscopic plastic strain are shown in Fig. 5(a) and (b). As it can be seen, the trace of the lenticular shaped tensile twins is apparent within the microstructure. Based on the optical microscopy measurements the twin area density in the aged sample has increased by a factor of ~ 2.7 over the non-aged sample. There is an analogy between decreasing the grain size [13] and introduction of precipitates into the microstructure. A higher number density of twins per grain boundary area has been observed in smaller grain sized materials [13] where according to Hall-Petch relation the yield strength is higher. Therefore, independent of the source of stress (grain boundary or precipitates) higher twin numbers correspond to higher stresses. In the presence of precipitates, formation of higher twin numbers to accommodate the applied stress might have been energetically more favorable compared to growth of individual twins. The twins in the aged sample (Fig. 5 (b)) seems to be thinner compared to the twins in non- aged sample (Fig. 5(a)). This may support the twin growth inhibition by precipitates.

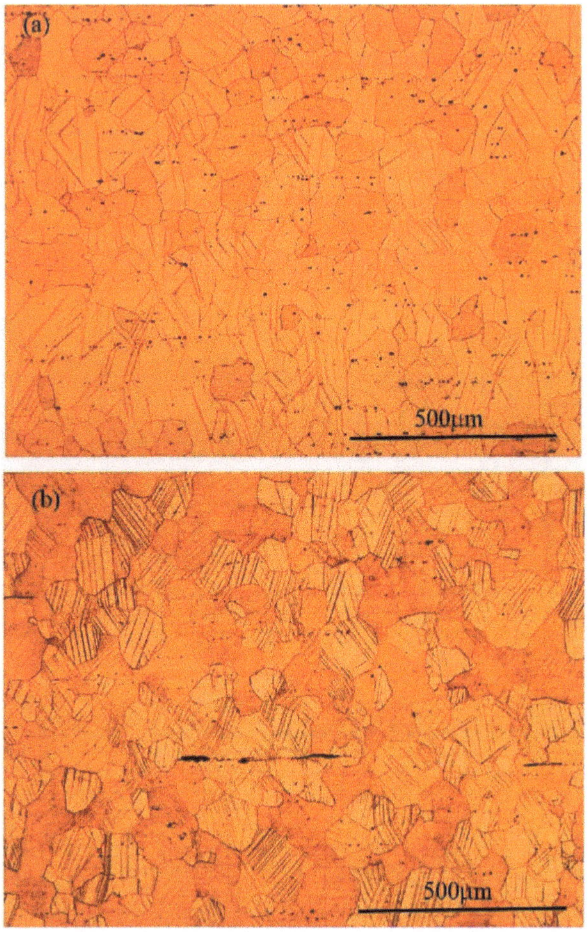

Fig. 5. Optical images of the deformed microstructures. a). non-aged; deformed to plastic strain of $\varepsilon_p = 0.98\%$. b). 24 hours aged deformed to plastic strain of $\varepsilon_p = 0.72\%$.

Other possibility for the increase in the number of twins is that the local stress concentrations around the precipitates leads to twin nucleation [14]. Although precipitate morphology is different, the current results are in close agreement with the observations in Mg-Zn alloys [15] but different from the twinning behavior in AZ91 alloy [5]. In the aged AZ91 alloy, an increase of two to three orders of magnitude in twin number density is expected over the non-aged sample [5]. In Mg-Zn alloy an increase by a factor of almost 2 has been verified [16].

Conclusion

The macroscopic twin dominated yield strength of Mg-6Sn-3Zn-0.04Na alloy increased by 150% following 24 hours of ageing. The strong initial texture induced by extrusion provides the opportunity to make use of diffraction techniques for quantifying the twinning activity. The in-situ synchrotron results showed that the CRSS for twinning has increased by ~20MPa in the presence of precipitates. The elastic lattice strain in the $(10\bar{1}0)$ parent reflection followed the elastic prediction of 45GPa in the stress range applied (nominal to -113MPa and -170 MPa for the non-aged and aged samples, respectively). Crystallographic reorientation due to twinning at higher applied stresses leads to an increase in the intensity of (0002) peak during compression testing along the extrusion direction. Although the twin reflections form in a more relaxed state with respect to their parent, they start to take load as the applied stress increases. In the presence of precipitates (aged sample) the twin area density has almost tripled following ~ 1% plastic strain.

Acknowledgement

The authors would like to thank the Australian Synchrotron as part of this research was undertaken on the Powder Diffraction Beamline at the Australian Synchrotron, Victoria, Australia.

References

[1] C.L. Mendis, C.J. Bettles, M.A. Gibson, C.R. Hutchinson, Materials Science and Engineering: A, 435–436 (2006) 163-171.

[2] M.A. Gibson, X. Fang, C.J. Bettles, C.R. Hutchinson, Scripta Materialia, 63 (2010) 899-902.

[3] J.D. Robson, N. Stanford, M.R. Barnett, Metall and Mat Trans A, 44 (2013) 2984-2995.

[4] J.D. Robson, N. Stanford, M.R. Barnett, Acta Materialia, 59 (2011) 1945-1956.

[5] N. Stanford, A.S. Taylor, P. Cizek, F. Siska, M. Ramajayam, M.R. Barnett, Scripta Materialia, 67 (2012) 704-707.

[6] S.R. Agnew, D.W. Brown, C.N. Tomé, Acta Materialia, 54 (2006) 4841-4852.

[7] C.C. Aydıner, J.V. Bernier, B. Clausen, U. Lienert, C.N. Tomé, D.W. Brown, Physical Review B, 80 (2009) 024113.

[8] S.R. Kada, P.A. Lynch, M.R. Barnett, Journal of Applied Crystallography, 48 (2015) 365-376.

[9] A.A. Coelho, J. Evans, I. Evans, A. Kern, S. Parsons, Powder Diffraction, 26 (2011) S22-S25.

[10] M.A. Gharghouri, G.C. Weatherly, J.D. Embury, J. Root, Philosophical Magazine A: Physics of Condensed Matter, Structure, Defects and Mechanical Properties, 79 (1999) 1671-1695.

[11] S.R. Kada, P.A. Lynch, M.R. Barnett, Materials science forum, 773-774 (2014).

[12] S.R. Kada, PhD Thesis, Deakin University, 2013, pp. 223.

[13] A. Ghaderi, M.R. Barnett, Acta Materialia, 59 (2011) 7824-7839.

[14] S.R. Agnew, R.P. Mulay, F.J. Polesak Iii, C.A. Calhoun, J.J. Bhattacharyya, B. Clausen, Acta Materialia, 61 (2013) 3769-3780.

[15] N. Stanford, M.R. Barnett, Materials Science and Engineering: A, 516 (2009) 226-234.

[16] J.D. Robson, N. Stanford, M.R. Barnett, Scripta Materialia, 63 (2010) 823-826.

Magnesium Technology 2016
Edited by: Alok Singh, Kiran Solanki, Michele V. Manuel, and Neale R. Neelameggham
TMS (The Minerals, Metals & Materials Society), 2016

First-principles Study of Diffusion Coefficients of Alloy Elements in Dilute Mg Alloys

Bi-Cheng Zhou[1], Shun-Li Shang[1], Yi Wang[1], Zi-Kui Liu[1]
[1]Department of Materials Science and Engineering, The Pennsylvania State University, University Park, PA 16802

Keywords: Solute Diffusion, First-principles, Mg alloys

Abstract

The equilibrium thermodynamic properties of Mg alloys are reasonably well understood, but the kinetics of these alloys is explored to a far lesser extent, especially diffusion coefficients of alloying elements in Mg. In the present work, first-principles calculations based on the Density Functional Theory (DFT) were used to calculate the dilute tracer diffusion coefficients in Mg for 61 elements, including rare earths, as a function of temperature using an 8-frequency model. An improved generalized gradient approximation of PBEsol was used in the present work, which is able to well describe both vacancy formation energies and vibrational properties. Systematic trends of diffusion activation energies of various solutes were analyzed. Remarkable agreements of the calculated results compared with available experimental data were obtained.

Introduction

Mg alloys have gained an increasing research attention because their low density, earth abundance, high specific strength, and good castability [1]. The majority of Mg alloys derives their mechanical properties from precipitation hardening [2], while the study of precipitation process demands accurate thermodynamic and kinetic (diffusion) data. Thermodynamics of Mg alloys has been extensively studied, and several comprehensive thermodynamic databases have been established [3]. However, the kinetics of Mg alloys has been studied to a far lesser extent, especially diffusion coefficients of various solutes in Mg. Due to the issues related to corrosion, oxidation, and contamination during sample preparation in diffusion measurements, few experimental data are available in the literature for diffusion coefficients of solutes in Mg [4]. In order to further understand the diffusion related properties of Mg alloys such as aging, creep, and corrosion, a better understanding of the atomic diffusion processes of solutes in Mg is desperately needed [5, 6].

Fortunately, it is now possible to calculate many aspects of diffusion [7, 8]. First-principles calculations based on the density functional theory (DFT) have shown great merit in terms of calculating diffusion coefficients, especially when experimental data are lacking [9, 10]. These calculations are usually coupled with the transition state theory (TST) under the harmonic or the quasi-harmonic approximations [8]. TST has become a practical tool in the context of DFT calculations when efficient algorithms for finding the minimum-energy path have been developed, such as the nudged elastic band (NEB) and the climb image nudged elastic band (CI-NEB) method [11].

Recently, Ganeshan et al. [10] in our group demonstrated using an 8-frequency model to calculate the diffusion coefficients of Al, Zn, Sn, and Ca in dilute hcp Mg. However, their calculated results compared with experimental data still need to be further improved, and especially more alloying elements need to be considered for Mg alloys. In the present study, we use first-principles calculations coupled with the TST and the 8-frequency model to calculate the dilute solute tracer diffusion coefficients for 61 substitutional elements in hcp Mg, as shown in Figure 1. The effects of different exchange-correlation (X-C) functionals on diffusion properties are examined. It is shown that the recently developed PBEsol X-C functional [12] yields better agreement with experimental data compared with the commonly used X-C functionals such as the local density approximation (LDA) and the generalized gradient approximation (GGA). The vibrational properties are derived from the quasi-harmonic Debye model [13, 14]. Therefore, we are able to calculate not only the migration barriers but also the temperature-dependent jump frequencies and the diffusion pre-factors, which are related to vibrational entropic contributions. Finally the dilute solute tracer diffusion coefficients in hcp Mg are calculated. The diffusion pre-factors and the activation energies are obtained by fitting the calculated diffusion coefficients to the Arrhenius-type diffusion equation.

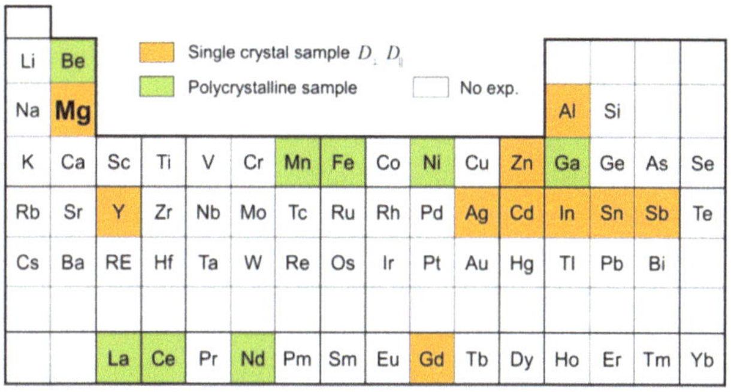

Figure 1. 61 alloying elements in dilute hcp Mg studied in the present work together with available experiments of diffusion data denoted in the periodic table. Note that the diffusion coefficients of Ba and K were not calculated because their direct migration barriers are vanishingly small.

Methodology

<u>Diffusion theory</u>

Note that solute diffusion in dilute alloys is also referred to impurity diffusion in the literature [15]. In the remainder of the paper, "solute" is synonymous with "impurity". For hcp lattice, the crystal anisotropy results in two unique solute jumps, one within the basal plane and the other between adjacent basal planes. These two solute jumps in hcp structure are illustrated in Figure 2(a). This leads to two distinctive diffusion tensors. The diffusion coefficient perpendicular to the *c*-axis, $D_\perp$, results from jumps between the adjacent basal planes and jumps within the basal plane; while the diffusion coefficient parallel to the *c*-axis, $D_\parallel$, attributes to jumps between the adjacent basal planes only. Correspondingly, the diffusion coefficients can be calculated using the following equations [16]:

$$D_{\perp} = \frac{1}{2} C_v a^2 \left(3 f_{Bx} \omega_X + f_{Ab} \omega'_X\right) \quad (1)$$

$$D_{\parallel} = \frac{3}{4} C_v c^2 f_{Az} \omega'_X \quad (2)$$

where a and c are lattice parameters of hcp lattice, C_v corresponds to vacancy concentration adjacent to the solute atom, ω_X and ω'_X are the solute-vacancy exchange jump frequencies within and out of the basal plane, respectively, f_{Bx} is the partial correlation factor for the solute jump within a basal plane, f_{Ab} is the partial correlation factor corresponding to the horizontal component of a solute jump into an adjacent basal plane, and f_{Az} is the partial correlation factor corresponding to the vertical component of a solute jump into the adjacent basal plane.

The vacancy concentration adjacent to the solute atom, C_v in Eqs. (1) and (2) in the dilute limit, is calculated by a Boltzmann relation:

$$C_v = \exp\left(-\frac{\Delta G_f}{k_B T}\right) \quad (3)$$

where $\Delta G_f = \Delta G_f^0 - \Delta G_b$ is the free energy of vacancy formation adjacent to the solute atom. ΔG_f^0 is the vacancy formation free energy in pure Mg without solute, as is calculated by

$$\Delta G_f^0 = G\left(\text{Mg}_{N-1}\text{Va}_1\right) - \frac{N-1}{N} G\left(\text{Mg}_N\right) \quad (4)$$

where Va indicates a vacancy, and N the number of lattice sites in the supercell. ΔG_b is the solute-vacancy binding free energy, defined as the free energy difference between a solute atom/vacancy pair and the two as isolated defects [17, 18]:

$$-\Delta G_b(\text{X}-\text{Va}) = G\left(\text{Mg}_{N-2}\text{X}_1\text{Va}_1\right) + G\left(\text{Mg}_N\right) - G\left(\text{Mg}_{N-1}\text{X}_1\right) - G\left(\text{Mg}_{N-1}\text{Va}_1\right) \quad (5)$$

where X represents a solute atom. The minus sign in front of ΔG_b in Eq. (6) is to keep the binding energy consistent with the convention in the literature such that favorable solute-vacancy binding is positive [19].

Based on Eyring's reaction rate theory [20] of activated complex, as applied by Wert and Zener [21] to solute diffusion in solids, the solute-vacancy exchange jump frequencies within and between the basal planes ω_X and ω'_X in their general form can be written as:

$$\omega = \frac{k_B T}{h} \exp\left(-\frac{\Delta G_m}{k_B T}\right) \quad (7)$$

where $\Delta G_m = G_{TS} - G_{IS}$ is the solute migration free energy barrier with G_{TS} being the free energy of the transition state (TS) and G_{IS} the free energy of the initial state (IS). A detailed derivation of Eq. (7) is given by Wimmer et al. [22].

Free energies ΔG_f and ΔG_m can be obtained from DFT calculations based on the quasi-harmonic approximations to calculate the 8 jump frequencies as shown in Figure 2(b) and 2(c), and the vacancy concentrations. We then implement them into the 8-frequency model, and subsequently obtain the diffusion coefficients from Eqs. (1) and (2). The calculated diffusion coefficients D in Eqs. (1) and (2) can be fitted into the conventional Arrhenius form:

$$D = D_0 \exp\left(-\frac{Q}{k_B T}\right) \quad (8)$$

where Q is the diffusion activation energy and D_0 the pre-exponential factor. The activation energy Q corresponds to the enthalpy of vacancy formation, the enthalpy of solute migration, and part of the contribution from correlation factors; the diffusion pre-exponential factor D_0 corresponds to the entropy of vacancy formation, the entropy of solute migration, the lattice parameters, and also part of the correlation factors.

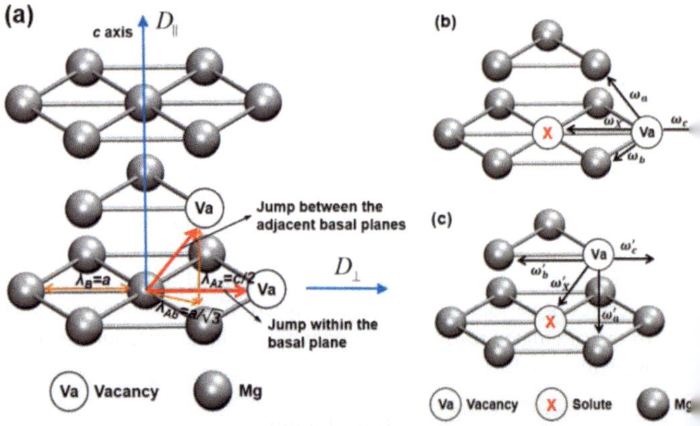

Figure 2. Left: (a) Illustration of vacancy-mediated diffusion jump components in an hcp lattice showing different jump distances (λ_B, λ_{Ab}, and λ_{Az}). The jump components along basal plane with jump distances λ_B and λ_{Ab} contribute to the diffusion coefficient $D_{\perp}$ ($\perp$ c-axis), while the jump component along distance λ_{Az} contributes to the diffusion coefficient $D_{\parallel}$ ($\parallel$ c-axis). Right: Illustration of the eight possible vacancy exchanges in an hcp lattice for vacancy and solute starting (b) within the basal plane and (c) between adjacent basal planes. ω_X, ω'_X are the jump frequencies for the solutes (X) and ω_a, ω_b, ω_c, ω'_a, ω'_b, ω'_c are the jump frequencies for the solvents (Mg).

Computational procedure

First-principles calculations based on DFT were employed to calculate the free energies needed in the diffusion equations and the 8-frequency model. The ion-electron interaction was described by the projector augmented plane-wave (PAW) method [23] and the X-C functional was described by an improved GGA of PBEsol [12], as implemented in the VASP 5.3.2 code [24]. The suitability of PBEsol for the present study is discussed in the following section. The recommended core configurations by VASP were adopted for each element in the present work. Due to the magnetic nature of V, Cr, Mn, Fe, Co, and Ni, first-principles calculations containing these elements were performed with spin polarization approach. An energy cut-off of 350 eV was used for the plane-wave expansion of the electronic wave functions.

After tests of supercell size, we chose a 64-atom supercell (4×4×2 conventional hcp unit cells) for most solutes, which was sufficient to isolate the vacancy and solute atom from the periodic images. For Ba, Bi, Ca, K, Pb, Sr, and all the rare earth elements, a 96-atom 4×4×3 supercell was used due to the large atomic size of these elements. All structural degrees of freedom (ionic coordinates, cell volume, and cell shape) in the supercell were fully relaxed via a conjugate gradient method to an energy convergence of 10^{-5}

eV/atom, followed by a final static calculation using the tetrahedron method with Blöchl corrections [25] with energy convergence of 10^{-6} eV/atom to get the accurate total energy. An 8×8×9 Γ-centered *k*-point mesh was used for the 64-atom supercell for the electronic integration in the Brillouin zone. For calculations using 96-atom supercells, a 5×5×4 Γ-centered *k*-point mesh was used in structural relaxation and a 7×7×7 Γ-centered *k*-point mesh in subsequent static calculations.

To determine the position of the saddle point and associated minimum-energy pathway during solute migration, calculations were performed employing the CI-NEB method [11] with 3 images. The two end-point structures, i.e. the initial and final structures, were fully relaxed first. The same supercell size and *k*-point settings were used in the CI-NEB calculations as described above. Such calculations were performed for all the solute (solvent)-vacancy exchange in the 8-frequency model for each investigated elements.

Vibrational contributions to the free energies were calculated using the quasi-harmonic approximations from phonon or Debye model. Helmholtz free energy of a configuration as a function volume (V) and temperature (T) can be expressed as [14]:

$$F(V,T) = E_0(V) + F_{vib}(V,T) + F_{el}(V,T) \quad (9)$$

where $E_0(V)$ is the static energy at 0 K without the zero-point vibrational energy, $F_{vib}(V,T)$ the vibrational contribution, and $F_{el}(V,T)$ the thermal-electronic contribution. Note that at zero pressure, the Helmholtz free energy is equal to the Gibbs free energy. In the present work, $E_0(V)$ was calculated via first-principles directly, and $F_{el}(V,T)$ obtained by integrating the electronic density of states [26]. The detailed implementation of quasi-harmonic Debye model can be found in [5].

Results and discussion

<u>Self-diffusion in Mg</u>

Figure 3 shows the calculated Mg self-diffusion coefficients using different X-C functionals of LDA, GGA (PW91), and PBEsol compared with available experiments [27-30]. The self-diffusion coefficients from PBEsol show the best agreement with available experimental values. According to the Arrhenius diffusion of Eq. (8), the slope of the diffusion coefficient line in Figure 3 corresponds to the activation energy Q and the intersect with vertical axis corresponds to the diffusion pre-factor D_0. The enthalpies of vacancy formation and migration mainly contribute to Q while the entropies of vacancy formation and migration mainly contribute to D_0.

Diffusion coefficients calculated using LDA yield a decent slope but a lower intersect, while the calculations using GGA give a good intersect but a lower slope compared with experiments. The reason is because LDA and GGA have different advantages in terms of getting accurate properties. LDA is most suitable for vacancy formation energy due to its surface error cancellation [31]. But LDA causes significant overbinding in metallic systems, which results in a lower D_0. On the contrary, although improper for vacancy formation energy, GGA is most suitable to obtain the equilibrium and vibrational properties for metallic systems [32]. This explains why the self-diffusion coefficients from LDA have a decent slope, i.e. the enthalpy-related Q value, and a good intersect from GGA, i.e. the entropy-related D_0 value. Since PBEsol is a revised GGA that improves equilibrium properties of densely packed solids and their surfaces [12], and possesses both advantages from LDA and GGA, it hence improves the predictions for both Q and D_0.

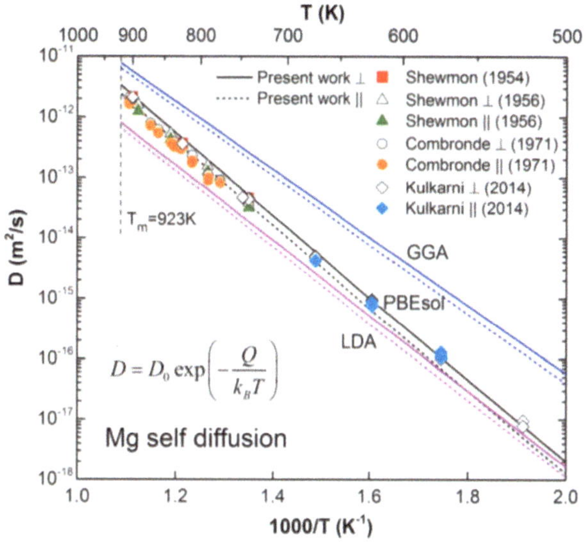

Figure 3. Predicted self-diffusion coefficients in Mg from different X-C functionals of LDA, GGA (PW91), and PBEsol compared with experimental data in the literature. Calculated results with PBEsol are from present work, results with GGA and LDA are from Ganeshan et al. [33]. Experimental data are taken from Shewmon [28], Combronde and Brebec [29], and Kulkarni et al. [30].

<u>Solute diffusion in Mg</u>

In many experiments, the measured diffusion coefficient data points were few and scattered, and the fitted values of activation energy Q and diffusion pre-factor D_0 have relatively large error bars. For solutes with more measurements available in the literature, such as Al [34-37], the present predictions agree remarkably well with experimental values within a factor of 3, as shown in Figure 4.

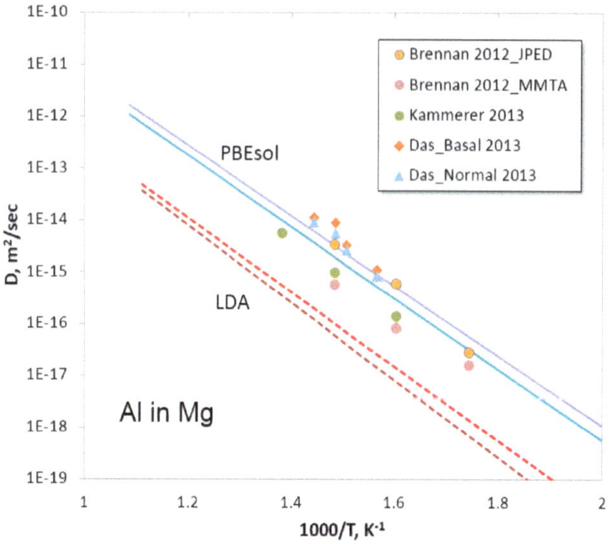

Figure 4. Predicted dilute solute tracer diffusion coefficients of Al. Results with LDA are from Ganeshan et al. [10]. Experimental

diffusion data are taken from Brennan et al. [34, 37], Kammerer et al. [36], and Das et al. [35].

As can be seen in Figure 5, the activation energy vs. atomic number curves show an inverse "V" shape, indicating the TM solutes have large activation energy. This may due to the strong interaction between solute atoms with *d* electrons and the Mg matrix. Most early TMs, such as Ti, V, Cr, Mn, Fe, Ni, Zr, and Hf, do not form favorable bonds with Mg. In binary Mg-X alloy systems with these elements, no ordered intermetallic compounds are experimentally observed. These alloying elements in hcp Mg also tend to have positive enthalpies of mixing, often indicating an energetic preference for phase separation and limited solid solubility. For rare earth elements, they consistently show very low activation energies, which indicate their diffusion in Mg is faster than most of the substitutional elements.

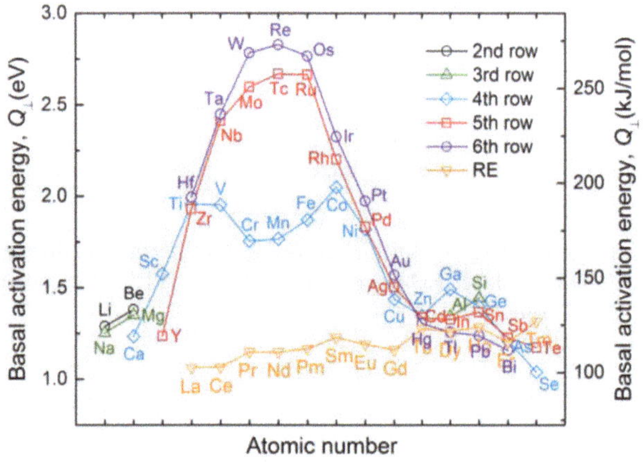

Figure 5. Calculated activation energies $Q_\perp$ of basal diffusion coefficients of solutes in Mg as a function of atomic number.

Figure 6 and 7 shows the calculated basal impurity diffusion coefficients of all the elements considered in the present work. As shown in Figure 6, As, Ca, Ga, Hg, Li, Na, Sb, Se, Sr, Te, Tl, and Zn are faster diffusers than Mg self-diffusion, while other solutes diffuse slower than Mg self-diffusion. Solutes which have high diffusion coefficients and low solid solubility, such as Ca, Na, Sb, and Sr, may serve as good alloying candidates for texture refinement since they have a high tendency to segregate to the grain boundaries and pin the grain boundary movement. Shown in Figure 7, rare earth elements have similar diffusion coefficients in Mg with relatively low diffusion activation energies.

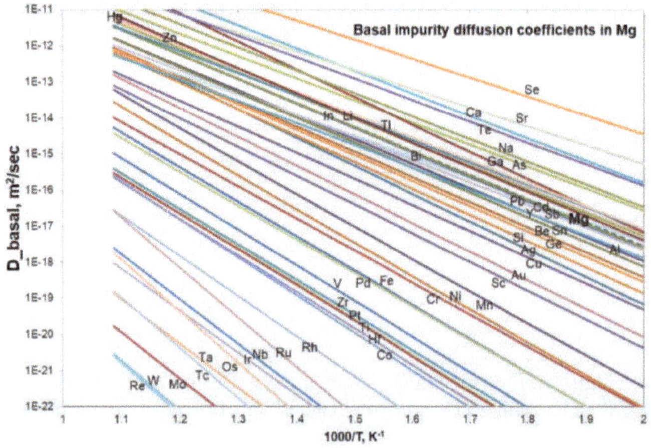

Figure 6. Predicted basal impurity diffusion coefficients $D_\perp$ of 47 non rare earth solutes in hcp Mg. The basal self-diffusion coefficient of Mg is plotted in a dashed line.

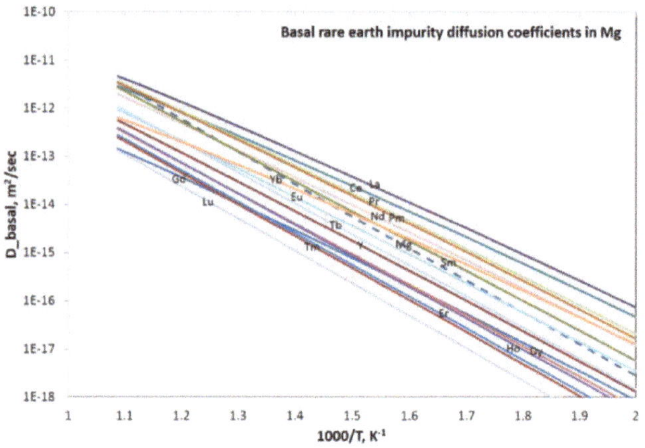

Figure 7. Predicted basal impurity diffusion coefficients $D_\perp$ of 15 rare earth solutes in hcp Mg. The basal self-diffusion coefficient of Mg is plotted in a dashed line.

Conclusion

As an effort for computational and data-driven development of advanced Mg-based alloys, we present a comprehensive study of dilute solute diffusion coefficients of 61 substitutional alloying elements in hcp Mg. The 8-frequency model for solute diffusion within an hcp lattice has been used to compute the correlation factors for different jumps. All the energies and vibrational quantities required to compute the jump frequencies and hence derive the solute diffusion coefficients from the 8-frequency model are calculated with the DFT based first-principles calculations in terms of the X-C functional of PBEsol and the quasi-harmonic Debye model. Saddle point configurations are predicted using the first-principles based CI-NEB method. It is found that:

(1) Compared with LDA and GGA, PBEsol is able to well describe both vacancy formation energies and vibrational properties, resulting in more accurate quantitative predictions of diffusion coefficients.
(2) The solute diffusion coefficients in hcp Mg are roughly inversely proportional to bulk modulus, i.e. their bonding to Mg. Transition metal elements with *d* electrons strongly interact with Mg and have very large diffusion activation energies, which cannot be explained by a simple strain model based on atomic size.
(3) The predicted solute diffusion coefficients in Mg compare remarkably well with available experiments in the literature. For most solutes basal diffusion coefficients are bigger than the non-basal ones.
(4) The present work lays the foundation of diffusion data for future rational design of novel casting and wrought Mg alloys. The calculated diffusion data can be used to develop CALPHAD-type diffusion mobility databases for multi-component Mg alloys [38]. The theoretical methodology used herein can be readily applied to solute diffusion in other hcp systems.

Acknowledgement

The present work was funded by the National Science Foundation (NSF) through Grant No. DMR-1006557. First-principles calculations were carried out partially on the LION clusters at the Pennsylvania State University supported by the Materials Simulation Center and the Research Computing and Cyberinfrastructure unit at the Pennsylvania State University, partially on the resources of NERSC supported by the Office of Science of the US Department of Energy under contract No. DE-AC02-05CH11231, and partially on the resources of XSEDE supported by NSF with Grant No. ACI-1053575. Calculations were also carried out on the CyberSTAR cluster funded by NSF through Grant No. OCI-0821527.

References

[1] B.L. Mordike, T. Ebert. Magnesium - properties - applications - potential, *Mater. Sci. Eng., A* 302 (2001) 37-45.
[2] J.F. Nie. Precipitation and hardening in magnesium alloys, *Metall. Mater. Trans. A* 43A (2012) 3891-3939.
[3] S.L. Shang, H. Zhang, S. Ganeshan, Z.K. Liu. The development and application of a thermodynamic database for magnesium alloys, *JOM* 60 (2008) 45-47.
[4] G. Neumann, C. Tuijn. Self-diffusion and impurity diffusion in pure metals: handbook of experimental data, 2009.
[5] B.C. Zhou, S.L. Shang, Y. Wang, Z.K. Liu. Diffusion coefficients of alloying elements in dilute Mg alloys: A comprehensive first-principles study, *Acta Mater.* (2015) In press.
[6] B.C. Zhou, S.L. Shang, Y. Wang, Z.K. Liu. Data set for diffusion coefficients of alloying elements in dilute Mg alloys from first-principles, *Data in Brief* (2015) Accepted.
[7] W.Y. Wang, B.C. Zhou, J.J. Han, H.Z. Fang, S.L. Shang, Y. Wang, X. Hui, Z.K. Liu. Prediction of diffusion coefficients in liquids and solids, *Defect and Diffusion Forum* 364 (2015) 182-191.
[8] M. Mantina, L.Q. Chen, Z.K. Liu. Predicting diffusion coefficients from first principles via Eyring's reaction rate theory, *Defect and Diffusion Forum* 294 (2009) 1-13.
[9] M. Mantina, Y. Wang, L.Q. Chen, Z.K. Liu, C. Wolverton. First principles impurity diffusion coefficients, *Acta Mater.* 57 (2009) 4102-4108.
[10] S. Ganeshan, L.G. Hector, Z.K. Liu. First-principles calculations of impurity diffusion coefficients in dilute Mg alloys using the 8-frequency model, *Acta Mater.* 59 (2011) 3214-3228.
[11] G. Henkelman, B.P. Uberuaga, H. Jonsson. A climbing image nudged elastic band method for finding saddle points and minimum energy paths, *J. Chem. Phys.* 113 (2000) 9901-9904.
[12] J.P. Perdew, A. Ruzsinszky, G.I. Csonka, O.A. Vydrov, G.E. Scuseria, L.A. Constantin, X.L. Zhou, K. Burke. Restoring the density-gradient expansion for exchange in solids and surfaces, *Phys. Rev. Lett.* 100 (2008) 136406.
[13] M.A. Blanco, E. Francisco, V. Luana. GIBBS: isothermal-isobaric thermodynamics of solids from energy curves using a quasi-harmonic Debye model, *Comput. Phys. Commun.* 158 (2004) 57-72.
[14] S.L. Shang, Y. Wang, D. Kim, Z.K. Liu. First-principles thermodynamics from phonon and Debye model: Application to Ni and Ni3Al, *Comput. Mater. Sci.* 47 (2010) 1040-1048.
[15] P.G. Shewmon. Diffusion in solids, McGraw-Hill New York, 1963.
[16] N.L. Peterson. Self-diffusion in pure metals, *J. Nucl. Mater.* 69 (1978) 3-37.
[17] D. Shin, C. Wolverton. First-principles study of solute-vacancy binding in magnesium, *Acta Mater.* 58 (2010) 531-540.
[18] J.E. Saal, C. Wolverton. Solute-vacancy binding of the rare earths in magnesium from first principles, *Acta Mater.* 60 (2012) 5151-5159.
[19] C. Wolverton. Solute-vacancy binding in aluminum, *Acta Mater.* 55 (2007) 5867-5872.
[20] H. Eyring. The activated complex in chemical reactions, *J. Chem. Phys.* 3 (1935) 107-115.
[21] C. Wert, C. Zener. Interstitial atomic diffusion coefficients, *Physical Review* 76 (1949) 1169-1175.
[22] E. Wimmer, W. Wolf, J. Sticht, P. Saxe, C.B. Geller, R. Najafabadi, G.A. Young. Temperature-dependent diffusion coefficients from ab initio computations: Hydrogen, deuterium, and tritium in nickel, *Physical Review B* 77 (2008) 134305.
[23] G. Kresse, D. Joubert. From ultrasoft pseudopotentials to the projector augmented-wave method, *Physical Review B* 59 (1999) 1758-1775.
[24] G. Kresse, J. Furthmuller. Efficient iterative schemes for ab initio total-energy calculations using a plane-wave basis set, *Physical Review B* 54 (1996) 11169-11186.
[25] P.E. Blochl, O. Jepsen, O.K. Andersen. Improved tetrahedron method for Brillouin-zone integrations, *Physical Review B* 49 (1994) 16223-16233.
[26] Y. Wang, Z.K. Liu, L.Q. Chen. Thermodynamic properties of Al, Ni, NiAl, and Ni3Al from first-principles calculations, *Acta Mater.* 52 (2004) 2665-2671.
[27] P.G. Shewmon, F.N. Rhines. Rate of self-diffusion in polycrystalline magnesium, *Trans. AIME* 200 (1954) 1021-1025.
[28] P.G. Shewmon. Self-diffusion in magnesium single crystals, *Trans. AIME* 206 (1956) 918-922.
[29] J. Combronde, G. Brebec. Anisotropy for self diffusion in magnesium, *Acta Metall.* 19 (1971) 1393-1399.
[30] N.S. Kulkarni, R.J.B. Warmack, B. Radhakrishnan, J.L. Hunter, Y. Sohn, K.R. Coffey, G.E. Murch, I.V. Belova. Overview of SIMS-based experimental studies of tracer diffusion in solids and application to Mg self-diffusion, *J. Phase Equilib. Diff.* 35 (2014) 762-778.
[31] C. Freysoldt, B. Grabowski, T. Hickel, J. Neugebauer, G. Kresse, A. Janotti, C.G. Van de Walle. First-principles calculations for point defects in solids, *Rev. Mod. Phys.* 86 (2014) 253-305.
[32] S.L. Shang, A. Saengdeejing, Z.G. Mei, D.E. Kim, H. Zhang, S. Ganeshan, Y. Wang, Z.K. Liu. First-principles calculations of pure elements: Equations of state and elastic stiffness constants, *Comput. Mater. Sci.* 48 (2010) 813-826.
[33] S. Ganeshan, L.G. Hector, Z.K. Liu. First-principles study of self-diffusion in hcp Mg and Zn, *Comput. Mater. Sci.* 50 (2010) 301-307.
[34] S. Brennan, A.P. Warren, K.R. Coffey, N. Kulkarni, P. Todd, M. Kilmov, Y. Sohn. Aluminum impurity diffusion in magnesium, *J. Phase Equilib. Diff.* 33 (2012) 121-125.
[35] S.K. Das, Y.M. Kim, T.K. Ha, R. Gauvin, I.H. Jung. Anisotropic diffusion behavior of Al in Mg: diffusion couple study using Mg single crystal, *Metall. Mater. Trans. A* 44A (2013) 2539-2547.
[36] C.C. Kammerer, N.S. Kulkarni, R.J. Warmack, Y.H. Sohn. Interdiffusion and impurity diffusion in polycrystalline Mg solid solution with Al or Zn, *J. Alloys Compd.* 617 (2014) 968-974.
[37] S. Brennan, K. Bermudez, N.S. Kulkarni, Y. Sohn. Interdiffusion in the Mg-Al System and Intrinsic Diffusion in β-Mg2Al3, *Metall. Mater. Trans. A* 43A (2012) 4043-4052.
[38] Z.L. Bryan, P. Alieninov, I.S. Berglund, M.V. Manuel. A diffusion mobility database for magnesium alloy development, *CALPHAD* 48 (2015) 123-130.

Magnesium Technology 2016
Edited by: Alok Singh, Kiran Solanki, Michele V. Manuel, and Neale R. Neelameggham
TMS (The Minerals, Metals & Materials Society), 2016

STUDY OF ZE10 MAGNESIUM ALLOY WELDED JOINTS PRODUCED WITH DISK LASER

Miroslav Sahul, Martin Sahul
Slovak University of Technology in Bratislava; Faculty of Materials Science and Technology in Trnava; Paulínska 16, 917 24 Trnava, Slovak Republic
miroslav.sahul@stuba.sk, martin.sahul@stuba.sk

Keywords: laser welding, disk laser, rare earth elements, ZE10 magnesium alloy, columnar crystals, equi-axed grains

Abstract

The paper deals with the study of weldability of magnesium alloy with the addition of rare earth elements. ZE10 magnesium alloy with 0.06 wt. % of La and 0.08 wt. % of Ce was suggested as base material. The thickness of ZE10 magnesium alloy was 1.65 mm. TruDisk 4002 laser with the power of 2.0 kW was used for production of welded joints. Bead on plate welds were produced with various welding parameters. Laser power ranged from 0.5 kW to 1.8 kW, welding speed was in the range from 30 mm/s to 90 mm/s, focusing of the laser beam was within the range of -3 to 3 mm. Shielding gases argon, helium and nitrogen were used for protection of the weld pool against the ambient atmosphere. Light and electron microscopy, microhardness measurements across the welded joints interface and tensile test were utilized to analyze the welded joints.

Introduction

Magnesium alloys are currently receiving great interest, especially in the automotive and aerospace industries mainly due to their high strength to weight ratio, good damping capacity, good castability and excellent recyclability. Magnesium and its alloys offer great potential to reduce the weight by replacing steel and aluminium for structural components [1-6]. The application of magnesium wrought products in vehicles is limited mainly due to poor formability of hexagonal close packed lattice. New magnesium wrought alloys have been developed in order to improve formability. The addition of rare earth elements is currently of particular interest [7]. Thus, it is necessary to concern with welding of magnesium alloys. Information on welding magnesium alloys with the addition of rare earth elements is limited.
Cao et al. investigated Nd:YAG laser welds of sand cast ZE41A-T5 magnesium alloy with the thickness of 2.0 mm. They found, that focal position greatly influenced the establishment of conduction or keyhole welding mode. Conduction welding was observed at 4.0×10^5 W/cm^2 and keyhole welding was reached at the power density of 1.5×10^6 W/cm^2. The fusion zone consisted of fine equi-axed grains formed through cellular growth [3].
Rao et al. studied friction stir spot welding of ZEK100 magnesium alloy containing rare earth elements. They found that overlapped welded joints produced with tool revolutions 1500 rpm and 0.2 mm tool shoulder plunge depth exhibited the highest lap shear strength [2].
Al Kazzaz et al. investigated application of laser beam for welding ZE41A-T5 magnesium alloy sand castings. The FZ contained fine equi-axed grains while the HAZ exhibited large grains, similar to the cast equi-axed structures in the base metal. They found, that no grain growth occurred in the HAZ. Authors observed defects such as sag and under fill which were associated due to the low viscosity and low surface tension of molten magnesium [4].
Dhari et al. concerned with CO_2 laser welding of AZ91 and WE43 magnesium alloys for automotive and airspace industries. Similarly, they observed grain refinement in the welded joint. Grain size in the WE43 base metal was 45 μm and 15 μm in the case of AZ91 magnesium alloy. The grain size in the welded joint reached only 5 μm. Grain refinement resulted in good mechanical properties of welded joints [5].
Palanivel et al. studied the influence of initial microstructure on the microstructural evolution and joint efficiency of WE43 alloy during friction stir welding [8]. They observed that FSW of WE43 magnesium alloy resulted in grain refinement, breakage and dissolution of second phase particles leading to homogenization.
The paper deals with the study of disk laser welded joints of ZE10 magnesium alloy with the addition of rare earth elements.

Experimental

ZE10 magnesium alloy with the addition of rare earth elements was suggested as experimental material. The thickness of ZE10 magnesium alloy is 1.65 mm [9]. The chemical composition of proposed base material is given in Tab. 1. The chemical composition was detected by means of EDX analysis. Area of base material subjected to EDX analysis is given in Fig. 1.

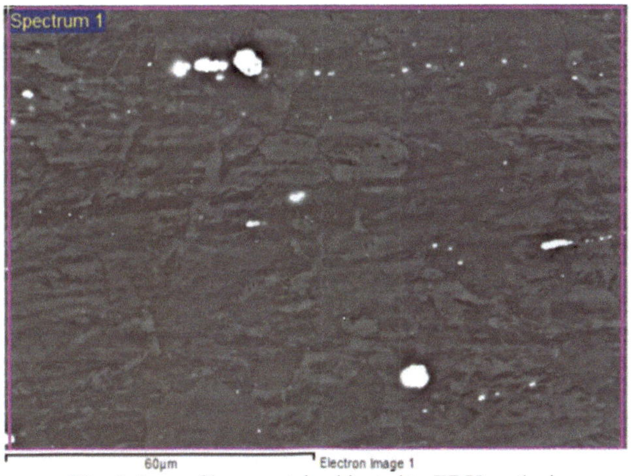

Fig. 1 Area of base metal subjected to EDX analysis

Table 1 Chemical composition of ZE10 magnesium alloy (wt. %)

Zn	Zr	La	Ce	Mg
1.15	0.26	0.06	0.08	98.45

TruDisk 4002 disk laser was used for production of bead on plate welds. The maximum power of laser TruDisk 4002 was 2.0 kW. The wavelength of laser radiation is λ = 1.03 μm. The laser radiation was transported from the laser source to the place of technological processing through laser light cable with the core diameter of Ø 400 μm. The spot size of the laser beam focused direct on the surface of magnesium alloy was Ø 400 μm. Laser power used for production of welded joints ranged from 0.5 kW to 1.8 kW, welding speed was in the interval from 30 mm/s to 90 mm/s and focusing of the laser beam was within the range of -3 to 3 mm.

Focal distance of the focusing lens was 200 mm. Focusing optics BEO D70 of TRUMPF Company was used. The optics was mounted on 6-axis Fanuc M710iC/50 robot. Argon, helium and nitrogen with the flow rate of 15 l/min and 22 l/min were used as shielding gases due to high affinity of magnesium to oxygen. After disk laser welding, the welded joints were cut off. Cross-sections were prepared for metallographic analysis by common procedure. 3 % Nital solution was used in order to reveal the macrostructure of welded joints. Etchant consisting of 10 ml acetic acid, 4.2 g picric acid, 10 ml water and 70 ml ethanol was used in order to reveal the microstructure of the magnesium alloy welded joints [10]. Light microscopy, electron microscopy with EDX analysis, microhardness measurements across welded joints interface and tensile testing were performed in order to analyze the properties of produced welded joints. Light microscope Neophot 32 was used for analysis of macrostructures and microstructures of welded joints. EDX analysis was conducted to more detailed study of weld metal. Scanning electron microscope JEOL JSM 7600 F fitted with EDX analysis detector X-max 50 mm^2 was used for EDX analysis. Buehler IndentaMet 1100 microhardness tester was used for Vickers microhardness measurements. The loading used during measurements was 100 g and dwell time 10 s. The distance between indents was 100 μm. Static tensile testing of welded joints was conducted according to STN EN 10002 standard on LabTest 5.250 SP1 – VM machine.

Results

Macrostructures of selected welded joints are given in Fig. 2 and 3. From the macrostructures, it is evident the influence of the focal position on the weld width. When the focal position was direct on the surface of magnesium alloy, the weld metal width was 1 372 μm.

Fig. 2 The macrostructure of ZE10 magnesium alloy welded joint (P = 1.1 kW, v = 60 mm/s, f = 0 mm, Ar flow rate 15 l/min)

When the focal position was set to - 1 mm below the surface of ZE10 magnesium alloy, the weld width was larger. Weld width increased to the value of 1 484 μm. It arises from the defocusing of the laser beam. The power density decreased in this case.

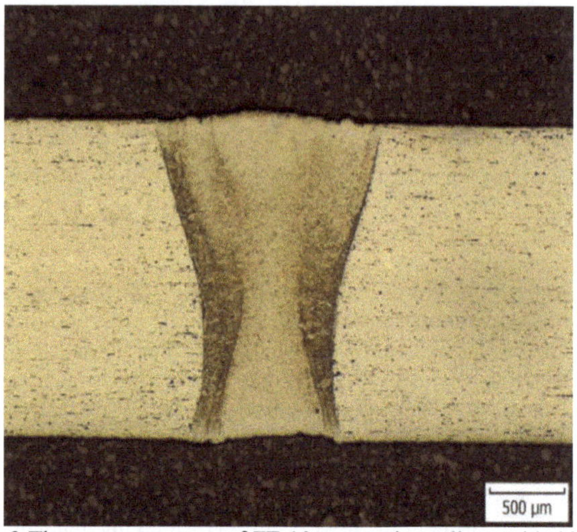

Fig. 3 The macrostructure of ZE 10 magnesium alloy welded joint (P = 1.1 kW, v = 60 mm/s, f = - 1 mm, Ar flow rate 15 l/min)

The influence of laser power on weld width is given in Fig. 4. As it arises from Fig. 4, the higher the laser power the wider the weld metal. The change of the weld metal width is associated with the increase of heat input. The maximum weld width was documented in the case of laser powers 1.2 and 1.3 kW.

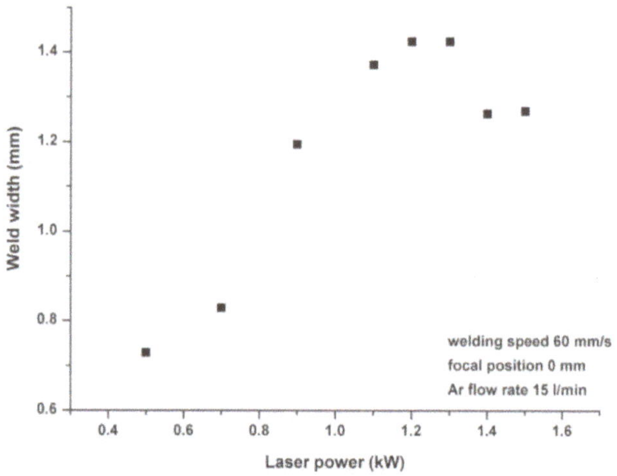

Fig. 4 Influence of laser power used for welding ZE10 magnesium alloy on weld width

Base metal - fusion zone interface is given in Fig. 5 (P = 1.1 kW, f = 0 mm, v = 50 mm/s and He gas flow rate of 22 l/min). Fusion zone consists of columnar crystals growing in the opposite direction to heat dissipation. Columnar crystals were observed at the fusion boundary. In this area, there is high thermal gradient. The columnar to equi-axed transition occurs when the movement of the columnar front is blocked by enough equi-axed grains formed in the liquid ahead of the columnar front [11]. Contrary, fine equi-axed grains were observed in the weld centre. This part of the weld metal is characteristic by grain refinement. The rapid cooling experienced during laser welding led to significant grain refinement in the fusion zone centre [11].

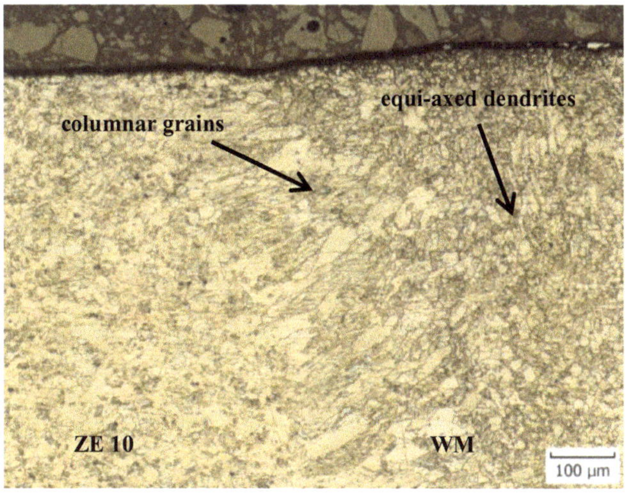

The influence of heat input on the grain size is documented in Fig. 6. The welding parameters used for production of welded joints were as follows: P = 1.1 kW, f = 0 mm and Ar = 15 l/min. The welding speeds were 40 mm/s and 90 mm/s. Microstructure of fusion zone of the welded joint produced with welding speed of 40 mm/s is given in Fig. 6a. As it arises from the microstructure, the grains in the centre of fusion zone were greater than that observed in the fusion zone of welded joint produced with the welding speed of 90 mm/s. It is related to the lower heat input. Grain refinement occurred in the second case, when higher welding speed was used. Grain refining is attributed to high welding speed, which is characteristic for laser beam welding, and thus high cooling rate of molten metal.

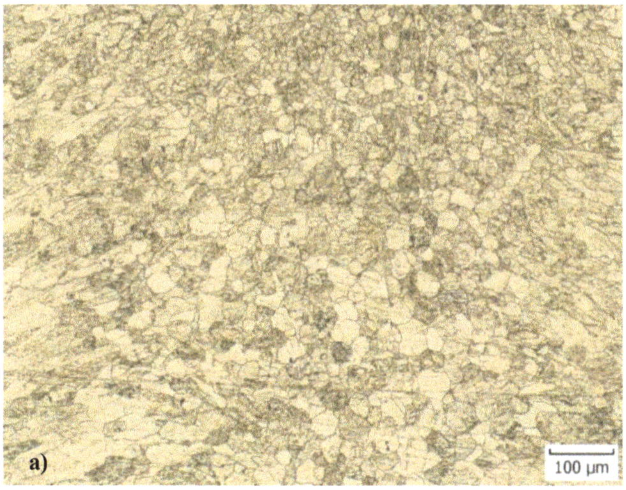

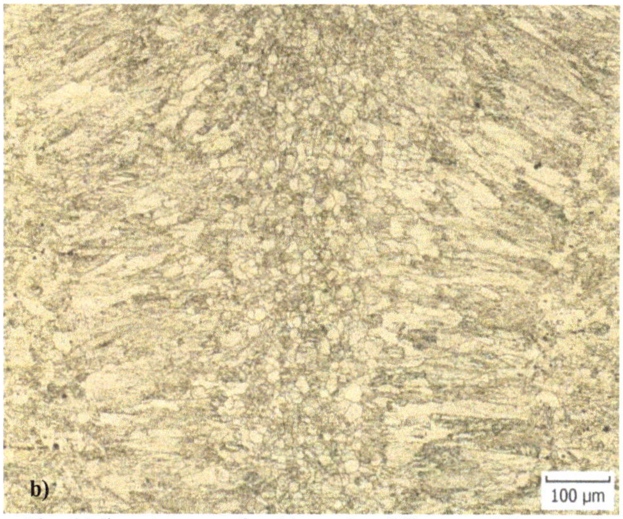

Fig. 6 Microstructure of weld metal at different welding speed (P = 1.1 kW, f = 0 mm and Ar = 15 l/min): a) v = 40 mm/s, b) v = 90 mm/s

The structure of weld metal recorded with scanning electron microscopy is given in Fig. 7. As mentioned above, ZE10 magnesium alloy contains rare earth elements (La and Ce). Fine bright particles were observed in weld metal grains. Results of EDX analysis are given in Tab. 2.

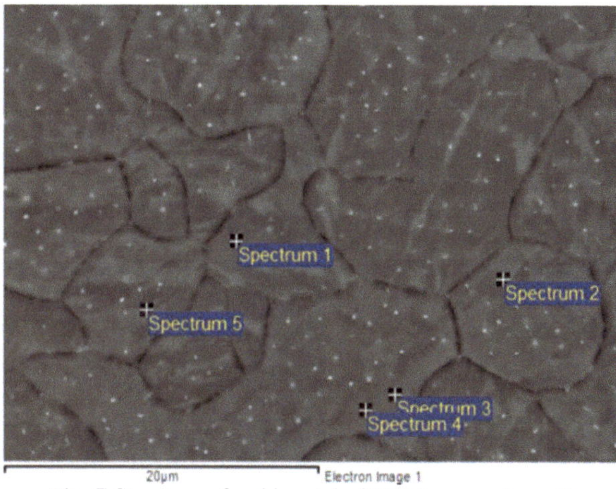

Fig. 7 Structure of weld metal of ZE10 magnesium alloy

As it follows from results given in Tab. 2, Spectrums 1 to 3 contain from 98.41 to 98.68 wt. % of magnesium. No presence of rare earth elements was recorded. The matrix is created of solid solution of alloying elements in magnesium. In the regions with the presence of bright particles, there were rare earths detected. In spectrum 4, there were 0.01 wt. % of La and 0.09 wt. % of Ce. In spectrum marked as 5, there was detected only Ce (0.04 wt. %). For more detailed analysis of bright particles, it will be necessary to carry out transmission electron microscopy.

Table 2 Chemical composition of locations marked as Spectrum 1 to 5 in weld metal (wt. %)

Spectrum No.	Zn	Zr	La	Ce	Mg
1	1.11	0.21	-	-	98.68
2	1.14	0.21	-	-	98.65
3	1.40	0.19	-	-	98.41
4	1.23	-	0.01	0.09	98.67
5	1.11	-	-	0.04	98.85

The influence of welding speed on the microhardness course across the welded joints is given in Fig. 8. The welded joints were produced at constant laser power of 1.1 kW, focal position of 0 mm and Ar shielding gas flow rate of 15 l/min. When welding with speed of 50 mm/s the welded joint was wider. It is due to higher heat input. Calculated heat input was 22 J/mm. The microhardness of base metal averaged 52 HV0.1 and the microhardness of the weld metal averaged 45 HV0.1. Slight drop of microhardness was documented in this case. When welding with speed of 90 mm/s, the heat input extensively decreased. Increasing the welding speed caused the reduction of heat input to about 12 J/mm. The higher welding speed and thus higher cooling rate resulted in grain refinement in the centre of the fusion zone. The grain refinement was also supported by zirconium. The content of zirconium in the base metal was 0.26 wt. %. The microhardness in the fusion zone averaged in the case of lower heat input about 50 HV0.1.

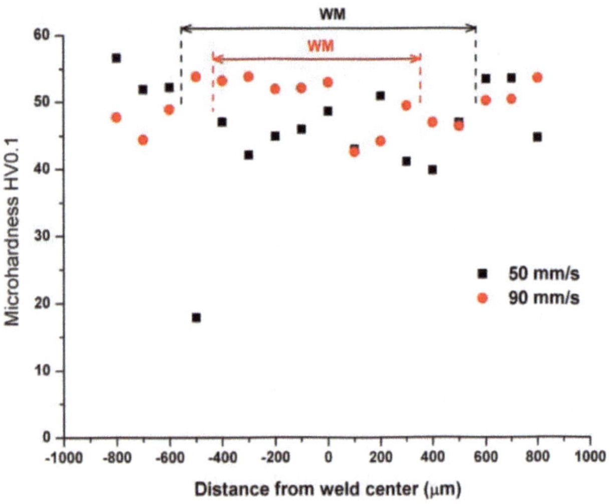

Fig. 8 Course of microhardness across base metal - weld metal - base metal interface

The influence of laser power on tensile force is given in Fig. 9. Other welding parameters were kept constant (v = 60 mm/s, f = 0 mm and Ar flow rate 15 l/min). The tensile force at which the welded joint was broken was 4 686 N, when welding with the power of 1.1 kW. When laser power increased to 1.2 kW, tensile force necessary to break welded joint increased to 4 996 N. Slight decrease of tensile load was recorded when laser power was set to 1.3 kW. Increase of the laser power to 1.4 kW caused increase of the tensile force measured during tensile testing.

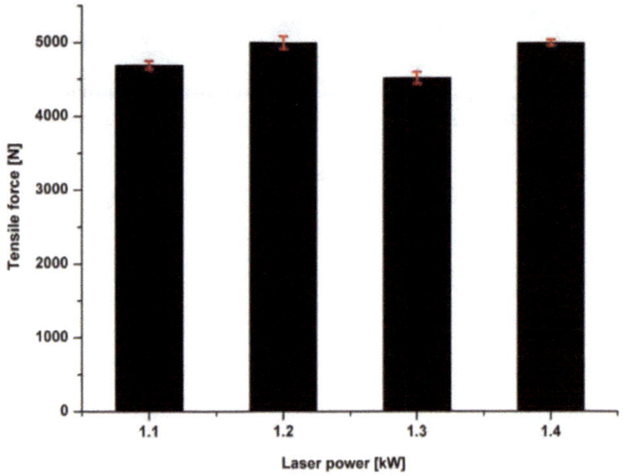

Fig. 9 Influence of laser power on tensile force

Conclusion

The paper focused on the study of properties of welded joints of rare earth containing magnesium alloy. ZE10 magnesium alloy was proposed as material to be welded. The magnesium alloy contains < 1.0 wt. % of rare earth elements (La, Ce). Following welding parameters: laser power of 0.9 kW, welding speed 60 mm/s, focal position 0 mm and Ar flow rate of 15 l/min were not suitable for full penetration of ZE10 magnesium alloy sheet. In this case weld bead reinforcement was observed. The full penetration was documented when laser power increased to 1.1 kW. When focal position changed to + 1 and + 2 mm above the surface of ZE10 magnesium alloy, extensive bead reinforcement was observed. This state was observed, when laser beam was focused above the surface of base metal. Furthermore, it was found that the higher the laser power the larger the weld width. It was attributed to increased heat input. Contrary, the higher the welding speed the narrower was the weld bead. The weld metal was created by columnar and equi-axed grains. Columnar grains were detected at the fusion boundary, where is high thermal gradient. Contrary, fine equi-axed grains were observed in the weld centre. This part of the weld metal is characteristic by grain refinement. Slight porosity was observed in some welded joints. Porosity belongs to the main problems when welding magnesium alloys. The reason could be in this case high welding speed, which caused that the gases present in molten weld metal were entrapped in this area during solidification stage. Drop of microhardness in weld metal was recorded during microhardness measurements. The base material was in as-rolled condition and the decrease of microhardness could be attributed to the recrystallization, which occurred during welding process. Increase of the welding speed to 90 mm/s resulted in increasing the average weld metal microhardness. Bright particles present in weld metal were detected by electron microscopy. Presence of rare earth elements was recorded in that location. The highest tensile force was measured in the case of welded joint produced with the parameters: P = 1.1 kW, v = 60 mm/s, f = + 3 mm and Ar flow rate of 15 l/min.

Acknowledgement

The paper was prepared under the support of The Ministry of Education, Science, Research and Sport of the Slovak Republic Grant Agency VEGA and Slovak Academy of Sciences, projects No. **1/0876/15** and **1/1010/16**.

The paper was prepared also under the support of the project: Centre for development and application of advanced diagnostic methods in processing of metallic and non-metallic materials, **ITMS:26220120014**, supported by the Research & Development Operational Programme funded by the ERDF.

References

[1] C. W. TAN, L. Q. LI, Y. B. CHEN, A. M. NASIRI, AND Y. ZHOU: Microstructural Characteristics and Mechanical Properties of Fiber Laser Welded Brazed Mg Alloy Stainless Steel Joint. In Welding Journal 2014, vol. 93, p. 399-409.

[2] H. M. RAO, R. I. RODRIGUEZ, J. B. JORDON, M. E. BARKEY, Y. B. GUO, H. BADARINARAYAN, W. YUAN: Friction Stir Spot Welding of Rare-Earth Containing ZEK100 Magnesium Alloy Sheets. In Materials and Design 2014, vol. 56, p. 750-754.

[3] X. CAO, M. XIAO, M. JAHAZI, J. P. IMMARIGEON: Continuous Wave Nd:YAG Laser Welding of Sand Cast ZE41A-T5 Magnesium Alloy. In Materials and Manufacturing Processes 2005, vol. 20, p. 987-1004. ISSN 1042-6914.

[4] H. Al KAZZAZ, X. CAO, M. JAHAZI, M. MEDRAJ: Reliability of Laser Welding Process for ZE41A-T5 Magnesium Alloy Sand Castings. In Materials Transactions 2008, vol. 49, No. 4, p. 774-781.

[5] M. DHAHRI, J. MASSE, J. MATHIEU, G. BARREAU, and M. AUTRIC: Laser Welding of AZ91 and WE43 Magnesium Alloys for Automotive and Aerospace Industries. In Advanced Engineering Materials 2001, vol. 3, No. 7, p. 504-507.

[6] M. WAHBA, S. KATAYAMA: Laser Welding of Magnesium Alloys. In Transactions of JWRI 2011, vol. 41, No. 1, p. 11-23.

[7] X. NIU, T. SKSZEK, M. FABISCHEK, A. ZAK: Low Temperature Warm Forming of Magnesium ZEK100 Sheets for Automotive Applications. In Materials Science Forum 2014, vol. 783-786, p. 431-436.

[8] S. PALANIVEL, R. S. MISHRA, B. DAVIS, R. DeLORME, K. J. DOHERTY K. C. Cho: Effect of Initial Microstructure on the Microstructural Evolution and Joint Efficiency of a WE43 Alloy During Friction Stir Welding. In TMS (The Minerals, Metals & Materials Society), 2013, p. 253-261.

[9] M. ŽITŇANSKÁ: Disk Laser Welding of ZE10 Magnesium Alloy. Trnava: MTF STU, 2015, 84 p.

[10] T. KRAMÁR, P. VONDROUŠ, M. KOLAŘÍKOVÁ, K. KOVANDA, L. KOLAŘÍK, M. ONDRUŠKA: Resistance Spot Welding of Magnesium Alloy AZ 61. In MM Science Journal 2015, p. 596-599.

[11] L. LIU: Welding and Joining of Magnesium Alloys. Cambridge: Woodhead Publishing Limited, 2010. ISBN 978-1-84569-692-4.

Magnesium Technology 2016
Edited by: Alok Singh, Kiran Solanki, Michele V. Manuel, and Neale R. Neelameggham
TMS (The Minerals, Metals & Materials Society), 2016

SIMILAR AND DISSIMILAR ULTRASONIC SPOT WELDING OF A RARE-EARTH CONTAINING ZEK100 MAGNESIUM ALLOY

A. Macwan[1], D.L. Chen[1]
[1]Department of Mechanical and Industrial Engineering, Ryerson University, 350 Victoria Street, Toronto, Ontario M5B 2K3, Canada

Keywords: Magnesium alloy, Aluminum alloy, Ultrasonic spot welding, Microstructure, Tensile strength

Abstract

Lightweight magnesium (Mg) alloys are increasingly used in the transportation industries to reduce the weight of vehicles due to their high strength-to-weight ratio. The applications of Mg alloys inevitably involve welding and joining. Ultrasonic spot welding (USW), an emerging and promising solid-state joining technology, can be suitably applied to join Mg alloys. In this study a low rare-earth containing Mg alloy, ZEK100 with improved formability and corrosion resistance, was welded similarly (ZEK100-ZEK100) and dissimilarly (ZEK100-Al5754) at varying levels of welding energy. It was observed that the interface grain size increased in the similar welding, while an intermetallic diffusion layer formed in the dissimilar welding and its thickness increased with increasing welding energy. The tensile lap shear strength first increased, reached its optimum value, and then decreased with increasing welding energy. The strength of dissimilar welds was about 75% of that of the magnesium similar welds.

Introduction

Automotive and aerospace industries are moving towards a lightweight design to meet stringent requirements of fuel efficiency and diminish anthropogenic climate-changing, environment-damaging, costly and human death-causing emissions [1,2]. Mg alloys with a high strength-to-weight ratio are attractive for light-weighting applications. Although the wrought (extruded, rolled) Mg alloys such as AZ31 and AZ61 show better mechanical properties than the cast counterparts, their applications are limited by their poor room temperature formability due to the strong deformation texture, which leads to the tension-compression yield asymmetry and mechanical anisotropy [3,4]. This is related to the hexagonal closed-packed crystal structure of Mg alloys with a nearly ideal *c/a* ratio (1.624), which gives limited deformation modes (i.e., basal slip and twining) and strong textures at room temperature [3,5]. Additionally, Mg alloys have a poor corrosion resistance compared with Al alloys [6-9]. Recent studies showed that the addition of rare-earth (RE) elements to Mg alloys along with Zn can soften or randomize the texture and improve their room temperature formability [3,10,11]. Moreover, it can also enhance the corrosion resistance [12,13]. Therefore, some new Mg alloys are being developed with the minimum RE elements to reduce alloy cost and achieve better mechanical properties. One such Mg alloy is ZEK100, which contains very small (0.2 wt.%) RE element (Nd). Several recent studies showed that this alloy exhibited superior room temperature formability [10,14,15], and as a result it is being considered as a promising candidate for structural and closure components in automotive and aerospace sectors. These applications unavoidably involve similar and dissimilar joining of ZEK100 Mg alloy to manufacture multi-material vehicle body structures and parts.

In the past decade, many studies have been done on the similar (Mg-Mg) and dissimilar (Mg-Al) joining using advanced high efficiency ultrasonic spot welding (USW) [16-18]. For example, Patel et al. [18] studied ultrasonic spot welded (USWed) AZ31 Mg alloy joints and showed that the grain size near the weld interface region increased with increasing welding energy, and developed a Hall-Petch type relationship between grain size and microhardness. Panteli et al. [17] have studied the growth behavior of Mg-Al IMCs under various levels of welding energy (or welding time) and clamping pressure in USWed AZ31-Al6111 dissimilar joints, and reported that IMCs growth rate was over twice than that under static condition. However, the feasibility of this joining technique for ZEK100-ZEK100 similar and ZEK100-Al5754 dissimilar joints is unknown. Hence, the purpose of the present study was to examine the feasibility of joining ZEK100-ZEK100 and ZEK100-Al5754 sheets and identify the effect of weld energy on the microstructure and mechanical properties of these joints.

Experimental details

The materials used in the present study were 1.5 mm and 2.0 mm thick sheets of ZEK100-O Mg alloy supplied by Magna International and University of Waterloo, Canada, and 1.5 mm thick sheets of Al5754-O Al alloy supplied by General Motors, USA. The chemical composition of both alloys are listed in Table I. The welding was performed using a 2.5 kW dual-wedge reed Sonobond MH2016 HP USW system operated in an energy mode at 20 kHz with 8 mm × 5 mm flat serrated welding tips having nine parallel teeth to ensure good grasping of the sheets during welding. Prior to welding, the surfaces of the sheets were slightly ground using 120 grit sand paper, and then cleaned with acetone followed by an ultrasonic cleaning using alcohol to remove lubricant and debris.

Table I. Composition (wt.%) of the alloys used in this study.

Material	Zn	Zr	Nd	Mn	Mg	Al
ZEK100	1.3	0.25	0.2	0.01	Bal.	-
Al5754	-	0.22	-	0.63	3.42	Bal.

Different levels of welding energy from 500 to 2000 J for ZEK100-ZEK100 and 250 J to 1500 J for ZEK100-Al5754, with a constant power setting of 2 kW, an impedance setting of 8, and a clamping pressure of 0.4 MPa, were applied. Here, the welding energy (Q) is determined by the level of power (P) and weld time (t), where $Q \sim P \times t$, e.g., a welding energy of 1500 J at 2 kW is equivalent to ~0.75 s. The welded joints were sectioned across their center, parallel to the direction of vibration using a low speed diamond cutter and then cold-mounted in epoxy and mechanically polished using abrasive papers, diamond paste and colloidal silica. The microstructural features were revealed using an etchant of acetic picral solution (i.e., 4.2 g picric acid, 10 ml acetic acid, 10 ml H_2O and 70 ml ethanol). The etched samples were examined using an optical microscope (OM) equipped with an image analysis system and the grain sizes were measured using a linear intercept method on twenty images taken at a magnification of 1000× at the weld interface. The polished microstructure samples of ZEK100-Al5754 dissimilar joints were examined with a JSM-6380LV scanning electron microscope equipped with Oxford EDS, EBSD, and 3D surface/fractographic analysis capacities.

To identify the optimum welding condition, the tensile lap shear tests were conducted to measure the lap shear failure load using a fully computerized United testing machine in air at a constant crosshead speed of 1 mm/min at room temperature. The tensile test coupons were welded using 15 mm × 80 mm strips with the weld located at the center of a 20 mm overlap. The joints were obtained by a transverse relative displacement between the sheets, where the vibration direction was perpendicular to the rolling direction.

Results and discussion

During USW the near-interface weld region defined as a nugget zone (NZ) underwent a significant microstructural changes due to the generation of frictional heat and plastic deformation. Fig. 1 shows the grain size of base metal (BM) and NZ at a welding energy of 1500 J and 2000 J. The BM showed a fine equiaxed and recrystallized grains with an average grain size of 5.7±0.9 μm as shown in Fig. 1(a). The relatively fine grain structure of the BM was due to alloying elements, Nd and Zr, which provided a resistance to grain growth [3,10]. The average grain size at 1500 J and 2000 J was 11.8±1.4 μm and 13.6±1.8 μm, respectively. The NZ experienced a high degree of plastic deformation and temperature rise due to frictional heating, resulting in the formation of relatively coarser grains compared with the BM, which was characterized by partial recrystallization of elongated/deformed grains and the coarsening of original equiaxed gains. Furthermore, it is seen from Fig. 1(a) and (b) that the average grain size of NZ at 2000 J was larger than at 1000 J, indicating that the welding energy had a significant effect on the NZ microstructure.

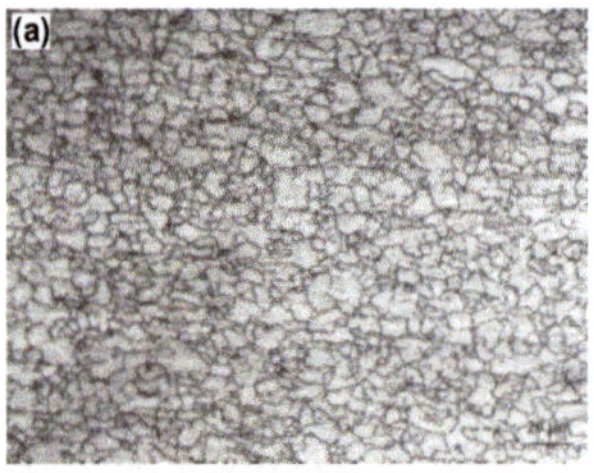

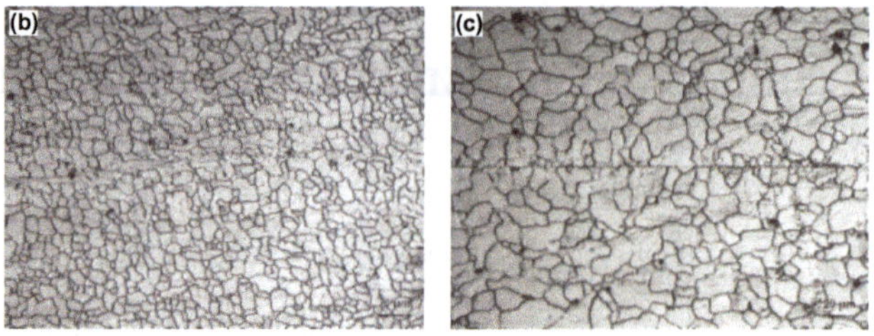

Fig. 1 Optical micrographs of (a) ZEK100 base metal, and samples welded at a welding energy of (b) 1500 J, and (c) 2000 J.

Fig. 2 shows a change of the average NZ grain size with welding energy. A linear increase of the grain size with increasing welding energy was observed. Previous studies showed that the increase in welding energy during USW increased temperature at the weld interface [19]. Besides, the increasing weld energy also increased total weld time when the welding power was kept constant. Therefore, a higher temperature at the weld interface for a longer time would promote grain growth, leading to a coarser grain size.

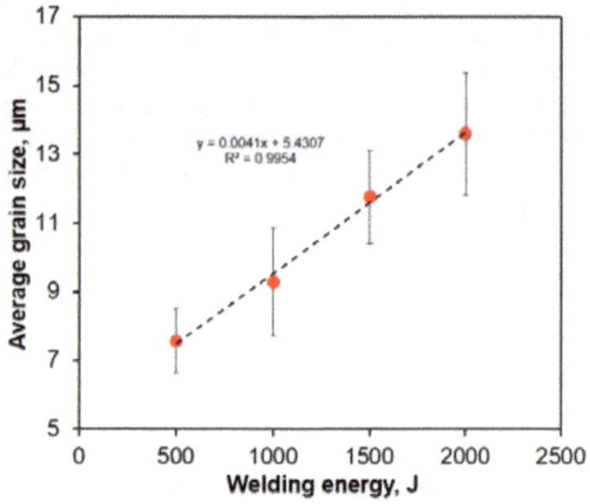

Fig. 2 Average grain size in NZ as a function of welding energy.

The most common bonding mechanism during dissimilar USW is interdiffusion. During USW, the oxide film will first be broken at asperities allowing a metal-to-metal contact, then the interdiffusion will occur allowing the formation of intermetallic products under sufficiently rapid kinetics reaction at the interface. Fig. 3 shows typical microstructures of the NZ of the USWed ZEK100-Al5754 joints at a welding energy of 500 J and 1000 J, respectively. It is seen that the interdiffusion caused the formation of a non-uniform diffusion layer across the interface. With increasing welding energy from 500 J to 1000 J the thickness of the diffusion layer increased. This was attributed to the increase in the interface temperature [19] and longer weld times, which caused increasing diffusion of atoms across the interface. To identify the interface diffusion layer, EDS element mapping and point analysis were performed.

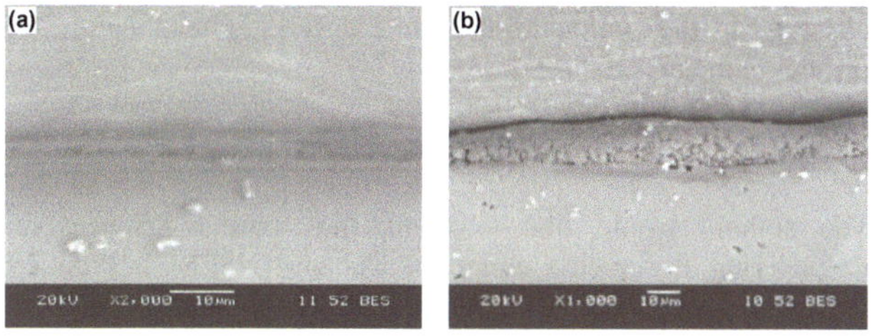

Fig. 3 Typical SEM images showing the interface microstructure of ZEK100-Al5754 dissimilar joint welded at a welding energy of (a) 500 J and (b) 1000 J.

Fig. 4 shows the results of EDS elemental mapping at the weld interface of a sample welded at 1000 J. It is seen that the interface diffusion layer was composed of mainly Mg and Al. EDS point analysis showed that the interface layer (in at.%) was approximately 60.4 % Mg, 39.3% Al, and 0.4 % Zn. By referring to the Al-Mg phase diagram [20], the interface diffusion layer should consist of a eutectic structure of Mg solid solution and $Al_{12}Mg_{17}$ phase via a eutectic reaction ($L \rightarrow Mg + Al_{12}Mg_{17}$). In our earlier studies [6], it was presented that the $Al_{12}Mg_{17}$ phase had a hardness of about 275 HV, which was significantly higher than the ZEK100 BM (60 HV). This suggested that $Al_{12}Mg_{17}$ IMC was brittle in nature and could have a detrimental effect on the overall strength of the joints under both static and dynamic loads [16,17]. Therefore, an optimum thickness of interface layer is required to achieve greater strength.

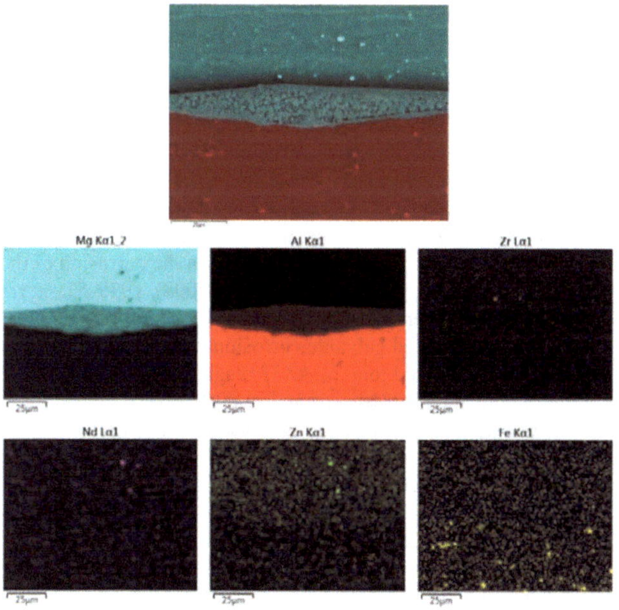

Fig. 4 EDS elemental mapping at the Al-Mg weld interface of a sample welded at a welding energy of 1000 J.

Fig. 5(a) shows the tensile lap shear strength of similar ZEK100-ZEK100 and dissimilar ZEK100-Al5754 joints as a function of welding energy at a constant power of 2 kW and constant clamping pressure of 0.4 MPa. It is seen that in both cases the tensile lap shear strength first increased, reached a maximum and then decreased with increasing welding energy. There was an increase in temperature at the weld interface with increasing welding energy [19], which facilitated increase in the plastic deformation and therefore, the bonding via diffusion, metallurgical adhesion, and mechanical interlocking. The average maximum strength were about 72 MPa and 55 MPa (calculated as the maximum tensile lap shear load divided by the nugget area of 8×5 mm^2), for similar and dissimilar joints, respectively. The relatively low strength of dissimilar joints was attributed to the presence of a brittle IMC $Al_{12}Mg_{17}$ in the interface layer, which reduced the plasticity and strength of the joint. After achieving the maximum strength, further rise in welding energy reduced the strength of both similar and dissimilar joints. The underlying mechanism for this was the presence of a higher stress concentration at the edge of the nugget in the similar joint, and the additional thickening of interface layer with a brittle IMC in the dissimilar joint. It was interesting to observe that a maximum strength was achieved at a relatively low welding energy of 500 J in the case of dissimilar joint compared with the similar joint where it was achieved at a welding energy of 1500 J. This could be due to a rapid formation of $Al_{12}Mg_{17}$ containing interface layer, where an optimum thickness of about 5 μm to 10 μm was quickly achieved at a lower welding energy, as reported in [17]. The strength of both similar and dissimilar joints was similar at 500 J as seen in Fig. 5(a). The average maximum tensile lap shear strength of dissimilar ZEK100-Al5754 joint compared with that of the similar ZEK100-ZEK100 and Al5754-Al5754 joints is shown in Fig. 5(b). It is seen that ~78% of the ZEK100-ZEK100 joint strength and ~55% of the Al5754-Al5754 joint strength were achieved in the ZEK100-Al5754 dissimilar joint.

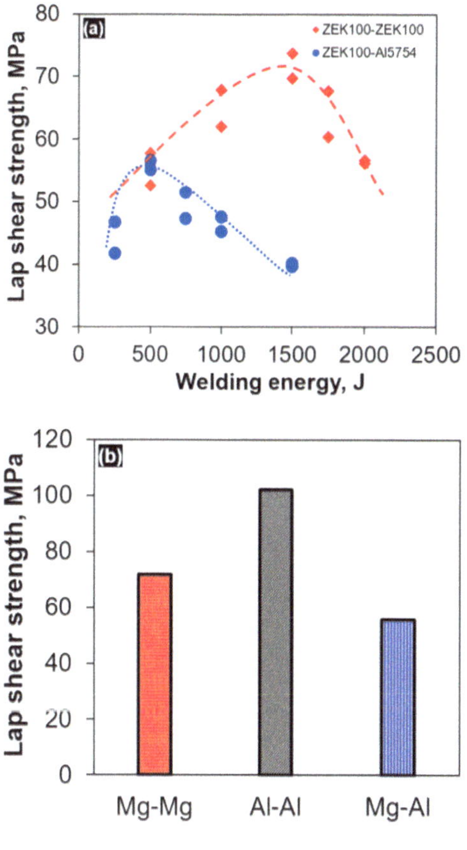

Fig. 5 (a) The tensile lap shear strength as a function of welding energy, and (b) a comparison of the maximum tensile lap shear strength of the Mg-Al dissimilar joint with that of the Mg-Mg and Al-Al similar joints.

Fig. 6 shows typical failure modes observed during the tensile lap shear testing of USWed ZEK100-ZEK100 joint at a welding energy of 1500 J and 2000 J (Fig. 6(a) and (b)) and ZEK100-Al5754 joint at a welding energy of 500 J and 1500 J (Fig. 6(c) and (d)). The similar joint showed two failure modes of interfacial (500 J, 1000 J, 1500 J) and nugget pull-out (1750 J, 2000 J), while the dissimilar joint exhibited only interfacial failure mode (250 J, 500 J, 750J, 1000 J, 1500J). In the case of similar welding two phenomena of (1) interfacial bonding strength and (2) stress concentration at the nugget edge played a role, while in the dissimilar welding, the brittleness of interface layer was responsible for the failure. Nugget pull-out was associated with the significant plastic deformation at the edge of the nugget under even a moderate clamping pressure at high temperatures, which led to a higher stress concentration at the nugget edge. Evidence from the fracture surfaces (Fig. 6) confirmed that microbonding first occurred in the areas with a higher pressure. Similarly, the interface diffusion layer thickness was larger at the nugget edge due to the relatively high diffusion during dissimilar joining of ZEK100 to Al5754.

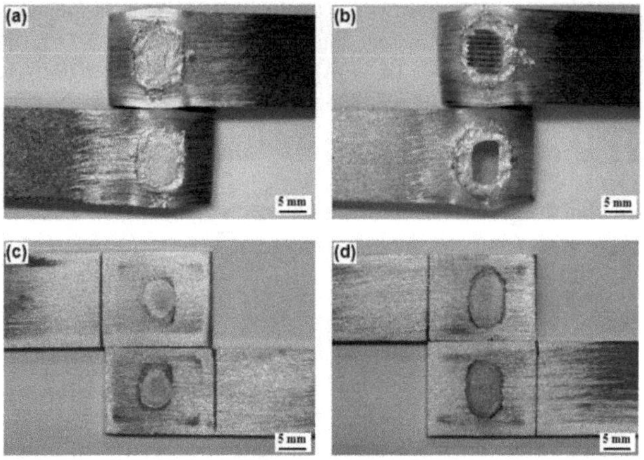

Fig. 6 Macroscopic images of the failed tensile lap shear samples of ZEK100-ZEK100 at a welding energy of (a) 1500 J and (b) 2000 J, and ZEK100-Al5754 at a welding energy of (c) 500 J and (d) 1500 J.

Conclusions

High power USW was used to join a low rare-earth containing ZEK100 Mg alloy at different levels of welding energy in both similar (ZEK100-ZEK100) and dissimilar (ZEK100-Al5754) configurations. The welding energy had a significant effect on the interface microstructure. The interface grain size increased via dynamic recrystallization in the similar welding, while the formation and thickening of interface diffusion layer containing brittle IMC $Al_{12}Mg_{17}$ occurred in the dissimilar joint with increasing welding energy. The tensile lap shear strength first increased, reached a maximum, and then decreased with increasing welding energy. About 78% of ZEK100-ZEK100 joint strength and ~55% of Al5754-Al5754 joint strength were achieved in the ZEK100-Al5754 dissimilar joint via USW. Two different modes of interfacial failure and nugget pull-out were observed in the similar joints, while only interfacial failure was present in the dissimilar joints.

Acknowledgements

The authors would like to thank the Natural Sciences and Engineering Research Council of Canada (NSERC), Premier's Research Excellence Award, NSERC-DAS Award, AUTO21 Network of Centers of Excellence, and Ryerson Research Chair program for providing financial support. The authors would like to thank Dr. T. Skszek and Dr. X.P. Niu (Magna International Inc.) Professor M. Worswick (University of Waterloo), and Professor Alan A. Luo (Ohio State University) for supplying test materials. The assistance of Q. Li, A. Machin, J. Amankrah, and R. Churaman in performing the experiments is gratefully acknowledged.

References

1. A.I. Cooper, "Cooperative carbon capture," *Nature*, 519 (7543) (2015), 294-5.
2. M. Jakob and J. Hilaire, "Unburnable fossil-fuel reserves," *Nature*, 517(7533) (2015), 150-2.
3. F.A. Mirza, D.L. Chen, D.J. Li, and X.Q. Zeng, "Low cycle fatigue of a rare-earth containing extruded magnesium alloy," *Mater. Sci. Eng. A*, 575 (2013), 65-73.
4. J. Bohlen, M.R. Nürnberg, J.W. Senn, D. Letzig, S.R. Agnew, "The texture and anisotropy of magnesium–zinc–rare earth alloy sheets," *Acta Mater.*, 55(6) (2007), 2101-12.
5. F.A. Mirza and D.L. Chen, "Fatigue of magnesium alloys," in Aerospace Materials Handbook, ed. S. Zhang and D.L. Zhao (CRC Press/Taylor & Francis Group, Boca Raton, FL, 2013), 647-98.
6. A. Macwan, X.Q. Jiang, C. Li, and D.L. Chen, "Effect of annealing on interface microstructures and tensile properties of rolled Al/Mg/Al tri-layer clad sheets," *Mater. Sci. Eng. A*, 587 (2013), 344-51.
7. M.C. Zhao, M. Liu, G. Song, and A. Atrens: Corros, "Influence of the β-phase morphology on the corrosion of the Mg Alloy AZ91," *Corros. Sci.*, 50(7) (2008), 1939-53.
8. C. Liu, D.L. Chen, S. Bhole, X. Cao, and M. Jahazi, "Polishing-assisted galvanic corrosion in the dissimilar friction stir welded joint of AZ31 magnesium alloy to 2024 aluminum alloy," *Mater. Charact.*, 60(5) (2009), 370-6.
9. A. Macwan, X.Q. Jiang, and D.L. Chen, "Interfacial characterization of dissimilar joints between Al/Mg/Al tri-layered clad sheet to high-strength low-alloy steel," *JOM*, 67(7) (2015), 1468-77.
10. F. Mokdad and D.L. Chen, "Strain-controlled low cycle fatigue properties of a rare-earth containing ZEK100 magnesium alloy," *Mater. Des.*, 67 (2015), 436-47.
11. J.B. Jordon, J.B. Gibson, M.F. Horstemeyer, H. El Kadiri, A.A. Luo, "Effect of twinning, slip, and inclusions on the fatigue anisotropy of extrusion-textured AZ61 magnesium alloy," *Mater. Sci. Eng. A*, 528(22-23) (2011), 6860-71.
12. J.E. Gray and B. Luan, "Protective coatings on magnesium and its alloys - a critical review," *J. Alloy Compd.*, 336 (2002), 88-113.
13. M. Yamasaki, N. Hayashi, S. Izumi, and Y. Kawamura, "Corrosion behavior of rapidly solidified Mg-Zn-rare earth element alloys in NaCl solution," *Corros. Sci.*, 49 (2007), 255-62.

14. H.M. Rao, R.I. Rodriguez, J.B. Jordon, M.E. Barkey, Y.B. Guo, H. Badarinarayan, and W. Yuan, "Friction stir spot welding of rare-earth containing ZEK100 magnesium alloy sheets," *Mater. Des.*, 56 (2014), 750-4.
15. J. Min and J. Lin, Anelastic behaviour and phenomenological modeling of Mg ZEK100-O alloy sheet under cyclic tensile loading-unloading," *Mater. Sci. Eng. A*, 561 (2013), 174-82.
16. V.K. Patel, S.D. Bhole, and D.L. Chen, "Microstructure and mechanical properties of dissimilar welded Mg-Al joints by ultrasonic spot welding technique," *Sci. Technol. Weld. Join.*, 17(3) (2012), 202-6.
17. A. Panteli, Y.C. Chen, D. Strong, X. Zhang, and P.B. Prangnell, "Optimization of aluminum-to-magnesium ultrasonic spot welding," *JOM*, 64(3) (2012), 414-20.
18. V.K. Patel, S.D. Bhole, and D.L. Chen, "Influence of ultrasonic spot welding on microstructure in a magnesium alloy," *Scr. Mater.*, 65(10) (2011), 911-4.
19. A. Macwan and D.L. Chen, "Microstructure and mechanical properties of ultrasonic spot welded copper-to-magnesium alloy joints," *Mater. Des.*, 84 (2015), 261-9.
20. J.L. Murray, "The Al-Mg (Aluminum-Magnesium) system," *Bull. Alloy Phase Diagr.*, 3 (1982), 60-74.

Magnesium Technology 2016
Edited by: Alok Singh, Kiran Solanki, Michele V. Manuel, and Neale R. Neelameggham
TMS (The Minerals, Metals & Materials Society), 2016

Effect of Filler Wires on Cracking along Edges of Magnesium Welds

Tao Yuan[1], Xiao Chai[2] and Sindo Kou[3]
[1]School of Materials Science and Engineering, Tianjin University, Tianjin, China 300072
[2]Novelis Global Research and Technology Center, 195 Vaughn Road, Kennesaw, GA 30144
[3]Department of Materials Science and Engineering, University of Wisconsin, Madison, WI 53706

Keywords: Welding, Mg Alloys, Liquation Cracking, Partially Melted Zone, Scheil Equation

Abstract

Since Mg alloys often have a relatively low eutectic temperature T_E, they are susceptible to intergranular liquid formation and hence cracking in the region of the workpiece along the weld edge, where temperature exceeds T_E during welding. To evaluate the effect of the filler wires on weld-edge cracking in butt welding of AZ91 Mg alloy, the circular-patch weld test was used. An AZ91 Mg circular patch was fitted inside a circular hole in an AZ91 Mg workpiece and butt welded to it. Mg filler wires AZ31 Mg and AZ92 Mg were used for welding. Temperature (T) was plotted against fraction solid (f_S) for the workpiece and the weld. It was found that the susceptibility to cracking can be determined by comparing these two T-f_S curves.

Introduction

Mg alloys are very similar to Al alloys, many of which have been reported to be susceptible to liquation [1-5] and liquation cracking [6-9]. Eutectic particles can exist in Mg alloys, either as composite-like α–β eutectic particles or divorced β eutectic particles, where α is the Mg-rich phase and β is $Mg_{17}Al_{12}$. During welding of Mg alloys, the region of the workpiece next to the fusion zone (the weld bead) is heated to above the eutectic temperature T_E. Due to rapid heating in welding, these particles do not have enough time to dissolve completely in the matrix by solid-state diffusion. Thus, upon reaching T_E the remaining α–β eutectic particles can melt or the eutectic reaction $\alpha + \beta \rightarrow L$ can occur to form liquid L. This liquid formation is called liquation, and the region of the workpiece in which liquation occurs during welding is called the partially melted zone (PMZ) [1]. The mechanism of the remaining intermetallic particles reacting with α to form liquid was called "constitutional liquation" by Pepe and Savage, who first reported the phenomena [10, 11]. The liquid can form along grain boundaries (and within grains as well) to weaken bonding between grains in the PMZ.

Upon solidification the weld pool contracts and pulls the PMZ. This induces tension in the PMZ and cracking can occur along the liquated grain boundaries in the PMZ. This is called liquation cracking, and it occurs along the weld edge in the PMZ. Mg alloys are susceptible to liquation cracking because they tend to have a rather low T_E, e.g., 437°C in binary Mg-Al alloys and 340°C for binary Mg-Zn alloys. Furthermore, when a substantial amount of β-$Mg_{17}Al_{12}$ is present, a wide PMZ can form with much liquid separating grains. In fact, liquation and liquation cracking were encountered in friction stir welding (FSW), either spot welding of Mg alloys [12] or seam welding of Mg alloys [13] or seam welding of Mg alloys to Al alloys [14-16] even though FSW is considered as a solid-state welding process. Recently, the susceptibility of Mg alloys to liquation cracking was evaluated by bead-on-plate circular welding [17]. The susceptibility to liquation cracking in butt welding of dissimilar Mg alloys was evaluated by circular-patch welding [18]. In the present study circular-patch welding was used to study the effect of the filler wire on butt welding of Mg alloys.

Experimental Procedure

The compositions of the workpiece material and filler wires are listed in Table 1 [19, 20]. The circular-patch welding test was used to evaluate the susceptibility of Mg welds to cracking along the weld edge. The specimen consisted of two pieces, that is, the workpiece as the outer piece and the circular patch as the inner piece as shown in Fig. 1. The specimen was 3.2 mm thick but only 1.6 mm thick in the patch and the surrounding area in the workpiece [18]. Welding was conducted by gas-metal arc welding (GMAW), which is also called metal-inert gas (MIG) welding. A process controller called controlled short circuiting (CSC) [21] was used to suppress spatter, which can easily occur in MIG welding of Mg alloys [22, 23].

The welds were cut in the transverse direction, polished and etched in order to determine the volumes of the workpiece, patch and filler in the weld. They were then first etched for 3 s in a solution of 3 ml HCl (concentrated) + 100 ml ethanol and then etched for another 3 s in a solution of 5 ml acetic acid + 5 g picric acid + 10 ml distilled water +100 ml ethanol [17]. After etching, transverse macrographs of the welds were taken using a digital camera with a close-up lens and bellows. The composition of the weld was determined from the volumes of the workpiece, patch and filler in the weld and their compositions following the procedure described previously [17].

Results and Discussion

Before discussing the experimental results, the criterion proposed by Kou and co-workers [24, 25] for the susceptibility of Al welds to liquation cracking is described as follows. The interaction between the mushy zone weld and the partially melted zone during welding is illustrated in Fig. 2. Based on the phase diagram (Fig. 2a) and quenching of the weld pool and its surrounding area during welding, Kou and Le [26] established the semisolid microstructure around the weld pool (Fig. 2b). Based on this microstructure, Kou and coworkers [24, 25] proposed the criterion illustrated in Fig. 2c and d. The reasoning here is that, if PMZ develops strength later than mushy zone does, cracking can occur in the former. The criterion for the susceptibility to cracking along the weld edge can be written as follows:

$$\text{Susceptible if } \textbf{\textit{weld}}\ \boldsymbol{f_S} > \textbf{\textit{workpiece}}\ \boldsymbol{f_S} \qquad (1)$$

The susceptibility to liquation cracking is significant when the weld f_S is significantly higher than the workpiece f_S, for instance, by 0.05. Eqn. (1) applies after some strength has developed in the semisolid, for instance, fraction solid $f_S > 0.3$. According to Flemings [27], significant strength develops during solidification after about $f_S > 0.3$.

Fig. 3 shows the weld made with AZ91 Mg as the workpiece and the patch and AZ31 Mg as the filler wire. As shown in Fig. 3a, cracking occurs along the outer edge of the weld. As shown in Table 1, the dilution is 39.37%, that is, the filler wire is diluted by the workpiece and the patch such that the weld consists of 39.37% of AZ91 Mg and 60.63% of the filler wire AZ31 Mg. Thus, the weld composition is Mg-5.36Al-0.88Zn-0.44Mn. Based on their compositions, the T-f_S curves were calculated for the workpiece and the weld using the thermodynamic software Pandat [28] and Mg database PanMagnesium [29] of CompuTherm, LLC in Madison, WI. As shown in Fig. 3b, the weld f_S is significantly greater than the workpiece f_S up to about $f_S = 0.92$ as indicated by the region encircled by the dotted oval. The curves suggest a significant susceptibility to cracking. Beyond 0.92 the weld f_S becomes slightly lower than the workpiece f_S, but this is too late to keep cracking from occurring. Thus, the crack susceptibility predicted by the T-f_S curves in Fig. 3b is consistent with the cracking along the outer edge in Fig. 3a.

Fig. 4 shows the weld made with AZ91 Mg as the workpiece and the patch and AZ92 Mg as the filler wire. As shown in Fig. 4a, no cracking occurs along the outer edge of the weld. As shown in Table 1, the dilution is 39.14%, that is, the weld consists of 39.14% of AZ91 Mg and 60.86% of the AZ92 Mg filler wire. Thus, the weld composition is Mg-9.0Al-1.49Zn-0.26Mn. As shown in Fig. 4b, the weld f_S is slightly less than the workpiece f_S as indicated by the region encircled by the dotted oval. The curves suggest the absence of a significant susceptibility to cracking. Thus, absence of crack susceptibility predicted by the T-f_S curves in Fig. 4b is consistent with the absence of weld-edge cracking in Fig. 4a.

Conclusions

1. Weld-edge cracking can occur in butt welding of AZ91 Mg, depending on the filler wire and dilution used.
2. The susceptibility to weld-edge cracking in AZ91 Mg can be predicted by comparing the T-f_S curves of the workpiece and the weld.

Acknowledgements

This work was supported by the National Science Foundation under Grant No. IIP-1034695, the American Welding Society Foundation (the AWS Foundation Fellowship Program for Xiao Chai), and the University of Wisconsin Foundation through the Industry/University Collaborative Research Center (I/UCRC) for Integrated Materials Joining Science for Energy Applications. The authors would like to thank: 1) CompuTherm, LLC in Madison, WI and Professor Rainer Schmid-Fetzer, Clausthal University of Technology at Clausthal-Zellerfeld, Germany for the software package Pandat and the database PanMagnesium; 2) Tom Kurilich of US Magnesium, LLC in Salt Lake City, Utah for donating 45 Kg (100 lbs) of pure Mg ingots.

Table 1 Nominal compositions of materials used for welding [19, 20]

	Al (wt %)	Zn (wt%)	Mn (wt %)	Mg (wt %)
Workpiece, patch				
AZ91	9.0	0.7	0.2	balance
Filler wire				
AZ31	3.0	1.0	0.6	balance
AZ92	9.0	2.0	0.3	balance
Weld				
With AZ31 filler (39.37 % Dilution)	5.36	0.88	0.44	balance
With AZ92 filler (39.14 % Dilution)	9.0	1.49	0.26	balance

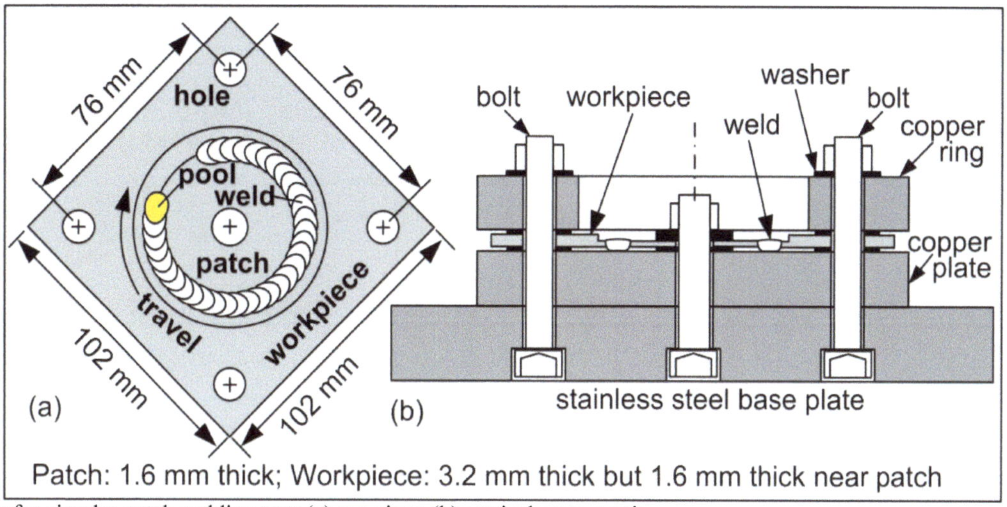

Fig. 1 Apparatus for circular-patch welding test: (a) top view; (b) vertical cross-section.

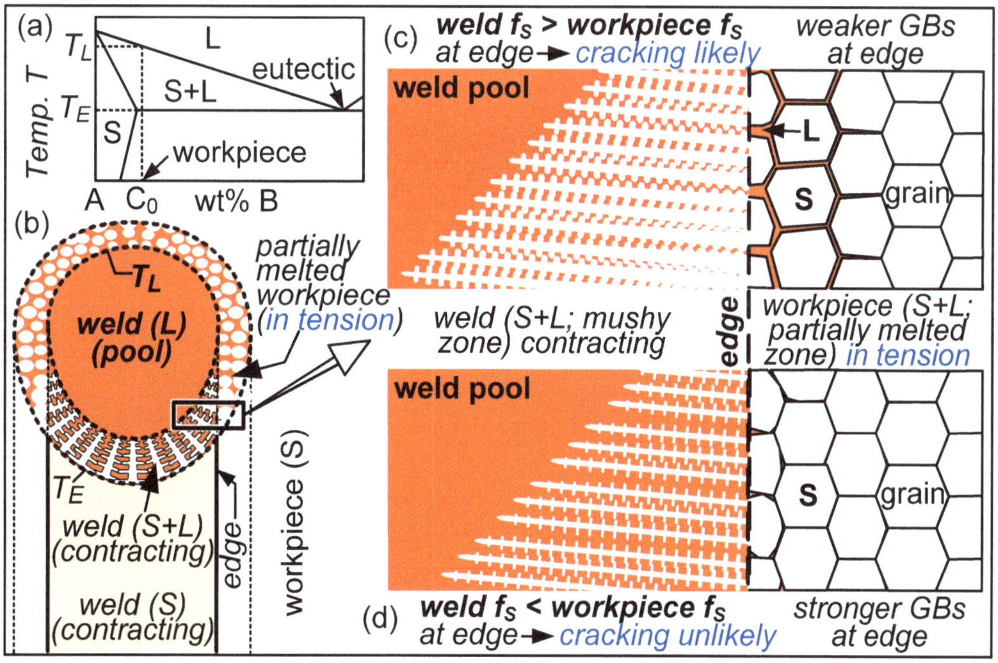

Fig. 2 Susceptibility to cracking along weld edge: (a) phase diagram showing alloy C_o (workpiece); (b) microstructure around weld pool in welding of alloy C_o; (c) cracking likely in welding an alloy; (d) cracking unlikely in welding another alloy with another filler wire.

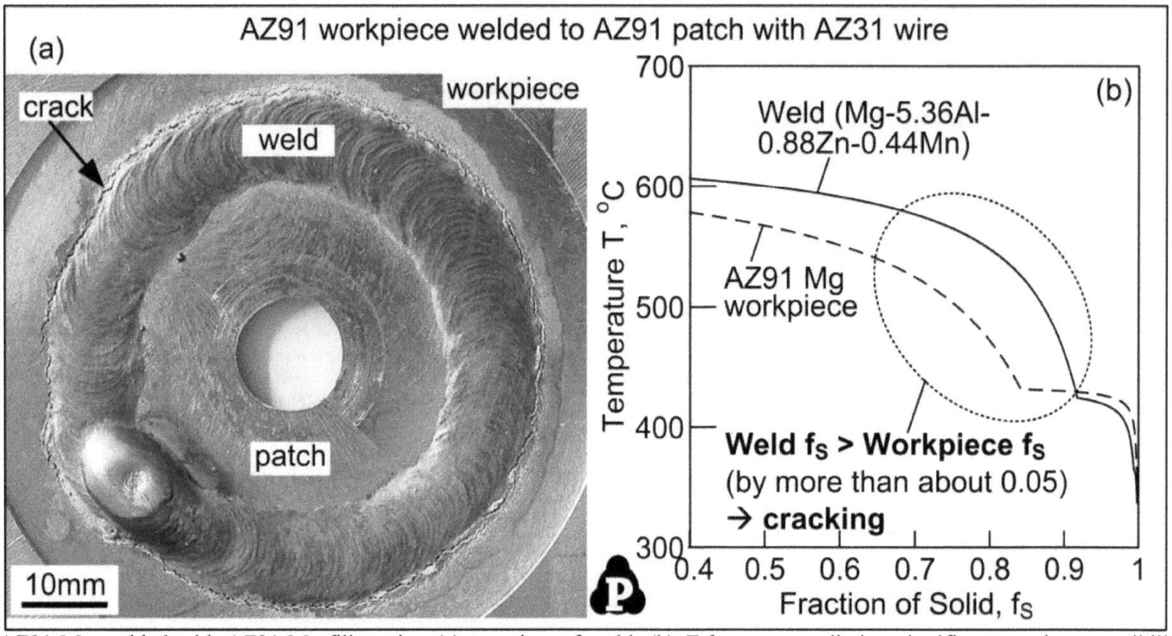

Fig. 3 AZ91 Mg welded with AZ31 Mg filler wire: (a) top view of weld; (b) T-f_S curves predicting significant crack susceptibility. T-f_S curves were calculated using Pandat [28] and PanMgnesium [29] of CompuTherm LLC.

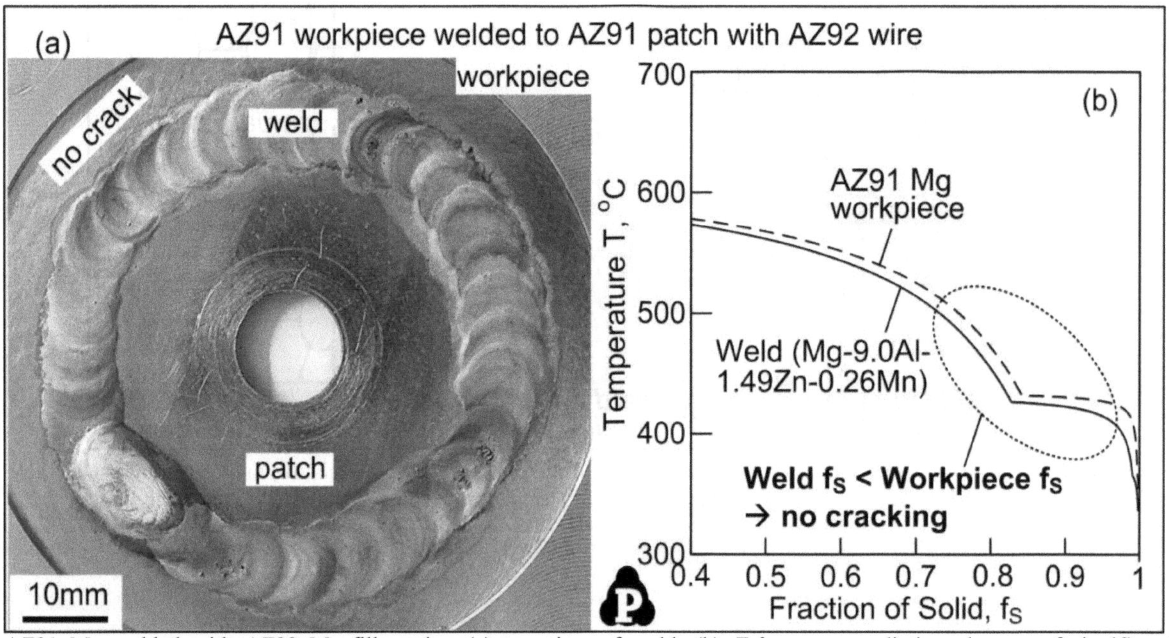

Fig. 4 AZ91 Mg welded with AZ92 Mg filler wire: (a) top view of weld; (b) T-f_S curves predicting absence of significant crack susceptibility. T-f_S curves were calculated using Pandat [28] and PanMgnesium [29] of CompuTherm LLC.

References

1. S. Kou, *Welding Metallurgy, 2nd edition* (Hoboken, NJ: John Wiley & Sons, 2003), 122-141, 97-121, and 301-340.
2. C. Huang, and S. Kou, "Partially Melted Zone in Aluminum Welds - Liquation Mechanism and Directional Solidification," *Welding Journal*, 79 (2000), 113s-120s.
3. C. Huang, and S. Kou, "Partially Melted Zone in Aluminum Welds - Planar and Celluar Solidification," *Welding Journal*, 80 (2001), 46s-53s.
4. C. Huang, and S. Kou, "Partially Melted Zone in Aluminum Welds - Solute Segregation and Mechanical Behavior," *Welding Journal*, 80 (2001), 9s-17s.
5. C. Huang and S. Kou, "Liquation Mechanisms in Multicomponent Aluminum Alloys during Welding," *Welding Journal*, 81 (2002), 211s to 222s.
6. G. Cao and S. Kou, "Liquation Cracking in Full-Penetration Al-Si Welds," *Welding Journal*, 84 (2005), 63s-71s.
7. C. Huang and S. Kou, "Liquation Cracking in Full-Penetration Al-Cu Welds," *Welding Journal*, 83 (2004), 50s-58s.
8. C. Huang and S. Kou, "Liquation Cracking in Full-Penetration Al-Mg-Si Welds," *Welding Journal*, 83 (2004), 111s-122s.
9. C. Huang, G. Cao and S. Kou, "Liquation Cracking in Partial-Penetration Aluminum Welds: Assessing Tendencies to Liquate, Crack and Backfill," *Science and Technology of Welding and Joining*, 9 (2004), 1-9.
10. J. J. Pepe and W. F. Savage, "Effects of Constitutional Liquation in 18-Ni Managing Steel Weldments," *Welding Journal*, 46 (1967), 411s-422s.
11. J. J. Pepe and W. F. Savage, "Weld Heat-affected Zone of The 18Ni Maraging Steels," *Welding Journal*, 49 (1970), 545s-553s.
12. Y. K. Yang et al., "Liquation of Mg Alloys in Friction-Stir Spot Welding," *Welding Journal*, 87 (2008), 167s-177s.
13. D. C. Wagner et al., "Liquation Cracking in Arc and Friction-Stir Welding of Mg-Zn Alloys", *Metallurgical and Materials Transactions A*, 46 (2015), 315-327.
14. V. Firouzdor and S. Kou, "Al-to-Mg Friction Stir Butt Welding: Effect of Material Position, Travel Speed and Rotation Speed," *Metall. and Mater. Trans.*, 41A (11) (2010), 2914-2935.
15. V. Firouzdor and S. Kou, "Formation of Liquid and Intermetallics in Al-to-Mg Friction Stir Welding," *Metallurgical and Materials Transactions A*, 41 (12) (2010), 3238-3251.
16. V. Firouzdor and S. Kou, "Al-to-Mg Friction Stir Welding: Effect of Positions of Al and Mg with Respect to the Welding Tool," *Welding Journal*, 88 (2009), 213s-224s.
17. X. Chai, T. Yuan and S. Kou, "Liquation and Liquation Cracking in Partially Melted Zones of Mg Welds," *Welding Journal*, accepted in July 2015.
18. T. Yuan, X. Chai, Z. Luo and S. Kou, "Predicting susceptibility of magnesium alloys to weld-edge cracking," *Acta Materialia*, 90 (2015), 242–251.
19. Magnesium Alloys, Products Literature, Washington Alloys, Rancho Cucamonga, California 91730, http://www.weldingwire.com/Images/Interior/documentlibrary/magnesium%20alloys.pdf.
20. Electron AZ31B Sheet, Plate & Coil, Data Sheet 482, Magnesium Electron, Manchester, England.
21. CSC-MIG Weld Process Controller, Jetline Engineering, Irvine, CA 92618, http://pdf.directindustry.com/pdf/jet-line-engineering/csc-mig-weld-process-controller/27577-54392.html.
22. D.C. Wagner, Y.K. Yang, S. Kou, "Spatter and Porosity in Gas-Metal Arc Welding of Magnesium Alloys: Mechanisms and Elimination," *Welding Journal* 92 (2013), 347s-362s.
23. X. Chai, Y.K. Yang, B.E. Carlson, S. Kou, "Gas Metal Arc Welding of Magnesium Alloys: Oxide Films, High Crowns, and Fingers," *Welding Journal* 94 (2015), 16s-33s.
24. G. Cao and S. Kou, "Predicting and Reducing Liquation-Cracking Susceptibility Based on Temperature vs. Fraction Solid," *Welding Journal*, 85 (2006), 9s to 18s.
25. S. Kou, "Fluid Flow and Solidification in Welding: Three Decades of Fundamental Research at the University of Wisconsin", *Welding Journal*, 91 (2012), 287s-302s.
26. S. Kou, and Y. Le, "The Effect of Quenching on the Solidification Structure and Transformation Behavior of Stainless Steel Welds," *Metall. Trans.* 13A (1982), 1141.
27. M. C. Flemings, *Solidification Processing* (New York, NY: McGraw-Hill Book Co., 1974), Appendix B.
28. Pandat, Phase Diagram Calculation Software Package for Multicomponent Systems, Computherm LLC, Madison, WI, 2013.
29. PanMg, Thermodynamic Database for Mg Alloys, Computherm LLC, Madison, WI, 2013.

Magnesium Technology 2016

Magnesium-Rare Earth Alloys

Magnesium Technology 2016
Edited by: Alok Singh, Kiran Solanki, Michele V. Manuel, and Neale R. Neelameggham
TMS (The Minerals, Metals & Materials Society), 2016

HOT TEARING IN MAGNESIUM-RARE EARTH ALLOYS

Mark Easton[1], Serge Gavras[2], Mark Gibson[1,2,3], Suming Zhu[1], Jian-Feng Nie[2], Trevor Abbott[1,4]

[1]School of Aerospace, Mechanical and Manufacturing Engineering, RMIT University, Carlton, VIC 3053, Australia
[2]Department of Materials Science and Engineering, Monash University, Clayton, VIC 3800, Australia
[3]CSIRO Manufacturing Flagship, Clayton, VIC 3168, Australia
[4]Magontec Ltd, St Potts Point NSW 2011, Australia

Keywords: Magnesium alloys, hot tearing, solidification processing, casting

Abstract

Magnesium-rare earth based alloys generally show good creep resistance at elevated temperatures and are consequently very promising candidates for powertrain applications. However, for magnesium-rare earth alloys to be used in high pressure die-cast components, they also need to have sufficient die-castability. This paper analyses the data from a series of investigations into binary and ternary magnesium-rare earth (RE) based alloys with the REs including La, Ce, Nd, Y and Gd, with an aim to identify alloy compositions that are castable. It is found that Mg-La based alloys are least susceptible to hot tearing whilst alloys containing Nd, Y or Gd tend to have very high hot tearing susceptibility. The hot tearing susceptibility of Mg-RE-Zn and Mg-Al-RE alloys is also addressed.

Introduction

Improving the properties of high-pressure die-cast (HPDC) alloys at elevated temperatures has been a major focus of research with the aim of extending the use of such alloys to hotter parts of the engine [1-4]. Alloys containing rare earths (REs) as the primary alloying element are some of the most creep resistant alloys, and the addition of REs to Mg-Al alloys also leads to substantial improvements in elevated temperature properties [2, 5].

The most common RE elements found in misch-metal are Ce, La and Nd with Ce and La being relatively cheap (~3 times the cost of Mg) and have consequently been the major focus of the current authors [6-8]. Y and Gd are also common alloying elements in ultra-high performance Mg-based alloys [9] because of their exceptional precipitation hardening properties [10, 11], particularly in combination with other alloying elements such as Zn [12, 13].

One of the most difficult aspects of alloy design is to be able to combine the triage of performance, processability and cost. This is particularly difficult for Mg-RE based alloys. It has been found that whilst different REs have similar effects on the as-cast strength and ductility of the alloys, they have widely different effects on the creep resistance, precipitation hardening and castability. Elements such as Nd, Gd and Y that tend to have the highest mechanical properties often have problems with castability/hot tearing [14]. Whilst alloys with La tend not to hot tear [8] but have the lowest creep resistance of all the alloys [7].

The purpose of this paper is to bring together a range of published and unpublished work on the effect of different RE elements on hot tearing resistance to understand their effect on one another. Also, the role of other elements in mitigating or not the hot tearing resistance in Mg-RE systems will be discussed. The reason for choosing hot tearing resistance as a measurement of castability is that it can be easily quantified. Other approaches to a more complete understanding of castability have been attempted elsewhere but require substantial campaigns [15, 16].

Evaluation of Hot Tearing

Substantial effort has gone into the measurement of hot tearing. The most simple tests involve a dog-bone sample with constrained ends [17, 18]. This has been further developed into castings with multiple done bones of different lengths relating to different magnitudes of shrinkage stress and consequent tear propensities [19]. A number of researchers have developed hot tearing tests further by instrumenting a constrained bar to measure forces developed during solidification or contraction [19-22] or performing in-situ strain measurements using neutron diffraction [23].

The current authors have found that the casting into tensile samples in a high-pressure die-casting machine provided a simple way to assess the hot tearing resistance of alloys. This had two distinct advantages: that hot tearing was assessed on alloys under processing conditions and with similar microstructures to production as it is known that processing conditions are critically important even to phase selection in these alloys [24, 25], and that a large number of samples can be made easily which gives some statistical confidence on the assessment.

The rating scale developed to assess the hot tearing susceptibility (HTS) of the castings was [8]:
0 = No cracks/dimples in the sample observed by the naked eye.
1 = Small observable dimple/s that require close examination to see.
2 = Dimple/s easily seen by the naked eye.
3 = Crack/s observed on the surface of the sample, but the sample was not broken.
4 = Sample broken.
Examples of these ratings can be seen in Figure 1 and at least 30 samples were assessed for each alloy. The average of these ratings is known as the hot tearing index (HTI).

The other area of interest is to be able to predict hot tearing using a criteria or a model. There have been quite comprehensive models of hot tearing based on the physics of solidification and the materials properties [26, 27]. However, there is also a strong history of developing models based on relatively simple criteria such as the Clyne-Davies model [25], which continues to be used to explain the effect of composition variations on hot tear susceptibility [28, 29]. This model predicts that the hot tear susceptibility is related to the 'time' which the alloy spends where shrinkage stress cannot be relaxed between fraction solids of 0.9 and 0.99 compared with the time spent in the region where stresses can be relaxed by interdenditic feeding between 0.4 and 0.9 fraction solid. Whilst it can predict the alloys effects on hot tearing it does not work so well in predicting the effect of processing parameters [30]. Most recently, a parameter that considers the gradient of the T-(fs)$^{1/2}$ has further been proposed in a similar vein and looks like a promising approach [31, 32].

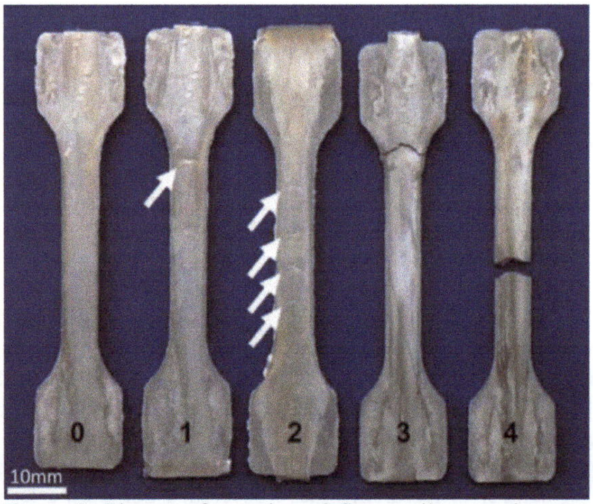

Figure 1. Picture of a series of flat tensile test pieces each of which is representative of one of the categories in the hot tearing scale defined above (the reference number corresponding to the scale is indicated at the bottom of each). The white arrows indicate the surface dimples that are associated with internal cracking [8].

Some of the current authors [8] have used a similar indicator (S_{ht2}) of hot tearing susceptibility from the integral of the temperature-fraction solid curve from the coherency (the temperature when feeding becomes interdendritic) to the coalescence (the temperature at which the metal has solidified sufficiently to act as a solid) points (Eqn. 1),

$$S_{ht2} = \int_{T_0}^{T_{co}} f_s(T)\, dT \qquad (1)$$

where $f_s(T)$ is the fraction solid at temperature T_0 is the temperature at which coherency occurs, T_{co} is the temperature at which coalescence has been achieved. A sensitivity analysis was performed and fit was found choosing T_o at a fraction solid of 0.7 and T_{co} at 0.98 were appropriate. This will also be used in this paper.

Typically, alloys have a Λ-curve with alloy composition where the pure metal shows little hot tearing there is an increase to peak hot tear susceptibility and then a reduction. Whilst the models described above predict this, it has been known for a long time that solidification path models do not incorporate all the factors that affect hot tearing. For example, systems with low temperature eutectics usually, but do not always, have higher hot tearing susceptibilities [18]. Furthermore it has been recently realized that the morphology and the eutectic intermetallic phases have a large effect on the hot tearing susceptibility of alloys [33]. These simple criteria (and even the complicated models) currently do not adequately incorporate the effect of eutectic morphology in the hot tearing susceptibility and this should be a key aspect of future research.

It should also be noted that whilst the hot tearing indicators discussed here do not include the substantial effect that the process parameters have on the hot tear susceptibility [27, 34]. This is of particular concern in high-pressure die-casting where solidification conditions are highly dynamic.

Whilst having this in mind, the predictions of such simple criteria can provide substantial insight into the mechanisms operating in hot tearing. They can predict the effect of solidification path, which can explain many results or point to the importance of eutectic morphology when that has a greater influence. Hence the combination of measurement and predictions using criteria are of considerable use in understanding hot tearing.

Properties of Mg-RE Systems

To begin to understand the differences in the properties of different RE elements in Mg, a basic understanding of their phase diagrams is required (Table **I**).

Table I. Important features obtained from the Mg binary phase diagrams [35]. Data for Mg-Al and Mg-Zn is provided for comparison.

Rare Earth	Maximum Solid Solubility, c_{ss} (wt.%)	Eutectic Composition, c_{eut} (wt.%)	Eutectic Temperature (°C)
Al	12.6	33.3	437
Zn	6.2	51.3	340
Y	12.02	26.3	565
La	0.23	16.5	612
Ce	0.74	20.5	590
Nd	3.6	33.0	552
Gd	23.5	40.4	542

Mg-La alloys have the shortest solidification ranges, with the highest eutectic temperature. Mg-Ce has a lower eutectic temperature whilst Mg-Nd, Mg-Y and Mg-Gd all have eutectic temperatures approximately 100°C below the melting point of Mg.

The Mg-Nd, Mg-Y and Mg-Gd alloys have lower eutectic temperatures and have greater solid solubilities and significant positive slopes on their solvus lines. Consequently, these are more able to precipitation harden and consequently have better high temperature properties than Mg-La and Mg-Ce alloys [7].

Hot Tearing of Binary and Ternary Mg-RE Alloys

The hot tearing behavior of Mg-La, Mg-Ce, and Mg-Nd alloys has been reported recently (Figure **2**) [8]. The samples were cast using a 250 tonne Toshiba cold chamber HPDC machine, with a melt temperature of 740°C. The ram velocity reached a maximum of 2.25 m/s, with a maximum pressure of almost 120 MPa and a biscuit size of 40 mm. These show almost no hot tearing susceptibility for Mg-La alloys across a wide range of alloy concentrations. Mg-Ce alloys showed some hot tear susceptibility with a Λ-curve peak at around 0.5-1.0% Ce. The Mg-Nd alloys showed a very high HTS at concentrations of approximately 1wt% and above and it was only at greater than 7wt% Nd that the susceptibility decreased again. The hot tearing susceptibility of Mg-Gd and Mg-Y has been investigated by others [36, 37] and was also found to be high.

Given the critical importance of castability in developing a commercially useful alloy, it was decided to investigate using Mg-La alloys as a base, under the same casting conditions as the binary alloys. Whilst the creep properties of Mg-La alloys are not as good as other RE elements [7], they are substantially better than Mg-Al alloys and provide a similar castable base alloy. Then alloying elements that were likely to provide improvements to properties, i.e. Nd, Y, and Gd were added [14].

Currently of these systems, the only one with available thermodynamic database is the Mg-La-Nd system [38] and consequently the analysis will focus on this system (Figure **3**). There are two key solidification reactions:

1. a monovariant Mg(l)->Mg(s)+$Mg_{12}RE$ where La and Nd are in solid solution as RE at 590-610°C, and
2. Mg(l)->Mg(s)+$Mg_{41}RE_5$, where there is limited solubility of La in the intermetallic phase [38, 39].

In actuality, an eutectic containing $Mg_{12}RE$ is observed with little evidence of $Mg_{41}RE_5$ [14], so it may be that the final reaction is not reached.

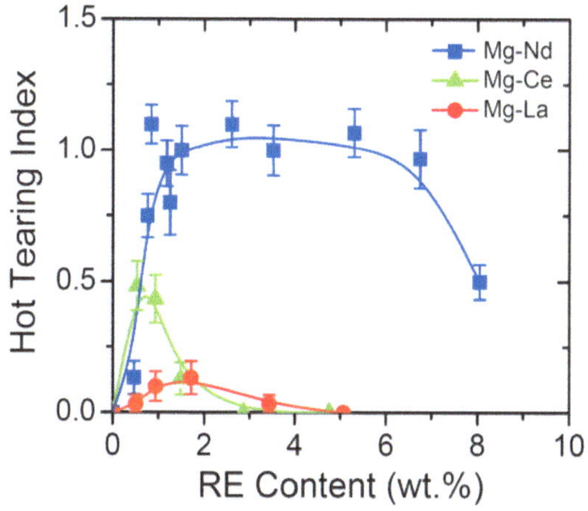

(a)

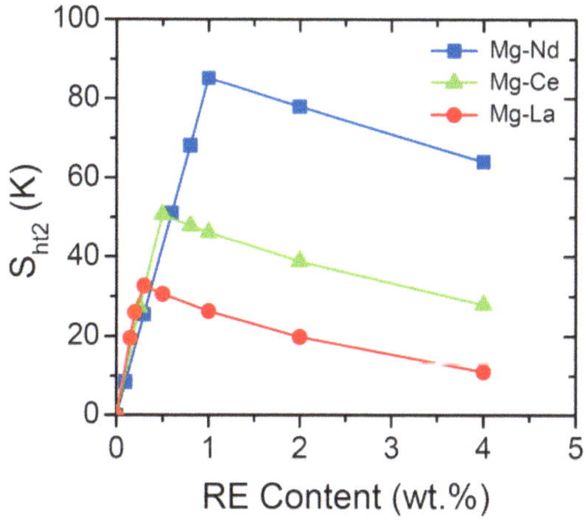

(b)

Figure 2. (a) Hot tearing index of binary Mg-La, Mg-Ce and Mg-Nd alloys. (b) Predictions of hot tearing susceptibility using Equation 1.

There are two key results that can be observed here (Figure **3**(b)): one is that the peak HTI is lower than in the binary Mg-Nd alloys (Figure **2**(a), and the other is that the increase to the peak may be less rapid. Hence it does appear that using a Mg-La base alloy is reducing the severity of the hot tearing in Mg-Nd based alloys, potentially allowing for additions to be made to improve the properties of the alloys.

Gd and Y were found to behave similarly to Nd with substantial HTS. Additions increased the HTI at lower addition levels but also the HTI reduced more rapidly at higher concentrations (Figure **4**). Hence it seems that the most likely option (not considering properties) for a castable high performance alloy is to add small additions of Nd to an Mg-La base alloy.

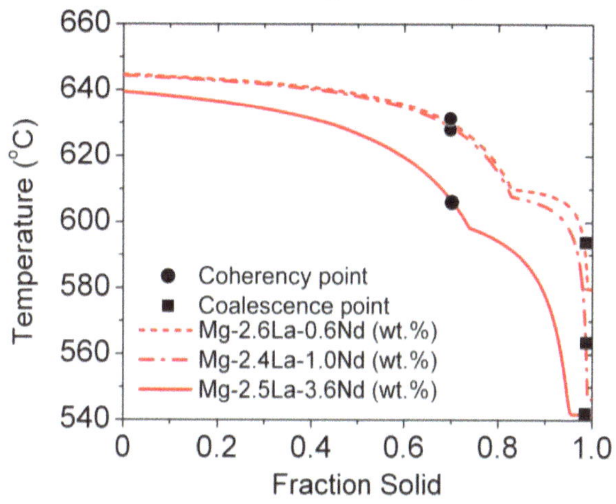

(a)

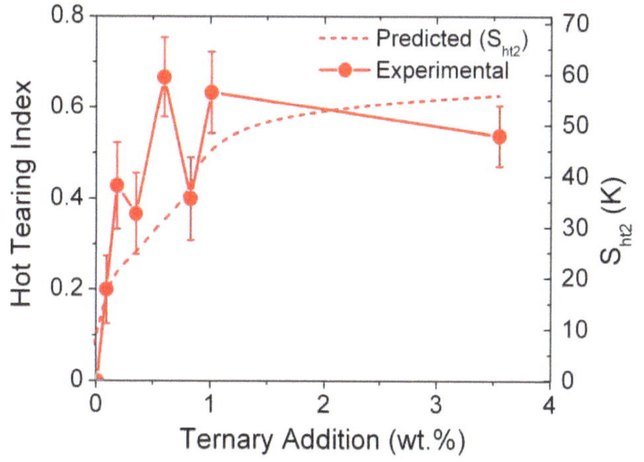

(b)

Figure 3. (a) Predictions of the solidification path of Mg-La-Nd alloys with increasing Nd content using Pandat. (b) Associated hot tearing susceptibility observations and predictions for Mg-2.5La-Nd alloys. Data from [14].

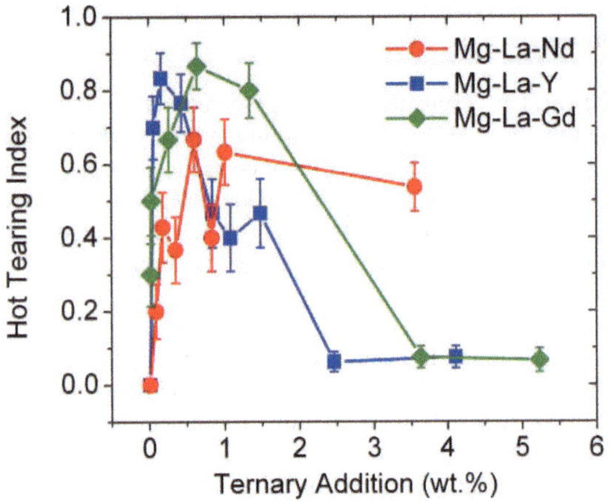

Figure 4. A comparison of the hot tearing index (HTI) for Mg-2.5wt.% La alloys with the addition of Nd, Y and Gd. Data from [14]

The Effect of Zn Additions on Hot Tearing of Mg-RE Alloys

High performance Mg-RE alloys and especially Mg-Nd alloys have improved age-hardening response with the addition of Zn [13]. Hence it is also important to consider whether Zn may affect the castability of these alloys. Hence Zn additions were made to a Mg-RE base alloy. The base alloy contained ~1.5wt% Nd, 0.5%Ce, 0.3%La. Hence it was already a relatively hot tear prone alloy. What was found was that the HTI increased dramatically up to an HTI above 2 (Figure **5**). Hence it can be seen as has been documented elsewhere that Zn additions can have a highly detrimental effect on the HTS of Mg alloys [29, 40-43]. This is not surprising because of the very low eutectic temperature (Table **I**). Hence the addition of Zn needs to be limited to less than 0.5%.

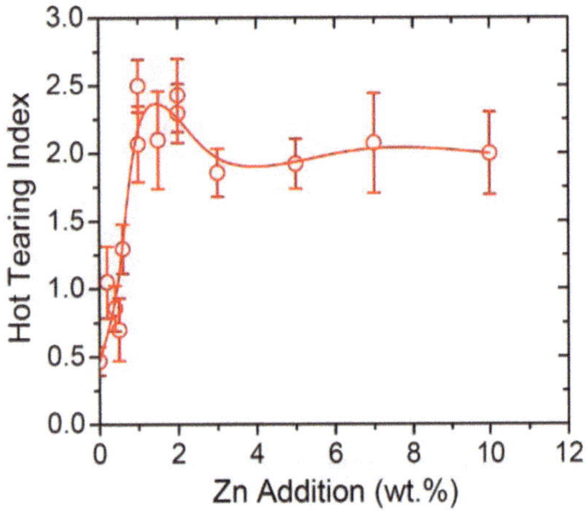

Figure 5. The hot tearing index for a 1.5wt% Nd, 0.5wt% Ce, 0.3wt% La alloy with Zn additions.

The above work has been instrumental in developing the formulation of a high-pressure die-castable, high performance magnesium alloy, AM-HP2plus. The composition is approximately 1.6%La, 1.0Ce, 1.0Nd, 0.4Zn, 0.1Y, 0.05Al which was found to have a HTI of approximately 0.2±0.03 under optimum casting conditions. This alloy has exceptional creep performance at 175°C, better than other proposed commercial, or near commercial Mg alloys, and aluminium die casting alloy A380 [5].

The Effect of RE additions on Hot Tearing of Mg-Al Alloys

Another approach to using RE additions to improve the properties of Mg alloys, has been to add them to Mg-Al alloys or Mg-Zn-Al alloys to improve elevated temperature properties [44, 45]. Whilst these alloys do not have as high creep performance as the Mg-RE based alloys, Mg-Al-RE alloys do have good creep performance and an excellent combination of strength and ductility [5].

Primarily this has led to the development of the alloy AE44. In the 1990s, AE42 was considered to be the benchmark creep-resistant HPDC alloy, but it became apparent that it was susceptible to hot tearing [46], hence further development of this alloy system was required.

Figure **6** shows a box die cast using a cold chamber diecasting machine performed by Hydro Magnesium. Hot tearing and other defects were quantified by counting the numbers of defects within the ribs of each casting. The series employed an Mg-4%Al alloy as the base and progressively added REs in the form of misch metal. Hot tearing severity increased with RE additions up to 2% RE then fell to very low levels by **4**% RE. AE44 has been cast into the part shown in Figure **6** **[**5, 40**]** and did not show any evidence of cracking.

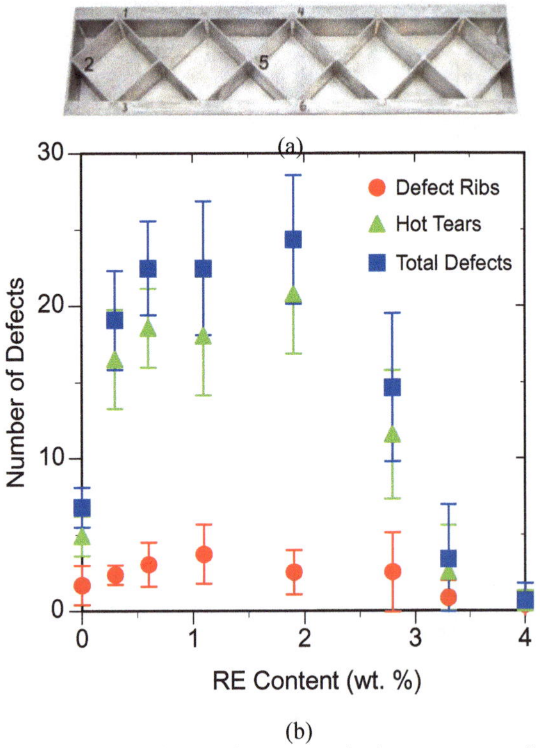

Figure 6. (a) Casting used to assess the hot tear susceptibility of the Mg-4%Al alloys with various additions of RE. (b) A plot of the effect of RE additions (as Ce-rich misch-metal) against the propensity of hot tears [47].

Discussion

Hot tearing susceptibility is a key parameter that can make alloys unprocessible particularly in high-pressure die-casting and castability issues are of particular concern in higher performance alloys that often contain REs.

Whilst Mg-Al alloys have been found to have a hot tearing Λ-curve with a peak at around 1wt%, the current authors have never observed hot tearing in these alloys in high-pressure die-casting. It is quite likely this is at least partly due to the poor creep properties meaning that significant creep can occur during casting to accommodate shrinkage stress, particularly during intensification.

In Mg-RE based alloys, the Mg-La system showed almost no hot tearing primarily due to the lower solidification range (Table **I**). Whilst this system has the poorest high temperature properties of the REs [7] it can provide a base alloy with a significant amount of eutectic to reduce the hot tearing susceptibility of the alloy when the higher performance alloying elements such as Nd, Y and Gd are added. This approach can make RE-rich magnesium alloys high-pressure die-castable and can have exceptional properties [5]. Even then, there is still some susceptibility to hot tearing, although it is only minor. It is clear, however, that Nd-rich alloys will be almost impossible to high-pressure die-cast although they can be acceptable gravity cast alloys.

Zn-additions can have significant benefits to the age hardening behavior in Mg-RE alloys, they are particularly prone to hot tearing because of the particularly low eutectic temperature in that system (Figure **5**), although small amounts <0.5% can be tolerated. However, sometimes the combination of Zn and REs can be beneficial to hot tearing resistance if the low temperature eutectic can be avoided [28].

Whilst more hot tear resistant alloys can be developed by understanding the roles of individual REs, it is apparent that the best castability option is still for a Mg-Al alloy with RE additions. These alloys have an excellent combination of strength and ductility although the high temperature creep properties are not as good as Mg-RE based alloys [5].

Although the intermetallic phases involved in the Mg-Al-RE and Mg-RE systems are different, the general behavior of a peak in hot tearing severity is observed in both types of systems with the addition of extra REs reducing the hot tear susceptibility. A similar behavior is observed in Mg-Al-Ca alloys [48] where increasing additions of Ca leads to a reduction in hot tearing to a Mg-4Al alloy. It is also likely that the addition of Al will decrease the HTS as in the case of the Mg-Al-Sr system [49].

Further work is required to be able to use a hot tearing criterion such as Equation 1 to investigate the role of the solidification path on the hot tearing behavior. There is already some information available on the Mg-Al-Ce system [50] to which recent additions to include other REs [51] will provide the basis for such a study.

Conclusions

Substantial advances have been made in understanding the role of rare earths on the hot tearing behavior of magnesium alloys. Rare earth elements have been found to have very different effects on the hot tear susceptibility, with peaks in HTS at about 0.5-2% and the height of the peak being related to the solidification range. Mg-La alloys were found to have the lowest hot tear susceptibility and it was found that the addition of higher performing rare earths such as Nd, Gd and Y were less susceptible to hot tearing than with binary alloys. Zn additions were also found to increase the hot tear susceptibility. It was found that Mg-RE based alloys with relatively low hot tear resistance could be developed for use in high-pressure die-casting. However, alloying Mg-Al with RE additions provide the lowest susceptibility to hot tearing. Results from Hydro Magnesium showed that AE44 has substantially lower HTS than AE42. Further work is required to more fully characterize hot tearing in the Mg-Al-RE system.

Acknowledgements

The authors acknowledge funding from ARC Linkage grant LP130100828. Much of this work was performed with the CAST Co-operative Research Centre, which was established under, and funded in part by the Australian governments co-operative research centre program. Mr. Andrew Yob and Ms. Maya Gershenzon from CSIRO are acknowledged for producing many of the samples studied.

References

1. A. Luo and M.Ö. Pekgüleryüz, "Review Cast Magnesium Alloys for Elevated Temperature Applications." *Journal of Materials Science*, 29 (1994), 5259-5271.
2. A. Luo, "Recent Mg alloy development for elevated temperature applications." *International Materials Reviews*, 49(1) (2004), 13-30.
3. M.Ö. Pekgüleryüz and M. Celikin, "Creep resistance in magnesium alloys." *International Materials Reviews*, 55(4) (2010), 197-217.
4. A. Luo, "Magnesium: current and potential automotive applications." *JOM*, 54(2) (2002), 42-48.
5. S.M. Zhu, et al., "Evaluation of Magnesium Die-casting Alloys for Elevated Temperature Applications: Microstructure, Tensile Properties and Creep Resistance." *Metallurgical & Materials Transactions A*, 46(8) (2015), 3543-3554.
6. T.L. Chia, et al., "The effect of alloy composition on the microstructure and tensile properties of binary Mg-rare earth alloys." *Intermetallics*, 17 (2009), 481-490.
7. S.M. Zhu, et al., "The relationship between microstructure and creep resistance in die-cast magnesium-rare earth alloys." *Scripta Materialia*, 63(7) (2010), 698-703.
8. M.A. Easton, et al., "An a priori hot tearing indicator applied to die-cast magnesium-rare earth alloys." *Metallurgical & Materials Transactions A*, 45A(7) (2014), 3586-3595.
9. C. Antion, et al., "Hardening precipitation in a Mg-4Y-3RE alloy." *Acta Materialia*, 51(18) (2003), 5335-5348.
10. P.J. Apps, et al., "Precipitation reactions in magnesium-rare earth alloys containing yttrium, gadolinium or dysprosium." *Scripta Materialia*, 48 (2003), 1023-1028.
11. C. Antion, et al., "Hardening precipitation and mechanical properties in new Mg–Mn–Y–Gd alloys." *International Journal of Materials Research*, 99 (2008), 168-177.
12. J.F. Nie, et al., "Enhanced age hardening response and creep resistance of Mg–Gd alloys containing Zn." *Scripta Materialia*, 53(9) (2005), 1049-1053.

13. D. Choudhuri, et al., "Role of Zn in enhancing the creep resistance of Mg–RE alloys." *Scripta Materialia*, 86 (2014), 32-35.
14. S. Gavras, et al., "Microstructure and property evaluation of high-pressure die-cast Mg–La–rare earth (Nd, Y or Gd) alloys." *Journal of Alloys and Compounds*, 597 (2014), 21-29.
15. M.Ö. Pekgüleryüz and P. Vermette, "Developing castability index for magnesium die casting alloys." *International Journal of Cast Metals Research*, 22(5) (2009), 357-366.
16. K. Strobel, et al., "Evaluation of the castability of high pressure die cast magnesium based alloys." *International Journal of Cast Metals Research*, 23(2) (2010), 81-91.
17. R.A. Dodd, et al., "Hot tearing of magnesium casting alloys." *AFS Transactions*, 65 (1957), 110-117.
18. R.A. Rossenberg, et al., "Nonferrous binary alloys hot tearing." *AFS Transactions*, 68 (1960), 518-528.
19. G. Cao and S. Kau, "Hot cracking of binary Mg-Al alloy castings." *Materials Science & Engineering A*, 417 (2006), 230-238.
20. S. Instone, et al., "New apparatus for characterising tensile strength development and hot cracking in the mushy zone." *International Journal of Cast Metals Research*, 12 (2000), 441-456.
21. D.G. Eskin, et al., "Contraction of aluminum alloys during and after solidification." *Metallurgical and Materials Transactions A*, 35A(4) (2004), 1325-1335.
22. M.G. Pokorny, et al., "Simulation of stresses during casting of binary magnesium-aluminum alloys." *Metallurgical & Materials Transactions A*, 41A(12) (2010), 3196-3207.
23. L. Bichler and C. Ravindran, "New developments in assessing hot tearing in magnesium alloy castings." *Materials and Design*, 31 (2010), S17-S23.
24. M.A. Easton, et al., "The role of crystallography and thermodynamics on phase selection in binary magnesium–rare earth (Ce or Nd) alloys." *Acta Materialia*, 60 (2012), 4420-4430.
25. T.W. Clyne and G.J. Davies, "The influence of composition on solidification cracking susceptibility in binary alloy systems." *British Foundryman*, 74 (1981), 65-73.
26. M. Rappaz, et al., "A new hot-tearing criterion." *Metallurgical and Materials Transactions A*, 30A (1999), 449-455.
27. M. M'Hamdi, et al., "TearSim: A two-phase model addresssing hot tearing formation during aluminum direct chill casting." *Metallurgical and Materials Transactions A*, 37A(10) (2006), 3069-3083.
28. P. Gunde, et al., "Influence of yttrium additions on the hot tearing susceptibility of magnesium–zinc alloys." *Materials Science and Engineering A*, 527 (2010), 7074-7079.
29. L. Zhou, et al., "Influence of composition on hot tearing in binary Mg–Zn alloys." *International Journal of Cast Metals Research*, 24(3/4) (2011), 170-176.
30. Suyitno, et al., "Hot tearing criteria evaluation for direct-chill casting of an Al-4.5 Pct Cu alloy." *Metallurgical and Materials Transactions A*, 36A(6) (2005), 1537-1546.
31. S. Kou, "A criterion for cracking during solidification." *Acta Materialia*, 88 (2015), 366-374.
32. J. Liu and S. Kou, "Effect of diffusion on susceptibility to cracking during solidification." *Acta Materialia*, 100 (2015), 359-368.
33. L. Sweet, et al., "Hot tear susceptibility of Al-Mg-Si-Fe alloys with varying iron contents." *Metallurgical & Materials Transactions A*, 44A(12) (2013), 5396-5407.
34. H. Huang, et al., "Effect of pouring and mold temperatures on hot tearing susceptibility of AZ91D and Mg–3Nd–0.2Zn–Zr Mg alloys " *Transactions of the Nonferrous Metals Society of China*, 24 (2014), 922-929.
35. L.L. Rokhlin, "Magnesium alloys containing rare earth metals." 2003, New York, USA: Taylor & Francis.
36. A. Srinivasan, et al., "Hot tearing characteristics of binary Mg-Gd castings." *Metallurgical & Materials Transactions A*, 44(5) (2013), 2285-2298.
37. Z.G. Wang, et al., "Hot tearing susceptibility of binary Mg–Y alloy castings." *Materials and Design*, 47 (2013), 90-100.
38. J. Gröbner, et al., "Phase analysis of Mg-La-Nd and Mg-La-Ce alloys." *Intermetallics*, 28 (2012), 92-101.
39. R. Schmid-Fetzer, et al., *Thermodynamics of phase formation in Mg-La-Ce-Nd alloys*, in *Magnesium Technology 2013*, N. Hort, et al., Editors. 2013, TMS (The Minerals, Metals & Materials Society): San Antonio, Texas. p. 243-248.
40. W. Xiao, et al., "Casting defects and mechanical properties of high pressure die cast Mg-Zn-Al-RE alloys." *Advanced Engineering Materials*, 14(1-2) (2012), 68-76.
41. M.A. Easton, et al., *An assessment of high pressure die cast Mg-Zn-Al alloys*, in *Magnesium Technology 2008*, M.O. Pekguleryuz, et al., Editors. 2008, The Metals, Minerals and Materials Society: New Orleans LA. p. 323-328.
42. G. Foerster, *Improved magnesium die casting alloys*, in *Transactions of the 8th SDCE International Die Casting Congress*. 1975: Detroit. p. paper G-T75-112.
43. L. Zhou, et al., "Prediction of hot tearing susceptibility for Mg-Zn-(Al) Alloys." *Advanced Materials Research*, 509 (2012), 138-146.
44. G. Pettersen, et al., "Microstructure of a pressure die cast magnesium-4wt.percent aluminium alloy modified with rare earth additions." *Materials Science & Engineering A*, 207(1) (1996), 115-120.
45. P. Bakke and H. Westengen, "Die casting for high performance - focus on alloy development." *Advanced Engineering Materials*, 5(12) (2003), 879-885.
46. L. Bichler, et al., "Onset of hot tearing in AE42 magnesium alloy." *Canadian Metallurgical Quarterly*, 48(1) (2009), 81-90.
47. P. Bakke, "unpublished research." (2006), unpublished research.
48. G. Cao and S. Kou, "Hot tearing of ternary Mg-Al-Ca alloy castings." *Metallurgical & Materials Transactions A*, 37A(12) (2006), 3647-3663.
49. G. Cao, et al., "Onset of hot tearing in ternary Mg-Al-Sr alloy castings." *Metallurgical & Materials Transactions A*, 41A(8) (2010), 2139-2150.
50. J. Gröbner, et al., "Thermodynamic modeling of Al-Ce-Mg phase equilibria couple with key experiments." *Intermetallics*, 10 (2002), 415-422.
51. L. Jin, et al., "Al–Mg–RE (RE = La, Ce, Pr, Nd, Sm) systems: Thermodynamic evaluations and optimizations coupled with key experiments and Miedema's model estimations." *Journal of Chemical Thermodynamics*, 58 (2013), 166-195.

Magnesium Technology 2016
Edited by: Alok Singh, Kiran Solanki, Michele V. Manuel, and Neale R. Neelameggham
TMS (The Minerals, Metals & Materials Society), 2016

HOT TEARING SUSCEPTIBILITY OF Mg-5Nd-xZn ALLOYS

Francesco D'Elia, Domonkos Tolnai, Chamini Lakshi Mendis, Norbert Hort
Magnesium Innovation Centre (MagIC), Helmholtz Zentrum Geesthacht, Max Planck Strasse, 1, Geesthacht, 21502, Germany

Keywords: Magnesium alloys, hot tearing, solidification, microstructure

Abstract

Magnesium-neodymium-zinc (Mg-Nd-Zn) alloys are promising candidates as creep resistant alloys. Further, Nd is a rare earth (RE) addition with lower solid solubility and a relatively lower cost. Hence, the use of such alloys may result in a feasible and cost effective alternative for enhancing Mg alloy use in high temperature applications. Nevertheless, studies on the castability of Mg-Nd-Zn alloys are lacking. As such, the aim of this research was to investigate the hot tearing susceptibility of Mg-5Nd-xZn (x = 0, 3, 5, 7 wt%) alloys during permanent mold casting. Specifically, a constrained-rod casting mold equipped with a load cell was used to characterize hot tearing severity and determine the onset temperature of hot tearing. The onset solid fraction of hot tearing was subsequently determined via thermodynamic software. The results suggest that hot tearing severity increased initially with addition of Zn (up to 5 wt%), but then decreased with further addition to 7 wt%. This was likely attributed to both the low onset solid fraction of hot tearing (i.e. 0.5) recorded for this alloy, which enabled enhanced feeding and opportunity to heal developing hot tears, as well as the divorced eutectic structure observed which may have facilitated late stage feeding of eutectic liquid and hence limit the alloy's susceptibility to hot tearing.

Introduction

Magnesium alloys are widely used in transport and consumer electronic applications to reduce the weight of components. In the past, automotive use of Mg was generally limited to a few select applications such as instrument panels, steering wheels and valve covers [1]. Nowadays, with the continuous development of Mg-RE alloy systems resulting in improved creep resistance and high strength at elevated temperatures, automotive use of Mg is increasing towards powertrain applications and further advancements are desired for future automobiles. However, one limiting factor remains the high cost of such RE elements.

Addition of Zn to Mg-RE alloys has recently been shown to reduce the amount of RE required for optimum alloy strength and ductility during hot rolling [2]. Such reduction in RE content significantly lowers the cost of these alloys, thereby making them more attractive for industrial applications. Moreover, Nd is a RE addition with lower solid solubility and a relatively lower cost [3]. Hence, the use of Mg-Nd-Zn alloys may result in a feasible and cost effective alternative for enhancing Mg alloy use in high temperature applications. Preliminary studies [4-6] have attempted to characterize the phase formation, microstructure development and mechanical properties of Mg-Nd-Zn alloys. Such alloys have been found to possess optimum hardness and strength after heat treatment due to the formation of plate-shaped GP zones and precipitates on prismatic planes of the Mg matrix [6]. Nevertheless, studies on the castability of such alloys are lacking. Castability of a new alloy must be at an acceptable level, as this is the major processing route for production of Mg alloy components [1]. As such, this research aims to characterize the hot tearing behavior of this alloy system.

Hot tearing is a common and severe defect occurring in alloys at late stages of solidification. Reviews carried out by Sigworth [7], Eskin [8] and Li and Apelian [9] detail the many theories and mechanisms attributed to the formation of such defects. It has been suggested that hot tearing is a complex phenomenon with many factors influencing their formation. An improvement in the understanding of such factors will help to develop methods towards preventing the formation of such defects in Mg alloys and thereby enhancing the use of Mg in industrial applications.

This research aims to characterize the hot tearing behavior of four Mg-5Nd-xZn (x = 0, 3, 5, 7 wt%) alloys during permanent mold casting. The load cell method was used to determine the onset temperature of hot tearing. Such a technique has shown success in investigating the hot tearing behavior of many Mg alloy systems [10-13]. Pandat thermodynamic software was used to estimate the onset solid fraction of hot tearing. Further, thermal analysis was carried out to understand the role of alloy solidification on hot tearing. Finally, scanning electron microscopy was used to characterize the microstructure of the alloys.

Experimental Procedure

Melting and Casting

Casting experiments were carried out at the Magnesium Innovation Centre (MagIC) at Helmholtz-Zentrum Geesthacht in Germany. The Mg-5Nd-xZn (x = 0, 3, 5, 7 wt%) alloys were prepared from pure (99.95%) Mg ingots with additions of pure (99.95%) Nd and pure (99.9999%) Zn. The alloys were cast into billets, which were then used for hot tearing experiments. All castings were made with virgin ingot material (i.e. no recycled alloy was used).

The alloys (in billet form) were melted under a protective atmosphere consisting of Ar + SF_6 gas in an electric resistance furnace. Approximately 350 g of the alloys were melted in a stainless steel crucible for each casting trial. Each alloy was poured with a superheat of 100 °C into a mold preheated to 250 °C. The resulting pouring temperatures were 740 °C, 730 °C, 730 °C and 725 °C for the Mg-5Nd, 3Zn-, 5Zn- and 7Zn-contaning alloys, respectively. Two repeat casting trials were carried out for each alloy.

Hot Tearing Apparatus

The hot tearing apparatus consisted of a constrained-rod casting (CRC) steel mold and a contraction force measurement system with a load cell, as shown in Figure 1. Details of the apparatus are provided elsewhere [14]. The load cell was connected to a data acquisition unit. The mold was comprised of two parts: a vertical sprue and a 148 mm long horizontal circular rod. The circular rod was slightly tapered (from 12.5 mm to 10 mm diameter) to minimize the effect of friction from the mold way to the solidifying melt [12,13]. The principal function of the apparatus was to monitor the evolution of hot tearing in the casting using the measurement of contraction force in conjunction with the temperature during solidification.

Mold temperature was monitored via three K-type thermocouples inserted along three locations of the mold: one at the sprue-rod junction (T1) and two (T2, T3) along the horizontal rod (Figure 1b). The thermocouple at the sprue-rod junction was inserted into the casting cavity (to monitor temperature of the melt during solidification) while the remaining two thermocouples were not placed in contact with the casting rod, but strictly used to monitor the mold temperature. The method for which the mold was heated is described elsewhere [14]. The thermocouples were connected to a data acquisition unit. Upon pouring of the alloys the temperature and force data were collected and used to generate force-time and temperature-time plots, which enabled the determination of the onset of hot tearing.

The mold was coated with boron nitride to prevent corrosion of the steel mold into the Mg melt. The coating was removed and reapplied after each casting experiment. Two repeats were carried out for each casting trial to ensure repeatability of results. The reproducibility of the present apparatus is well documented and reported elsewhere [15].

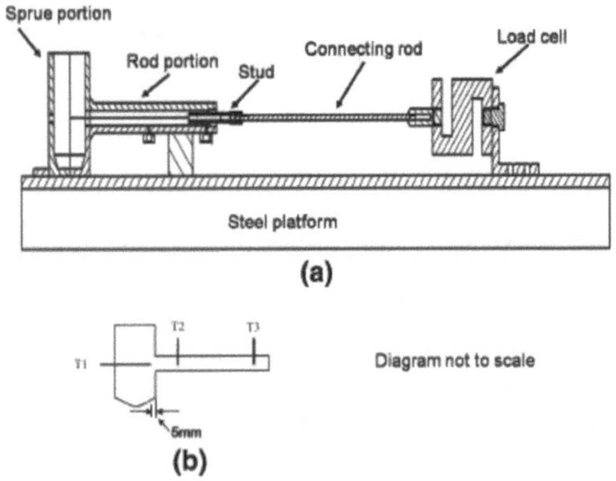

Figure 1. Schematic diagram showing a) CRC mold apparatus and b) location of thermocouples.

Microscopy

Representative samples were sectioned from the castings and subsequently mounted and ground using 500, 800, 1200 and 2500 grit SiC papers. The samples were then polished with 1 μm diamond suspension. The polished samples were viewed under a Tescan scanning electron microscope (SEM) to characterize the morphology and distribution of second phases.

Results and Discussion

Observation of Casting Surfaces

Upon completion of solidification, the castings were removed from the mold and examined for hot tears. The casting surfaces of the investigated alloys are illustrated in Figure 2. A fine tear was present on the casting surface of the Mg-Nd alloy. Addition of 3 wt% Zn resulted in a significant increase in hot tearing severity, as shown in Figure 2b. In the case of the Mg-5Nd-5Zn alloy (Figure 2c), the hot tear was so severe that it resulted in the complete separation of the horizontal rod from the sprue. Finally, further addition of Zn to 7 wt%, was seen to significantly reduce the susceptibility of the alloy to hot tearing, as shown in Figure 2d.

Onset of Hot Tearing

The force-time and temperature-time curves are presented in Figure 3 for the investigated Mg alloys. The thermocouple at the sprue-rod junction, T1, was used for temperature measurements. At the beginning of each trial, increased temperature in the temperature-time curve was coupled with a simultaneous increase in compressive load in the force-time curve for each condition. This was likely attributed to the filling of the casting cavity by molten metal. Initially, as the molten alloy entered the casting cavity and flowed toward the load cell, the pressure against the load cell generated a compressive load. Upon filling the casting cavity, solidification commenced and the casting began to contract with partial restriction (from the steel stud attached to the load cell), thereby changing the path of the force curve towards the opposite direction (tensile magnitude). The load continued towards a tensile state until a relief occurred. This relief in contraction force was attributed to the formation of a hot tear. Following this relief, the load again increased towards tension as the casting continued to contract. In some cases, tensile load magnitude was not obtained during solidification, as a result of the initial high magnitude of compression induced by the incoming melt on the load cell.

The hot tears present on the casting surfaces (Figure 2) were confirmed by the dip in the force-time curves, as shown in Figure 3. In the case of the Mg-5Nd alloy (Figure 3a), two hot tearing initiation points were observed. This suggested that two separate hot tears occurred during solidification. The corresponding initiation temperatures for the first and second tear were 602 °C and 559 °C, respectively. The onset solid fractions of hot tearing were calculated via Pandat thermal database software. The plots of solid fraction vs. temperature are given in Figure 4 for each investigated alloy. In the case of the Mg-5Nd alloy, the corresponding onset solid fractions of hot tearing were 0.77 and 0.86 for the first and second tear, respectively. Hot tears are known to typically occur within the solid fraction range of 0.85-0.95 [8]. However, solid fractions as low as 0.78 have been reported [10], thus the results are consistent with literature.

The large tear present on the casting surface of the Mg-5Nd-3Zn alloy (Figure 2b) was in agreement with the force-time curve (Figure 3b), which depicted a significant drop in force magnitude. Since hot tearing is a stress-relieving process, the size of the dip in the force curve can be directly related to the propagation of the hot tear. The hot tear present on the surface of this alloy was significantly larger than that of the Mg-5Nd alloy, and thus and greater drop in force magnitude was observed. The hot tear initiation temperature was recorded at 592 °C. This corresponded to a fraction solid of 0.7.

In the case of the Mg-5Nd-5Zn alloy, the severe hot tear that resulted in the complete separation of the horizontal rod from the sprue was consistent with the force curve (Figure 2c). An initial hot tear was seen to initiate at 576 °C and propagated for some time, as no significant increase in load was observed. From there, the compressive force magnitude was seen to decrease towards tension until a second tear occurred at 492 °C. This tear was severe enough to result in the complete separation of the horizontal rod from the sprue, as from that instant there was further development of force magnitude (i.e. the force curve remained stagnant). Such a trend is typical of the bar contracting freely (i.e. without restriction from the sprue). The corresponding solid fractions of hot tearing were 0.69 and 0.9 for the two tears, respectively.

Finally, the fine tear seen on the casting surface of the Mg-5Nd-7Zn alloy resulted in a minor drop in the force-time curve, as shown in Figure 3d. In comparison to the 3Zn- and 5Zn-containing alloys, the magnitude of tear propagation (i.e. force drop or stagnant force development) was significantly lower. The hot tear was found to occur at 592 °C at an estimated solid fraction of 0.5. Thus, the tear was found to occur quite early during solidification. In turn, this likely enabled more time for the initiated hot tear to heal by liquid metal feeding during the remainder of solidification, thereby resulting in a less severe tear for this alloy in comparison to the 3Zn- and 5Zn-containing alloys.

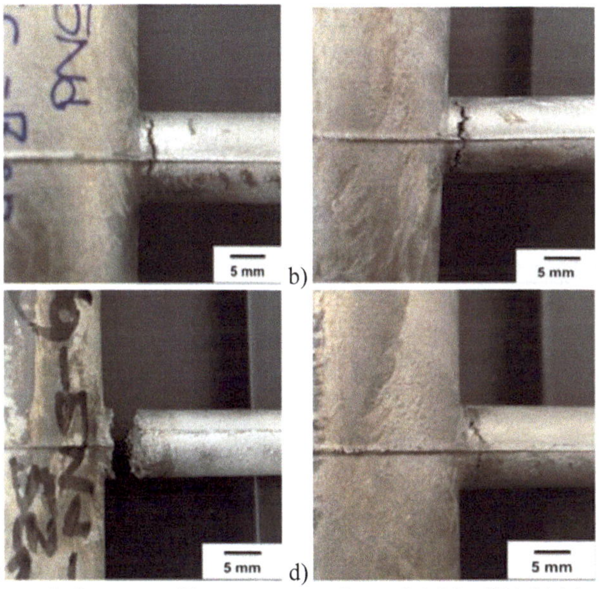

Figure 2. Presence of hot tears on surface of a) Mg-5Nd, b) Mg-5Nd-3Zn, c) Mg-5Nd-5Zn and d) Mg-5Nd-7Zn alloy castings.

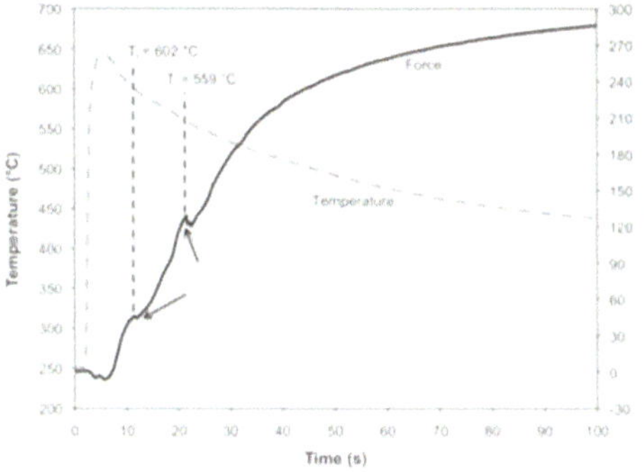

a)

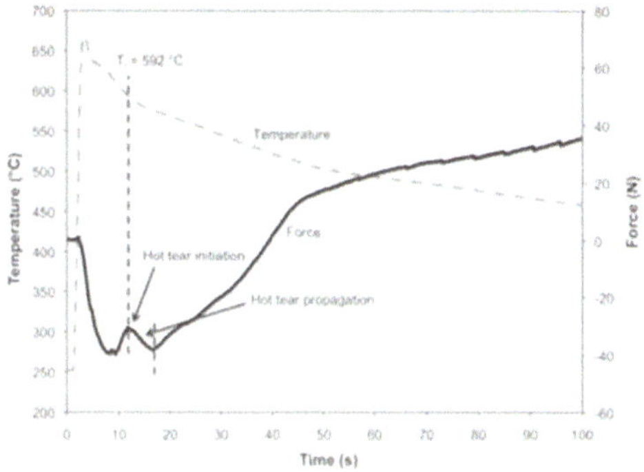

b)

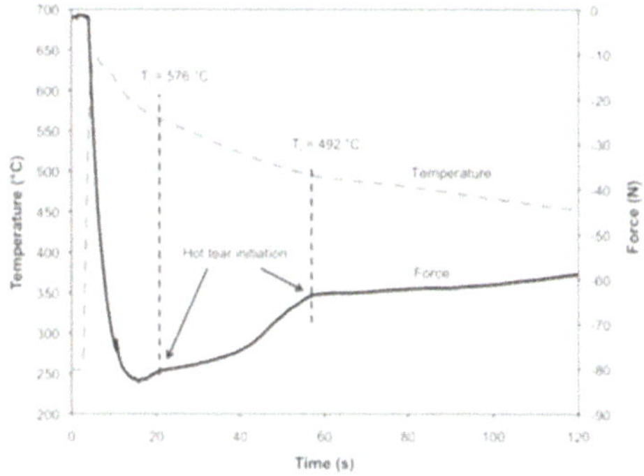

c)

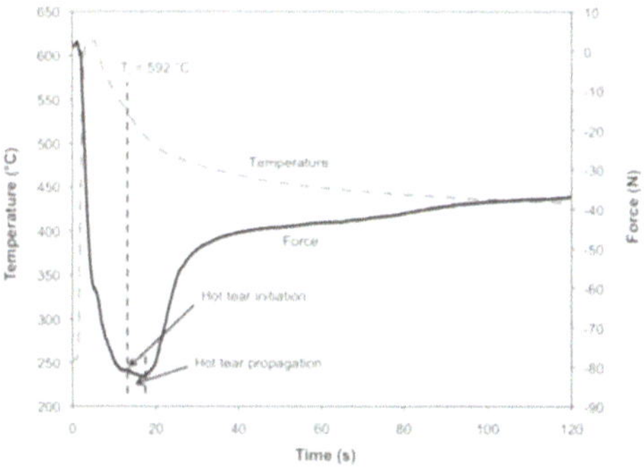

d)

Figure 3. Force-temperature-time curves for the a) Mg-5Nd, b) Mg-5Nd-3Zn, c) Mg-5Nd-5Zn and d) Mg-5Nd-7Zn alloys.

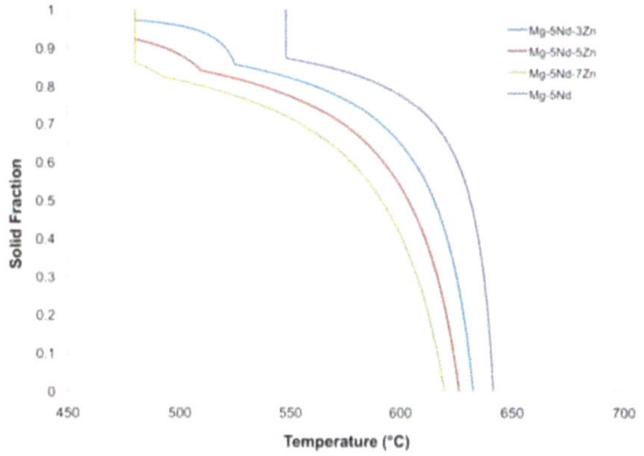

Figure 4. Solid fraction development for the Mg-5Nd-xZn alloys.

Thermal Analysis

Conventional thermal analysis was carried out from the temperature-time (cooling) curves of each alloy. The first derivative of each cooling curve was taken and used to determine the freezing range, solidification time and the onset temperature of dendrite coherency. The onset temperature of dendrite coherency was determined from methods readily available in literature [16, 17]. The results of thermal analysis are summarized in Table I for the four Mg alloys. Further, differences in temperature (T) and time (t) between the points of interest are given in Table II. A discussion on the time and temperature intervals is provided and related to casting feedability.

Table I. Thermal analysis data for Mg-5Nd-xZn castings.

Alloy	Nucleation		Coherency		Solidification	
	T_N (°C)	t_N (s)	T_C (°C)	t_C (s)	T_S (°C)	t_S (s)
5Nd	641	1.4	628	4.1	522	32.4
5Nd-3Zn	632	2.9	616	5.6	496	53.2
5Nd-5Zn	628	2.5	610	5.2	490	56.5
5Nd-7Zn	624	2.4	610	3.8	480	63.1

Table II. Temperature and time differences between points of interest.

Alloy	$t_C - t_N$ (s)	$T_N - T_S$ (°C)	$(T_N - T_S)_S$ (°C)	$t_S - t_N$ (s)
5Nd	2.7	119	93	31.0
5Nd-3Zn	2.7	136	153	51.3
5Nd-5Zn	2.7	138	146	54.0
5Nd-7Zn	1.4	144	140	60.7

The duration of mass feeding of liquid metal was represented by the time difference between nucleation and coherency (i.e. $t_c - t_N$). The results suggest that minimal time was available for mass feeding in each alloy. This is consistent with the poor hot tearing resistance observed for all alloys (Figure 2). Further, the time for mass feeding was virtually identical for the Mg-5Nd, Mg-5Nd-3Zn and Mg-5Nd-5Zn alloys. In the case of the Mg-5Nd-7Zn alloy, however, the time interval for mass feeding was found to decrease to almost half with respect to the other alloys. Such an early onset of dendrite coherency in this alloy, may suggest the reason for which the hot tear occurred at such an early stage of solidification (0.5 solid fraction) with respect to the other alloys. Stress and strain begin to build within a casting after the dendrites become coherent and thus, an earlier development of stress and stain, likely may have lead to an earlier initiation of hot tearing in the Mg-5Nd-7Zn alloy.

The temperature interval, $T_N - T_S$, represents the freezing range of the alloys, while the time interval $t_S - t_N$, represents the solidification time. The freezing calculated via Scheil using Pandat thermodynamic software is represented as, $(T_N - T_S)_S$. A good agreement between the actual experimental values and those obtained thermodynamically was observed. Further, the results suggest that both freezing range and solidification time increased with the addition of Zn.

The increase in freezing range and solidification time was likely the reason for which the hot tearing severity was increased for the 3Zn- and 5Zn-containing alloys with respect to the Mg-5Nd binary alloy. It is well established that an increase in freezing range (and solidification time) often results in increased hot tearing susceptibility [8, 18]. However, in the case of the 7Zn-containing alloy, a further increase in freezing range and solidification time as well as the lowest time interval of mass feeding was not consistent with the reduced hot tearing severity observed for this alloy (Figure 2). It is possible that the early initiation of hot tearing in this alloy enabled more time for (interdendritic) liquid metal feeding, as mentioned earlier. Another possible reason for this occurrence can be described by considering the work of Clyne and Davies [18]. The authors found that for many alloy systems, there reaches an amount of solute that results in a peak severity of hot tearing. Further increases in solute beyond the peak content, typically results in subsequent decreases in hot tearing severity. This trend is referred to as the lambda (λ) curve due to its shape. Thus, it may be the case that the addition of 7 wt% Zn to the binary Mg-5Nd alloy may be past the critical amount of Zn required for severe hot tearing. This may be due to an increased amount of eutectic liquid available for feeding (and subsequent healing of developing tears) at late stages of solidification as a result of the higher amount of solute. The microstructure of each alloy is presented in the following section.

Microscopy

Representative SEM micrographs are shown in Figure 5 for the investigated Mg alloys. Examination of the micrographs illustrates coarse dendrite morphology with distribution of second phases along interdendritic regions for each alloy. The second phase present in each alloy is likely either the Mg_3Nd phase. It has been shown that Mg-Nd casting alloys tend to contain $Mg_{12}Nd$ at relatively slow cooling rates or Mg_3Nd if the cooling rate is particularly rapid, as in gravity die casting [19]. Further inspection of the micrographs suggests that the addition of Zn did not significantly increase the amount or affect the distribution of second phases in the microstructure. Thus, it cannot be suggested that a greater amount of eutectic liquid was present for healing the developing hot tear in the Mg-5Nd-7Zn alloy. The alloys were therefore examined at higher magnification in order to see whether changes in the morphology of the second phases occurred with additions of Zn.

a) 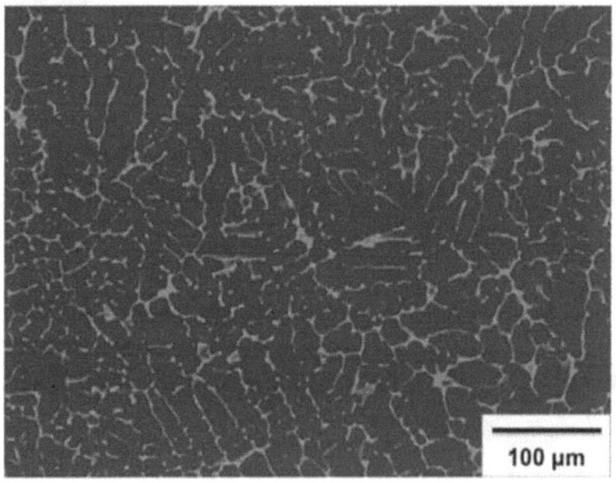

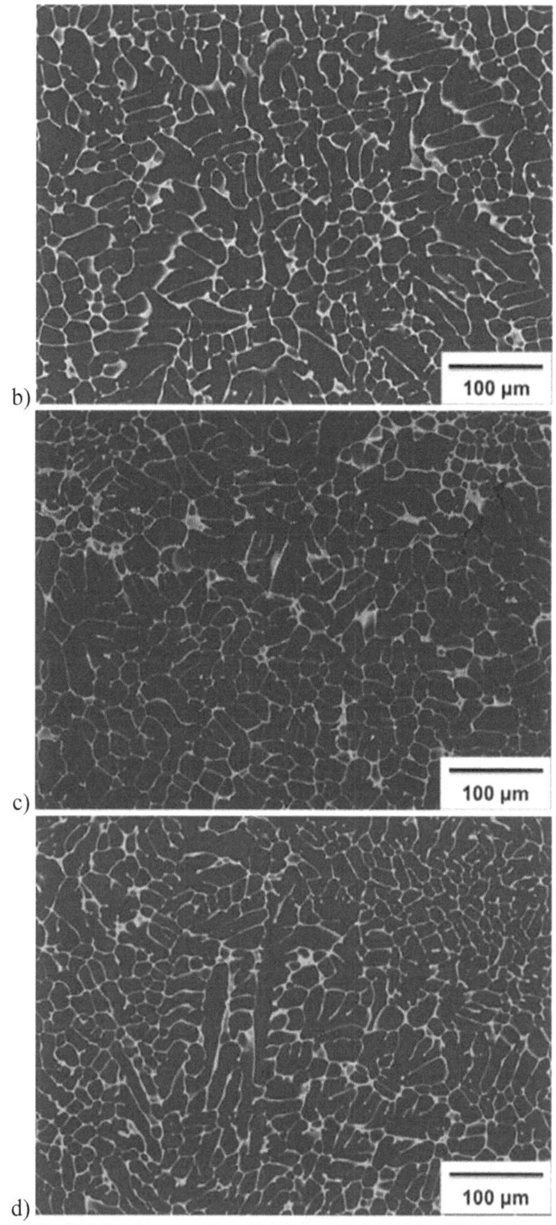

Figure 5. SEM micrographs of a) Mg-5Nd, b) Mg-5Nd-3Zn, c) Mg-5Nd-5Zn and d) Mg-5Nd-7Zn.

Three representative micrographs at higher magnification are illustrated in Figure 6 a, b and c for the Mg-5Nd, Mg-5Nd-5Zn and Mg-5Nd-7Zn alloys, respectively. In the case of the Mg-5Nd alloy, a fully divorced eutectic structure of the $Mg_{12}Nd$ phase was observed as separate 'islands' of this phase was present along the Mg matrix. In contrast, a significant change in the morphology of the intermetallic was observed in the Mg-5Nd-5Zn alloy, as the structure appeared more plate-like. This plate-like structure was likely more brittle and thus more prone to hot tear formation. This was likely a significant factor affecting the alloy's resistance to hot tearing. Finally, in the case of the Mg-5Nd-7Zn alloy, shown in Figure 6c, the intermetallic appeared to become more divorced once again. The reason for this occurrence is not completely understood and subject to further investigation. Nevertheless, such larger 'islands' of intermetallic may suggest improved feeding through larger channels at late stages of solidification, and thus an overall improved resistance to hot tearing.

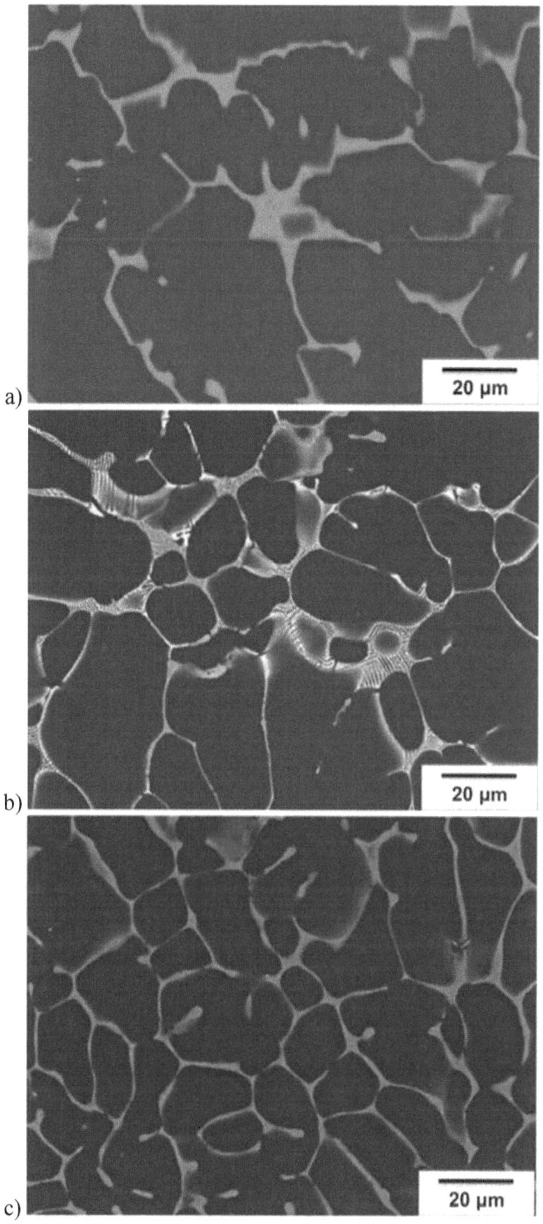

Figure 6. SEM micrographs of a) Mg-5Nd, b) Mg-5Nd-5Zn and c) Mg-5Nd-7Zn at higher magnification.

Conclusions

The hot tearing susceptibility of Mg-5Nd-xZn (x=0, 3, 5, 7) alloys was investigated during permanent mold casting. The following conclusions can be drawn from this study:

1. Hot tearing was found to occur in all investigated alloys. Addition of Zn up to 5 wt% significantly increased the hot tearing severity. However, further addition of Zn to 7 wt% resulted in a decrease in hot tearing severity.
2. The onset temperature of hot tearing was obtained from the force-time and temperature-time curves generated from the load cell. The hot tears for all alloys, with

exception of the Mg-5Nd-7Zn alloy, were seen to occur at solid fractions ranging from 0.7-0.9. In the case of the Mg-5Nd-7Zn alloy, the onset solid fraction of hot tearing was recorded at 0.5.

3. Thermal analysis revealed an increase in freezing range and solidification time with addition of Zn. Such results are consistent with the increase in hot tearing severity observed for the Mg-5Nd-3Zn and Mg-5Nd-5Zn alloys with respect to the binary Mg-5Nd alloy. However, conflicting results were seen for the Mg-5Nd-7Zn alloy, as it was found to have the largest freezing range and longest solidification time but the lowest susceptibility to hot tearing of the three Zn-containing Mg-5Nd alloys.
4. Scanning electron microscopy illustrated a significant transformation in eutectic structure from fully divorced for the Mg-5Nd alloy to a more plate-like brittle structure for the Mg-5Nd-5Zn alloy. Such a brittle structure likely increased the alloy's susceptibility to hot tearing, thereby making it the most prone of the alloys investigated. In the case of the Mg-5Nd-7Zn alloy, however, the eutectic structure was again found to be more divorced, which in turn may have facilitated late stage feeding of eutectic liquid and thereby limit the alloy's susceptibility to hot tearing.

Acknowledgements

The authors are thankful to Mr. G. Meister of MagIC at Helmholtz Zentrum Geesthacht for help with casting experiments. One of the authors (F. D'Elia) acknowledges the financial support of the Alexander von Humboldt Foundation.

References

1. M.O. Pekguleryuz, A.A. Kaya, "Creep Resistant Magnesium Alloys for Powertrain Applications," *Advanced Engineering Materials,* 5 (12), 2003, 866-878.
2. T. Itoi et al., "Microstructure and Mechanical Properties of Mg-Zn-Y Alloy Sheet Prepared by Hot Rolling," *Materials Science and Engineering A,* 560 (2013), 216-223.
3. A.A. Nayeb-Hashemi, J. B. Clark Binary, *Phase Diagrams of Magnesium Alloys* (Materials Park, Ohio: ASM International, 1988).
4. H.Y. Qi et al., "Thermodynamic Description of the Mg-Nd-Zn Ternary system," *Journal of Alloys and Compounds*, 509 (2011), 3274-3281.
5. H. Xu et al., "Experimental Determination of the Phase Equilibria Mg-Nd-Zn System at 320 °C," *Journal of Alloys and Compounds*, 603 (2014), 100-110.
6. R. Wilson et al., "Precipitation Hardening in Mg-3wt%Nd(-Zn) Casting Alloys," *Materials Science Forum*, 419-422 (2003), 267-272.
7. G.K. Sigworth, "Hot Tearing of Metals," *AFS Transactions*, 104 (1996), 1053-1062.
8. D.G. Eskin, Suyitno, L. Katgerman, "Mechanical Properties in the Semi-solid State and Hot Tearing of Aluminum Alloys," *Progress in Materials Science*, 49 (2004), 629-711.
9. S. Li and D. Apelian, "Hot Tearing of Aluminum Alloys: A Critical Literature Review," *International Journal of Metalcasting*, 5 (2011), 23-40.
10. G. Cao, S. Kou, "Hot Tearing of Ternary Mg-Al-Ca Alloys," *Metallurgical and Materials Transactions A*, 37 (2006), 3647-3663.
11. G. Cao, I. Haygood, S. Kou, "Hot Tearing Susceptibility of Ternary Mg-Al-Sr Alloy Castings," *Metallurgical and Materials Transactions A*, 41 (2010), 2139-2150.
12. Z. Wang et al., "Hot Tearing Susceptibility of Binary Mg-Y Alloy Castings," *Materials and Design*, 47 (2013), 90-100.
13. A. Srinivasan et al., "Hot Tearing Characteristics of Binary Mg-Gd Alloy Castings," *Metallurgical and Materials Transactions A*, 44 (2013), 2285-2298.
14. F. D'Elia et al., "Hot Tearing Mechanisms of B206 Aluminum-Copper Alloy," *Materials and Design*, 64 (2014), 44-55.
15. Z.S. Zhen et al., "Hot Tearing Behaviour of Binary Mg-1Al Alloy Using a Contraction Force Measuring Method," *Internation Journal of Cast Metal Research*, 22 (2009), 331-334.
16. L. Bäckerud, G. Chai, J. Tamminen, *Solidification of Aluminum Alloys, Volume 2: Foundry Alloys* (Stockholm: AFS/Skanaluminum, 1990).
17. F. D'Elia, C. Ravindran, D. Sediako, "Interplay Among Solidification, Microstructure, Residual Strain and Hot Tearing in B206 Aluminum Alloy," *Materials Science and Engineering A*, 624 (2015), 169-180.
18. T.W. Clyne, G.J Davies, "The Influence of Composition on Solidification Cracking in Binary Alloy Systems," *British Foundryman*, 74 (1981), 65-73.
19. M.A. Easton et al., "The Role of Crystallography and Thermodynamics on Phase Selection in Binary Magnesium-Rare Earth (Ce or Nd) Alloys," *Acta Materialia*, 60 (2012), 4420-4430.

Magnesium Technology 2016
Edited by: Alok Singh, Kiran Solanki, Michele V. Manuel, and Neale R. Neelameggham
TMS (The Minerals, Metals & Materials Society), 2016

SOLID SOLUTION STRENGTHENING IN Mg–Gd ALLOYS

Yuling Xu, Zheng Ren, Yuanding Huang, Karl Kainer, Norbert Hort

MagIC-Magnesium Innovation Centre, Helmholtz-Zentrum Geesthacht,
Max-Planck-Str. 1, 21502 Geesthacht, Germany

Keywords: Mg-Gd alloy, Solid solution strengthening, Mechanical properties.

Abstract

In order to improve mechanical properties and formability of magnesium alloys, a range of new rare earth (RE) containing alloys was developed. However, the strengthening mechanisms of RE alloys are still unclear to a certain extent. In the present work numerous Mg-Gd and Mg-Gd-0.5Zr alloys were prepared. The effects of solute Gd atoms on hardness and yield stress were investigated with its contents in the range from 0.32 to 2.57 at. %. The hardness linearly increases with the increment of Gd content. The ultimate tensile strength and tensile yield strength increase obviously after adding Gd element. The maximum values of elongation were observed for Mg10Gd and Mg2GdZr alloys, respectively. After the extraction of grain size strengthening, the yield strength of Mg-Gd alloys increases linearly with c^n, where c is the Gd concentration and n=1/2 or 2/3.

1. Introduction

As a lightest structure material, magnesium alloys are becoming one of the key engineering materials, because they have a high specific strength among all the structural alloys used in industry [1, 2]. However, the present applications of magnesium alloys are still limited. The main obstacles are their low mechanical properties both at room and high temperatures. For those reasons, large number of new Mg alloys has been developed to improve the mechanical properties. Mg-REs (rare earth elements) alloys attract tremendous attentions due to their excellent creep resistance and mechanical properties [3-6].

The strengthening effects of RE in Mg alloys have been explained by two mechanisms: solid solution strengthening and precipitation hardening [7-9]. Among all RE elements, Gd has a high solid solubility in Mg, with a maximum solubility 23.49 wt. % at the eutectic temperature. Previous work indicated that gadolinium shows supersaturate solid solution in Mg alloys during the solidification [10-16]. In this work, the solid solution strengthening on the hardness and yield strength for polycrystalline Mg-Gd and Mg-Gd-Zr alloys were studied.

2. Experimental

A number of Mg-Gd, Mg-Gd-Zr alloys were investigated. Permanent mould direct chill casting was used to prepare 8 alloys with nominal chemical compositions listed in Table 1. High-purity Mg (MEL, UK, 99.94 wt. %) was molten in a mild steel crucible under a protective atmosphere (Ar + 2% SF6). Pure Gd (Grirem, China, 99.5 wt.%) and Mg-33.3wt.%Zr master alloy were added at a melt temperature of 750 °C. The melt was stirred at 200 rpm for 20 min. After that, the melt was cast using a directly chilled permanent mould casting method. Firstly, the melt was poured into a mould preheated at 500 °C. The filled mould was held at 670 °C for 30 min under the protective gas. Then the whole steel crucible with the melt was lowered into cooling water at a rate of 10 mm/s. When the bottom of steel crucible touched the water, it was stopped for 10 second. When the melt was fully immersed, the solidification finished. The ingot was cylinder, with a size of Φ120 mm × 200 mm. The chemical compositions of all alloys were analyzed by X-ray fluorescence (XRF) (Table 1). For all alloys, a solution treatment (T4) was performed at 530 °C for 6 to 8h, followed by water quenching.

In order to investigate the influence of Gd content on the mechanical properties, Vickers hardness was measured. The specimens for hardness tests were prepared by mounting and grinding with silicon carbide emery paper up to 2500 grit. The hardness measurement was carried out using a Vickers hardness testing machine (KARL FRANK GMBH) with a load of 5 kg and a dwell time of 10 seconds. An average of 10 measurements was made for each specimen. Tensile tests were performed at room temperature using a Zwick 050 machine (Zwick GmbH & Co., KG, Ulm, Germany) according to DIN EN 10002. The tensile specimens with a gauge length of 30 mm, a diameter of 6 mm and threaded heads were used. All tension tests were done under a strain rate of 1×10^{-3} s^{-1}. At least three specimens were tested under each condition.

The metallographic specimens for microstructural observations were etched in a solution of 8 g picric acid, 5 mL acetic acid, 10 mL distilled water and 100 mL ethanol after mechanical polishing. Microstructures were characterized using light microscope (Reichert-Jung MeF3) with a digital camera attachment. A Zeiss Ultra 55 (Carl Zeiss GmbH, Oberkochen, Germany) scanning electron microscope (SEM) equipped with energy dispersive X-ray (EDX) analyzer was further used to observe the microstructure at an accelerative voltage of 20 kV. EDX was used to analyze the compositions of different phases with a minimum live time of 50 seconds. The grain size was measured using linear intercept method. At least 200 grains were counted for the alloys with various Gd contents.

3. Results and discussion

3.1 Solid solution effects on hardness

Fig. 1 shows the effects of elements Gd and Zr on the grain size of the alloys. It is seen that there has little influence on the grain size of Mg-Gd binary alloys by adding Gd element. For Mg-Gd-Zr alloys it is obvious that the grain size decreases with the increase of Gd content. Compared with Mg-Gd binary alloys, the grain sizes of the alloys with Zr drop more drastically.

Table I Chemical composition, grain size and solution treatment conditions of the alloys.

Alloys	Gd content (wt. %) [at. %]	Zr content (wt. %)	Average grain size (μm)	Solution treatment
Pure Mg	0.00 [0.00]	0.00	709.25 ± 455.97	530°C / 6h
Mg-5Gd	5.07 [0.81]	0.00	989.00 ± 464.40	530°C / 6h
Mg-10Gd	10.04 [1.69]	0.00	892.45 ± 467.95	530°C / 8h
Mg-15Gd	14.57 [2.65]	0.00	446.35 ± 213.86	530°C / 8h
Mg-2Gd-Zr	2.02 [0.33]	0.39	363.55 ± 161.44	530°C / 6h
Mg-5Gd-Zr	4.95 [0.80]	0.20	407.16 ± 250.72	530°C / 6h
Mg-10Gd-Zr	10.04 [1.56]	0.29	128.64 ± 52.02	530°C / 8h
Mg-15Gd-Zr	11.95 [2.05]	0.22	98.30 ± 39.81	530°C / 8h

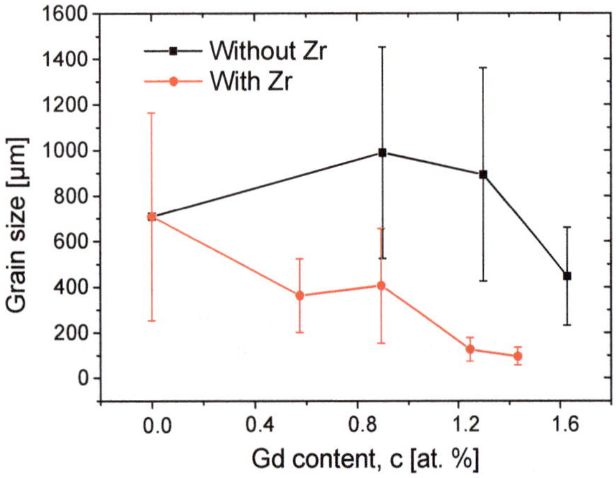

Fig. 1 Grain size of solid solution treated Mg-Gd alloys with or without Zr.

The microstructures of the as-cast and the T4 treatment Mg-2.65 at. % Gd binary alloys are shown in Fig. 2. It reveals that the as-cast alloy is mainly composed of continuous equiaxed dendrites. The distribution of Gd in the as-cast alloy is different in the different zones. The line distribution of Gd was analyzed, which is shown in Fig. 2b. Gd element segregates mainly at dendritic boundaries. The second phases were disappeared at the grain boundaries after the solution treatment (Fig. 2c). Only few small precipitates distribute inside the grains. This is the case for all studied alloys after solution treatment.

The main purpose of adding Zr element is to refine the grains of Mg-Gd alloys. Since its low concentration, the influence of Zr on solid strengthening can be neglected.

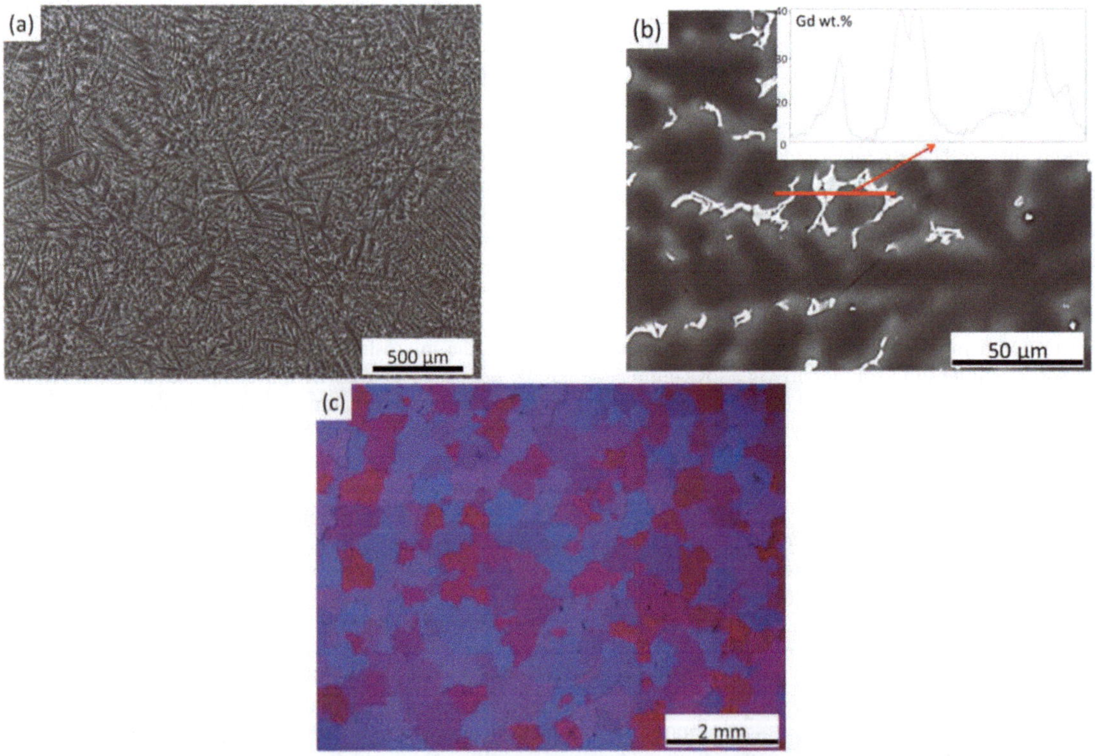

Fig. 2 Microstructures of Mg-2.65 at % Gd alloy: (a) and (b) SEM photos of as-cast alloy; (c) optical microstructure of solution treatment alloy.

As shown in Fig.3, it is obvious that the hardness of the experimental alloys increases monotonically with Gd content. The Mg-Gd alloys with or without Zr have a similar linear relationship. The hardness (H_{v5}) has the relation with the Gd content (c) according to the following equations for the alloys without Zr and with Zr, respectively:

$$H_{v5}\left(\text{kg mm}^{-2}\right) = 25.03 + 23.42\, c_{Gd}\ (\text{at. \%})\quad (R^2 = 0.97) \tag{1}$$

$$H_{v5}\left(\text{kg mm}^{-2}\right) = 25.99 + 25.23\, c_{Gd}\ (\text{at. \%})\quad (R^2 = 0.99) \tag{2}$$

Fig. 3 and Eq. (1), (2) indicate a near linear increase of hardness with solute concentration of Gd. The linear slope is 23.42 kg·mm^{-2}/at. % Gd for the alloys without Zr and 25.23 kg·mm^{-2}/at. % Gd for the alloys with Zr , respectively, which are much higher than that reported for Mg-Al alloys ($\approx$3.3 kg mm^{-2}/at. % Al) [17], Mg-Zn alloys ($\approx$9 kg mm^{-2}/at. % Zn) [18], Mg-Y alloys ($\approx$13.23 kg mm^{-2}/at. % Y) [19] and Mg-Sn alloys ($\approx$ 6.88 kg mm^{-2}/at. % Al) [20]. It is worth noting that the slope of Mg-Gd alloys ($\approx$14 kg mm^{-2}/at. % Gd) reported by Gao which is much lower than this study[12].

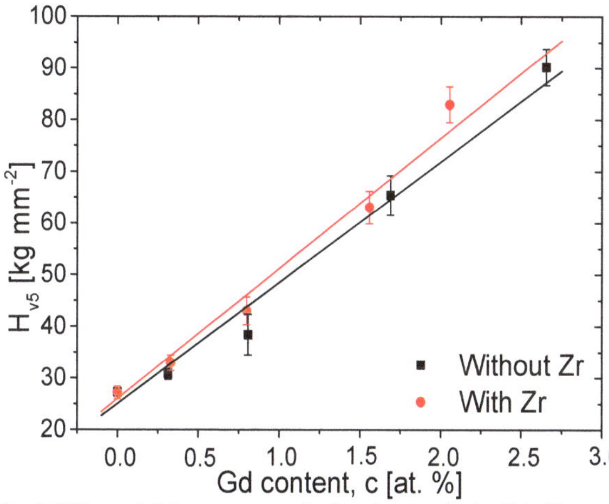

Fig. 3 Effect of Gd content on the hardness of Mg-Gd alloys.

3.2 Solid solution effects on yield strength

Fig.4 shows the typical strain-stress curves of Mg-Gd alloys with or without Zr after solid solution treatment. The corresponding values of tensile yield strength (TYS), ultimate tensile strength (UTS) and elongation are shown in Fig. 5. It can be seen that UTS and TYS increase obviously after adding Gd element, especially the TYS. A maximum value for the elongation is observed for Mg10Gd alloy. After alloying with Zr both UTS and TYS are also improved with increasing Gd content. The maximum elongation is obtained for the alloy Mg2GdZr.

The strengthening effects in the Mg-Gd alloys originate from solid solution and grain size strengthening effects. The grain size strengthening effect, defined as $\Delta\sigma_g$, can be expressed by Hall-Petch law:

$$\Delta\sigma_g = \sigma_0 + kd^{-1/2} \tag{3}$$

where σ_0 and k are parameters determined in the polycrystalline material and d is the grain size.

The k-value for Mg-Gd alloy was obtained in the literature [12]. For Mg-Gd alloys it has been suggested that k = 188 MPa/µm. With $\sigma_0 \approx 11$ MPa for pure Mg and d values from Table 1, the strengthening contribution from grain size can be calculated with Eq. (3). The effect of solid solution strengthening $\Delta\sigma_s$ can be calculated as follows:

$$\Delta\sigma_s = \sigma_{0.2} - \Delta\sigma_g = \sigma_{0.2} - (\sigma_0 + kd^{-1/2}) \tag{4}$$

where $\sigma_{0.2}$ is the 0.2% yield strength in Fig. 5.

Fleischer[21] and Labusch[22] reported that the strength in metallic materials is related to the concentration of solute atoms. The relationship between yield strength and solute concentration has the following equations:

$$\sigma_{y_s} = \sigma_{y_0} + Z_F G \xi_F^{3/2} c^{1/2} \tag{5}$$

$$\sigma_{y_s} = \sigma_{y_0} + Z_L G \xi_L^{4/3} c^{2/3} \tag{6}$$

where σ_{y_0} is the yield stress of pure magnesium; Z_F and Z_L are constants; ξ_F and ξ_L are different linear combination of the size misfit parameter and modulus misfit parameter, respectively; G is the shear modulus of pure magnesium; c is the atomic concentration of solute. The data points in Fig. 6 have been plotted following Fleischer and Labusch´s equations with n = 1/2 and 2/3. The best fitting lines show that both the alloys without (Fig. 6(a)) and with (Fig. 6(b)) Zr follow Eq. (5) and Eq. (6) well. The slopes of fitting lines for Mg-Gd alloys are 223.4 and 335.2 for n = 1/2 and 2/3 (R^2 = 0.999 and 0.999), respectively. For Mg-Gd-Zr alloys the slopes are 445.9 and 718.6 for n = 1/2 and 2/3 (R^2 = 0.957 and 0.937), respectively.

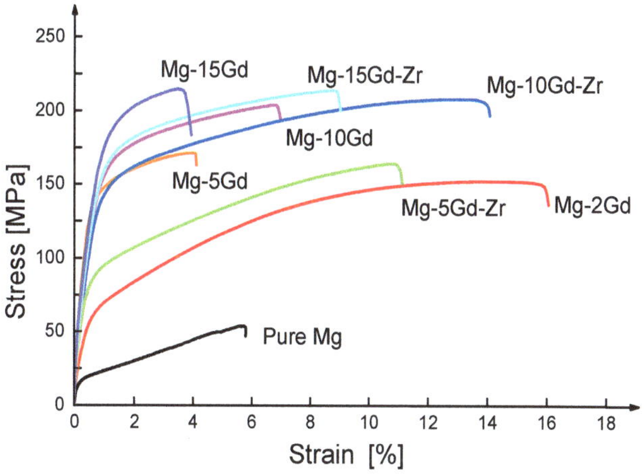

Fig. 4. Typical flow curves of Mg-Gd alloys (T4) with or without Zr.

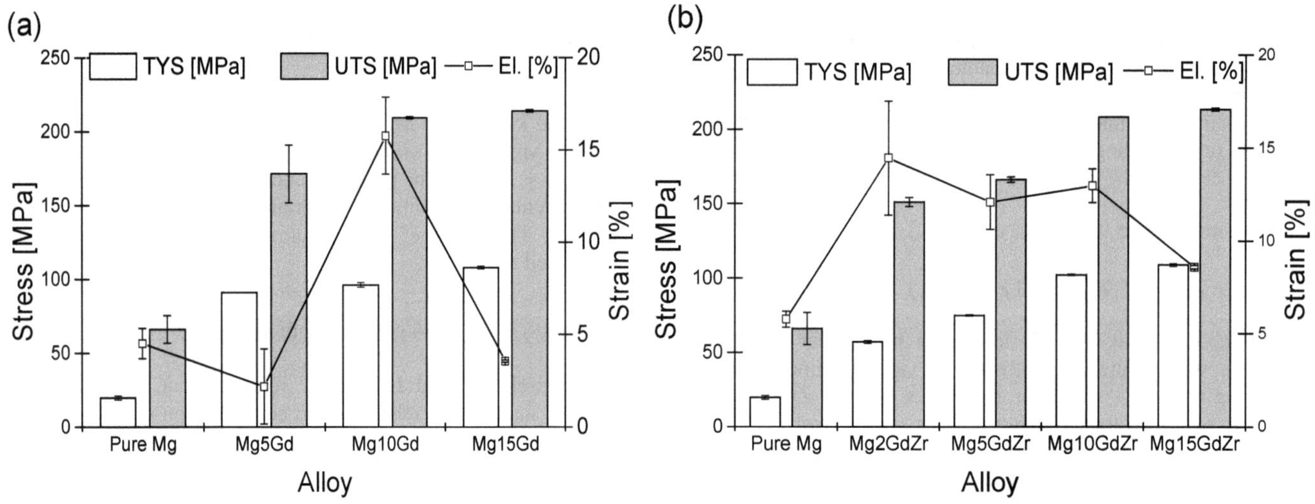

Fig. 5 Tensile properties of Mg-Gd alloys without (a) and with (b) Zr.

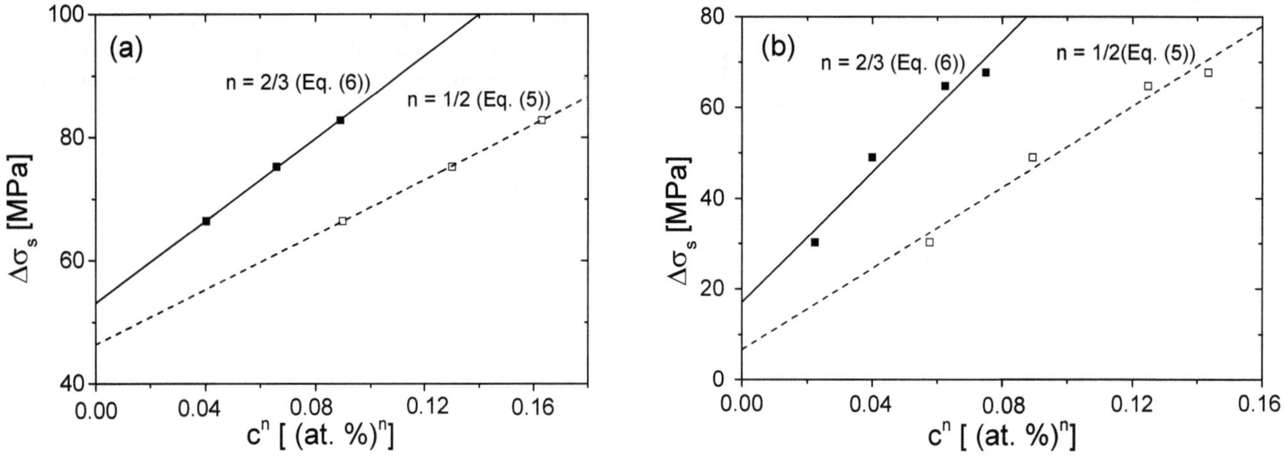

Fig. 6. $\Delta\sigma_s$ of alloys without (a) and with (b) Zr as a function of c^n for given values of n

4. Conclusions

Solid solution effects on the hardness and yield stress have been investigated for Mg-Gd and Mg-Gd-Zr alloys with Gd contents between 0.32 and 2.57 at. %

The hardness linearly increases with increasing Gd content with a slope of $\approx$ 23.42 ~ 25.23 kg·mm^{-2}/at. % Gd for alloys without and with Zr addition. After correcting for grain size strengthening, the yield strength also increases linearly with c^n, where c is the atom concentration of Gd, and n = 1/2 and 2/3.

Reference

1. K.U. Kainer, *Magnesium Alloys and Technology* (Weinheim, Germany: Wiley-VCH GmbH, 2003), 1-23.

2. L.L. Rikhlin, *Magnesium alloys containing rear earth metals: Structure and properties* (Taylor and Francis, London, 2003), 43-45.

3. H. Somekawa and C. A. Schuh. "Effect of solid solution elements on nanoindentation hardness, rate dependence, and incipient plasticity in fine grained magnesium alloys," *Acta Mater.*, 59 (2011), 7554-7563.

4. K.U. Kainer et al., "Status of the Development of Creep Resistant Magnesium Materials for Automotive Applications," *Mater Sci Forum*, 2009, 638-642.

5. X. Gao et al., "Microstructure Evolution in a Mg-15Gd-0.5Zr (wt.%) Alloy during Isothermal Aging at 250°C," *Mater. Sci. Eng. A*, 431 (2006), 322-327.

6. J. Yan et al., "Microstructure and mechanical properties in cast magnesium-neodymium binary alloys". *Mater. Sci. Eng A*, 476 (2008), 366-371.

7. M. Suzuki et al., "Effects of zinc on creep strength and deformation substructures in Mg-Y alloy," *Mater. Sci. Eng*

A, 387 (2004), 706-709.

8. K. Maruyama, M Suzuki and H. Sato, "Creep strength of magnesium-based alloys," *Metall. Mater. Trans. A*, 33 (2002), 875-882.

9. M. Suzuki et al., "Creep behavior and deformation microstructures of Mg-Y alloys at 550 K," *Mater Sci Eng A*, 252 (1998), 248-255.

10. F. Witte et al., "Degradable biomaterials based on magnesium corrosion," *Curr. Opin. Solid State Mater. Sci.* 12 (2008), 63-72.

11. A.A. Nayeb-Hashemi and J.B. Clark. *Phase Diagram of Binary Magnesium Alloys* (Metal Park, OH, USA: ASM International, 1988).

12. L. Gao, R.S. Chen and E.H. Han, "Effects of rare-earth elements Gd and Y on the solid solution strengthening of Mg alloys," *J. Alloys Compd.*, 481 (2009), 379-384.

13. B.Y. Tang et al., "First-principles investigation of the structural and mechanical properties of phase in Mg-Gd alloy system," *J. Alloys Compd.*, 492 (2010), 416-420.

14. Q. Peng, N. Ma and H. Li, "Gadolinium solubility and precipitate identification in Mg-Gd binary alloy," *J. Rare Earths*, 30 (2012), 1064-1068.

15. N. Hort et al., "Magnesium alloys as implant materials-Principles of property design for Mg-RE alloys," *Acta Biomater*, 6 (2010),1714-1725.

16. N. Stanford et al., "The effect of Gd on the recrystallization, texture and deformation behavior of magnesium-based alloys," *Acta Mater.*, 58 (2010), 6773-6783.

17. C.H. Caceres and D.M. Rovera, "Solid solution strengthening in concentrated Mg-Al alloys," *J. Light Metals*, 1 (2001), 151-156.

18. C.H. Caceres and A. Blake. "The Strength of concentrated Mg-Zn solid solutions," *Physica Status Solidi A*, 194 (2002), 147-158.

19. L. Gao, R.S. Chen and E.H. Han. "Solid solution strengthening behaviors in binary Mg-Y single phase alloys," *J. Alloys Compd.*, 472 (2009), 234-240.

20. B.Q. Shia, R.S. Chen and W. Ke. "Solid solution strengthening in polycrystals of Mg-Sn binary alloys," *J. Alloys Compd.*, 509 (2011), 3357-3362.

21. R.L. Fleischer, "Solution hardening," *Acta Metall.*, 9 (1961), 996-1000.

22. R. Labusch, "A Statistical theory of solid solution hardening," *Phys. Stat. Sol. A*, 41 (1970), 659-664.

Magnesium Technology 2016
Edited by: Alok Singh, Kiran Solanki, Michele V. Manuel, and Neale R. Neelameggham
TMS (The Minerals, Metals & Materials Society), 2016

EFFECTS OF HOMOGENIZATION ON MICROSTRUCTURE AND MECHANICAL PROPERTIES OF A ZE20 Mg ALLOY PROCESSED BY INDIRECT EXTRUSION

Z. McClelland[1,3,4], B. Li[2], S.J. Horstemeyer[1], M.F. Horstemeyer[1,3], A.L. Oppedal[1]
[1]Center for Advanced Vehicular Systems, Mississippi State University, Mississippi State, MS 39762
[2]Department of Chemical and Materials Engineering, University of Nevada, Reno, NV 89557
[3]Department of Mechanical Engineering, Mississippi State University, Mississippi State, 39762
[4]Engineer Research and Development Center Vicksburg, MS, 39180

Keywords: magnesium – rare earth alloy, extrusion, texture, EBSD, deformation twinning

Abstract

The effects of billet homogenization on final grain size, texture, and mechanical properties after indirect extrusion of a ZE20 magnesium (Mg) alloy were studied. Using a laboratory scale indirect extrusion process, both as-cast, and homogenized ZE20 billets were extruded into solid rods with an extrusion ratio of 25, at 454°C and at several ram speeds. Electron backscatter diffraction (EBSD) was used to characterize texture and grain size of the extruded material. Energy-dispersive X-ray spectroscopy (EDS) was used to characterize the effect of homogenization on elemental segregation. To determine the effects of homogenization and extrusion on mechanical properties, tension and compression tests were performed at room temperature and quasi-static test conditions. In tension, the homogenized extruded material exhibited a 29% elongation to failure, a 5% increase over the as-cast extruded material that obtained a 24% elongation to failure. The texture of the homogenized extruded material was also significantly weaker than that of the as-cast extruded materials.

Introduction

The desire to decrease weight in the automotive industry for improved fuel efficiency has led to the study of magnesium (Mg) and its alloys as a viable replacement for aluminum (Al) and steel in some applications [1-2]. Extrusions are particularly attractive for parts such as the subframes and bumper beams, which can be extruded in one step to manufacture the part. However, the limited ductility of Mg alloys at lower temperatures currently do not allow it to be a suitable replacement economically. The limited ductility or extrudability requires Mg alloys to be formed at lower speeds or higher temperatures, requiring more time and energy to manufacture Mg parts.

Low ductility in Mg alloys is attributed to the limited number of deformation modes available, due to the hexagonal close packed (HCP) structure of Mg. Based on the von Mises criterion for uniform deformation, five independent slip systems are required [3]. At room temperature only basal slip $\{0001\}< 11\bar{2}0 >$ has a low enough critical resolved shear stress (CRSS) to be activated, giving only two slip systems to accommodate strain. Non-basal slip in pure Mg has been shown to require up to 100 times the energy to be activated than basal slip [3]. A current avenue to increase the number of available slip systems is alloying Mg. Koike et al. [4] showed an AZ31 Mg alloy deformed by non-basal slip in 40% of the sample when tested in tension at room temperature. Koike et al. [4] have also shown that the CRSS for pyramidal $\{11\bar{2}2\} < \bar{1}\bar{1}23 >$ and prismatic slip $\{1\bar{1}00\} < 11\bar{2}0 >$ were substantially lowered to 2.5 and 2.2 times what is required for basal slip.

One recent focus of research to increase the ductility of Mg is the addition of rare earth elements (RE)[5-8]. Rare earth elements are added in either a misch-metal form or pure RE elements. Misch-metal is the combination of several RE elements and can provide a lower cost alternative to adding pure RE elements for alloying. The addition of RE elements to Mg extrusion alloys has been shown to lower the texture intensities as well as change the orientation of the produced texture [9-11]. Mishra et al. [12] showed that the addition of 0.2% Cerium (Ce) to Mg not only decreased the texture intensity, but also increased the room temperature ductility of the extruded alloy significantly. These improvements were attributed to smaller grain size, changed slip distribution, and a texture favorable for basal dislocation activity [12]. Mackenzie et al. [11] found that a ZE10 alloy, containing less than 1% Zn and 0.2% Ce, underwent a 45° reorientation during recrystallization of the texture producing a split "non-basal" texture.

A ZE20 Mg alloy containing 2% Zn and 0.3% Ce in wt % has shown particular promise for the application of extrusions due to a decreased extrusion texture and increased ductility [12-14]. In the present study, an in-house built, indirect, extrusion set up was used to extrude cast and homogenized ZE20 Mg alloy at various ram speeds. The effect of homogenization of the cast microstructure on the final extruded microstructure as well as mechanical properties was studied using scanning electron microscopy (SEM), electron backscatter diffraction (EBSD), energy dispersive x-ray spectroscopy (EDS), and mechanical testing. The purpose of this study was to better understand the microstructural and mechanical response of an extruded ZE20 Mg alloy.

Experimental Methods

The ZE20 material for this study was supplied by the United States Automotive Materials Partnership (USAMP) in the form of cast ingots containing 2% Zn and 0.3% Ce (all values for the present study are given in wt. %). Small sections of a 4 inch tall by 6 inch diameter cast billet were cut in two directions to obtain the initial microstructure. The directions were designated normal direction (ND) and radial direction (RD).

Billets from the ZE20 cast ingot were machined to dimensions of 31.75 mm diameter and 25.4 mm height using a 33.02 mm hole saw followed by a mechanical lathe. Some of the cast billets were homogenized at 450°C for 5 hours in an argon atmosphere and then quenched in water. The billets were extruded at 454°C using a laboratory scale, indirect extrusion fixture and an Instron 8850 test frame with a max load capacity of 298 kN. Each 31.75 mm diameter billet was extruded into a 6.35 mm solid rod extrudate, an extrusion ratio of 25. The as-cast samples were extruded at 10, 20, 30, and 40 mm/min ram speed and the homogenized billets were extruded at the upper and lower bounds of 10 and 40 mm/min.

After extrusion, tension and compression samples for both the as-cast extruded and the homogenized extruded materials were machined from the middle section of the extrudate. Compression samples were cut to a diameter of 6.35 mm and 9.53 mm in height and the cylindrical tension samples were machined to a gauge length of 6.5 mm with a diameter of 3.2 mm. Tension and compression tests were performed at room temperature and 0.001 s^{-1} strain rate using an Instron 5869 dual column mechanical test frame.

Scanning electron microscopy (SEM), electron backscatter diffraction (EBSD), and energy-dispersive X-ray spectroscopy (EDS) was performed to characterize the microstructure. Sections of the as-cast extruded and homogenized extruded materials were cut along the extrusion direction (ED) as well as the radial direction (RD). Metallographic samples were cold mounted and each sample for EDS was grinded using 1200, 2400, and then 4000 grit SiC paper followed by polishing with Struers Mol, Nap, and diamond suspension. The last polishing step consisted of using a suspension of 0.05 micron alumina in ethylene glycol with a Struers Chem cloth. The samples were polished for 45 min using a force of 2.5 N and 15 RPM. After polishing, the samples were analyzed for chemical content using a Zeiss Supra 40 FEG scanning electron microscope with an EDS detector. Line scans were taken in various locations with a step size of 0.05 µm. The area fraction of Mg-Zn-Ce intermetalics was then quantified using traditional image processing techniques of analyzing differences in grey scale over multiple SEM images.

Samples for EBSD were polished with 1200, 2000, and 4000 grit SiC paper and then electropolished using a Struers LectroPol-5 with Struers C1 (160 mg sodium thiocyanate, 800 ml ethanol,80 ml ethylene glycol monobutyl ether, and 20 ml distilled water solution) at 20 volts for 20 seconds. The electropolish procedure was repeated until the desired surface quality was obtained. After electropolishing, the samples were etched using an Acetic-Nital solution (40 ml distilled water, 120 ml ethanol, 30 ml acetic acid, and 10 ml nitric acid) for 5 seconds to increase pattern indexing quality. A Zeiss Supra 40 FEG SEM operating at 20 kV and 120 µm aperture with an EDAX Hikari camera was used to capture EBSD data. EBSD scans were taken with an area of 400 × 400 µm and a step size of 0.3 µm. EBSD data was analyzed using TSL OIM Analysis 6 software to generate IPF maps and pole figures.

Results and Discussion

The initial microstructure of the as-cast and homogenized ZE20 was studied using SEM. SEM observations revealed a large dendritic microstructure in the as-cast material with an average grain size of 526 µm. Figure 1a shows a SEM micrograph of the cast material. The darker sections of the image are the dendrite arms and the lighter areas are the interdendritic regions. Using EDS, the white precipitates along the grain boundaries were found to consist of an Mg-Zn-Ce composition. Figure 1b shows the SEM image of the homogenized cast material. The segregation that causes the contrast between the dendrites and the interdendritic regions (Figure 1a) was eliminated by the homogenization.

Figure 2a shows an EDS line scan at the boundary of the homogenized cast material. As the scan crossed over the precipitate, the amount of both Ce and Zn increased. Figure 2b shows a larger EDS line scan of the as-cast material. Within the grain, the scan revealed only Mg and Zn. As the scan crossed the precipitates of the grain boundaries, there was a spike in Ce indicating that Ce is only present in the precipitates since the solubility of Ce in solid Mg is close to zero at room temperature.

EDS was also used to determine the distribution of Ce before and after the homogenization process. The difference in area of the Mg-Zn-Ce intermetallics between the as-cast and homogenized materials was calculated to better understand the effect of the homogenization heat treatment on the microstructure of the cast ZE20. The original area of the Mg-Zn-Ce in the as-cast material was found to be 1.4%, and the area of the homogenized material was found to be 1.1%. The slight change to the percent area was expected due to the limited amount of solubility of Ce in Mg. According to the phase diagram, Ce only has a 0.06% solubility in Mg at 500°C [15]. Zinc however has a 6.2% solubility in Mg at a eutectic temperature of 340°C, allowing Zn to dissolve more readily into the matrix and provide solid solution strengthening [10].

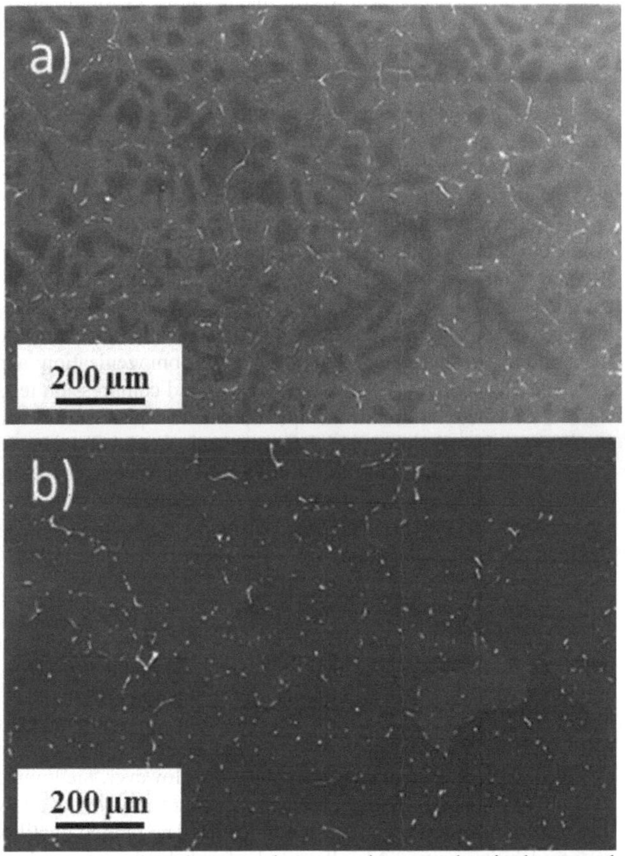

Figure 1. Scanning electron microscope images taken in the normal direction of a) as-cast and b) homogenized materials. Both images show a large segregation of precipitates at the grain boundaries. Scale bars are 200 µm.

Mechanical testing of the as-cast and homogenized materials was performed in tension to better understand the ductility of the material and can be seen in Figure 3. The as-cast material exhibited a higher yield stress of 90 MPa compared to the homogenized material at 60 MPa. The homogenized material exhibited a higher elongation to failure with an increase of approximately 5% total strain compared to the as-cast material. The as-cast and homogenized samples both displayed a large amount of scatter in the elongation to failure from test to test; this amount of scatter is expected with a non-homogenous cast microstructure. The decrease in strength after homogenization could be due to the slight decrease in the volume fraction of precipitates.

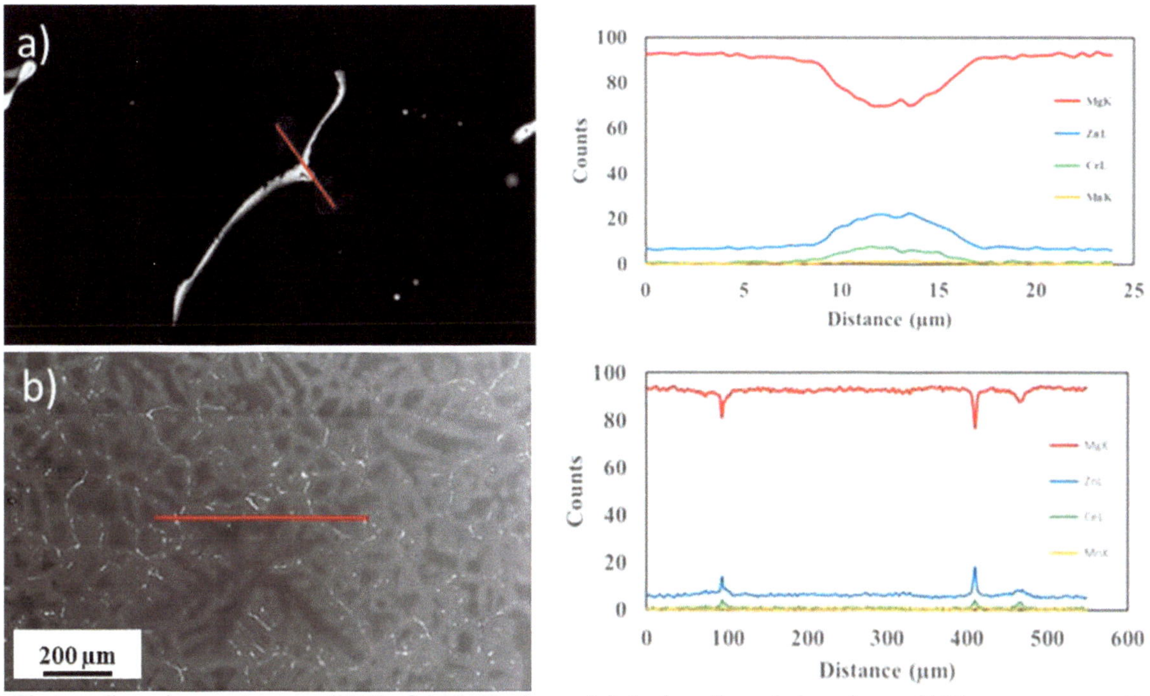

Figure 2. Electron dispersive spectroscopy (EDS) of a) Mg-Zn-Ce precipitate along the grain boundary and b) line scan across entire grain showing spikes in zinc and cerium along the interdendritic region. Elements in line scan map are Mg, Zn, Ce, and Mn from top to bottom.

As-cast and homogenized extruded ZE20

The maximum load data from the Instron 8850 during extrusion can be seen in Table 1. The maximum load data indicated that as the ram speed increased, the load required to perform the extrusion increased. The as-cast extruded sample required a max load of 165 kN when extruded at 10 mm/min and 202 kN when extruded at 40 mm/min. There was also a noticeable increase in maximum required load between the as-cast extruded and the homogenized extruded materials. The homogenized extruded materials required approximately 45 kN more force when extruded at 10 mm/min and 60 kN more when extruded at 40 mm/min. The requirement of a higher load seems counter intuitive considering the yield stress of the homogenized material was lower than that of the as-cast. The answer to this is found in the microstructural analysis of the extruded materials. The homogenized extruded material produced a significantly smaller grain size than the as-cast extruded material suggesting a change in the recrystallization kinetics that contribute to the increase in overall load required to perform the extrusion.

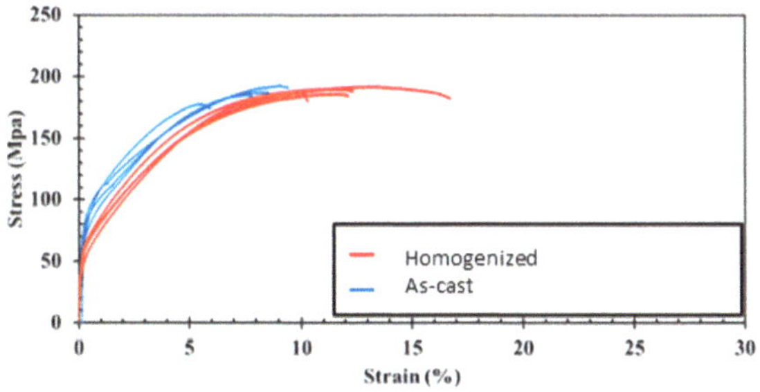

Figure 3. Plot of engineering stress vs. engineering strain behavior of the as-cast and homogenized materials tested in tension at room temperature and 0.001 s^{-1} strain rate

To better understand the resulting microstructure and texture of the as-cast extruded and homogenized extruded ZE20 at 10 and 40 mm/min ram speeds, EBSD scans were taken halfway through the extrudate in both ED and RD. The EBSD scans revealed a homogenized microstructure with an average grain size of 24 µm in ED and 23 µm in RD over the 2 different extrusions for the as-cast extruded material. EBSD of the homogenized, extruded material revealed a homogenous microstructure with an average grain size of approximately 17 µm and 19 µm for the 10 and 40 mm/min samples, respectively. Recrystallization was studied using the EBSD data and grain orientation spread (GOS) analysis[18]. All of the extrudates were fully recrystallized when extruded at 454˚C suggesting that at the time of cooling, grain nucleation and growth had already been completed. The 40 mm/min as-cast extruded sample displayed the largest grain size with an average diameter of 26 µm. As the ram speed increased, the local temperature at the bearing increased. The increase in temperature led to rapid grain growth of the recrystallized grains, changing the overall average grain size. The as-cast extruded sample with the lowest ram speed, 10 mm/min, displayed remnants of extension twins.

Table 1. Maximum load values to perform extrusions of as-cast and homogenized materials

	Extrusion Speed (mm/min)	Max Load (kN)
As-Cast	10	165
	40	202
Homogenized	10	211
	40	264

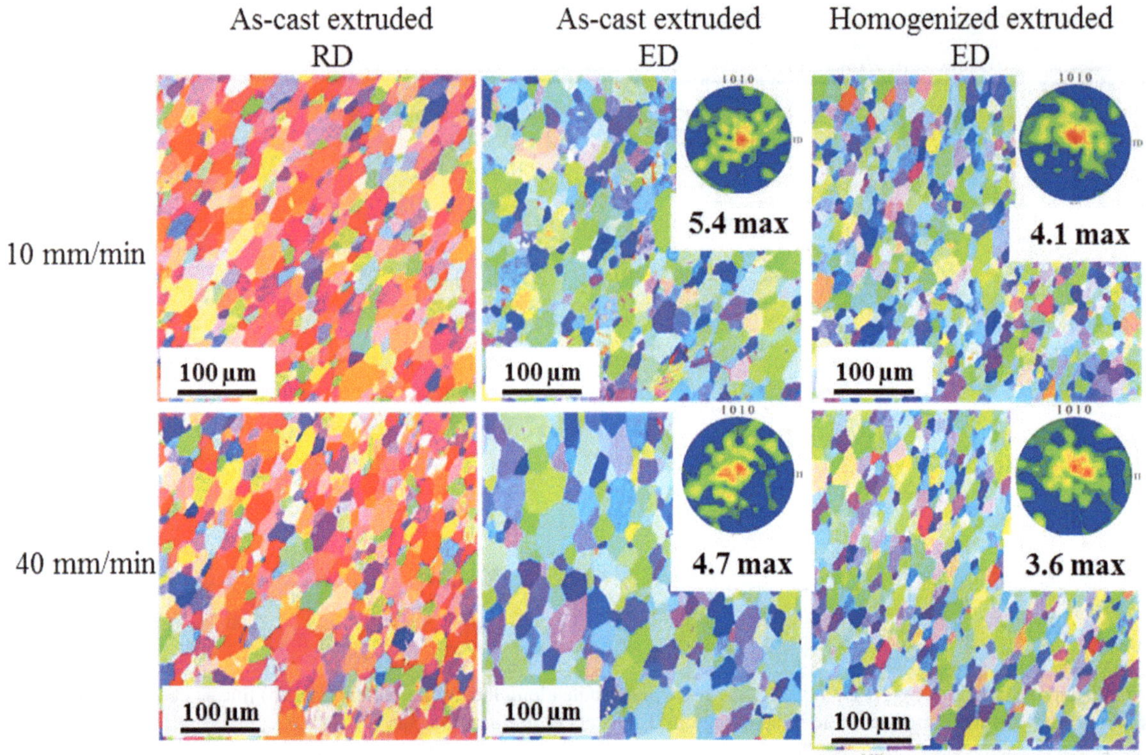

Figure 4. Electron backscatter diffraction comparison of the as-cast extruded materials in RD (Left) and ED (Center) and homogenized extruded materials in ED (Right) for the 10 and 40 mm/min. Scale bar for each image is 100 μm.

The extruded material exhibited the expected prismatic texture that is typical of extruded Mg alloys with the < c >-axis aligned perpendicular to the extrusion direction. The as-cast, extruded 10 mm/min sample was found to have the strongest texture intensity with a value of 5.4 multiples of random density (MRD).

Yu et al. (2013) showed that as the extrusion speed was increased the texture intensity was decreased due to the increased rate of recrystallization. The present data shows a slight decrease in overall texture strength when comparing the 10 mm/min as-cast, extruded sample and the 40 mm/min sample. The maximum texture intensities of the homogenized, extruded material were at least a point lower in both the 10 and 40 mm/min extrudates compared to the as-cast, extruded material. The weakening of the texture can be attributed to the reduced size of the dynamically recrystallized grains in the 40 mm/min sample with a more random orientation. A decrease in texture would suggest an increase in ductility based on the findings of Stanford et al. [6]. They showed that as the texture intensity decreased the amount of ductility decreased for a commercial M1 Mg alloy with Ce rich misch metal added.

Mechanical responses

Mechanical testing in both tension and compression revealed a yield and UTS strength asymmetry between the two loading directions. Tension and compression testing were performed along the extrusion direction. The compression tests exhibited a maximum failure stress of 425 MPa for the as-cast, extruded material. Both as-cast and homogenized, extruded samples displayed very similar stress vs. strain curves with typical strain hardening when tested in compression in which twinning dominates the initial deformation. Sigmoidal shaped curves, as seen when tested in compression, are indicative of extension twinning which was verified using EBSD and can be seen in the pole figures of Figure 5. The as-cast, extruded samples tested in tension reached an elongation to failure of 23% whereas, the homogenized, extruded samples reached an elongation of 29%. The higher ductility can be explained by the decrease in texture intensity in the homogenized extruded materials. The decreased texture intensity allows for more deformation modes to be activated, readily allowing for slip and increasing the deformation allowed before failure. Slightly after yield, the material exhibits behavior like a twinning hump representative of extension twinning with a fair amount of strain hardening, but not as significantly as the compression samples. The twinning hump is a slight inflection in the strain hardening region of the stress strain curve and has been observed occasionally when tested in tension along the extrusion direction at elevated temperatures [19], [20]. The ultimate tensile strength (UTS) of the as-cast, extruded 40 mm/min sample was found to be 235 MPa when tested in tension.

When tested in tension, the homogenized extruded materials displayed slightly different behaviors between the 10 and 40 mm/min samples. The 10 mm/min sample had a slightly higher strain hardening rate than that of the 40 mm/min sample. The 40 mm/min sample had a slightly higher elongation to failure at 29% compared to the 27% of the 10 mm/min sample. These differences can be attributed to the change in both grain size and texture of the homogenized, extruded materials between the 10 and 40 mm/min extrusions. The 10 mm/min sample has a lower grain size and a higher texture intensity when compared to the 40 mm/min sample. The decreased grain size increased the amount of strength seen in the 10 mm/min sample and the increased texture decreased the amount of ductility the material exhibited. The increased texture also effected the "twinning hump" by having more grains favorable for extension twinning; therefore, increasing the strain hardening rate seen in the 10 mm/min sample.

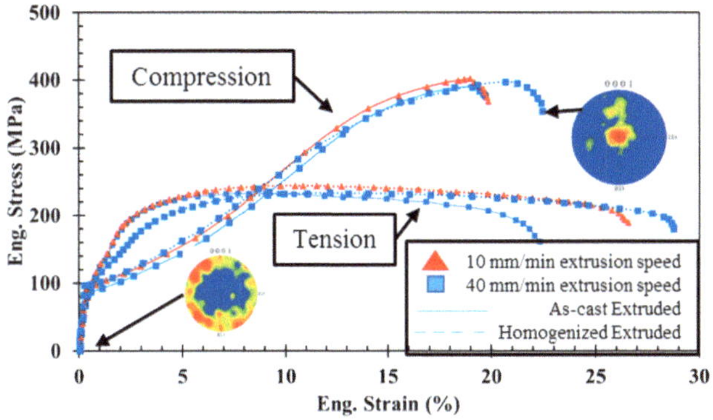

Figure 5. Stress vs. strain of extruded ZE20 in tension and compression along the extrusion direction at 0.001 s^{-1} strain rate at room temperature. The starting texture is the prismatic texture typical of extrusion. The "sigmoidal" shape during compression is typical of strong twinning activity in the deformation behavior. Dashed lines represent homogenized extruded material.

Conclusions

Slight mechanical and microstructural differences were observed between the as-cast and homogenized materials. The homogenization eliminated the micro-segregation during casting, resulting in a higher elongation to failure. The homogenized material also displayed a lower yield than that of the original as-cast material. When the two materials were extruded the homogenized extruded material required a higher load, by at least 50 kN, to perform the extrusion at the same parameters over the as-cast material. The increase in required load was found to be caused by a decreased grain size in the homogenized extruded material. The as-cast extruded and homogenized extruded materials both exhibited an increase in grain size as the ram speed increased. The increase in grain size can be attributed to the increase in temperature at the bearing due to the increase in extrusion speed.

The homogenized extruded material displayed a decrease in both texture intensity and grain size as the extrusion ram speed increased. The texture change is most likely due to the reduced size of the DRX grains. The homogenized extruded materials displayed approximately 5 % more elongation to failure than the as-cast extruded materials due to the change in texture intensities allowing for more deformation modes to accommodate the strain.

Traditional Mg alloys that have been extruded typically display a low ductility. Lou et al. (2011) reported only 17% elongation to failure of an extruded AZ31 Mg alloy [10]. However, Mishra et al. (2008) reported 31% elongation to failure of an Mg-Ce alloy [12]. The elongation to failure found in the present study matches well with that reported by Luo et al..

Acknowledgments

This material is based upon work supported by the Department of Energy National Energy Technology Laboratory under Award Number DE-EE0005660, with the United States Automotive Partnership LLC (USAMP). This report was prepared as an account of work sponsored by an agency of the United States Government. Neither the United States Government nor any agency thereof, nor any of their employees, makes any warranty, express or implied, or assumes any legal liability or responsibility for the accuracy, completeness, or usefulness of any information, apparatus, product, or process disclosed, or represents that its use would not infringe privately owned rights. Reference herein to any specific commercial product, process, or service by trade name, trademark, manufacturer, or otherwise does not necessarily constitute or imply its endorsement, recommendation, or favoring by the United States Government or any agency thereof. The views and opinions of authors expressed herein do not necessarily state or reflect those of the United States Government or any agency thereof.

References

[1] M. K. Kulekci, "Magnesium and its alloys applications in automotive industry," *Int. J. Adv. Manuf. Technol.*, vol. 39, no. 9–10, pp. 851–865, Nov. 2008.

[2] J. Hirsch and T. Al-Samman, "Superior light metals by texture engineering: Optimized aluminum and magnesium alloys for automotive applications," *Acta Mater.*, vol. 61, no. 3, pp. 818–843, Feb. 2013.

[3] H. Wang, "Constitutive Modeling of Hexagonal Close Packed Polycrystals," McMaster University, 2011.

[4] J. Koike, T. Kobayashi, T. Mukai, H. Watanabe, M. Suzuki, K. Maruyama, and K. Higashi, "The activity of non-basal slip systems and dynamic recovery at room temperature in fine-grained AZ31B magnesium alloys," *Acta Mater.*, vol. 51, no. 7, pp. 2055–2065, Apr. 2003.

[5] N. Stanford and M. R. Barnett, "The origin of 'rare earth' texture development in extruded Mg-based alloys and its effect on tensile ductility," *Mater. Sci. Eng. A*, vol. 496, no. 1–2, pp. 399–408, Nov. 2008.

[6] N. Stanford, D. Atwell, A. Beer, C. Davies, and M. R. Barnett, "Effect of microalloying with rare-earth elements on the texture of extruded magnesium-based alloys," *Scr. Mater.*, vol. 59, no. 7, pp. 772–775, Oct. 2008.

[7] T. Al-Samman and X. Li, "Sheet texture modification in magnesium-based alloys by selective rare earth alloying," *Mater. Sci. Eng. A*, vol. 528, no. 10–11, pp. 3809–3822, Apr. 2011.

[8] T. Wu, L. Jin, W. X. Wu, L. Gao, J. Wang, Z. Y. Zhang, and J. Dong, "Improved ductility of Mg–Zn–Ce alloy by hot pack-rolling," *Mater. Sci. Eng. A*, vol. 584, pp. 97–102, Nov. 2013.

[9] J. Bohlen, M. R. Nürnberg, J. W. Senn, D. Letzig, and S. R. Agnew, "The texture and anisotropy of magnesium–zinc–rare earth alloy sheets," *Acta Mater.*, vol. 55, no. 6, pp. 2101–2112, Apr. 2007.

[10] A. A. Luo, R. K. Mishra, and A. K. Sachdev, "High-ductility magnesium–zinc–cerium extrusion alloys," *Scr. Mater.*, vol. 64, no. 5, pp. 410 413, Mar. 2011.

[11] L. Mackenzie and M. Pekguleryuz, "The recrystallization and texture of magnesium–zinc–cerium alloys," *Scr. Mater.*, vol. 59, no. 6, pp. 665–668, Sep. 2008.

[12] R. K. Mishra, A. K. Gupta, P. R. Rao, A. K. Sachdev, A. M. Kumar, and A. A. Luo, "Influence of cerium on the

texture and ductility of magnesium extrusions," *Scr. Mater.*, vol. 59, no. 5, pp. 562–565, Sep. 2008.

[13] S. H. Park, H. Yu, J. H. Bae, C. D. Yim, and B. S. You, "Microstructural evolution of indirect-extruded ZK60 alloy by adding Ce," *J. Alloys Compd.*, vol. 545, pp. 139–143, Dec. 2012.

[14] T. Laser, C. Hartig, M. R. Nürnberg, D. Letzig, and R. Bormann, "The influence of calcium and cerium mischmetal on the microstructural evolution of Mg–3Al–1Zn during extrusion and resulting mechanical properties," *Acta Mater.*, vol. 56, no. 12, pp. 2791–2798, Jul. 2008.

[15] V. Pavlyuk, B. Marciniak, and E. Różycka-Sokołowska, "The isothermal section of the phase diagram of Ce–Mg–Zn ternary system at 470 K," *Intermetallics*, vol. 20, no. 1, pp. 8–15, Jan. 2012.

[16] A. Jain, O. Duygulu, D. W. Brown, C. N. Tomé, and S. R. Agnew, "Grain size effects on the tensile properties and deformation mechanisms of a magnesium alloy, AZ31B, sheet," *Mater. Sci. Eng. A*, vol. 486, no. 1–2, pp. 545–555, Jul. 2008.

[17] E. O. Hall, "The Deformation and Ageing of Mild Steel: III Discussion of Results," *Proc. Phys. Soc. Sect. B*, vol. 64, no. 9, pp. 747–753, Sep. 1951.

[18] W. Polkowski, P. Jóźwik, and Z. Bojar, "EBSD and X-ray diffraction study on the recrystallization of cold rolled Ni3Al based intermetallic alloy," *J. Alloys Compd.*, vol. 614, pp. 226–233, Nov. 2014.

[19] S. H. Park, B. S. You, R. K. Mishra, and A. K. Sachdev, "Effects of extrusion parameters on the microstructure and mechanical properties of Mg–Zn–(Mn)–Ce/Gd alloys," *Mater. Sci. Eng. A*, vol. 598, pp. 396–406, Mar. 2014.

[20] L. Li, J. Zhou, and J. Duszczyk, "Determination of a constitutive relationship for AZ31B magnesium alloy and validation through comparison between simulated and real extrusion," *J. Mater. Process. Technol.*, vol. 172, no. 3, pp. 372–380, Mar. 2006.

Magnesium Technology 2016
Edited by: Alok Singh, Kiran Solanki, Michele V. Manuel, and Neale R. Neelameggham
TMS (The Minerals, Metals & Materials Society), 2016

AGE-HARDENING OF DUAL PHASE Mg-Sc ALLOY AT 573 K

Yukiko Ogawa, Daisuke Ando, Yuji Sutou, Junichi Koike

Department of Materials Science, Tohoku University;
6-6-11-1016, Aoba, Aramaki, Aoba-ku, Sendai, 980-8579, Japan

Keywords: Magnesium alloy, Age hardening, bcc/hcp dual phase

Abstract

In this study, we investigated the effect of aging treatment at 573 K on hardness in Mg-Sc alloy with bcc/hcp dual phase prepared by annealing at 873 K. It was found that hardness abruptly increased after some incubation time, and then reached steady state level. Moreover, by further aging, the hardness increased again. It was confirmed by SEM observation and XRD that the initial drastic hardening was due to the precipitation of α phase. Furthermore, XRD results suggested that the secondary hardening was attributed to the formation of MgSc phase with ordered B2 structure.

Introduction

Mg alloys have been attracting attention for next generation structural materials because of its high specific strength. However, conventional Mg alloys have been used only in a limited field because of their poor formability. This is due to the anisotropic hcp structure of Mg alloys. Thus, the approach introducing bcc structure which can deform isotropically in hcp matrix has been suggested.

Among binary Mg-X systems, only Mg-Li and Mg-Sc alloys have a bcc structure in a Mg rich composition region. Mg-Li alloys with bcc/hcp dual-phase is reported to show a high elongation [1]. However, Mg-Li alloys shows low tensile strength and poor corrosion resistance [2,3]. Meanwhile, recently our group has found that Mg-Sc alloy with bcc (β) / hcp (α) dual-phase shows much higher tensile strength than Mg-Li alloys [4]. Moreover, β/α dual phase boundary of Mg-Sc system shifts towards a higher Mg content with increasing temperature [4-7]. This indicates that the microstructure of Mg-Sc alloy can be controlled by aging treatment after annealing or solution treatment at high temperature. In fact, our group achieved a high hardness of over 230 Hv by aging treatment after solution treatment, attributing to the precipitation of a fine α phase in β single-phase [8].

In this report, we present the aging effect on hardness in β/α dual phase Mg-Sc alloy at 573 K.

Experimental

The Mg-Sc alloys were prepared by induction melting, where melted alloy was cooled in crucible. The nominal composition of Sc was 20 at.%. The obtained samples were annealed at 873 K for 1 h followed by water quenching to obtain an α+β dual phase. Finally, samples were aged at 573 K in air. The hardness of aged samples was measured by Vickers hardness tester under a load of 1 kg. The microstructure of samples after aging was observed by scanning electron microscopy (SEM), where samples were mechanically polished and then chemically etched before SEM observation. The crystal structure of samples was determined by XRD measured with Cu-Kα.

Results and discussion

Figure 1 shows age hardening behavior at 573 K. The hardness drastically increases up to around 120 Hv after some incubation time. And then, the hardness reaches steady state level. After aging for 18000 s, the hardness increases again and reaches a maximum value of 128 Hv.

The microstructure of samples observed by SEM are shown in Fig. 2. Fig. 2(a) shows as-annealed sample at 873 K. Dark contrast region is α phase and bright contrast region is β phase, where black color regions are cavities. Fig. 2(b), (c) and (d) show microstructures of samples aged at 573 K for 300 s, 600 s and 90000 s, respectively. In the aged samples, fine precipitates appear in the β phase and their volume fraction and size increase with increasing aging time. It can be seen that the hardness dramatically increases with increasing the number of fine precipitates.

Figure 3 shows XRD patterns of the as-annealed sample and samples aged for 600 s and 90000 s. In Fig. 3, white circles denote α phase, while full circle denotes β phase. The as-annealed sample has sharp peaks derived from α and β phases. For the sample aged for 600 s, the intensity of α peak at about 32° becomes stronger than that of the as-annealed sample. Combining with SEM observation, the fine precipitates in β phase after aging seem to be α phase. Meanwhile, for the sample aged for 90000 s, small peaks at 24.8° as indicated by star mark and at 34.7° as indicated by grayed square were clearly observed. According to reported phase diagram, there is ordered bcc (B2)-MgSc phase below 723 K [5-7]. The small peak at 24.8° is well agreed with the peak position of 100 B2-MgSc phase. And the small peak at 34.7° is considered to be 110 B2-MgSc phase. Therefore, the secondary hardening is suggested to be due to the formation of B2-MgSc phase. However, the existence of B2-MgSc phase has been not yet confirmed by TEM observation. Further study will be needed in order to understand the mechanism of the secondary hardening of Mg-Sc alloy.

Conclusion

The age hardening of Mg-Sc alloy with bcc/hcp dual phase alloy was confirmed at 573 K. The hardness drastically increased up to 116 Hv after aging for 600 s and then reached steady state level. Moreover, the hardness further increased after aging for 90000 s. From SEM and XRD results, the initial hardening was due to the precipitation of α phase. It was suggested by XRD that the secondary hardening may be due to the precipitation of B2-MgSc phase.

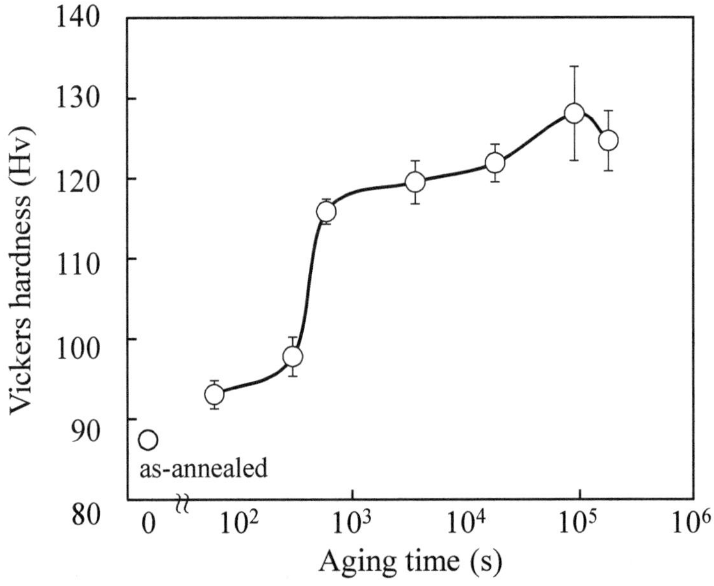

Figure. 1 Age hardening behavior at 573 K.

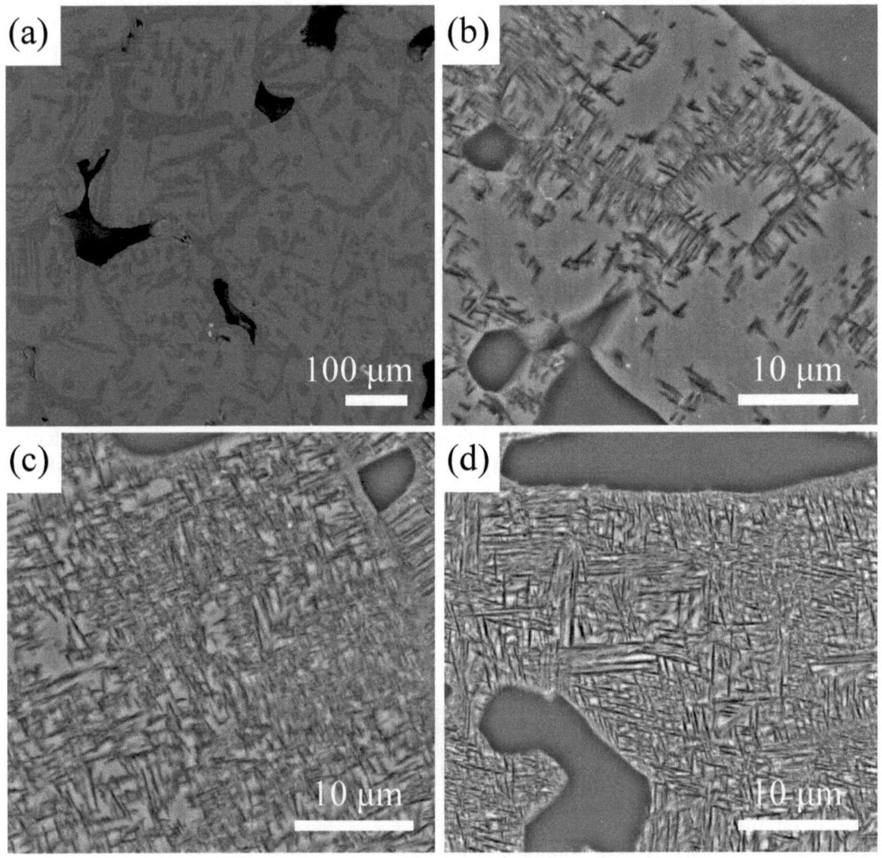

Figure. 2 Microstructure of samples aged for (a) 0 s (as-annealed), (b) 300 s, (c) 600 s, (d) 90000 s.

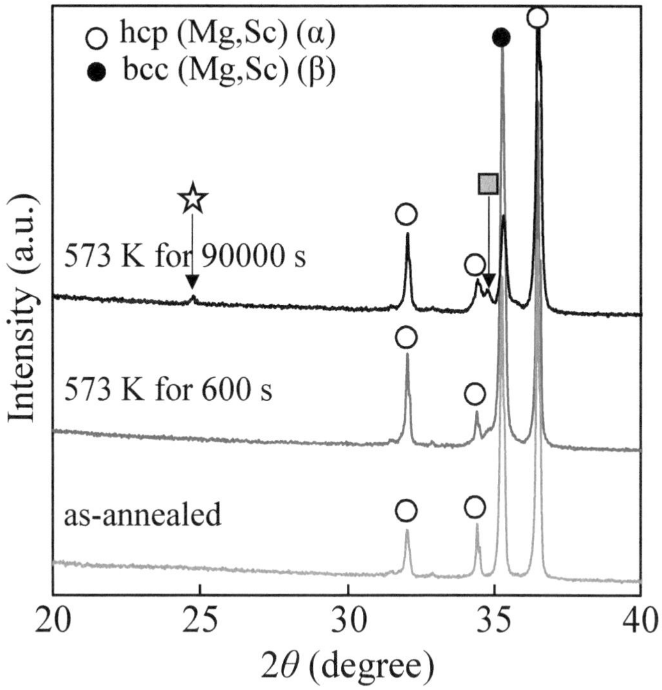

Figure. 3 XRD patterns of as-annealed samples and the samples aged for 600s and 90000 s.

Acknowledgement

This work was supported by KAKENHI (15H05549).

References

[1] G. González-Doncel. et al., "The use of foil metallurgy processing to achieve ultrafine grained Mg-9Li laminates and Mg-9Li-5B$_4$C particulate composites", *Journal of Materials Science*, 25 (1990), 4535-4540.
[2] T.C. Chang et al., "Mechanical properties and microstructures of various Mg-Li alloys", *Materials Letters*, 60 (2006), 3272-3276.
[3] C. Zhang et al., "Electrochemical characterization of the corrosion of a Mg-Li Alloy", *Materials Letters*, 62 (2008), 2177-2180.
[4] Y. Ogawa et al., "Determination of hcp/bcc boundaries and mechanical properties of dual phase alloy in binary Mg-Sc system", *Proceedings of the International Conference on Solid-Solid Phase Transformations in Inorganic Materials 2015*, (2015), 419-420.
[5] B.J. Beaudry and A.H. Daane, "A study of the scandium-magnesium system from 0 to 60 at. % scandium", *Journal of the Less-Common Metals*, I8 (1969), 305-308.
[6] A.A. Nayeb-Hashemi and J.B. Clark, "The Mg-Sc (Magnesium-Scandium) System", *Bull. Alloy Phase Diagrams*, 7(6) (1988), 574-577.
[7] H. Okamoto, "Mg-Sc (Magnesium-Scandium)", *Journal of Phase Equilibria and Diffusion*, 30(6) (2009), 660-661.
[8] D. Ando et al., "Age-hardening effect by phase transformation of High Sc containing Mg alloy", *Materials Letters*, 161 (2015) 5-8.

Magnesium Technology 2016
Edited by: Alok Singh, Kiran Solanki, Michele V. Manuel, and Neale R. Neelameggham
TMS (The Minerals, Metals & Materials Society), 2016

The Structure of β" and β' in an Aged Mg-Nd Alloy

Ellen L. S. Solomon[1], Emmanuelle A. Marquis[1]
[1]University of Michigan, 2300 Hayward Street, Ann Arbor, MI, 48109, USA

Keywords: Mg alloy, precipitation, electron microscopy

Abstract

The structures of the β" and β' phases in a Mg-2.35 wt. % Nd alloy aged at 250°C for 3 min, 15 min and 2 hr were investigated using high-angle annular dark field scanning transmission electron microscopy. Both phases consist of different arrangements of Nd atoms including triangular structures and hexagonal rings. In the $[0001]_{Mg}$ orientation, some β' precipitates also exhibit unexpected $D0_{19}$ stacking, which is explained by a projection artifact of stacking defects visible in the $\langle 11\text{-}20\rangle_{Mg}$ zone axes. The β' phase appears to nucleate from preexisting β" chains. The β' phase is coherent and contains the highest strain along the $\langle 11\text{-}20\rangle_{Mg}$ directions, which agrees with the morphology of the precipitates being thinnest along the $\langle 11\text{-}20\rangle_{Mg}$ directions.

Introduction

Magnesium rare earth (RE) alloys perform exceptionally well as structural materials in automotive and aerospace components where the operating temperatures can reach up to 250°C. These alloys achieve good mechanical properties due to the formation of precipitates with habit planes parallel to the prismatic planes of the Mg matrix during ageing [1-6], which act as barriers to dislocation motion [1]. Nevertheless, the mechanisms by which precipitates and dislocations interact and the optimum phase characteristics remain to be established.

Among the Mg-RE alloys, those based on the Mg-Nd system are well known for their significant hardening responses [2, 3] and a substantial amount of research on Mg-Nd-Y-Zr-RE alloys has focused on the precipitation sequence in WE alloys such as WE43 and WE54 [4-8]. However, this alloy system is very complex and contains some inhomogeneity in the solute distribution, making it difficult to quantify precipitate evolution. While Zr is often added for grain refinement and Nd and Y for precipitation hardening, the precise contributions of each alloying element to the precipitation processes are unclear. The hardening contribution of the precipitates is also difficult to assess. To gain a more fundamental understanding of the precipitation and deformation mechanisms in these alloys, a simpler binary Mg-Nd alloy that exhibits similar precipitate phases as the more complex WE alloys is considered.

The precipitation sequence is generally assumed to involve the successive formation of the following metastable phases from the supersaturated solid solution: G. P. zones or β" (Mg_3Nd), orthorhombic β' (Mg_7Nd), hcp β_2, fcc β_1 (Mg_3Nd), bct β ($Mg_{12}Nd$), and tetragonal β_e ($Mg_{41}Nd_5$) [4, 9-12]. These studies have revealed important information about the general structures, morphologies and compositions of the main precipitating phases in Mg-Nd alloys. Yet much work remains to quantify the precipitation kinetics and strengthening mechanisms. While most of what we know about the appearance of these precipitate phases have been obtained via transmission electron microscopy imaging along the $[0001]_{Mg}$ zone axis, accurate interpretation of these complex structures requires analysis in multiple orientations. The present work focuses on the precipitate phases present at peak hardness with a discussion on the formation mechanism of β' and the different structural features observed within β" and β' precipitates.

Experimental Methods

A cast Mg-2.35 wt.% (0.40 at.%) Nd alloy was provided by Magnesium Elektron North America. The alloy composition was determined by inductively coupled plasma-mass spectrometry (ICP-MS). Samples were encapsulated in quartz tubes under an argon and sulfur hexafluoride atmosphere, solution treated at 560°C for 24 hours, quenched in water and aged at 250°C for times ranging between 3 min and 128 hr. Hardness measurements were obtained using a Vickers microhardness indenter with a load of 100 grams and dwell time of 15 seconds. Specimens for transmission electron microscopy were prepared by punching 3 mm discs from heat treated foils, ground to a thickness of 0.10 mm and twin-jelectropolished in a solution of 10.6 g lithium chloride, 22.32 g magnesium perchlorate, 1000 ml methanol and 200 ml 2-butoxyethanol at -55°C. High-angle annular dark field (HAADF)- scanning transmission electron microscopy (STEM) images were obtained using a JEOL 2100F microscope operated at 200 kV with a collection angle of 52 mrad and a double-corrected JEOL 3100 microscope operated at 300 kV with a collection angle of 74 mrad.

Experimental Results and Discussion

The ageing response of the Mg-2.35 wt. % Nd alloy at 250°C (Fig. 1) shows that the alloy reaches its peak hardness between 7 and 15 min of ageing, where after the hardness gradually decreases. Using conventional and scanning transmission electron microscopy, the phases present at these different stages of ageing were determined.

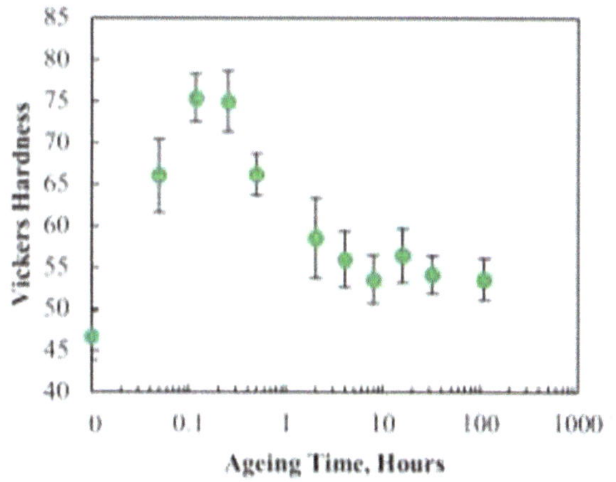

Fig. 1: Hardness curve for a Mg-2.35 wt. % Nd alloy at 250°C.

The β' phase appears to be the hardening phase in this alloy. Indeed, at 3 min of ageing, the microstructure of Mg-Nd contains β" and β' (Fig. 2a), while at the hardness peak, after 15 min of ageing, the microstructure contains primarily larger β' with residual β" (Fig. 2b). At 2 hr, where hardness has dropped significantly, the β" phase has completely dissolved and both the β' and $β_1$ phases are present, as shown in Fig. 2c. After 4 hr, β' has completely dissolved and the microstructure contains $β_1$ and β phases.

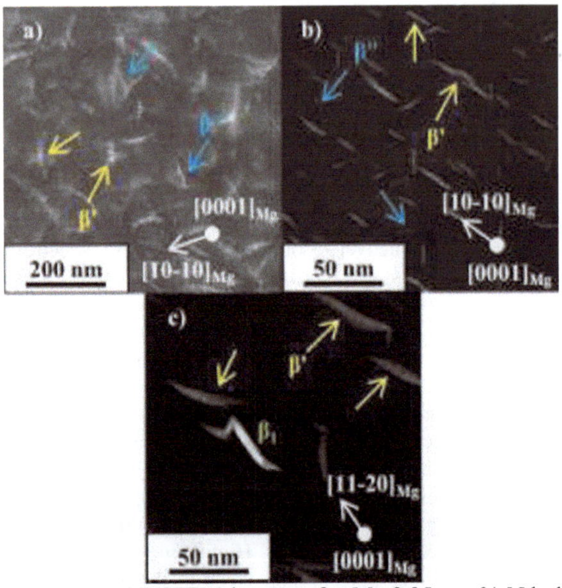

Figure 2: HAADF-STEM images of a Mg-2.35 wt. % Nd alloy age treated for a) 3 min, b) 15 min and c) 2 hr at 250°C with the incident beam parallel to the $[0001]_{Mg}$ direction. β", β' and $β_1$ phases are identified.

In the peak aged condition, the β" and β' phases consist of (0001) columns of Nd atoms arranged in variations of a triangular pattern, as illustrated in Fig. 3 and Fig. 4. The β" phase exists as either individual hexagonal rings, V or N shapes with a $D0_{19}$-type ordering, as was previously reported in [9]. The V and N shapes are arranged in chains extending in the $<11\text{-}20>_{Mg}$-type directions.

The β' phase is also made up of these triangular units yet arranged in a different pattern (Fig. 5). The triangular units of β' form long chains that are aligned in parallel rows of zigzag chains extending along the $<10\text{-}10>_{Mg}$-type directions, consistent with the Mg_7Nd stoichiometry.

The interface between the β' precipitates and the matrix is coherent and no interfacial dislocation was observed. The average spacing between the Nd columns along the $[11\text{-}20]_{Mg}$ direction measured from five β' precipitates similar in size to that shown in Fig. 6 is 0.72±0.02 nm yielding an estimated strain between 9 and 15% between the β' phase and matrix (Fig. 6). The average spacings between the $\{010\}_{β'}$ and $\{001\}_{Mg}$ planes along the $<10\text{-}10>_{Mg}$ and $[0001]_{Mg}$ directions were measured at 1.1±0.01 nm and 0.51±0.01 nm, which match closely to four $\{1\text{-}100\}_{Mg}$ plane spacings ($\{1\text{-}100\}_{Mg}$ plane spacing = 0.278 nm) and one $(0001)_{Mg}$ plane spacing (0.521 nm), resulting in little (<2%) to no strain in these directions. This would suggest that the morphology of these β' precipitates is controlled by strain minimization as they are longest along the $[0001]_{Mg}$ direction and shortest along a $<11\text{-}20>_{Mg}$ direction.

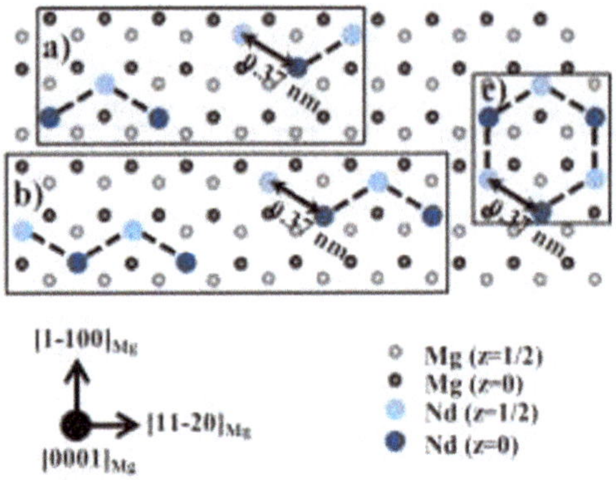

Figure 3: Schematic diagram showing the atomic structure of the β" phase in the a) V-shaped & b) N-shaped and c) hexagonal ring arrangements. The structures were created using VESTA [13].

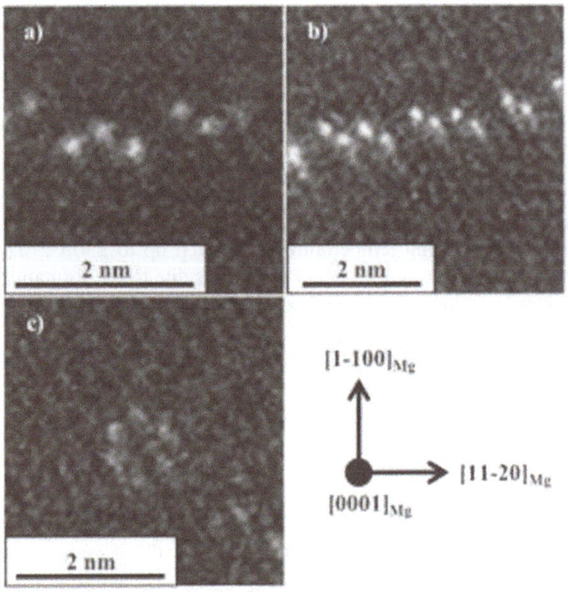

Figure 4: HAADF-STEM images of a 2.35 wt. % Nd alloy age treated for 15 min at 250°C with the incident beam parallel to the $[0001]_{Mg}$ zone axis showing the a) V-shaped & b) N-shaped arrangements and c) hexagonal ring of the β" phase.

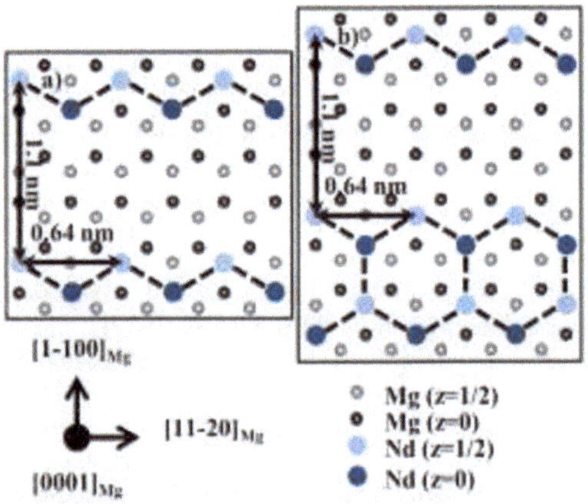

Figure 5: Schematic diagram showing the atomic structure of the β' phase containing a) only zigzag chains and b) a zigzag chain and row of hexagonal rings. The structures were created using VESTA [13].

On closer inspection, it appears that the structure of β' is more complex than previously reported. In addition to the rows of zigzag chains, the β' phase contains rows of hexagonal rings (Fig. 5b and Fig. 6). The number of rows of hexagonal rings to the number of zigzag chains varies from precipitate to precipitate as illustrated in Figs. 6a and b. The numbers of hexagonal rings and zigzag chains locally determines the concentration of the β' precipitates. The composition of precipitates containing only chains would be 12.5 at. % Nd (Mg_7Nd) while the composition of precipitates containing only rings would be 16.7 at. % Nd. Most precipitates exhibit both structures in different proportions. The Nd concentration therefore varies between 12.5 and 16.7 at. % Nd from precipitate to precipitate and locally varies within the same precipitate.

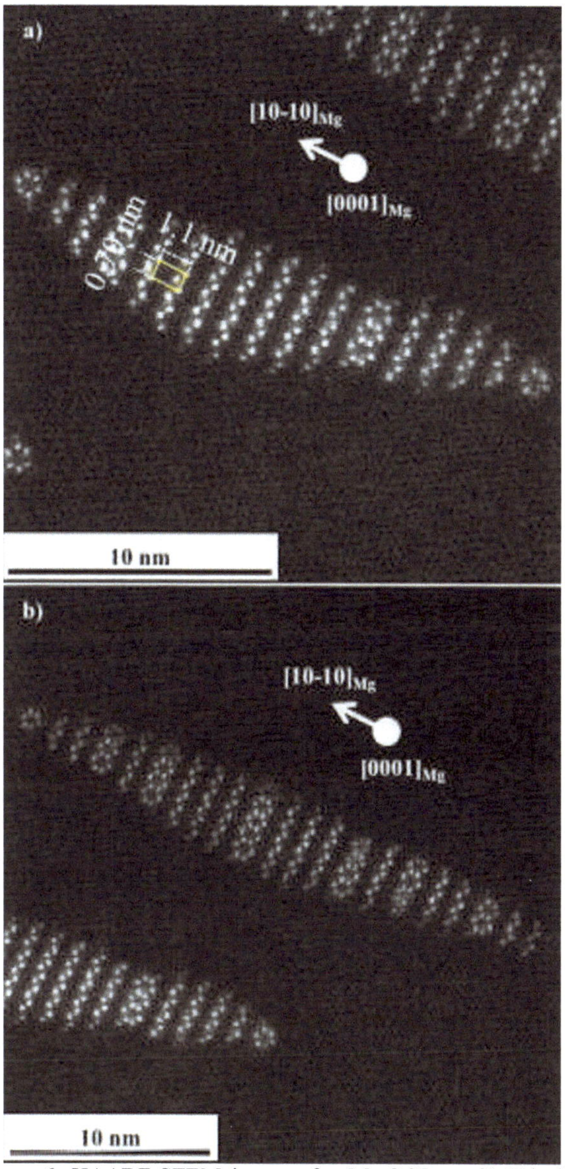

Figure 6: HAADF-STEM image of a Mg-2.35 wt. % Nd alloy aged for 2 hr at 250°C with the incident beam parallel to the [11-20]$_{Mg}$ zone axis.

Besides the hexagonal rings and zigzag chains, Nd arrangements consistent with the $D0_{19}$ structure were also observed within some β' precipitates (Fig. 7). These regions are often "blurry", suggesting that parts of these structures may be located at a different depth within the sample compared to the other regions of the precipitate. Stacking errors leading to shifts in the positions of the chains are also observed in many β' precipitates (Fig. 7).

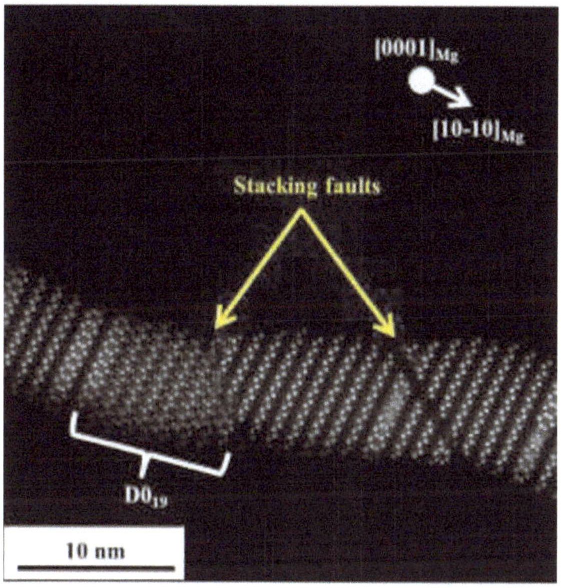

Figure 7: HAADF-STEM image of a Mg-2.35 wt. % Nd alloy aged for 2 hr at 250°C with the incident beam parallel to the [0001]$_{Mg}$ zone axis showing a region of the $D0_{19}$ structure and stacking faults.

To elucidate the nature of the "blurry" regions, imaging of β' precipitates was performed along the [11-20]$_{Mg}$ zone axis. In this orientation, the hexagonal rings appear as 2 zigzag chains spaced 0.55±0.02 nm apart. The β' precipitate shown in Fig. 8a shows that a single hexagonal ring does not necessarily extend the entire depth along the [0001]$_{Mg}$ direction. For instance, there are three segments of hexagonal rings shown in Fig. 8a (individual hexagonal rings are shown in the yellow boxes) and where one segment ends as a hexagonal ring another segment is shifted by 0.55±0.02 nm along a <10-10>$_{Mg}$ direction. The discontinuity of hexagonal ring segments results in the appearance of a $D0_{19}$-like structure that contains blurry contrast when viewed along the [0001]$_{Mg}$ direction. Fig. 8b also illustrates stacking errors on zigzag chains, which may be consistent with the stacking faults shown in Fig. 7 when viewed along a <11-20>$_{Mg}$ zone axis. The complexity and high density of stacking defects present in β' precipitates makes it difficult to interpret the STEM images. Therefore, it is important when studying these complex structures with TEM to consider their appearance in multiple orientations.

Finally, due to the rapid precipitation process in Mg-Nd at 250°C, the β" precipitates are only seen up to 15 min. Yet from the observations at earlier times, some of the β' precipitates are connected to β" chains (Fig. 2a & b and Fig. 9) and contain features such as hexagonal rings, which are also seen in β". It is speculated that the β' phase forms from the preexisting β" by nucleating at the ends of the β" chains. Interestingly, there is a hexagonal ring consistently located at the connection point between the β" and β' phases (Fig. 9), which may facilitate the formation of β'. As the β' grows with ageing, some precipitates come into close proximity to each other and lead to stacking defects. Although stacking defects can result from dislocations, these have not been observed in the vicinity of the precipitates.

Future work will involve examination of the aged alloy after it has undergone deformation to understand dislocation interactions with precipitates and the effect of deformation on the precipitate structures.

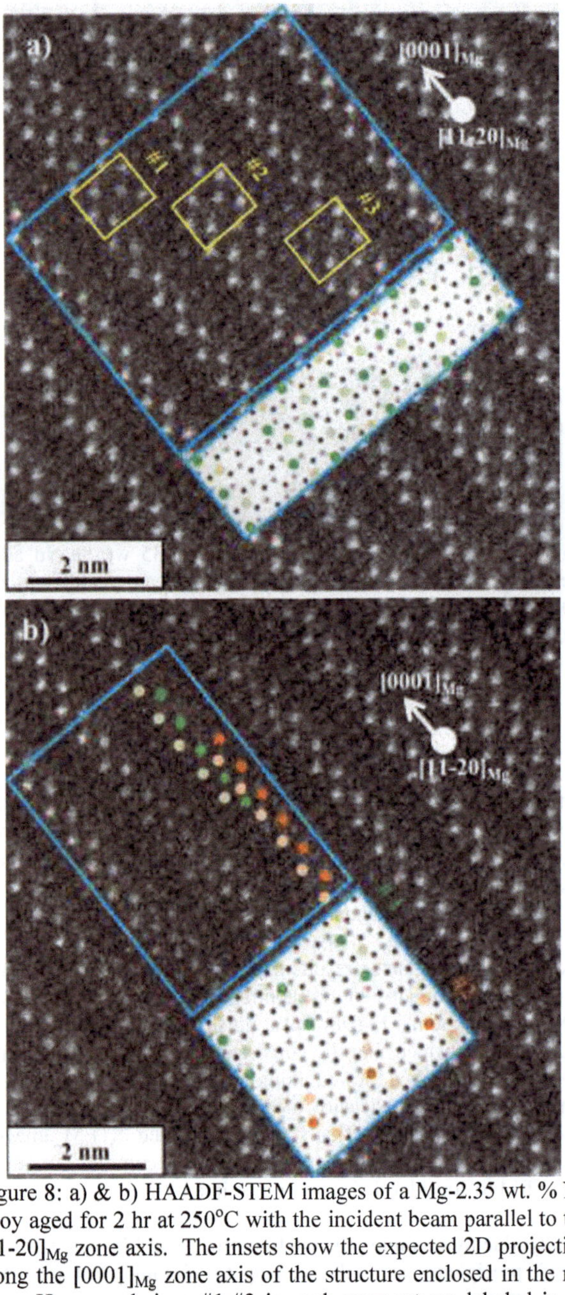

Figure 8: a) & b) HAADF-STEM images of a Mg-2.35 wt. % Nd alloy aged for 2 hr at 250°C with the incident beam parallel to the $[11\text{-}20]_{Mg}$ zone axis. The insets show the expected 2D projection along the $[0001]_{Mg}$ zone axis of the structure enclosed in the red box. Hexagonal rings #1-#3 in each segment are labeled in a). Precipitates #1 & #2 are labeled in b). The Mg lattice was created using VESTA [13].

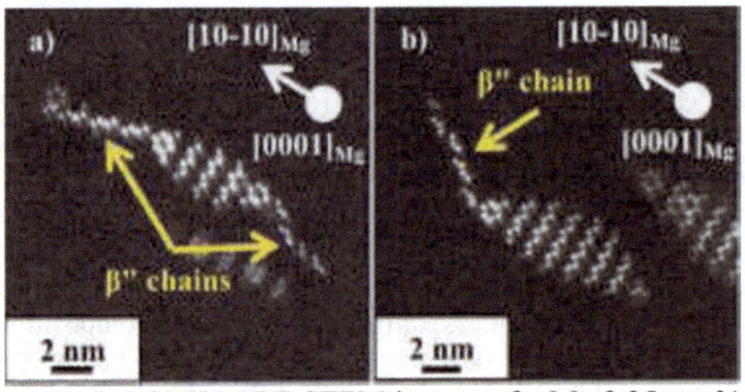

Figure 9: a) & b) HAADF-STEM images of a Mg-2.35 wt. % Nd alloy age treated for 15 min at 250°C with the incident beam parallel to the $[0001]_{Mg}$ direction.

Conclusions

In the present work, the structures of β" and β', which form in the early to peak stages of ageing, were investigated using HAADF-STEM. Observations of their internal structures show that both phases consist of similar triangular units arranged in different patterns. Due to their similar structural arrangements and location with respect to each other, it seems that β' nucleates from preexisting β" chains. The structure of β' was shown to be more complex than just the Mg_7Nd structure and contains different structural features resulting in local composition variations as well as stacking defects.

Acknowledgements

The authors thank B. Davis (MEL) for providing the alloy. This work was supported by the Department of Energy – Basic Energy science program [grant award number DE-SC0008637]; the University of Michigan College of Engineering (access to microscopy facilities), and the University of Michigan Rackham Graduate School (fellowship for E. S.).

References

[1] B.L. Mordike, K.U. and Kainer, Magnesium alloys and their applications, MAT INFO/Werkstoff-Informationsgesellschaft, Frankfurt, 1998.
[2] M.O. Pekguleryuz, M.M. Avedesian, J. Jpn. Inst. Light Met., 42 (1992) 679.
[3] B.L. Mordike, Mater. Sci. Eng. A, 324 (2002) 103-112.
[4] H. Karimzadeh, PhD Thesis, University of Manchester, (1985).
[5] J.F. Nie, B.C. and Muddle, Scripta Mater., 40 (1999) 1089.
[6] J.F. Nie, B.C. Muddle, Acta Mater., 48 (2000) 1691.
[7] M. Ahmed, G.W. Lorimer, P. Lyon, R. and Pilkington, in: Magnesium Alloys and Their Applications, DGM Informationsgesellschaft, Garmisch Partenkirchen, Germany, 1992, pp. 301.
[8] G. Lorimer, R. Azari-Khosroshahi, M. and Ahmed, Proc. Int. Conf. of Solid-Solid Phase Transform., 12 (1999) 185.
[9] K. Saito, K. and Hiraga, Mater. Trans., 52 (2011) 1860.
[10] T.J. Pike, B. Noble, J. Less-Common Met., 30 (1973) 63.
[11] M. Hisa, J.C. Barry, G.L. Dunlop, Philos. Mag. A, 82 (2002) 497.
[12] Y.M. Zhu, H. Liu, Z. Xu, Y. Wang, J.F. Nie, Acta. Mater., 83 (2015) 239.
[13] K. Momma, F. Izumi, J. Appl. Crystallogr., 44 (2011).

Magnesium Technology 2016

LPSO Alloys and Composites

Magnesium Technology 2016
Edited by: Alok Singh, Kiran Solanki, Michele V. Manuel, and Neale R. Neelameggham
TMS (The Minerals, Metals & Materials Society), 2016

SOLID SOLUTION HARDENING IN Mg-Gd-TM (TM=Ag, Zn and Zr) ALLOYS: AN INTEGRATED DENSITY FUNCTIONAL THEORY AND ELECTRON WORK FUNCTION STUDY

William Yi Wang,[1, 4, *] Shun Li Shang,[1] Yi Wang,[1] Hongyeun Kim,[1] Kristopher A. Darling,[2] Laszlo J. Kecskes,[2] Suveen N. Mathaudhu,[3] Xi Dong Hui,[4] and Zi-Kui Liu [1]

[1] Department of Materials Science and Engineering, The Pennsylvania State University, University Park, PA 16802, USA

[2] U.S. Army Research Laboratory, Weapons and Materials Research Directorate, Aberdeen Proving Ground, MD 21005, USA

[3] Department of Mechanical Engineering, University of California, Riverside, CA 92521, USA

[4] State Key Laboratory for Advanced Metals and Materials, University of Science and Technology Beijing, Beijing 100083, China

* Corresponding author. William Yi Wang (Tel.: +1-814-863-9957, e-mail: yuw129@psu.edu)

Keywords: Long Periodic Stacking Ordered Structures (LPSOs); Bonding Charge Density; Hardness; Mg alloys.

Abstract

The present work [1] aims to reveal the effects of solute atoms (TM=Ag, Zn and Zr) on the age-hardening of Mg-Gd-based alloys via the density functional theory and electron work function (EWF) approaches. Based on the electronic structures of LPSOs (including 6H, 10H, 14H, 18R and 24R) [2], the 10H LPSO phases of Mg-Gd-TM alloys are selected as the model case due to the improved strength and ductility such long periodic stacking ordered precipitates (LPSOs) offer. The CALPHAD-modeling method is applied to predict the EWF in the ternary Mg-Gd-TM alloys. The obtained EWFs of these Mg alloys match well with previous experimental and theoretical results. Moreover, the variation of EWF in the ternary Mg-Gd-TM alloys is attributed to the structure contribution (i.e., the formation of FCC-type fault layers) and the chemical effect of solute atoms (i.e., electron redistributions characterized by bonding charge density – $\Delta\rho$ [3-5]). Comparisons of electron redistributions caused by mechanical and chemical contributions of solute atoms posit correlations between EWF and the formation energy of LPSO, which is critical to yield a predictive mesoscale or phenomenological model for age-hardening of Mg. It is found that the interfacial energy of 10H LPSO is decreased significantly with the addition of Zn and Zr, indicating the plasticity of 10H LPSO will be increased in the Mg-Gd-Zr and Mg-Gd-Zn alloys. The enhanced electrons along the basal plane caused by atomic clusters of Gd-TM suggest that the bond strength is improved along basal plane, while the reduced electrons in the prismatic and pyramidal planes indicate the bond strengths are weakened along prismatic and pyramidal planes. The EWF and hardness of Mg-Gd-TM (TM= Ag, Zn and Zr) alloy are also correlated, revealing that the EWF variations of ternary Mg-Gd-TM alloys are attributed to not only the mechanical contribution caused by lattice distortion but also the chemical effect of solute atoms. The attractive combination of physical ($\Delta\rho$ and EWF) and mechanical properties provides a new insight into studying the solid solution hardening behaviors of Mg-RE alloys.

References:

[1] Wang WY, Shang SL, Wang Y, Kim HY, Darling KA, Kecskes LJ, Mathaudhu SN, Hui XD, Liu Z-K. Solid Solution Hardening in Mg-Gd-TM (TM=Ag, Zn and Zr) Alloys: An Integrated Density Functional Theory and Electron Work Function Study. JOM 67 (2015) 2433-2441.

[2] Wang WY, Shang SL, Wang Y, Darling KA, Kecskes LJ, Mathaudhu SN, Hui XD, Liu Z-K. Electronic structures of long periodic stacking order structures in Mg: A first-principles study. J. Alloy. Compd. 586 (2014) 656-662.

[3] Nakashima PNH, Smith AE, Etheridge J, Muddle BC. The Bonding Electron Density in Aluminum. Science 331 (2011) 1583-1586.

[4] Wang WY, Shang SL, Wang Y, Darling KA, Mathaudhu SN, Hui XD, Liu ZK. Electron localization morphology of the stacking faults in Mg: A first-principles study. Chem. Phys. Lett. 551 (2012) 121-125.

[5] Wang Y, Wang WY, Chen L-Q, Liu Z-K. Bonding charge density from atomic perturbations. J. Comput. Chem. 36 (2015) 1008-1014.

Graphic abstract

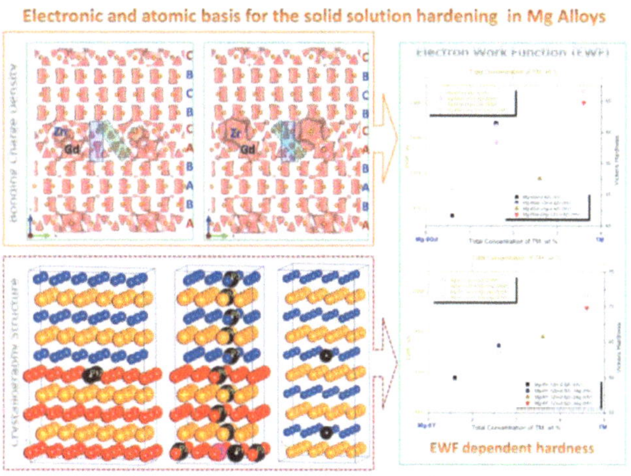

Note: These pictures are reproduced from Ref. [1].

Magnesium Technology 2016
Edited by: Alok Singh, Kiran Solanki, Michele V. Manuel, and Neale R. Neelameggham
TMS (The Minerals, Metals & Materials Society), 2016

MICROSTRUCTURE AND MECHANICAL PROPERTIES NEW MAGNESIUM-ZINC-GADOLINIUM ALLOYS

Sankaranarayanan Seetharaman [1], Sravya Tekumalla [1], Bhavesh Lalwani [2], Hardik Patel [2], Nguyen Quy Bau [1] and Manoj Gupta [1]
[1] Department of Mechanical Engineering, National University of Singapore, 9 Engineering Drive 1, Singapore 117576
[2] Department of Mechanical Engineering, National Institute of Technology, Surathkal, Karnataka, India 575025

Keywords: Mg-Zn-Gd alloy, Microstructure, Mechanical Properties

Abstract

Magnesium based materials are effective for structural/component weight reduction in automotive applications. However, their real time applications are limited because of their inadequate mechanical properties, especially the absolute strength and creep resistance. In this regard, the formation of thermally stable ternary compounds is believed to positively influence the properties of Mg-Zn-RE alloys. In this study, new Mg alloys containing Zn and Gd (Mg-2.0Zn-0.5Gd and Mg-3.4Zn-0.8Gd, in at.%) were developed using disintegrated melt deposition technique followed by hot extrusion. The developed alloys were investigated for their microstructural and mechanical properties in hot-extruded conditions. The mechanical properties examined under indentation, tension and compression loads indicated improved mechanical performance due to Zn and Gd addition. The observed mechanical properties are presented using structure-property relationship.

Introduction

The extremely lightweight of magnesium with density (1.74 g/cc) only slightly higher than that of plastics makes it an attractive choice for structural weight savings in automobiles [1]. Besides being light, magnesium (Mg) also possesses excellent machinability, good castability, and damping capacity. However, the adaptation of Mg and its alloys in critical engineering applications is largely limited by their inherent inadequacies such as low absolute strength, ductility, creep resistance and corrosion resistance [2]. Existing literature illustrates that the judicious selection and addition of alloying elements aids in overcoming the above mentioned limitations [3-5].

Despite considerable efforts made thus far, adoption of magnesium alloys in critical engineering applications remains limited. One of the important reasons for such underutilization is that there are limited magnesium alloys available for the designers. Given the limited choices available, the widely used AZ and AM series alloys exhibit some inadequate properties such as creep-resistance and strength loss at elevated temperatures due to the poor thermal stability of $Mg_{17}Al_{12}$ phase [3,4]. Hence, it becomes increasingly essential to develop new Mg alloys that can provide improved strength, creep and corrosion resistance on a cost-effective basis [5].

While it is well known that the addition of Zn improves the strength of Mg without reducing its ductility, the rare earth addition has been found beneficial in improving the thermal stability, creep and room temperature / elevated temperature mechanical properties of magnesium alloys [6,7]. In this context, the Mg-Zn-RE alloys are characterized by high-strength and creep resistance for automotive and aerospace applications [8]. Le et al. [9] investigated the properties of RE containing Mg-2Zn-0.4RE alloys and reported that the Y containing alloy exhibit the highest ductility. However, the available literature suggests that the Mg alloys containing Y are associated with processing difficulties and hence, it is essential to look for non-yttrium containing alloys [10]. Apart from Y, Gd and Nd additions are also believed to positively influence the mechanical properties [6-12]. Previous investigations demonstrate excellent mechanical properties of Mg-Gd alloys both at room temperature and at elevated temperature [6-12]. In a recent study, Wu et al. [13] reported superior ductility for rolled Mg-Gd-Zn alloy sheets due to the texture weakening effects of Gd. Recently, Srinivasan et al. [14] investigated the mechanical properties of Mg-Gd-Zn alloys and Mg-10Gd-xZn alloys (x = 2, 6) exhibited good strength due to the presence of 14H-type LPSO phases in the matrix. It was also reported in this study that those alloys containing lower Gd exhibited good ductility due to the lower fraction of LPSO. Given the positive benefits of Zn and Gd alloying addition, an attempt is made in this research work to develop new Mg-Zn-Gd alloys using disintegrated melt deposition (DMD) technique followed by hot extrusion. The microstructural and tensile properties of the developed alloys were studied in detail. Structure–property relationships were invoked in order to comprehend the properties exhibited by the developed alloys.

Experimental

<u>Synthesis of Mg-Zn-Gd alloy</u>

Mg-2.0Zn-0.5Gd and Mg-3.4Zn-0.8Gd (at.%) alloys required for the current study were synthesized using disintegrated melt deposition (DMD) technique. It involves heating of Mg turnings (> 99.9% purity, supplied by ACROS Organics, New Jersey, USA) together with zinc powder (> 99.9% purity, supplied by Phoenix Scientific Industries, UK) and Mg-30Gd master alloy ingots (99.9% purity, supplied by Sunrelier Metal Company, China) in a graphite crucible to 750 °C in an electrical resistance furnace under inert argon gas protective atmosphere. The superheated melt was then stirred (at 465 rpm for 5 min) and bottom-poured into a copper mold (after disintegration by two jets of argon gas oriented normal to the melt stream). Following deposition, an ingot of 40 mm diameter was obtained. The obtained ingot was then soaked at 400 °C for 5 h and hot extruded at 350 °C into rods of 8 mm diameter.

<u>Characterization of Mg-Zn-Gd alloys</u>

<u>Microstructural Characterization:</u> Microstructural studies were conducted on the polished samples in a direction perpendicular to the extrusion direction in order to determine the average matrix grain size and the presence and distribution of second phases. For this purpose, a small section (8 to 10 mm in length) was cut from

the extruded rod and the ends were then ground using 600 and 1200 grit size sand paper in order to remove the large surface scratches and also to produce a flat surface. Once all visible surface scratches and cracks were removed, the sample was polished using a polishing disc with 5 micron diamond slurry, followed by 1 micron diamond slurry and lastly by 0.3 micron alumina slurry. The surface of the polished samples was then etched (to make the grain boundaries visible) with citral (4.2 g citric acid monohydrate in 100 ml of water) and observed under an Olympus optical microscope. Using selected optical micrographs, a total of 120–150 grains were selected to calculate the grain size using the Scion Image analysis software. In order to study the morphology and distribution of secondary phases, the as-polished samples were etched with acetic picral (10 ml acetic acid, 10 ml water, 70ml picral (95% ethanol + 5% 2,4,6-Tri-nitro-phenol)) and observed in a field emission scanning electron microscope (FESEM-S4300, Hitachi Ltd., Tokyo, Japan, SEI mode).

Mechanical Properties: The mechanical properties of the developed Mg-Al-Er alloy were evaluated under indentation, tension and compression loads. Microhardness measurements were carried out on the as-polished samples of developed Mg-Zn-Gd alloys using a Matsuzawa MXT 50 automatic digital Microhardness tester in accordance with ASTM standard E3 84-99. To determine the tensile and compressive properties, standard ASTM test methods E8/E8M-13a and E9/89a were followed. A fully automated servo-hydraulic mechanical testing machine was used for this purpose. The fractured surfaces were then studied using the Hitachi S-4300 FESEM in order to understand the fracture characteristics.

Results and Discussion

Microstructure

Visual observation of the cast alloy ingots and the extruded rods clearly indicate the absence of any macro surface defects. This confirms the suitability of processing parameters used during melting, casting and extrusion processes. The microstructural properties of the developed Mg-Zn-Gd alloys are studied in terms of: (a) second phase precipitates and (b) grain morphology. Fig. 1 shows the distribution of second phases in Mg-2.0Zn-0.5Gd and Mg-3.4Zn-0.8Gd alloys. It shows the presence of fine second phase particles distributed uniformly throughout the samples. Careful observation of the SEM micrographs suggests that there are more than one type of intermetallics present in the developed Mg-Zn-Gd alloys. The results of EDX analysis (Fig. 1c) conducted on the bright intermetallic phases (as marked in Fig. 1b) shows dominant peaks corresponding to Mg and Gd phases in Mg-3.4Zn-0.8Gd. This confirms that the bright intermetallic phases are Mg-Gd based Mg_5Gd intermetallics and the fine intermetallics distributed throughout the sample surface are $MgZn_2$. From the SEM images, it should also be noted that the size of second phase precipitates in Mg-3.4Zn-0.8Gd is generally larger than that of those observed in Mg-2.0Zn-0.5Gd.

The grain characteristics of Mg-2.0Zn-0.5Gd and Mg-3.4Zn-0.8Gd alloys are shown in Fig. 2. It reveals nearly equiaxed grains in both the alloys, thus indicating complete recrystallization of Mg-matrix during extrusion [15,16]. Using Scion Image processing software, the average grain size and aspect ratio of the developed alloys were calculated and the results are presented in Table 1. It shows that the developed alloys exhibit a fine equiaxed microstructure with average grain size less than 10 μm which could be attributed to the efficacy of $MgZn_2$ and Mg_5Gd intermetallic phases in inhibiting the matrix grain growth during (extrusion associated) recrystallization [15,16].

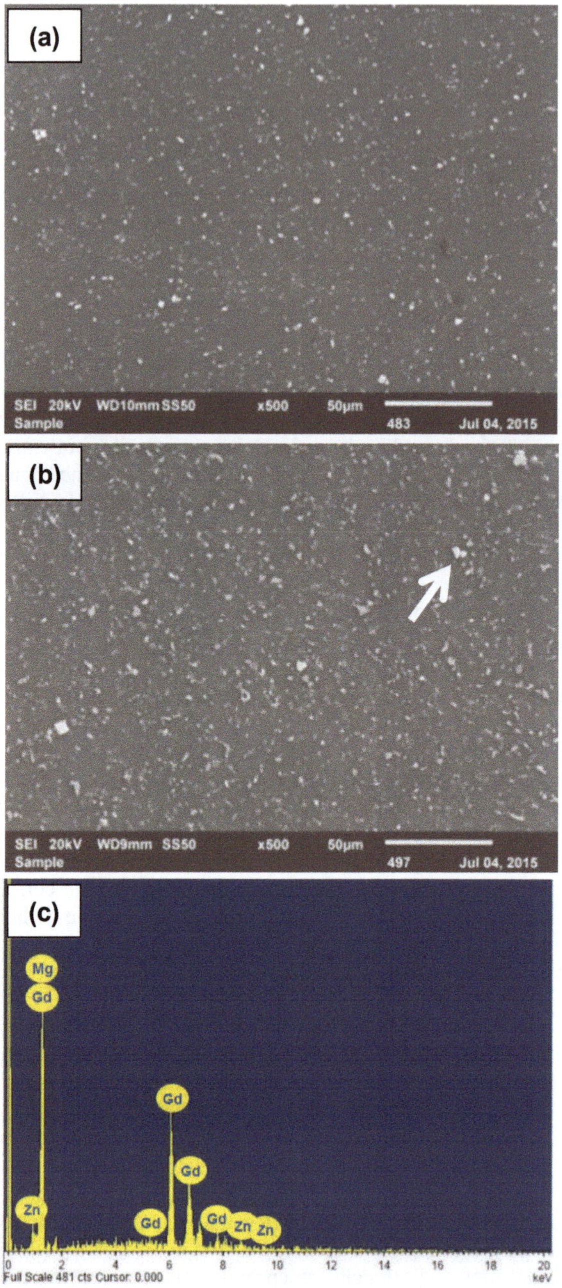

Fig. 1 SEM micrographs showing the distribution of second phases in Mg-2.0Zn-0.5Gd (a) and Mg-3.4Zn-0.8Gd (b) alloys. EDX spectrum showing the compositional details of intermetallic phases (marked 'b') in Mg-3.4Zn-0.8Gd (c).

Table 1 - Results of Microstructural Measurements

Materials	Elemental Composition [wt.%]		Grain size	Aspect ratio
	Zn	Gd	[μm]	
Mg-2.0Zn-0.5Gd	5.07	3.05	7±3	1.62 ± 0.2
Mg-3.4Zn-0.8Gd	8.31	4.7	4±2	1.53 ± 0.2

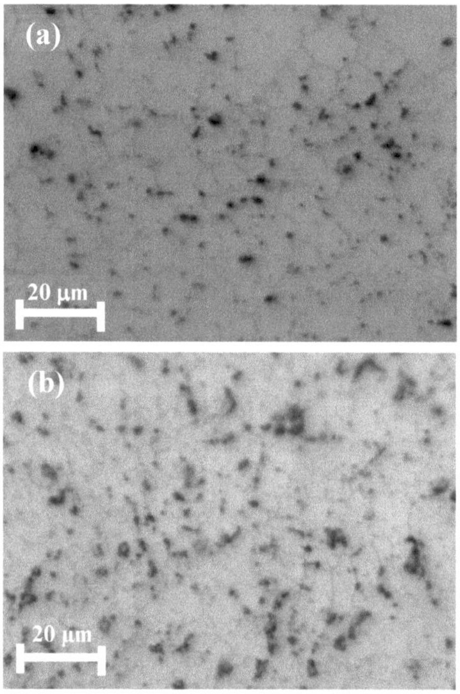

Fig. 2 Optical micrographs showing the grain characteristics of Mg-2.0Zn-0.5Gd (a) and Mg-3.4Zn-0.8Gd (b) alloys.

Mechanical behavior

The results of mechanical property measurements conducted on Mg-2.0Zn-0.5Gd and Mg-3.4Zn-0.8Gd alloys are summarized in Table 2. For comparison purpose, the properties of similar magnesium alloys are also listed. From the micro-hardness readings, it can be seen that the Mg-3.4Zn-0.8Gd alloy showed better indentation resistance (121±6 Hv) when compared to Mg-2.0Zn-0.5Gd (105±4 Hv) and the noticeable improvement can be ascribed to the increased amount of second phase precipitates (Fig. 1) imparting resistance to indentation [17,18].

Table 2 highlights the benefits of Zn and Gd alloying addition on the tensile properties of Mg. It is interesting to note that the tensile failure strain as high as 24% was obtained for Mg-2.0Zn-0.5Gd alloy. The results further highlight that the increasing amount of Zn and Gd addition enhances the yield strength and ultimate tensile strength of Mg-3.4Zn-0.8Gd alloy. However, the tensile failure strain was adversely affected.

Table 2 - Results of Microhardness and Tensile testing

Materials	Micro-hardness [H_v]	Tensile Properties		
		Yield Strength [MPa]	Ultimate Strength [MPa]	Failure Strain [%]
Mg-2.0Zn-0.5Gd	105±4	126 ± 7	258 ± 7	23.6 ± 2.1
Mg-3.4Zn-0.8Gd	121±6	169 ± 5	286 ± 7	16.1 ± 3.0
Mg-2Zn-0.4Gd [9]	-	125	220	26
Mg-2Zn-0.4Ce [9]	-	190	255	18
Mg-2Zn-0.4Y [9]	-	160	240	30
Mg-2Zn-0.4Nd [9]	-	175	245	28

The results of compression tests conducted on the extruded samples of Mg-2.0Zn-0.5Gd and Mg-3.4Zn-0.8Gd alloys are listed in Table 3. Similar to tensile test results, compression test results also show an increase in the yield and ultimate strength due to increasing Zn and Gd amounts.

Table 3 - Results of Compression Tests

Materials	Compressive Properties		
	Yield Strength [MPa]	Ultimate Strength [MPa]	Failure Strain [%]
Mg-2.0Zn-0.5Gd	130 ± 1	476 ± 4	22.0 ± 0.7
Mg-3.4Zn-0.8Gd	163 ± 4	494 ± 2	17.6 ± 0.3

Based on the microstructural analysis; it is evident that the addition of Zn and Gd alloying elements to Mg has resulted in a fine equi-axed microstructure. Hence, the strength properties improvement (under both tensile and compressive loads) due to increasing Zn and Gd addition can be attributed to the following strengthening mechanisms [17,18]:

(i) Strengthening due to the grain size reduction and formation of refined and equiaxed grains,
(ii) Strengthening from the reasonably distributed hard grain boundary second phases ($MgZn_2$ and Mg_5Gd) that impede the stress transfer across the grain boundaries.

The high ductility obtained due to Zn and Gd addition could be attributed to the activation of non-basal slip systems as suggested by Koike et al. [19] wherein, the non-basal slip activation resulting larger deformation strain (elongation) was observed in the case of fine-grained Mg-alloys.

The microscopic features of the tension and compression test fractured samples are shown in Figs. 3&4.

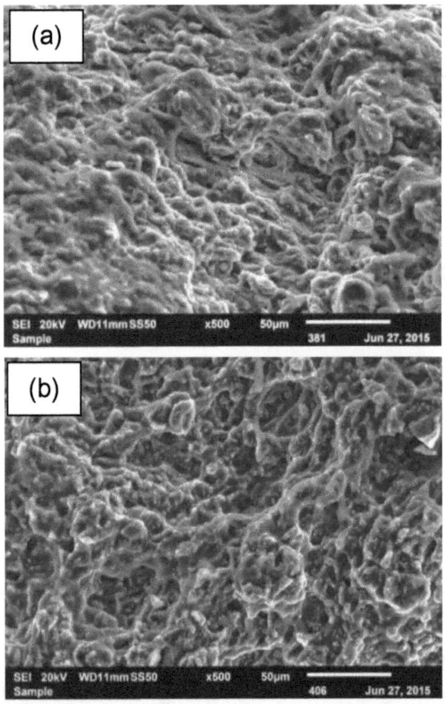

Fig. 3 SEM micrographs showing the morphology of tensile test fractured Mg-2.0Zn-0.5Gd (a) and Mg-3.4Zn-0.8Gd (b) alloys.

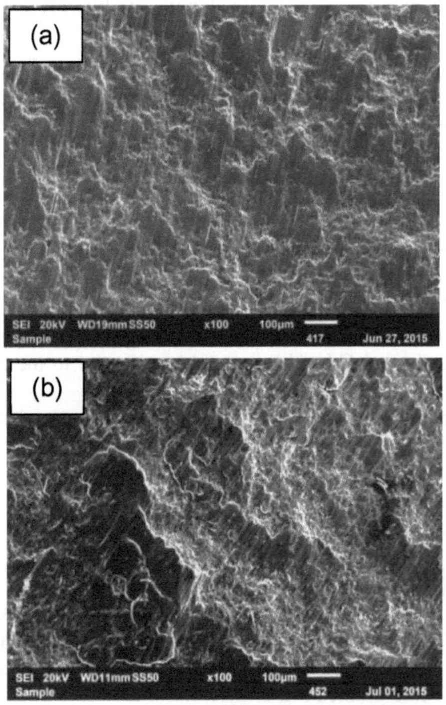

Fig. 4 SEM micrographs showing the morphology of compression test fractured Mg-2.0Zn-0.5Gd (a) and Mg-3.4Zn-0.8Gd (b) alloys.

While the tensile fractographs (Fig. 3(a-b) reveals prominent ductile features due to plastic deformation, the compression test fractured samples showed wavy pattern. Commonly, cleavage features which are indicative of the inability to plastic deformation are observed for HCP magnesium materials that are failed under tensile loads. [20]. However, in the present case, the developed Mg-Zn-Gd alloys show prominent ductile features representing extensive plastic deformation, thus corroborating with the tensile test results. The wavy pattern seen for compression test fractured samples (Fig. 4(a-b)) correspond to shear mode fracture behavior as commonly observed for magnesium materials.

Conclusions

In the present work, new Mg-alloys containing Zn and Gd were produced using the disintegrated melt deposition technique followed by hot extrusion. From the microstructural and mechanical property characterization, the following conclusions can be drawn:

1. The addition of Zn and Gd to Mg resulted in Mg-2.0Zn-0.5Gd and Mg-3.4Zn-0.8Gd alloys with refined microstructure (~ 10μm average grain size) containing evenly distributed $MgZn_2$ and Mg_5Gd intermetallic phases.
2. The developed Mg-Zn-Gd alloys displayed superior mechanical properties under indentation, tension and compression loads.
3. Increasing amount of Zn and Gd addition enhanced the micro-hardness, yield strength and ultimate strength (both tensile and compressive) of Mg-3.4Zn-0.8Gd alloy when compared to Mg-2.0Zn-0.5Gd. However, the tensile and compressive ductility were negatively affected.

Acknowledgement

The financial support of Ministry of Education Academic Research Funding (WBS# R-265-000-498-112) is gratefully acknowledged.

References

1. Luo, A.A., Magnesium: current and potential automotive applications. JOM 2002, 54, (2), 42-48.
2. Avedesian, M. M.; Baker, H., ASM specialty handbook: magnesium and magnesium alloys. ASM international 1999, 274.
3. Bohlen, J.; Yi, S.; Letzig, D.; Kainer, K. U., Effect of rare earth elements on the microstructure and texture development in magnesium–manganese alloys during extrusion. Mater. Sci. Eng. A 2010, 527, (26), 7092-7098.
4. Bettles, C.; Gibson, M.; Zhu, S.-M., Microstructure and mechanical behaviour of an elevated temperature Mg-rare earth based alloy. Mater. Sci. Eng. A 2009, 505, (1), 6-12.
5. Zhu, S.-M.; Gibson, M.; Easton, M.; Nie, J., The relationship between microstructure and creep resistance in die-cast magnesium–rare earth alloys. Scripta Materialia 2010, 63, (7), 698-703.
6. Andrzej Kielbus. "Microstructure and Properties of Elektron 21 Magnesium Alloy," Magnesium Alloys - Design, Processing and Properties, ed Frank Czerwinski (InTech Publishers, 2011), 281-296.

7. He, S.M.; Zeng, X.O.; Peng, L.M.; Gao, X.; Nie, J.F., Ding W.J., Precipitation in a Mg–10Gd–3Y–0.4Zr (wt.%) alloy during isothermal ageing at 250°C, J. Alloys Compd. 2006, 421, 309–313.
8. Rokhlin, L; Nikitina, N.; Dobatkina, T, Solid-state phase equilibria in the Mg corner of the Mg-Gd-Sm phase diagram, J. Alloys Compd. 1996, 239, 209-213.
9. Le, Q.-C.; Zhang, Z.-Q.; Shao, Z.-W.; Cui, J.-Z.; Xie, Y. Microstructures and mechanical properties of Mg-2%Zn-0.4%Re alloys. Trans. Nonferrous Metals Soc. China 2010, 20, s352–s356.
10. Lorimer, G.; Apps, P.; Karimzadech, H.; King, J., Improving the performance of Mg-Rare Earth alloys by the use of Gd or Dy additions. Mater. Sci. Forum 2003, 419-422, 279-284.
11. Smola, B.; Stulikova, I.; von Buch, F.; Mordike, B., Structural aspects of high performance Mg alloy design, Mater. Sci. Eng. A 2002, 324, 113-117.
12. Hu, Y.B.; Deng, J.; Zhao, C.; Pan, F.S.; Peng, J., Microstructure and mechanical properties of Mg–Gd–Zr alloys with low gadolinium contents, J. Mater. Sci. 2011, 46(17) 5838-5846.
13. Wu, D.; Chen, R.S.; Han, E.H. Excellent room-temperature ductility and formability of rolled Mg–Gd–Zn alloy sheets. J. Alloy. Compd. 2011, 509, 2856–2863.
14. Srinivasan, A.; Huang, Y.; Mendis, C.L.; Blawert, C.; Kainer, K.U.; Hort, N. Investigations on microstructures, mechanical and corrosion properties of Mg–Gd–Zn alloys. Mater. Sci. Eng. A 2014, 595, 224–234.
15. Davies, R.; Randle, V.; Marshall, G., Continuous recrystallization—related phenomena in a commercial Al–Fe–Si alloy. Acta materialia 1998, 46, (17), 6021-6032.
16. Humphreys, F. J.; Hatherly, M., Recrystallization and related annealing phenomena. Elsevier: 1995.
17. Callister, W. D. and Rethwisch, D. G., Materials science and engineering: An Introduction. (John Wiley & Sons. NJ, USA: 2007).
18. Dieter, G. E., Mechanical metallurgy. (McGraw-Hill International, 1981).
19. Koike, J.; Kobayashi, T.; Mukai, T.; Watanabe, H.; Suzuki, M.; Maruyama, K.; Higashi, K., The activity of non-basal slip systems and dynamic recovery at room temperature in fine-grained AZ31B magnesium alloys. Acta Mater. 2003, 51, (7), 2055-2065.
20. Gupta, M.; Sharon, N. M. L., Magnesium, magnesium alloys, and magnesium composites. John Wiley & Sons. NJ, USA: 2011.

Magnesium Technology 2016
Edited by: Alok Singh, Kiran Solanki, Michele V. Manuel, and Neale R. Neelameggham
TMS (The Minerals, Metals & Materials Society), 2016

EFFECTS OF ALLOYING ELEMENTS ON MICROSTRUCTURES AND MECHANICAL PROPERTIES OF Mg-Gd-Zn-Ca ALLOYS

Hyunkyu Lim[1], Youngkyun Kim[1], Bonghwan Kim[1], Daeguen Kim[2], Young-Ok Yoon[1], Shae K. Kim[1]
[1] Korea Institute of Industrial Technology; 156 Gaetbeol-ro, Yeonsu-gu, Incheon, 21999, South Korea
[2] GI tech; 133 Asan valley nam-ro 110beon-gil, Dunpo-myeon, Asan, 31409, South Korea

Keywords: Extrusion, Strength, Ignition-resistance

Abstract

For the development of high strength extruded magnesium alloys with ignition-resistance, we had investigated the effect of addition of yttrium into Mg-Gd-Zn-Ca alloy. The addition of yttrium increased the volume fraction and thermal stability of the LPSO-like phases. The addition of yttrium also increased the tensile strength by about 70 MPa but decreased the ignition temperature by about 150 °C. In addition, the effect of the extrusion conditions on the mechanical properties of alloys has been discussed.

Introduction

Recently, magnesium alloys are recognized as alternative to steel and aluminum alloys to reduce the weight of structural materials [1]. In spite of the attractive properties of magnesium alloys such as low density, high specific strength (ratio of strength to density), excellent castability, good machinability and so on [2], magnesium alloys have some drawbacks which should be overcome in order to apply to structural parts. Some drawbacks of magnesium alloys might be low absolute strength, limited formability, low corrosion resistance and lower ignition temperature. Therefore, most researches on the development of magnesium alloys have been focused on finding the solutions for these drawbacks. From this point of view, Mg-RE-TM alloys with LPSO (Long-Period Stacking Ordered) phase are very attractive because these alloys exhibited very high strength and good heat resistance [3-5]. According to Kawamura *et al.*, there are two types of LPSO Mg alloys: one is as-cast alloys and the other is soaked alloys [6]. It was reported that the Mg-Gd-Zn alloy, which is one of the soaked alloys, exhibited high strength and large elongation among other tested alloys. Furthermore, many researchers have developed new magnesium alloys with the ultra-high strength. Li *et al.* [5] reported that Mg-12Gd-3Y-0.6Zr (wt.%) alloy, which was produced by conventional extrusion plus hydrostatic extrusion and T5 exhibited tensile yield strength (TYS) of 453 MPa, ultimate tensile strength (UTS) of 510 MPa, and elongation to fracture of 4 %. Honma *et al.* [7] reported that the extruded and peak-aged Mg-10Gd-5.6Y-1.6Zn-0.5Zr (wt.%) alloy exhibited high TYS of 473 MPa, UTS of 542 MPa, and elongation of 8 %. However, it is very hard to extrude these alloys at industrial scale due to the high strength of these as-cast billets.
On the other hand, to use magnesium alloys for the aircraft components, the ignition resistance is important. Although magnesium alloys in the molten state can be protected from coming into contact with air by the usage of protective gases such as SF_6 and SO_2 gases [1], it is impossible to protect the parts with protective gases. Many researches have been done to improve the ignition resistance of magnesium alloys with the addition of rare earth elements, metallic calcium and beryllium [8-9]. There are a few methods of alloying calcium into magnesium alloys; one is the use of metallic calcium and the other is the reaction of CaO on the surface of molten magnesium alloys. Kim *et al.* [10, 11] reported that the addition of CaO to magnesium alloys provides an effect similar to that of adding metallic calcium to magnesium because CaO is reduced when it reacts with the liquid magnesium. This finding is important from the viewpoint of industry because CaO is more cost effective and less hazardous to handle than metallic calcium.
According to our previous study, the Ca added Mg-Gd-Zn alloys exhibited high ignition temperatures. In addition, the addition of Ca in Mg-Gd-Zn alloy formed stacking faults in as-cast state and after heat treatment. LPSO phase coexisted with stacking faults in Mg-Gd-Zn-Ca alloys. However, the strength of extruded alloys were lower than those of other reported alloys. Therefore, the main task of this study was to investigate the effects of alloying elements on microstructures and mechanical properties of Mg-Gd-Zn-Ca alloys for improving the strength of Mg-Gd-Zn-Ca alloys. In this study, addition of Ca to tested alloys was conducted by making Mg-Ca master alloys using CaO powders and pure magnesium.

Table 1. Nominal compositions of tested alloys

Alloy designation	Chemical compositions (wt.%)				
	Mg	Gd	Y	Zn	Ca*
VZO5107	93.3	5	-	1	0.7
VWZO51107	92.3	5	1	1	0.7
VWZO52107	91.3	5	2	1	0.7
VWZO53107	90.3	5	3	1	0.7

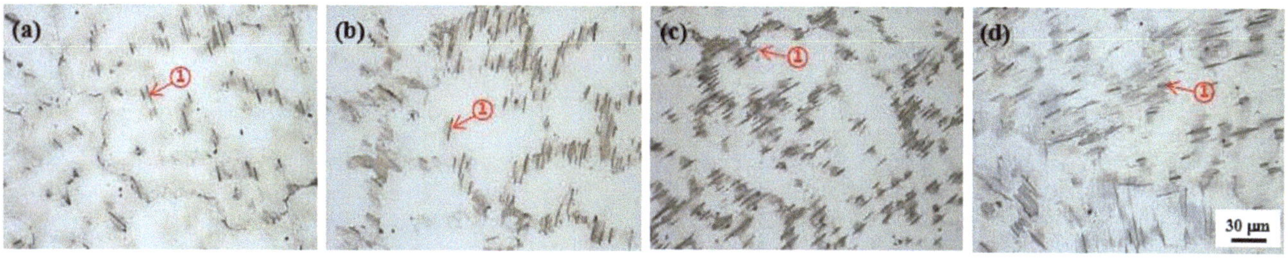

Fig. 1. Optical microstructures of as-cast (a) VZO5107, (b) VWZO51107, (c) VWZO52107, and (d) VWZO53107 alloys

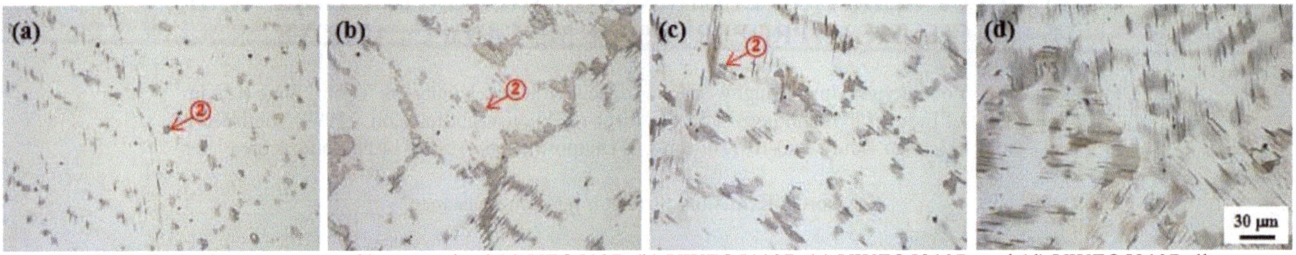

Fig. 2. Optical microstructures of homogenized (a) VZO5107, (b) VWZO51107, (c) VWZO52107, and (d) VWZO53107 alloys

Experimental procedure

Tested alloys in this study were designed as in Table 1. We used "O" as a code letter to be distinguished from alloys with metallic calcium whose code letter is "X". The as-cast billets were prepared by melting commercial purity Mg and Zn, Mg-40Gd, Mg-25Y and Mg-4.16Ca master alloys in an electric resistance furnace using a steel mold under the protection of the mixed gas of CO_2 and SF_6. The Mg-Ca master alloy was fabricated by EMK Co. Ltd. (Chungcheongbuk-do, Republic of Korea) using CaO powders and pure magnesium. After holding for 30 min at 690 °C, the melt was poured into a cylindrical steel mold with an inside diameter of 50 mm and a length of 200 mm. The cast billets were homogenized at 450 °C for 24 h. The indirect extrusions of the homogenized billets were carried out at 300 °C, 350 °C, and 400 °C with the extrusion ratio of 20 or 25. Before the extrusion, billets were preheated at extrusion temperatures for 40 min.

Uniaxial tensile tests were performed on a dog-bone specimens (specimen gauge length of 25 mm) machined from as-extruded samples at room temperature with the constant crosshead speed of 1.0 mm/min. DTA analyses were performed under dry air atmospheres to measure the ignition temperatures of tested alloys. The sample weight for DTA test was about 7~10 mg and the heating rate was 10 °C/min. For an analysis of the microstructural characteristics of specimens, an optical microscope (OM) and a scanning electron microscope (SEM) were used and chemical compositions of phases were analyzed by an Energy Dispersive Spectroscopy (EDS).

Results and Discussion

Microstructures and Mechanical Properties

The microstructures of as-cast alloys were shown in Fig. 1. All alloys consisted of α-Mg and the secondary solidified phases. These phases seemed to be LPSO phases and, according to our previous study [12], the addition of Ca in Mg-Gd-Zn alloy formed stacking faults in as-cast state. In addition, black needle-type phase (① in Fig. 1) was a Ca-bearing phase, but the exact composition of the black needle-type phases in Fig. 1 could not be measured unambiguously by SEM and EDS analysis due to the large interaction volume of the beam. With increasing yttrium content, the volume fraction of LPSO-like phases increased. Fig. 2 shows the microstructures of alloys homogenized at 450 °C for 24 h. As seen in Fig. 2, some rounded phases (② in Fig. 2) could be observed. These phases might be formed during the homogenizing process by the migration of atoms. However, with increasing yttrium content, it was hard to observe these rounded phases. That means, the addition of yttrium enhanced the thermal stability of the secondary solidified phases (LPSO-like phases). Fig. 3 shows SEM images of as-cast and homogenized VWZO52107 alloys. The results of EDS were listed in Table 2. Compared with the EDS result of LPSO-like phase (Fig. 3 (b) ①), more amount of alloying elements were detected in the rounded phase (Fig. 3 (b) ②) by EDS.

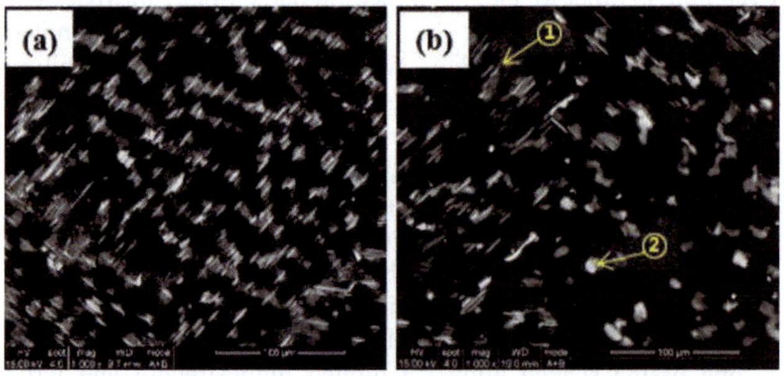

Fig. 3. SEM images of (a) as-cast and (b) homogenized VWZO52107 alloy

Table 2. EDS results of phases in Fig. 3 (b)

Marks	Elements (wt.%)				
	Mg	Gd	Y	Zn	Ca
①	77.51	13.79	4.32	3.51	0.86
②	64.75	15.56	5.80	5.64	8.25

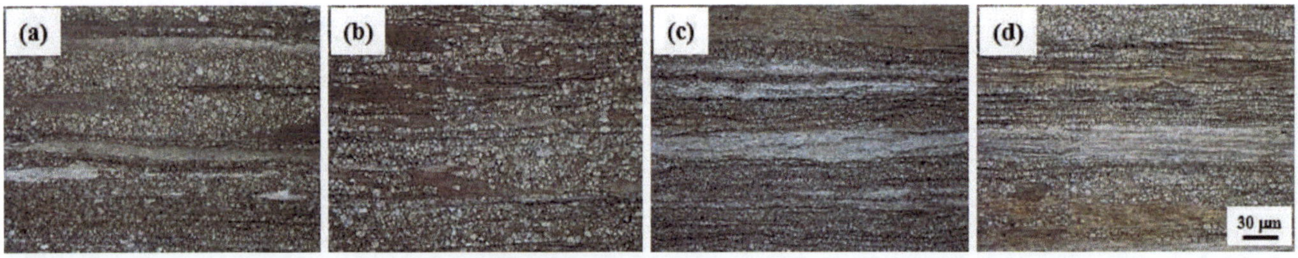

Fig. 4. Optical microstructures of as-extruded (a) VZO5107, (b) VWZO51107, (c) VWZO52107, and (d) VWZO53107 specimens (Extrusion temperature: 400 °C and extrusion ratio: 25)

Table 3. Mechanical properties of as-extruded alloys

Alloy	TYS (MPa)	UTS (MPa)	El. (%)
VZO5107	221+/-4.2	277+/-4.8	14.3+/-2.5
VWZO51107	238+/-3.1	299+/-5.9	13.4+/-1.8
VWZO52107	296+/-3.8	339+/-6.1	12.0+/-2.1
VWZO53107	294+/-4.6	341+/-5.6	11.7+/-1.5

The mechanical properties of as-extruded alloys, which were fabricated at 400 °C with the extrusion ratio of 25, are listed in Table 3. The microstructures of as-extruded alloys were shown in Fig. 4. With increasing yttrium content, the strength increased, but elongation decreased. The reason for increase of strength might be the grain refinement and the increment of the volume fraction of LPSO-like phases. From the results of tensile test, it could be seen that the addition of more than 2 % of Y is effective for increasing the strength of Mg-5Gd-1Zn-0.7Ca alloy.

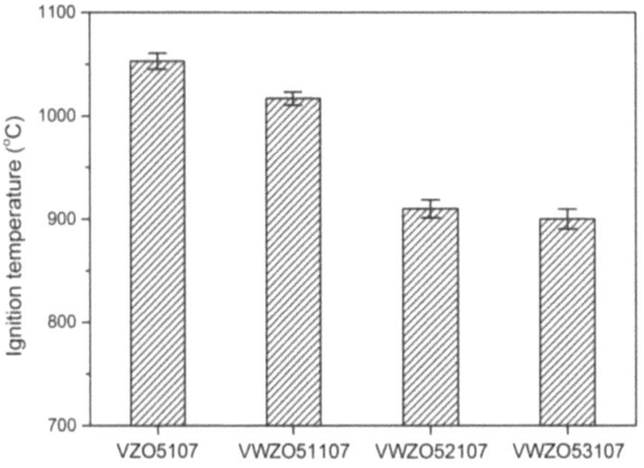

Fig. 5. Ignition temperatures of tested alloys measured by DTA

Ignition Temperature

Fig. 5 showed the results of DTA for alloys tested under dry air atmosphere. The ignition temperatures of VZO5107, VWZO51107, VWZO52107, and VWZO53107 alloys were 1053.2 °C, 1017.0 °C, 910.3 °C, and 899.6 °C, respectively. From the results of DTA, it could be noticed that the increase of Y content weakened ignition resistance of these alloys. According to Ravi Kumar *et al.*, during oxidation under an ambient atmosphere, an yttrium rich oxide could quickly form on the surface of samples, so the yttrium rich oxide could prevent the ignition of the specimens at high temperature [13]. However, in this alloy system, the formation of the yttrium rich oxide might break the dense and complex oxides (MgO, CaO, and etc.) layer on the surface of specimens because of the large value of the Pilling-Bedworth ratio of the yttrium oxide, which is about 1.13. But it should be needed more study on the reason of the decrease of ignition temperature by the adding of Y in this alloy system.

Effect of Extrusion Conditions

The results of tensile test at room temperature for as-extruded VWZO52107 specimens, which were prepared with various extrusion conditions, were shown in Fig. 6. Fig. 7 shows the microstructures of as-extruded specimens with various extrusion conditions. With decreasing extrusion temperature, the strength increased due to the fine recrystallized grains. These fine recrystallized grains might be caused by lower friction heat between dies and billet. Generally, the friction heat between dies and billet influence the microstructure and the surface quality of extruded specimens. Therefore, the friction heat might be reduced with decreasing extrusion temperature. On the other hand, when the specimens were extruded at the same temperature, the extrusion ratio had an effect on mechanical properties. With increasing the extrusion ratio, the strength and elongation increased simultaneously. Also, it might be because of the fine recrystallized grains.

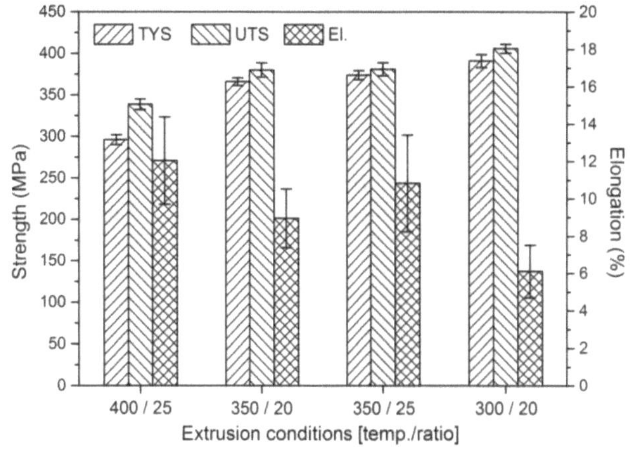

Fig. 6. Mechanical properties of as-extruded VWZO52107 specimen with various extrusion conditions

Conclusion

In the present study, microstructures, ignition temperatures, and mechanical properties of yttrium added Mg-Gd-Zn-Ca alloys were investigated. The conclusions are as follows:

(1) The addition of the yttrium into Mg-Gd-Zn-Ca alloys increased the volume fraction of LPSO-like phases and enhanced the thermal stability of LPSO-like phases.

(2) With increasing yttrium content, the strength increased, but elongation decreased. The increase of strength resulted from the fine grain size and the increment of the volume fraction of LPSO-like phases.

(3) The addition of yttrium into Mg-Gd-Zn-Ca alloys decreased the ignition-resistance of alloys due to formation of yttrium rich oxide with the large value of the Pilling-Bedworth ratio.

(4) In Mg-Gd-Y-Zn-Ca alloy system, with decreasing extrusion temperature, the strength increased and with increasing the extrusion ratio, the strength and elongation increased simultaneously.

References

[1] H. Watarai, "Trend of research and development for magnesium alloys", *Quarterly Review (NISTEP)*, 18 (2006) 84-97.

[2] Avedesian, H. Baker, eds., *Magnesium and Magnesium Alloys*, ASM International, Metals Park, 1999.
[3] J.F. Nie, X. Gao, and S.M. Zhu, "Enhanced age hardening response and creep resistance of Mg-Gd alloys containing Zn", *Scripta Mater*, 53 (2005) 1049-1053.

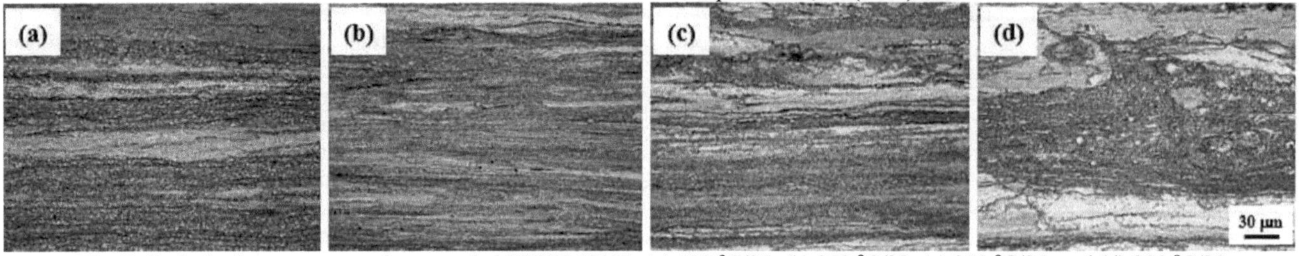

Fig. 7. Optical microstructures of as-extruded VWZO52107; (a) 400 °C/25, (b) 350 °C/20, (c) 350 °C/25, and (d) 300 °C/20

[4] T. Honma et al., "Effect of Zn additions on the age-hardening of Mg-2.0Gd-1.2Y-0.2Zr alloys", *Acta Mater*, 55 (2005) 4137-4150.
[5] X. Li, C. Liu, and T. Al-Samman, "Microstructure and mechanical properties of Mg-12Gd-3Y-0.6Zr alloy upon conventional and hydrostatic extrusion", *Mater Lett*, 65 (2011) 1726-1729.
[6] Y. Kawamura and M. Yamasaki, "Formation and Mechanical Properties of $Mg_{97}Zn_1RE_2$ Alloys with Long-Period Stacking Ordered Structure", *Mater Trans*, 48 (2007) 2986-2992.
[7] T. Homma, N. Kunito, and S. Kamado, "Fabrication of extraordinary high-strength magnesium alloy by hot extrusion", *Script Mater*, 61 (2009) 644-647.
[8] X. Zeng et al., "Behavior of surface oxidation on molten Mg-9Al-0.5Zn-0.3Be alloy", *Mat Sci Eng A*, 301 (2001) 154-161.
[9] Z. Xiaoqin et al., "Study of ignition proof magnesium alloy with beryllium and rare earth additions", *Scripta Mater*, 43 (2000) 403-409.
[10] J.K. Lee and S.K. Kim, "Effect of CaO Addition on the Ignition Resistance of Mg-Al Alloys", *Mater Trans*, 52 (2011) 1483-1488.
[11] J.K. Lee and S.K. Kim, in: Proceedings of 18th International Conference on Processing and Fabrication and Advanced Materials (PFAM18-2009), Sendai, December 2009, PFAM18, Japan.
[12] H.K Lim, "Investigation of Microstructures and Mechanical Properties of Ca Added Mg-Gd-Zn Alloys" (Paper presented at the 10th International Conference on Magnesium Alloys and Their Application, Jeju, Korea, 14 October 2015)
[13] N.V. Ravi Kumar et al., "Effect of alloying elements on the ignition resistance of magnesium alloys", *Scripta Mater*, 49 (2003) 225-230.

Magnesium Technology 2016
Edited by: Alok Singh, Kiran Solanki, Michele V. Manuel, and Neale R. Neelameggham
TMS (The Minerals, Metals & Materials Society), 2016

CREEP OF A Mg-Zn-Y ALLOY AT ELEVATED TEMPERATURES

Weiwei Hu, Zhiqing Yang*, Jianfang Liu, Hengqiang Ye
Shenyang National Laboratory for Materials Science, Institute of Metal Research, Chinese Academy of Science; Shenyang, 110016, China
*E-mail: yangzq@imr.ac.cn

Keywords: Mg alloy, Strength, Creep, Long-periodic stacking ordered phase, Dislocation, Stacking fault

Abstract

Mg alloys containing long-period stacking ordered (LPSO) phases have superior strength at elevated temperatures, which makes them potential creep resistant materials. We have studied the creep behavior of a $Mg_{97}Zn_1Y_2$ (at.%) alloy. No creep strain was detected for the $Mg_{97}Zn_1Y_2$ alloy at 150 ℃ under a tensile stress of 70MPa in 100h. The creep strains were measured to be about 0.01% and 0.28% at 200 ℃ and 250 ℃, respectively, under a tensile stress of 70MPa in 100h, showing excellent creep resistance. Fracture occurred with a creep strain of 7.55% after 42h of tensile creep under a tensile stress of 70MPa at 300 ℃. The microstructures were characterized by transmission electron microscopy and scanning-transmission electron microscopy techniques, in order to understand the creep deformation behavior and microstructural origin of the excellent creep resistance of this alloy. LPSO phase was found playing an important role in the alloy's creep resistance. Generation and motion of basal "**a**" dislocations led to bending of LPSO phase. No voids were formed at LPSO/Mg interfaces. Suzuki segregation occurred widely along stacking faults resulting from dissociation of either "**a**" or "**a + c**" dislocations in the Mg matrix, which hindered dislocation motion and thus played an important role in strengthening the Mg grains that are softer than the LPSO phases. This segregation should have considerable contribution to the alloy's excellent creep resistance at elevated temperatures. Growth of stacking faults on basal planes in Mg can effectively restrict motion of non-basal dislocations. The excellent creep resistance of the $Mg_{97}Zn_1Y_2$ (at.%) alloy at elevated temperatures is intimately associated with the LPSO strengthening phase, wide stacking faults with Zn and Y segregation formed during solidification, and dynamic Suzuki segregation along nanometer-sized stacking faults produced by dislocation dissociation.

Introduction

Reduction in vehicle weight can play an important role in reducing fuel consumption and CO_2 emission [1,2]. Mg alloys are attractive materials for the transportation industry due to their low density, high specific strength, excellent damping and ease of recycling [1]. Mg alloys have been used to make automotive components, such as steering wheels, instrument panels, valve covers and et al. There are also increasing demands for heat-resistant Mg alloys for powertrain applications. The powertrain applications require materials with good creep resistance under stresses in the range of 50-70 MPa at elevated temperatures as high as 200 ℃ or above [3]. However, the relatively low strength and poor ductility of Mg alloys limit their widespread engineering applications. Improving strength and ductility is a central goal for widespread application of Mg alloys. In addition, development of Mg alloys with good creep resistance at elevated temperatures is of general interest and importance [2]. Great efforts have been made in alloy design and processing in order to improve their mechanical properties. Mg alloys with LPSO structures are found to show good strength at both ambient and elevated temperatures. A tensile yield strength of ~ 600 MPa and elongation of ~ 5.0% were achieved by rapidly solidified powder metallurgy processing of a $Mg_{97}Zn_1Y_2$ (at.%) alloy [4]. Hot-extruded $Mg_{97}Y_2Cu_1$ (at.%) alloy exhibited tensile strength and elongation of 377 MPa and 8.1% at ambient temperature, and 344 MPa and 16.3% at 200 ℃ [5]. The high tensile strength at 200 ℃ in Mg alloys strengthened by LPSO structures has not been achieved in Mg alloys consisting of other types of secondary strengthening phases. For example, the tensile strength of extruded AZ31B alloy decreased significantly from about 300 MPa to about 100 MPa when raising the temperature from ~25 ℃ to 200 ℃ [5].

Creep properties are of great importance for the application of Mg alloys at elevated temperatures. However, creep, especially tensile creep of Mg alloys with LPSO has not been investigated comprehensively [6]. The aim of our study is to reveal the microscopic creep behavior under tensile stress and the creep-resistance mechanisms of Mg alloys with LPSO structures, by using a $Mg_{97}Zn_1Y_2$ (at.%) alloy as the model material. The superior mechanical properties of this kind of Mg alloys should not be solely attributed to the LPSO structures distributed along grain boundaries. The Mg matrix should also play a role in the high strength, especially at elevated temperatures, since a highly softened Mg matrix is expected to result in a remarkable decrease in the material's strength. Microstructures of both the LPSO structures and Mg grains in crept samples were characterized in order to get some in-depth understanding of the creep-resistance mechanisms.

Experiments

A $Mg_{97}Zn_1Y_2$ (at.%) alloy was prepared using a high frequency induction furnace under argon atmosphere protection. The alloy's yield strength is about 90 MPa at 300 ℃ [7]. Creep tests were performed under a tensile stress of 70 MPa at temperatures ranging from 150 ℃ to 300 ℃. The samples were protected using argon atmosphere during the creep tests. Samples were held at the given temperature for at least five minutes before starting the creep tests. The creep test was ended at a creep time of 100 h if no fracture occurred. The crept samples were cooled down to room temperature after the creep tests. A Tecnai G2 F30 transmission electron microscope (TEM) and a Titan 60-300 scanning TEM (STEM) were used to characterize the lattice defect microstructures [8,9]. Samples for electron microscopy characterization were prepared by standard ion milling.

Results and discussion

The as-cast $Mg_{97}Zn_1Y_2$ (at.%) alloy has a duplex microstructure with LPSO phases distributed along grain boundaries of Mg dendrite matrix. Furthermore, TEM and STEM investigations show that there are high density stacking faults (SFs) and nanometer thick LPSO plates in Mg grains, as shown in Figure 1. Those straight SFs were formed during solidification, and most of them go across the whole Mg grains. The SFs are much brighter than the Mg matrix in the high-angle annular dark-field (HAADF) STEM image, demonstrating the segregation of Zn/Y atoms along those growth SFs [8-10].

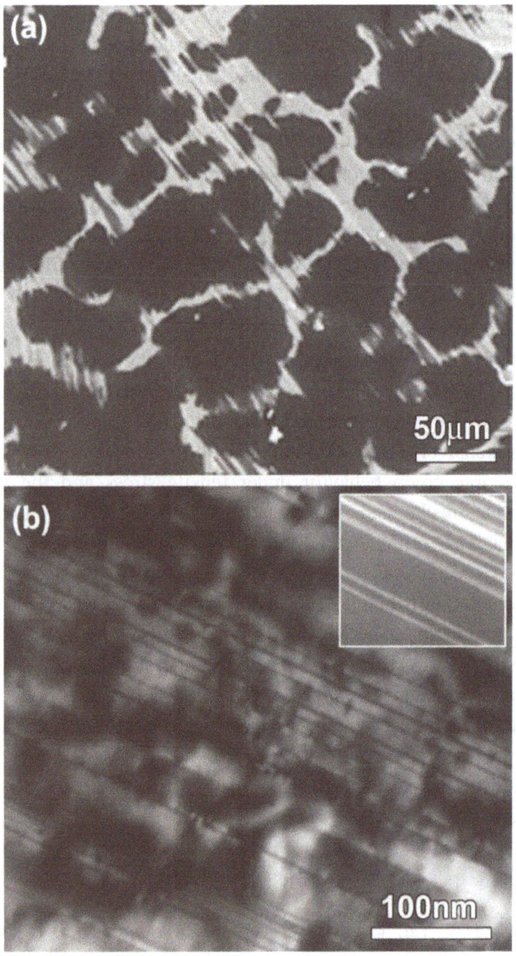

Figure 1. (a) A back-scattered scanning electron microscopy image showing the duplex microstructure of the as-cast alloy. (b) A TEM image showing SFs in Mg grains. The upper-right inset in (b) is a STEM image demonstrating Zn/Y segregation along SFs.

Figure 2 shows curves of creep strain as a function of creep time under a tensile stress of 70 MPa at different temperatures. No creep strain was detected for the $Mg_{97}Zn_1Y_2$ alloy at 150 ℃ after tensile creep for 100h. The creep strains were measured to be about 0.01% and 0.28% at 200 ℃ and 250 ℃, respectively, after tensile creep for 100 h. Those tensile creep tests demonstrate excellent creep resistance of the $Mg_{97}Zn_1Y_2$ (at.%) alloy at temperatures as high as 250 ℃. Fracture occurred with a creep strain of 7.2% after 42 h of tensile creep under a tensile stress of 70 MPa at 300 ℃. Data fitting shows that creep at 200 ℃ is in the first stage during the whole tensile test, while steady-state creep, i.e., the second stage, started after the sample subjected to the tensile load for about 50 h, when the temperature was increased to 250 ℃. At 300 ℃, the first and second creep stages lasted for about 2.5 h and 17.5 h, respectively; the sample was then subjected to accelerated creep deformation (the third stage of creep), and fracture happened about 42 h. The creep rates were calculated to be 5.05×10^{-9} s^{-1} and 1.58×10^{-7} s^{-1} for the steady-state creep stage at 250 ℃ and 300 ℃, respectively. The obtained tensile creep rates are similar to those for tests under a compressive stress of 70 MPa [6].

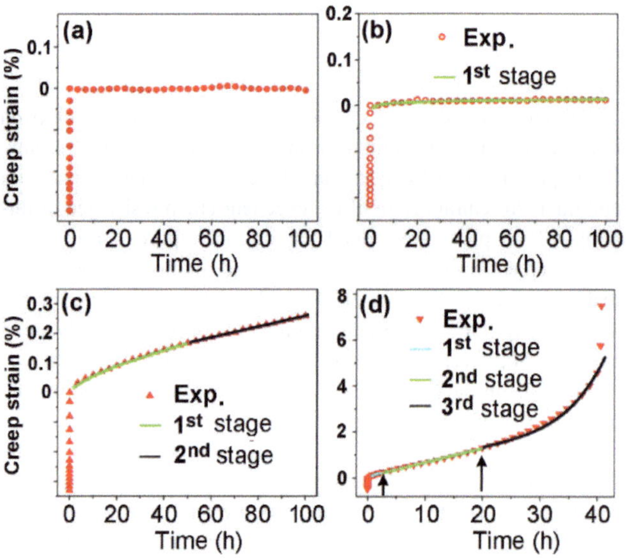

Figure 2. Creep curves of the Mg alloy under a tensile stress of 70 MPa, at temperatures of (a) 150 ℃, (b) 200 ℃, (c) 250 ℃, (d) 300 ℃, respectively. Arrows in (d) point to the ending the 1st and 2nd stages of creep.

Since no creep took place for the sample tested at 150 ℃, we did not perform detailed TEM and STEM investigations on those crept samples. Two-beam bright-field TEM imaging was carried out, in order to check the nature of dislocations that were activated in Mg grains during the creep tests. Figure 3 show series of two-beam bright-field TEM images recorded from Mg grains in samples creep-tested at 200 ℃, 250 ℃ and 300 ℃, respectively. Three diffraction vectors, (0002), $(01\bar{1}0)$ and $(01\bar{1}1)$ were chosen in the two-beam imaging characterization of the same region of interest, so that we can identify both "**a**" and "**a** + **c**" dislocations. Systematic analyses showed that generation and motion of "**a**" dislocations played an important role in creep strain of most Mg grains at 200 ℃ and 250 ℃, as shown in Figures 3a-c and 3d-f, respectively. The dislocation density in Mg grains in the sample tested at 250 ℃ is slightly higher than that

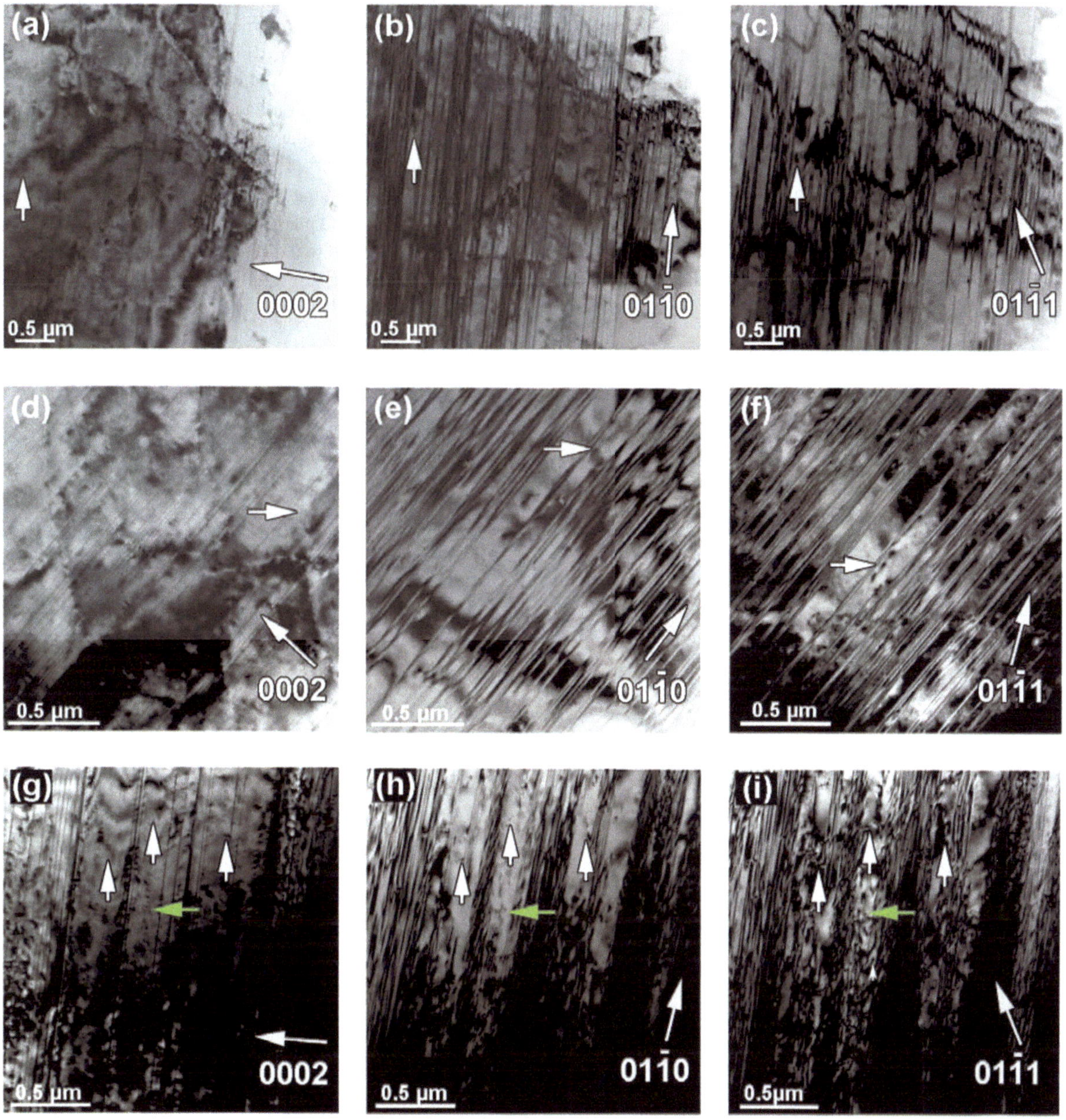

Figure 3. TEM images recorded creep curves of the Mg alloy at (a-c) 200℃, (d-f) 250℃, (g-i) 300℃. The three diffraction vectors used in recording the two-beam TEM images and their direction are given by a long arrow just above them in each image. White and green arrows point to "**a**" and "**a** + **c**" dislocations, respectively.

tested at 200 ℃. Both "**a**" and "**a** + **c**" dislocations were observed in Mg grains in the sample crept at 300 ℃, as indicated by white and green arrows in Figures 3g-i, respectively. The presence of high-density SFs in Mg grains produced complex fringes in the two-beam TEM images, which makes the imaging of dislocations difficult using diffraction vectors of $(01\bar{1}0)$ and $(01\bar{1}1)$, as shown in Figure 3.

STEM imaging is another technique that can provide microstructural information of materials down to an atomic scale. Additionally, STEM can provide atomic-scale chemical information of materials [8-10], which is very useful for investigations on microstructural evolution involving changes in microchemistry. Figure 4 shows low-magnification HAADF-STEM images recorded along the $\langle 11\bar{2}0\rangle$ zone axis for samples crept at 200℃, 250 ℃ and 300℃. Although motion of "**a**" dislocations played a predominant role in most Mg grains in samples crept at 200 ℃, evidence of motion of "**a** + **c**" dislocations was observed in a small number of Mg grains, as shown in Figure 4a. It is clearly seen that several SFs and thin

LPSO plates in Mg were broken, producing steps in the [0001] direction, as indicated by arrows in Figure 4a. This is clear evidence for local non-basal shear in Mg grain due to the generation and motion of "**a** + **c**" dislocations. Furthermore, small regions showing higher brightness appeared in Mg, which should be attributed to local segregation of Zn/Y atoms. Since there is no such local segregation of Zn/Y atoms present in the as-cast materials, it should be formed during the creep deformation at 200 ℃. Cottrell introduce the concept of "atmospheres" in 1949, in order to explain the serrated deformation behaviour of steels [11]. The elastic interaction between dislocations and solute atoms may result in diffusion of solute atoms toward dislocations, and form solute segregation along dislocations, which releases partly the elastic strain of those lattice defects. The solute atoms and dislocations pin each other, which strengthens the materials. "**a** + **c**" dislocations have a high elastic strain energy, which can thus play an important role in attracting Zn/Y atoms to form Cottrell atmospheres during the creep deformation. Little plastic deformation was observed in LPSO phases in the sample tested at 200 ℃. Slight kinking was observed occasionally at a few positions in the LPSO phases, as shown in Figure 4b. The kinking of LPSO is attributed to the generation and rearrangement of basal "**a**" dislocations [12]. In addition, a fraction of Mg in thin Mg slices sandwiched between LPSO thick plates diffused away

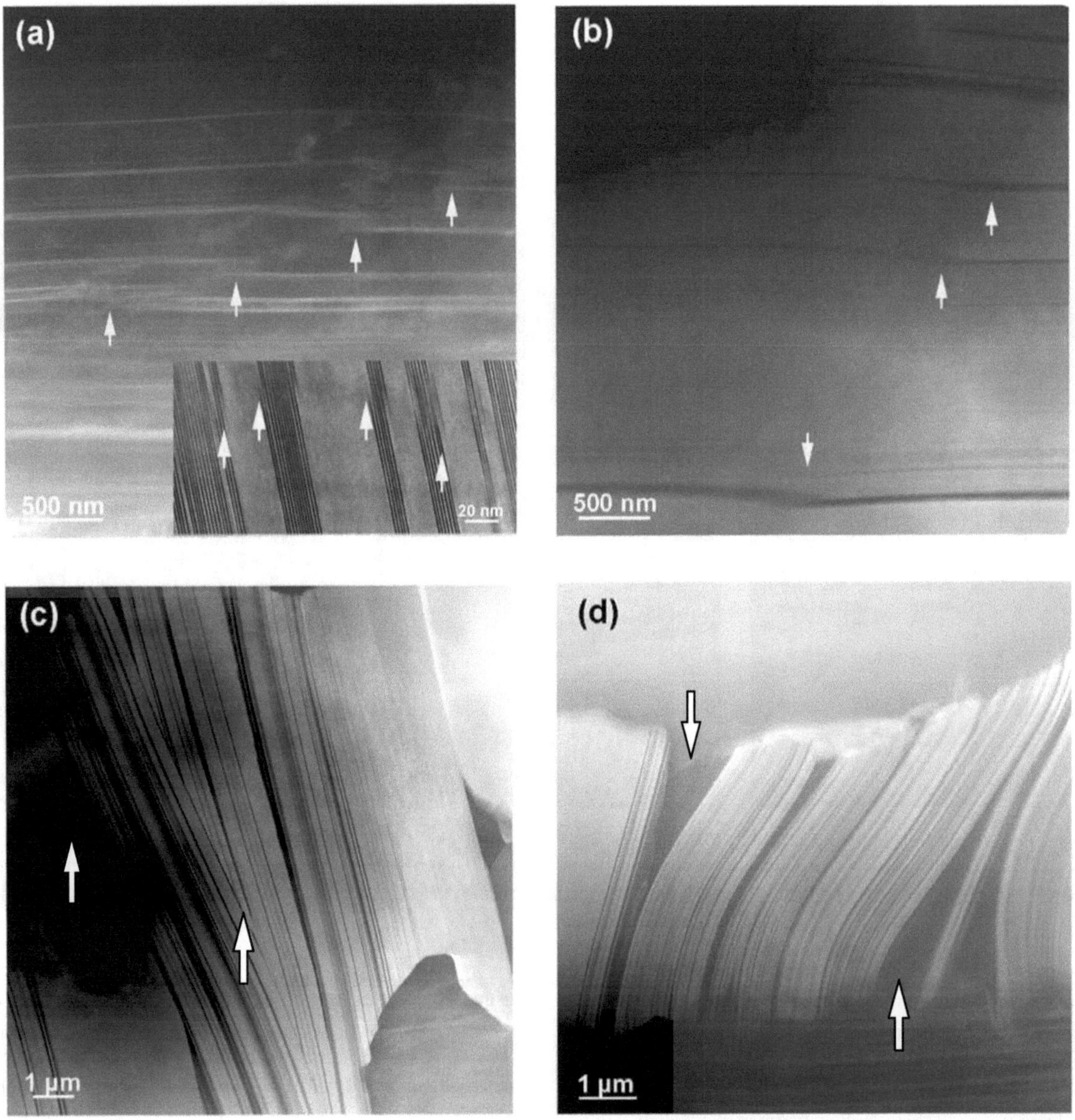

Figure 4. Low-magnification HAADF-STEM images for crept samples. (a) An image showing defects in Mg at 200 ℃. The bottom-right inset is a two-beam TEM image showing local steps along SFs. Small steps on both sides of the SFs or thin LPSO plates were produced by the cut of "**a** + **c**" dislocations. (b) An image showing deformation of LPSO at 200 ℃. (c) An image showing both basal slide and slight bending of LPSO at 250 ℃. (d) An image showing obvious bending and delaminating of LPSO at 300 ℃.

under the applied stress during creep deformation at 200 ℃, resulting in local decrease in their thickness.

The deformation microstructure in Mg grains in the sample tested at 250 ℃ is more or less similar to those in the sample tested at 200 ℃, according to both TEM and STEM observations. Figure 4c is a low-magnification HAADF-STEM image for LPSO in samples crept at 250 ℃, showing a higher level of plastic deformation of LPSO with respect to samples tested at 200 ℃. Those LPSO plates indicated by arrows in Figure 4c are wedge-shaped, demonstrating large shear deformation along their basal planes due to generation and glide of a large number of basal "**a**" dislocations in those LPSO plates. This suggests that basal "**a**" dislocations still play an important role in the creep deformation of this alloy at 250 ℃. Obvious bending of LPSO was observed in samples crept at 300 ℃, as shown in Figure 4d. Lattice bending is due to the presence of high density of basal "**a**" dislocations. In addition, thermally driven diffusion of Mg was able to accommodate to bending of the LPSO plates during the creep deformation at 300 ℃. So, no voids or cracks were formed at interfaces between Mg thin slices and LPSO plates due to the obvious bending of LPSO plates, as indicated by arrows in Figure 4d.

Figure 5a is a low-magnification HAADF-STEM image for Mg grains in samples crept at 300 ℃. Some of the long basal growth SFs and thin LPSO plates were subjected to non-basal shear due to glide of "**a** + **c**" dislocations, which is similar to that shown in Figure 4a. Furthermore, there are many bright lines that are shorter than 100 nm and show brightness lower than the growth SFs or thin LPSO plates in the crept Mg grain, as indicated by arrows in Figure 5a, indicating local segregation of Zn/Y atoms on basal planes in Mg. Those shorter should be newly formed during the creep deformation, since no such features were observed in the as-cast $Mg_{97}Zn_1Y_2$ (at.%) alloy, as shown in Figure 1b. Figure 5b shows a high-resolution HAADF-STEM image for a region containing one growth SF and one LPSO plate. There is a step of two atomic layers high along the growth SF and each sub-layer with local face-centered cubic (FCC) stacking in the LPSO plate, as indicated by the black lines in Figure 5b. The two atomic-layer steps were produced by the cut of an "**a** + **c**" dislocation. Severe lattice distortion can be observed at those steps. Those basal growth SFs with Zn/Y segregation and thin LPSO plates in Mg grains have little resistance to glide of basal "a" dislocations. However, they can hinder the movement of "**a** + **c**" dislocations, which benefits the creep resistance at elevated temperatures when non-basal slip systems are activated. It has been demonstrated that formation of a small fraction of chemically ordered structures/phases can improve materials' strength at elevated temperatures, which thus benefits their creep resistance [13-16].

Figure 5c is a high-resolution HAADF-STEM image for one short newly formed bright line with Zn/Y segregation next to a thin LPSO plate containing three FCC sub-layers. White lines outline a Burger's circuit around the region with Zn/Y segregation, showing a closure failure of the Burger's vector of a 60° basal "**a**"

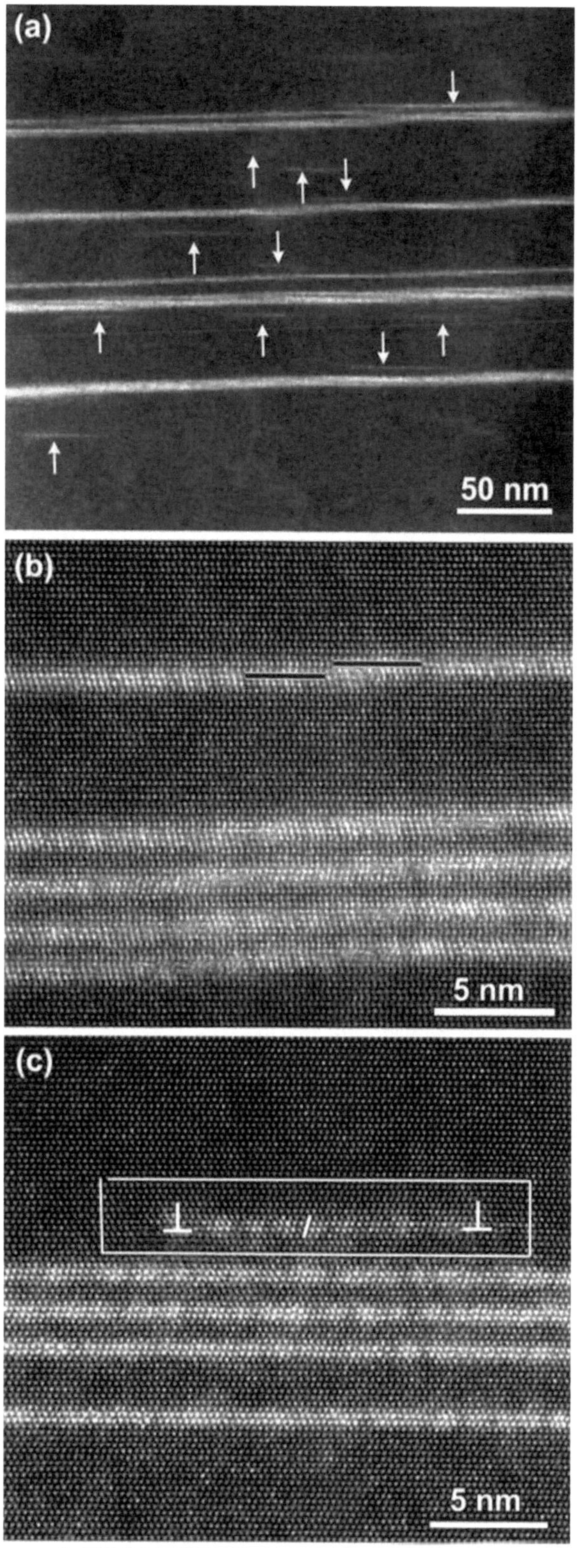

Figure 5. HAADF-STEM imaging of creep microstructures in Mg grains in samples crept at 300 ℃. (a) Low-magnification image showing growth SFs and newly formed nanometer-wide structures on basal Mg planes. (b, c) High-resolution HAADF-STEM images showing atomic structures of the non-basal shear of growth SFs and a thin LPSO plate, and Suzuki segregation with the dissociation of a 60° basal "**a**" dislocation.

dislocation. Close analyses found that the 60° basal "**a**" dislocation was dissociated into two Shockley partials and produced one SF in between, as indicated by "⊥" and a short bar respectively in Figure 5c [10,12]. Therefore, Suzuki segregation took place during the creep deformation at 300 ℃, instead of the formation of Cottrell atmospheres during creep at 200 ℃ (Figure 4a). The occurrence of Suzuki segregation of Zn/Y atoms requires relocation of Zn/Y atoms along moving dislocations or in the vicinity of SFs. It is most likely that diffusion of Zn/Y atoms is not fast enough to realize their relocation at 200 ℃, while the thermally driven diffusion becomes fast enough with increasing the creep temperature to 300 ℃. It should be noted that Suzuki segregation was also observed in samples subjected to compressive deformation with a strain rate of 1×10^{-3} s^{-1} which is much faster than the tensile creep rate at 70 MPa [10]. The Suzuki segregation can pin the dissociated dislocations, as well as act like tiny precipitates or GP zones impeding movement of other dislocations. Therefore, both Cottrell atmospheres (Figure 4a) and Suzuki segregation (Figure 5a and c) make contributions to the excellent creep resistance of the $Mg_{97}Zn_1Y_2$ (at.%) alloy.

Severe kinking of LPSO was widely observed in the $Mg_{97}Zn_1Y_2$ (at.%) alloy subjected to compressive creep with stepwise loading: the load was increased to a new value after a steady-state creep rate was established in each step, and the creep rate was increased to 10^{-4}-10^{-3} s^{-1} finally [6]. The tensile creep of the $Mg_{97}Zn_1Y_2$ (at.%) alloy at 200 ℃ and 250 ℃ are in the first and second stage of creep deformation, respectively; and the final creep strains are only 0.01% and 0.28%, respectively. No obvious plastic deformation of the LPSO phase would be expected at such low creep strains, since the LPSO is much harder than the Mg matrix [7]. The steady-state creep rate at 300 ℃ is only 1.58×10^{-7} s^{-1} in our tensile creep experiment, which is three to four orders of magnitude lower than the final creep rate during compressive creep [6]. The low creep rate and/or creep strain are believed responsible for the lack of kinking of the LPSO phases in our crept samples.

Conclusions

In summary, we studied the tensile creep behavior of the $Mg_{97}Zn_1Y_2$ (at.%) alloy under an applied stress of 70 MPa. No creep deformation was detected at 150 ℃, after 100 h of tensile loading. The creep strain was measured to be 0.01% and 0.28%, respectively, after 100 h tensile loading at 200 ℃ and 250 ℃. The sample was broken with a creep strain of 7.55% at 300 ℃. Therefore, the $Mg_{97}Zn_1Y_2$ (at.%) alloy shows excellent creep resistance at temperatures as high as 200 ℃.

Both "**a**" and "**a** + **c**" dislocations were activated during tensile creep at 200 ℃ and above. Growth SFs and thin LPSO plates in Mg grains can hinder the glide of "**a** + **c**" dislocations which were activated during creep at elevated temperatures, so they should have contributions to the creep resistance of the $Mg_{97}Zn_1Y_2$ (at.%) alloy. Local solute segregation, in the form of either Cottrell atmospheres or Suzuki segregation, was observed in crept samples. The solute atmospheres may have higher pinning effects on dislocations compared with randomly distributed individual solute atoms, which can play an important role in improving the creep resistance of the material. Lattice bending, instead of kinking was often observed in the LPSO in tensile crept samples.

Acknowledgement

The research was financially supported by the NSFC (51171189, 51371178 and 51390473), Liaoning Province (201501701) and SYNL (2015RP17).

References

[1] M. M. Avedesian, H. Baker. Magnesium and magnesium alloys. Maerials Park (OH): ASM international, 1999.

[2] M. Bamberger, G. Dehm. Annu. Rev. Mater. Res. 38, 2008, 505.

[3] M. O. Pecguleryuz, A. A. Kaya. Adv. Eng. Mater. **5**, 2003, 866.

[4] A. Inoue et al. J. Mater. Res. **16**, 2001, 1894.

[5] Y. Kawamura et al. Scripta Mater. **55**, 2006, 453.

[6] G. Garces et al. Mater. Sci. Eng. A 539, 2012, 48.

[7] X. H. Shao et al. Acta Mater. **58**, 2010, 4760.

[8] S. J. Pennycook. Ultramicroscopy **30**, 1989, 58.

[9] S. J. Pennycook. Annu. Rev. Mater. Sci. **22**, 1992, 171.

[10] Z. Q. Yang et al. Acta Mater. **61**, 2013, 350.

[11] A. H. Cottrell, B. A. Bilby. Proc. Phys. Soc. Lond. Sec. A **62**, 1949, 49.

[12] J. P. Hirth, J. Lothe. Theory of dislocations. New York: Wiley, 1982.

[13] J. Weertman. J. Appl. Phys. 28, 1957, 362.

[14] J. Weertman. J. Appl. Phys. 28, 1957, 1185.

[15] R. Srinivasan et al. Phys. Rev. Lett. 102, 2009, 086101.

[16] R. W. Cahn. Mater. Sci. Eng. A 324, 2002, 1.

Magnesium Technology 2016
Edited by: Alok Singh, Kiran Solanki, Michele V. Manuel, and Neale R. Neelameggham
TMS (The Minerals, Metals & Materials Society), 2016

AN INSIGHT INTO USE OF HOLLOW FLY ASH PARTICLES ON THE PROPERTIES OF MAGNESIUM

Gururaj Parande [1], Vyasaraj Manakari [1], Manoj Gupta [1,*]

[1]Department of Mechanical Engineering, National University of Singapore, 9 Engineering Drive 1, Singapore - 117576
mpegm@nus.edu.sg

Keywords: Magnesium, Fly ash cenospheres, Syntactic foams

In the recent years, magnesium based materials has received wide acceptability and attention owing to their attractive properties like high specific strength, high specific stiffness, low density, excellent machinability, castability, and good damping properties. Therefore, they have become a promising choice for applications in the automotive, transport, aerospace and electronic packaging industries [1, 2]. Magnesium is also a biocompatible material and is an important constituent of the bone. The volume of scientific publications on the research of magnesium as a biomaterial has increased exponentially in the past 8-10 years. Further, when compared to other light metals, the Young's modulus of magnesium based materials (40–45 GPa) is comparable to that of natural bone (3–20 GPa). This property of magnesium materials assists in mitigation of stress shielding effects with possibility to eliminate secondary surgery for the implant removal especially for clinical applications [3]. However, costs incurred during the synthesis of these materials are a major limitation. Fly ash cenospheres are alumino-silicate particles which are mainly a by-product of coal combustion in thermal power plants with an extremely low density of about 0.4-0.8g/cc [4]. The incorporation of fly ash particles into magnesium is an encouraging option for light-weight applications as the replacement of magnesium matrix by fly ash particles can further decrease the density of the composite and also the overall cost of raw materials [5]. The addition of hollow particles like cenospheres into polymeric or metallic matrices results in a special type of closed cell foams known as syntactic foams. For the past decade, a lot of research has been carried out mainly on aluminium syntactic foams and researchers have reported enhanced compressive strength and modulus, isotropic behaviour, high energy absorption and attractive strength to weight ratio [6, 7]. Moreover, the addition of fly ash cenosphere particles in composites is an effective way to reduce the carbon emissions and thereby finding applications in automotive, aerospace and military applications. The effect of fly ash cenosphere particles addition in magnesium based alloys and composites have become an encouraging area for researchers for the past 10 years. As magnesium is expanding into more critical structural applications, there is a need to tailor the properties of magnesium syntactic foams for potential use in lightweight energy absorbing components. This paper focuses on magnesium matrix syntactic foams which contain porosity in the form of hollow fillers and a review of our recent findings in the study of magnesium syntactic foams synthesized by disintegrated melt deposition technique.

In past, researchers have studied the effect of fly cenospheres on AZ91, ZC63 magnesium alloy systems and reported intriguing results on the behaviour of the magnesium syntactic foams [4, 8, 9]. However, the influence of fly ash particles on the behaviour of pure magnesium has not been studied in the available literature. In a recent study Bau *et al.* successfully incorporated industrial waste cenosphere particles into magnesium to enhance its overall performance. The goal of the this study was to examine the effect of hollow fly ash cenosphere particles on the microstructural and mechanical properties of pure magnesium synthesized by disintegrated melt deposition (DMD) technique. DMD is a one-of-a-kind technique which brings together the cost effectiveness of the conventional casting process and the technological potential and scientific innovativeness of the spray processes. DMD utilises lower impinging gas jet velocities and higher superheat temperatures to produce bulk composite materials. The synthesis of magnesium composites through DMD technique is thoroughly explained in a previous study [10]. Composites containing 5, 10 and 15 weight percentage of cenospheres were fabricated. The cast ingots obtained from DMD were secondary processed using hot extrusion with an extrusion ratio of 20:25:1.

The addition of cenosphere particles led to a significant reduction (23%) in density of the composites in comparison to the monolithic magnesium from 1.74 g/cc to 1.42 g/cc. The densities exhibited by the synthesized composites could be directly compared to those of polymer based composites such as PE-30% fiber glass composite. However, the measured density of the composites was marginally higher than the calculated density by using Rule of Mixtures. This is attributed to the filling of some fly ash particles with Mg matrix. The hardness of the magnesium matrix was found to increase with the addition of fly ash cenospheres. It is observed that the composite foam containing 15 weight% cenosphere was almost three times harder than pure magnesium. This increase in the hardness can be attributed to the presence of harder cenosphere walls in the matrix and a higher constraint to the localized matrix deformation during indentation due to the presence of fly ash particles. The results on the study of coefficient of thermal expansion (CTE) showed that the CTE of composite foams was significantly reduced due to the increasing presence of cenosphere particles. These results suggest that composite foams are more dimensionally stable with respect to temperature. The reasons for this reduction in CTE values can be attributed to the presence of cenospheres, which contains Al_2O_3 and SiO_2 with low CTE values of $7.4 \times 10^{-6}K^{-1}$ and $0.54 \times 10^{-6}K^{-1}$.

In most of the available studies [4, 8, 9], the strength of syntactic foams is found to be nearly equal to that of matrix. The composites synthesized by DMD technique in this study showed an increase in 0.2% compressive yield strength from 70 MPa to 100, 130 and 150 MPa when the amount of cenosphere was increased from 5 to 15 weight percent. A similar trend was seen in ultimate compressive strength with a maximum strength of 370 MPa which means 106% increment. The significant increase in 0.2% compressive yield strength and ultimate compressive strength of composite foams can primarily be attributed to the presence of reasonably well distributed harder cenosphere particles and secondary phases, load transfer from matrix to reinforcement/second phases and coefficient of thermal expansion mismatch between the matrix and reinforcement/second phases, and Orowan strengthening mechanism [9]. The energy absorption capabilities of syntactic foams are mainly related to the collapse plateau behaviour wherein sequential crushing of microballoons takes place and the material absorbs energy without any significant change in the strength. More than 247% increment in energy absorption was achieved with only 5 weight% addition of cenospheres. These results reflect the superior capability of cenospheres reinforced Mg syntactic foams to absorb significant extent of energy during the scenarios of impact and collisions. Further, it can be observed that all the composites which had higher strength than the matrix didn't show any stress-plateau region and densification trend. This may indicate the fracture of microballoons during fabrication of composites. Fractured cenospheres become higher density solid ceramic particles and result in increased composite density. The tensile properties obtained in this study showed that the ultimate tensile strength increased with addition of 5 weight% cenosphere, but had no effect with the further addition of cenosphere particles. This can be attributed to the presence of reasonably well distributed harder cenosphere particles and the formation of hard secondary phases. Microstructural evaluation showed the presence of secondary phase particles like Mg_2Si, MgO and Al_2O_3. The lack of continued increase in the tensile strength with the increased addition of cenospheres can be attributed to the micro-void formation at the uneven surface of hollow particles.

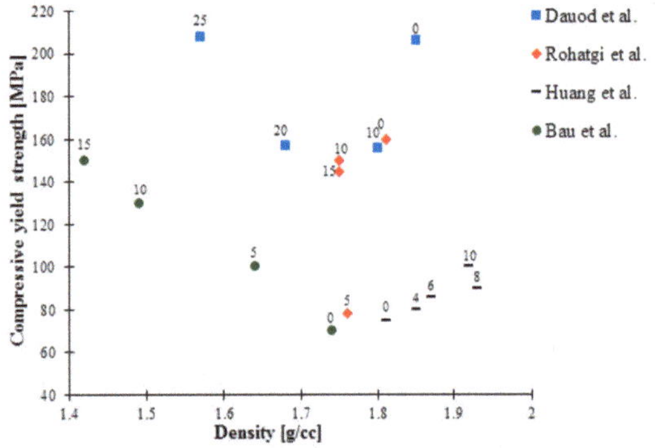

* Note that the weight fraction of reinforcements is labelled with respect to the corresponding data points.

Figure 1. Compressive yield strength of cenosphere reinforced syntactic foams plotted against density

The weight saving advantages of the cenosphere reinforced syntactic foams can be realised by plotting the values of compressive yield strength of the composites with respect to density. Some of the compressive properties presented here are extracted from the published graphs to the best possible accuracy. It can be observed from Figure 1 that the syntactic foam properties can be tailored over a wide range of strength values. Low density syntactic foams can effectively replace the matrix alloys to achieve weight saving in load bearing applications.

Therefore, depending on the specific requirements of the applications the best material can be selected. Also, optimizing the processing techniques used in the production of the composites and better control over the cenosphere microballoon cracking is necessary before these materials can see widespread use. The mechanical and physical properties of the syntactic foams mainly depend on the matrix porosity and the native integrity of the hollow particles. The properties of syntactic foams are also found to be enhanced by heat treatment [11].The strength and modulus of syntactic foams can be further enhanced by secondary processes like rolling, forging and extrusion. However, the presence of large amount of porosities in the matrix may lead to fracture of microballoons when secondary processing is carried out. Extensive studies on the optimization of the secondary processing techniques is required to meet the challenges in enabling manufacture of large near-net shaped parts of syntactic foams.

References

[1] Gupta M, Sharon NML. Magnesium, magnesium alloys, and magnesium composites: John Wiley & Sons; 2011.

[2] Habibi MK, Joshi SP, Gupta M. Hierarchical magnesium nano-composites for enhanced mechanical response. Acta Materialia. 2010;58:6104-14.

[3] Gupta M, Meenashisundaram GK. Insight Into Designing Biocompatible Magnesium Alloys and Composites: Processing, Mechanical and Corrosion Characteristics: Springer; 2015.

[4] Rohatgi P, Daoud A, Schultz B, Puri T. Microstructure and mechanical behavior of die casting AZ91D-Fly ash cenosphere composites. Composites Part A: Applied Science and Manufacturing. 2009;40:883-96.

[5] Rohatgi P, Guo R, Keshavaram B, Golden D. Cast aluminum, fly ash composites for engineering applications. American Foundrymen's Society, Inc(USA). 1996:575-9.

[6] Gupta N, Luong DD, Cho K. Magnesium matrix composite foams—density, mechanical properties, and applications. Metals. 2012;2:238-52.

[7] Rohatgi PK, Gupta N, Schultz BF, Luong DD. The synthesis, compressive properties, and applications of metal matrix syntactic foams. JOM. 2011;63:36-42.

[8] Daoud A, El-Khair MA, Abdel-Aziz M, Rohatgi P. Fabrication, microstructure and compressive behavior of ZC63 Mg–microballoon foam composites. Composites Science and Technology. 2007;67:1842-53.

[9] Huang Z-Q, Yu S-R, Li M-Q. Microstructures and compressive properties of AZ91D/fly-ash cenospheres composites. Transactions of Nonferrous Metals Society of China. 2010;20:s458-s62.

[10] Gupta M, Wong W. Magnesium-based nanocomposites: Lightweight materials of the future. Materials Characterization. 2015;105:30-46.

[11] Gupta N, Rohatgi PK. Metal Matrix Syntactic Foams: Processing, Microstructure, Properties and Applications: Destech Publications Incorporated; 2014.

Magnesium Technology 2016
Edited by: Alok Singh, Kiran Solanki, Michele V. Manuel, and Neale R. Neelameggham
TMS (The Minerals, Metals & Materials Society), 2016

ROLE OF SiC IN GRAIN REFINEMENT OF ALUMINUM-FREE Mg-Zn ALLOYS

Jian Gu, Yuanding Huang, Karl Ulrich Kainer, Norbert Hort
Magnesium Innovation Centre, Helmholtz-Zentrum Geesthacht, Max-Planck-Str. 1, D-21502 Geesthacht, Germany

Keywords: Magnesium alloys, Grain refinement, SiC inoculation, Solidification

Abstract

The addition of SiC particles effectively refined the grains of a range of aluminum-free Mg-Zn alloys. Simultaneously, the corresponding mechanism of grain refinement has also been discussed. In the present work the effect of SiC on the microstructure of as-cast Mg-Zn alloy was investigated using optical and scanning electron microscopy. The results show that the primary alloying element Zn segregates at the grain boundaries and the phase Mg_7Zn_3 locates inside Mg matrix of Mg-Zn grains. The efficiency of SiC as a grain refiner can be seen with 45% average grain size reduction of Mg-3Zn alloy at 0.3 wt% SiC addition, indicating an existence of close relationship between the contents of SiC and Zn to affect the grain size. The interaction between SiC and the impurities such as Mn and Fe in Mg-Zn alloys is suggested to explain the grain refinement.

Introduction

Mg alloys can be generally classified into two broad groups: Al-free and Al-bearing, depending upon whether they are alloyed with Al [1]. For Al-free Mg alloys, Zr is an extremely effective nucleant but an expensive one. Reducing the consumption of Zr and hence the cost of grain refinement or finding a substitute for it is of great interest. Carbon (C) inoculation using various systems, such as graphite, paraffin wax, lampblack, organic compounds, carbides and bubbling the melt with carbonaceous gases, was known to be the most popular grain refining treatment for Mg-Al type alloys [1-3]. However, a common view that C inoculation is only effective to Mg alloys that contain more than 2 wt% Al is well accepted [4]. Especially following the idea of the work by Qian et al. [5], no grain refinement was observed in Mg-3Zn alloy after 0.6 wt% of C addition in the form of Nucleant 5000 at 750 °C.

Interestingly, Cao et al. [6, 7] successfully prepared Mg-Zn/SiC nano-composites by ultrasonic cavitation-based solidification processing. Their results showed that the grain size of Mg–Zn alloy was reduced considerably by the addition of SiC nanoparticles. Unfortunately, they focused on the castability and properties of those alloys. They only attributed the grain refinement to SiC nanoparticles served as nuclei for heterogeneous nucleation, but no direct evidence was obtained [6]. Both the physical and chemical properties are totally different when comparing SiC with other C-containing sources, and therefore the viewpoint of C inoculation only effective to Al-bearing Mg alloys may be debatable.

In addition, SiC served as an effective grain refiner has been widely investigated in Mg alloys containing Zn. No poisoning grain refining effect of Zn was found and different grain refinement mechanisms were proposed. Luo [8] studied the heterogeneous nucleation and grain refinement in cast $AZ91/SiC_p$ alloys. A number of SiC particles were found within the primary Mg grains suggesting a possible heterogeneous nucleation mechanism for grain refinement. Luo [9] also found that some of the SiC particles were pushed away to the grain boundaries. Therefore, the grain refinement effect of SiC may also be attributed to the reduced growth rate of the primary phase since the presence of SiC particles around the growing Mg crystals may create a diffusion barrier. Inem et al. [10, 11] investigated the nucleation and crystallographic orientation of both the Mg matrix and the eutectic $Mg(ZnCu)_2$ phase at the β-SiC particle surface in ZC63 and ZC71 Mg matrix composites. They found that the eutectic nucleates at the particle surface with an identical crystallographic orientation, but no distinct crystallographic orientation of α-Mg with SiC particle has been resolved.

Therefore, the purpose of this study is to investigate the grain refining effect of SiC in Mg-Zn alloys. Special attention will be paid to the discussion of the grain refinement mechanism during the solidification process, which may give new insights into the grain refinement mechanism of C-inoculation in Al-free Mg alloys.

Experimental Procedures

Mg–Zn alloys with the contents of 1%, 3% and 6% Zn made from commercial purity Mg and Zn inoculated with different contents of SiC were prepared. The nominal compositions of the tested alloys are given in Table I (all compositions are given in wt% unless specified). SiC particles (supplied by Alfa Aesar GmbH & Co KG, Germany) with the average size of 2 μm were used as refiner. SEM image of the micron-SiC particles is shown in Figure 1. Mg and Zn ingots were melted at 700 °C in an electrical resistance furnace using a mild steel crucible under a protective gas mixture of high pure Ar + 0.2% SF_6. The melt was manually stirred for 2 min and then the melt surface was skimmed. SiC particles preheated to 500 °C under Ar atmosphere were added into the melt directly. After that the melt was stirred vigorously at 100 rpm for 5 min to ensure good dispersion of the particles and then held at 700 °C for 15 min before casting. Each ingot was cast by pouring the melt into a steel mold preheated to 200 °C with a diameter of 70 mm at the bottom and 80 mm at the top and a height of 250 mm.

Figure 1. SEM image of the micron-SiC particles.

Table I. Nominal compositions of the alloys investigated

Alloy	Composition (wt%)		
	Mg	Zn	SiC
Mg-1Zn	Bal.	1	-
Mg-3Zn	Bal.	3	-
Mg-6Zn	Bal.	6	-
Mg-1Zn-0.3SiC	Bal.	1	0.3
Mg-3Zn-0.1SiC	Bal.	3	0.1
Mg-3Zn-0.2SiC	Bal.	3	0.2
Mg-3Zn-0.3SiC	Bal.	3	0.3
Mg-3Zn-0.5SiC	Bal.	3	0.5
Mg-3Zn-1.0SiC	Bal.	3	1.0
Mg-3Zn-10SiC	Bal.	3	10
Mg-6Zn-0.3SiC	Bal.	6	0.3

Metallographic samples were transversally sectioned from the same position of 20 mm from the bottom of the castings, and prepared according to a standard procedure. Samples for optical microscopy were etched with a solution of 8 g picric acid, 5 ml acetic acid, 10 ml distilled water, and 100 ml ethanol. The average grain size was measured by the linear intercept method from the micrographs taken using polarized light in an optical microscope. A Zeiss Ultra 55 (Carl Zeiss GmbH, Oberkochen, Germany) scanning electron microscope (SEM) equipped with energy dispersive spectroscopy (EDS) was also used to observe the microstructures of selected samples.

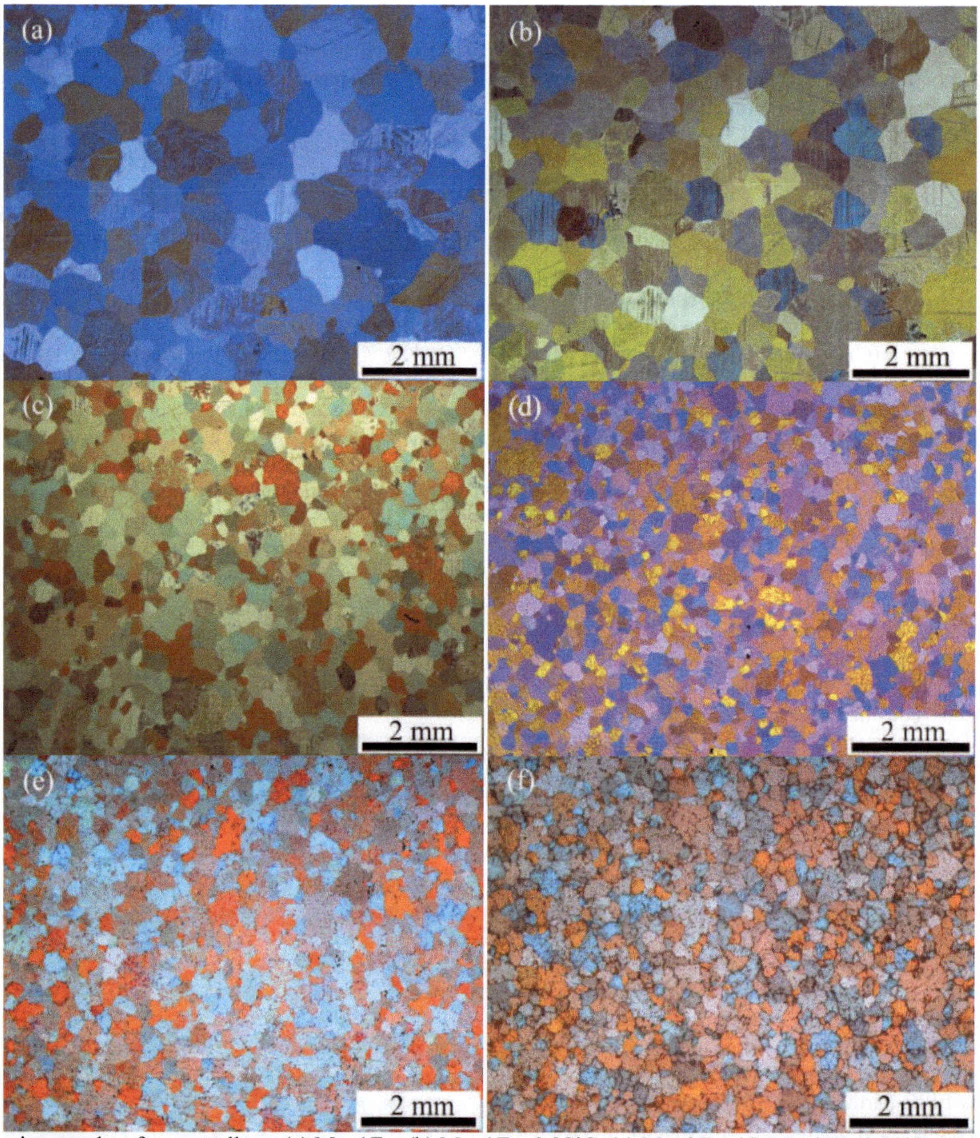

Figure 2. Optical micrographs of as-cast alloys: (a) Mg–1Zn; (b) Mg–1Zn–0.3SiC; (c) Mg–3Zn; (d) Mg–3Zn–0.3SiC; (e) Mg–6Zn; (f) Mg–6Zn–0.3SiC.

Results and Discussion

Figure 2 shows the changes in microstructure of as-cast Mg-*x*Zn (x = 1, 3, 6) alloys prior to and after the addition of 0.3% SiC. The grain size of the Mg-*x*Zn alloys without refiner decreased with the increase of Zn contents. It is known that solute element plays a significant role in controlling the growth of the nucleated grains and in subsequent nucleation [12, 13]. The growth restriction factor (GRF), Q, was used as a measurement of the effect of solute on grain refinement. The value of Q can be calculated with equation (1) [14]:

$$Q = \sum_i m_i C_{0,i}(K_i - 1) \qquad (1)$$

where m_i is the slope of the liquidus line, k_i is the distribution coefficient, and $C_{0,i}$ is the initial concentration of element i.
Higher value of $m(k-1)$ means higher value of GRF and stronger segregating power, consequently, a better grain refinement can be obtained. For each wt % of Zn it is 5.31, which is a little bit higher than that of Al (4.32) in Mg [14, 15]. When Zn was added, Zn generates constitutional undercooling in a diffusion layer ahead of the advancing solid/liquid interface, which restricts grain growth since the diffusion of the solute occurs slowly, thus limiting the rate of crystal growth. In addition, further nucleation occurs in front of the interface because nuclei in the melt are more likely to survive and be activated in the constitutionally undercooled zone. Therefore, the strong segregation ability of Zn in Mg could be the main reason of grain refinement of Mg with the addition of Zn.
The grain size of Mg-Zn alloys with the addition of 0.3% SiC decreased in all cases. Notably, it decreased significantly from 330 ± 89 μm for Mg–3Zn alloy to 180 ± 45 μm for Mg–3Zn–0.3SiC alloy. The alloys with lower (1%) or higher (6%) Zn contents, however, have poor grain refinement with SiC addition, the grain size decreased from 624 ± 118 μm to 540 ± 103 μm for Mg-1Zn alloy and from 242 ± 41 μm to 192 ± 35 μm for Mg-6Zn alloy, respectively. In order to assess the combining effects of Q of the solutes in a melt and the potency of nucleant particles on the evolution of a grain structure, the following equation (2) was suggested [1]:

$$d = a + b/Q \qquad (2)$$

where d is the grain size, the intercept, a, corresponds to the number of particles that actually nucleate grains at infinite values of Q, and the slope, b, is related to the potency of the nucleant particles. It should be noted that the only element that makes a significant contribution to the Q value of the alloy in this case is Zn. Based on the equation (2) and previous work, it is known that alloys with lower contents of solutes are much more effectively grain refined by the addition of an effective nucleant than that with higher contents of solutes [15-17], however, contrary result obtained in Mg-1Zn-0.3SiC alloy indicates that proper content of Zn, e.g. 3%, is essential for the grain refinement process during solidification in this case.
Figure 3 shows the changes in average grain size of as-cast Mg-3Zn alloy inoculated by different contents of SiC. It clearly shows a slight decrease of grain size from 330 ± 89 to 285 ± 72 μm with the increase of SiC contents from 0 to 0.2%. Then, a sharp decrease of grain size from 285 ± 72 to 180 ± 45 μm was obtained with increasing of SiC contents from 0.2% to 0.3%. After that, the grain size remains stable with the increase of SiC contents, even up to 10% SiC.

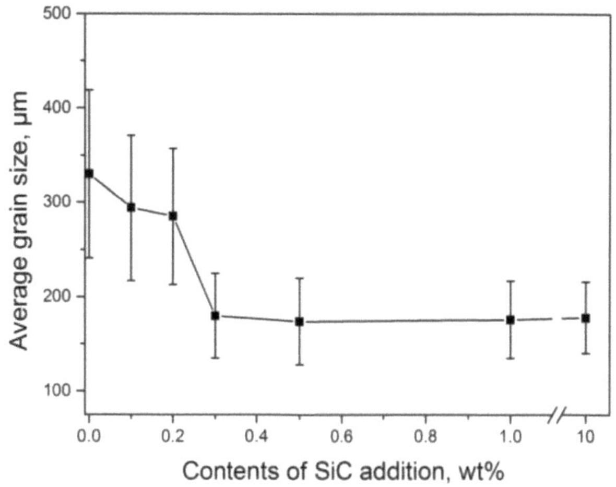

Figure 3. Changes in average grain size of Mg-3Zn alloy inoculated by different contents of SiC.

Figure 4(a) shows the optical micrograph of as-cast Mg-3Zn-10SiC alloy. Comparing it with Mg-3Zn-0.3SiC alloy (Figure 2(d)), more black dots confirmed as SiC clusters can be found, indicating that SiC is not well distributed in Mg-Zn alloy with high content of SiC. In addition, when 10% SiC was added into the melt, part of the added SiC floating on the melt surface can be found before casting even after vigorously stirring. This might be due to the poor wettability between SiC and the melt or an increase in viscosity with higher SiC concentration. Hence, the effective efficiency of grain refinement effect is reduced. Conversely, when the content of SiC was less than 1%, especially lower than 0.5%, SiC can be added into the melt well and distributed uniformly within the matrix. Former research work also shows that the proportion of typically active particles served as the nucleation substrates in Mg–Al alloys with SiC is very low [15]. Therefore, the grain refinement effect could be saturated after the addition of 0.5% SiC in this case.
Figure 4(b) shows the SEM microstructure of Mg-3Zn-10SiC alloy. The corresponding quantitative analysis of the areas by EDS is given in Table II. In backscattered electron image, SiC cluster was confirmed in area 1, which is agreed with the result of optical microstructure in Figure 4(a). Elements of both Si and C were detected in areas 2, 3 and 4. These particles are likely to be SiC as they have similar contrast and morphology to that of area 1. Zn segregation was easily found along the grain boundaries in all casting samples, like the area showed in 6. The large amounts of Mg_7Zn_3 phases with bright white contrast located inside Mg matrix. The Mg/Zn ratio in area 5 is approximately 2.8, which is close to that in Mg_7Zn_3 phase. SiC particles located inside grains (areas 3 and 4) could be some evidence of nucleation substrates for the heterogeneous nucleation of primary Mg. Besides heterogeneous nucleation, the rate of grain growth also determines the final solidification structure. Area 2 gives the evidence that some SiC particles were pushed away and then located around the Mg-Zn grain boundaries. The presence of SiC particles around the growing Mg-Zn grains during solidification may create diffusion barriers to grain growth due to the particle/solidification front interaction, leading to a reduced grain growth rate for the primary α-Mg phase. However, the ratio of those SiC particles located inside grains or located around the grain boundaries in Mg-Zn alloys is very few, even in alloys inoculated with 10% SiC, meaning that it is not the main factor to affect the grain refining performance in this case.

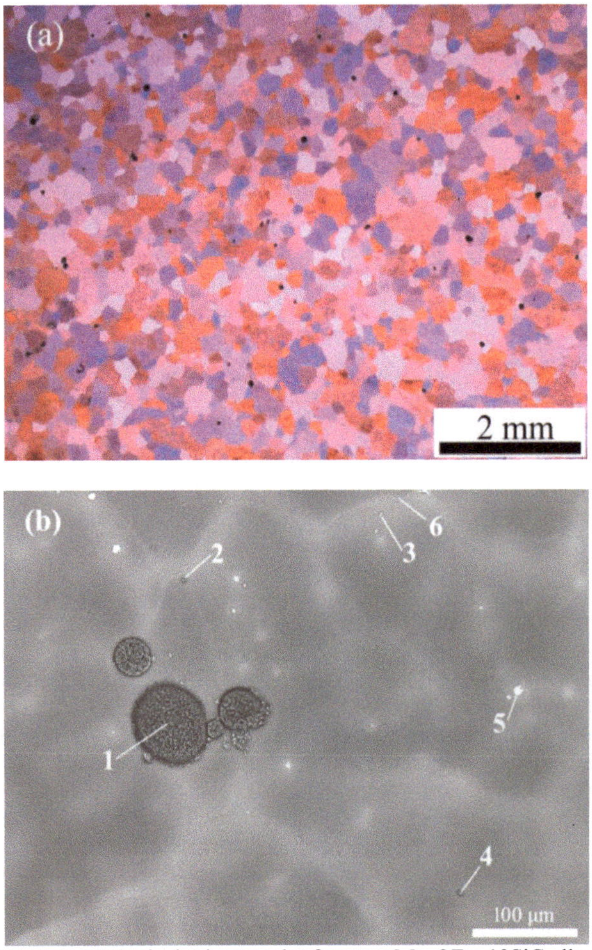

Figure 4. (a) Optical micrograph of as-cast Mg–3Zn–10SiC alloy; (b) Backscattered electron SEM image of Mg–3Zn–10SiC alloy.

Table II. Quantitative EDS results of the areas as indicated in Figure 4(b)

Area	Composition (at%)				
	Mg	Zn	Si	C	O
1	34.7	1.0	32.7	26.9	4.7
2	38.9	0.5	1.4	49.4	9.8
3	87.9	1.0	1.9	7.4	1.8
4	78.2	0.5	7.1	8.0	6.1
5	74.1	25.9	–	–	–
6	97.4	2.6	–	–	–

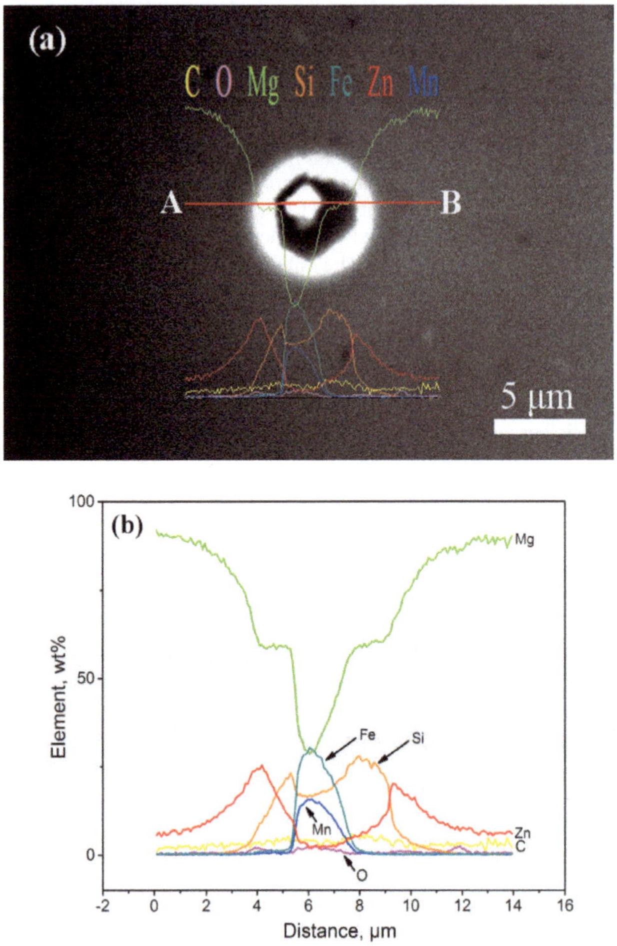

Figure 5. (a) Typical EDS line-scan analysis of a nucleating site in Mg-3Zn-0.3SiC alloy; (b) corresponding line-scan analysis curves along line A-B in Figure 5(a) for clarity.

Interestingly, in the present study a common microstructure indicating a possible nucleating site can be found in all the Mg-Zn alloys inoculated by SiC. Figure 5(a) shows this typical feature obtained in Mg-3Zn-0.3SiC alloy detected by EDS line-scan analysis. In backscattered electron image, the black part enriched with Si between both the inside white and outside white part can be confirmed as a SiC particle, although the amount of C contained in the particle is difficult to be assessed due to the limitation of the available technique. This particle has even darker contrast than that of Mg matrix suggesting that its average atomic number is less than that of Mg matrix. The inside white part was enriched with Fe and Mn, indicating that SiC is likely to associate with Fe and Mn impurities to serve as a grain nuclei. As the commercial purity Mg was used in this study, Fe and Mn impurities are inevitably existed [18, 19]. In addition, it is well known that the grain size of Mg–Al alloys has a complicated relationship with the presence of impurity elements such as Fe, Mn, and/or C. Du et al. [20] found that the refining efficiency could be further obtained by the combination of Fe and C in the grain refinement of Mg-3Al alloy. In this case, C containing SiC particle might be partly reacted with Fe and Mn impurities to form Fe-Mn-Si-(C) rich intermetallic phase served as a nucleating site, which is attributed to the grain refinement. Moreover, the SiC was wrapped by the outside white part, which is confirmed as the Zn enriched area, giving the evidence that Zn has close relationship with SiC and plays an important role for grain refinement of Mg-Zn alloys inoculated by SiC. Further detailed analyses by transmission electron microscopy (TEM) and focused ion beam (FIB) will be carried on in order to obtain more convincing evidences to illustrate this issue.

Conclusion

In summary, Mg-Zn alloys can be readily grain refined by SiC inoculation. Grain refining efficiency with SiC inoculation is not significant at both the lower and higher Zn contents. The proper content of Zn, e.g. 3%, is essential for the grain refinement process. SiC contents varied from 0.3% to 0.5% are the optimized condition for the grain refinement of Mg-3Zn alloys. Further addition of SiC leads to the aggregation of this particle, and the grain size cannot be further refined. The average grain size reduction of Mg-3Zn alloy with 0.3% SiC addition reaches to 45%. The primary alloying element Zn segregates at the grain boundaries and the phase Mg_7Zn_3 locates inside Mg matrix of Mg-Zn grains. The interaction between SiC and the impurities such as Mn and Fe might mainly contribute to the grain refinement.

Acknowledgements

The authors would like to thank Mr. Gert Wiese and Mr. Günter Meister for their technical supports. Financial support from China Scholarship Council (CSC) for this work is greatly appreciated.

References

1. D.H. StJohn, et al., "Grain Refinement of Magnesium Alloys", Metallurgical and Materials Transactions A, 36 (2005), 1669-1679.

2. E. Yano, et al., "Effect of Carbon Powder on Grain Refinement of an AZ91E Magnesium Alloy", Materials Transactions, 44 (2003), 107-110.

3. D. Vinotha et al., "Grain Refining Mechanisms in Magnesium Alloys – an Overview", Transactions of the Indian Institute of Metals, 62(6) (2009), 521-532.

4. D.H. StJohn, et al., "Grain Refinement of Magnesium Alloys: A Review of Recent Research, Theoretical Developments, and Their Application", Metallurgical and Materials Transactions A, 44 (2012), 2935-2949.

5. M. Qian and P. Cao, "Discussions on Grain Refinement of Magnesium Alloys by Carbon Inoculation", Scripta Materialia, 52 (2005), 415–419.

6. G. Cao et al., "Mg–6Zn/1.5%SiC Nanocomposites Fabricated by Ultrasonic Cavitation-Based Solidification Processing", Journal of Materials Science, 43 (2008), 5521-5526.

7. G. Cao et al., "Tensile Properties and Microstructure of SiC Nanoparticle–Reinforced Mg-4Zn Alloy Fabricated by Ultrasonic Cavitation–Based Solidification", Metallurgical and Materials Transactions A, 39 (2008), 880-886.

8. A. Luo, "Development of Matrix Grain Structure During the Solidification of a Mg(AZ91)/SiC_P Composite", Scripta Metallurgica Et Materialia, 31 (1994), 1253-1258.

9. A. Luo, "Heterogeneous Nucleation and Grain Refinement in Cast Mg(AZ91)/SiC_p Metal Matrix Composites", Canadian Metallurgical Quarterly, 35 (1996), 375-383.

10. B. Inem and G. Pollard, "Interface Structure and Fractography of a Magnesium-Alloy, Metal-Matrix Composite Reinforced with SiC Particles", Journal of Materials Science, 28 (1993), 4427-4434.

11. B. Inem, "Crystallography of the Second Phase/SiC Particles Interface, Nucleation of the Second Phase at β-SiC and Its Effect on Interfacial Bonding, Elastic Properties and Ductility of Magnesium Matrix Composites", Journal of Materials Science, 30 (1995), 5763-5769.

12. M.A. Easton and D.H. StJohn, "Grain Refinement of Aluminum Alloys: Part I. the Nucleant and Solute Paradigms—a Review of the Literature", Metallurgical and Materials Transactions A, 30 (1999), 1613-1623.

13. M.A. Easton and D.H. StJohn, "Grain Refinement of Aluminum Alloys: Part II. Confirmation of, and a Mechanism for, the Solute Paradigm", Metallurgical and Materials Transactions A, 30 (1999), 1625-1633.

14. Y.C. Lee, A.K. Dahle, and D.H. StJohn, "The Role of Solute in Grain Refinement of Magnesium", Metallurgical and Materials Transactions A, 31 (2000), 2895-2906.

15. M.A. Easton et al., "Grain Refinement of Mg–Al(–Mn) Alloys by SiC Additions", Scripta Materialia, 55 (2006), 379-382.

16. Y. Liu, X. Liu and X. Bian, "Grain Refinement of Mg–Al Alloys with Al_4C_3–SiC/Al Master Alloy" Materials Letters, 58 (2004), 1282-1287.

17. M.A. Easton and D.H. StJohn, "An analysis of the Relationship between Grain Size, Solute Content, and the Potency and Number Density of Nucleant Particles", Metallurgical and Materials Transactions A, 36 (2005), 1911-1920.

18. J.D. Hanawalt, C.E. Nelson, and G.E. Holdeman. US Patent 2, 267, 862, (1940).

19. P. Cao, M. Qian, and D.H. StJohn, "Effect of Manganese on Grain Refinement of Mg–Al Based Alloys", Scripta Materialia, 54 (2006), 1853-1858.

20. J. Du et al., "Effect of Iron and/or Carbon on the Grain Refinement of Mg-3Al Alloy", Materials Transactions, 48 (2007), 2903-2908.

Magnesium Technology 2016
Edited by: Alok Singh, Kiran Solanki, Michele V. Manuel, and Neale R. Neelameggham
TMS (The Minerals, Metals & Materials Society), 2016

HOT DEFORMATION AND PROCESSING MAP IN AN Mg-Zn-Mn-Y ALLOY

N. Tahreen[1], D.F. Zhang[2,3], F.S. Pan[2,3,4], X.Q. Jiang[4,5], D.Y. Li[6], D.L. Chen[1]
[1]Department of Mechanical and Industrial Engineering, Ryerson University, Toronto, Ontario M5B 2K3, Canada
[2]College of Materials Science and Engineering, Chongqing University, Chongqing 400045, China
[3]National Engineering Research Center for Magnesium Alloys, Chongqing University, Chongqing 400044, China
[4]Advanced Materials Research Center, Chongqing Academy of Science and Technology, Chongqing 401123, China
[5]Faculty of Materials and Energy, Southwest University, Chongqing 400715, China
[6]Department of Chemical and Materials Engineering, University of Alberta, Alberta T6G 2V4, Canada

Keywords: Magnesium alloy, Hot deformation, Processing map, LPSO phase.

Abstract

In recent years Mg-Zn-Y series alloys have drawn significant interest due to their superior mechanical properties. The addition of yttrium to the Mg-Zn alloy leads to the formation of a special phase ($Mg_{12}YZn$) with a long-period stacking ordered (LPSO) structure, which plays an important strengthening role. The main objective of this study was to identify the deformation mechanisms of LPSO phase in an as-extruded yttrium-containing magnesium alloy. On the other hand, the successful manufacturing of magnesium alloy components requires the optimal hot working parameters. The processing map is an effective tool which can be used to pinpoint the optimal processing regions and the instability regions which should be avoided. Another objective of this study was thus to evaluate the hot deformation behavior of a Mg-Zn-Mn-Y alloy via isothermal compression testing at different temperatures and strain rates. The processing maps were constructed and the optimal hot processing conditions were identified.

Introduction

The much talked concept of "lightweighting" which is infusing on the horizon of the auto industry, inspires researchers to develop new wrought magnesium alloys with superior properties [1]. For extensive applications in the transportation industries, the improvement on strength, thermal stability and formability of magnesium alloys is required. There is a persistent effort to improve the high temperature performance of wrought magnesium alloys, where one of possibilities being explored is the addition of rare-earth (RE) metals. In recent years, Mg-Zn-Y series alloys have shown great potential due to their excellent mechanical properties at both room and elevated temperatures [2]. The addition of Y to the Mg-Zn alloy leads to the formation of a unique phase with long-period stacking ordered (LPSO) structure [3]. This phase is highly beneficial as it contributes considerably to alloy strengthening because of the deflection of deformation twins and the formation of kink bands [4]. The LPSO phase has a (0001) basal plane which is the same as that in magnesium, but its stacking periodicity is lengthened to 18-fold or 14-fold along the *c*-axis [5,6]. Apart from its exceptional crystal structure, LPSO phase containing magnesium alloys exhibit a significantly enhanced strength and ductility [4,7-8]. The credit is associated with the factors such as matrix grain refinement, the development of a weaker texture and the formation of kink bands within the LPSO phase [4,9-10]. For example, a Mg-1Zn-2Y alloy processed via rapidly solidified powder metallurgy was reported to reach a tensile yield strength of 610 MPa and ductility of 5% at room temperature [11].

Due to limited slip systems, forming of magnesium alloys at room temperature is challenging [12]. At elevated temperatures, their workability increases as additional slip systems start functioning. However, the elevated temperature inevitably carries some concerns such as oxidation, energy consumption, grain growth and particle coarsening [13,14]. It is, therefore, critical to optimize the processing window to secure the desired post-deformation properties. It has been widely recognized that the processing map is an effective tool for optimizing hot working/processing parameters [15,16]. By using the processing maps, safe processing parameters related to the occurrence of dynamic recrystallization can be determined [16]. The mechanical properties of an as-extruded ZM31+6Y alloy with LPSO phase have been studied in our earlier study [17,18]. To the authors' knowledge no information about optimum hot working parameters of the as-extruded Mg-3Zn-1Mn-6Y alloy is available in the open literature. Therefore, the present study was aimed to identify the hot deformation constraints of the alloy based on isothermal compression tests. The processing maps of the alloy are constructed with a view to analyze and optimize the hot working parameters. Moreover, the deformation mechanisms within the LPSO phase are discussed by microstructural examinations.

Experimental

The material used in the present study was an as-extruded ZM31+6Y magnesium alloy with a nominal composition (wt.%) of 3 Zn, 1 Mn, and 6 Y. The alloy was extruded at 380°C at an extrusion ratio of 25. Cylindrical specimens with a diameter of 5 mm and a height of 8 mm were machined from the as-extruded rod with the compression axis parallel to the extrusion direction (ED). Compression tests were conducted at temperatures of 300°C, 350°C and 400°C and strain rates of 0.001, 0.01, 0.1 and 1 s^{-1}, using a computerized United tensile testing machine equipped with an environmental chamber having a temperature accuracy of ±5°C. The samples were first heated up to the desired temperatures and held for 300 s prior to compression. Two samples were tested in each case. After hot deformation the samples were quenched in water immediately to retain the deformed microstructures. In evaluating the stress-strain curves, to acquire the actual deformation amount of test samples the machine deformation was eliminated using a calibration curve at each temperature. The samples for microstructural examination were cut parallel to the compression axis using a slow diamond cutter, cold mounted, ground with SiC abrasive papers and polished with diamond paste and colloidal silica. Microstructures were observed via scanning electron microscope (JSM-6380LV) equipped with Oxford energy dispersive X-ray spectroscopy (EDS), after etching with an acetic picral solution containing 4.2 g picric acid, 10 ml acetic acid, 10 ml water, and 70 ml ethanol.

Results and Discussion

Fig. 1 shows a typical SEM micrograph of alloy ZM31+6Y with larger and fiber-shaped precipitates of LPSO phase, which were laying along the extrusion direction. The average grain size of the as-extruded alloy was estimated to be ~5.1 μm.

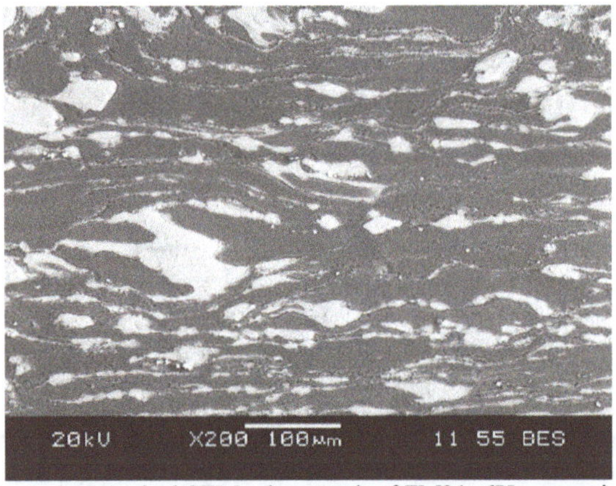

Figure 1. A typical SEM micrograph of ZM31+6Y magnesium alloy in the as-extruded condition.

The typical true stress-true strain curves of as-extruded ZM31+6Y alloy obtained at different deformation temperatures in the range of 300-400°C and strain rates of 0.001-0.1 s^{-1} are shown in Fig. 2.

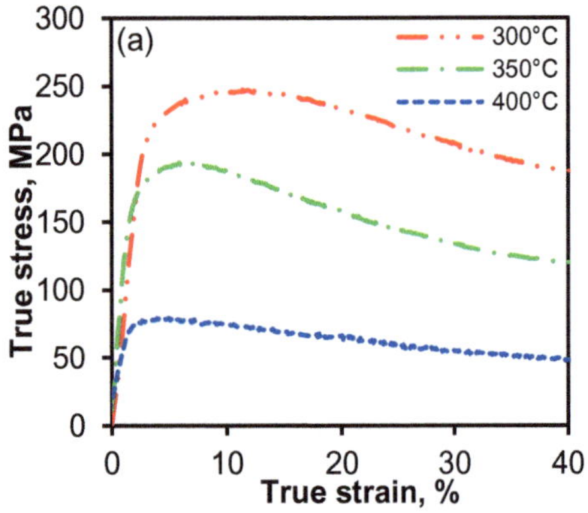

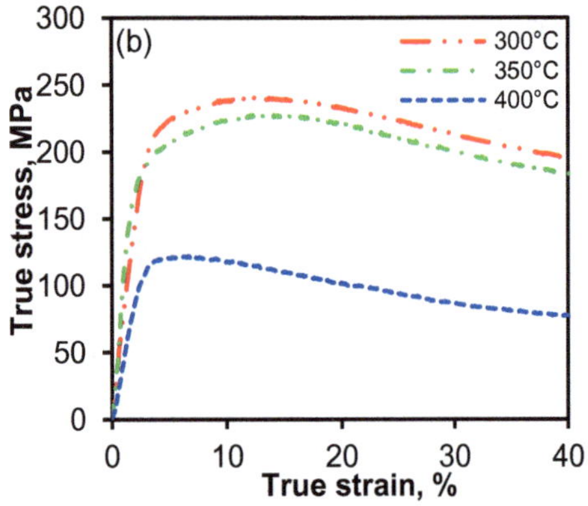

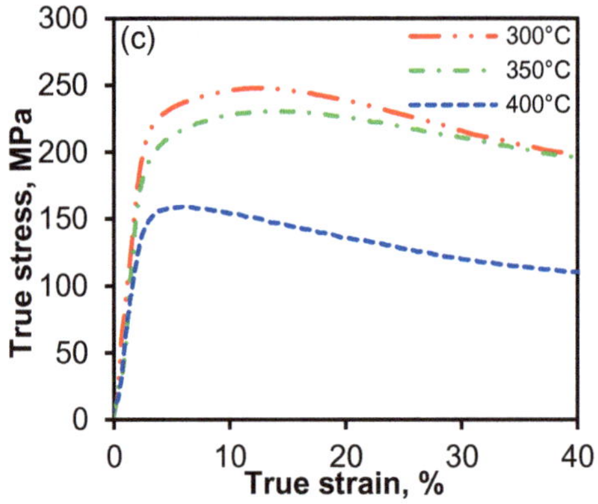

Figure 2. Compressive true stress-true strain curves of ZM31+6Y alloy deformed at different temperatures at strain rates of (a) 0.001 s^{-1}, (b) 0.01 s^{-1}, and (c) 0.1 s^{-1} [17].

At higher temperatures the material exhibited less significant hardening followed by minor flow softening leading to a quasi-steady state. In contrast, at lower temperatures the alloy displayed greater hardening followed by larger flow softening and shear fracture at 45° with respect to the compression axis (or ED). It is also noted that the peak flow stress was reliant on the deformation temperature and strain rate. At a given strain rate, the flow stress decreased with increasing temperature (Fig. 2). This was due to the enhanced thermal activation. The kinetic energy of metal atoms accumulated which boosted the movement of dislocations at higher temperatures. The higher the strain rate, the greater the flow stress at a given temperature. This is because there was not enough time for energy accumulation and dislocation annihilation at higher strain rates [13,17].

Processing map has been regarded as an effective tool for the optimization of hot working process for a diversity of materials including magnesium alloys. The concept of processing map was developed on the basis of dynamic material model (DMM) [19]. According to this model, the workpiece dissipates power during hot deformation. The total power P (per unit volume) absorbed by the workpiece during plastic flow can be expressed as a sum of two paired functions [20],

$$P = G + J = \int_0^{\dot{\varepsilon}} \sigma d\dot{\varepsilon} + \int_0^{\sigma} \dot{\varepsilon} d\sigma \,, \tag{1}$$

where G represents the power dissipated by plastic deformation, most of which is converted into heat, and J represents the dissipation through microstructural changes such as dynamic recovery (DRV), dynamic recrystallization (DRX), deformation twinning, phase transformation as well as crack propagation. At a certain temperature and strain, the splitting of power between J and G is expressed as,

$$\left(\frac{\partial J}{\partial G}\right)_{T,\dot{\varepsilon}} = \left(\frac{\partial \ln \sigma}{\partial \ln \dot{\varepsilon}}\right)_{T,\dot{\varepsilon}} = m \,, \tag{2}$$

where m is the strain rate sensitivity of the material which varies with temperature and strain rate. The value of J may be normalized with respect to J_{max} to obtain a dimensionless parameter called the efficiency of power dissipation defined as,

$$\eta = \frac{J}{J_{\max}} = \frac{2m}{m+1}. \quad (3)$$

The variation of η with temperature and strain rate reflects the characteristics of power dissipation which creates a power dissipation map. The map is presented as a contour plot of efficiency variation in the temperature-strain rate field where different domains reflect different deformation mechanisms. A continuum instability criterion developed on the basis of extremum principles of irreversible thermodynamics, which is given by [21],

$$\xi(\dot{\varepsilon}) = \frac{\partial \ln(m / m+1)}{\partial \ln \dot{\varepsilon}} + m < 0. \quad (4)$$

It is used to designate the regions of flow instability. The instability parameter $\xi(\dot{\varepsilon})$ is evaluated as a function of temperature and strain rate and is plotted to obtain an instability map where ξ is negative. It is superimposed upon the power dissipation map to build a processing map.

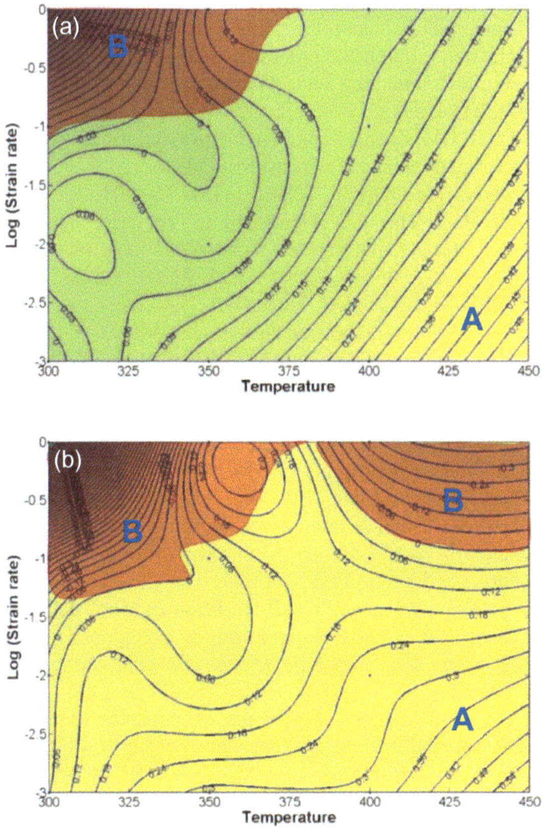

Figure 3. Processing maps of the ZM31+6Y alloy at strains of (a) ε= 0.1 and (b) ε= 0.4.

The 2D processing maps of as-extruded ZM31+6Y alloy developed at strains of 0.1 and 0.4 are shown in Fig. 3, where the contour values denote the efficiency of power dissipation and the shaded brown areas stand for the regions of unstable flow. The processing map at strain amounts of 10% and 40% could be allotted into two different domains: (1) Domain A indicates a stable region where fairly high values of efficiency of power dissipation were seen, which corresponds to the samples deformed at relatively higher temperatures and lower strain rates. It lay in a temperature range of 325-450°C and a strain rate range of 0.001-0.1 s^{-1}, with a peak efficiency of ~48% occurring at 425°C and 0.003 s^{-1}. (ii) Domain B corresponds to the samples deformed in the lower temperature and higher strain rate conditions in which very low or negative values of efficiency of power dissipation were obtained. This is often referred to as instability region.

As can be seen from Fig. 3 the contour of power efficiency varies with increasing strain. Generally, dissipation efficiency is connected with microstructure evolution mechanisms where higher values represent DRX and relatively lower values represent DRV [21,22]. DRX is a desired beneficial phenomenon in hot deformation since it gives good workability to the material by reducing flow localization and simultaneous softening. It is known that domains with a higher efficiency of power dissipation commonly represent optimum processing conditions. Since a higher efficiency value was observed in the ranges range of 350-450°C and 0.001-0.03 s^{-1} it could be inferred as the favorable condition for hot processing of ZM31+6Y alloy.

Domain B is characterized by unstable material flow where the instability parameter ξ is estimated as negative. Domain B is only visible in the high strain rate and low temperature region at a strain of 0.1. With increasing strain Domain B extends and also appears in the high temperature regime, which suggests that the hot-workability of the alloy in the high strain rate region became poor as the strain amount increased. Hence, it became increasingly difficult to deform the alloy at higher strain rates. Microstructures within the instability regions were often characterized by local deformation bands, twins and micro-cracks [15]. When a material

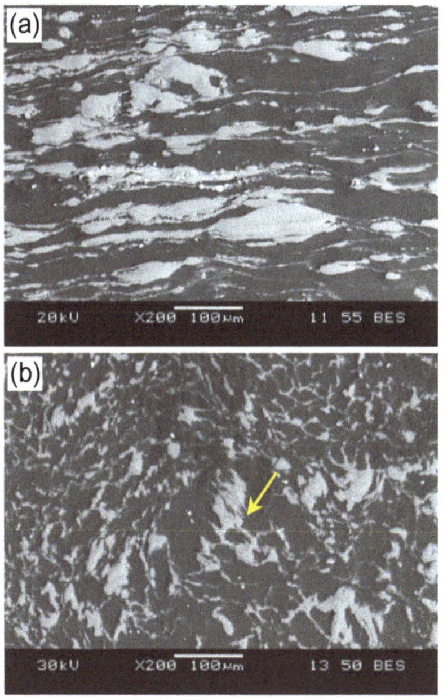

Figure 4. Typical microstructures of the ZM31+6Y alloy deformed at (a) 300°C and 1 s^{-1}, and (b) 400°C and 0.001 s^{-1}.

is deformed in the instability domain, it is not easy to achieve the desired mechanical properties. Thus, it is important to avoid such processing conditions during hot deformation.

Fig. 4 shows the SEM micrographs of ZM31+6Y alloy deformed at 300°C/1 s^{-1} and 400°C/0.001 s^{-1}. At 300°C/1 s^{-1}, LPSO phase did not exhibit notable alteration in their shape and distribution. In contrast, the LPSO phase experienced extensive distortion during compressive deformation at 400°C/0.001 s^{-1}. The alignment of the LPSO phase was significantly altered and deformation kink bands (indicated by a yellow arrow) appeared within some LPSO phases. Moreover, no micro-crack or debonding between LPSO phase and Mg matrix was detected even after compression up to a high strain amount (50%). The phase could accommodate a large compressive strain within itself without inducing crack when compressed along ED at elevated temperatures. This suggests superior formability of the LPSO phase at elevated temperatures.

Conclusions

The processing map of as-extruded ZM31+6Y alloy showed a maximum efficiency of power dissipation of 48% and the optimum processing parameters of 350-450°C and 0.001-0.03 s^{-1}.

During compression at elevated temperatures the LPSO phase exhibited a high degree of deformability, thus accommodating a large compressive strain without cracking.

Acknowledgements

The authors would like to thank the Natural Sciences and Engineering Research Council of Canada (NSERC), AUTO21 Network of Centres of Excellence, Premier's Research Excellence Award, NSERC-DAS Award, Canada Foundation for Innovation, and Ryerson Research Chair program for providing financial support. The authors also thank Ministry of Science and Technology of China (2011DFR50950-05, 2014DFG52810), Chongqing Science and Technology Commission for the financial support (CSTC2009BA4045 and CSTC2011gjhz50001) and National Great Theoretic Research Project of China (2013CB632200). The authors would also like to thank Messrs. A. Machin, Q. Li, C. Ma, J. Amankrah and R. Churaman for easy access to the laboratory facilities of Ryerson University and their assistance in the experiments.

References

1. T.A. Schaedler, A.J. Jacobsen and W.B. Carter, "Toward lighter, stiffer materials," *Science*, 341 (2013), 1181-1182.
2. X.H. Shao, Z.Q. Yang and X.L. Ma, "Strengthening and toughening mechanisms in Mg-Zn-Y alloy with a long period stacking ordered structure," *Acta Materialia*, 58 (2010), 4760-4771.
3. J. Gröbner, A. Kozlov, X.Y. Fang, J. Geng, J.F. Nie, R. Schmid-Fetzer, "Phase equilibria and transformations in ternary Mg-rich Mg-Y-Zn alloys," *Acta Materialia*, 60 (2012), 5948-5962.
4. M. Yamasaki, K. Hashimoto, K. Hagihara and Y. Kawamura, "Effect of multimodal microstructure evolution on mechanical properties of Mg-Zn-Y extruded alloy," *Acta Materialia*, 59 (2011), 3646-3658.
5. M. Yamasaki, K. Hagihara, S.-I. Inoue, J.P. Hadorn, Y. Kawamura, "Crystallographic classification of kink bands in an extruded Mg-Zn-Y alloy using intragranular misorientation axis analysis," *Acta Materialia*, 61 (2013), 2065-2076.
6. L.B. Tong, X.H. Li and H.J. Zhang, "Effect of long period stacking ordered phase on the microstructure, texture and mechanical properties of extruded Mg-Y-Zn alloy," *Materials Science and Engineering: A*, 563 (2013), 177-183.
7. M. Yamasaki, T. Anan, S. Yoshimoto, Y. Kawamura, "Mechanical properties of warm-extruded Mg-Zn-Gd alloy with coherent 14H long periodic stacking ordered structure precipitate," *Scripta Materialia*, 53 (2005), 799-803.
8. T. Itoi, K. Takahashi, H. Moriyama, M. Hirohashi, "A high-strength Mg-Ni-Y alloy sheet with a long-period ordered phase prepared by hot-rolling," *Scripta Materialia*, 59 (2008), 1155-1158.
9. K. Hagihara, N. Yokotani and Y. Umakoshi, "Plastic deformation behavior of $Mg_{12}YZn$ with 18R long-period stacking ordered structure," *Intermetallics*, 18 (2010), 267-276.
10. E. Oñorbe, G. Garcés, P. Pérez and P. Adeva, "Effect of the LPSO volume fraction on the microstructure and mechanical properties of Mg-Y_{2X}-Zn_X alloys," *Journal of Materials Science*, 47 (2012), 1085-1093.
11. Y. Kawamura, K. Hayashi, A. Inoue and T. Masumoto, "Rapidly solidified powder metallurgy $Mg_{97}Zn_1Y_2$ Alloys with excellent tensile yield strength above 600 MPa," *Materials Transactions*, 42 (2001), 1171-1174.
12. J. Hirsch and T. Al-Samman, "Superior light metals by texture engineering: Optimized aluminum and magnesium alloys for automotive applications," *Acta Materialia*, 61 (2013), 818-843.
13. N. Tahreen, D.F. Zhang, F.S. Pan, X.Q. Jiang, D.Y. Li and D.L. Chen, "Hot deformation and processing map of an as-extruded Mg-Zn-Mn-Y alloy containing I and W phases," *Materials & Design*, 87 (2015), 245-255.
14. L. Li and X.M. Zhang, "Hot compression deformation behavior and processing parameters of a cast Mg-Gd-Y-Zr alloy," *Materials Science and Engineering: A*, 528 (2011), 1396-1401.
15. Y. Xu, L.X. Hu, T.Q. Deng and L. Ye, "Hot deformation behavior and processing map of as-cast AZ61 magnesium alloy," *Materials Science and Engineering: A*, 559 (2013), 528-533.
16. Q. Chen, X.S. Xia, B.G. Yuan, D.Y. Shu, Z. Zhao, and J.C. Han, "Hot workfability behavior of as-cast Mg-Zn-Y-Zr alloy," *Materials Science and Engineering: A*, 593 (2014), 38-47.
17. N. Tahreen, D.F. Zhang, F.S. Pan, X.Q. Jiang, C. Li, D.Y. Li and D.L. Chen, "Characterization of hot deformation behavior of an extruded Mg-Zn-Mn-Y alloy containing LPSO phase," *Journal of Alloys and Compounds*, 644 (2015), 814-823.
18. N. Tahreen, D.F. Zhang, F.S. Pan, X.Q. Jiang, C. Li, D.Y. Li and D.L. Chen, "Hot deformation and work hardening behavior of an extruded Mg-Zn-Mn-Y alloy," *Journal of Materials Science and Technology*, 2015. (accepted)
19. O. Sivakesavam and Y.V.R.K. Prasad, "Hot deformation behaviour of as-cast Mg-2Zn-1Mn alloy in compression: a study with processing map," *Materials Science and Engineering: A*, 362 (2003), 118-124.
20. J.Q. Li, J. Liu and Z.S. Cui, "Characterization of hot deformation behavior of extruded ZK60 magnesium alloy using 3D processing maps," *Materials & Design*, 56 (2014), 889-897.
21. Y.V.R.K. Prasad and T. Seshacharyulu, "Processing maps for hot working of titanium alloys," *Materials Science and Engineering: A*, 243 (1998), 82-88.
22. M.H. Wang, L. Huang, M.L. Chen and Y.L. Wang, " Processing map and hot working mechanisms of Cu-Ag alloy in hot compression process," *Journal of Central South University*, 22 (2015), 821-828.

Magnesium Technology 2016

Twinning and Plasticity

Magnesium Technology 2016
Edited by: Alok Singh, Kiran Solanki, Michele V. Manuel, and Neale R. Neelameggham
TMS (The Minerals, Metals & Materials Society), 2016

WHAT IS IN A STRAIN HARDENING "PLATEAU"?

Sean R. Agnew[1], Christopher A. Calhoun[1], and Jishnu J. Bhattacharyya[1]

[1] Department of Materials Science and Engineering, University of Virginia, Charlottesville, Virginia, 22904, USA

Keywords: Twinning, strain localization, EBSD

Abstract

A yield plateau occurs in some Mg alloys during compressive deformation that has been ascribed to the localization of twinning. In fine-grained AZ31, this plateau was explained by a region of large twin volume fraction nucleating in a small band and propagating, similar to a "Lüders band" like phenomena. Once the band traverses the entire gauge length, the sample begins to strain harden. Similarly, ZK60 samples exhibit the same Lüders like phenomena during extrusion direction compression as confirmed by digital image correlation. However, the band is not sufficient to fully explain the plateau in the stress-strain curve. Postmortem electron backscattered diffraction (EBSD) reveals the twin structure evolution through this plateau. It is found that twinning occurs in large elongated grains before spreading to the finer grains.

Introduction

Twinning in magnesium single crystals and textured polycrystals is often characterized as giving rise to a "sigmoidal" or S-shaped flow curve. However, although the initial low strain hardening rate regime is commonly referred to as a "strain hardening plateau", there frequently is significant hardening present [1-3]. Under such circumstances, the term "plateau" is a misnomer. On the other hand, some recent studies on Mg alloys have also shown a true plateau during compression associated with the onset of twinning during loading along the rolling or extruded directions of wrought products. In fact, Barnett et al. [4] have shown that the prominence of a yield plateau is strongly correlated with the grain size. Fine-grained samples exhibit plateau (and even a yield point peak followed by a period of negative strain hardening rate), while coarse-grained samples do not. Furthermore, a sort of Lüders phenomenon has been associated with this yield plateau [5,6]. Recently, digital image correlation (DIC) studies of Mg alloy deformation have confirmed this effect [7]. To this end, the current study aims at investigating the strain hardening behaviour of ZK60 alloy during twinning deformation and revisit the general consensus about the twinning "plateau."

The currently examined material has a sort of duplex microstructure comprised of large, unrecrystallized grains as well as finer, presumably recrystallized grains [8]. It is of interest to consider the effect of such a complex microstructure on the propagation of twinning through the material. For example, does twinning occur first in large grains due to a sort of Hall-Petch effect? Will material in which only a fraction of the volume is fine-grained exhibit the Lüders phenomenon? During the course of the work a number of other interesting effects became apparent, such as the pre-existence of twin boundaries in the microstructure prior to deformation and the impact of detwinning on the ex-situ characterization of twin volume fractions.

Experimental Procedure

Material:

Alloy ZK60A was received in the T5 temper as a thick (~ 61 mm) extruded plate. The nominal composition in wt. % is 4.8-6.2 Zn, 0.45 min. Zr, and balance Mg. After machining into the compression sample geometries described below, all of the samples were solution heat treated at 400 °C for 1.5 hours followed by a water quench and aging for 24 hours at 175 °C to create peaked aged, ZK60A-T6.

Compression testing:

Compression samples were EDM cut from the extruded plated to have their loading axis along the extrusion direction. Samples used for in-situ neutron diffraction and ex-situ DIC measurements were cylindrical and were 12.5 mm diameter x 27.5 mm tall. Samples for the ex-situ EBSD study were 9.75 x 9.75 x 19.5 mm. The in-situ straining procedure is described below. The ex-situ compression studies were performed within a conventional screw driven universal testing system with Teflon tape used as a lubricant between the sample an platens at a strain rate of 10^{-3} s^{-1}. After testing samples to failure, specific samples were strained under uniaxial compression to three levels of plastic strain: 0.007, 0.02 and 0.04. On the 0.02 sample, digital image correlation (DIC) was used to monitor any potential strain localization.

Ex-situ microstructure and texture characterization:

The initial microstructure was examined optically after solution heat treatment using conventional mechanical polishing and an acetal-picral etchant. Subsequently, electron backscattered diffraction (EBSD) measurements were carried out on the undeformed, peak-aged material and specimens deformed up to 0.7, 2, 4.2 and 8.8 % strain, in order to investigate the microstructure and texture evolution. The samples were cold-mounted in epoxy, mechanically ground down to 1200-grit SiC paper, polished with 3 and 1 μm oil-based diamond paste, and final polished with 0.06 μm colloidal silica. After this mechanical polishing procedure, the samples were ion polished using a Hitachi IM4000 to remove the fine-scale deformation layer which remains even after fine polishing and facilitate indexing.

EBSD was performed within a FEI Quanta 650 FEG scanning electron microscope equipped with Oxford /HKL system with a Nordys EBSD detector. The electron microscope was operated at 20 kV, with a working distance of 15-20 mm at a magnification of 500X. The step size and the spot size during EBSD data collection were 1 μm and 5.5 (corresponding to a beam current of

11 nA), respectively. Additionally, some more scans were performed using a finer step size (0.2-0.3 μm) to resolved finer details of the twins interacting with the grain boundaries. The fraction of points successfully indexed was 80–95%. Band contrast and inverse pole figure (IPF) maps of different directions were constructed using HKL Channel 5 software to analyze the microstructure evolution post deformation. The twin boundaries were identified based on their misorientation: 86° ± 5° about $\langle 11\bar{2}0\rangle$ axis for extension twins, 56° ± 5° $\langle 11\bar{2}0\rangle$ for contraction twins and 38° ± 5° about $\langle 11\bar{2}0\rangle$ axis for double twins. It is to be noted that no "clean-up" procedure based on extrapolation of data from nearest neighbors was carried out because such extrapolation increases the chance of introducing artifacts in the data. The twin volume fraction (TVF) values were obtained according to the method described by Jain and Agnew [9]. A "cut-off radius" of 40° tilt from the compression axis was used to determine the twin volume fraction.

In-situ neutron diffraction compression testing:

The compression tests were performed on a custom built Instron (Norwood, MA) load frame in the flight path of the time-of-flight neutron beam line, Spectrometer for Materials Research at Temperature and Stress (SMARTS), at the Manuel Lujan Jr. Neutron Scattering Center, (LANSCE, Los Alamos, NM). The tests were carried out at room temperature in position control at a constant velocity 0.01 mm/s which corresponds to an initial strain rate of 5×10^{-4}/s. An extensometer was attached to the sample using rubber bands to measure the strain. The samples were lubricated using molybdenum disulfide (MoS_2) grease.

A detailed description of SMARTS is given elsewhere [10]. Only a short description is provided here. The loading axis is oriented at 45° relative to the incident neutron beam. Two detector banks are situated at ±90°, making it possible to simultaneously collect two complete diffraction patterns, one parallel and one perpendicular to the applied load. Since the in-situ testing setup is horizontal, the compression sample was preloaded to a stress of ~5 MPa. The neutron diffraction measurements were carried out at several points along the stress-strain curve. The load frame was held at a constant position during diffraction pattern collection, and the hold time was varied to obtain a constant neutron count at the detectors (governed by the source current and the material neutron absorption and scattering cross sections). For this material, the collection time is approximately 10 to 12 minutes per pattern.

The results of this internal strain analysis were presented in a previous Mg Technology Symposium paper [8]. More important for the present work was the in-situ measurement of texture. The texture evolution was assessed via axial distributions of {00.2} pole figure intensity determined by simultaneous Rietveld fitting of the entire diffraction spectra collected in both the parallel and transverse detector banks using General Structural Analysis System (GSAS) [11]. This approach employs a finite series of spherical harmonics description of texture, and assumes radial symmetry. Because the texture of the initial material was confirmed to have near axis-symmetry by ex situ x-ray diffraction (see below), and because the strain path during in-situ deformation is axis-symmetric, the assumption of radial symmetry is warranted for all of the in situ data. These axial distributions are used to assess the volume fraction of twins as a function of deformation, as described in the results section below.

Results and Discussion

Microstructure:

Optical metallography (Fig. 1) of the solution heat treated material reveal a "duplex" microstructure of large, elongated and unrecrystallized grains surrounded by finer, equiaxed and recrystallized grains, which is typical of this Zr-bearing alloy. The added Zr helps to refine the microstructure during both the original casting (inoculation) and subsequent hot working processes. The low magnification of Fig. 1 better conveys the extent to which the structure is comprised of two very distinct grain sizes and morphologies.

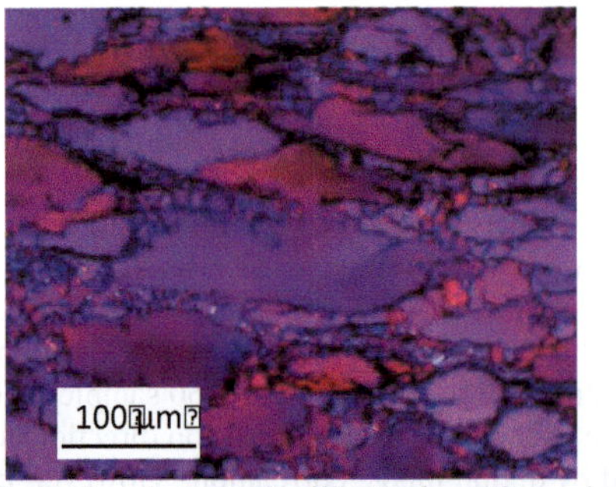

Figure 1: Optical micrograph illustrating the microstructure which results from a solutionizing heat treatment after extrusion (adapted from [8]). Note that the extrusion direction is horizontal.

Figure 2 presents an extrusion direction (ED) IPF color map of the as-received material and the IPF color legend for all subsequent maps. In this particular map, the microstructure is viewed along the ED showing that the grains are nearly equiaxed, from this perspective, whereas the large grains are highly elongated when viewed along the transverse direction (see Figs. 1 & 3).

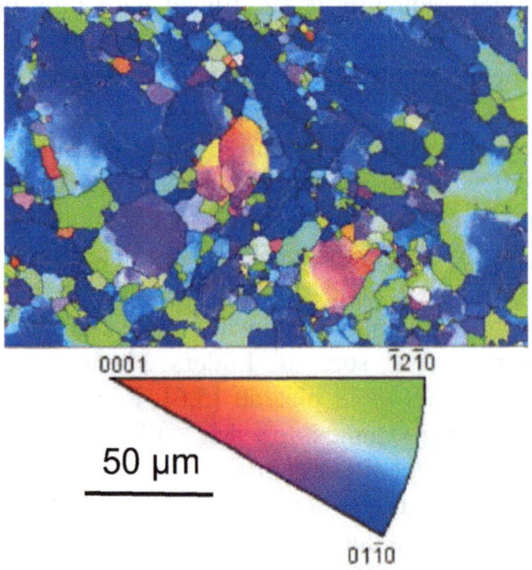

Figure 2: IPF map of the heat treated microstructure looking down the extrusion direction with the IPF taken with respect to the transverse direction.

Also shown in Figure 3 is the fact that there is a significant number of grain boundaries which have the twin orientation relationship, 86° ± 5° about $\langle 11\bar{2}0\rangle$, in the undeformed, as heat-treated condition. However, the morphology of the regions surrounded by these boundaries is not consistent with deformation twins induced by small strains or that of annealing twins (e.g., in fcc metals). Rather, they appear to be remnants of the thermomechanical processing which survived the solution heat treatment. Figure 4 presents a boundary misorientation distribution, which further emphasizes the fact that this material contains a large number of twin boundaries, even in the initial undeformed, heat-treated condition.

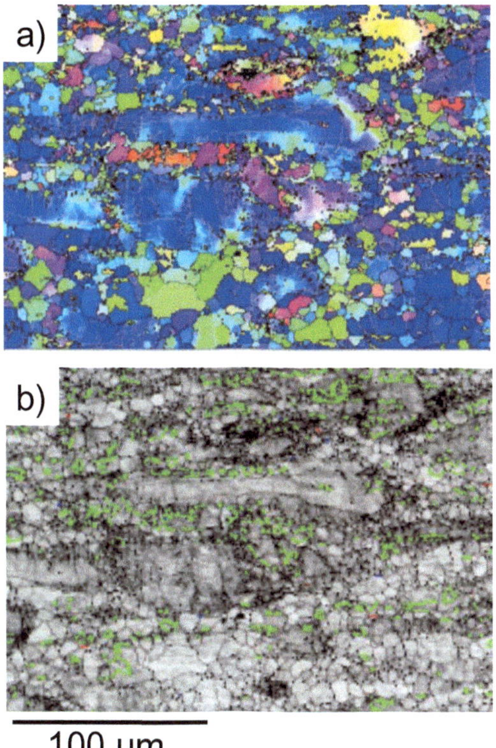

Figure 3: Undeformed microstructure viewed along transverse direction presented as a) an ED-IPF map and b) a band contrast map highlighting the presence of boundaries with the characteristic geometry of extension twins (green), contraction twins (red), and double twins (blue).

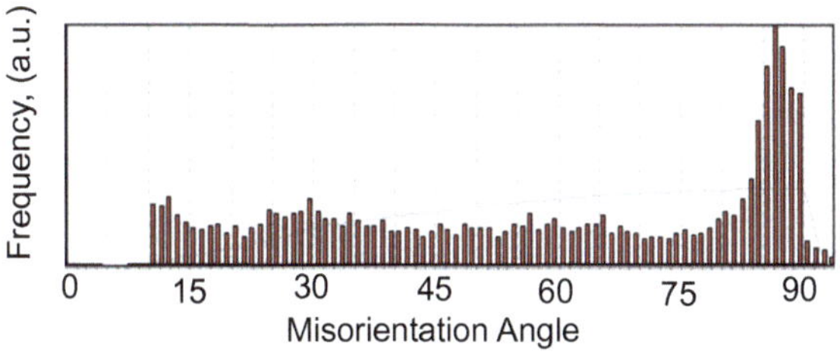

Figure 4: Boundary misorientation distribution showing a preference for boundaries with near 90° misorientation in the undeformed, heat treated condition.

The initial texture measured by EBSD is consistent with that previously measured by X-ray diffraction [8]. Even though the material was obtained in the form of a rectangular extruded plate, Figure 5 reveals a <10.0> fiber parallel to the extrusion axis, which is typical of axis-symmetric (i.e. round) extrusions. The texture from the region where the mechanical test samples were excised does not show the strong <00.1> component parallel to plate normal direction. Rather, it is nearly radially symmetric, a texture commonly referred to as the rod texture [12]. For thoroughness, the texture was measured twice on two different pieces of material from the region of testing and nearly identical textures were observed.

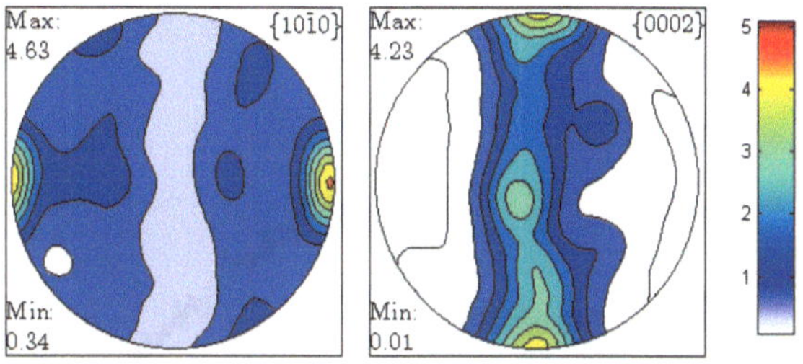

Figure 5: {00.2} and {10.0} pole figures of extruded and heat treated alloy ZK60A-T6 showing a prominent <10.0> fiber || extrusion axis (horizontal).

<u>Mechanical behaviour</u>

Figure 6 shows the full true stress-true strain response as well as the stress-strain locations where tests were interrupted to allow for ex-situ characterization of the microstructure by EBSD. The strain hardening behaviour is distinctly S-shaped, characteristic of {10.2} extension twinning as evidenced by neutron diffraction studies from the same material [8]. As the twinning mode exhausts its capacity to accommodate strain, the material begins to rapidly harden until a stress level is obtained at which hard dislocation glide mechanisms can occur within the twins. In ZK60A, the "plateau" saturates at roughly 5% compressive strain, which is similar to the response reported in Hazeli et al. [7] for alloy AZ31B with a grain size of ~30 μm. Note that the unload from ~10% strain exhibits a strong curvature indicative of a Bauschinger effect, even before fully unloading.

A plot of the strain hardening rate, as a function of the applied compressive strain, also reveals the distinct strain hardening regime with a lower initial strain hardening rate regime, up to 5% strain. However, this plot makes it clear that the hardening rate is distinctly non-zero, having a value of about 1.2 GPa ~ E/37.5, given the measured value of Young's modulus of 45 GPa. Therefore, even this lower strain rate regime should *not* be considered as a low, in general. It appears low only because of the exceptionally rapid hardening which follows. Consider that the initially high strain hardening rate exhibited by fcc polycrystals is observed to be in the range of E/50 [13]. Also of interest is the fact that the strain hardening rate *after* the yield plateau was observed to tend toward a value between 1.3-1.5 GPa independent of grain size in the studies of Barnett et al. [4-6]. Given that the strain is largely accommodated by twinning, we can conclude that there is a significant strain hardening of the twinning mode itself.

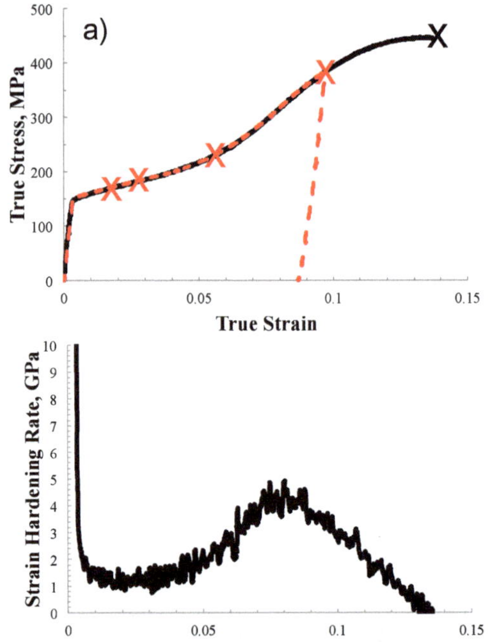

Figure 6: Compression response of ZK60 alloy along ED (a) flow curve showing the locations of tests interrupted for ex-situ characterization (b) strain hardening rate as a function of strain.

DIC measurements revealed the strain to be uniform within the sample with the exception of a slight localization due friction with the platens, despite the use of Teflon lubricant. Figure 7 shows the DIC measured strain field on the surface when the sample was at ~1% strain. At this strain level, it is clearly seen that the strain is non-uniform. This non-uniformity began at the onset of plastic deformation, when a region of roughly 0.7% higher strain than the surrounding areas, appeared at the bottom of the sample and then grew until consumed the entire sample. It is noted that this region did not move in a clear Lüders front, rather it has a wavy appearance. The strong localization seen in AZ31B (a band with 1.2% strain greater than the surroundings) [7] was not observed during the present study.

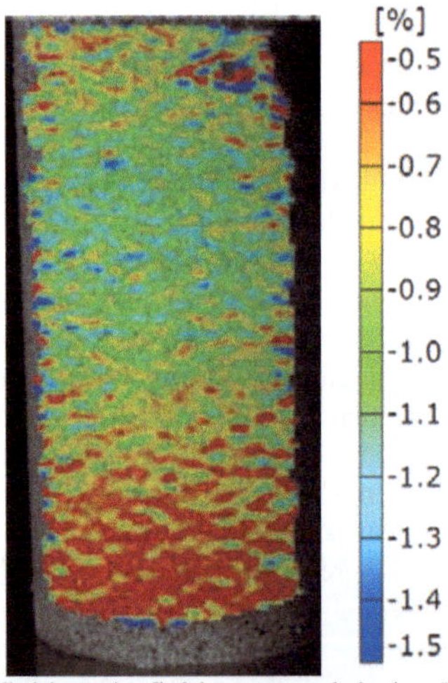

Figure 7: Full field strain field measured during ED compression showing strain localization but not distinct Lüders band structure.

Ex-situ EBSD

The microstructure post deformation was characterized using EBSD. Figure 8 shows the microstructure after 0.7% strain. It is clearly seen that there are a large number of thin, lenticular twins which were not present in the as heat-treated condition (Figs. 2 & 3). Figures 8a and b present the microstructure with two different IPF colouring schemes, TD and ED, respectively. The ED-IPF map reveals the twins more clearly, as they tend to be red on a blue or green background. The TD-IPF map provides more clarity regarding the grain orientation, since many of the grains have a similar color in the ED-IPF map. The hypothesis that the large grains will undergo twinning initially does not seem to hold. Though there is a higher number of twins within individual large grains, there are many small grains with twins in them.

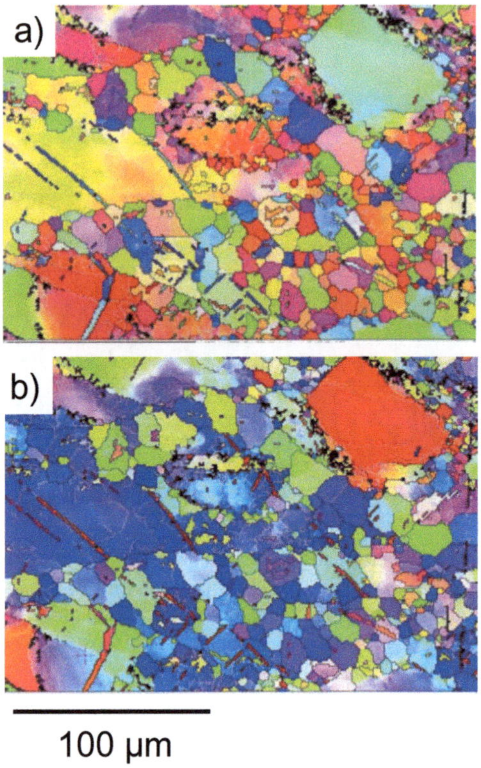

Figure 8: IPF map of the sample tested to 0.7% strain. With IPF oriented (a) along the transverse direction and (b) along the rolling direction.

Figure 9 shows the deformation microstructure after 2.0% strain. It is clearly seen that there is a large number of twins in the large grains, at this stage in the deformation. However, it is also clearly seen that twins are present in both the large and the small grains. It is noted that twins in the small grains frequently appear to have nucleated off twins in surrounding grains. Notice also what appears to be continuous twin across multiple grains, what has been termed 'autocatalytic twinning' [14]. Finally, it is noted that the twins appear with a characteristic spacing within the large grains. In Figure 10, obtained from the sample deformed to 4.1% strain, there are some grains in which the majority of the area appears to be twinned. Note that many grains appear predominantly red in the ED-IPF coloured map, whereas most of the grains were originally blue or green in this type of map. The characteristic spacing of the twins is quite obvious at this strain.

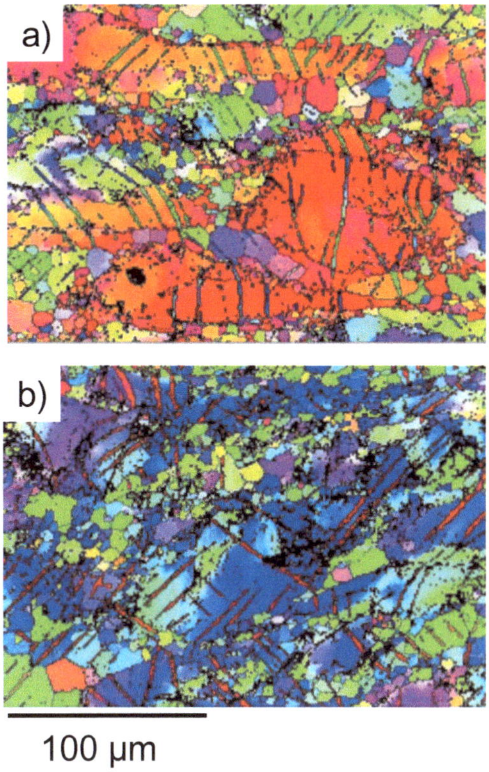

Figure 9: IPF map of the sample tested to 2.0% strain, with IPF coloring oriented along the (a) TD and (b) ED, respectively.

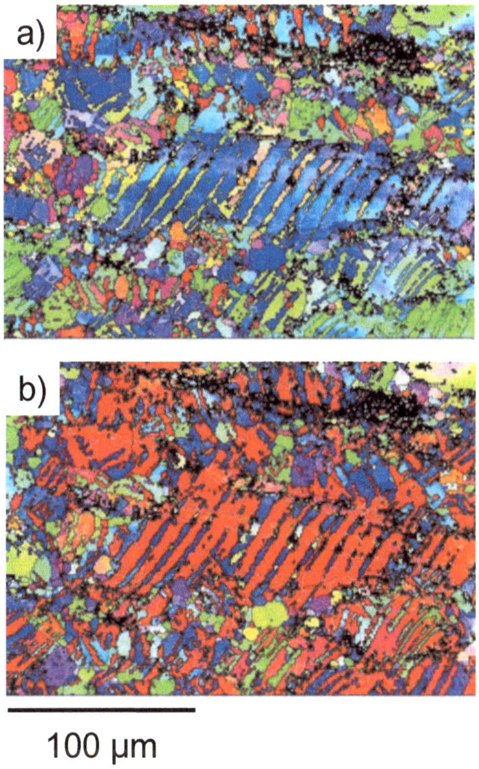

Figure 10: IPF map of the sample tested to 4.1% strain, with IPF coloring oriented along the (a) TD and (b) ED, respectively.

Figure 11 shows the appearance of the microstructure after 8.8% strain, and the majority of the microstructure has adopted the twin orientation. Note the preponderance of red orientations in the ED-IPF map. Only small regions of the microstructure remain which have orientations with colours close to blue or green. Because the orientation of the matrix grains and the twins is so distinct, the area (volume) fraction of each can be computed, employing the EBSD data [9]. Figure 12 presents the results of this assessment, along with in-situ estimates of the twin volume fraction (TVF) based upon neutron diffraction data per ref. [15]. The discrepancy between the in-situ and ex-situ data estimates of TVF can be reconciled by the fact that detwinning occurs during unloading of the sample [16]. In fact, if one assumes detwinning accounts for all of the reverse plastic strain during unload, one estimates a decrease in TVF of 11%, close to the level of discrepancy.

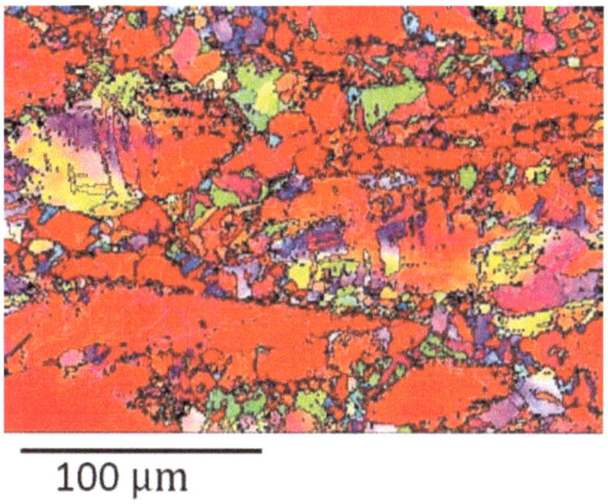

Figure 11: IPF map of the sample tested to 8.8% strain, with IPF coloring oriented along the ED.

One puzzling feature is the rapid accumulation of TVF at low plastic strains. This is distinct from previous studies, e.g., [15], in which the initial plastic strain was clearly accommodated by dislocation glide, even when the macroscale yield was controlled by twinning. See Figure 12 for a comparison of the TVF obtained on the present ZK60A sample and that of two AZ31B samples published previously [15]. Notably, ZK60A samples of various heat treatment conditions [8] all exhibit the same unusual rapid initial twinning. This result is readily explained if one recalls that there were many twin boundaries present in the material prior to deformation. Thus, rather than having to nucleate twins, the twin boundaries only have to migrate through the grains. It has recently been emphasized that twin migration in Mg is quite easy [17]. Thus, the initial twinning in this microstructure bears some similarity to detwinning and retwinning observed recently [18].

Conclusions

Based on the measured stress-strain response, full field strain measurements (DIC), ex-situ microstructural characterization, and in-situ neutron diffraction, it is concluded that:

- A partially recrystallized microstructure is observed in peak-aged ZK60A-T6 with large unrecrystallized grains surrounded by small recrystallized grains.
- The region of "low" strain hardening, often termed as the twinning plateau, is not low in the general sense indicating significant hardening during twin dominated deformation.
- The Lüders bands seen in fine-grained AZ31B and ZM20 are not observed in the present ZK60A-T6 and the strain localization is far less, on a macroscopic scale.

- Though it was hypothesized that twinning may occur first in the large grains and only subsequently grow into the small recrystallized grains that surround them, the initial data do not appear to bear this out.
- It is suggested that pre-existing twin boundaries within the microstructure permit early, rapid twinning in ZK60A because no nucleation of such twins is required.
- The possibility of detwinning during unloading should make researchers cautious of employing ex-situ measurements of twin volume fractions.

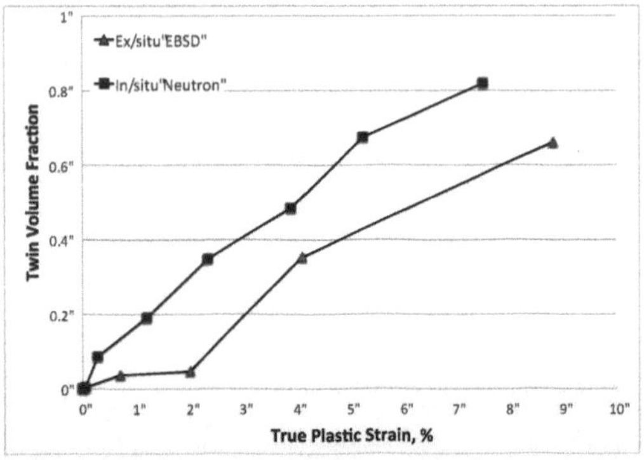

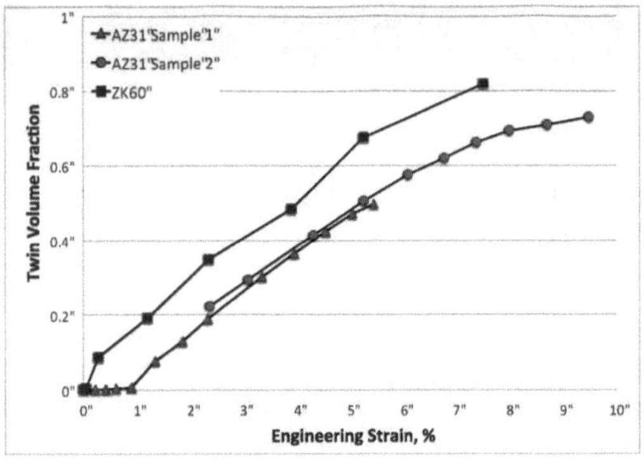

Figure 12. A comparison of twin volume fraction measured by in-situ neutron diffraction and EBSD as a function of plastic strain as well as a comparison of the present in-situ neutron data with previous data obtained on AZ31 [15].

Acknowledgements

The authors would like to thank the NSF, Grant CMMI 1235259, and the U.S. ARO, under contract number W911NF-12-1-0455 monitored by Drs. Mathaudhu and Stepp. The views and conclusions contained in this document are those of the authors and should not be interpreted as representing the official policies, either expressed or implied, of the Army Research Laboratory or the U.S. government. The U.S. government is authorized to reproduce and distribute reprints for government purposes notwithstanding any copyright notation hereon.

References

[1] Chapuis, A. & Driver, J. H. (2011). Temperature dependency of slip and twinning in plane strain compressed magnesium single crystals. *Acta Materialia,* 59(5) 1986-1994.

[2] Xin, Y., Wang, M., Zeng, Z., Nie, M., & Liu, Q. (2012). Strengthening and toughening of magnesium alloy by {10–12} extension twins. *Scripta Materialia, 66*(1), 25-28.

[3] Ma, Q., El Kadiri, H., Oppedal, A. L., Baird, J. C., Li, B., Horstemeyer, M. F., & Vogel, S. C. (2012). Twinning effects in a rod-textured AM30 Magnesium alloy. International Journal of Plasticity, 29, 60-76.

[4] Barnett, M. R., Keshavarz, Z., Beer, A. G., & Atwell, D. (2004). Influence of grain size on the compressive deformation of wrought Mg–3Al–1Zn. *Acta materialia, 52*(17), 5093-5103.

[5] Muránsky, O., Barnett, M. R., Luzin, V., & Vogel, S. (2010). On the correlation between deformation twinning and Lüders-like deformation in an extruded Mg alloy: In situ neutron diffraction and EPSC. 4 modelling. *Materials Science and Engineering: A, 527*(6), 1383-1394.

[6] Barnett, M. R., Nave, M. D., & Ghaderi, A. (2012). Yield point elongation due to twinning in a magnesium alloy. *Acta materialia, 60*(4), 1433-1443.

[7] Hazeli, K., Cuadra, J., Vanniamparambil, P. A., & Kontsos, A. (2013). In situ identification of twin-related bands near yielding in a magnesium alloy. *Scripta Materialia, 68*(1), 83-86.

[8] Agnew, S. R., Calhoun, C. A., & Clausen, B. (2014). In-situ Neutron diffraction study of aging in alloy ZK60A. *Magnesium Technology 2014*, 387-394.

[9] Jain, A., & Agnew, S. R. (2007). Modeling the temperature dependent effect of twinning on the behavior of magnesium alloy AZ31B sheet. Materials Science and Engineering: A, 462(1), 29-36.

[10] Bourke, M. A. M., Dunand, D. C., & Ustundag, E., "SMARTS–a spectrometer for strain measurement in engineering materials." *Applied Physics A* 74 (2002), S1707-S1709.

[11] Von Dreele, R. B., "Quantitative Texture Analysis by Rietveld Refinement" *J. Applied Crystallography* 30 (4) (1997), 517–25

[12] Turner, P. A., & Tomé, C. N. (1994). A study of residual stresses in Zircaloy-2 with rod texture. *Acta metallurgica et Materialia, 42*(12), 4143-4153.

[13] Kocks, U. F., & Mecking, H. (2003). Physics and phenomenology of strain hardening: the FCC case. *Progress in materials science, 48*(3), 171-273.

[14] Ecob, N., & Ralph, B. (1983). The effect of grain size on deformation twinning in a textured zinc alloy. *Journal of Materials Science, 18*(8), 2419-2429.

[15] Clausen, B., Tomé, C. N., Brown, D. W., & Agnew, S. R. (2008). Reorientation and stress relaxation due to twinning: modeling and experimental characterization for Mg. *Acta Materialia, 56*(11), 2456-2468.

[16] Wu, L., Jain, A., Brown, D. W., Stoica, G. M., Agnew, S. R., Clausen, B., & Liaw, P. K. (2008). Twinning–detwinning behavior during the strain-controlled low-cycle fatigue testing of a wrought magnesium alloy, ZK60A. *Acta Materialia, 56*(4), 688-695.

[17] El Kadiri, H., Barrett, C. D., Wang, J., & Tomé, C. N. (2015). Why are twins profuse in magnesium?. *Acta Materialia, 85*, 354-361.

[18] Lou, X. Y., Li, M., Boger, R. K., Agnew, S. R., & Wagoner, R. H. (2007). Hardening evolution of AZ31B Mg sheet. *International Journal of Plasticity, 23*(1), 44-86.

Magnesium Technology 2016
Edited by: Alok Singh, Kiran Solanki, Michele V. Manuel, and Neale R. Neelameggham
TMS (The Minerals, Metals & Materials Society), 2016

ASYMMETRIC GROWTH OF TENSILE TWINS IN MAGNESIUM

Zhe Li[1], Chengliang Li[2], Ben Xu[1], Wei Liu[1]
[1]School of Materials Science and Engineering, Tsinghua University, Beijing, China
[2]China Nuclear Power Engineering Co., Ltd, Shenzhen 518172, China

Keywords: Twinning, twin growth, stress distribution, molecular dynamics, magnesium

Abstract

Molecular dynamics studies were carried out to simulate the growth of a single {-1012} tensile twin and two adjacent {-1012} twins. For the single twin, transverse propagation is faster than thickening. For the two adjacent twins, one twin's growth is predominant over the other when they are 45° to each other. These phenomena are attributed to the interaction of stress field around PB interfaces and twinning planes. The asymmetric stress distribution leads to the twins' asymmetric growth.

Introduction

Magnesium alloys attract great scientific and industrial interests since they have several desirable properties such as high specific strength, natural abundance and recyclability. However, the low formability near the room temperature has limited their application. The nucleation, transverse propagation and thickening of twinning, which is always seen as the sequence of deformation, plays an important role in the plastic deformation of magnesium alloys [1]. Focusing on the nucleation and propagation, several twinning mechanisms, such as the pole mechanism [2] and the slip dislocation mechanism [3], have been proposed. In recent years, some experiments and simulations demonstrated that twinning interfaces are inherently defective with kink-like steps and curvature [4] and the experimental observations on Mg showed that the twinning boundaries of the {10-12}<-1011> twin model deviate significantly from the (10-12) plane [5,6].These phenomena cannot be explained by the traditional point that the twinning dislocations are strictly bounded in the twinning planes. The atomic shuffling mechanism seems reasonable to explain this phenomenon as it permits the twinning dislocations to migrate from the twinning plane [5,7,8]. However, the shuffling-dominated mechanism has led to some debate [9,10].

With regard to twin growth, the mechanism of thickening, that is, the motion perpendicular to the twinning plane, has attracted much attention [8,11,12]. Interestingly, the transverse propagation of the twin, i.e. its extension in the twinning direction, received less attention.

By mimicking the analytical method of Eshelby's inclusion problem[13], Xu created a volume containing a twinned subdomain and simulated the nucleus' growth process. He noted that transverse propagation is faster than thickening and a specific interface named PB (prismatic/basal) interface, which has been observed in several experiments [14,15], plays an important role in the growth of the twin nucleus [16]. However, the reason for the asymmetric growth was not studied. Although the growth of the twins is strongly dependent on the stress distribution around the twins, the existing study of this topic [17] is nevertheless not enough, thus in the following parts it will be discussed.

In this study, we create twin nuclei bordered by conjugate {-1012} twinning planes. With molecular dynamics (MD), we simulate the growth process of a single nucleus and two adjacent nuclei. In order to explain the phenomena observed in the process, the stress distribution around the twins is illustrated.

Simulation Methods

Twin growth in magnesium was studied using MD. The embedded atom method (EAM) potential developed by Liu et al. [18] was used to simulate interatomic interaction.

One hypothetical twin nucleus as a volume containing a twinned subdomain was considered as shown in Figure 1. Horizontal planes

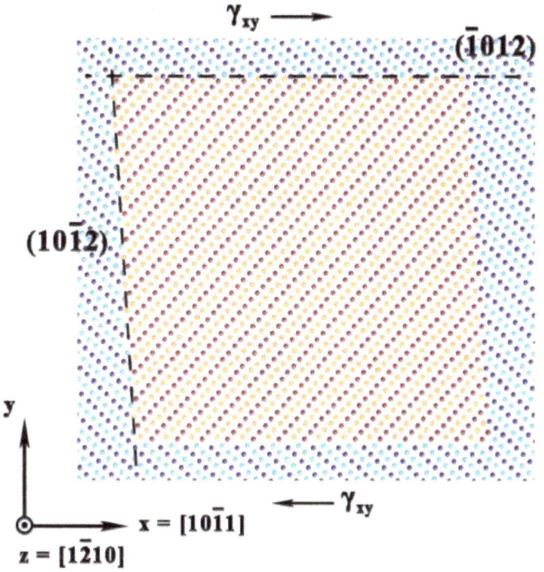

Figure 1. The volume containing a twinned subdomain. The two dashed lines represent the (-1012) and (10-12) planes. Different colors represent different planes parallel to basal planes.

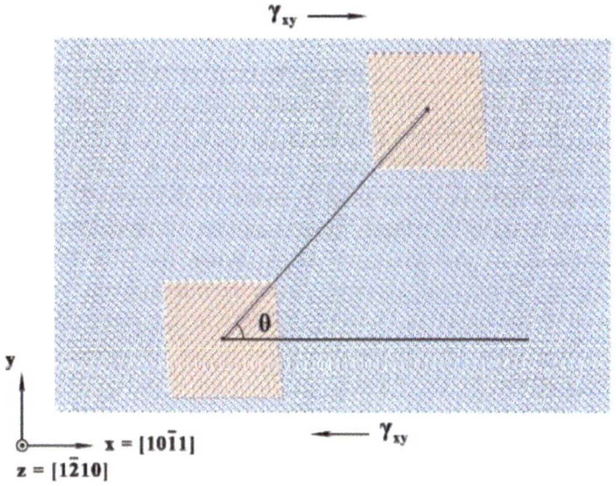

Figure 2. The volume containing two adjacent nuclei. The angle between the connecting line of two nuclei's center and the [10-11] direction is 45° (θ = 45°).

are (-1012) twinning planes with the x-axis along the [10-11] twinning direction. The z-axis perpendicular to the plane of the figure is the [1-210] direction. A second twinning plane, the (10-12) plane, is almost perpendicular to the horizontal one (the angle between the two {-1012} planes is 93.7°, with c/a=1.623 for Mg). A region bounded by (-1012) and (10-12) planes as illustrated in Figure 1 was initially cut out. Mirror symmetry with respect to the (-1012) and (10-12) planes was then applied to the atoms inside the cut-out region. In this way, a nucleus bordered by (-1012) and (10-12) twinning planes was created. A volume containing two adjacent nuclei is shown in Figure 2. The angle between the connecting line of two nuclei's center and the [10-11] direction is 45°. The coordinate system in Figure 2 is the same as that in Figure 1.

In the following study, for the single nucleus and the two adjacent nuclei, periodic boundary conditions were used in all directions in a cell of dimensions 83.71 × 41.77 × 0.96 nm^3 containing 145200 atoms. Note that if no strain is applied to the simulation cell, the twinned region is unstable, it shrinks and disappears to minimize the energy. However, if the initial nucleus is sufficiently large and a proper value of shear strain (γ_{xy}) is applied along [10-11] direction in the (-1012) planes, the nucleus may be stabilized. Here, the initial nucleus is stabilized by application of 5% shear to the whole volume in advance. The way to apply shear strain is to change the x-coordinate with the increasing time as shown in equation 1. x and y is the x-coordinate and y-coordinate of the atom at some time respectively. x' is the atom's x-coordinate at the next moment. ε_{xy} is the shear strain rate. dt is the time of applying shear strain.

$$x' = x + \varepsilon_{xy} * y * dt \qquad (1)$$

Results

The Growth Process of A Single Twin

To simulate the growth process, the nucleus shown in Figure 1 was applied an additional 1.5% shear strain. The growth process was simulated by MD at 0K with a constant strain rate (ε_{xy}=0.00001/ps). This low temperature was chosen in order to avoid influence of other thermally activated processes, such as nucleation of dislocations. Some snapshots of the process are shown in Figure 3. At the initial stage, shown in Figure 3a-b, the nucleus is bordered by conjugate {-1012} twinning planes and interfaces at ~45° from the horizontal twinning planes named PB interfaces in some references [14,15,16].

When shear strain is increased, the twin grows both vertically (thickening in the y-direction) and laterally (transverse propagation in the x-direction). Transverse propagation is faster than thickening, which has been proved by several simulations [16,19,20,21]. In the lateral direction, the twin front is made of the (10-12) twinning plane and two PB interfaces. Lateral motion of the twin occurs by the nucleation and glide of steps along the PB interfaces. Thickening of the twin also occurs via the nucleation and glide of steps along the upper and lower (-1012) twinning planes. Steps are visible in Figure 3d-f. It should be noted that the growth process does not depend on the initial shape of the nucleus. For instance, an initial hexagonal nucleus bordered by (-1012) twinning planes, prismatic planes and basal planes also grows in a similar way [16].

The Growth Process of Two Adjacent Twins

The growth process of two adjacent twins was also studied in this paper. Unexpected growth process was observed when the two

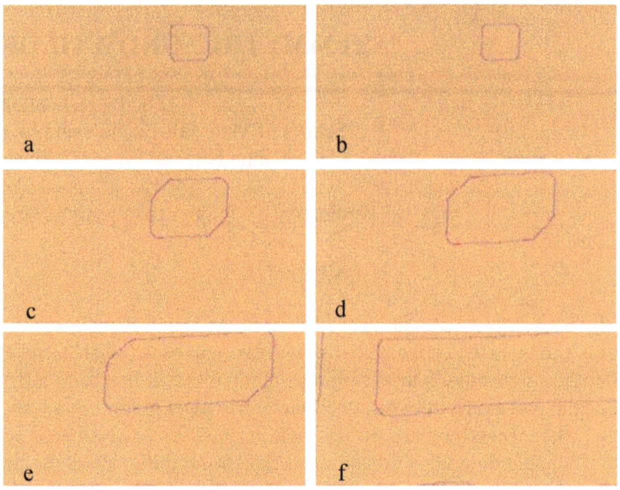

Figure 3. Growth process of a single twin simulated by MD at 0K. The outline of the twin area is plotted out by different color, which represent the different coordination number of these atoms.

twins were 45° to each other shown in Figure 2. The nuclei were applied an additional 3% shear strain in the whole simulation. The growth process, as shown in Figure 4, was simulated by MD at 0K, using a constant strain rate (ε_{xy}=0.00001/ps). Figure 4a-b show that the nuclei are bordered by conjugate {-1012} twinning planes and PB interfaces at the initial stage. With increasing deformation, shown in Figure 4c-f, the growth of one twin expanding laterally is predominant over the other, while the other one remain unchanged at the latter stage (Figure 4e-f).

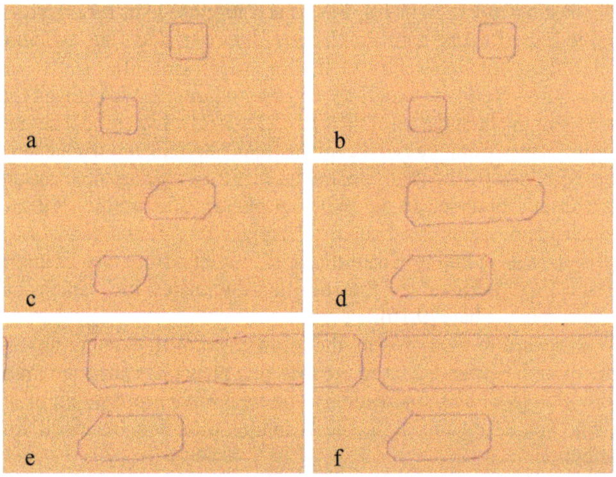

Figure 4. Growth process of two adjacent twins which are 45° to each other. The process was simulated by MD at 0K. The outline of the twin area is plotted out by different color, which represent the different coordination number of these atoms.

Discussion

The growth of {-1012} tensile twins was simulated by MD. For the single twin, transverse propagation is faster than thickening. For the two adjacent twins which are 45° to each other, one twin's growth is predominant over the other. These phenomena are attributed to the stress distribution around the twin nuclei.

The Stress Distribution of The Single Twin

The shear stress (τ_{xy}) distribution during single twin's growth is shown in Figure 5. There is a one-to-one correspondence between Figure 5 and Figure 3. At the initial stage, the stress around PB interfaces and twinning planes is lower than other part, as shown in Figure 5a-b. This distribution is similar to the stress distribution of the stiff rectangular inclusion in Ref. [22]. With the increasing shear strain, the interaction of stress field around the two (-1012) twinning planes and the PB interfaces leads to a lower stress value on the thickening direction, which results in the difficulty of thickening (Figure 5c-f).

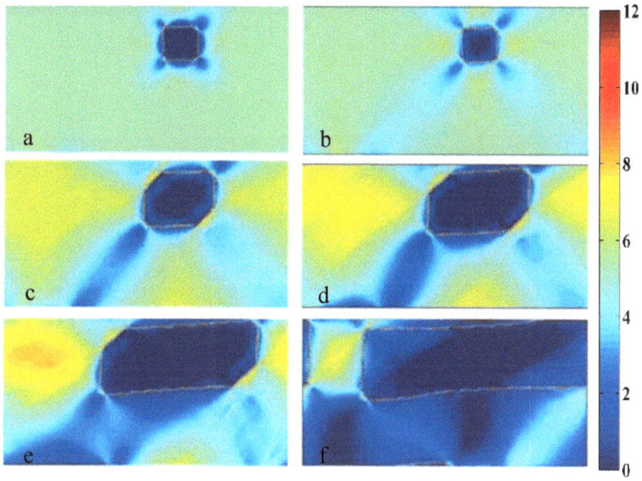

Figure 5. The shear stress (τ_{xy}) distribution during the single twin's growth process. Color scale shows stress value (the unit is MPa), from blue at low value to red at high value.

The Stress Distribution of The Two Adjacent Twins

In Figure 6, the shear stress (τ_{xy}) distribution during the growth process of the two adjacent twins which are 45° to each other is shown. There is a one-to-one correspondence between Figure 6 and Figure 4. As shown in Figure 6a-b, the stress around PB interfaces and twinning planes is lower than other parts at the initial stage. When shear strain is increased, the interaction of stress field around the (-1012) twinning planes and the PB interfaces leads to a lower stress value on the thickening direction, which results in the difficulty of thickening for the two twins (Figure 6b-f). With further increasing shear strain, one twin's growth stops due to the lower stress value around it as shown in Figure 6e-f.

Conclusion

We simulated the growth of {-1012} tensile twins by MD. For a single twin, transverse propagation is faster than thickening. For the two adjacent twins, one twin's growth is predominant over the other when they are 45° to each other. These phenomena are attributed to the interaction of stress field around PB interfaces and twinning planes. When shear strain is applied to the twins, stress around these planes is lower than other parts. In conclusion, the asymmetric stress distribution leads to the twins' asymmetric growth.

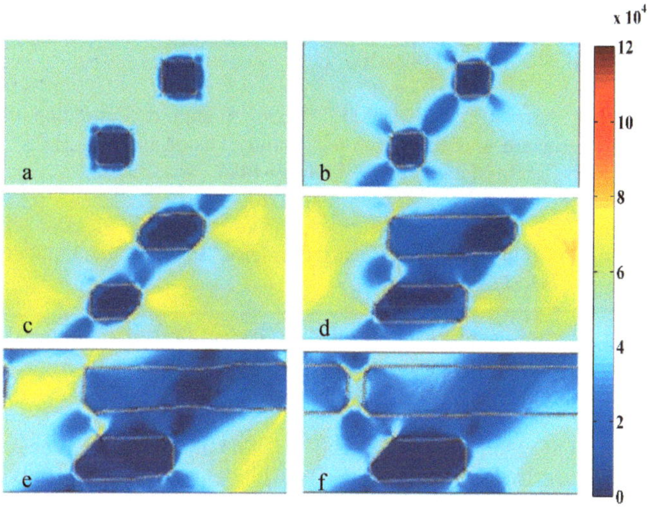

Figure 6. The shear stress (τ_{xy}) distribution during the growth process of two adjacent twins which are 45° to each other. Color scale shows stress value (the unit is MPa), from blue at low value to red at high value.

References

1. J.W. Christian and S. Mahajan, "Deformation Twinning," Prog. Mater. Sci., 39 (1995), 1-157.

2. N. Thompson and D.J. Millard, "Twin Formation in Cadmium," Philos. Mag., 43 (1952), 422-440.

3. S. Mendelson, "Dislocation Dissociations in Hcp Metals," Appl. Phys., 41 (1970), 1893-1910.

4. Y.M. Wang, F. Sansoz, T.L. Grange, R.T. Ott, J. Marian, T.W. B. Jr and A.V. Hamza, "Defective Twin Boundaries in Nanotwinned Metals," Nat. Mater., 12 (2013), 697-702.

5. X.Y. Zhang, B. Li, X.L. Wu, Y.T. Zhu, Q. Ma, Q. Liu, P.T. Wang and M.F. Horstemeyer, "Twin Boundaries Showing Very Large Deviations from the Twinning Plane," Scr. Mater., 67 (2012), 862-865.

6. D. Zhang, B. Zheng, Y. Zhou, S. Mahajan and E.J. Lavernia, "Prism Stacking Faults Observed Contiguous to a {10-12} Twin in a Mg-Y Alloy," Scr. Mater., 76 (2014), 61-64.

7. J. Wang, Q. Yu, Y. Jiang, and I.J. Beyerlein, "Twinning-associated Boundaries in Hexagonal Close-packed Metals," Jom, 66 (2013), 95-101.

8. B. Li and E. Ma, "Atomic Shuffling Dominated Mechanism for Deformation Twinning in Magnesium," Phys. Rev. Lett., 103 (2009), 035503.

9. A. Serra, D.J. Bacon and R.C. Pond, "Comment on 'Atomic Shuffling Dominated Mechanism For Deformation Twinning In Magnesium'," Phys. Rev. Lett., 104 (2010), 029603.

10. B. Li and E. Ma, "Li and Ma Reply," Phys. Rev. Lett., 104 (2010), 029604.

11. A. Serra, D.J. Bacon and R.C. Pond, "The Crystallography and Core Structure of Twinning Dislocations in H.C.P. Metals," Acta Metall., 36 (1988), 3183-3203.

12. A. Serra, R.C. Pond and D.J. Bacon, "Computer Simulation of the Structure and Mobility of Twinning Dislocations in H.C.P. Metals," Acta Metall. 39 (1991), 1469-1480.

13. J.D. Eshelby, "The Determination of the Elastic Field of an Ellipsoidal Inclusion, and Related Problems," Proc. R. Soc. London A, 241 (1957), 376-396.

14. Q. Sun, X.Y. Zhang, Y. Ren, J. Tu and Q. Liu, "Interfacial Structure of {-1012} Twin Tip in Deformed Magnesium Alloy," Mater. Sci., 90 (2014), 41-44.

15. B.M. Morrow, E.K.Cerreta, R.J. McCabe and C.N. Tomé, "Toward Understanding Twin–Twin Interactions in Hcp Metals: Utilizing Multiscale Techniques to Characterize Deformation Mechanisms in Magnesium," Mater. Sci. Eng. A, 613 (2014), 365-371.

16. B. Xu, L. Capolungo and D. Rodney, "On the Importance of Prismatic/Basal Interfaces in the Growth of (-1012) Twins in Hexagonal Close Packed Crystals," Scr. Mater., 68 (2013), 901-905.

17. M.A. Kumar, A.K. Kanjarla, S.R. Niezgoda, R.A. Lebensohn and C.N. Tomé, "Numerical Study of the Stress State of a Deformation Twin in Magnesium," Acta Mater., 84 (2015), 349-358.

18. X.Y. Liu, J.B. Adams, F. Ercolessi and J.A. Moriarty, "EAM Potential for Magnesium from Quantum Mechanical Forces," Modell. Simul. Mater. Sci. Eng., 4 (1996), 293-303.

19. I.J. Beyerlein, R.J. McCabe and C.N. Tomé, "Effect of Microstructure on the Nucleation of Deformation Twins in Polycrystalline High-purity Magnesium: a Multi-scale Modeling Study," J. Mech. Phys. Solids, 59 (2011), 988-1003.

20. L. Leclercq, L. Capolungo and D. Rodney, "Atomic-Scale Comparison Between {-1101} and {-1102} Twin Growth Mechanisms in Magnesium," Mater. Res. Lett., 2 (2014), 152-159.

21. Z.P. Pi, Q.H. Fang, B. Liu, H. Feng, Y. Liu , Y.W. Liu and P.H. Wen, "A Phase Field Study Focuses on the Transverse Propagation of Deformation Twinning for Hexagonal-closed Packed Crystals," Int. J. Plast., 76 (2016), 130-146.

22. D. Misseroni, F.D. Corso, S. Shahzad and D. Bigoni, "Stress Concentration Near Stiff Inclusions: Validation of Rigid Inclusion Model and Boundary Layers by Means of Photoelasticity," Eng. Fract. Mech., 121 (2014), 87-95.

Magnesium Technology 2016
Edited by: Alok Singh, Kiran Solanki, Michele V. Manuel, and Neale R. Neelameggham
TMS (The Minerals, Metals & Materials Society), 2016

NON-DISLOCATION BASED ROOM TEMPERATURE PLASTIC DEFORMATION MECHANISM IN MAGNESIUM

Bo-Yu Liu[1], Zhi-Wei Shan[1], Evan Ma[1,2]
[1]Center for Advancing Materials Performance from the Nanoscale (CAMP-Nano) & Hysitron Applied Research center in China (HARCC), State Key Laboratory for Mechanical Behavior of Materials, Xi'an Jiaotong University, Xi'an 710049, P.R. China
[2]Materials Science and Engineering, Johns Hopkins University, Baltimore, MD 21218, USA

Keywords: Magnesium, TEM, Twin, Basal-Prismatic interface, Unit-cell-reconstruction

Abstract

Dislocation and deformation twinning are traditionally known to be plasticity carriers of crystalline materials at room temperature. By using in-situ TEM mechanical testing technique, here we report that the plasticity of a specially orientated single crystal magnesium can be mediated neither by dislocation nor by twinning, but through a non-dislocation based process, termed as unit-cell-reconstruction. After the reconstruction, a ~7% strain is produced. The newly formed grain and its parent grain are separated by a boundary that mainly consisted of basal-prismatic interfaces. Such boundary can migrate back and forth under a cyclic loading and therefore produce a reversible plastic deformation. The reported novel mechanism may have important implications for the alloy design of magnesium.

Introduction

When external stress that is applied on the crystalline materials exceeds the so-called yield stress, plastic deformation will occur. Such process is traditionally known to be carried out through the slip of lattice dislocations or through the deformation twinning that accomplished by successively slip of twinning dislocations [1]. The dislocation slip and deformation twining produce shear strain that makes the material elongated or shorten. Both the dislocation activity and deformation twinning are restricted on a specific crystallographic plane. For dislocation slip, such plane is named as slip plane, which is usually a low index close-packed plane. For deformation twinning, such plane is named as twinning plane, which serves as the slip plane for the twinning dislocations. The twin boundary that separates the matrix volume and the twinned volume must be parallel to the twinning plane [2].

For magnesium and its alloys, the number of dislocation slip systems is limited. Only basal slip can be easily activated. Therefore, several deformation twinning modes play important role to accommodate the plastic strain. Among those twinning modes, the twinning on $\{10\bar{1}2\}$ plane is the most common one [3]. However, some unusual phenomena are difficult to understand by the above classical twinning theory based on the twinning dislocations. Examples include the deviation of twin boundaries from the twinning plane [4, 5], the easy de-twinning process and the reversibility of this twinning mode [6-8].

Recently, twinning mechanism based on non-dislocation process attracts strong research interesting [9-11]. By performing the in-situ TEM mechanical testing on micro/nano-scaled magnesium pillars or dog-bone shaped tensile samples, we reported a new plastic deformation mechanism, which is different from the traditional dislocation-based process [12, 13]. This new mechanism can be used to interpret the above unusual experimental phenomena.

Methods

A FEI Focus Ion Beam (FIB, Helios 600 Dual Beam system) was used to fabricate the tested pillar samples and dog-bone samples. The preparation of dog-bone shape sample consisted of two steps. First, a thin lamella was fabricated by a top-down ion beam. In order to minimize the taper, each side of the lamella was over-tilted by 2 more degree. Second, the lamella was shaped by ion beam from a front view to form a 'dog-bone' shape. In situ compression and tension testing was performed with a Hysitron PI-95 TEM PicoIndentor inside JEOL 2100F transmission electron microscope operating at 200 keV. All tension and compression tests were run under the displacement control mode. Figure 1 shows the experimental setup.

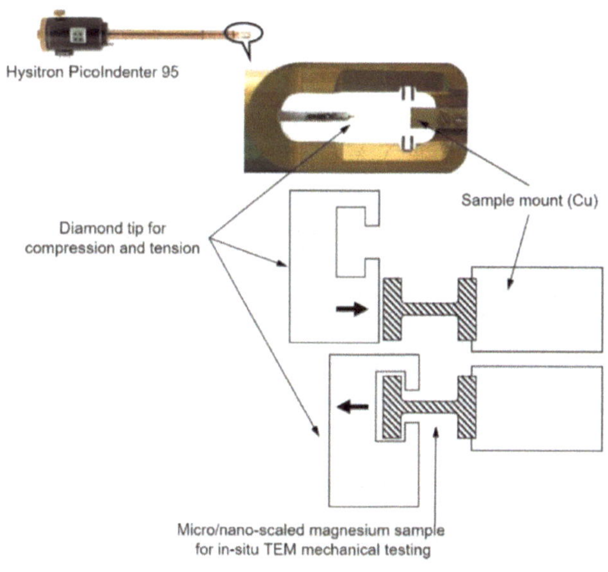

Figure 1. Experimental setup of the in-situ testing.

Results and discussions

Sixteen samples were tested, including 8 compression tests and 8 tension tests. The compression loading was along the axial direction of the pillar and the dog-bone sample, which were set to be the $[1\bar{1}00]$ orientation. The tension loading was along the axial direction of the dog-bone sample, which was set to be the $[0001]$ orientation. The electron beam was set to be along the $[11\bar{2}0]$

orientation. The geometric configurations are shown in Figure 2a and 2b.

Under the above loading condition, a new grain would be generated after the stress reached a critical value. The orientation relation between the new grain and the matrix grain was akin to that of the $\{10\bar{1}2\}$ deformation twining. The c-axis and the basal plane in the new grain are perpendicular to those in the matrix grain, respectively. However, the boundary between the new grain and the matrix grain significantly deviated from the $\{10\bar{1}2\}$ twinning plane. We defined the angle between the boundary and the loading direction as α. If the boundary is parallel to the twinning plane, it would appear to be edge-on when observed along the $[11\bar{2}0]$ orientation and α would be 43.15 degree under compression or 46.85 degree under tension. However, the measured values were in the range of 4 degree to 89 degree [12].

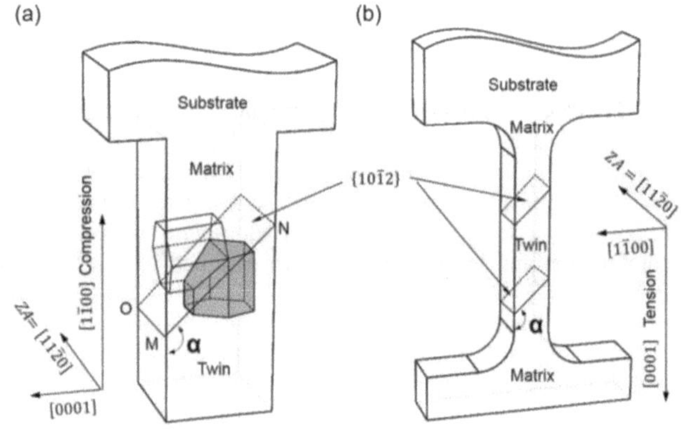

Figure 2. Loading condition and geometry configuration in pillar sample (a) and dog-bone shape sample (b).

One pillar sample was thinned down to about 100 nm by using FIB. After that, the boundary morphology between the new grain and its matrix was characterized by high resolution TEM (FEI Titan), as shown in Figure 3. The viewing direction was set to be$[11\bar{2}0]$. The basal plane in the matrix, and in the new grain, is outlined by white solid lines, as well as the orientation of the $\{10\bar{1}2\}$ plane. In Figure 3a, the boundary is almost parallel to the $\{10\bar{1}2\}$ plane. However, the boundary is fuzzy rather than a sharp line, which means the boundary is not edge-on along the observation direction. In Figure 3b, the boundary is almost perpendicular to the basal plane in matrix. Figure 3c-3e show boundaries with zigzag morphology [12]. Figure 3f shows a boundary with significant width. Fast Fourier Transformation finds that the broad boundary area is actually the overlap of the matrix and the new grain, which is the reason for the blurred boundary morphology in Figure 3a-3e. The above results point out that the boundary is not a crystallographic plane, and the boundary can possess an irregular morphology in 3D space, which is consistent with the observation in bulk experiments about the $\{10\bar{1}2\}$ twin boundaries deviating from the $\{10\bar{1}2\}$ plane in magnesium and its alloys.

In order to reveal the basic component of such boundary, Cs-corrected imaging was performed (ARM200F) under STEM mode. Figure 4a shows one typical example. The viewing direction was set to be$[11\bar{2}0]$. The new grain and its matrix are bonded by several steps. Part of the interface is parallel to the basal plane in new grain and the prismatic plane in matrix, while the other part is parallel to the basal plane in matrix and the prismatic plane in new grain. Therefore, such interfaces are termed basal-prismatic (BP) interfaces.

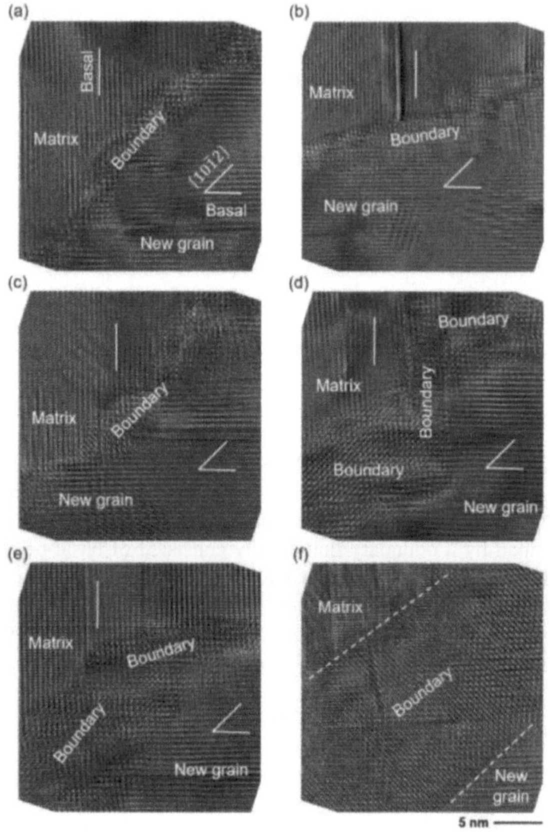

Figure 3. High resolution TEM images of the boundary. The basal plane in matrix and new grain, and the $\{10\bar{1}2\}$ plane are outlined by white solid lines.

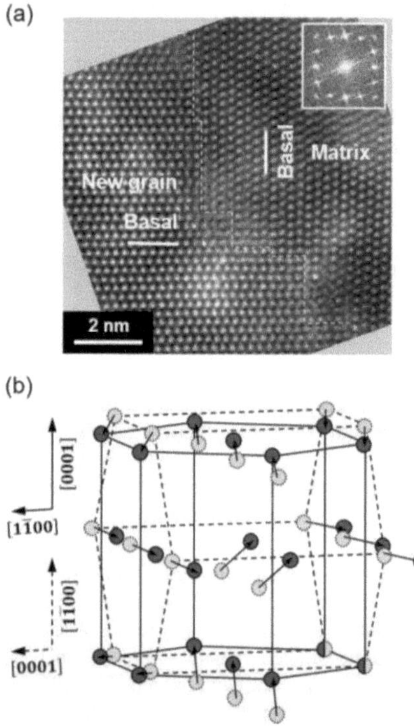

Figure 4. (a) Atomic scale image showing the basal-prismatic interface. (b) One possible route for the unit-cell-reconstruction. The matrix HCP cell is outlined by dashed lines, while the new HCP cell is outlined by solid lines [12, 13].

The BP interfaces can migrate through the conversion of basal plane and prismatic plane. One possible route is shown in Figure 4b [13]. Light gray spheres with dashed circles represent atoms in matrix. Dashed lines connecting those atoms form the original HCP structure. Dark gray spheres with solid circles represent atoms in new grain. Solid lines connecting those atoms form the new HCP structure. The original HCP cell can become the new HCP cell merely by the local movement of atoms, without involving dislocation activities. Therefore, such local atomic rearrangement is termed unit-cell-reconstruction. After such lattice reconstruction, the new basal plane is parallel to the original prismatic plane, and the new prismatic plane is parallel to the original basal plane. As a result, BP interfaces are created.

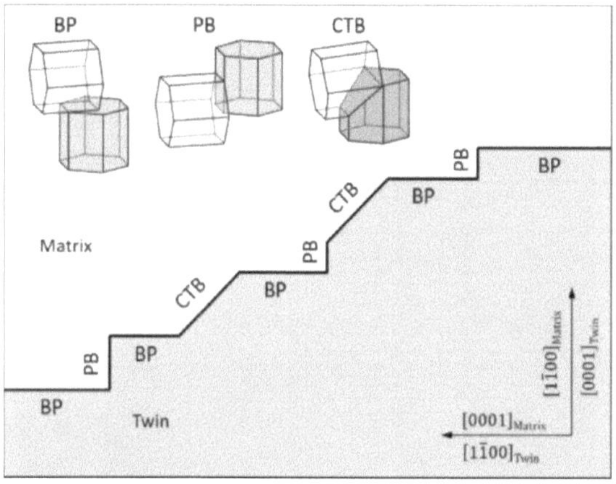

Figure 5. Schematic illustration showing a boundary consisting of segments of BP interfaces and coherent twin boundaries.

Note that the BP and PB interfaces are perpendicular to each other, Therefore, they can compose a boundary with arbitrary morphology in 3D space, as schematically shown in Figure 5.

During cyclic loading, such a BP boundary can migrate back and forth. The migration speed of the boundary is found to be closely related to the loading stress and the crystal structure, which will be discussed in detail elsewhere.

Summary

The newly found plastic deformation mechanism, unit-cell-reconstruction, is different from the dislocation slip and deformation twinning. First of all, it is the local atom rearrangement that dominates the plastic deformation instead of the gliding of lattice dislocations or twinning dislocations. Second, the boundary that separates the new grain and its matrix grain is not a crystallographic mirror plane. Third, both dislocation slip and deformation twinning produce shear strain, while the unit-cell-reconstruction doesn't have to involve shear strain. Details regarding the mechanical response of unit-cell-reconstruction will be discussed in future.

Acknowledgements

The authors acknowledge the support provided by grants from NSFC (50925104, 11132006, 51231005, 51201127 and 51321003), 973 Program of China (2010CB631003), and the support from the 111 Project of China (B06025). We also thank Dr. Jian Wang, Dr. Bin Li, and Dr. Ju Li for discussion. The authors declare no competing financial interests.

References

[1] Weertman J, Weertman JR. Elementary Dislocation Theory: Macmillan; 1964.
[2] Christian JW, Mahajan S. Deformation twinning. Progress in Materials Science. 1995;39:1-157.
[3] Yoo MH. Slip, twinning, and Fracture in hexagonal close-packed metals. Metallurgical Transactions a-Physical Metallurgy and Materials Science. 1981;12:409-18.
[4] Zhang XY, Li B, Wu XL, Zhu YT, Ma Q, Liu Q, et al. Twin boundaries showing very large deviations from the twinning plane. Scr Mater. 2012;67:862-5.
[5] Partridge PG, Roberts E. Formation + behaviour of incoherent twin boundaries in hexagonal metals. Acta Metallurgica. 1964;12:1205-&.
[6] Wu L, Jain A, Brown DW, Stoica GM, Agnew SR, Clausen B, et al. Twinning-detwinning behavior during the strain-controlled low-cycle fatigue testing of a wrought magnesium alloy, ZK60A. Acta Materialia. 2008;56:688-95.
[7] Wang J, Liu L, Tomé CN, Mao SX, Gong SK. Twinning and de-twinning via glide and climb of twinning dislocations along serrated coherent twin boundaries in hexagonal-close-packed metals. Materials Research Letters. 2013;1:81-8.
[8] Li B, McClelland Z, Horstemeyer SJ, Aslam I, Wang PT, Horstemeyer MF. Time dependent springback of a magnesium alloy. Mater Des. 2014:575-80.
[9] Wang J, Yadav SK, Hirth JP, Tomé CN, Beyerlein IJ. Pure-shuffle nucleation of deformation twins in hexagonal-close-packed metals. Materials Research Letters. 2013;1:126-32.
[10] Liu BY, Li B, Shan ZW. Twin boundary migration creating zero shear strain: in-situ tem observations and atomistic simulations. Magnesium Technology 2013. 2013:107-11.
[11] Li B, Ma E. Atomic shuffling dominated mechanism for deformation twinning in magnesium. Physical Review Letters. 2009;103:035503.
[12] Liu B-Y, Wan L, Wang J, Ma E, Shan Z-W. Terrace-like morphology of the boundary created through basal-prismatic transformation in magnesium. Scr Mater. 2015;100:86-9.
[13] Liu B-Y, Wang J, Li B, Lu L, Zhang XY, Shan ZW, et al. Twinning-like lattice reorientation without a crystallographic twinning plane. Nat Commun. 2014;5:3297.

Magnesium Technology 2016
Edited by: Alok Singh, Kiran Solanki, Michele V. Manuel, and Neale R. Neelameggham
TMS (The Minerals, Metals & Materials Society), 2016

INVESTIGATION OF THE PLASTIC FLOW FIELD IN MAGNESIUM ALLOY AZ31B IN THREE ORIENTATIONS FOR EMPIRICAL PENETRATION MODELS

Tyrone L. Jones[1], John P. Riegel, III[2], Christopher S. Meredith[1], Kris Darling [1], Jim Catalano [1] and Anthony Roberts[1]

[1] *US Army Research Laboratory, APG, MD 21005 USA*
[2]*R3 Technology, Inc., 7324 Fountain Spring Ct., Springfield, VA 22150 USA*

Keywords: AZ31B-H24, Ballistic Analysis, Constitutive Modeling

Abstract

Experimental evidence to the extent of the plastic flow field in magnesium targets after ballistic impact has been lacking. The work described in this paper will include the ballistic characterization of the Magnesium alloy AZ31B-H24 in three orientations: normal direction; transverse direction; and rolling direction. Semi-infinite impacts from penetrators in each of the three directions will be shown. The targets were sectioned and machined using electrical discharge machining in preparation for polishing and etching to determine the extent of plastic flow that can be seen. The authors will investigate the applicability of Effective Flow Stress in the similitude-based empirical model and the Walker-Anderson Penetration Model for small arms penetration prediction. The results will be included in this paper.

Introduction

Magnesium (Mg) is the lightest structural metal available. Historically, it has not been perceived to have properties suitable for armor applications. However over the last decade, the US Army Research Laboratory (ARL) has been examining Magnesium (Mg) alloys for ballistic applications [1]. This work has improved the understanding of Mg properties, especially strength and ductility, important factors in the application of Mg in lightweight armors. This paper continues the effort, exploring the ballistic performance of Mg as a function of the plate direction. Mg is known for showing significant variation in properties as a function of the loading direction [2]. However, in a search of the literature, the authors were unable to find work that carefully examined the ballistic performance as a function of plate direction. The directions evaluated are the rolling also known as longitudinal direction (RD or LD), normal direction (ND), and transverse direction (TD) as shown in Figure 1 [3].

For this work, the authors fired 7.62-mm APM2 projectiles against blocks of Mg AZ31B-H24 cut from 76.2-mm (3") thick plates. In addition to conducting terminal ballistics experiments, the quasi-static and dynamic material properties of the Mg were determined with standard tests and Split-Hopkinson Pressure Bar (SHPB) experiments in each of the three directions of the plate.

The goal of this effort was to develop experimental data comparing performance of AZ31B-H24 in the various orientations. A secondary goal was to acquire static and dynamic properties in each of the directions for use in empirical, analytical, and numerical modeling of penetration.

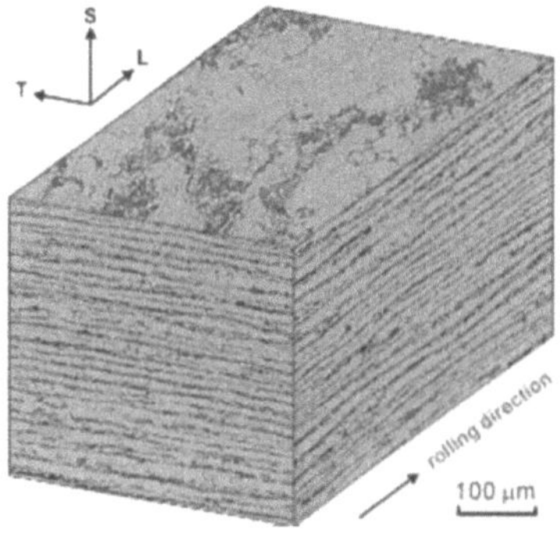

Figure 1. Banded microstructure "T" (TD), "L" (RD or LD) and "S" (ND).

Magnesium Alloy AZ31B-H24

The Mg plate was procured from Magnesium-Elektron in Madison, IL. It is from the same lot of material used in a prior project at ARL, enabling this effort to build on the past work [4]. The material was produced in accordance with ASTM-B90-98. The certification provided by Magnesium-Elektron included static properties in the longitudinal (rolling) direction. The Ultimate Strength was reported as 38.0 ksi and the Yield Strength was reported as 24.5 ksi. That corresponds to SI units of 262 MPa and 169 MPa. The elongation was quoted as 9.5%. The texture of the plate was measured using XRD and has typical texture where the c-axis is strongly aligned with the plate normal direction.

ASTM quasi-static tensile samples were prepared by traditional machining dog bone specimens in three orthogonal directions, those being normal, transverse and along the rolling direction of the plate. The prepared samples were tested in an Instron, load framed at a rate of 2.0 mm/min, using digital image correlation (VIC 2D) to remove the machine compliance and accurately track the straining of the samples during testing. In addition to the tensile samples, 1-cm blocks of the plate were removed and mounted as to allow imaging via optical microscopy. The samples were prepared such that the microstructure could be viewed along the three orthogonal directions of the plate (normal, transverse and rolling).

Further samples were prepared for high rate mechanical compression studies. Cubic specimens with dimensions of 6.35-mm (0.25") were machined via electrical discharge machining with the RD, TD and ND tracked and marked on each specimen. High strain rate compression experiments were carried out using a conventional Split-Hopkinson

Pressure Bar. The bars are 1841.5-mm (72") long and 12.7-mm (0.5") in diameter and are made out of AA7075-T6.

Microscopy

Optical micrograph, Figure 2 a-c provides the as-rolled microstructures. In general the microstructures are consistent with being deformed, that is, a high density of twins are present in the microstructure. However, the degree to which twins are revealed by the etching process appear to differ. More twins can be observed in rolling and transverse directions as compared to the normal direction. The average grain size in all directions appears to be less than 30 microns, however isolated grain as large as 80- 160 microns were observed in some orientations. The average twin spacing was observed to be less than 4 microns in all orientations.

High Rate Compression

Three experiments were performed in each direction for repeatability and all samples were loaded to failure (except for Tests 14 and 15). Table 1 summarizes the high strain rate experiments.

Table 1. High Strain Rate Experiments

Test #	Orientation Direction	Nominal Strain Rate (1/s)
5	Normal	4100
6	Normal	2800
7	Normal	4000
8	Normal	3200
9	Normal	3000
10	Normal	4000
14*	Normal	1000
15*	Normal	1300
17	Rolling	3500
18	Rolling	3500
19	Rolling	3500
20	Transverse	3500
21	Transverse	3600
22	Transverse	3500

*Sample didn't fail during test.

Taking the Normal direction as a starting point, Figure 3 shows the results of the SHPB experiments. There are a couple of things to note. First, the lower strain rates produce lower peak stresses. Second, the specimens failed at lower strains for the tests at lower strain rates. Such a behavior indicates that for larger, slower projectile impacts, the Mg will look weaker and will fail earlier than when impacted by smaller, faster projectiles. The shape of the stress-strain curve is indicative of slip-dominated deformation of the material [5].

Figure 4 shows the SHPB results for the RD specimens. For the strain rate of approximately 3500 s^{-1} a clear step can be seen at approximately 100 MPa. After about 5% strain, the peak stress climbs to approximately 450 MPA as compared to slightly less than 400 MPa for the normal direction. This step is indicative of tension twinning which accommodates the tensile strains along the c-axis aligned with the ND [5].

The behavior in the TD mirrors that found in the RD because both loading directions are perpendicular to the strong c-axis texture aligned with the ND, Figure 5. The deformation mechanism is via the same tension twinning. The ductility in this loading direction is slightly less than the RD.

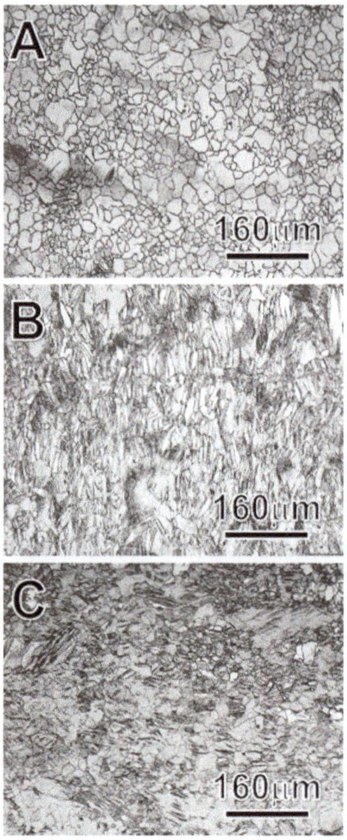

Figure 2. Optical micrographs of (A) Normal, (B) Rolling and (C) Transverse directions.

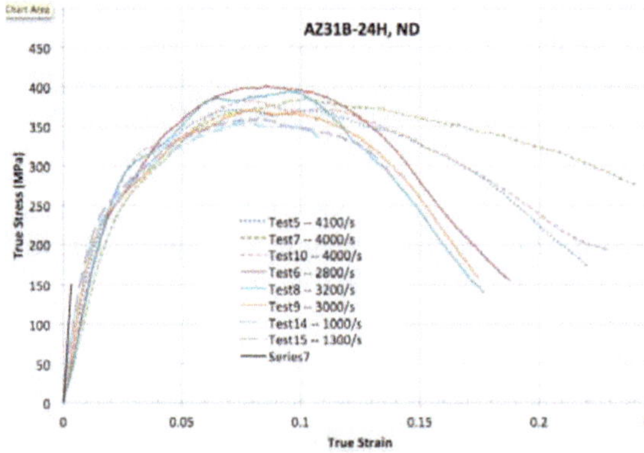

Figure 3. Dynamic compression of AZ31B-H24 in the ND.

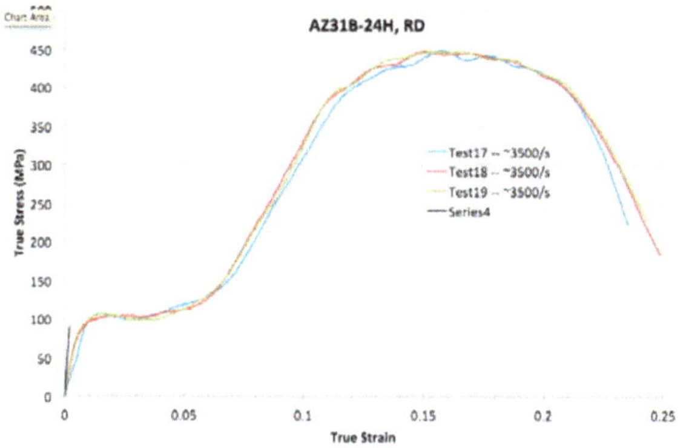

Figure 4. Dynamic compression of AZ31B-H24 in the RD.

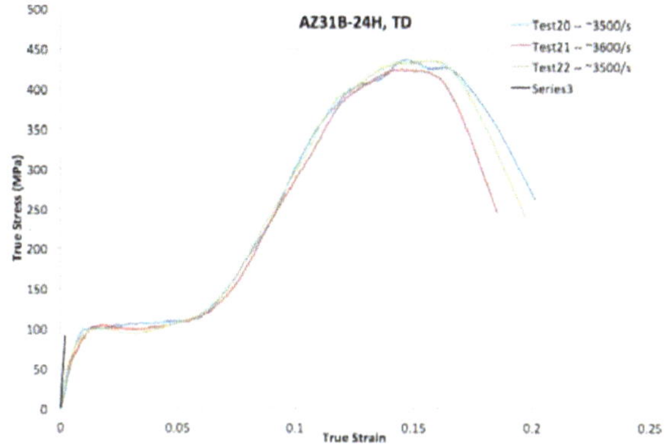

Figure 5. Dynamic compression of AZ31B-H24 in the TD.

Figure 6 shows the results for all three directions. It is obvious that the rolling and transverse directions behave similarly. They are significantly different than the normal direction. Assuming that Mg used in an armor configuration is aligned such that a normal impact on the armor is entering the Mg normal to the rolling direction, this behavior implies that performance differences may exist as a function of the projectile size, impact velocity, and impact obliquity.

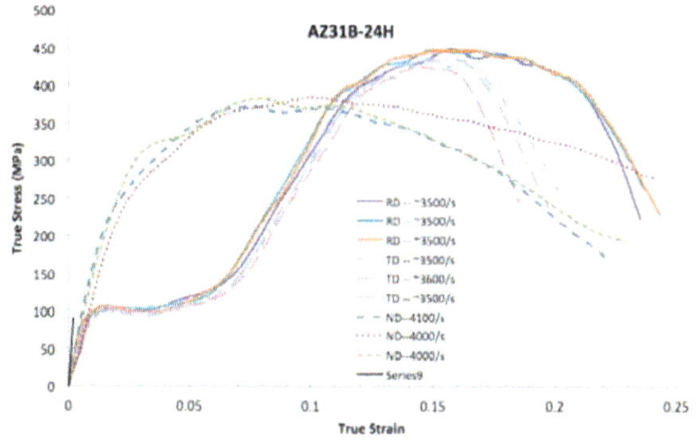

Figure 6. Dynamic compression in the three directions.

Quasi-static Tensile Analysis

Figure 7 shows the results for all three directions. The first noticeable difference is in the relative activities between slip and twinning. That is, under tension, only samples pulled along the normal direction showed that twinning was activated, while the other two directions (i.e. rolling and transverse) displayed deformation mainly controlled by dislocation slip. When compared to compression data the exact opposite relationship with orientation and operative deformation mechanisms is observed. These relationships between tension and compression have been reported before in the work of H. Wang et al. [6] and SR Agnew et al. [7]. Additionally, while it is expected that some asymmetry will exist between compression and tension data for AZ31B [8], a significant increase in the flow stress directly related to the loading rate can be observed in samples tested dynamically, as compared to samples tested quasi-statically.

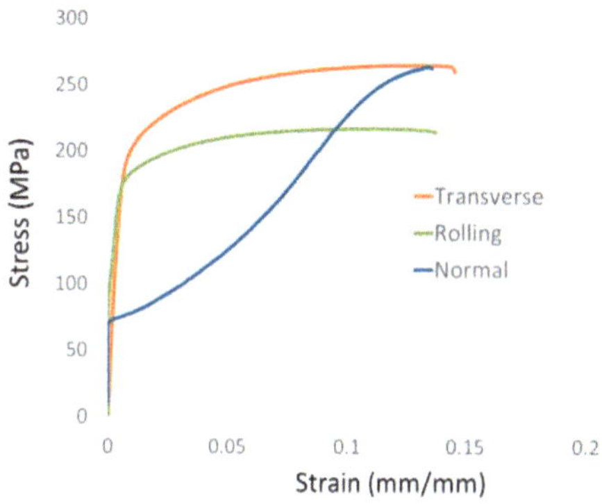

Figure 7. Quasi-Static Tension Data of AZ31B-H24 in the three directions.

Table 2 provides a summary of the average quasi-static tension properties measured for the three orientations. It is obvious from Figure 7 that the rolling and transverse directions behave similarly and that their response is dramatically different than that of the normal direction. Upon comparison, samples tested in the transverse direction have slightly higher strength properties i.e. (yield and ultimate tensile strengths) than those obtained in the rolling direction. This is also consistent with the work of SR Agnew et al. [7] for this particular alloy. Such a response (transverse/rolling) is indicative of dislocation mediated deformation, where once the strain hardening capacity is exhausted necking and ultimately failure ensues. In comparison, the normal direction initially undergoes twinning at low strains. As the capacity to further orient the grains in the normal direction reduces, dislocation slip becomes progressively more dominant, ultimately controlling the deformation response at higher strain values.

Table 2. Quasi-static Tension Properties

AZ31B	Yield Stress (MPa)	Ultimate Stress (MPa)
Transverse	175	263
Rolling	165	223
Normal	67	255

*Rate of 2.0 mm/min

Ballistic Experimental Results

The specific ballistic threats used to test the magnesium alloy plate samples were the 7.62-mm APM2 armor-piercing projectile [9], shown in Figure 8. The projectile was fired at each face of the Mg AZ31B-H24 targets. The size of the targets were 152.4-mm x 152.4-mm x 76.2-mm thick.

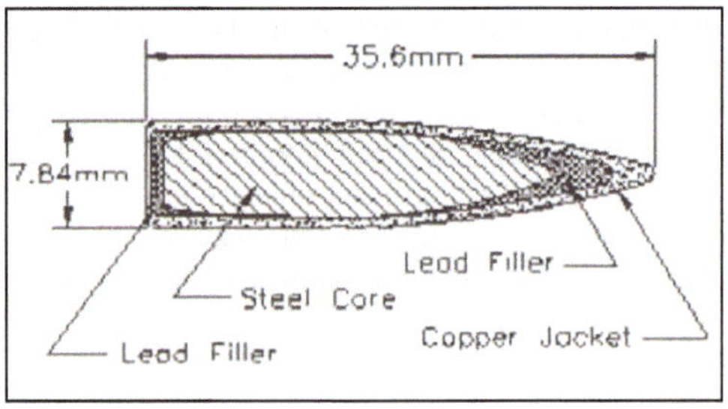

Figure 8. Sketch of 7.62-mm APM2 projectile.

Target impact with a total yaw less than ½ degree was desired. The pitch and yaw for each shot were determined by using orthogonal flash x-rays. The total yaw values obtained were typically between 1 and 2 degrees. The potential degradation in penetration as a function of yaw has been previously discussed by Riegel [10] and by Yaziv, Rozenberg, and Riegel [11]. In short, the critical yaw angle is a function of the penetration crater diameter relative to the projectile diameter. For rigid penetration, the ratio of the crater diameter to the projectile diameter goes to 1. That translates to a critical total yaw of 0 degrees. In practice, the amount of degradation depends on the material properties of the target, the shape of the projectile nose, and other factors, making it impossible to assume a simple relationship between penetration obtained with an unyawed projectile and that of a yawed projectile. Figure 9 illustrates the setup of the ballistic experiments.

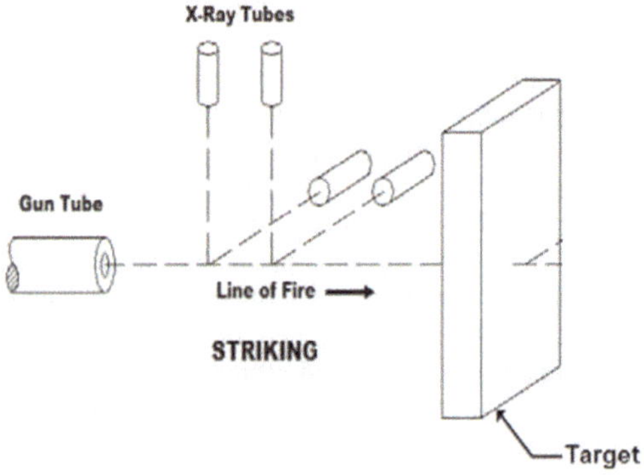

Figure 9. Setup of the ballistic experiments.

The distance from the muzzle of the gun barrel for the 7.62 APM2 projectile and the target face was 4.52-m (15').
The primary velocity measurement tool for the 7.62 APM2 projectile was two sets of orthogonal 150kV flash x-rays at the strike face and residual face of the target. The spacing of the striking face x-rays were 254-mm (10") from center-to-center. The spacing from the center of the closest striking x-ray and the target face was 381-mm (15"). The flash x-rays capture orthogonal images of the projectile at two times. The images are spatially calibrated to enable the distance traveled during the time interval between x-ray flashes to be determined. Analysis of the x-rays images includes the projectile orientation relative to the shot line in addition to locating the center of mass of the projectile in each image. This data permits the determination of the projectile velocity as well as the projectile pitch and yaw. The ballistic launch view is shown in Figure 10. The experimental target fixture is shown in Figure 11.

Figure 10. Ballistic Launch View

Three Oehler Model 57 infrared screens near the muzzle of the gun barrel were used as a backup system for the velocity measurements. These instruments optically detect the passage of a projectile across a plane that is normal to the shot line. The separation between the light planes, couple with the time that the light is interrupted, permits the projectile velocity to be computed. These devices do not provide a method for determining pitch or yaw and they are interrupted by the leading edge of the projectile. Therefore, the computed velocity of a severely tumbling projectile would be subject to errors associated with the distance between the leading edge and the center of mass. In these tests, all pitch and yaw values were less than 3 degrees. Therefore, there was negligible difference between the times computed using the light screens and those computed using the x-rays.

Figure 11. Experimental Target Fixture

Semi-infinite penetration occurs when the projectile is fired into a target where the thickness is such that the rear surface is not deformed. The target is sectioned and the penetration normal to the original impact face is measured. For each shot, the impact velocity, projectile pitch, projectile yaw, and depth of penetration in to the AZ31B-H24 targets were recorded. This first set of data was intended to be at a constant velocity of 457m/s to capture the rigid mode of the penetrator. In practice, the impact velocities varied from 440 m/s to 473 m/s. The desired limit for total yaw was 0.5 degrees. The measured total yaw values ranged from 0.93 to 1.85 degrees. Table 3 shows the data for this series of shots. The table also includes the penetration value estimated by WAPEN [11] using an Effective Flow Stress (EFS) of 250 MPa for all three material orientations. While the agreement between WAPEN and the experimental results is quite good, it is unclear that the selected value of Effective Flow Stress is the best value to use.

Table 3. Experimental data and comparison to WAPEN.

Shot ID	Dir	V (m/s)	Pen (mm)	WAPEN (mm)	Yaw
12684	LD	444	21.5	24.5	1.28
12685	LD	458	26.2	26.0	0.93
12686	TD	473	27.7	27.7	1.39
12687	TD	440	23.5	24.0	1.22
12688	ND	466	24.3	26.9	1.85

In Figure 12 the penetration is plotted against the impact velocity. It is interesting to note the difference in slope between the TD direction shots and the LD shots. The highest penetration LD shot impacted at a total yaw of 0.93. The lower penetration LD shot had a yaw of 1.28 and the two TD shots had yaws of 1.28 and 1.22. There are not sufficient data to draw conclusions regarding the role that yaw plays.

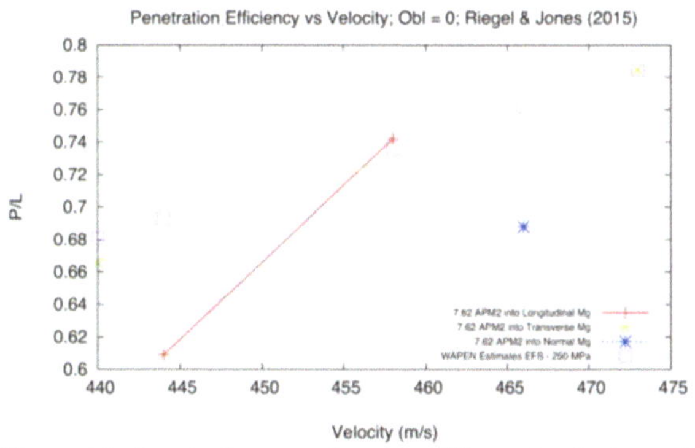

Figure 12. Penetration as a function of impact velocity and target material orientation.

Figure 13 shows the sectioned target for shot #12688, fired in the ND orientation followed by Figure 14, which shows an outline of the target crater and projectile core profiles. The outlines more clearly indicate that the crater is not symmetric, with a larger crater in the yaw direction. The sectioned target for shot #12685 is shown in Figure 15. That shot was fired in the LD direction and was at the lowest yaw, 0.93 degrees, obtained during the series. Figure 15 illustrates the jacket is being stripped in a more symmetric manner than seen in shot #12688, and shows that the core has rebounded from the point of maximum penetration. The targets were sectioned along the shot line so the depth of penetration could be more clearly visualized.

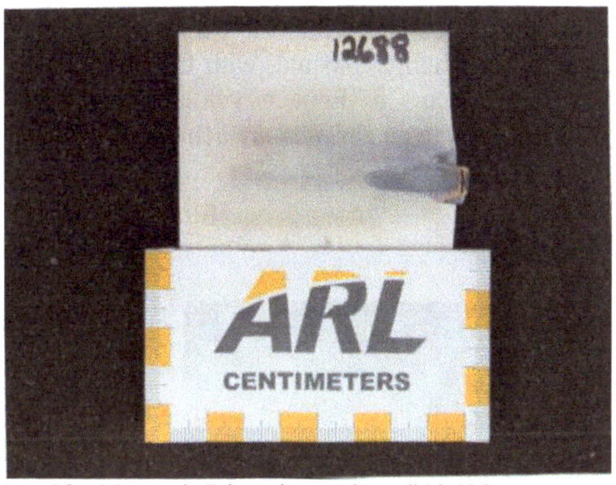

Figure 13. Normal Direction, shot #12688, yaw = 1.85 degrees.

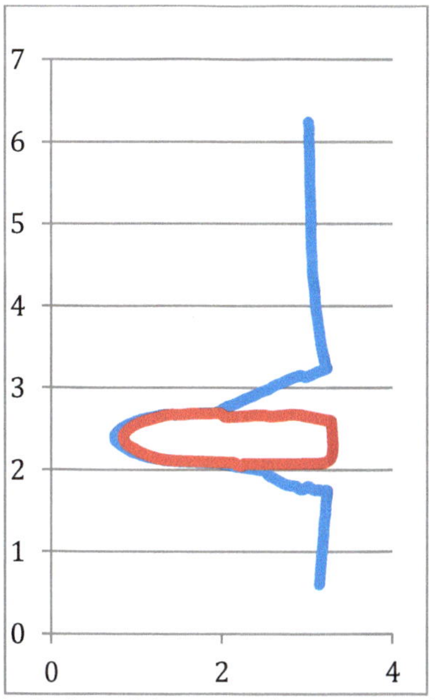

Figure 14: Outline of the core and target crater from shot #12688 with 1.85 degrees of yaw. Note the non-symmetric crater.

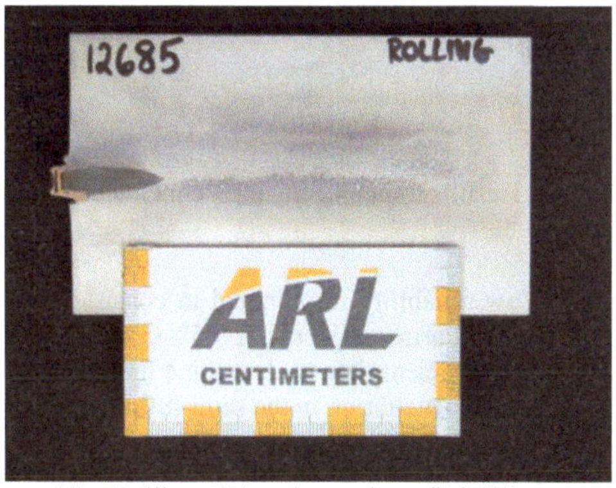

Figure 15. Rolling Direction, shot #12685, yaw 0.93 degrees.

Figure 16 is an etched optical image, indicating the depth of plastic deformation or flow. It can be seen that the region is relatively constant extending beyond the projectile. As future work proceeds, similar images will be created for different impact velocities and target material orientation. It is expected that the extent of the flow field will remain reasonably constant, with the possibility that a change may occur if impact velocities sufficient to transition to eroding penetration are achieved.

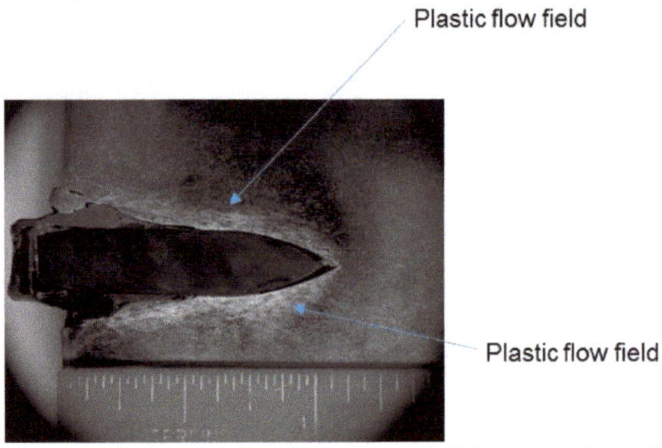

Figure 16. Penetration into Normal Direction channel showing jacket stripping and core rebound.

Figure 17 is the view of the penetration produced when the Mg was shot in the transverse direction.

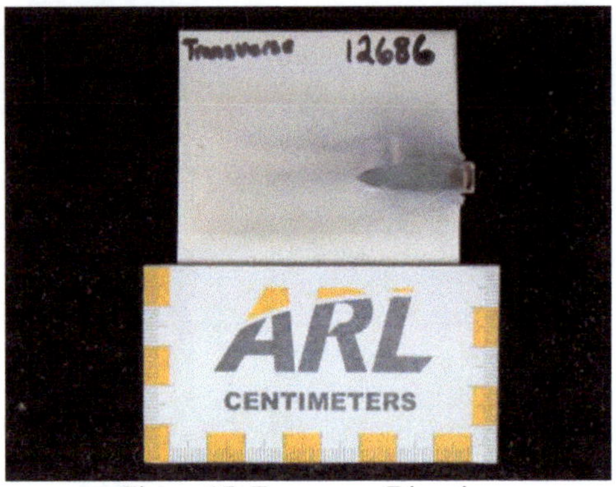

Figure 17. Transverse Direction

Figure 17 illustrates the jacket is being stripped in a more symmetric manner than seen in shot #12688, and shows that the core has rebounded from the point of maximum penetration.

Conclusions

From our results, the following conclusions can be deduced:

1. Yaw reliability was limited in consistently firing a small arms projectile at 0.5° or less. There was not sufficient data to determine the effect of yaw on the penetration in the semi-infinite blocks.
2. The shots fired here were intended to be replicated at the same velocity to establish the repeatability of the shots and determine the difference in penetration as a function of the material orientation. There was slightly more scatter than desired in the impact velocities, but the preliminary results do indicate different penetration. Future shots will be needed to determine if it is reasonable to approximate the target resistance in each direction with a single value of EFS or if values must be defined for each direction.
3. Impacts at higher velocities are needed to capture the spectrum of failure from the projectile into the semi-infinite Mg plate. The higher velocity regimes will also provide data regarding how yaw effects penetration at those velocities.
4. Using an EFS of 250 MPa in the WAPEN model works well for this limited set of data. The value of 250 MPa seems reasonable based on the quasi-static values of yield and ultimate, but appears low based on the SHPB data. Tests over a wider velocity regime will be required.

Improvements to the experimental methodology or the use of a gas gun should be implemented to control the variability of the yaw. Capturing the plastic flow fields in the microstructural analysis at the penetration zone will provide the extent of the material damage and will be the basis of future work.

References

1. T.L. Jones, R.D. DeLorme, M.S. Burkins, and W.A. Gooch, Ballistic Evaluation of Magnesium Alloy AZ31B, ARL-TR-4077, U.S. Army Research Laboratory: Aberdeen Proving Ground, MD, April 2007.
2. B. Powers, A. Dwivedi, T. Walter, C. Williams, C. Meredith, J. Ligda, B. Schuster, and D. Casem, AMX602 – Fundamental penetration Experiments & Basic Mechanical Characterization, Proceedings of AMX602 Interest Group Workshop, Aberdeen Proving Ground, MD, October 2014.
3.http://cml.postech.ac.kr/2008/Steel_Microstructure/SM2.html.
4. T.L. Jones, Development of a Ballistic Specification for Magnesium Alloy AZ31B, ARL-TR-4664, U.S. Army Research Laboratory: Aberdeen Proving Ground, MD, December 2008.
5. A. Jain, and S.R. Agnew, Mat. Sci. Eng. A, Vol 462 (1-2) 2007, pg. 29-36.
6. H. Wang, B. Raeisinia, P.D. Wu, S.R. Agnew, and C.N. Tomé, Int. J. Solids and Struct. Vol 47 (21) 2010, pg. 2905-2917.
7. S.R. Agnew, and O. Duygulu, Int. J. Plasticity 2005 Vol 21 (6) 2005, pg. 1161-1193.
8. M. Al-Maharbi, I. Karaman, I.J. Beyerlein, D.C. Foley, K.T. Hartwig, L.J. Kecskes, and SN Mathaudhu, MSEA Vol 528 2011, pg. 7616-7627.
9. T.L. Jones, et al, Development of a Ballistic Specification for Magnesium Alloy AZ31B, ARL-TR-4664, US Army Research Laboratory: Aberdeen Proving Ground, MD, December 2008.
10. J.P. Riegel III, "The Analysis and Reporting of V50 – Should we do Better?", Proceedings of the 27th International Symposium on Ballistics, April 2013.
11. D. Yaziv, Z. Rosenbery, and J.P. Riegel III, "Penetration Capability of Yawed Long Rod Penetrators", Proceedings of the 12th International Symposium on Ballistics, Oct 29 – Nov 1, San Antonio, TX, 1990.

Magnesium Technology 2016
Edited by: Alok Singh, Kiran Solanki, Michele V. Manuel, and Neale R. Neelameggham
TMS (The Minerals, Metals & Materials Society), 2016

Deformation behavior of Mg single crystals compressed along c-axis

Kelvin Y. Xie[1], Zafir Alam[1], Alexander Caffee[1], Kevin J. Hemker[1]
[1]Department of Mechanical Engineering, Johns Hopkins University, Baltimore, MD, 21218, USA

Keywords: Magnesium; single crystal; plastic deformation; pyramidal slip

Abstract

Pyramidal I <c+a> slip ({10-11}<11-23>) was observed to be the dominant slip mode in magnesium single crystals compressed quasi-statically along c-axis at room temperature. Transmission electron microscopy (TEM) observations revealed the <c+a> dislocation character and structure: some <c+a> dislocations are long, straight and aligned with the basal plane, while others are curved and are not contained within the basal plane.

Introduction

Magnesium (Mg) and its alloys have attracted tremendous interest in both the material science research community and industry because of their low density and high specific strength. However, the limited formability and overall deformation behavior of Mg is complicated due to the low crystal symmetry of its hexagonal close-packed (HCP) crystal structure, and a clear understanding of the active deformation mechanisms for Mg is still being developed. Multiple dislocation-based slip systems (basal, prismatic and pyramidal slip) and twining modes (extension and contraction twins) are possible and the operative mechanisms have been shown to depend on the loading conditions of the material. Among basal (0 0 0 1), prismatic {0 -1 1 0} and pyramidal ({1 0 -1 1} and {1 1 -2 2}) slip systems, basal <a> slip is by far the easiest to activate but pyramidal <c+a> slip is of particular importance because it provides the additional independent slip systems needed for general plasticity in polycrystalline materials. Compressing single crystals provides a way of suppressing basal slip and studying the non-basal slip systems that include <c> component glide. It has been reported that plastic deformation of Mg crystals compressed along the c-axis is mainly carried by <c+a> pyramidal slip [1-5]. A few contraction twins were also reported, but these twins were only observed occasionally and it was determined that they had little or no overall contribution to deformation [4, 6]. The importance of pyramidal slip has also been illustrated in samples initially oriented away from c-axis compression, but in which extension twins can form and consume the matrix during deformation. This reorients the grains by 86° so that the fully twinned grains are then subjected to c-axis compression and undergo pyramidal slip [7].

To date, experimental investigations of pyramidal <c+a> dislocations in Mg have been quite limited. A few investigators have studied polycrystalline Mg with micro-pillar compression and attributed c-axis deformation to the activation of pyramidal II glide, but closer inspection of these studies indicate that many investigators simply referenced the 1974 work of Obara et al. [1] without conducting additional analyses, see for example [2, 4, 5]. The conclusion for pyramidal II glides was recently called into question [8, 9] and the issue as to whether <c+a> dislocations preferentially glide on {10-11} pyramidal I or on {1-211} pyramidal II planes remains and open question. A second and related question involves the character and overall structure of <c+a> dislocations. Answers to these questions are needed to moderate the dominance of basal glide and to guide and provide benchmarks for molecular dynamics (MD) simulations that have predicted various stable dislocation core configurations for <c+a> edge and screw dislocations and estimated the relative mobility of these dislocations on pyramidal I and II glide planes [6, 10, 11]. In this manuscript, we will address these two scientific questions experimentally with slip trace analysis and TEM observations.

Materials and Methods

Compression samples with dimensions of 6 mm x 6.5 mm x 14 mm were fabricated from a 99.999% pure bulk [0 0 0 1] Mg single crystal (Metals Crystals and Oxides Ltd, UK). Laue diffraction confirmed that the misorientation of the single crystal is within 1° along the compression direction. The bulk material was oriented and electrical discharge machined such that the compression samples have perpendicular surfaces parallel to the (0 -1 1 0) and (2 -1 -1 0) planes. The EDM machining parameters were adjusted to low power, low water pressure and low feed rate in order to minimize prior deformation. The cut surfaces were then chemically polished using 10% nitric acid in water solution to remove the EDM damaged layer. Quasi-static compression tests were carried out on a stroke-controlled MTS machine at a strain rate of 10^{-4} s^{-1}. Compressing the single crystals along the c-axis suppresses both basal and prismatic slip, allowing the pyramidal slip to be isolated and investigated in detail. Compressed samples were immediately imaged using a confocal microscope (Keyence 3D laser scanning microscope) and the confocal images where used to perform detailed slip trace analyses. TEM specimens were obtained by slicing the deformed samples using a diamond wire saw. The slices were then chemically polished using 10% nitric acid in water solution to remove the surface damage layer and to reduce the foil thickness to ~150 μm. To create electron-transparent regions, the foils were electropolished in 10% nitric acid in methanol solution at -40°C. To avoid introducing new dislocations in the TEM foils, no mechanical polishing was involved in the specimen preparation. TEM observations were carried out using a Tecnai12 TEM operating at 100 kV.

Results and Discussions

Slip trace Analysis. A total of 4 samples were compressed in this study. A typical engineering stress and strain curve is plotted in Figure 1. The shape of the curve and the stress levels are similar to the observations reported by Obara *et al.* and Syed *et al.*: no apparent yield point, high flow stress up to 300 MPa before failure and much higher strain hardening than pure FCC and BCC single crystals [1, 4, 12]. The yield strength at 0.2% strain offset is estimated to be ~135 MPa, similar to that reported by Syed *et al.* [4].

All 4 compressed samples developed slip traces. After 3% deformation a high density of inclined and a few horizontal slip traces (parallel to the basal planes) were observed. Most inclined slip traces on the (0 -1 1 0) surface were found to reside at an angle of ~55° to the horizontal or basal plane (an example is provided in Fig. 2a). The calculated slip trace angles projecting on the (0 -1 1 0) surface are 58° for pyramidal I slip and 39° and

0° for pyramidal II slip. The experimental measurement is close to the calculated pyramidal I slip and quite different from the pyramidal II slip, indicating prevalent pyramidal I slip. In addition, we also examined the slip traces on the (2 -1 -1 0) surface (Fig. 2b). Most of the slip trace angles were observe to range from 50° to 60°. The calculated angles for pyramidal I slip trace on this surface are 43° and 62°. MD simulations suggested the screw component for <c+a> dislocations are much more mobile than the edge component [11]. Screw component can cross slip from one pyramidal I plane to another pyramidal I plane. Depending on the likelihood and frequency of cross slip, the slip trace angle can range from 43° and 62°. The horizontal slip traces in Figure 2 should be interpreted with care as some basal slips likely took place during the deformation. Although the compression samples were aligned in a way that basal and prismatic slip are restrained, the CRSS of basal slip is extremely low (~0.5 MPa) [13, 14] and slight misalignment or vibration during the compression tests could activate some basal slip.

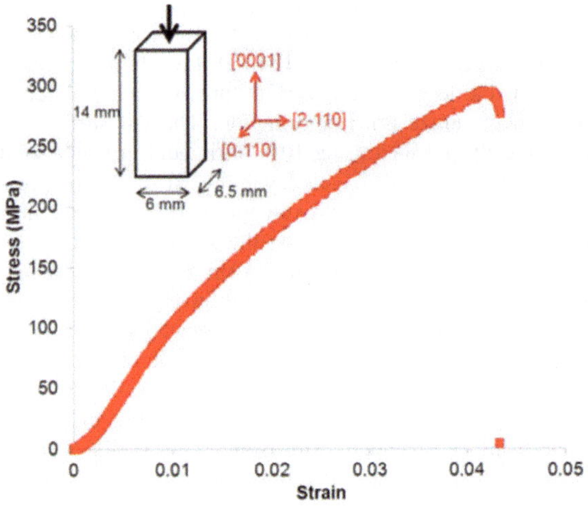

Figure 1: A typical engineering stress-strain curve Mg single crystal compressed along c-axis at room temperature.

We note that two pyramidal II planes do intersect the (0 -1 1 0) front surface at an angle of 58° from the basal plane, but that these (-2 1 1 2)<- 2 1 1 -3> and (2 -1- 1 2)<2 -1 -1 -3> slip systems have no out-of-plane component on the (0 -1 1 0) surface. The Burgers vectors for these systems lie in (0 -1 1 0) plane, and no slip traces should be observed on the (0 -1 1 0) surface for these two slip systems. Moreover, if pyramidal II slip was active other pyramidal II slip systems would have resulted in prevalent 39° slip traces on the (0 -1 1 0) surface, which was not observed. These observations all point to and confirm the conclusion that pyramidal I slip is favored over pyramidal II slip in c-axis compressed Mg single crystals.

The prevalence of pyramidal I slip can be attributed to its relatively lower CRSS compared to that of pyramidal II slip. This proposition is supported by recent molecular dynamics (MD) simulations showing that <c+a> dislocation glide on the {1 0 -1 1} pyramidal I planes is favored due to low Peierls stresses [8, 11, 15]. MD simulations by Tang *et al.* suggest that <c+a> dislocations are nucleated and glide on pyramidal I planes and that these dislocations may be able to cross slip to pyramidal II planes but only at very high stresses (stresses realized in some MD simulations but likely not achievable in experiments). [11].

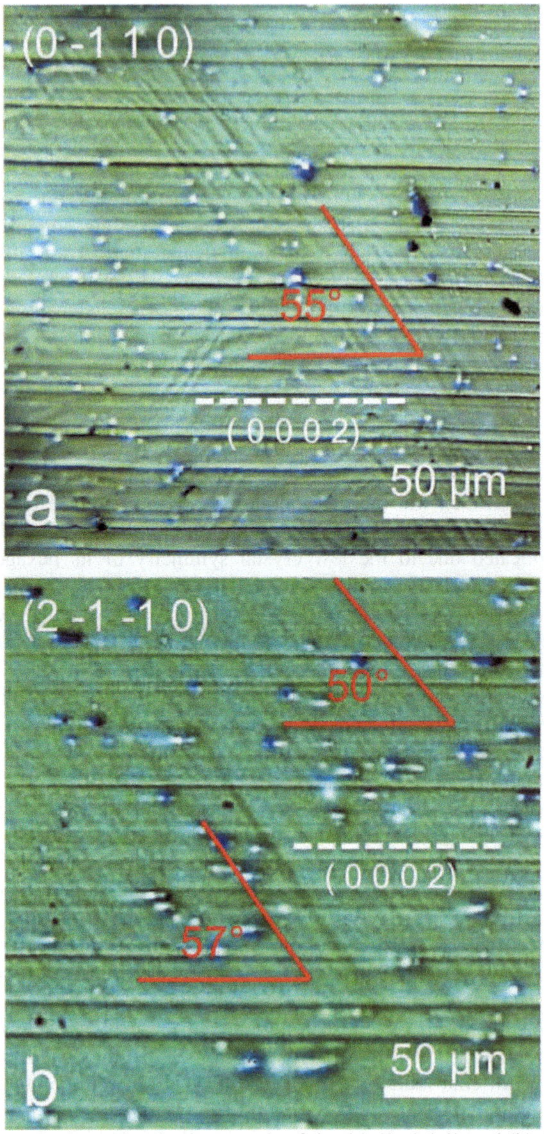

Figure 2: A confocal micrograph of the (a) (0 -1 1 0) and (b) (2 -1 -1 0) faces of a 3% deformed Mg single crystal. The contrast from pyramidal slip traces is very weak and the images were purposely colored to enhance the contrast.

TEM Analysis. In order to reveal the shape and character of the dislocations, we prepare TEM thin foils parallel to the (0 -1 1 0) plane and employed double-tilt experiments and weak beam dark field (WBDF) techniques to image the <a> and <c> components of the Burgers vector. Two-beam diffraction conditions with diffraction vectors <$\mathbf{g}_{hkl}$> that caused the <a> or <c> components to go invisible, e.g. <a> dislocations are invisible for $\mathbf{g}_{hkl}$= [0 0 0 2] and <c> dislocations were invisible for $\mathbf{g}_{hkl}$= [1 1 -2 0]. If a dislocation shows up for both diffraction conditions, it is then a <c+a> dislocation. A typical example of dislocation structures in the c-axis compressed Mg single crystals is shown in Figure 3. The nanoscale contamination (marked by "*") was used as a reference for the tilting experiment so that the dislocations in Figures 3a and 3b are the same and can be directly correlated.

Many dislocations were observed to show up in both <a> and <c> conditions, indicating these are indeed <c+a> dislocations. Some of these <c+a> dislocations appear long and straight, and lie parallel to the basal plane (highlighted by the dashed boxes). Others are curved and do not align with the basal plane (highlighted by the solid line boxes). In addition to linear dislocations, numerous dislocation loops (speckles in Fig. 3) were also observed in the deformed samples. The origin of these loops is currently unknown, but they were not observed in undeformed foils and are related to the deformation of the crystal. These loops may have nucleated during the deformation, or what is more likely is that they may have been left behind by pencil-like glide of the <c+a> dislocations that involves cross-slip and local pinching off of glissile dislocations. Whatever the source of these loops, they are believed to impede the movement of mobile <c+a> dislocations and may contribute to the high strain hardening rate observed in Figure 1 [16].

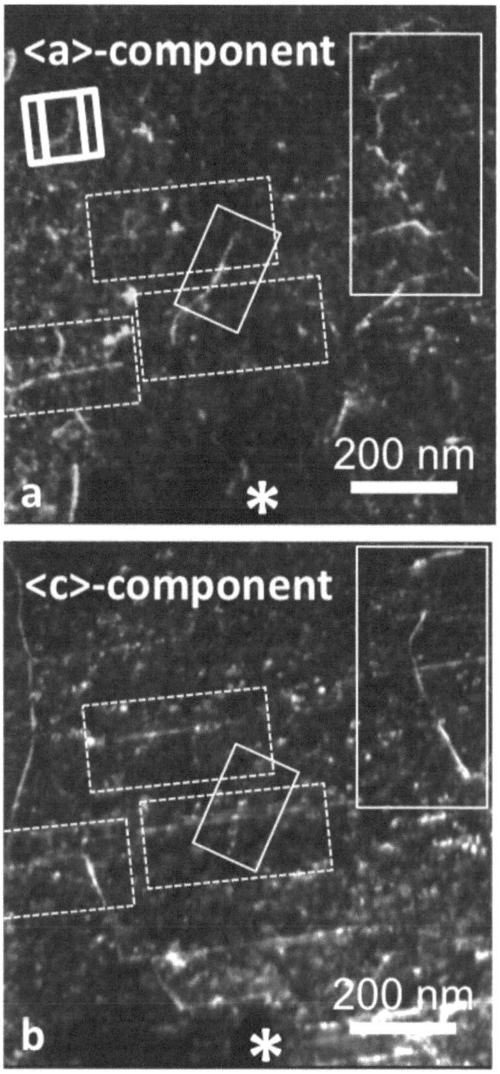

Figure 3: WBDF images showing (a) <a> component and <c> component of dislocations in the same area of the TEM foil. <c+a> dislocations are highlighted by boxes.

Summary and Conclusions

In summary, we have gathered experimental evidence to demonstrate that pyramidal I <c+a> slip is the dominant slip mode that accommodates c-axis deformation in Mg. The structure and character of <c+a> dislocations was observed to be complicated: some long, straight, dislocations were aligned parallel to the basal plane, while others were curved and lying outside the basal plane. The observation of loops, possibly formed by pencil glide, may be used to explain the high strain hardening that was recorded in the compression experiments.

Acknowledgement

This research was sponsored by the Army Research Laboratory and was accomplished under Cooperative Agreement Number W911NF-12-2-0022. The views and conclusions contained in this document are those of the authors and should not be interpreted as representing the official policies, either expressed or implied, of the Army Research Laboratory or the U.S. Government. The U.S. Government is authorized to reproduce and distribute reprints for Government purposes notwithstanding any copyright notation herein.

References

[1] T. Obara, H. Yoshinga, and S. Morozumi, Acta Metallurgica **21**, 845 (1973).
[2] C. M. Byer, B. Li, B. Cao, and K. Ramesh, Scripta Materialia **62**, 536 (2010).
[3] C. M. Byer and K. Ramesh, Acta Materialia **61**, 3808 (2013).
[4] B. Syed, J. Geng, R. Mishra, and K. Kumar, Scripta Materialia **67**, 700 (2012).
[5] E. Lilleodden, Scripta Materialia **62**, 532 (2010).
[6] S. Sandlöbes, S. Zaefferer, I. Schestakow, S. Yi, and R. Gonzalez-Martinez, Acta Mater. **59**, 429 (2011).
[7] N. Dixit, K. Y. Xie, K. J. Hemker, and K. Ramesh, Acta Materialia **87**, 56 (2015).
[8] H. Fan and J. A. El-Awady, Materials Science and Engineering: A **644**, 318 (2015).
[9] K. Y. Xie, Z. Alam, A. Caffee, and K. J. Hemker, Scripta Materialia **(accepted)** (2015).
[10] T. Nogaret, W. A. Curtin, J. A. Yasi, L. G. Hector, and D. R. Trinkle, Acta Mater. **58**, 4332 (2010).
[11] Y. Tang and J. A. El-Awady, Acta Mater. **71**, 319 (2014).
[12] T. J. Balk, K. J. Hemker, and L. P. Kubin, Scripta Materialia **56**, 389 (2007).
[13] H. Yoshinaga and R. Horiuchi, Trans. JIM **4**, 1 (1963).
[14] H. Yoshinaga and R. Horiuchï, Transactions of the Japan Institute of Metals **5**, 14 (1963).
[15] T. Nogaret, W. Curtin, J. Yasi, L. Hector, and D. Trinkle, Acta Materialia **58**, 4332 (2010).
[16] J. Geng, M. F. Chisholm, R. Mishra, and K. Kumar, Philosophical Magazine Letters **94**, 377 (2014).

THE USE OF ACOUSTIC EMISSION AND NEUTRON DIFFRACTION TO REVEAL THE ACTIVE DEFORMATION MECHANISMS IN POLYCRYSTALLINE MAGNESIUM AND COMPARISON TO THEORETICAL MODELING

Jan Čapek[1], Kristián Máthis[1], Tomáš Krajňák[1]
[1]Charles University in Prague, Department of Physics of Materials; Ke Karlovu 5; Prague 2, 121 16, Czech Republic

Keywords: Magnesium, Nondestructive testing, Acoustic emission, Neutron diffraction

Abstract

A detailed analysis of the loading mode dependence of the deformation mechanisms in randomly textured cast magnesium is presented. An elasto-plastic self-consistent model (EPSC) is used to model the evolution of the dislocation structure and twinning. The results are quantitatively compared with experimental data obtained by in-situ neutron diffraction (ND) and acoustic emission (AE) methods. Both EPSC calculations and ND line profile analysis show an increased activity of prismatic slip with increasing strain and a loading mode dependence of the activity of the second-order pyramidal slip. The AE measurements and the modeling indicate different amount of nucleated twin variants and twinned volume in tension and compression, respectively.

Introduction

The application of magnesium alloys as lightweight structural elements in the transportation industry has been significantly increased in the last two decades. However, there are still many issues, as e.g. limited room temperature formability or asymmetric response on the strain path changing, which hinder the wider usage of magnesium alloys. The specific deformation behavior of the hexagonal closed packed (hcp) structure of magnesium is usually in the background of these problems[1].

The $(0001)\langle 11\bar{2}0\rangle$ basal slip is the most easily activated system at room temperature, followed by $\{10\bar{1}0\}\langle 11\bar{2}0\rangle$ prismatic and first-order $\{10\bar{1}1\}\langle 11\bar{2}0\rangle$ pyramidal slip systems. All of these systems provide deformation within the basal plane and offer not more than four independent slip systems. Thus, the von Mises criterion of five independent slip systems for homogeneous deformation is not fulfilled and a deformation along the c-axis is not possible. The slip in the second-order ($\{11\bar{2}2\}\langle 11\bar{2}3\rangle$) pyramidal system would satisfy the von Mises criterion. However, the activation of the $< c + a >$ dislocations is rather difficult owing to the high critical resolved shear stress (CRSS)[2, 3].

Thus, deformation twinning is a preferred mechanism that accommodates deformation out of the basal plane. In magnesium, the $\{10\bar{1}2\}$-type extension twinning, associated with extension along the c-axis and reorientation of the lattice by 86.3° and $\{10\bar{1}1\}$-type compression twinning, resulting in contraction along the c-axis and tilting of the lattice by 56° are the most often reported and studied twinning mechanisms[4].

The activation of above listed mechanisms is influenced by many microstructural features (initial texture, alloying content, grain size)[5, 6] and the experimental condition (testing temperature, strain rate, strain path)[6, 7]. The key role of basal slip on magnesium plasticity is well known. Nevertheless, the activation and importance of other slip systems is not fully understood as well as the influence of twinning on plastic flow [8, 9].

Some theoretical models have been developed for studies of deformation of HCP structured alloys. The atomistic simulation can describe the fundamental mechanisms; other models use finite elements method, or self-consistent approach[8, 9]. Despite the theoretical models have been improved a lot and provide new insight to this topic, the experimental proof of theoretical results is important. Besides the *ex-situ* experimental techniques (optical light, electron scanning and transmission microscopy) which are limited to relatively small volume, the *in-situ* testing (neutron diffraction, acoustic emission) became popular in the last several years[10-12]. The main advantage of the *in-situ* approach is that it provides statistically relevant information from entire volume of the specimen.

Experimental procedures

Pure Mg, binary Mg–2 wt.% Al and Mg–9 wt.% Al (further referred as Mg2Al and Mg9Al) were used for the experiments. The neutron diffraction measurements did not reveal any preferential orientation of specimens[13]. The testing was carried out using cylindrical specimens with a diameter of 9 mm and gauge length of 20 mm. Tensile and compression testing were carried out using a horizontal 250 kN load frame at a strain rate of $10^{-3}s^{-1}$ in strain control mode. In order to collect ND data with good enough statistics, the tests were stopped at predefined strain levels (0.1, 0.5, 1, 2, 3, 4, 5, 6%) for approx. 60 min. In-situ neutron diffraction measurements were carried out at the SMARTS engineering instrument in the Lujan Neutron Scattering Center[14]. The acoustic emission measurement was done separately at the Department of Physics of Material, Charles University in Prague.

The neutron diffraction patterns were evaluated using the Convolutional Multiple Whole Profile (CMWP) fitting (see [15] for details) method. The output of CMWP fitting procedure were the dislocation density (ρ), and the parameters q_1 and q_2 which can be used for the determination of the fractions of dislocations in the different slip system families. A recent statistical method (adaptive sequential *k*-means – ASK) worked out by Pomponi and Vinogradov[16] was successfully applied for the determination of the dominant deformation mechanisms from the AE signal at the various stages of deformation.

Results and discussion

Raw AE signals recorded during tension and compression in the threshold-less regime (waveform streaming) are presented in Fig. 1. Using the ASK analysis we can determine the time evolution of AE source mechanisms (Fig.2) which can be compared with the modeling results. In agreement with our previous work, the ASK analysis shows limited twin nucleation after the yield point in compression while twinning nucleation is present during the entire test in tension[17].

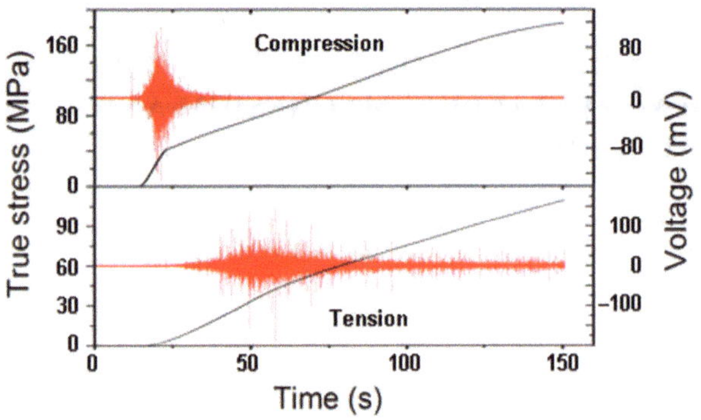

Figure 1. AE waveform stream and the corresponding monotonic stress-strain curves

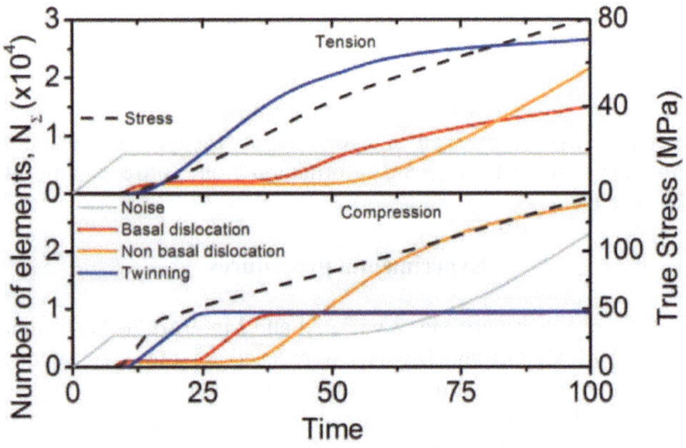

Figure 2. Strain evolution of cumulative number of elements in the AE clusters assigned to noise (grey line), basal dislocations (red line), non-basal dislocations (orange line) and twinning (blue line). The black line represents the experimental stress strain curves, measured concurrently with AE.

The influence of Al content on the mechanical properties together with the AE count rates (i.e. number of crossing of threshold level per second) corresponding to the stress-strain curves are presented in Fig. 3. In agreement with the literature data, the strength of alloys increased with the increasing Al content[18].

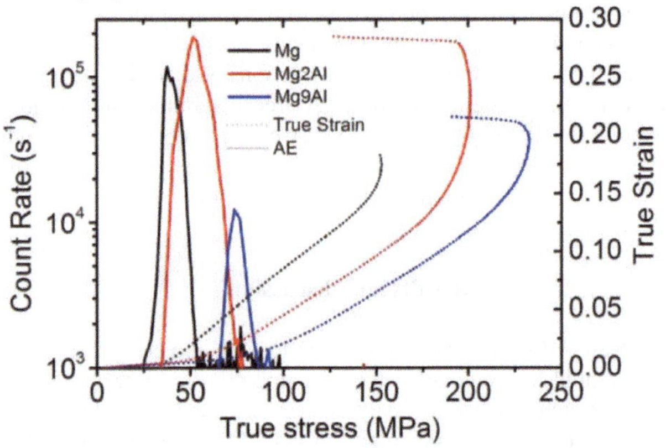

Figure 3. The true stress – true strain curves and AE count rates as measured in compression for various Al concentration.

It is obvious that the Mg9Al specimen has the lowest AE response, whereas the pure Mg and Mg2Al behave similarly. The characteristic peak in the vicinity of the yield point can be attributed to synergic effect of twin nucleation and massive dislocation motion. It can be seen that the peak values of the count rate are shifted towards higher stresses with increasing Al content, which is a clear sign that the twin nucleation stress depends on the alloying content.

From the changes of peaks intensities we can determine the twin volume fraction (TVF)[8]. The change of the integrated intensity of peaks favorably oriented for twinning is directly proportional to twinned volume fraction in these grain families. Figure 4 shows, similar twinned volume of Mg and Mg2Al and lower twinned volume for Mg9Al.

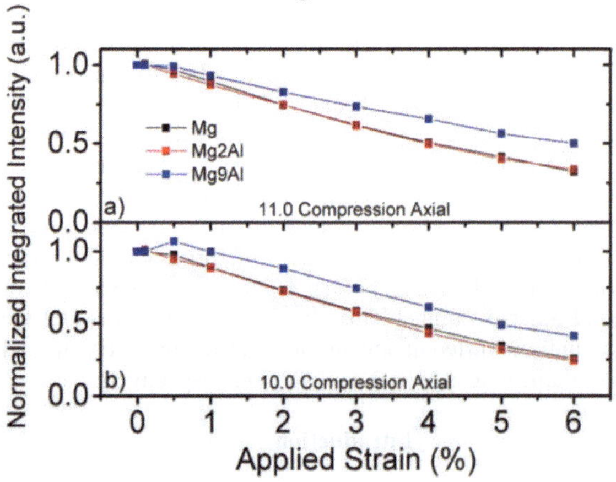

Figure 4. Change of the integrated intensity due to the twinning in grains favorably oriented for twinning a) 11.0, b) 10.0

The evaluation of the overall twin volume fraction can be performed by a 2-bank Rietveld refinement assuming an axisymmetric texture[8]. Figure 5 shows the twin volume fraction as a function of the composition and applied strain. It is obvious that the strain dependence of the TVF for pure Mg and Mg-2%Al is the same within the experimental error, which fits well with the AE data. In contrast, the TVF for Mg-9%Al is smaller, which indicates that there is another significant strain accommodation mechanism besides the twinning. Similar dependence of twinned volume on Al content was observed in the same set of specimens by Nagarajan[19] using micrograph analysis.

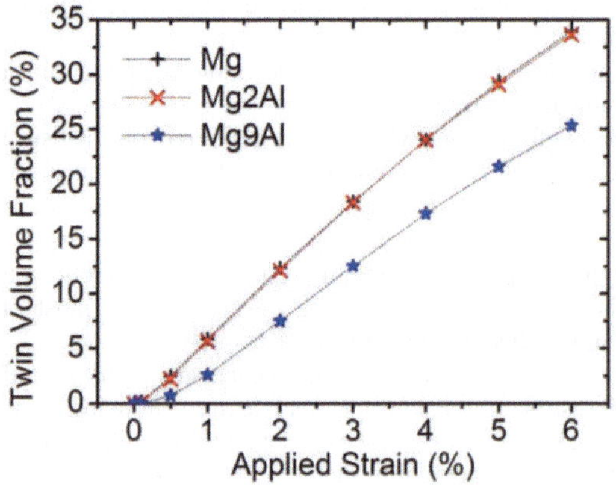

Figure 5. Evolution of twin volume fraction for all examined Al concentrations as a function of strain.

Using CMWP, we were able to determine dislocation densities and types. For Mg9Al alloys at higher strain level the fraction of <c+a> dislocations significantly increases, whereas for Mg2Al and pure Mg the increment is very small Fig.6. Such a large difference can be given by restricted twin growth in Mg9Al, shown in Fig.5. Since the strain cannot be accommodated by twinning, an alternative mechanism is required for continuing the plastic deformation, which is realized in the form of <c+a>-slip.

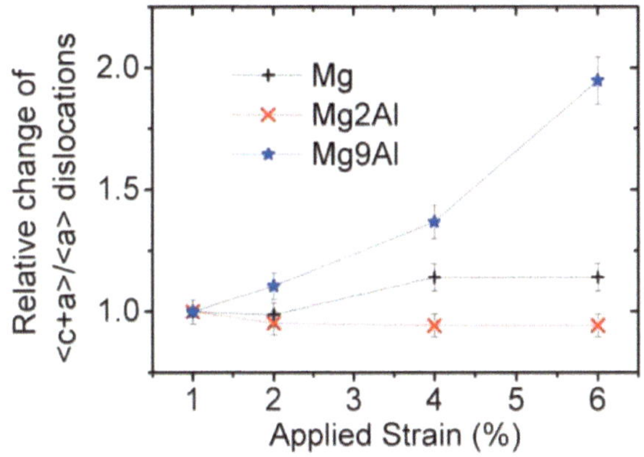

Figure 6. Relative change of the ratio of <c+a> and <a> dislocations as a function of the applied strain and Al concentration.

Conclusions

The loading mode dependence of deformation mechanisms in a pure, randomly oriented magnesium polycrystal and the influence of solute content on the deformation mechanisms were investigated at room temperature using various in-situ experimental (AE, ND). The maximum applied strain was 6%. It can be concluded that the particular results form a complementary dataset, which helps in getting a more detailed description of the processes of the plastic deformation.

Generally, the novel potential of AE and ND experimental methods was revealed:

AE – using a proper analysis of raw AE waveform streaming dataset, high-time resolution information about the dynamics of active deformation processes can be derived. It is shown that the method is capable to make difference not only between dislocation slip and twinning AE events, but also between basal and non-basal slip events.

ND – in combination with CMWP evaluation procedure, quasi in-situ information can be obtained about the fraction of dislocations in various slip systems. The evaluation method is capable for revealing the evolution of density of <c+a>, as well as basal and non-basal <a> dislocations with applied stress, which can be hardly realized with such a good statistics using other experimental methods.

Acknowledgement

The authors are grateful for the financial support of the Czech Science Foundation under the contract 14-36566G. JC acknowledges the support from the Grant Agency of Charles University, Faculty of Mathematics and Physics under the contract Nr. 251715.

References

1. J. Hirsch and T. Al-Samman, "Superior light metals by texture engineering: Optimized aluminum and magnesium alloys for automotive applications". *Acta Materialia*, 2013. **61**(3): p. 818-843.

2. A. Chapuis and J. H. Driver, "Temperature dependency of slip and twinning in plane strain compressed magnesium single crystals". *Acta Materialia*, 2011. **59**(5): p. 1986-1994.

3. M. M. Avedesian and H. Baker, *Magnesium and Magnesium Alloys (ASM Specialty Handbook)*1999: ASM International.

4. J. W. Christian and S. Mahajan, "Deformation Twinning". *Progress in Materials Science*, 1995. **39**(1-2): p. 1-157.

5. P. Dobron, et al., "Grain size effects on deformation twinning in an extruded magnesium alloy tested in compression". *Scripta Materialia*, 2011. **65**(5): p. 424-427.

6. C. H. Caceres and A. H. Blake, *Solute and temperature effects on the strain hardening behaviour of Mg-Zn solid solutions*, in *Materials Structure & Micromechanics of Fracture V*, P. Sandera, Editor 2008. p. 45-50.

7. K. Máthis, et al., "Investigating deformation processes in AM60 magnesium alloy using the acoustic emission technique". *Acta Materialia*, 2006. **54**(20): p. 5361-5366.

8. B. Clausen, et al., "Reorientation and stress relaxation due to twinning: Modeling and experimental characterization for Mg". *Acta Materialia*, 2008. **56**(11): p. 2456-2468.

9. S. R. Agnew, D. W. Brown, and C. N. Tome, "Validating a polycrystal model for the elastoplastic response of magnesium alloy AZ31 using in situ neutron diffraction". *Acta Materialia*, 2006. **54**(18): p. 4841-4852.

10. M. A. Gharghouri, et al., "Study of the mechanical properties of Mg-7.7at.% Al by in-situ neutron diffraction". *Philosophical Magazine a-Physics of Condensed Matter Structure Defects and Mechanical Properties*, 1999. **79**(7): p. 1671-1695.

11. T. Ungar, "Micro structural parameters from X-ray diffraction peak broadening". *Scripta Materialia*, 2004. **51**(8): p. 777-781.

12. O. Muransky, et al., "Investigation of deformation twinning in a fine-grained and coarse-grained ZM20 Mg alloy: Combined in situ neutron diffraction and acoustic emission". *Acta Materialia*, 2010. **58**(5): p. 1503-1517.

13. K. Mathis, et al., "Investigation of the dependence of deformation mechanisms on solute content in polycrystalline Mg-Al magnesium alloys by neutron diffraction and acoustic emission". *Journal of Alloys and Compounds*, 2015. **642**: p. 185-191.

14. M. A. M. Bourke, D. C. Dunand, and E. Ustundag, "SMARTS - A spectrometer for strain measurement in engineering materials". *Applied Physics A: Materials Science and Processing*, 2002. **74**(SUPPL.II): p. S1707-S1709.

15. K. Mathis, et al., "Effect of the loading mode on the evolution of the deformation mechanisms in randomly textured magnesium polycrystals - Comparison of experimental and modeling results". *International Journal of Plasticity*, 2015. **72**: p. 127-150.

16. E. Pomponi and A. Vinogradov, "A real-time approach to acoustic emission clustering". *Mechanical Systems and Signal Processing*, 2013. **40**(2): p. 791-804.

17. J. Capek, et al., "Study of the loading mode dependence of the twinning in random textured cast magnesium by acoustic emission and neutron diffraction methods". *Materials Science and Engineering a-Structural Materials Properties Microstructure and Processing*, 2014. **602**: p. 25-32.

18. C. H. Cáceres and D. M. Rovera, "Solid solution strengthening in concentrated Mg–Al alloys". *Journal of Light Metals*, 2001. **1**(3): p. 151-156.

19. D. Nagarajan, C. H. Caceres, and J. R. Griffiths, "Anelastic Phenomena in Mg-Al Alloys". *Acta Physica Polonica A*, 2012. **122**(3): p. 501-504.

Magnesium Technology 2016
Edited by: Alok Singh, Kiran Solanki, Michele V. Manuel, and Neale R. Neelameggham
TMS (The Minerals, Metals & Materials Society), 2016

STRAIN RATE DEPENDENT DEFORMATION AND FAILURE PROCESS OF MAGNESIUM FOAMS

Peifeng Li
School of Mechanical and Aerospace Engineering, Nanyang Technological University, 50 Nanyang Avenue, 639798 Singapore

Keywords: Magnesium foams, Strain rate, Deformation, Failure, Dynamic behaviour.

Abstract

Lightweight closed-cell magnesium foams have the potential to dissipate impact energy in automotive and aerospace structures which are subjected to ballistic and impact loads. This work investigated the strain rate dependent compressive deformation and failure process of closed-cell Mg foams fabricated by the direct foaming process. The split-Hopkinson pressure bar technique with high speed imaging was used to characterise the dynamic response which was then compared to the measured quasi-static behaviour. The shape of the dynamic stress–strain curve is similar to that at low strain rates. However, both the peak and plateau stresses increase with the strain rate. It was found that under quasi-static compression the damage evolution in the foam causes several big fragments, while at high strain rates the extensive cracking in the foam results in a large quantity of small fragments, thus dissipating more impact energy.

Introduction

Recently there is an increasing interest in using lightweight magnesium foams in automotive and aerospace sectors where weight reduction is a critical design issue. The direct foaming process is one of the effective manufacturing techniques to fabricate closed-cell Mg foams with controllable unit cell sizes and morphologies [1,2]. Closed-cell Mg foams have the potential to dissipate impact energy in automotive and aerospace structures which are subjected to ballistic and impact loads. The high speed applications require the understanding of compressive deformation and failure of Mg foams at various strain rates.

Most studies of the compressive properties of Mg foams have focused on quasi-static conditions (e.g., at the strain rate of $0.001\ s^{-1}$) [3-5]. These reports quantitatively characterised the stress–strain response and energy dissipation capacity. Extensive research has also been carried out to investigate the effect of various factors such as unit cell size, porosity and cell wall properties on the bulk mechanical behaviour [6,7]. It was found that the strength of Mg foams increases with the reduced unit cell size [7]. However, there is a paucity of studies in the literature on the dynamic properties of Mg foams in spite of their high strain rate applications [8,9]. Moreover, this limited literature [8,9] focused on the open-cell but not closed-cell Mg foams.

The aim of this study was to investigate the strain rate dependent compressive behaviour of closed-cell Mg foams fabricated by the direct foaming process. The microstructure of unit cell walls was first examined in scanning electron microscope (SEM) and optical microscope. Uniaxial compression experiments in the INSTRON and split-Hopkinson pressure bar (SHPB) system were conducted to characterise the stress–strain response, deformation, failure and energy dissipation of the foam at the low and high rates, respectively. The strain rate effect on the compressive behaviour was finally analysed.

Materials and Microstructure

Fabrication of Mg Foams

The direct foaming process was developed to fabricate the bulk Mg foams. The processing steps are detailed as follows.

Melting and Thickening. The Mg ingot was molten in a crucible. The Ca particles (3–5 wt%) were added to prevent rapid oxidation and burning. The viscosity of the molten Mg was then increased by adding the SiC particles (3–5 wt%).

Foaming and Stirring. After the dispersion of $CaCO_3$ (3–5 wt%) particles coated with a SiO_2 layer, the molten Mg was stirred for 30 seconds to achieve the more uniform foamed structure.

Holding and Solidification. The foamed Mg was held at the specific temperature for 1–2 min and finally solidified/cooled in the air.

Figure 1 illustrates a typical cylindrical specimen of the diameter 18 mm and the length 18 mm for uniaxial compression tests, which was cut from the bulk Mg foam. The porosity (65–70%) of the foam was determined by controlling the foaming parameters such as the temperature, stirring time and holding time.

Figure 1. A cylindrical Mg foam specimen fabricated by the direct foaming process.

Microstructure of Mg Foams

The SEM examination on Mg foams reveals that the majority of the unit cells are closed and spherical (see Figure 2(a)). The cell size ranges from a few hundred microns up to approximately 2 mm. It was also observed that interconnect channels occur between some unit cells. The unit cell structure is homogeneous as a result of the improved thermal stability of foaming agents ($CaCO_3$) that were coated with SiO_2 particles [1,2,10].

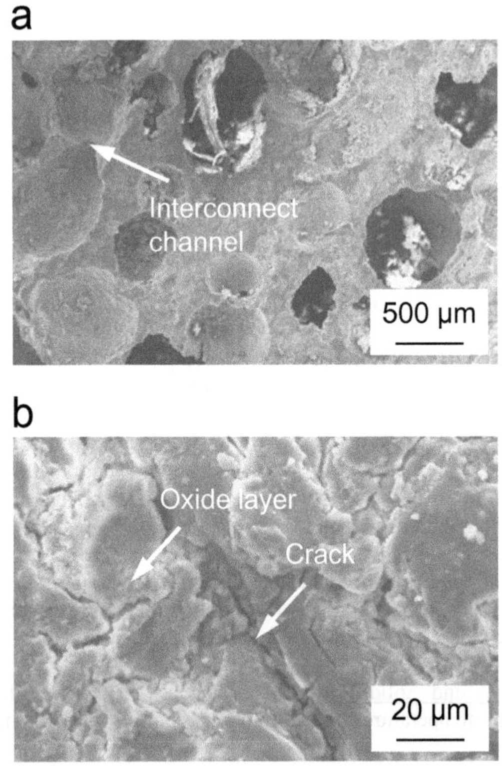

Figure 2. SEM images of the microstructure of Mg foams showing (a) the unit cell topology and morphology and (b) the surface feature of cell walls.

In the Mg foam, a layer of oxide forms on the surface of unit cell walls (Figure 2(b)). The oxide layer breaks off, leaving the cracks on the cell surface. The sectioned unit cell walls were polished and examined in optical microscopy. As shown in Figure 3, ceramic compounds distribute in the Mg unit cell walls. The energy dispersive x-ray (EDX) microanalysis reveals that Mg, Ca, Si and O are the main elements in the compound. The formation of oxide layers on the surface and ceramic compounds in unit cells results from the addition of ceramic particles (e.g., Ca, SiC and $CaCO_3$ with SiO_2 layers) in the molten Mg during the fabrication. These brittle phases in the cell walls can significantly affect the deformation and failure process of the bulk Mg foam.

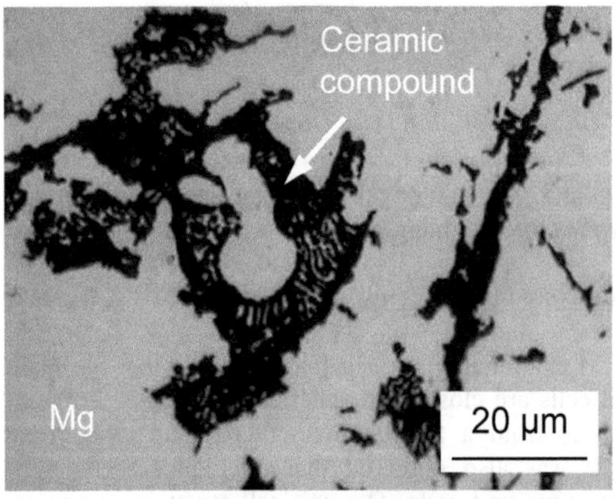

Figure 3. Optical microscopic image showing the ceramic compounds in unit cell walls of Mg foams.

Uniaxial Compression Experiments

Uniaxial compression experiments were conducted on the cylindrical Mg foam specimens at the quasi-static ($0.001\ s^{-1}$) and dynamic ($1000\ s^{-1}$) strain rates. The specimen aspect ratio of 1:1 (the diameter 18 mm and the length 18 mm) was selected to achieve uniform deformation [6,11]. The specimens were lubricated to reduce the interfacial friction.

Quasi-static compression tests of Mg foams were performed in an INSTRON mechanical testing machine. As shown in Figure 4, the cross head was controlled at a constant velocity of $0.018\ mm\ s^{-1}$, thus the equivalent strain rate of $0.001\ s^{-1}$. The applied displacement and resultant forces were recorded by the machine, which were then used to calculate the stress (σ) and strain (ε) histories of the specimen.

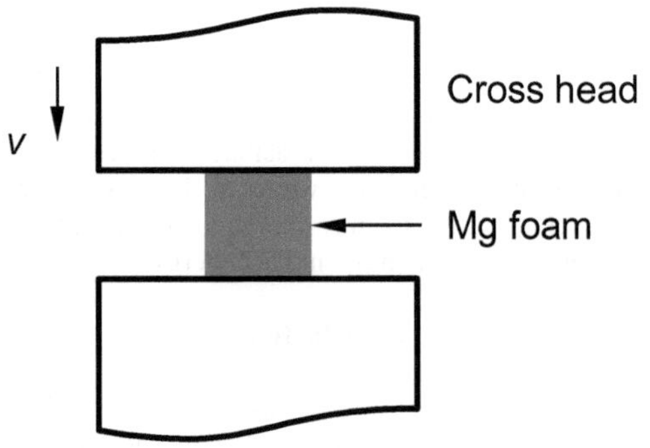

Figure 4. The schematic of INSTRON loading unit to measure the quasi-static compressive behaviour of Mg foams.

The dynamic compression experiments were conducted in an in-house split-Hopkinson pressure bar system as schematically shown in Figure 5. The striker, input and output bars of 20 mm in diameter were all made of aluminium alloys. The respective length of the input and output bars is 1200 mm. The strain gauges mounted approximately in the middle of the input and output bars recorded the incident (ε_i), reflected (ε_r) and transmitted (ε_t) strain waves (Figure 5). On the assumption of axial uniformity of specimen stress and strain fields ($\varepsilon_i + \varepsilon_r = \varepsilon_t$), the stress and strain histories of the specimen can be calculated as follows:

$$\sigma(t) = \frac{A_0 E_0}{A} \varepsilon_t(t) \tag{1}$$

$$\varepsilon(t) = \frac{2c_0}{l} \int_0^t \left[\varepsilon_i(\tau) - \varepsilon_t(\tau)\right] d\tau \tag{2}$$

where A_0 and A are the cross-sectional areas of the bars and the specimen respectively, E_0 is the Young's modulus of the bars, c_0 is the wave velocity in the bars, and t is the time. In the SHPB tests, the strain rate of $1000\ s^{-1}$ was controlled by the initial velocity of the striker which was determined by the pressure in the chamber.

The compressive deformation and failure process of a foam specimen was captured using various imaging systems with framing rates ranging from 0.125 to 210,000 frames s^{-1} in the experiments [12,13].

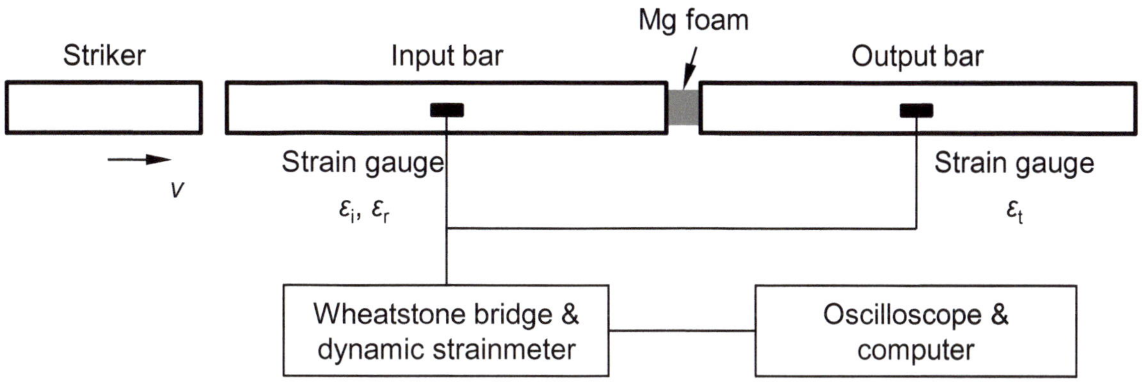

Figure 5. The schematic of the split-Hopkinson pressure bar (SHPB) system to measure the dynamic compressive behaviour of Mg foams.

Results and Discussion

Stress–strain Response

Figure 6 shows the representative stress–strain curves of the Mg foams at the quasi-static (0.001 s^{-1}) and dynamic (1000 s^{-1}) strain rates. Similar to most of other foams [10,12,14-17], the stress–strain curves can be divided into the elastic, plateau and densification regions for both the rates.

An initial, approximately linear, region corresponding to the elastic behaviour of the foam is observed in the stress–strain curve until the peak stress is reached (Figure 6). A drop in stress then occurs, suggesting the start of the failure of unit cell walls. The peak stress represents the facture resistance of the foam.

A nearly constant plateau region following the stress drop is due to the progressive collapse of unit cells (Figure 6). The plateau region thus indicates the energy dissipation capacity of the Mg foam; and the plateau stress can be defined as the average stress in this region. When a significant portion of unit cells fails, further loads lead to densification of the bulk foam, as illustrated by the abrupt increase of stress in the curve (Figure 6).

The shape of the dynamic stress–strain curve is similar to that at quasi-static rates (Figure 6). However, both the peak and plateau stresses increase with the strain rate. Thus the Mg foam subjected to impact loads can dissipate more energy than the foam under quasi-static loads.

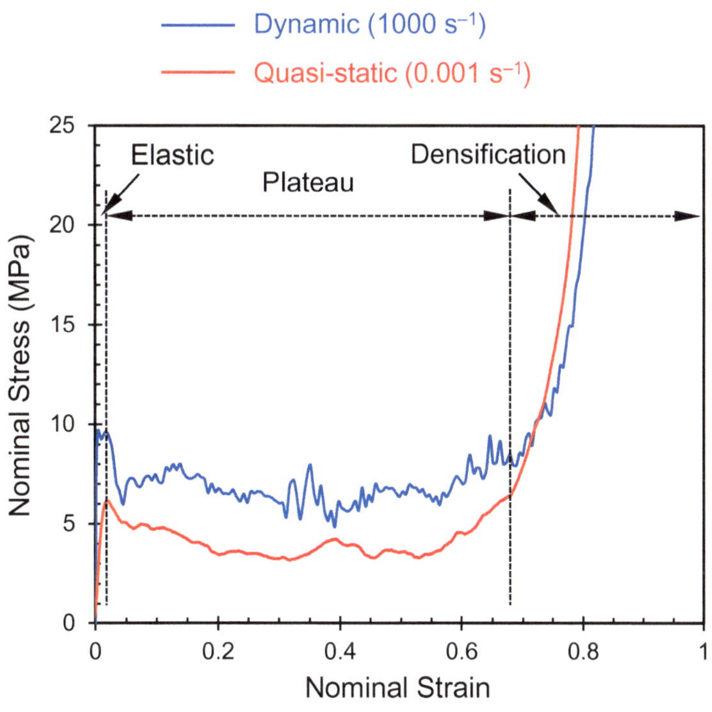

Figure 6. Representative stress–strain curves of Mg foams at quasi-static (0.001 s^{-1}) and dynamic (1000 s^{-1}) strain rates. The curves can be divided into the elastic, plateau and densification regions.

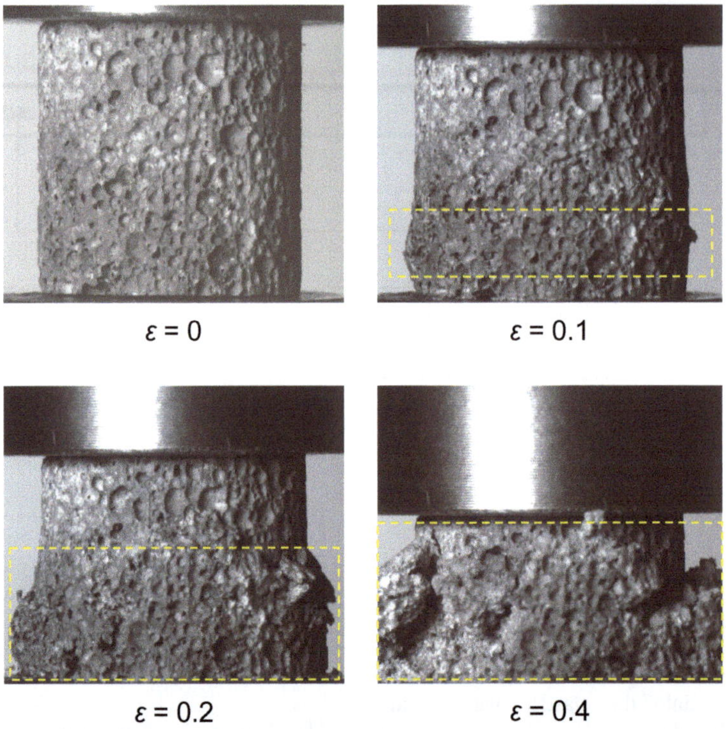

Figure 7. Deformation sequence of Mg foams in quasi-static compression experiments at the strain rate 0.001 s^{-1}.

Quasi-static Deformation and Failure

Figure 7 illustrates the recorded deformation series of the Mg foam in quasi-static compression. The failure of unit cells initiates at either side of the foam specimen. As indicated by the box with a dashed border, a damage band occurs almost in the transverse plane. The band expands as the foam is compressed. At the late stage of compression (e.g., the strain ε = 0.4 prior to densification), the edge of the foam flakes away, subsequently leaving behind several large fragments.

The unit cell size is much bigger than the wall thickness in the foam with the porosity of 65–70% (refer to Figure 2(a)). Thus most of the cell walls are subjected to bending instead of axial tension or compression [18,19]. Under the compression, the strength of unit cell walls increases owing to the hardening properties of Mg. At the same time, cracks may form in the cell walls and propagate along the ceramic compounds. Both the hardening and cracking behaviour in unit cells result in the plateau region of the bulk behaviour of the foam.

Dynamic Deformation and Failure

In high rate tests, shear bands arise in the specimen near either the input or output bar side (Figure 8). The shear band evolves along the diagonal (40–60° to the loading direction) towards the centre of the foam, and becomes more noticeable after the stress drop (refer to the strain ε = 0.1 in Figure 8). Catastrophic cracking occurring along the shear band at dynamic rates can contribute to the more substantial stress drop, compared to that at low loading rates (Figure 6).

The dynamic compression on the Mg foam gives rise to fragmentation. After the SHPB tests, the fragments are small but the quantity is large. This suggests that relative to quasi-static compression, more cracks form in the unit cell walls of the Mg foam under dynamic loads, which cause the foam to fracture into more fragments and thus dissipate more energy.

Discussion on Strain Rate Sensitivity

The strain rate sensitivity of the bulk behaviour of Mg foams, such as the peak and plateau stresses, can be attributed to a number of microscopic factors as follows.

1. The rate dependency of Mg and the alloy can influence the bulk rate sensitivity of the foam [20-22].
2. At high strain rate, more cracks evolve in unit cell walls, causing more fragments and dissipating more energy.
3. In the Mg foam with a high porosity that is loaded dynamically, there is a possible inertia effect manifested in the initiation of high order buckling modes of the slender cell walls before extensive deformation and fracture.

Conclusions

The compressive properties of Mg foams fabricated by the direct foaming process were experimentally characterised at the low and high strain rates. The following conclusions can be drawn.

1. Both the peak and plateau stresses of the Mg foam increase with the strain rate.
2. Under quasi-static compression, a damage band evolves in the transverse plane of the foam and causes several big fragments. However, at high strain rates, shear bands arise along the diagonal in the foam and cause the foam to fracture into a large quantity of small fragments, thus dissipating more impact energy.

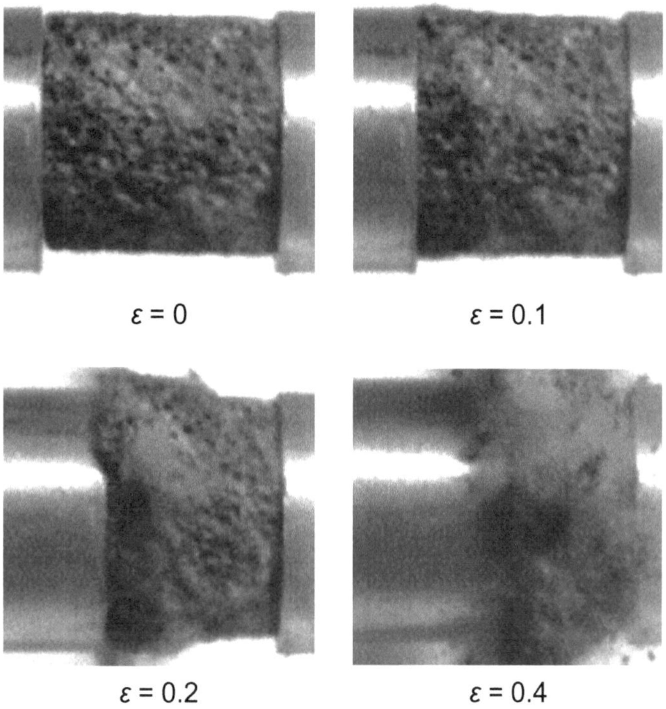

Figure 8. Deformation sequence of Mg foams in the SHPB compression experiments at the strain rate 1000 s^{-1}.

Acknowledgements

The author gratefully acknowledges the financial support of Academic Research Fund (AcRF) Tier 1 by Ministry of Education, Singapore.

References

[1] G.Q. Lu et al., "Preparation of closed-cell Mg foams using SiO_2-coated $CaCO_3$ as blowing agent in atmosphere," *Trans. Nonferrous Met. Soc. China.* 23 (6) (2013), 1832-1837.

[2] D.H. Yang, B.Y. Hur and S.R. Yang, "Study on fabrication and foaming mechanism of Mg foam using $CaCO_3$ as blowing agent," *J. Alloys Compd.* 461 (1-2) (2008), 221-227.

[3] J.O. Osorio-Hernandez et al., "Manufacturing of open-cell Mg foams by replication process and mechanical properties," *Mater. Des.* 64 (2014), 136-141.

[4] M. Tane et al., "Effect of crystallographic texture on mechanical properties in porous magnesium with oriented cylindrical pores," *Acta Mater.* 84 (2015), 80-94.

[5] D.H. Yang et al., "Compressive properties of cellular Mg foams fabricated by melt-foaming method," *Mater. Sci. Eng. A.* 527 (21-22) (2010), 5405-5409.

[6] X.C. Xia et al., "Effects of specimen aspect ratio on the compressive properties of Mg alloy foam," *Mater. Des.* 42 (2012), 32-36.

[7] Z.G. Xu et al., "Effects of cell size on quasi-static compressive properties of Mg alloy foams," *Mater. Des.* 34 (2012), 40-44.

[8] H. Kanahashi et al., "Experimental study for the improvement of crashworthiness in AZ91 magnesium foam controlling its microstructure," *Mater. Sci. Eng. A.* 308 (1-2) (2001), 283-287.

[9] T. Mukai et al., "Dynamic compressive behavior of an ultra-lightweight magnesium foam," *Scr. Mater.* 41 (4) (1999), 365-371.

[10] J. Banhart, "Metal foams: Production and stability," *Adv. Eng. Mater.* 8 (9) (2006), 781-794.

[11] P. Li, C.R. Siviour and N. Petrinic, "The effect of strain rate, specimen geometry and lubrication on responses of aluminium AA2024 in uniaxial compression experiments," *Exp. Mech.* 49 (4) (2009), 587-593.

[12] R. Huang and P. Li, "Elastic behaviour and failure mechanism in epoxy syntactic foams: The effect of glass microballoon volume fractions," *Compos. Part B.* 78 (2015), 401-408.

[13] P. Li, N. Petrinic and C.R. Siviour, "Quantification of impact energy dissipation capacity in metallic thin-walled hollow sphere foams using high speed photography," *J. Appl. Phys.* 110 (8) (2011), 083516.

[14] P. Li, "Constitutive and failure behaviour in selective laser melted stainless steel for microlattice structures," *Mater. Sci. Eng. A.* 622 (0) (2015), 114-120.

[15] P. Li et al., "Strain rate dependent compressive properties of glass microballoon epoxy syntactic foams," *Mater. Sci. Eng. A.* 515 (1-2) (2009), 19-25.

[16] P. Li et al., "Deformation behaviour of stainless steel microlattice structures by selective laser melting," *Mater. Sci. Eng. A.* 614 (2014), 116-121.

[17] P.J. Tan, S.R. Reid and J.J. Harrigan, "On the dynamic mechanical properties of open-cell metal foams - A re-assessment of the 'simple-shock theory'," *Int. J. Solids Struct.* 49 (19-20) (2012), 2744-2753.

[18] M.F. Ashby, "The mechanical properties of cellular solids," *Metall. Trans. A.* 14 (9) (1983), 1755-1769.

[19] L.J. Gibson and M.F. Ashby, "The mechanics of three-dimensional cellular materials," *Proc. R. Soc. Lond. A.* 382 (1782) (1982), 43-&.

[20] I.C. Choi et al., "Enhancement of strain-rate sensitivity and shear yield strength of a magnesium alloy processed by high-pressure torsion," *Scr. Mater.* 94 (2015), 44-47.

[21] S. Kurukuri et al., "Constitutive behavior of commercial grade ZEK100 magnesium alloy sheet over a wide range of strain rates," *Metall. Mater. Trans. A.* 45A (8) (2014), 3321-3337.

[22] W.Q. Song, P. Beggs and M. Easton, "Compressive strain-rate sensitivity of magnesium-aluminum die casting alloys," *Mater. Des.* 30 (3) (2009), 642-648.

Magnesium Technology 2016
Edited by: Alok Singh, Kiran Solanki, Michele V. Manuel, and Neale R. Neelameggham
TMS (The Minerals, Metals & Materials Society), 2016

EXPLORATION OF THIN-WALLED MAGNESIUM ALLOY TUBE EXTRUSION FOR IMPROVED CRASH PERFORMANCE

Robert W. Klein1, Bruce W. Williams2, Jonathan McKinley2, John R. Einhorn1, and Sean R. Agnew1
[1] Materials Science & Engineering, University of Virginia, 395 McCormick Road, Charlottesville, Virginia 22904-4745, USA [2] CanmetMATERIALS, Natural Resources Canada, 183 Longwood Road South, Hamilton, Ontario L8P 0A5, Canada

Keywords: extrusion, texture, Mg, ductility, crashworthiness

Abstract

Magnesium alloy tubes offer an attractive combination of strength and density. However, their poor crushing behavior under axial loads has prevented their use in crash-sensitive automotive structures such as crush rails. A combination of alloy selection, strain path design, and texture control were employed in an attempt to enhance the mechanical behavior of extruded Mg tubes. C-channel extrusion, porthole-die extrusion, and seamless tube extrusion over a mandrel were all explored. Mandrel extrusion was confirmed to produce tubes ideal for later axial crush tests, as there was no weld or extrusion seam to act as a source of weakness. Crush testing of the tubes revealed that conventional alloy, AZ31B, exhibited premature fracture prior to folding. Novel alloy, ZEK100, which exhibits slightly higher ductility than AZ31B, still exhibits poor folding behavior. The crush behavior of the Mg alloy tubes is discussed in terms of their crystallographic texture.

Introduction

Mg alloy components exhibit an appealing combination of strength and density for use in automobile construction. However, poor crash energy absorption due to low room temperature ductility, under complex loading conditions, prohibits use of Mg in particular automotive structures such as crush rails, e.g., [1]. Axial crush response is highly dependent upon the ductility of the material and can yield information regarding the acceptability of these components for use in automotive design.

Recent studies have helped develop the understanding of how texture in Mg components influences the mechanical properties of the material. In one instance, Chino et al. [2] and Orlov et al. [3] demonstrated that the texture of the material can influence the ductility of the material. Previous work suggests that different extrusion methods can yield a wide variety of resultant textures in Mg alloys [4] and mechanical properties [5]. The present study was initiated to determine if texture evolved during thin-walled extrusion of Mg alloys impacts the final axial crush performance of the material.

Two Mg alloys, AZ31B and ZEK100, were selected for investigation. Conventional alloy, AZ31B, containing 3wt%Al and 1wt% Zn is the most common Mg alloy used to fabricate extrusions. Alloy ZEK100, including 1wt% Zn, 0.2wt% rare earth elements (such as Ce or Nd), and 0.5% Zr is of interest because the inclusion of rare-earth elements has been shown to improve mechanical properties, such as ductility, e.g., [6].

Experimental methods

A vertical extrusion press at CanmetMATERIALS was utilized to extrude thin-walled Mg alloy components via three extrusion configurations: porthole-die, bridge-die, and conventional die. An example extrusion tooling for this system is depicted in Figure 1, in which a pre-extruded (AZ31B) or an as-cast (ZEK100) billet of 88.9 mm can be extruded into thin-walled components. The extrusions were all produced either at a rate of 254 mm/min, referred to as slow, or at a rate of 2032 mm/min, referred to as fast. Axial crush tests were performed on tube extruded at both rates, but all of the texture measurements and analysis reported in this paper were performed on samples extruded at the slower rate.

In this study, circular tubes with a wall-thickness of 2 mm and an outer diameter of 50 mm were extruded using a pierced billet and mandrel (Figure 1) which enforces an area reduction of 15:1 and results in a seamless extrusion product or using a circular bridge-die, often referred to as a porthole-die [7] (Figure 2), which enforces a higher extrusion ratio of 20:1 and creates a product with four extrusion seams. The focus of the study shifted towards the mandrel extrusion as previous studies revealed failures at the seams in tubes made using a bridge-die [7]. Further, technical difficulties, such as deformation of the bridges, with the extrusion of tubes through the bridge die made use of the mandrel expedient for the current study. Pierced billet extrusions were produced at temperatures of 350, 400, and 450 °C.

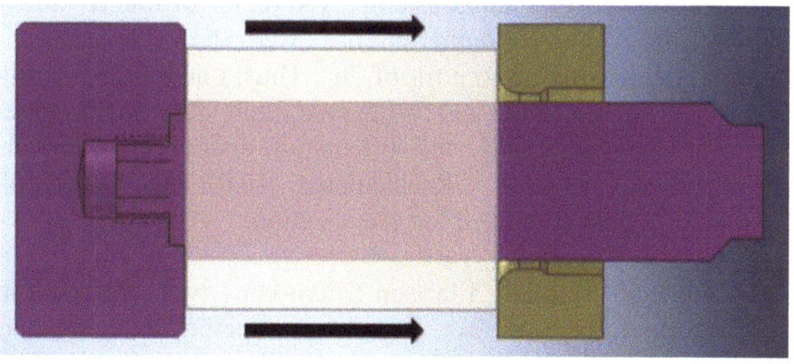

Figure 1 – Schematic design of the extrusion tooling to create a round-tube. The mandrel (purple) forces the billet (starting position shown in gray) through the die (gold).

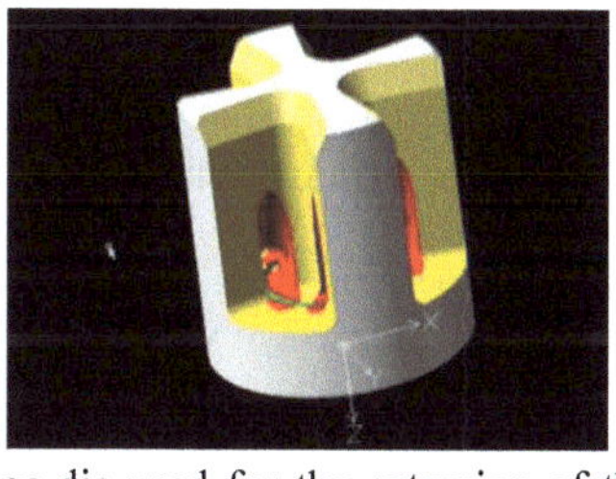

Figure 2 - Bridge-die used for the extrusion of thin-walled tube components visualized as a CAD (computer aided design) schematic. The creation of four seams is apparent from the visual.

C-channel (top-hat) shaped extrusions with a wall-thickness of 2 mm, width of 40 mm and height of 20 mm high were created using a conventional die from billets pre-heated to 400 °C. Notably, the extrusion process used to create these components enforced an area reduction of about 80:1. In Figure 3, a top-hat sample is visualized alongside a schematic showing how specimen were excised for texture measurement from the extruded component. Though the top-hat profiles have no extrusion seams, it should be noted that they must welded together to create a final square tube. Although texture data was obtained from samples of this geometry, no tube crush tests have been performed on this sample geometry.

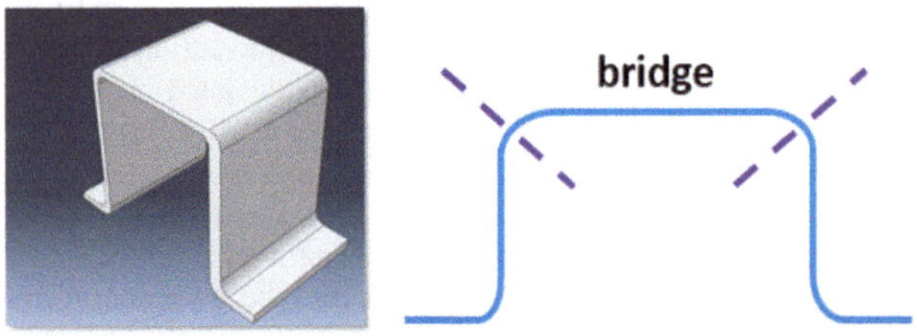

Figure 3 – Top-hat extrusion sample rendered as a CAD drawing. Specimens used in this paper are the "bridge" segment, which connects the two vertical portions. Dotted lines on the right show the cuts made to generate an investigated specimen.

Texture measurements were performed using x-ray diffraction (XRD) and electron backscatter diffraction (EBSD). Samples examined were in the as-cast, pre-extruded, thin-walled top-hat, and thin-walled extruded tube configurations. A PANalytical X'Pert Pro MPD diffractometer employing CuK_α radiation was used to experimentally measure the textures of Mg samples. Samples received standard metallographic preparation consisting of mechanical and final chemical polishing (using a 5% nital solution). Incident radiation was put through a point-focus collimator with 1 mm x 2 mm (Shulz reflection geometry to minimize defocusing effects). A parallel plate collimator with divergence of 0.04 radians was employed on the diffracted side of the machine. This configuration has been shown to result in minimal defocusing up to a tilt of 70°. Thus, uncorrected, partial (up to 70°) (10.0), (10.1) and (00.2) pole figure data was imported into MTEX, from which full orientation-distribution functions (ODF) were calculated. Recalculated (10.0) and (00.2) pole figures were then computed for visualization.

A Quanta 650 Scanning Electron Microscope was used with an accelerating voltage of 20kV in order to experimentally measure the texture of the Mg samples via EBSD. Samples received standard metallographic preparation up to colloidal silica polish followed by ion polishing using a Hitachi IM4000. EBSD scans were run with a 1μm step sizes over areas of 250μm x 160 μm. EBSD data was also imported into MTEX, from which full ODF's were calculated. ODF's from the EBSD data were also plotted as full pole figures for the (10.0) and (00.2) planes to allow for best comparison with the other texture results.

Axial crush tests were performed on several of the AZ31B and ZEK100 extrusions. The tube length was about 160 mm and three crush initiators were placed at about the mid-length of the tube in a triangular pattern, to initiate the location of folding and reduce the peak load. All tubes were tested in a 225 kN press at a quasi-static displacement rate of 1 mm/s. The experimental setup is shown in Figure 4.

Figure 4 – Experimental setup for an axial crush test being performed on a thin-walled tube component.

Results

The billets used in this study began as commercially extruded AZ31B and laboratory cast ZEK100, the textures of which were investigated using XRD. The recalculated pole figures obtained from the pre-extruded AZ31B billet are presented in Figure 5a, viewed down the axis of the extrusion. These pole figures show axisymmetry with two components, one with the (00.2) poles parallel and the other with (00.2) poles nearly perpendicular to the extrusion axis. The ZEK100 alloy was investigated via a representative specimen, which was cut from the center of the casting (Figure 5b). As expected, the overall texture of this as-cast material is weak. There is an unexpected feature of the texture as the (00.2) poles appear to have an off-axis preference.

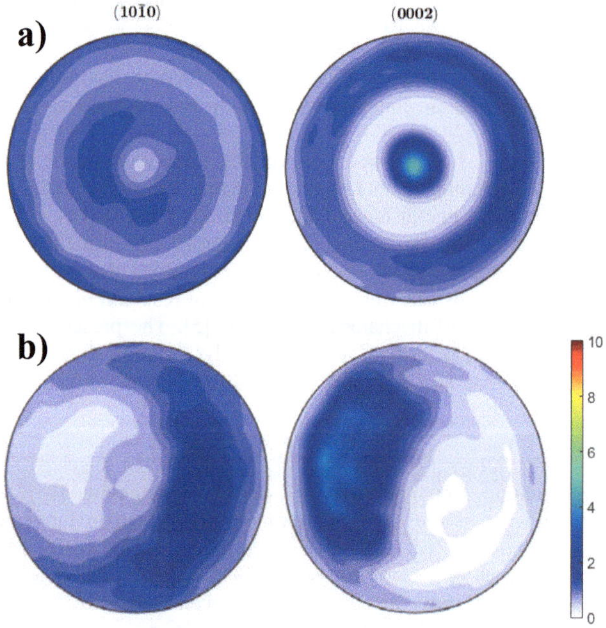

Figure 5 - Recalculated pole figures from the pre-extruded billet of AZ31B (a) and as-cast ingot of ZEK100 (b). Pole figures are viewed down the axis of the samples. (Note all pole figures presented in this paper employ the same contouring and color spectrum).

The texture of the thin-walled top-hat specimen of the AZ31B is presented in Figure 6. The texture of the outermost surface is presented in Figure 6a and that of the mid-plane of the "bridge" section (obtained by grinding through sufficient material) is shown in Figure 6b. The surface texture shows that the (00.2) poles lie near the normal direction (ND). For the mid-plane, (00.2) poles align closer to the normal direction, but also at angles tilted from the normal direction and towards the transverse direction.

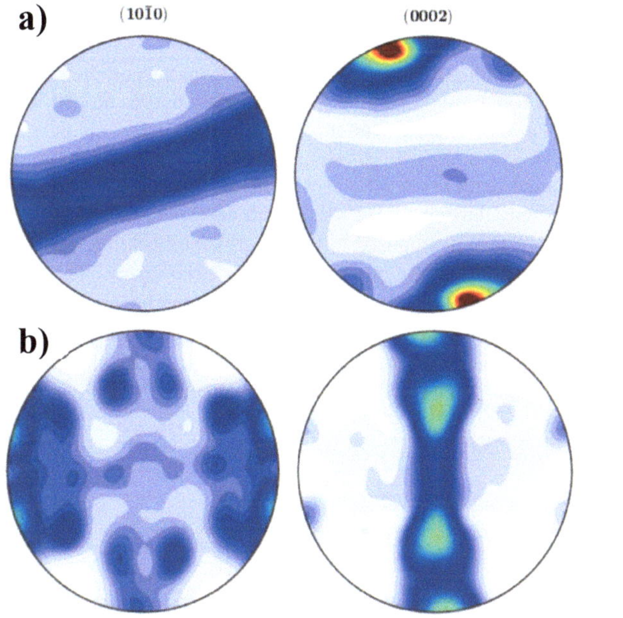

Figure 6 - Recalculated pole figures for the AZ31B top-hat specimen at the a) surface texture and b) mid-plane. Pole figures plotted with the ND pointing upwards, and the ED facing left.

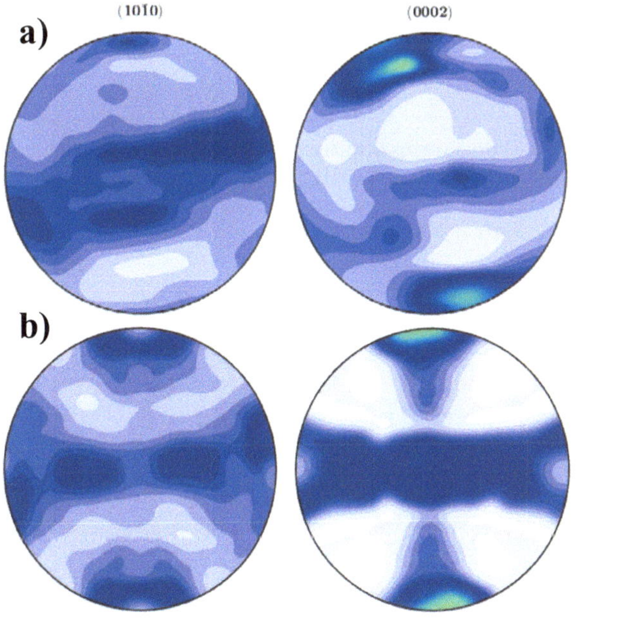

Figure 7 - Recalculated pole figures for the ZEK100 top-hat specimen at the a) surface texture and b) mid-plane. Pole figures plotted with the ND pointing upwards, and the ED facing left.

Similarly, the texture of the thin-walled top-hat specimen of the ZEK100 is presented in Figure 7. The texture of the outermost surface is presented in Figure 7a and that of the mid-plane of the "bridge" section (obtained by grinding through sufficient material) is shown in Figure 7b. The surface texture of the ZEK100 sample shows the (00.2) poles to be only slightly inclined from the ND, as observed for alloy AZ31B. In the mid-plane pole figures, the ZEK100 texture is shown to have a feature where the (00.2) poles are nearly perfectly aligned with the ND. The mid-plane (00.2) pole figure also features a band of preferred orientations sweeping from the extrusion direction (ED) to the transverse direction (TD).

A thin-walled tube specimen of the AZ31B made via the mandrel tooling was investigated using EBSD. Three EBSD scans were performed on this sample. The EBSD scans were taken at varying distances along the radial direction (RaD) on a plane normal to the tangential (hoop) direction (HD), as visualized in Figure 8. A thin-walled tube specimen of the ZEK100 alloy, also made by means of the mandrel tooling, was investigated using EBSD. Five EBSD scans were performed on this sample, at different locations along the radial direction of piece, on a plane normal to ED, as drawn schematically in Figure 9. The microstructure is shown as an inverse pole figure map for AZ31B (Figure 10a) and ZEK100 (Figure 10b). Note that the grain size is finer in the ZEK100. Notably, now strong variations in microstructure were observe along the tube radial direction in either case, though the AZ31B had a slightly finer grain size close to the inner surface tube.

Figure 8 - Schematic representation of site locations for EBSD performed on the thin-walled tube of AZ31B.

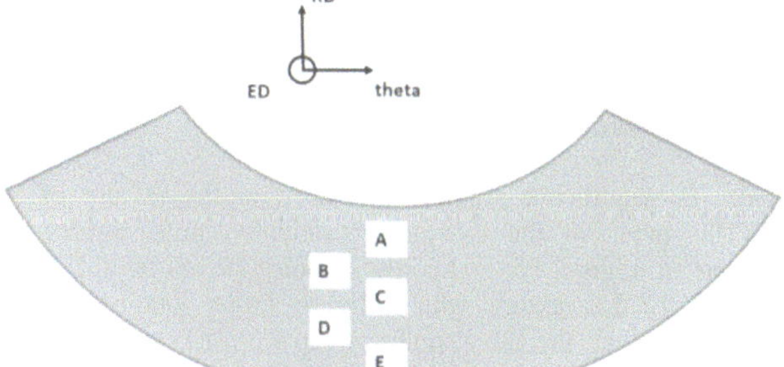

Figure 9 - Schematic representation of site locations for EBSD performed on the thin-walled tube of ZEK100.

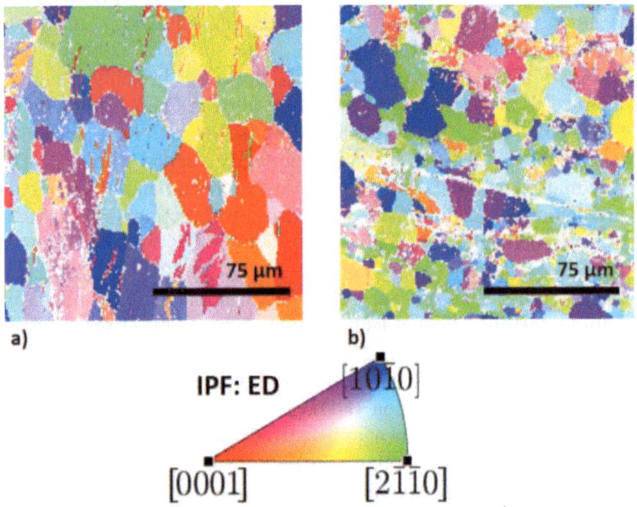

Figure 10 – Extrusion direction (ED) inverse pole figure maps showing the microstructures of a) AZ31B and b) ZEK100 extruded tubes.

As there was no significant texture gradient observed along the RaD for the AZ31B tube, a set of averaged, recalculated pole figures is featured in Figure 11. These pole figures show a preference for the c-axes of the grains to align close to the RaD, although they are tilted slightly aft of the precise RaD. Additionally, there is a weaker texture component with (00.2) poles parallel to the HD. The (10.0) recalculated pole figure exhibits a band of preferred orientations from the extrusion direction to theta, with highpoints along the ED and HD. Note also the 6-fold symmetry of the weak (10.0) peaks along the rim of the pole figure, which correspond to the (00.2) poles parallel to the HD (at the center of that pole figure).

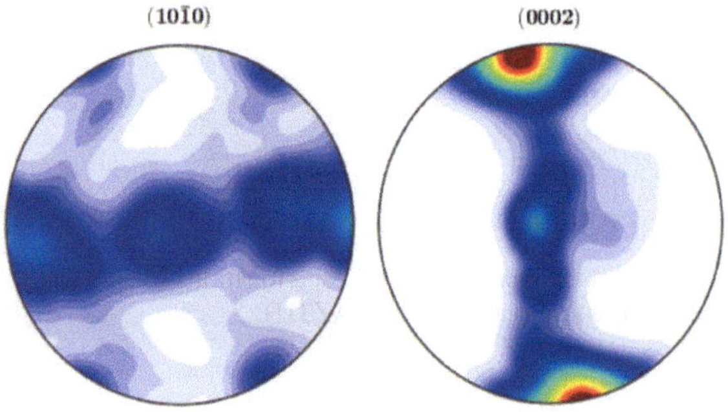

Figure 11 - Full pole figures recalculated from the complete ODF of an AZ31B thin-walled tube specimen. Pole figures plotted with the RaD points upwards, and the ED faces left.

Again, no significant gradient was seen in the texture of the ZEK100 tube, so the average recalculated pole figures are featured in Figure 12. The (00.2) poles lie along the radial direction for this texture, with a slightly larger rotation towards the ED. Further, the (10.0) poles sweep from ED to HD, with a slight preference to lie parallel to the ED and no preference for the HD which was observed in the case of AZ31B.

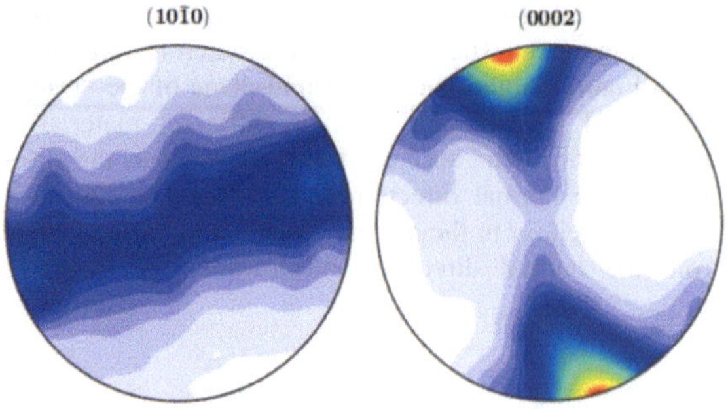

Figure 12 – Full pole figures recalculated from the complete ODF of a ZEK100 thin-walled tube specimen. Pole figures plotted with the RaD points upwards, and the ED faces left.

Subsequent axial crush tests were performed on the AZ31B and ZEK100 circular tube extrusions. In all cases, the tubes fractured on the first fold as opposed to the progressive buckling observed for steel and aluminum alloy tubes under axial crush. The force versus displacement curves for the AZ31B samples are shown in Figure 13 and for the ZEK100 samples in Figure 14. In Figure 15, the deformed states of all of the tested samples are shown.

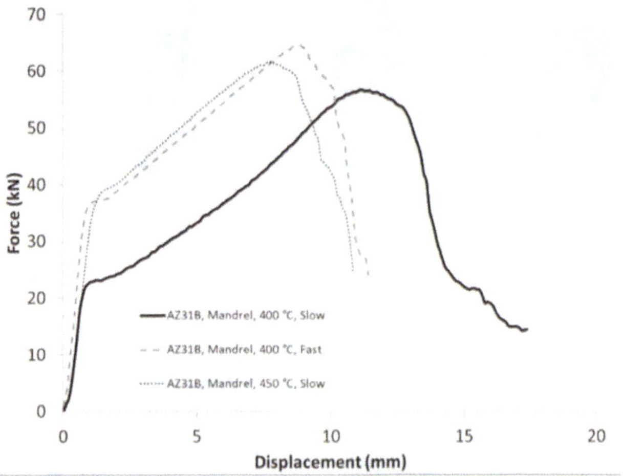

Figure 13 - Force versus displacement from quasi-static axial crush tests performed on AZ31B extrusions.

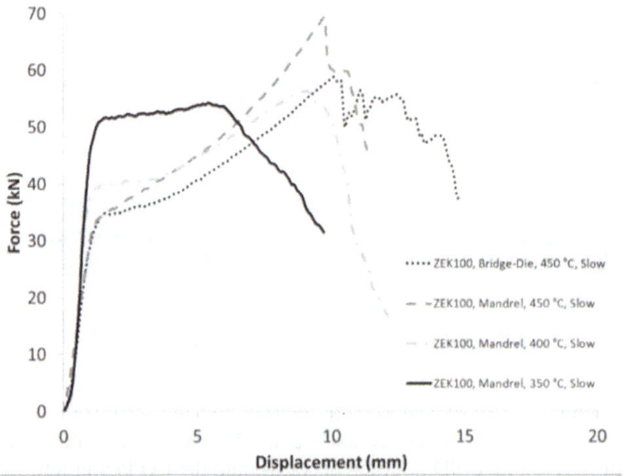

Figure 14 - Force versus displacement from quasi-static axial crush tests performed on ZEK100 extrusions.

The initial load at buckling for the ZEK100 tubes was greater than the peak load for the AZ31B tubes. The initial buckling load of about 51 kN for the ZEK100 tube extruded using the mandrel at 350 °C was higher than most of the other tubes for which the initial buckling load was between 34 to 40 kN. Interestingly, the buckling load of the slow AZ31B tube extruded using the mandrel at 400 °C was only about 22 kN. After initial buckling, the force response for all of the tubes was similar with an increasing load to about 60 kN. The only exception to this trend of increasing load was for the ZEK100 tube extruded using the mandrel at 350 °C in which the load increases only slightly to a displacement of 7 mm before decreasing.

a)

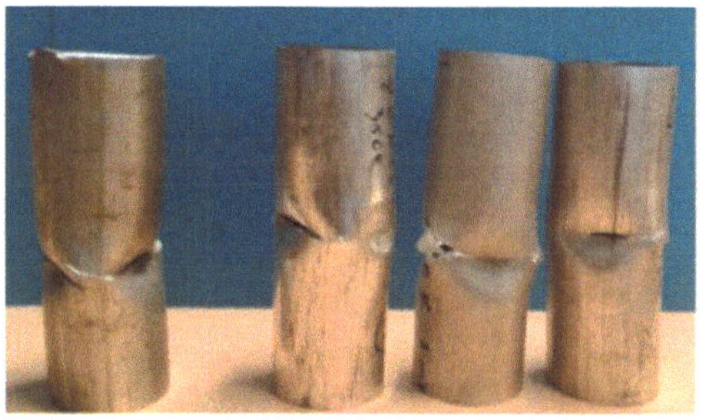

b)

Figure 15(a) - Axial crush profiles for AZ31B extrusion (left to right): 400 °C mandrel - slow, 400 °C mandrel – fast, 450°C mandrel – slow (b) ZEK100 (left to right: 450 °C bridge die - slow, 450 °C mandrel – slow, 400 °C mandrel – slow, 350 °C mandrel – slow

Discussion

The measured surface textures of the top-hat specimens of both AZ31B and ZEK100 are qualitatively similar, though the AZ31B texture is stronger. Pole figures for both alloys possess dyadic symmetry, and show the c-axes close to ND. The slight rotation of the c-axes from ND counter clockwise about the TD gives the impression of shear from friction, a notion supported by the absence of this feature in the mid-plane textures. The dyadic symmetry and near ND preference of the (00.2) poles correlates well with textures predicted by crystal plasticity simulations in a previous study [7].

Measured mid-plane textures for the top-hat specimens of AZ31B and ZEK100 are more dissimilar from each other. Both textures exhibit near orthotropic symmetry and have the c-axes of the grains closely aligned with the ND. The degree of symmetry suggests that the mid-planes experience pure shear deformation. A difference between the AZ31B and ZEK100 pole figures is the presence of lobes, tilted between ND and HD, in the (00.2) pole figure of the AZ31B texture measurement. The absence of these lobes in the ZEK100 pole figure could be explained by the rare-earth alloying elements impeding recrystallization, which is one explanation that has been invoked to explain the weaker textures sometimes observed in rare earth containing alloys [5].

The measured textures of the extruded tubes of both AZ31B and ZEK100 are also qualitatively similar, with the AZ31B texture appearing slightly stronger. While both of the textures show the c-axes close to the RaD, the AZ31B (00.2) pole figure contains an additional feature, a preferred orientation along the HD. Results from previous work suggest that this texture might be characteristic of deformation by plane strain compression due to its similarity to rolled Mg textures, and was most prominent in cases where the prismatic slip mechanism was dominant [7].

In general, the ZEK100 tubes exhibited only a slight improvement in the axial crush response compared to the AZ31B tubes. There are two possible explanations for the observed behavior. The first is the possibility that the ZEK100 tube has a higher overall strength due to the finer grain size. The other relates to the possibility that the distinct crystallographic texture (i.e., the slightly higher degree of tilt of the basal poles away from RaD) causes a specific increase in the compressive strength. Note that the hcp crystals compressed orthogonal to the c-axes are prone to undergo tensile twinning. By rotating the c-axes away from the RaD, twinning deformation is suppressed. This could have the effect of reducing the tension-compression asymmetry and improving the compressive ductility by blending a combination of slip and twin accommodated deformation, rather than relying on twin dominated deformation, which has been associated with premature compression failures of highly textured extrusions.

Conclusion

AZ31B and ZEK100 Mg alloys were extruded into thin-walled structures using three techniques: a bridge-die, a mandrel with a pierced billet, and a C-channel extrusion. The extrusion seams resulting from the bridge-die extrusion were a source of concern for weakness in the structure. The C-channel extrusion did not have an extrusion seam, but two C-channels would need to be welded together to form a tube, which was not carried out in the current work. This research investigated the mandrel extrusion in which a pierced billet was extruded into a thin walled tube with no extrusion weld seams. Texture analysis performed using EBSD showed that the range of textures produced in the thin-walled magnesium alloy extrusion is limited. The texture mainly indicated that the basal planes were aligned with their poles near the radial direction. Subsequent axial crush tests on AZ31B and ZEK100 circular extrusions were performed. In all cases, fracture initiated on the first fold and there was little energy absorption by the structure, much as observed by other researchers for similar thin-walled structures [9]. The ZEK100 tubes exhibited slightly higher strength than the AZ31B tubes. The ZEK100 tube extruded at 350 °C demonstrated the highest structural strength. Developing a better understanding of this behaviour and further efforts to improve the axial crush behavior of Mg alloy tubes is underway.

Acknowledgements

The authors would like to thank Dr. Amjad Javaid and the casting team at CanmetMATERIALS for casting the ZEK100 billets. The authors would also like to thank Lucian Blaga of CanmetMATERIALS for performing the extrusions. RWK and SRA would like to thank the National Science Foundation for their support of this research. Funding for this work was also provided by Natural Resources Canada through the Program of Energy Research and Development.

References

[1] J. L. Y. &. H. H. Yoon, " Investigation of deformation and collapse mechanism for magnesium tube in axial crushing test," *Journal of Mechanical Science and Technology,* vol. 27, no. 10, pp. 2917-2921, 10 2013.

[2] Y. Chino, K. Sassa and M. Mabuchi, "Enhancement of tensile ductility of magnesium alloy produced by torsion extrusion," *Scripta Materialia,* vol. 59, no. 4, pp. 399-402, 2008.

[3] D. Orlov, G. Raab, T. Lamark, M. Popov and Y. Estrin, "Improvement of mechanical properties of magnesium alloy ZK60 by integrated extrusion and equal channel angular processing," *Acta Materialia,* vol. 59, no. 1, pp. 375-385, 2011.

[4] S. Seipp, M. Wagner, K. Hockauf, I. Schneider, L. Meyer and M. Hockauf, "Microstructure, crystallographic texture and mechanical properties of the magnesium alloy AZ31B after different route of thermo-mechanical processing," *International Journal of Plasticity,* vol. 35, pp. 155-166, 2012.

[5] D. Li, V. Joshi, C. Lavender, M. Khaleel and S. Ahzi, "Yield asymmetry design of magnesium alloys by integrated computational materials engineering," *Computational Materials Science,* vol. 79, pp. 448-455, 2013.

[6] J. Bohlen, M. Nurnberg, J. Senn, D. Letzig and S. R. Agnew, "The texture and anisotropy of magnesium-zinc-rare earth alloy sheets," *Acta Materialia,* vol. 55, no. 6, pp. 2101-2112, 2007.

[7] G. Z. J. &. D. J. Liu, "FE analysis of metal flow and weld seam formation in a porthole die during the extrusion of a magnesium alloy into a square tube and the effect of ram speed on weld strength," *Journal of materials processing technology,* vol. 200, no. 1, pp. 185-198., 2008.

[8] B. W. Williams, S. R. Agnew, R. W. Klein and J. McKinley, "Development of Thin-Walled Magnesium Alloy Extrusions for Improved Crash Performance Based Upon Texture Control," in *Magnesium Technology 2015*, Hoboken, NJ, John Wiley & Sons, Inc., 2015.

[9] D. Steglich, X. Tian, J. Bohlen, S. Riekehr, N. Kashaev, K. U. Kainer and N. Huber, "Experimental and numerical crushing analysis of thin-walled magnesium profiles," *International Journal of Crashworthiness,* vol. 20, no. 2, 2015.

Magnesium Technology 2016
Edited by: Alok Singh, Kiran Solanki, Michele V. Manuel, and Neale R. Neelameggham
TMS (The Minerals, Metals & Materials Society), 2016

HIGH TEMPERATURE TENSILE BEHAVIORS AND DEFORMATION MECHANISMS OF Mg-x%Al ALLOYS

JI Jia-xing, BIAN Fu-bo, HE Min, NIU Tian-gang, QIAO Jun*
College of Materials Science and Metallurgy, the University of Science and Technology Liaoning, Anshan 114051, China

Keywords: magnesium alloys, high temperature deformation, deformation mechanism, stress exponent, creep

Abstract

Effects of Al and its content on high temperature tensile behaviors and deformation mechanisms of Mg-xAl (x=2, 3, 4, 6 wt. %) alloys were investigated. Elongation-to-failure tests were conducted at constant temperatures of 300, 350, 400, and 450 ℃, and constant strain rates of 10^{-2} and 10^{-3} s^{-1}. Strain-rate-change tests were conducted at the same temperatures and varying strain rate from 5×10^{-5} to 2×10^{-2} s^{-1}. Experimental results show that elongations of over 100% are achieved for each Mg-xAl alloy when deformed at 450 ℃ and 10^{-3} s^{-1}, which are ascribed to an evident drop of stress exponent due to Al addition. Particularly, the stress exponent decreases with an increase of Al content, indicating a potential variation of deformation mechanism. The mechanisms were analyzed and discussed in terms of activation energy, grain boundary sliding, and dislocation creep.

Introduction

Magnesium (Mg) alloy is so far the lightest metallic structural material with attracting features of high specific strength, high thermal conductivity, strong impacting-load carrying ability, and excellent corrosion resistance, etc. Mg alloys with great application prospect are replacing steels and aluminum (Al) alloys in fabricating structural parts in industries[1]. However, Mg alloys generally exhibit low plasticity and poor formability at room temperature because only two independent basal slip systems are active in the hexagonal close-packed crystalline structure, which is particularly a inhibitory factory for further processing of Mg alloy sheets[2, 3].

Plasticity of Mg alloys can be greatly improved at high temperatures, and even superplasticity can be achieved at appropriate deformation conditions. Grain boundary sliding (GBS) with a low stress exponent of $n\approx2$ is believed to be the dominate deformation mechanism for superplasticity of fine-grained Mg alloys deformed at low strain rates and high temperatures. GBS requires a fine average grain size of less than 10 μm which must be stable during high temperature deformation[4]. Coarse-grained Mg alloys can also present superplasticity at certain temperatures and strain rates[5, 6]. GBS may still be the dominate mechanism if the coarse grains were refined due to occurrence of dynamic recrystallization at high strain rates[7, 8]. On the other hand, prismatic and pyramidal slip systems can be thermally activated at high temperatures, and more active dislocation motions in these slip systems facilitate dislocation creep and plasticity. Dislocation creep includes solute-drag creep (SDC) and dislocation-climb creep (DCC). SDC could occur in a solid solution with a large size effect due to a large linear misfit actor between solute and solvent atoms, such as Al-Mg alloys[9]. The size effect could induce an elastic interaction between solute atoms and moving dislocations, and a dragging force which inhibits dislocation motion and improve plasticity at a stress exponent of $n\approx3$ and activation energy for diffusion of the solute atoms[10]. DCC is dislocation-climb controlled deformation through lattice diffusion or dislocation-pipe diffusion at a higher stress exponent of $n\geq5$ and an activation energy for lattice diffusion or pipe diffusion[11, 12].

As the most important alloying element in AZ Mg alloys, Al not only improves fluidity and castability of melted metals, but also improves room temperature strength through solid solution strengthening. Although superplasticity of AZ alloys has been extensively studied, effects of Al and its content on high temperature deformation behaviors and mechanisms have not been clarified, partially because of potential difficulties brought by other existing elements of Mn, Fe, and Zn. In this study, comparative experiments were conducted on pure Mg and Mg alloys with various Al contents, stress exponent and activation energy were calculated to distinguish deformation mechanisms, and the effects of Al and its content on mechanism evolution were discussed.

Experimental

The investigated Mg-xAl (x=2, 3, 4, 6 wt.%) ingots were made by melting pure Mg (99.98 wt.%) and pure Al (99.97 wt.%) in a resistant furnace. The contents of Al were measured using Inductive Coupled Plasma-Atomic Emission Spectroscopy (ICP-AES) after homogeneous annealing. Then, the 20 mm thick ingots were hot rolled into 3 mm thick sheets with a total reduction rate of 85%, and a 10 minute heating at 400 ℃ was applied between two passes. The rolled sheets were annealed again at 400 ℃ for 2 hours to ensure a fully recrystallized structure, which is expected to be stable during high temperature deformation under applied conditions.

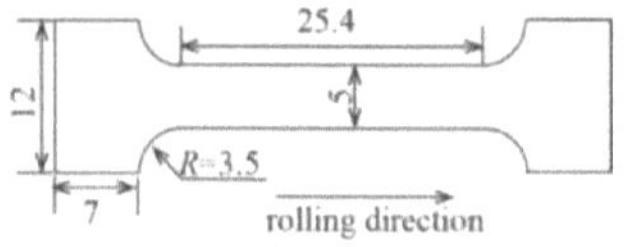

Fig. 1 Specimen geometry for high temperature tensile test (unit: mm)

Specimens for high temperature tensile tests were prepared with geometries shown in Fig.1 and longitudinal direction parallel to the rolling direction. High temperature elongation-to-failure (EF) tests and strain-rate-change (SRC) tests were conducted using a UTM-5305 testing machine equipped with a three-zone resistance furnace. EF tests were conducted at constant temperatures of 300, 350, 400, and 450℃ and constant strain rates of 10^{-2} and 10^{-3} s^{-1} until the specimen fractured, then elongations in percentage were measured. SRC tests were conducted at constant temperatures of 300, 350, 400, and 450 ℃, and varying strain rates from 5×10^{-5} to 2×10^{-2} s^{-1}. In a typical SRC test, a pre strain of 10% was firstly applied at a low strain rate to stabilize the tensile system and microstructure, followed by 9 steps of tension at increasing strain rates. A pair of data of force and displacement was taken at each step to calculate flow stress and true strain rate, and furtherly stress exponent and activation energy.

Metallographic specimens before and after deformation were prepared through standard grinding, polishing, and etching procedure. The applied etching solution was made of 5 g picronitric acid, 25 ml glacial acetic acid, 100 ml absolute ethyl alcohol, and 10 ml deionized water. Microstructures were observed under a JVCTK-350EG optical microscopy.

Results and Discussion

Microstructure

The Mg-xAl (x=2, 3, 4, 6 wt.%) alloys before the EF tests have similar microstructures of equiaxial grains, as shown in Fig. 2(a-d), respectively. The equiaxial grains were kept after the EF tests at 450 ℃ and 10^{-3} s^{-1}, while obvious grain growth occurred, as shown in Fig. 2(e-h), respectively. The average grain sizes of the alloys before and after the EF tests are listed in Table 1. The grain growth indicates that dynamic recrystallization and corresponding grain refinement are not likely to occur at 450 ℃ and 10^{-3} s^{-1}. The microstructures at other deformation conditions show the similar behavior.

Elongation-To-Failure (EF) Test

A representative true stress-true strain curve of Mg-2Al during EF tests at 10^{-2} s^{-1} is shown in Fig. 3(a). The peak stresses at higher temperatures of 400 and 450 ℃ are much lower than those at 300 and 350 ℃, while steady-state strain and elongation increase with the increase of temperature. Similar behaviors are observed for the other three alloys, and as expected, the steady-state strain and elongation of each alloy improve at the lower strain rate of 10^{-3} s^{-1}. Elongations of over 100% can be achieved for each alloy at 450 ℃ and 10^{-3} s^{-1}.

Effects of Al content on the tensile behaviors at 300 ℃ and 10^{-2} s^{-1} are shown in Fig. 3(b). Tensile strength increases with the increase of Al content (x=2, 3, 4, 6 wt.%) due to solid solution strengthening. Correspondingly, the Mg alloys with higher Al content exhibit lower elongations, which is also observed at other temperatures.

Strain-Rate-Change (SRC) Test

High temperature deformation of metallic materials generally follows the constitutive equation[13]:

$$\dot{\varepsilon} = A\left(\frac{Gb}{kT}\right)\left(\frac{\sigma}{E}\right)^n\left(\frac{b}{d}\right)^p D \qquad (1)$$

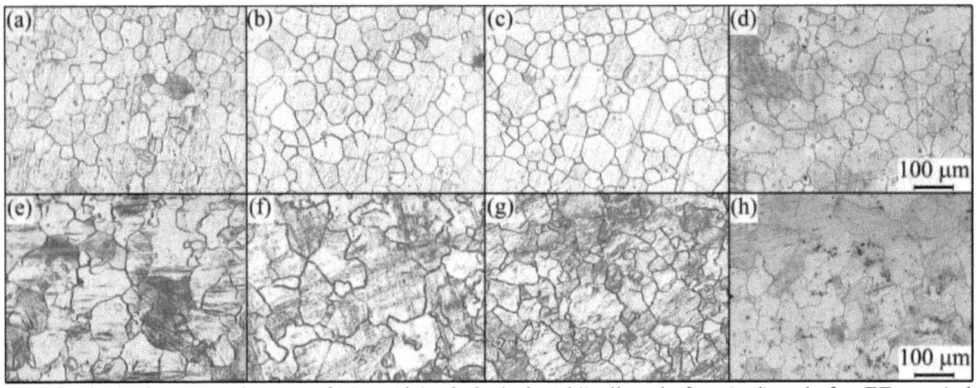

Fig. 2 Optical microscopy images of Mg-xAl (x=2, 3, 4, 6 wt.%) alloys before (a-d) and after EF tests(e-h) at 450 °C and 10^{-3} s^{-1}, respectively

Table 1 Average grain size before and after tensile tests at 450 ℃ and 10^{-3} s^{-1} (unit: μm)

	Mg-2Al	Mg-3Al	Mg-4Al	Mg-6Al
before	40.76±4.41	46.58±3.81	42.29±3.55	43.88±3.12
after	54.56±3.87	65.82±7.01	49.67±4.77	55.51±4.87

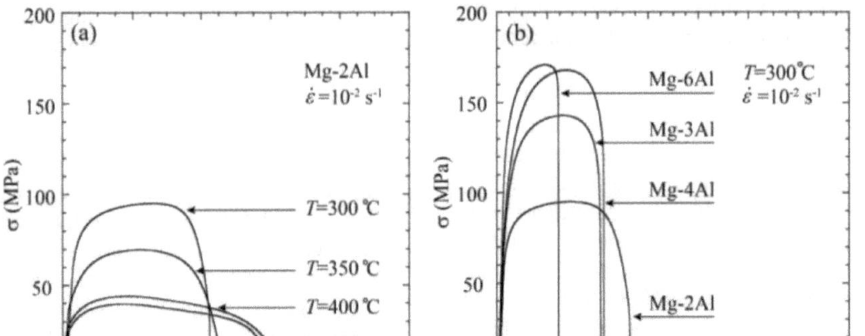

Fig. 3 True stress-true strain curves of EF tests: (a) Mg-2Al at 10^{-2} s^{-1} and different temperatures; (b) Mg-xAl at 300 ℃ and 10^{-2} s^{-1}

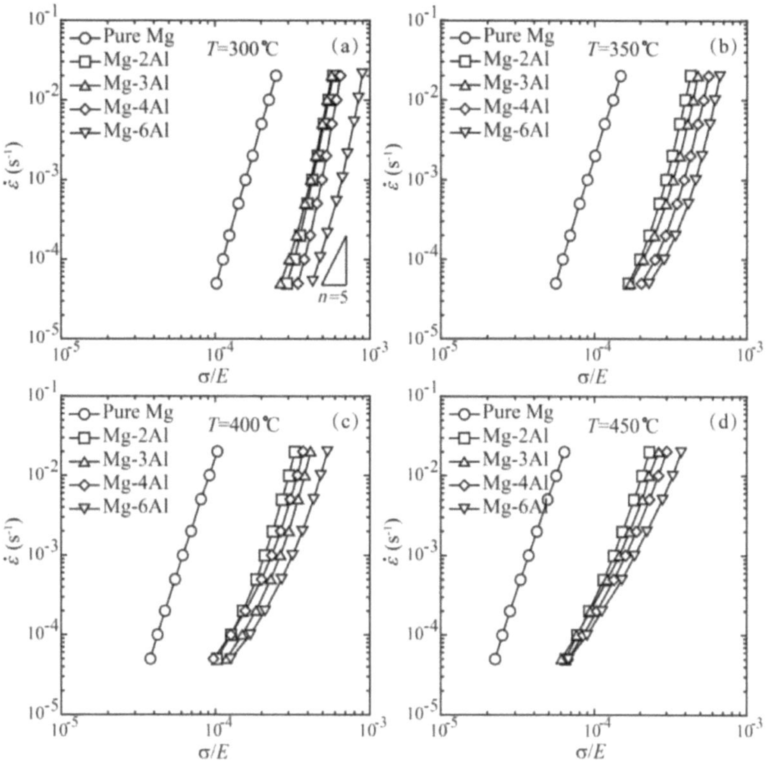

Fig. 4 SRC test data plotted as true strain rate versus modulus-compensated true stress

$\dot{\varepsilon}$ is the true strain rate; A is a constant related to the material, G is the shear modulus, b is the Burgers vector, k is the Boltzmann constant, T is the absolute temperature, σ is the flow stress, E is the elastic modulus, n is the stress exponent, d is the average grain size, p is the grain-size exponent (p=0 for coarse grain), D is the diffusivity, $D=D_0 \exp[-Q/(RT)]$, where D_0 is the pre-exponential factor, Q is the activation energy for creep, and R is the gas constant.

True stress-true strain curves of SRC tests are shown in dual logarithmic scales in Fig.4, where the true stress is compensated by the dynamic elastic modulus. According to Equation (1), the slope of each curve in Fig.4 is approximately taken as stress exponent n. The pure Mg exhibits a stable deformation mechanism with a nearly constant slope or stress exponent under the applied temperatures and strain rates. However, stress exponents of the Mg-xAl alloys increase with the increase of strain rate. Particularly, the stress exponents decrease with the increase of Al content at a constant temperature and a constant strain rate, which indicates that the addition of Al in Mg matrix may affect deformation mechanism and plasticity.

Tensile behaviors and deformation mechanism of the investigated alloys were evaluated and further analyzed by compensating the true strain rates in Fig.5 with diffusivity D, i.e. $\dot{\varepsilon}/D$, so that one master curve of $\dot{\varepsilon}/D$ versus σ/E integrating data at the four temperatures was built for each alloy, as shown in Fig. 5. The true strain rates of pure Mg were compensated by the self-diffusion coefficient with $D_0=1.0\times10^{-4}$ m^{-2} s^{-1} and Q=135 kJ·mol^{-1}[14], while the true strain rates of Mg-xAl were compensated by diffusion coefficient of Al in Mg with $D_0=1.2\times10^{-3}$ m^{-2}·s^{-1} and Q=143±10 kJ·mol^{-1}[15]. Pure Mg exhibits a stable stress exponent of $n\approx5.09$ under the applied temperatures and strain rates, as shown in Fig. 5. However, the stress exponents of the Mg-xAl alloys increases with the increase of $\dot{\varepsilon}/D$. Three regions can be classified according to the specific stress exponents in Fig.5, high n region, intermediate n region, and low n region, as shown in Table 2. The n value decreases with the increasing Al content in each region. The n values of the intermediate and low n regions are lower than $n\approx5.09$ of pure Mg. A low stress exponent is generally beneficial to plasticity. Strain rate sensitivity, the inverse of stress exponent, i.e. $m=1/n$, is an important parameter to evaluate deformation stability at high temperatures[16]. A higher m value generally indicates that the material has a stronger ability to inhibit necking, enlarge steady state deformation region, and enhance elongation. Therefore, higher elongations are expected for the intermediate and low n regions. In addition, a varying stress exponent usually indicates a varying deformation mechanism because different mechanisms have specific characteristic stress exponent[11]. Therefore, addition of Al not only improves the high temperature strength, but reduces stress exponent and improves plasticity by introducing new deformation mechanisms.

Deformation Mechanism

Both fine-grained superplasticity and dislocation creep could be deformation mechanism of high temperature deformation. Fine-grained superplasticity usually operates by grain boundary sliding in a microstructure with fine grain size (<10 μm), and exhibits a low stress exponent of $n\approx2$. However, the investigated Mg-xAl alloys have relatively large grain size, as shown in Table 1. Moreover, even the lowest stress exponent in the low n region is close to 3, i.e. $n\approx2.98$ for Mg-6Al, as shown in Table 2. Therefore, fine-grained superplasticity is not likely the dominate deformation mechanism.

Dislocation motion and plasticity can be significantly improved at high temperatures due to thermally activated prismatic and pyramidal slip systems in Mg alloys[17]. Dislocation motion is a sequential process of slip and climb, and

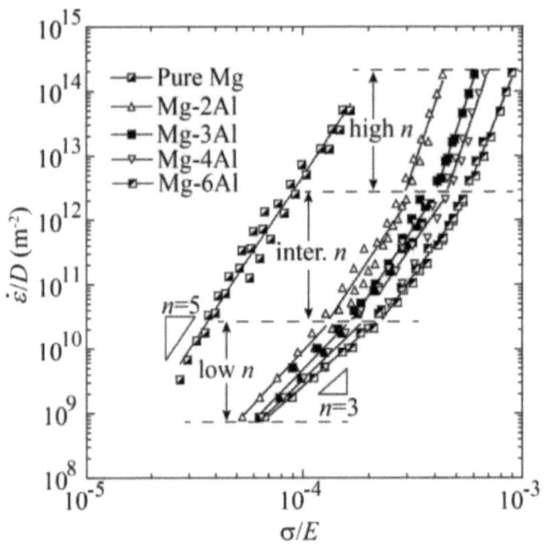

Fig. 5 SRC test data plotted as diffusivity-compensated true strain rate versus modulus-compensated true stress

Table 2 Calculated stress exponents of the three regions for Mg-xAl alloys

	$\dot{\varepsilon}/D$ (m^{-2})		
	Low *n* region 8.91×10^{8} ~2.65×10^{10}	Intermediate *n* region 2.65×10^{10}~2.88×10^{12}	High *n* region 2.88×10^{12} ~1.93×10^{14}
Mg-2Al	3.70	5.09	10.12
Mg-3Al	3.52	5.07	10.09
Mg-4Al	3.28	4.93	9.94
Mg-6Al	2.98	4.82	8.54

the slower one controls deformation rate. Dislocation climb is generally the slower one because it is a diffusional process through lattice or dislocation pipe. Therefore, dislocation-climb creep is the dominate mechanism in most pure metals[18]. The stress exponent of the investigated pure Mg is about 5.09, which is close to the characteristic value for dislocation-climb creep. Therefore, dislocation-climb creep should be the dominate mechanism in pure Mg.

In solid solutions, solute-drag creep could occur if dislocation slip were inhibited by a dragging force of solute-atom atmosphere interacting with the stress field of dislocations[5]. According to the Mg-Al binary phase diagram, β phase of $Mg_{17}Al_{12}$ gradually dissolves into the matrix with an increase of temperature[19]. In the low $\dot{\varepsilon}/D$ region, both of the high diffusivity at high temperatures and the low dislocation slip rate at low strain rates facilitate solute-drag creep, and the stress exponent is close to the characteristic value of solute-drag creep, $n\approx3$. In the intermediate $\dot{\varepsilon}/D$ region, deformation mechanism is transferring to dislocation-climb creep at lower temperatures and higher strain rates, and the stress exponent increases to $n\approx5$ at a lower diffusivity of solute atoms and higher dislocation slip rate. Dislocation creep gradually loses control on deformation with decreasing temperature and increasing strain rate in the high $\dot{\varepsilon}/D$ region, where the deformation is featured by power low breakdown, unstability, and fracture.

According to the solute drag theory, an increase of Al content facilitates solute-drag effect, improves its contribution to strain rate, and causes a drop of stress exponent. Therefore, the stress exponent in the low $\dot{\varepsilon}/D$ region decreases with an increase of Al content. It should be noted that several deformation mechanism could contribute simultaneously to high temperature deformation. For example, when dislocation-climb creep dominates at relatively low temperatures and high strain rates, dislocation slip rate may be still so low in certain slip systems that solute-drag creep becomes the local dominate mechanism, vice versa. Therefore, the *n* value decreases with an increase of Al content even in the intermediate and high $\dot{\varepsilon}/D$ regions in Fig.5.

Activation energy for deformation should be considered to further analyze deformation mechanism and plasticity of the investigated Mg-xAl alloys[20]. Activation energy for creep was calculated using the equation[21],

$$Q = nR\left.\frac{\partial(\ln\sigma)}{\partial(1/T)}\right|_{\dot{\varepsilon}} = nRk \tag{2}$$

where *n* is the stress exponent, *R* is the gas constant, σ is the peak stress in an EF test, *T* is the absolute temperature, and $\dot{\varepsilon}$ is the true strain rate. According to the EF test results at 10^{-3} s^{-1}, the Mg-xAl alloys exhibit large steady state deformation and enhanced elongation at 400 and 450 ℃, which approximately gives 1.78×10^{10} m^{-2} < $\dot{\varepsilon}_0/D$ <1.04×10^{11} m^{-2}, ranging from low to intermediate *n* regions. Data in this range in Fig.5 were linear regressed to calculate the *n* values in Equation (2), as shown in Fig. 6(a). EF data of σ at different temperatures were linear regressed to calculate *k* values in Equation (2), as shown in Fig. 6(b). The calculated activation energies using the *n* and *k* values are listed in Table 3.

According to Table 3, the activation energy when 1.78×10^{10} m^{-2} < $\dot{\varepsilon}_0/D$ <1.04×10^{11} m^{-2} is stable with the increase of Al content. The calculated activation energy is similar to both the activation energy for Al atom diffusion in Mg matrix, Q_{sol}=(143±10) kJ·mol^{-1}, and the activation energy for lattice diffusion of Mg, Q_{sd} =136 kJ·mol^{-1}[22]. Therefore, dislocation creep is the dominate mechanism responsible for the enhanced

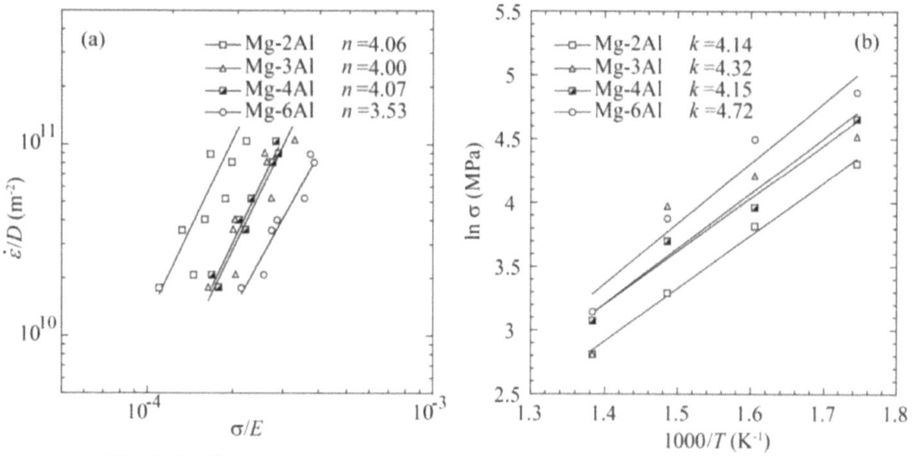

Fig. 6 The linear-regressed n and k values for activation energy calculation

Table 3 Creep activation energy of Mg-xAl alloys

material	$\dot{\varepsilon}$ (s^{-1})	n	k	Q (kJ/mol)
Mg-2Al		4.06	4.14	139.75
Mg-3Al	10^{-3}	4.00	4.32	143.67
Mg-4Al		4.07	4.15	140.43
Mg-6Al		3.53	4.72	138.52

elongations in the range of 1.78×10^{10} m^{-2} < $\dot{\varepsilon}_0/D$ <1.04×10^{11} m^{-2}, i.e. dislocation motions play a dominate role in plastic deformation, while grain boundary sliding only works as a coordinate mechanism. Both solute-drag creep and dislocation-climb creep could dominate as dislocation creep. However, solute-drag creep should contribute more to the enhanced plasticity based on the analysis above.

Conclusions

(1) The investigated coarse-grained Mg-xAl (x=2, 3, 4, 6 wt.%) alloys exhibit excellent plasticity at 450 ℃ and 10^{-3} s^{-1} with consistent tensile elongations of over 100%. Dislocation creep, particularly solute-drag creep, is responsible for the enhanced plasticity.

(2) Under the applied temperatures and strain rates, dislocation-climb creep is the dominate deformation mechanism of pure Mg, Al addition not only significantly improves strength, but also transfers the deformation mechanism to solute-drag creep. Stress exponent decreases with an increasing Al content at a constant strain rate due to the solute-drag effect.

(3) The stress exponent of each alloy increases with the increase of diffusivity-compensated strain rate until power law breaks down. The deformation mechanism transfers from solute-drag creep to dislocation-climb creep under the applied temperatures and strain rates.

Acknowledgements

The authors acknowledge the financial support of the Nationals Science Foundation of China (50801034), and the Foundation for "Ten-Hundred-Thousand" High-end Talents Introduction Project in Liaoning Province, China (2006207).

References

1. M. Suzuki et al.,"Creep behavior and deformation microstructures of Mg–Y alloys at 550 K,"Materials Science and Engineering: A, 252 (2) (1998), 248-255.
2. M. Suzuki et al.,"Strengthening effect of Zn in heat resistant Mg–Y–Zn solid solution alloys,"Scripta Materialia, 48 (8) (2003), 997-1002.
3. N. Stanford, K. Sotoudeh and P. S. Bate,"Deformation mechanisms and plastic anisotropy in magnesium alloy AZ31,"Acta Materialia, 59 (12) (2011), 4866-4874.
4. J. M. Boileau et al.,"Superplastic Response of Continuously Cast AZ31B Magnesium Sheet Alloys,"Journal of Materials Engineering and Performance, 19 (4) (2010), 467-480.
5. O. D. Sherby and E. M. Taleff,"Influence of grain size, solute atoms and second-phase particles on creep behavior of polycrystalline solids,"Materials Science & Engineering A, 322 (1) (2002), 89-99.
6. J. J. Blandin et al.,"Effect of thermal treatments on mechanical behaviour of thixoformed magnesium alloy,"Materials Science and Technology, 18 (3) (2002), 333-340.
7. Y. H. Wei et al.,"Current Status and DevelopmentTrends of Study on Superplastic Magnesium Alloys,"Materials Review, (09) (2002), 20-23.
8. S. Q. Zhu,"An exploratory study on the principle of the fabrication of ultrafine grained magnesium sheets using medium-high strain rate rolling technique and the related fundamental research"(Ph.D.thesis, Hunan University, 2012).
9. R. Fernández and G. González-Doncel,"A unified description of solid solution creep strengthening in Al–Mg alloys,"Materials Science and Engineering: A, 550 (2012), 320-324.
10. F. R. N. Nabarro,"Distribution of solute atoms round a moving dislocation,"Materials Science and Engineering: A, 400–401 (2005), 22-24.
11. S. L. Zhang et al.,"Creep mechanism of Mg-5.5Zn-2Gd-0.6Zr cast alloy "Transactions of Nonferrous Metals Society of China, (02) (2012), 331-336.

12. H. Somekawa et al.,"Dislocation creep behavior in Mg–Al–Zn alloys,"Materials Science and Engineering: A, 407 (1–2) (2005), 53-61.
13. Z. X. Wang, X. F. Liu and J. X. Xie,"Constitutive relationship of hot deformation of AZ91 magneesium alloy,"Acta Metallurgica Sinica, (11) (2008), 1378-1383.
14. H. J. Frost and M. F. Ashby.Deformation mechanism maps: the plasticity and creep of metals and ceramics (Oxford,OXON: Pergamon Press Inc,1984), 224-225.
15. S. S. Vagarali and T. G. Langdon,"Deformation mechanisms in h.c.p. metals at elevated temperatures--II. Creep behavior of a Mg-0.8% Al solid solution alloy,"Acta Metallurgica, 30 (6) (1982), 1157-1170.
16. H. Watanabe et al.,"Deformation mechanism in a coarse-grained Mg-Al-Zn alloy at elevated temperatures,"International Journal of Plasticity, 17 (3) (2001), 387-397.
17. Y. M. Zhan et al.,"Superplastic Deformation in Fined-grained Materials and Its study in magnesium Alloys,"Light Alloy Fabrication Technology, (08) (2003), 6-10.
18. S. Spigarelli et al.,"Analysis of the creep behaviour of a thixoformed AZ91 magnesium alloy,"Materials Science & Engineering A, 289 (1) (2000), 539-543.
19. Y. S. Wang et al.,"Models for casting solidification of Mg-9Al Alloy,"Acta Metallurgica Sinica., (05) (2002), 539-543.
20. Y. Gao,"Microstructure, Properties and Creep behavior of Mg-Gd-Y-Zn-Zr alloys"(Ph.D.thesis, Shanghai Jiao Tong University, 2009).
21. H. Q. Chen et al.,"Hot deformation mechanism and microstructure evolution of TC11 titanium alloy in β field,"Transactions of Nonferrous Metals Society of China, 18 (5) (2008), 1021-1027.
22. P. Zhang,"Creep behavior of the die-cast Mg–Al alloy AS21,"Scripta Materialia, 52 (4) (2005), 277-282.

Magnesium Technology 2016

Texture and Formability

Magnesium Technology 2016
Edited by: Alok Singh, Kiran Solanki, Michele V. Manuel, and Neale R. Neelameggham
TMS (The Minerals, Metals & Materials Society), 2016

IN-SITU EBSD OBSERVATIONS OF RECRYSTALLIZATION AND TEXTURE EVOLUTION IN ROLLED Mg-2Zn-xCe (wt.%)

Ajith Chakkedath[1], David Hernández Escobar[2], Jan Bohlen[3], Sangbong Yi[3], Dietmar Letzig[3], Carl Boehlert[1,4]

[1]Department of Chemical Engineering and Materials Science, Michigan State University, East Lansing, Michigan 48824-1226, USA
[2]Technical University of Madrid, Calle Ramiro de Maeztu, 7, 28040 Madrid, Spain
[3]Magnesium Innovation Centre MagIC, Helmholtz-Zentrum Geesthacht, Max-Planck-Str. 1, D-21502 Geesthacht, Germany
[4]IMDEA Materials Institute, C/Eric Kandel 2, 28906-Getafe, Madrid, Spain

Introduction

Wrought magnesium (Mg) alloys are attractive for automotive applications where light-weighting is critical. One of the major challenges with conventional Mg alloys, such as Mg-3Al-1Zn(wt.%), is the strong texture formation after rolling and subsequent annealing, which results in anisotropic mechanical properties and poor cold formability [1]. It was observed that relatively weak texture can be obtained by dilute rare-earth additions such as Ce [2]. However, the mechanisms responsible for the weak texture formation during annealing are not clear. The goal of this study was to investigate recrystallization and texture evolution during annealing in a Ce containing Mg alloy system using in-situ techniques.

Experimental methods

Mg-2Zn-0.2Ce(wt.%) and Mg-2Zn-0.6Ce(wt.%) alloy sheets, ~1mm thick, were studied in the as-rolled condition. For electron backscatter diffraction (EBSD) analysis, ~10mm wide and ~15mm long samples were cut from the sheet material using a diamond saw. One of the sample surfaces was mechanically polished and then electropolished using a mixture of 30% Nitric acid and 70% Methanol as electrolyte. The electrolyte temperature was kept below -25°C and a voltage of ~12V was used.

EBSD analysis was performed while the sample was kept at a desired temperature. The specimen was heated using a 6mm diameter tungsten-based heating element placed beneath the sample. EBSD maps of the same area of ~100μm x ~125μm was collected at 25°C, 150°C, 200°C, 225°C, 250°C, 275°C, 300°C, and 325°C. Each scan took ~45-55 minutes at a 0.5μm stepsize. The specimen temperature was kept constant within ±3°C during the scan. A thermocouple was spot welded to the sample surface to monitor the temperature.

Results

Both as-rolled materials exhibited a basal texture in which the c-axis tends to align perpendicular to the rolling direction. The Mg-2Zn-0.2Ce(wt.%) sheet exhibited a stronger texture, with a maximum intensity of ~10 times random in 0001 pole figure along the normal direction (ND), compared to the Mg-2Zn-0.6Ce(wt.%) sheet, which exhibited a maximum intensity of ~5 times random.

In Mg-2Zn-0.2Ce(wt.%), new grains started to appear at ~200°C (See Figure 1). At ~300°C, a completely recrystallized microstructure was formed. Grain boundaries with orientation relationships corresponding to $\{10\bar{1}2\}$ extension twinning (86° about $<11\bar{2}0>$), $\{10\bar{1}1\}$ contraction twinning (56° about $<11\bar{2}0>$), and $(10\bar{1}2)$-$(01\bar{1}2)$ extension double twin (60° about $<10\bar{1}0>$) were observed. This was expected to be due to the recovery and growth of the twins formed during the rolling process. Among the misorientation relationships observed between the newly formed high angle grain boundaries, rotation axis about $<10\bar{1}0>$, $<11\bar{2}0>$, and $<10\bar{1}1>$ were the most prevalent. It is noted that rotation axes about other <hki0> and <hkil> were also observed.

In the completely recrystallized microstructure, the fraction of high angle grain boundaries with a rotation axis about $<10\bar{1}0>$, $<11\bar{2}0>$, and $<10\bar{1}1>$ in Mg-2Zn-0.2Ce(wt.%) and Mg-2Zn-0.6Ce(wt.%) were similar with ~16% for each case. However, a significantly lower fraction (~6%) of grain boundaries with rotation axis about $<10\bar{1}1>$ was observed in Mg-3Al-1Zn(wt.%).

References

[1] Boehlert, C. J., et al. Acta Materialia 60.4 (2012): 1889-1904.
[2] Yi, Sangbong, et al. Acta Materialia 58.2 (2010): 592-605.

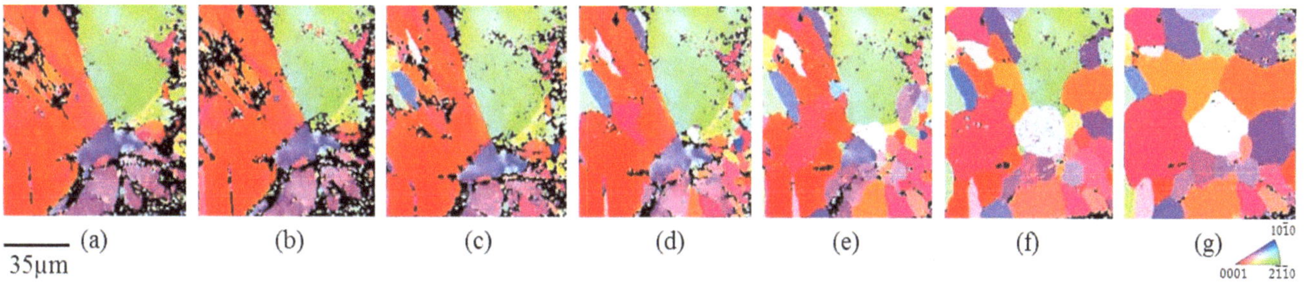

Figure 1. The EBSD IPF map in the normal direction of the same area analyzed for Mg-2Zn-0.2Ce(wt.%) captured at (a) 25°C (b) 150°C (c) 200°C (d) 225°C (e) 250°C (f) 275°C (f) 300°C and (g) 325°C. The rolling direction is horizontal.

Magnesium Technology 2016
Edited by: Alok Singh, Kiran Solanki, Michele V. Manuel, and Neale R. Neelameggham
TMS (The Minerals, Metals & Materials Society), 2016

NON-BASAL TEXTURE EVOLUTION DURING ANNEALING OF COLD-DEFORMED MAGNESIUM ALLOY

Abu Syed Humaun Kabir[1], Jing Su[1], In-Ho Jung[1], and Stephen Yue[1]

[1]Department of Mining and Materials Engineering, McGill University,
3610 University Street, Montreal, Quebec, Canada H3A 0C5

Keywords: Dynamic recrystallization, Annealing, Texture randomization

Abstract

In this study, effects of statically and dynamically formed precipitates on the microstructure and texture evolutions were investigated after two step deformations and subsequent post deformation annealing of Mg-3Al-3Sn (wt.%) alloy. At first, the alloy was compressed at low temperature and at a relatively high strain rate to simulate the rolling behavior. Annealing at moderate temperature recrystallizes the microstructure and forms static precipitation. Further deformation was performed on the recrystallized microstructures with the presence of precipitation followed by a subsequent annealing at the deformation temperature. Microstructures were characterized by optical and scanning electron microscopes and macro- and micro-texture were measured by X-ray diffraction (XRD) and Electron Back-Scattered Diffraction (EBSD) techniques, respectively. Fine recrystallized grains were obtained after annealing due to the presence of large volume fraction of precipitates. Also, the texture was weakened significantly after the final annealing step probably due a mechanism that involves the precipitates.

Introduction

Magnesium alloys are attractive materials for weight reduction applications in automotive industries due to their highest specific strength among the structural metals [1]. However, magnesium wrought products show poor formability at lower temperatures due to the insufficient number of operating slip systems related to its hexagonal closed packed structure. Grain refinement is one of the most significant approach to improve the strength as well as ductility of wrought magnesium products [2]. Significant increase in yield strength has been reported by reducing the grain size for extruded AZ31 [3]. During the hot deformation of Mg-alloys, dynamic recrystallization (DRX) takes place which itself is an important grain refining mechanism. During post deformation annealing, further grain growth or grain coarsening may take place that may decline the mechanical properties of the final products.

Strength and formability of wrought magnesium alloys largely depend not only on final grain size but also on deformation and recrystallization textures depending on the deformation conditions [4]. The basal texture is almost unavoidable for the typical industrial wrought magnesium alloys and processing techniques due to the lowest critical resolved shear stress (CRSS) of basal slip system [1]. The potential to improve the formability of magnesium alloys by modifying its texture has been reported number of times [4-5]. Several studies have also reported texture weakening of Mg alloys during post deformation annealing [5-7]. The effects of precipitates on texture weakening in magnesium alloys have been attributed mostly to the particle stimulated nucleation (PSN) mechanism [8-10]. Fine precipitates at the grain boundary may also affect the recrystallization textures. Recently, Xia et al. [11] reported a weaken texture due to deformation in the presence of grain boundary precipitation in an Mg-Gd-Y-Nd-Zr alloy. According to the authors, precipitates pinned the grain boundaries preventing the usual rotation of grains towards the basal orientation. The main precipitates for Mg-Al-Sn alloys has been reported to be Mg_2Sn when deformed or annealed at moderate to high temperature [12]. Mg_2Sn is incoherent in nature with the matrix and therefore, a strong pinning pressure is expected from this alloy system [12].
In this study, efforts have been taken to understand the effects of fine precipitates on the recrystallized grain size and texture evolution during deformation as well as during the post deformation annealing.

Experimental Procedures

Mg-3wt% Al- 3wt% Sn (AT33) alloy was produced with commercially pure Mg, Al and Sn melted using an induction furnace (20 kW, 5 kHz) in a graphite crucible and cast in a copper mold preheated to 250°C. The cast ingots were homogenized for 24 hours at 440°C to fully dissolve the as-cast precipitates. Compression samples of 11.4 mm in height and 7.6 mm in diameter (h/d ratio of 1.5) were machined from the homogenized bar. The uniaxial compression tests were performed using a servo-hydraulic materials testing system (MTS) with 100 kN capacity, coupled with a radiant furnace. At first, isothermal compression testing was performed at 120°C for a strain of 0.35 with a relatively high strain rate of $1.0s^{-1}$ in order to simulate the rolling behaviour. The specimen was annealed at 250°C for 60 minutes followed by water quenching. The specimen was further deformed at 250°C for a strain of 0.5 with a strain rate of $0.01s^{-1}$ followed by a second annealing at 250°C for 30 minutes. Specimens for metallurgical characterization were cut along the compression axis. Polished specimens were etched in a fresh solution of 10 ml acetic acid, 4.2 gm picric acid, 10 ml water and 70 ml alcohol for 3-8 seconds. Microstructural examination was carried out using Nikon Epiphot 200 optical microscope and Hitachi SU3500 scanning electron microscope. Grain size was measured by a linear intercept method according to the ASTM standard E112. The bulk texture was assessed by XRD using a Bruker D8 diffractometer (Co Kα source). Micro-texture was studied by electron backscattered diffraction (EBSD) technique in a Hitachi SU-3500 SEM operating at 20 kV with an Oxford EBSD detector.

Results and Discussions

Figure 1 shows the optical and back scattered electron (BSE) images of the specimen compressed at 120°C. The optical microstructure shows an expected heavily twinned structure. In the cold deformed Mg-alloys, deformation progressed by forming basal slip and twins, which have lower critical resolved shear stresses (CRSS) compared to prismatic and pyramidal slip systems. The SEM image shows that no precipitates formed, as the precipitation kinetics is very slow at this low temperature. For this particular alloy, significantly large fraction of dynamic or strain induced precipitates formed during deformation within the temperature range of 250-350°C, which in fact was termed by the authors as recrystallization induced precipitates as the precipitates always formed on the recrystallized grain boundaries [12].

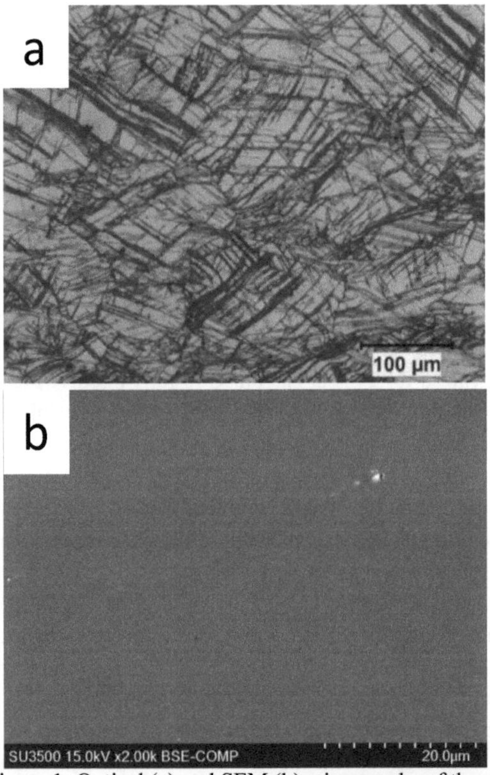

Figure 1: Optical (a) and SEM (b) micrographs of the as-compressed condition

Figure 2 shows the optical and BSE SEM images of the cold-compressed specimen after annealing at 250°C for 60 minutes. The optical microstructure shows a complete static recrystallization with an average grain size of 8.5 µm. Cold deformed alloy contains large number of dislocations in the substructure, which act as driving force for inducing the static recrystallization (SRX) during subsequent annealing. Also, large fraction of twins creates large number of twin intersections which are high in strain concentrations and therefore, preferential sites for SRX nucleation [13]. The BSE image indicates the formation of static precipitates on the SRXed grain boundaries. This indicates that the precipitates form after the static recrystallization of grains. This behavior somehow contradicts the precipitation behavior in micro-alloyed steel where the precipitation of niobium carbo-nitrides takes place on dislocations and low angle grain boundaries effectively stopping recrystallization [14].

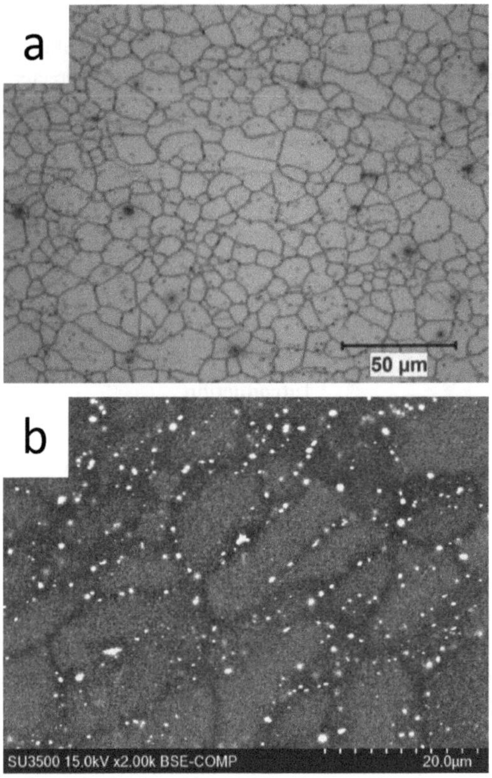

Figure 2: Optical (a) and SEM (b) micrographs after annealing at 250°C for 60 minutes

As reported before, the annealed specimen was further deformed at 250°C for a strain of 0.5 with a strain rate of $0.01s^{-1}$. Figure 3 shows the microstructures after deformation, where the optical image (Fig. 3-a) shows a significant reduction in recrystallized grain size (~3.5 µm) due to dynamic recrystallization. DRX grains size is a function of initial grain size, where finer DRX grains are obtained with smaller initial grain size due to the fact that the smaller initial grain size provides large fraction of grain boundaries to undergo dynamic recrystallization [15]. Figure 3-b shows the BSE image of the specimen after the hot deformation. Dynamic precipitates formed on the new recrystallized grain boundaries as well as at the interior of the grains.

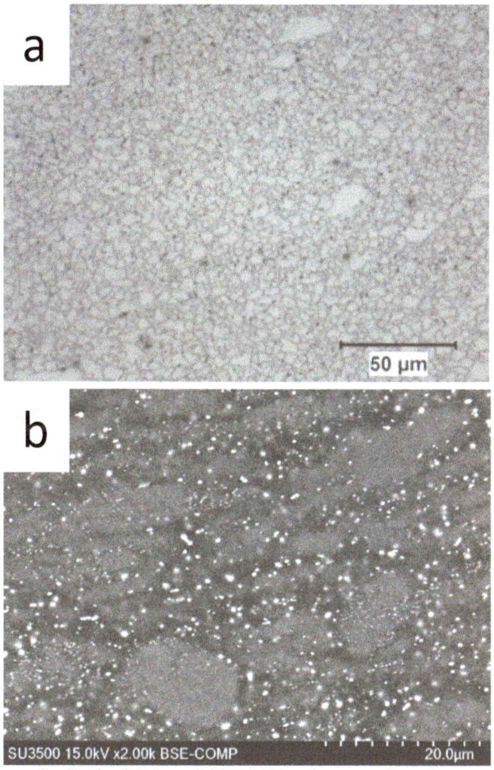

Figure 3: Optical (a) and SEM (b) micrographs after second deformation at 250°C at a strain rate of $0.01s^{-1}$

Figure 4 shows the optical and SEM images of the alloy annealed further at 250°C for 30 minutes. Grain size increases a little (from 3.5 to 4.1 µm) after the annealing for 30 minutes due to the strong precipitation pinning. Amount of precipitation also increases significantly after annealing.
Figure 5 shows the (00.2) pole figures of the alloy at different conditions. This macro-texture study reveals the formation of strong basal texture after compression with a maximum intensity of 3.2. It is well known that during deformation, the basal plane of the Mg-HCP lattice lies perpendicular to the deformation direction. Specimen after annealing for 60 min after cold deformation retains the basal texture but the texture intensity is lower from 3.2 to 2.8. Further deformation at 250°C, increases the intensity of the basal texture to 4.1. Annealing at 250°C after the second deformation significantly weakens the basal texture, where the intensity drops to 1.7. A non-basal texture component in the (00.2) pole figure along the ND direction is observed at this condition. To understand this new non-basal component, EBSD study was performed on this sample and is shown in Fig. 6. Low-angle grain boundaries (LAGB) (mis-orientation angle 3-15°) are marked out by thin white lines, while high-angle boundaries (HAGB) (mis-orientation angle more than 15°) are by black lines. The inverse pole figure map in Fig. 6-a indicates a significant fraction of non-basal grains in the structure. EBSD study on the previous conditions did not show the existence of these much non-basal grains. Figure 6-b shows the (00.1) pole figures of only the non-basal grains selected by the TSL-OIM software. The orientation of these non-basal grains is mostly towards the ND, which means texture weakening. Therefore, the micro-texture study by EBSD supports the results obtained from macro-texture.

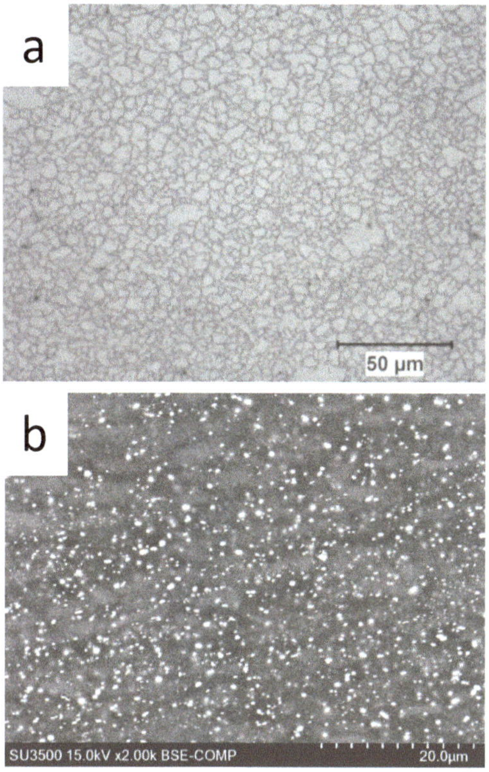

Figure 4: Optical (a) and SEM (b) micrographs after second annealing at 250°C for 30 minutes

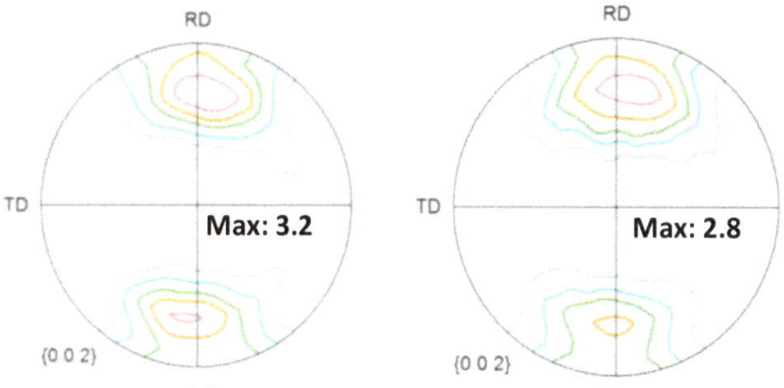

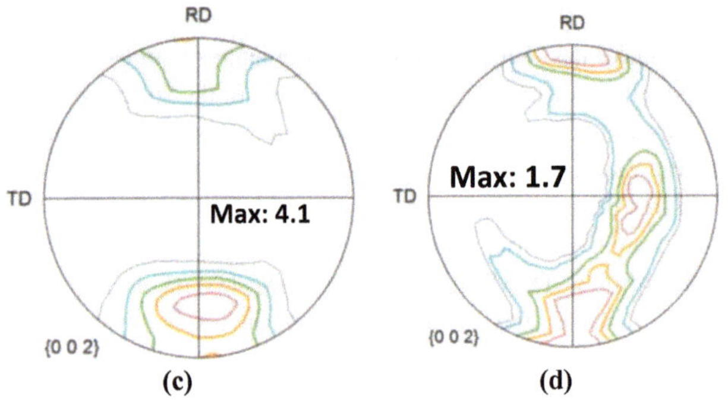

Figure 5: (00.2) pole figures of the alloy at all the conditions (a) as-compressed, 120°C, Strain rate 1.0/s, strain 0.35 (b) Annealed at 250°C for 60 minutes, (c) Recompressed, 250°C, strain rate 0.01/s, strain 0.5, (d) Annealed at 250°C for 30 minutes.

In this study, the amount of precipitates is the maximum at the final stage. The precipitates are identified as Mg_2Sn intermetallics with a size range of 70-200 nm. Generally, Mg_2Sn forms a non-coherent interface with the Mg-matrix, and it is well-known that dislocations cannot shear the non-coherent precipitates [16]. Therefore, a strong and effective pinning pressure is expected on the recrystallized grain boundaries that refined the final grain structure. Texture weakening mechanism after the final annealing step is still unknown. Aluminum in the solid solution may play a role in texture weakening through solute drag effect, but it has also been reported as a very slow mechanism [17]. Therefore, the precipitates are probably responsible for the texture weakening.

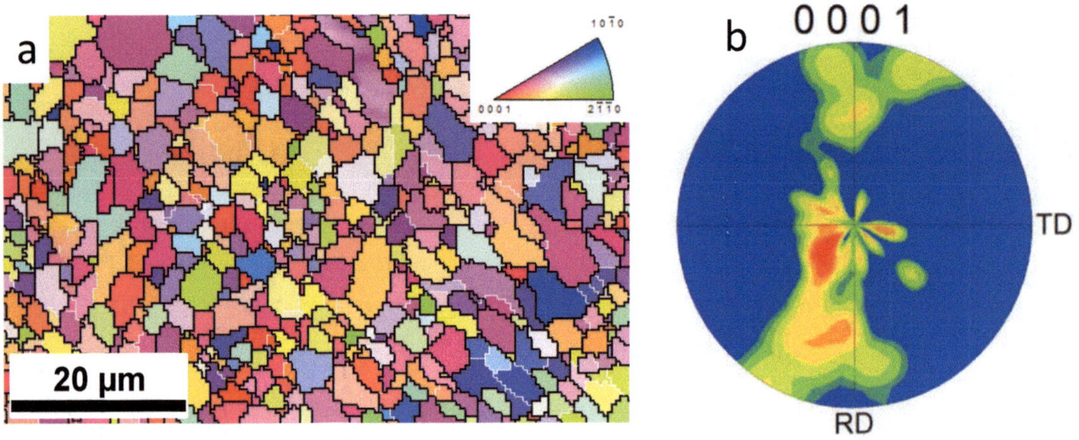

Figure 6: (a) Inverse pole figure mapping of the alloy at the final condition, (b) (0001) pole figure of the recrystallized grains

Summary

Mg-Al-Sn alloy was designed with the aim to form dynamic and static precipitates during hot deformation and annealing. The objective was to study the effect of precipitates on recrystallized microstructures and texture. The results indicate that static precipitates form after recrystallization on the SRX grain boundaries. Formation of large amount of precipitates significantly pins the grain boundary migration, resulting in steady grain sizes after annealing. Texture weakened significantly after two stages of annealing that may be related to the presence of significantly large fraction of precipitates on the recrystallized grain boundaries

Acknowledgements

Financial supports from General Motors and NSERC, Canada are highly acknowledged.

References

1. Huang, X., et al., "Influence of rolling temperature on static recrystallization behavior of AZ31 magnesium alloy," *Journal of Materials Science,* 47(11), (2012), 4561-4567.
2. Beer, A.G. and M.R. Barnett, "Microstructure evolution in hot worked and annealed magnesium alloy AZ31," *Materials Science and Engineering: A*,. 485(1-2), (2008), 318-324.
3. Barnett, M.R., et al., "Influence of grain size on the compressive deformation of wrought Mg-3Al-1Zn," *Acta Materialia*, 52(17), (2004), 5093-5103.
4. Sanjari, M., et al., "Promotion of texture weakening in magnesium by alloying and thermomechanical processing: (I) alloying," *Journal of Materials Science*, 49(3), (2014), 1408-1425.
5. Farzadfar, S.A., et al., "Texture weakening and static recrystallization in rolled Mg–2.9Y and Mg–2.9Zn solid solution alloys," *Journal of Materials Science*, 47(14), (2012), 5488-5500.

6. Huang, X., K. Suzuki, and Y. Chino, "Annealing behaviour of Mg–3Al–1Zn alloy sheet obtained by a combination of high-temperature rolling and subsequent warm rolling," *Journal of Alloys and Compounds*, 509(14), (2011), 4854-4860.
7. Huang, X., K. Suzuki, and Y. Chino, "Different annealing behaviours of warm rolled Mg–3Al–1Zn alloy sheets with dynamic recrystallized microstructure and deformation microstructure," *Materials Science and Engineering: A,* **560**(0), (2013), 232-240.
8. Ball, E.A. and P.B. Prangnell, "Tensile-compressive yield asymmetries in high strength wrought magnesium alloys," Scripta Metallurgica et Materialia 31(2), (1994), 111-116.
9. Bohlen, J., et al., "The texture and anisotropy of magnesium-zinc-rare earth alloy sheets," *Acta Materialia*, 55(6), (2007), 2101-2112.
10. Li, T., et al., "Dynamic precipitation during multi-axial forging of an Mg–7Gd–5Y–1Nd–0.5Zr alloy," *Journal of Magnesium and Alloys,* 1(1), (2013), 47-53.
11. Xia, X., et al., "Microstructure and texture of coarse-grained Mg–Gd–Y–Nd–Zr alloy after hot compression," *Materials & Design*, 44(0), (2013), 521-527.
12. Kabir, A.S.H., et al., "Effect of strain-induced precipitation on dynamic recrystallization in Mg–Al–Sn alloys," *Materials Science and Engineering: A*, 616(0), (2014), 252-259.
13. Yang, X., et al., "Annealing of a magnesium alloy AZ31 after interrupted cold deformation," *Materials & Design*, 36(0), (2012), 626-632.
14. Simielli, E.A., S. Yue, and J.J. Jonas, "Recrystallization kinetics of microalloyed steels deformed in the intercritical region," *Metallurgical Transactions A*, 23(2), (1992), 597-608.
15. Beer, A.G. and M.R. Barnett, "Influence of initial microstructure on the hot working flow stress of Mg–3Al–1Zn," *Materials Science and Engineering: A*, 423(1–2), (2006), 292-299.
16. Masoumi, M. and M. Pekguleryuz, "The influence of Sr on the microstructure and texture evolution of rolled Mg–1%Zn alloy," *Materials Science and Engineering: A*, 529(0), (2011), 207-214.
17. Ravi Kumar, N.V., et al., "Grain refinement in AZ91 magnesium alloy during thermomechanical processing," *Materials Science and Engineering A*, 359(1-2), (2003), 150-157.

On Modeling the Mechanical Behavior and Texture Evolution of Rolled AZ31 Mg for Complex Loadings Involving Strain Path Changes

Nitin Chandola[1], Crystal Pasiliao[2], Oana Cazacu[1]*, Benoit Revil-Baudard[1]
[1]University of Florida, Dept. of Mechanical and Aerospace Engineering,
REEF,1350 N. Poquito Road Shalimar, FL 32579, USA
[2]Air Force Research Laboratory, Munitions Directorate, Shalimar, FL 32579, USA

Keywords: Strain-path changes; Texture evolution; Visco-plastic Self Consistent Model; Twinning

Abstract

An accurate description of the deformation response of AZ31 Mg under changing strain paths requires consideration of its strong anisotropy and its evolution with accumulated plastic deformation. The general held belief is that without modeling de-twining it is impossible to describe the effect of the pre-strain on hardening behavior in low cycle compression-tension-compression and tension-compression-tension tests, respectively. In this paper, it is shown that using the viscoplastic self-consistent crystal plasticity model in conjunction with the pre-dominant twinning reorientation (PTR) scheme it is possible to accurately model the reorientation of the microstructure for such complex loadings. Comparison with recent data reported by Hama et al. [1, 2] shows that the stress-strain response is also very well predicted.

1. Introduction

The behaviour of AZ31 Mg under strain path changes, in particular the marked difference in response under cyclic loading depending on whether tension or compression is applied first, was experimentally investigated in Lou et al. [3], and most recently in Hama et al.[1, 2]. The very unusual work-hardening behaviour observed was attributed to twinning and de-twinning.

Starting from the lamellar grain model of Lebensohn ([4]), Tome and Kaschner ([5]) and Proust et al. (see [6]) developed a new description of twinning that accounts for twin shape and twin-parent interaction. In order to address the role of twinning during strain path changes in AZ31 Mg, the latter model was further extended in [6] to incorporate de-twinning, and implemented in the VPSC code. Very recently, the behaviour of AZ31 Mg during cyclic loadings was also modelled by Hama and collaborators using a crystal plasticity finite-element (CPFE) framework.

In this paper, we show that the behaviour of AZ31 Mg when subjected to strain-path changes can be accurately modelled if the reorientation of microstructure associated with deformation twinning is described using the predominant twin reorientation scheme (PTR). Specifically, it is shown that using the VPSC model in conjunction with the PTR scheme it is possible to predict with accuracy the influence of pre-strain on the work-hardening response of the material under cyclic tension-compression-tension loadings (T-C-T) as well as under cyclic compression-tension-compression (C-T-C) loadings. The predicted stress-strain response is compared with the very recent mechanical data reported in [1]. Furthermore, we present the predicted texture evolution for both loading scenarios. While there were no measured textures for comparison, Hama et al. presented in [1] the pictures of the microstructures observed with an optical microscope for a T-C-T test (no pre-strain) at different stages of the loading cycle. This allowed for verification of the predicted reorientation of the microstructure for cyclic loadings.

2. Single Crystal and Polycrystal Model

In the VPSC model a self-consistent approach is used to describe the interaction of a grain with its surroundings. Every grain is treated as an anisotropic ellipsoidal inclusion embedded in a homogeneous effective medium. Plastic deformation in each grain occurs by activation of both slip and twinning modes. The grain level constitutive behavior relating the Cauchy stress deviator and the viscoplastic strain rate **d** is:

$$\mathbf{d}(\mathbf{x})=\sum_{s}\mathrm{m}^{s}(\mathbf{x})\dot{\gamma}^{s}(\mathbf{x}),\quad \dot{\gamma}^{s}=\dot{\gamma}_{0}\left|\frac{\mathbf{m}^{s}:\boldsymbol{\sigma}'}{\tau_{C}^{s}}\right|^{1/n}\operatorname{sgn}\left(\mathbf{m}^{s}:\boldsymbol{\sigma}'\right), \tag{1}$$

where the sum is over all slip and twin systems, $\mathbf{m}^{s}$ is the Schmid tensor, $\dot{\gamma}^{s}$ is the shear rate on the system given by a power law with $\dot{\gamma}_0$, n, and τ_C being the reference slip rate, rate-sensitivity exponent, and resistance of the slip/twin system, s, respectively. A Voce-type hardening law is used to describe the evolution of the critical resolved shear stress as a function of accumulated shear strain:

$$\tau_{c}^{s}=\tau_{0}^{s}+\left(\tau_{1}^{s}+\theta_{1}^{s}\Gamma\right)\left(1-\exp\left(-\frac{\theta_{0}^{s}\Gamma}{\tau_{1}^{s}}\right)\right), \tag{2}$$

where τ_0^s , τ_1^s , θ_1^s , θ_0^s are the initial threshold stress, initial hardening rate, asymptotic hardening rate, and back extrapolated threshold stress, respectively. In addition, we allow for self and latent hardening using the coupling coefficients, which empirically account for the obstacles that new dislocations (or twins) with system s' create for the propagation of dislocations (or twins) on system *s*. The increase in the threshold stress is calculated as:

$$\Delta\tau_{o}^{s(r)}=\frac{d\tau^{*s(r)}}{d\Gamma^{(r)}}\sum_{k'}h^{ss'}\dot{\gamma}^{s'(r)}\Delta t, \tag{3}$$

The twinning contribution to texture development is accounted for by means of the PTR scheme (see [7]). In this scheme, within each grain the shear strain contributed by each twin system and the associated volume fraction in the grain are tracked. The sum over all the twin systems associated with a given twin mode and over all grains, represents the 'accumulated twin fraction', in the aggregate for the particular mode. The accumulated twin fraction for the aggregate, $\mathrm{V}^{\mathrm{acc,mode}}$ is calculated as,

$$\mathrm{V}^{\mathrm{acc,mode}}=\sum_{g}\sum_{t}\gamma^{g,t}/S^{t} \tag{4}$$

where, S^{t} (characteristic twin shear for the twinning mode) is a material parameter. Since it is not numerically feasible to consider each twinned fraction as a new orientation, a statistical approach is used to determine new orientations (for more details, see [7]).

At each incremental step, grains are fully re-oriented provided that certain conditions are fulfilled. Specifically, an effective twinned fraction $V^{eff,mode}$ associated with the fully reoriented grains for the given mode as well as a threshold value, $V^{th,mode}$ are defined. The threshold volume fraction $V^{th,mode}$ is defined as:

$$V^{th,mode} = A^{th1} + A^{th2}\frac{V^{eff,mode}}{V^{acc,mode}}, \qquad (5)$$

where, A^{th1} and A^{th2} are material parameters, $V^{eff,mode}$ is the effective twinned fraction and $V^{acc,mode}$ is the accumulated twin fraction. In any given grain, the twin system with the highest accumulated volume is identified. If the highest accumulated volume fraction, for the given mode is larger than the threshold value $V^{th,mode}$ then the grain is allowed to re-orient and $V^{eff,mode}$ and $V^{th,mode}$ are both updated. This process is repeated until either all grains are randomly checked or until the effective twin volume exceeds the accumulated twin volume. In the latter case reorientation by twinning is stopped and we proceed to the next deformation step.

3. Calibration of the Visco-plastic Self Consistent Model for AZ31 Mg

All the parameters of the VPSC model are related to the single crystal deformation behavior. Given that single crystal data are not available, the current practice is to fit the parameters associated to each deformation mode that is potentially active using macroscopic stress-strain curves and the initial texture of the polycrystalline material. In this work, the calibration is done using the room-temperature monotonic stress-strain curves in uniaxial tension, compression (strain rate of 10^{-4}/s) and shear reported by Khan et al. [8]. The initial texture of the material is determined from the entire EBSD scan (784x808 μm) to generate 2762 weighted orientations (see Figure 1). Note that the initial texture of the material is strongly basal with an almost equal fraction of grains having the c-axis slightly tilted away from the sheet normal towards +RD and –RD, respectively (around ±30^0).

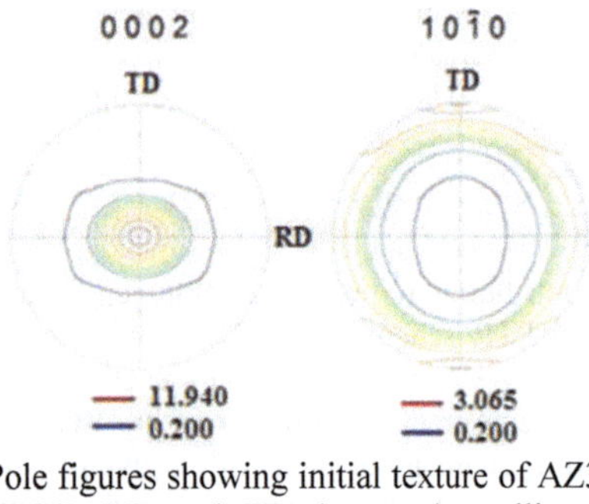

Figure 1. Pole figures showing initial texture of AZ31 Mg used as input in VPSC. RD and TD denote the rolling direction and transverse in-plane direction, respectively.

The step-by-step procedure that we use for identification of the VPSC model parameters is that proposed by Chandola et al [9]. The key idea is to isolate as much as possible the potential systems for accommodating the imposed plastic strain. First, the parameters associated to the basal and prismatic <a> slip systems are determined using the stress-strain curve in uniaxial tension along the rolling direction (RD). Since the c-axis of most grains is aligned with the normal direction or deviates by a very small angle from it (see Figure 1), the activation of pyramidal <c + a> slip system is expected in compression along the normal sheet direction (ND). From the uniaxial compression stress-strain response in ND (through thickness direction of sheet), we identify the parameters associated with this slip system. A characteristic ~86^0 reorientation of the basal poles during uniaxial compression along RD has been observed in experimental tests and is attributed to the activation of $\{10\bar{1}2\}\langle\bar{1}101\rangle$ extension twinning system [3,8]. Hence, from the uniaxial compression test along RD, we identify the parameters for $\{10\bar{1}2\}\langle\bar{1}101\rangle$ and further estimate the validity of the identification of the parameters for the <c + a> slip system obtained on the basis of the ND compression test alone. Table 1 gives the values of all the parameters of the model.

Table 1: VPSC parameters determined from monotonic test data

Mode	τ_0	τ_1	θ_0	θ_1
Basal, $\{0001\}\langle 1\bar{2}10\rangle$	17.5	5	3000	35
Prismatic <a>, $\{1\bar{1}00\}\langle 11\bar{2}0\rangle$	85	33	550	70
Pyramidal II <c + a>, $\{10\bar{1}1\}\langle\bar{1}\bar{1}23\rangle$	148	50	8500	0
Tensile Twinning, $\{10\bar{1}2\}\langle\bar{1}101\rangle$	52	0	0	0

Figure 2 shows the comparison between the only available experimental textures in uniaxial RD tension and RD compression test (reported in [8]) and the textures predicted by VPSC. We also verify that the predicted microstructure evolution contributing to plastic deformation corroborate with the observed hardening behavior for these simple monotonic loading paths (see predicted stress–strain curves and predicted texture evolution in Figures 3 and 4). In Figures 3, 4 the experimental data is represented by symbols while the VPSC predictions are represented by solid line.

For simulations of the monotonic response of AZ31 Mg for strain paths that were not used for identification, the reader is referred to [9], where the predicted material behavior in uniaxial tension and compression along 45^0 to the rolling direction in-plane and in TD as well as simple shear response in RD are presented (see [9]).

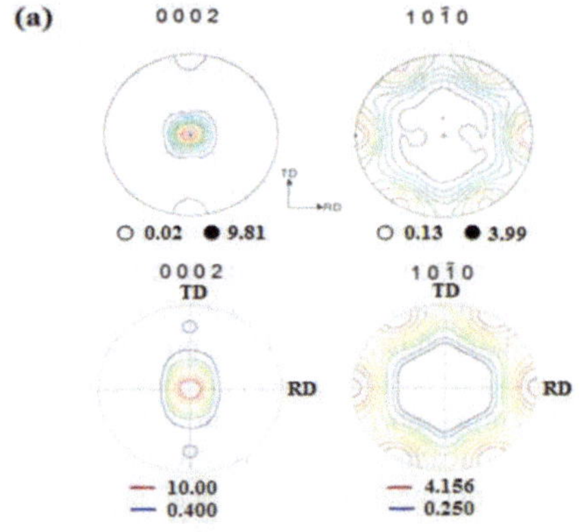

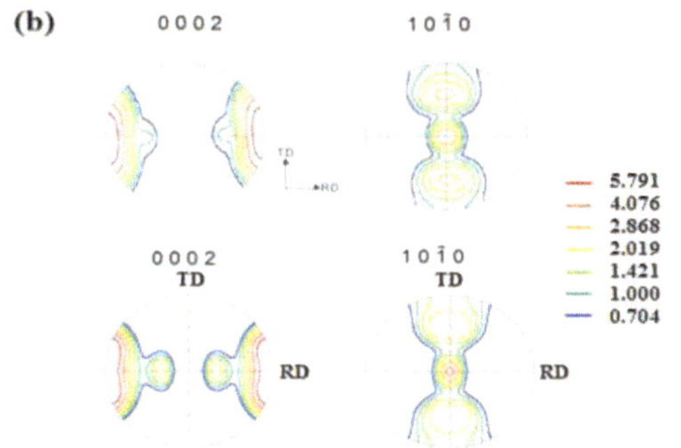

Figure 2. Pole figures showing measured texture (top) and predicted texture (bottom) in: (a) uniaxial tension along RD at 13% strain (failure) (b) uniaxial compression along RD at 8% strain.

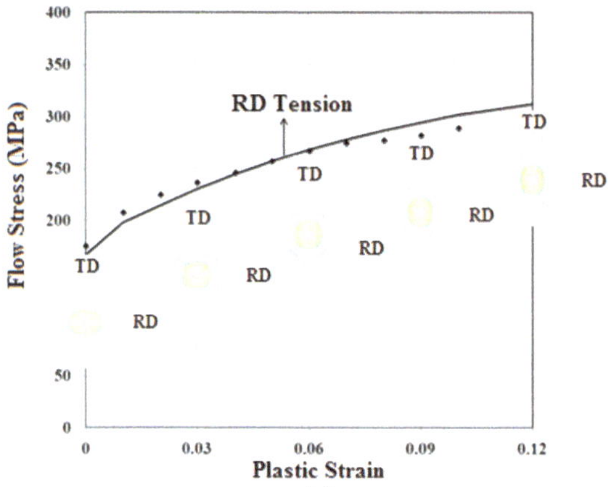

(a)

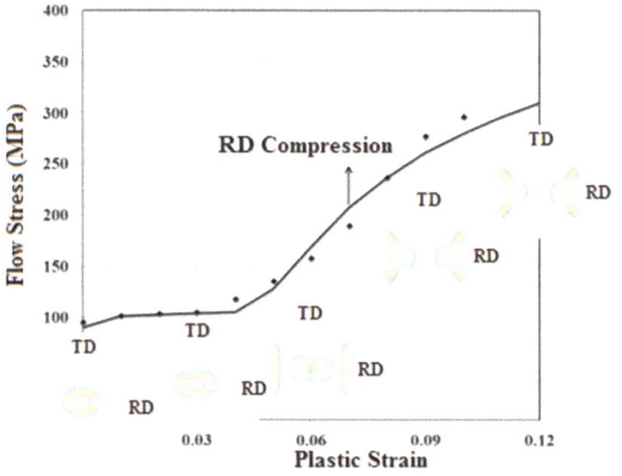

(b)

Figure 3. Stress-strain response and evolution of microstructure ((0 0 0 2) pole figure) predicted by the VPSC model in (a) RD tension, (b) RD compression

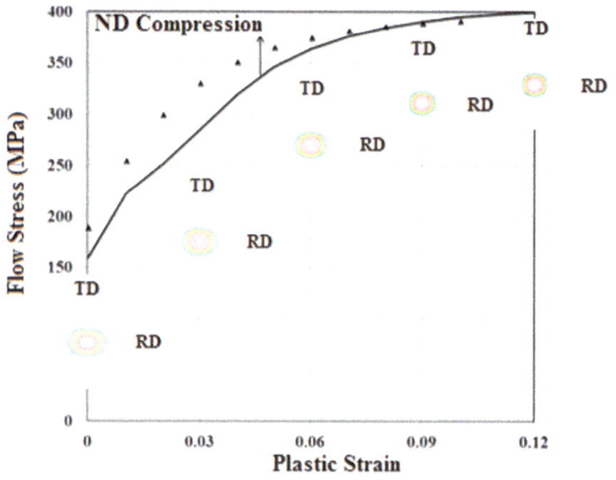

Figure 4. Stress-strain response and evolution of microstructure ((0 0 0 2) pole figure) predicted by the VPSC model in ND compression.

4. Predictions of the polycrystal model for changing strain paths

The major objective of this work is to investigate whether using the same set of parameters for VPSC it is possible to predict the mechanical response under loadings involving strain path changes of specimens of AZ31 Mg taken from rolled sheet of similar texture as that of Figure 1. For this purpose, we compare the simulation results with the experimental data recently reported by Hama and collaborators. As mentioned earlier, in [2] and [3] stress-strain curves obtained in cyclic tests starting with tension or compression were also presented. Here, we analyse tension-compression-tension (T-C-T) behavior, and respectively compression-tension-compression (C-T-C) behavior. The parameters S^t (see Eq. (4)), A^{th1} and A^{th2} used in the PTR scheme that will be used are: A^{th1}=0.7, A^{th2}=0.7 and S^t =0.105. The other model parameters are those given in Table 1. First, we simulate the stress-strain response for monotonic loading (uniaxial tension and uniaxial compression) and compare with those reported by Hama et al [1]. The agreement is very good (see Figure 5).

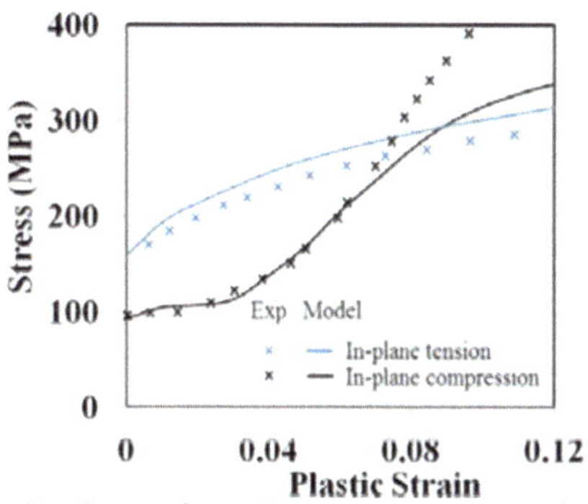

Figure 5. Comparison between experimental data from [2] (symbol) and VPSC model prediction (line) for uniaxial in-plane tension and in-plane compression tests.

Next, we compare our simulation results with the respective stress-strain response reported in Hama et al. ([1], [2]). We begin by simulating the mechanical response in tension-compression-tension observed in a cyclic test with strain amplitude of 4%. The imposed strain path consisted of uniaxial tension up to 2% strain (point A in Figure 6) followed by partial unloading to 1% strain (point B in Figure 6) then loading in compression to -2% strain (point D in Figure 6), then unloading, and reloading in tension (yielding in tension occurring at point E, see Figure 6) up to 2% strain.

The simulations show that under uniaxial tension, the initial texture is rotated such that the basal planes tend to align themselves along RD. Note also that there is no rotation of the c-axes of the grains. The predicted texture corresponding to 2% strain (point A) is shown in Figure 7(a). Unloading is considered elastic, with slope equal to the experimental Young modulus (~40 GPa, based on [1]), so the texture at points A and B are the same. During subsequent compression loading, the model predicts no texture evolution (compare textures at point A and point C), and almost constant hardening rate until tensile twinning activates (between points B and D) producing the typical kink in the stress-strain curve (sigmoidal shape). The texture at -2% strain (point D, see Figure 7(c)) also indicates that the c-axes of some of the grains are aligned with the compression direction (RD).

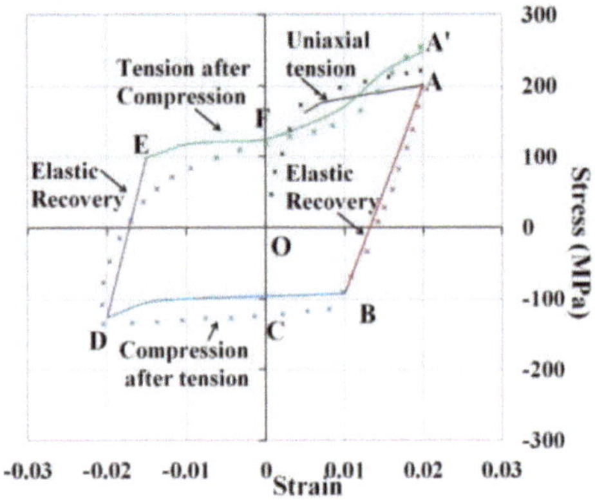

Figure 6. Comparison of the stress-strain response measured by [1] (symbol) against that predicted by VPSC model (lines) for AZ31 Mg material subjected to a tension-compression-tension test with 4% strain amplitude and no pre-strain.

Note that in the early stages of loading in tension following compression (i.e. up to point F, see Figure 7(d)), the model predicts no significant change in texture, the deformation being accommodated by slip. However, with increasing strain, tensile twinning is activated, thus inducing further rotation of the c-axes and the texture at 2% strain (point A') becomes basal (see Figure 7(e)).

Figure 8 shows the predicted stress-strain response in a cyclic test with strain amplitude of 6% in comparison with the experimental data reported in [1]. Note that at higher strain amplitude, the sigmoidal shape displayed in compression following tension is more pronounced. Also, more grains have their c-axes oriented along RD as it can be clearly seen by comparing the predicted texture at -2% shown in Figure 7(c) with the predicted texture at -3% shown in Figure 9(c)). Likewise, under uniaxial tension following compression the amount of twinning is larger as compared to the test at lower amplitude. Nevertheless, irrespective of the strain amplitude, the model predicts that a basal texture is recovered. It is to be noted that in [2] no texture measurements were reported, so no direct comparison with the predicted textures could be made. However, in [2] are given the photographs of the microstructures corresponding to the points A (strain of 3%), B (strain of 1.5%), C (strain of -0.75%), D (strain of -3%), E (strain of -1.5%), F (strain of 0.75%) and A' (strain of 3%) of Figure 9.

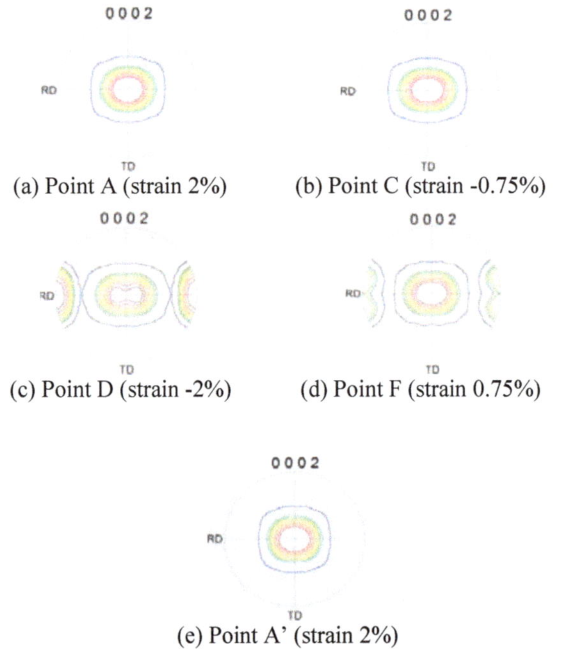

Figure 7. Predicted texture evolution in a tension-compression-tension (T-C-T) test with 4% strain amplitude and no pre-strain: (a) Texture at Point A; (b) Point C; (c) Point D; (d) Point F; (e) Point A'.

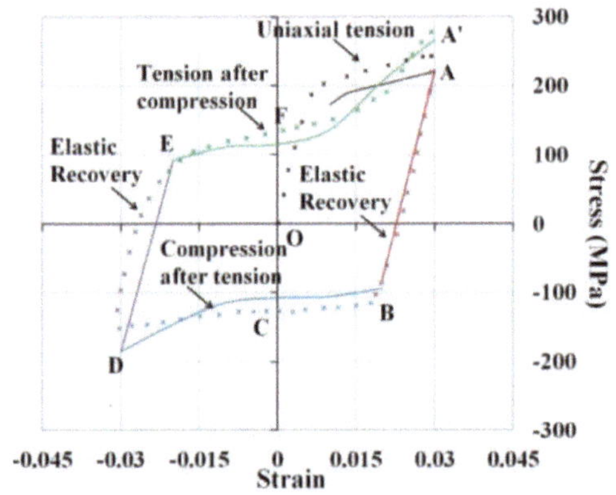

Figure 8. Comparison of stress-strain response measured by [1] (symbol) against that predicted by VPSC model (lines) for AZ31 Mg material subjected to a tension-compression-tension (T-C-T) test with 6% strain amplitude and no pre-strain.

Comparison between the initial microstructure and that at point A (i.e. after 3% loading in tension) indicates that the deformation is accommodated mainly by slip, as predicted by the model (see predicted texture at point A shown in Figure 9(a)). The data presented indicates that unloading does not affect the microstructure, so the hypothesis of elastic unloading is valid.

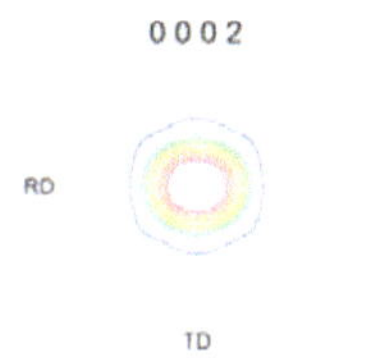

(a) Point A (strain 3%)

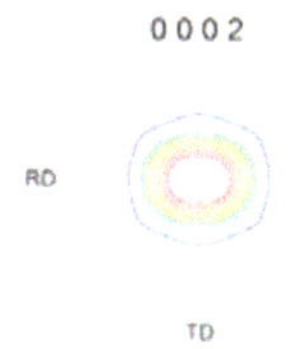

(b) Point C (strain -0.75%)

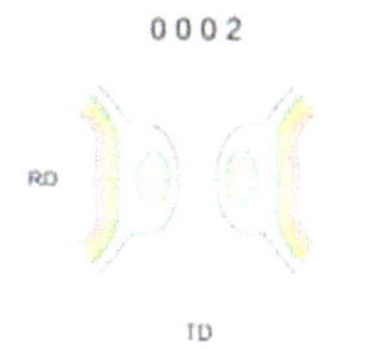

(c) Point D (strain -3%)

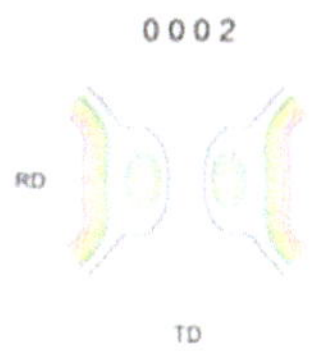

(d) Point F (strain 0.75%)

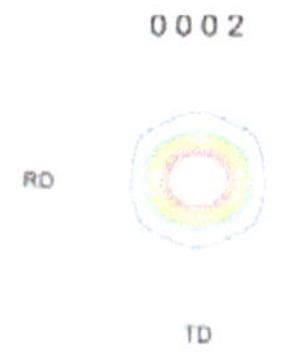

(e) Point A' (strain 3%)

Figure 9. Predicted texture evolution in a tension-compression-tension (T-C-T) test with 6% strain amplitude and no pre-strain: (a) Texture at Point A; (b) Point C; (c) Point D; (d) Point F; (e) Point A'.

The microstructure at point D, i.e. at the end of the compression stage, reported in [1] shows significant deformation twinning, which corroborates with the predicted texture at D, shown in Figure 9(c) (namely, rotation of the c-axes of most grains). The microstructure at point A', i.e. at the end of the T-C-T cycle, shows that the initial microstructure is recovered. This corroborates with the predicted texture at A' (see Figure 9(e)).

Next, we examine whether the model can predict the influence of pre-strain on the work-hardening response of the material under cyclic tension-compression loadings along RD. The simulation results are compared with the experimental stress-strain curves reported in [2]. Figure 10(a) shows the results obtained in a cyclic test with strain amplitude of 4% and a tensile pre-strain of 4%. The agreement between simulations and data is very good. Most importantly, comparison between the results for the same amplitude and 0% pre-strain (Figure 6) and the results when the tensile pre-strain was applied (Figure 10(a)) show that the overall tendencies remain the same. With or without tensile pre-strain, under tension loading followed by compression, and further tension (i.e. T-C-T), the material shows an increase in the rate of hardening during the second tensile loading, the stress-strain curve displaying a sigmoidal shape. On the other hand, both the simulation and experimental results obtained in a cyclic test consisting of compression followed by tension (C-T-C) with the same strain amplitude of 4% show that in tension after compression the stress-strain curve does not have a sigmoidal shape (see Figure 10(b)). However, the influence of amount of pre-strain in compression is very important. In a C-T-C test with compressive pre-strain of 2%, the rate of hardening in the tension stage is much higher than in the case when the pre-strain is larger, as evidenced by comparing the results shown in Figure 10(c) (compressive pre-strain of 2%) with those in Figure 10(b) (compressive pre-strain of 4%). This can be explained based on the texture evolution in RD compression predicted by the model. Indeed, in the CTC test shown in Figure 10(b), the tension stage followed a compressive loading to 6% while in the CTC test shown in Figure 10(c), the tension stage followed compression only to 4%.

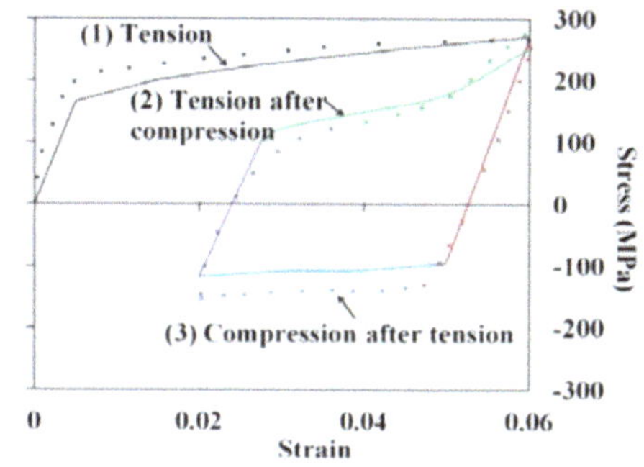

(a)

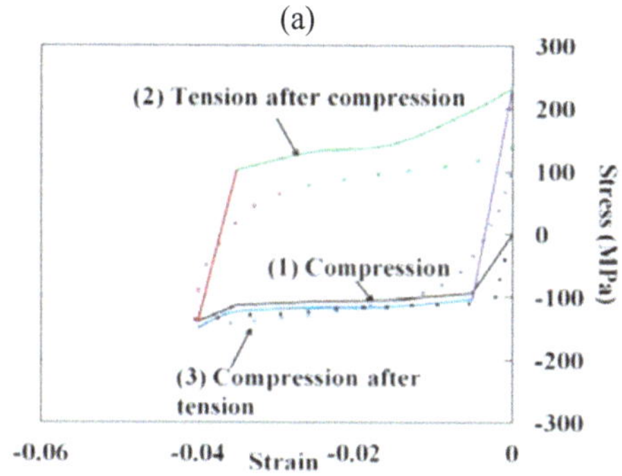

(b)

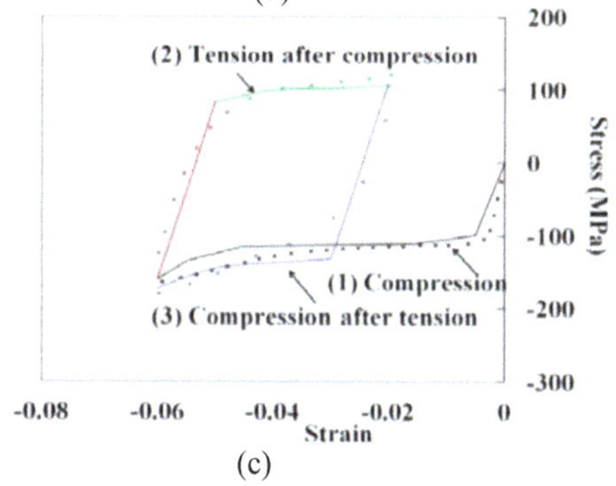

(c)

Figure 10. Comparison of stress-strain response measured by [3] (symbol) against predicted by VPSC model (lines) for tests with strain amplitude of 4% : (a) tension-compression-tension (T-C-T) with pre-strain of 4%, (b) compression-tension-compression (C-T-C) with pre-strain of -4%, and (c) compression-tension-compression (C-T-C) with pre-strain of -2%.

Comparison between the textures after 6% compression and that corresponding to 4% RD compression shows that in the latter much fewer grains have their c-axes reoriented than in the former case (see Figure 3 showing the predicted texture evolution during monotonic RD compression). Hence, the predicted difference in hardening behavior in tension following 4% compression, and -6% compression, respectively. Simulations of the behavior in CTC and respectively TCT tests with amplitude of 6% and pre-strains of 4%, -2%, -4% are shown in Figure 11. The agreement between model and data is also very good.

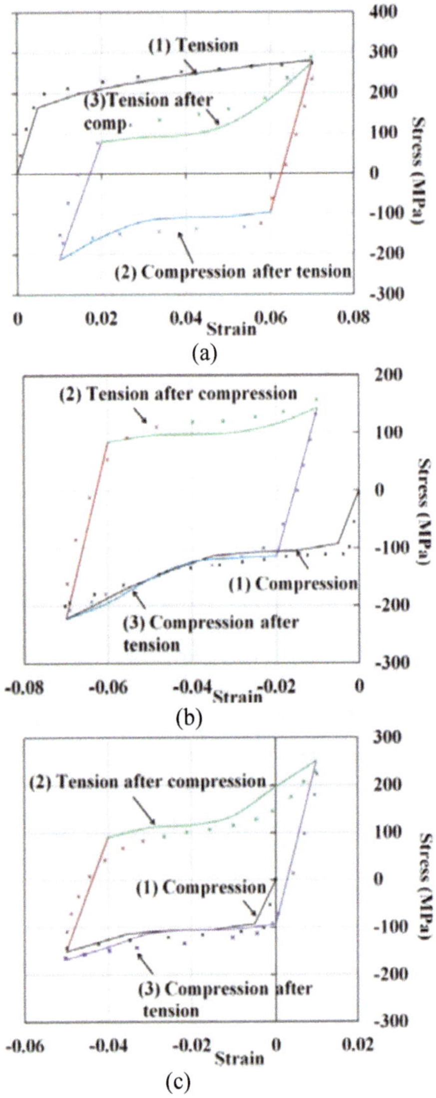

Figure 11. Comparison of stress-strain response measured by [2] (symbol) against that predicted by VPSC model (lines) tests with strain amplitude of 6% : (a) tension-compression-tension (T-C-T) test with pre-strain of 4%, (b) compression-tension-compression (C-T-C) test with pre-strain of -4% and (c) compression-tension-compression (C-T-C) test with pre-strain of -2%.

5. Conclusion

In this paper, it was investigated whether with the same set of parameters for all the slip and twin systems that were identified from simple monotonic tests, it is possible to predict the mechanical response of a thinner rolled AZ31 sheet (0.8 mm thickness) with similar initial texture for which room-temperature monotonic and cyclic tests were reported in [1, 2]. It was shown that if the slip and twinning activity in uniaxial loadings is properly identified with the simple PTR scheme for twinning re-orientation, it is possible to describe the difference in hardening behavior under strain-path changes, namely the difference in hardening between tension-compression-tension and compression-tension-compression loadings, respectively.

Since no texture measurements were given, a direct comparison of predicted and measured textures could not be made. However, partial verification of the overall trends in terms of microstructure evolution was done based on the photographs of the experimental microstructures.

The following conclusions can be drawn from the predicted texture evolution:

- With or without tensile pre-strain, under tension loading followed by compression, and further tension (i.e. T-C-T loading), the material shows an increase in the rate of hardening during the second tensile loading, the stress-strain curve displaying a sigmoidal shape.
- On the other hand, both the simulation and experimental results obtained in a cyclic test consisting of compression followed by tension (C-T-C) with the same strain amplitude show that in tension after compression the stress-strain curve may or may not have a sigmoidal shape. In this case, the influence of amount of pre-strain in compression is very important.

References

1. T., Hama, Y., Kariyazaki, N., Hosokawa, H., Fujimoto, H., Takuda, "Work-hardening behaviors of magnesium alloy sheet during in-plane cyclic loading," Mat. Sci. And Eng., A, 551 (2012), 209-217.

2. T., Hama, H., Nagao, Y., Kuchinomachi, H., Takuda, "Effect of pre-strain on work-hardening behavior of magnesium alloy sheets upon cyclic loading", Mat. Sci. Eng., 591, (2014), 69-77.

3. X.Y., Lou, M., Li, R.K., Boger, S.R., Agnew, R.H., Wagoner, "Hardening evolution of AZ31b Mg sheet," Int. J. Plast. 23 (2007), 44-86.

4. R., Lebensohn, "Modelling the role of local correlations in polycrystal plasticity using viscoplastic self-consistent schemes," Model. Sim. Mat. Sci. Eng., 7, (1999), 739.

5. C. N., Tomé, G. C., Kaschner, "Modeling texture, twinning and hardening evolution during deformation of hexagonal materials," Materials Science Forum, 495, (2005), 1001-1006.

6. G., Proust, C. N., Tomé, G.C., Kaschner, "Modeling texture, twinning and hardening evolution during deformation of hexagonal materials," Acta Mat., 55, (2007), 2137-2148.

7. R.A., Lebensohn, C.N., Tomé, „A self-consistent anisotropic approach for the simulation of plastic deformation and texture development of polycrystals: application to zirconium alloys," Acta Metall. Et. Mater. 41 (1993), 2611-2624.

8. A., Khan, A., Pandey, T., Gnaupel-Herold, R.K., Mishra, "Mechanical response and texture evolution of AZ31 alloy at large strains for different strain rates and temperatures," Int. J. Plast., 27, (2011), 688-706.

9. N., Chandola, R.K., Mishra, O., Cazacu, "Application of the VPSC model to the description of the stress-strain response and texture evolution in AZ31 Mg for various strain paths," J. Eng. Mat. Tech., 137(4), (2015), DOI: 10.1115/1.4030999.

Magnesium Technology 2016
Edited by: Alok Singh, Kiran Solanki, Michele V. Manuel, and Neale R. Neelameggham
TMS (The Minerals, Metals & Materials Society), 2016

FORMABILITY OF EXTRUDED MAGNESIUM ALLOY SHEETS WITH DIFFERENT TEXTURES

Jan Bohlen[1], Oliver Schlung[1], Sven Gall[2], Sören Müller[2], Dietmar Letzig[1]

[1]Magnesium Innovation Centre (MagIC), Helmholtz-Zentrum Geesthacht, Max-Planck Str. 1, D 21502 Geesthacht, Germany
[2]Extrusion Research & Development Center, TU Berlin, Gustav-Meyer-Allee 25, D 13355 Berlin, Germany

Keywords: extrusion, sheet, EBSD, Forming limits

Abstract

The rolling of magnesium sheets is carried out in a process consisting of multiple passes with low degree of deformation per pass. The low degree of formability of magnesium and its alloys makes it necessary to carry our rolling in multiple steps. Alternatively, it has been shown that the extrusion of flat and thin plates is feasible.
By applying this process on alloys AZ31, ZE10 and ME21, thin sheets with final gauge below 1 mm have been achieved. A microstructure and texture analysis as well as the mechanical properties of these sheets reveal distinct differences compared to a classically rolled counterpart.
A series of Nakajima tests has been used to investigate the formability of the sheets. Forming limit curves have been derived at elevated temperatures. Strain path related differences in the formability at elevated temperatures of such sheets are discussed based on an EBSD analysis of deformed samples for the alloys.

Introduction

Wrought magnesium alloys are in the spotlight for different application approaches in industrial environments. One focus at present is on magnesium sheets, which offers a broad field for applications as a semi-finished product [1, 2]. Such products are used in subsequent forming processes to produce shaped thin components. Thus, the formability and the resulting properties determine the applicability of such components. This motivates investigations on microstructure control during processing as well as texture development as two important characteristics that determine the formability of sheets and the properties of formed parts [3]. The general known limitation of magnesium and its alloys is the crystallization with a hexagonal close packed lattice structure and a resulting low formability due to a limitation in the number of active deformation mechanisms. Such a low formability [3, 4] has a significant impact on the process ability during massive deformation like rolling. In combination with strong textures, e.g. developing during rolling of sheets, unfavorable properties can result.
It has been shown that other production methods than the classical rolling process can help to overcome issues with the production of competitive magnesium sheets. Extrusion is such an alternative way to produce thin flat products as well as sheets. However, the maximum width of the extruded sheets is limited by the container diameter for direct extrusion and respectively by the inner diameter of the hollow stem for indirect extrusion. Usually the diameter of the surrounding profile circle is limited to 80 % of the container diameter. It has been shown that the extrusion of magnesium alloys even at high extrusion ratios is feasible if the processing parameters are adjusted. In this case an efficiency improvement can result from the fact that extrusion is also applied as a one-step method to the final product shape.

The aim of this study is to use three magnesium sheet alloys in the extruded condition and describe the microstructure and texture development. The properties of such sheets are investigated which especially includes the formability at elevated temperatures. The resulting microstructures after forming are exemplarily investigated in order to reveal the microstructure stability.

Experimental

Three magnesium alloys, AZ31, ZE10 and ME21 were used in form of billets for extrusion of thin sheets. The alloy composition is collected in Table 1.

Table 1: Chemical composition of the sheets of this study in wt.%; Rare earth (RE) consists of Ce, La and Nd; Mg balance

Alloy	Al	Zn	Mn	RE
AZ31	2.8	0.8	0.3	-
ZE10	-	1.5	-	0.6
ME21			1.7	1.3

At the Extrusion R&D Center a process chain was developed facilitating the extrusion of sheets with a width up to 170 % of the container diameter. The process consists of three process steps. At the beginning of the process an open tube profile is extruded, followed by an opening of the profile, which is finally flattened [5]. The extrusion trials were performed on the 8.32 MN horizontal press of the Extrusion R&D Center at the TU Berlin. The trials were carried out at process temperatures of 380°C for alloy AZ31, 450°C for alloy ZE10 and 420 °C for ME21, respectively (billet, container and die) with a ram speed of 1 mm/s. The extrusion ratio was 51:1. The as-extruded sheets were levelled and additionally annealed at 380°C for 60 min. All sheets were air-cooled after annealing. Sheets with a width of 180 mm and a thickness of 0.8 mm were achieved from the process.
Quasi-static tensile tests were conducted at ambient temperature and a constant strain rate of $10^{-3}s^{-1}$. Samples were cut by spark erosion in two orientations, extrusion direction (ED) and transverse direction (TD). A universal testing machine Zwick Z050 has been used with a 50 kN load cell to carry out such tests. Nakajima tests were conducted at two temperatures, 200°C and 350°C, in order to measure the formability of the sheets and to reveal a forming limit curve (FLC). The Nakajima test uses different specimen geometries in a common setup that all represent a different strain path. In the present study, specimen geometries according to Hasek [6] were tested. Sample geometries were scaled by 50% in order to fit the size of the sheets. The specimens were round blanks of 100mm diameter with semi-circular recesses on both sides with diameters of 0 (geometry I) to 40mm (geometry VII). A hemispherical punch (d=50mm) has been used to deform the clamped specimens until fracture. The punch speed was constant at 60mm/min. During the tests at 200°C a lubricant based on oiled PTFE-foil with a

thickness of 500 µm has been applied also using contact oil OKS 352. In case of tests at 350°C graphite foil with a thickness of 130 µm and contact oil OKS 230 has been used as lubricant system.

Local strain measurement during the Nakajima tests were realized by using an optical deformation measuring system, named ARAMIS™ (GOM). A statistical pattern based on graphite spray on the surface of the samples has been collected by two high resolution stereo cameras at a sampling rate of 12 Hz. Based on this the software allows local true strain data to be recalculated that were taken during the test. These data are used to determine the forming limit curves based on a time dependent strain inhomogeneity formed during the tests and identified as a point on the sample where a distinct increase in the strain rate occurs [7]. This linear "best fit method" identifies points of increasing thinning rates which are correlated to the respective major strain and minor strain to identify the point of the forming limit curve (FLC). At least 3 tests were carried out and averaged per geometry.

The microstructure and texture analysis was carried out by electron backscatter diffraction (EBSD). Standard metallographic sample preparation procedures were applied for sample preparation [8] followed by electropolishing in an AC2 solution (Struers™). Measurements of local orientation patterns were carried out using a a field emission gun scanning microscope (Zeiss, Ultra 55, EDAX/TSL) on longitudinal sections of the sheet samples. The accelerating voltage was 15 kV and the point to point step size was 0.3 µm. A software "TSL Orientation Imaging Microscopy Analysis" of EDAX© was employed to analyse the EBSD measurements.

Results

Fig. 1 collects the results of EBSD measurements on longitudinal sections of the extruded sheets used in this study. Orientation patterns are shown to reveal the microstructure. Pole figures are used to show the texture of the sheets. In all cases recrystallized microstructures are revealed with some distinct differences. In case of AZ31 (Fig. 1a) a bi-modular microstructure is found with a fraction of relatively small grains embedded into a structure of large grains. An average grain size of 16 µm is found. The texture is strong with a typical basal component, i.e. a basal texture, with a slightly broader angular distribution of basal planes to the ED that to TD. Prismatic planes, however, are aligned parallel to the normal direction (ND). This result compares well to conventionally rolled sheets of AZ31 [9]. In case of the ZE10 sheet (Fig. 1b) the grain structure is distinctly coarser with an average grain size of 24 µm. The texture is also strong with a tendency to align basal planes parallel to the sheet plane. Prismatic planes show a slight preference to be aligned perpendicular to ED. The texture of this sheet is significantly different compared to a conventionally rolled counterpart, where weak textures with no basal alignment are found and a broad angular spread of basal planes towards TD [9, 10]. For the ME21 sheet (Fig. 1c) a fine grained homogeneous microstructure is revealed with an average grain size of 9 µm. The spread of grain sizes is the smallest for the alloys investigated. A fine distribution of precipitates is hardly resolved in the orientation maps. The pole figures show two ED tilt peaks for the basal planes. In summary, a strong texture is found, where basal planes are not aligned parallel to the sheet plane. In case of this alloy series similar textures have been described for conventionally rolled sheets [11].

In an attempt to reveal insight into the microstructure and texture development during extrusion and annealing, different fractions of grains are looked at as a detail of the orientations patterns shown in Fig. 1. Results of these maps with separately looking at smaller or larger grains with respect to the average grain size are collected in Fig. 2. The applied software allows this to be carried out by using a grain size constraint. In case of AZ31 (Fig. 2a), a strong basal texture is verified for smaller grains (arbitrarily set to sizes below 10 µm). Band or necklace structures around larger grains may indicate a grain structure development due to discontinuous dynamic recrystallization [12] based on a grain nucleation and growth mechanism. However, larger grains, Fig. 2d, also exhibit other orientations with tilt to both, ED or TD as well as a component with c-axis parallel to TD. This component is well understood as a result of the extrusion shape. In case of ZE10 a very comparable result is found. The grain size spread throughout the microstructure is less-pronounced compared to AZ31 but smaller grains also tend to show a texture with basal alignment into the sheet plane and a tilt more towards ED (Fig. 2b). In larger grains no clear tendency is developed (Fig. 2e).

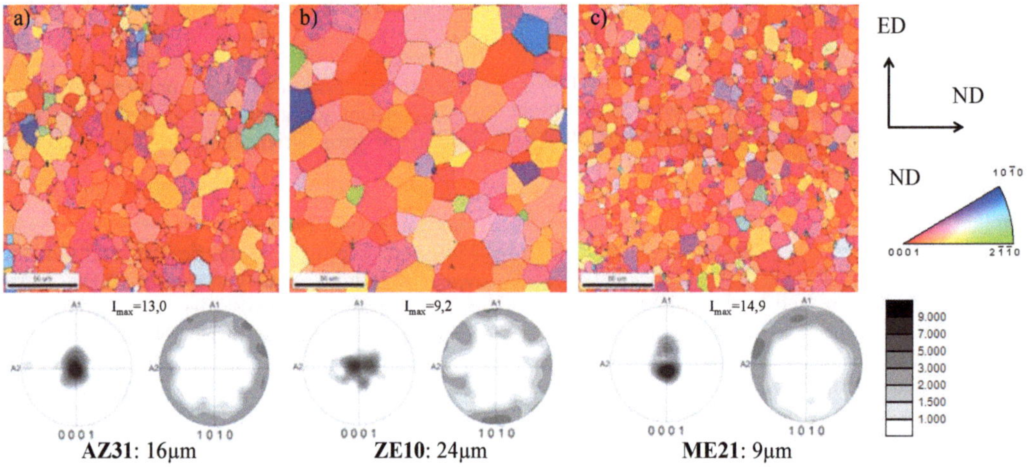

Fig. 1: Sections of orientation patterns and pole figures of EBSD measurements to present the microstructure and texture of the extruded sheets; a) AZ31, b) ZE10, c) ME21

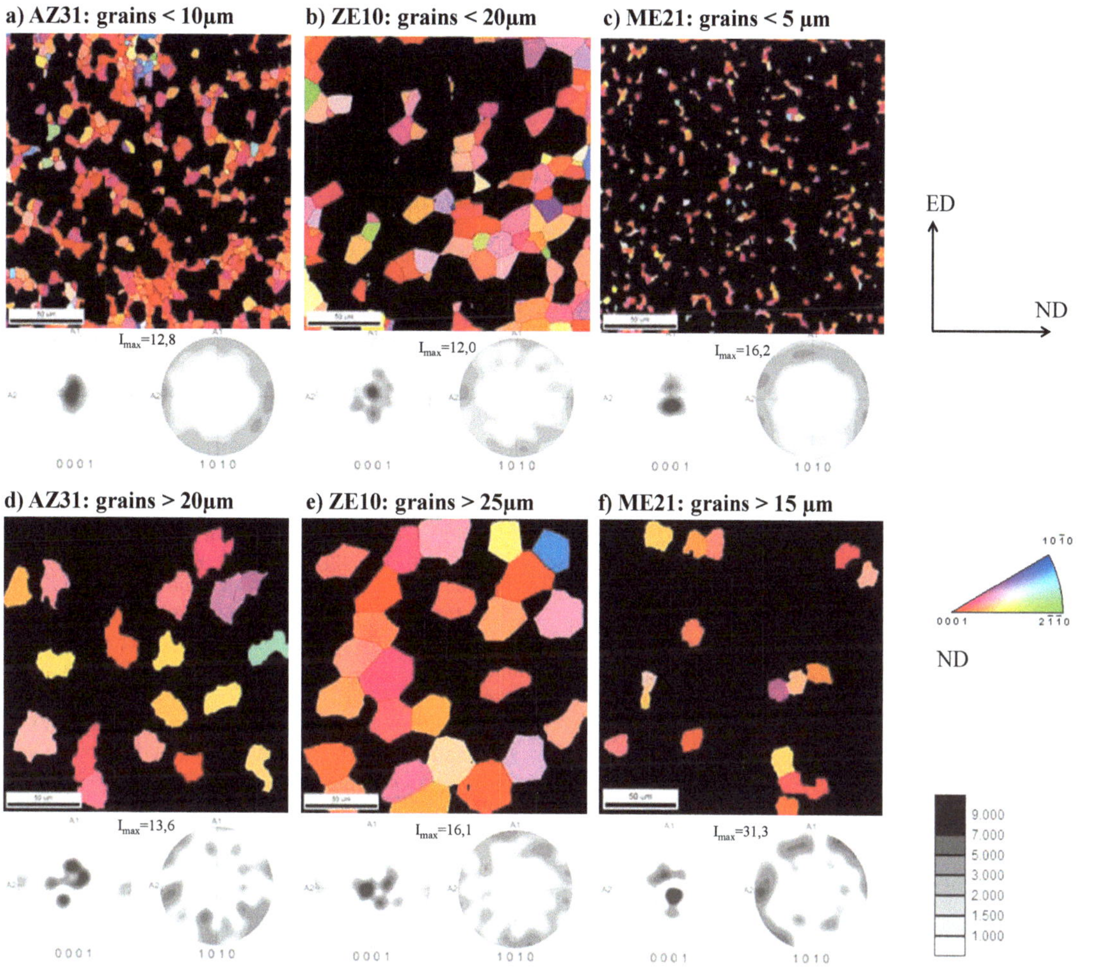

Fig. 2: Sections of orientation patterns and pole figures of EBSD measurements of selected grain fractions with different grain size; a) AZ31: grains < 10µm, b) ZE10: grains < 20µm , c) ME21: grains < 5µm, d) AZ31: grains > 20µm, e) ZE10: grains > 25µm , f) ME21: grains > 15µm

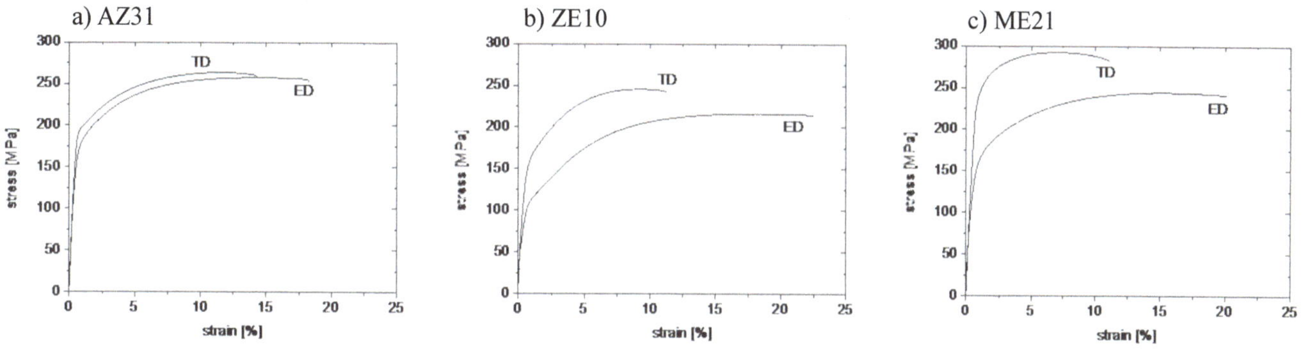

Fig. 3: Engineering stress-strain diagrams from tensile tests in ED and TD at ambient temperature and a constant strain rate of $10^{-3}s^{-1}$; a) AZ31, b) ZE10, c) ME21

In case of ME21 the texture is obviously not dependent on the grain size. In both fractions of grains, Figs. 2c and 2f, the split peaks with tilt towards ED are found. Also no conglomeration of grains is visible in form of bands or rings.

Fig. 3 collects stress-strain diagrams from uniaxial tensile tests at ambient temperature in two orientations, ED and TD. Table 2 collects the resulting mechanical properties of the sheets. All three alloys have in common that the resulting stresses are lower in ED than in TD and that the resulting fracture strains are higher in ED than in TD. Furthermore, the uniform elongation is closer to the fracture strain in TD indicating that flow localization occurs relatively early during testing in ED.

Table 2: Mechanical properties of the extruded sheets from tensile tests; TYS: tensile yield stress, UTS: ultimate tensile stress; standard deviation in brackets

Alloy		TYS [MPa]	UTS [MPa]	Uniform strain [%]	Fracture strain [%]
AZ31	ED	163 (±4)	261 (±4)	13.9 (±0.4)	18.7 (±2.0)
	TD	188 (±2)	262 (±1)	11.0 (±0.1)	14.0 (±0.6)
ZE10	ED	92 (±9)	218 (±1)	17.2 (±0.9)	21.5 (±1.6)
	TD	143 (±2)	246 (±1)	8.7 (±0.1)	11.0 (±0.3)
ME21	ED	131 (±6)	244 (±1)	14.4 (±0.5)	20.8 (±3.5)
	TD	228 (±3)	291 (±3)	6.3 (±0.1)	10.6 (±0.3)

The orientation dependence is lowest in case of AZ31 and highest in case of ME21. However, highest strength properties are observed for ME21 in TD whereas the TYS in ED is lower than for AZ31. The lowest strength properties are revealed for ZE10 in both directions. This is directly correlated to the coarse grained microstructure of this sheet. Vice versa, the finest grain size of ME21 corresponds to the highest stress values received, specifically in TD.

Fig. 4 shows the forming limit diagrams as a result of Nakajima tests for the three alloys at two different testing temperatures. Following the assumption that typical forming operations will be carried out at elevated temperatures, Fig. 4a shows results of tests at 200°C whereas Fig. 4b collects results of tests at a higher temperature of 350°C. Standard deviations show the scatter of results for each strain path. In case of the tests at 200°C comparably low strains are achieved for all three alloys on the right side of the FLC where positive minor strains are required. A limit of 0.2 as a major strain is not exceeded in all cases. This specifically indicates a distinct stretch forming limitation of the sheets. Only with increasing possibility of material flow along the minor strain direction (i.e. material flow along the width of the samples and moving towards the left of the diagram) higher strains are achieved. In this case ZE10 exhibits the highest achievable strains whereas AZ31 and ME21 compare quite well to each other. At 350°C, a distinctly higher temperature, the tendency that stretch forming operations (right side of the FLC) appear limited over forming operations with less significant sheet thinning (left side of the FLC) is principally maintained. However, the formability at all strain paths' increases with varied significance. The AZ31 sheet only shows a small increase in its overall formability, interestingly especially the stretch forming case on the outer right side (geometry VII) shows some tendency to increase. Obviously, at this high temperature the material starts to overcome a general limitation in formability.

In case of ZE10 and ME21 the changes are more significant. A visible increase along strain paths on the left side of the FLC in case of ZE10 can be compared to a distinct increase in case of ME21 allowing comparable forming results of the two alloys in this condition. Both alloys also do not overcome the principal flow restriction in stretch forming operations. However, ME21 shows strong temperature dependence whereas ZE10 shows a lower increase in formability. In all cases – with the exception of AZ31 tested at 350°C - there is no pronounced lower limit along with a plane strain path, i.e. zero material flow in the minor strain direction which is unusual compared to results from other forming curves on magnesium alloy sheets [3].

Fig. 5 exemplarily collects EBSD orientation patterns of deformed samples after testing to analyse the microstructure development. Each two samples are selected which refer to the most distinct strain paths, one (indicated as geometry I) from the so-called "deep-drawing case" which the lowest minor strain achieved (left side of the FLC) and one (indicated as geometry VII) which refers to the stretch forming condition with the highest minor strain achieved (right side of the FLC). Results including the textures for the two temperatures, 200°C and 350°C, are shown. In case of the samples tested in geometry I a distinct elongation of grains is found which refers to the underlying deformation process accommodated on a grain level in the microstructure. AZ31 shows small grains developing in this microstructure obviously as a result of dynamic recrystallization. At higher temperature the grain elongation is less pronounced but the development of small grains is enhanced. The texture consists of an enhanced spread of basal planes towards TD and a six peak symmetry of prismatic planes which has been associated with active prismatic slip [13] in uniaxially deformed samples. It is well understood that this is less pronounced at higher temperature. In case of the coarser grained ZE10 sheet and geometry I such signals of deformation are even stronger at both temperatures. Furthermore, deformation bands and twins can be identified especially after testing at 200°C. At the same temperature, recrystallization, i.e. the formation of a new fraction of small grains, appears to be delayed. The effect of grain elongation at 350°C is visibly more pronounced compared to AZ31 which is consistent with the higher strain achieved in this condition. This effect is even more distinct in case of ME21 at 350°C. At 200°C small grains appear to form in a microstructure not consisting of such strongly elongated grains in this finest grained alloy sheet. The texture is very comparable with the two other alloys.

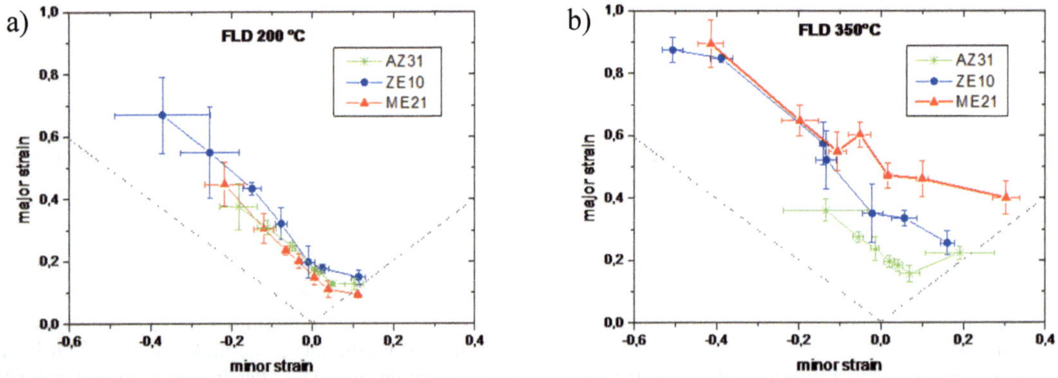

Fig. 4: Forming limit diagrams with forming limit curves of the three extruded sheet alloys with major strain direction parallel to RD; a) 200°C, b) 350°C

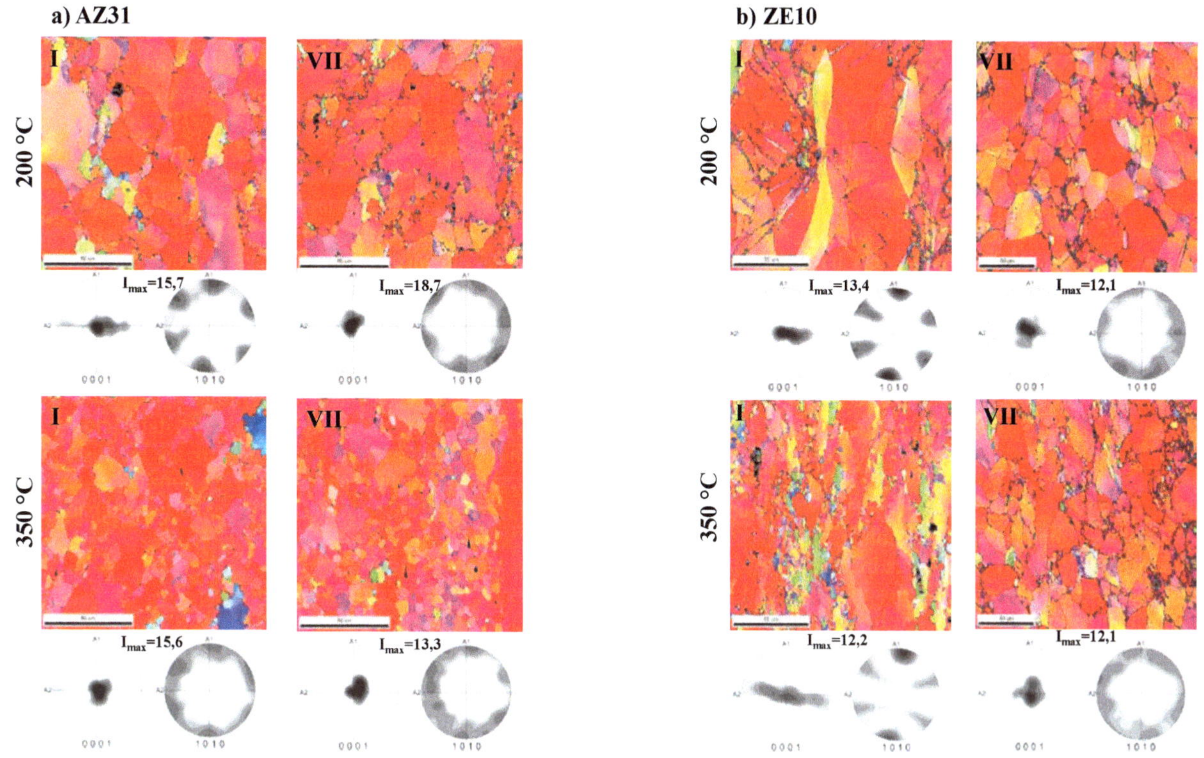

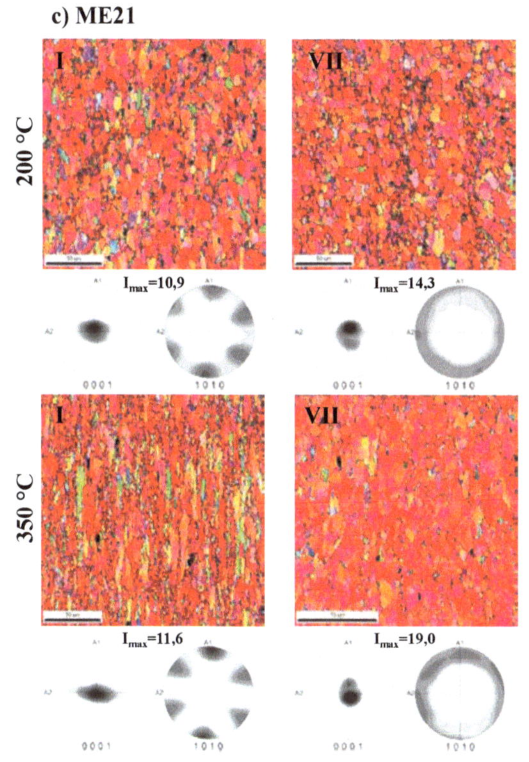

Fig. 5: EBSD orientation patterns of samples after Nakajima tests at two temperatures; geometry I refers to the test with highest negative minor strain, geometry VII refers to the test with the highest positive minor strain; colour legend and directions as in Fig. 1; a) AZ31, b) ZE10, c) ME21

In case of stretch forming, i.e. sample geometry VII, for AZ31 the beginning of recrystallization is seen to form necklace structures around deformed grains which is distinctly enhanced at higher temperature. The texture remains stronger at the lower forming temperature.

The distinct six peak symmetry of prismatic planes is not strongly visible anymore which is directly correlated to the biaxial stress condition. Again, a similar result is found for ZE10 with delayed development of the recrystallized grain structure at both temperatures. Only in case of ME21 a more homogeneous development of the microstructure is found. It is less temperature dependent and less texture influencing. If this is understood as a result of enhanced recrystallization a direct correlation to the improved formability of this sample condition applies.

Discussion

The microstructures of the sheets after the extrusion process show clear differences with respect to the texture development. In case of AZ31 a bimodal microstructure is seen correlated with a tendency to develop a basal texture. In case of ZE10 no such bimodality is seen any more but grain growth appears to be enhanced. Still, large grains of this alloy sheet exhibit orientations with a certain tilt to the TD which would be directly comparable to findings of conventionally rolled sheets of this alloy [10]. Smaller grains, however, also show a tendency to an ED split. In summary, the texture is strong unlike in the classically rolled counterpart.
A grain size dependence of the texture can be seen. Thus, growth enhancement or –vice versa – growth restriction – of grains with different orientations during dynamic recrystallization can be suggested, often used to explain the development of unusual sheet

textures in alloys that contain rare earth elements [14]. In case of ME21 generally grains remain smaller leading to the sheet with the finest grain size of this study. The spread of the grain size also is smaller compared to the other two alloys. No clear differences in the texture are observed for different grain sizes in this case leaving a pronounced ED split peak for the basal plane orientations of this sheet. Thus, more homogeneous recrystallization is revealed in this alloy.

Correlating the texture of the materials to the mechanical properties, it can be concluded that all sheets have a profound fraction of the microstructure with tilt or split peaks towards ED. Over small components with tilts towards or into TD this preferential tilt to ED can be directly correlated to lower yield stresses and higher ductility along ED rather that along TD if a preferential activation of basal slip is suggested to influence the mechanical behaviour [10, 15]. Furthermore, grain size effects are likely to explain the different stress levels of the alloy. The coarse grained ZE10 sheet exhibits the lowest stress properties in both directions, ED and TD. The finest grained ME21 sheet which also contains the highest amount of non-solute elements (i.e. precipitates) shows highest stress properties in TD but not in ED where a pronounced tilt of basal planes helps to understand lower stress properties due to activation of slip. In summary, a direct correlation of microstructure, i.e. grain size and texture, and the mechanical properties can be drawn. Furthermore, the lowest anisotropy of all mechanical properties in case of AZ31 is correlated to the strong and simple alignment of basal planes parallel to the sheet plane. Thus, texture weakening correlated with the development of asymmetric orientation distributions is also concurrent to the development of the anisotropy of mechanical properties.

In case of the forming behaviour the same considerations may apply as the uniform strains in ED show a comparable tendency to the forming behaviour along strain paths with distinct negative minor strain (left side of the FLC). In such cases – like in the tensile tests – the material flow is freely arranged between width reduction and thickness reduction of the sample. However, towards the right side of the FLC, namely in case of stretch forming operations, material flow is limited to the thickness reduction which is not enhanced if textures are strong and basal planes are aligned parallel to the sheet plane. As all three sheets exhibit a tendency to develop such strong textures the formability along such strain paths remains limited. Increased spread of basal planes, i.e. a potentially easier activation of basal slip, can then result in a distinct increase of the formability in cases at higher temperatures, where also other deformation mechanisms are easier activated. Furthermore, in Fig. 5 dynamic recrystallization to different extent can be understood as an active softening mechanism during deformation which allows higher strain in case of ME21 to be reached whereas recrystallization appears to be delayed in case of ZE10. In case of AZ31 it only seems to influence the stretch forming strain path (geometry VII) at the highest temperature.

If the resulting microstructures are seen as a microstructure development step during forming of parts from magnesium sheets it becomes very clear that the result aim towards changes of the mechanical properties of the sheets after forming. In case of AZ31 and ZE10 the test range of this study leads to deformed microstructures with distinct inhomogeneity. Only in case of ME21 the fine grained homogeneous grain structure is maintained.

Conclusions

Extrusion of magnesium sheets has been used as a feasible method to process different magnesium alloys to thin sheets with thicknesses even lower than 1 mm. The microstructure development leads to fully recrystallized grain structures but also to a stronger alignment of basal planes in the sheet plane unlike in some conventionally rolled magnesium sheet alloys (especially those that contain rare earth elements). This is consistent with a change in the balance of grain fractions developing due to deformation and recrystallization during the massive deformation process.

A strong alignment of basal planes limits the stretch formability of such sheets whereas deep drawing operations with negative minor strain can be high at elevated temperature. There are distinct differences between the alloys. Enhanced recrystallization, e.g. in case of fine grained ME21, also leads to an increase in the formability. Microstructure changes during forming have to be considered regarding the resulting properties of formed parts.

Acknowledgement

The authors appreciate financial support of this work by the Deutsche Forschungsgemeinschaft (Grant. BO 2461/3-1).

References

1. J. Hirsch, T. Al-Samman, Acta Mater. 61 (2013) 818 – 843.
2. H. Friedrich, S. Schumann, J. Mat. Process. Techn. 117 (2001), p. 276-281.
3. L. Stutz, J. Bohlen, G. Kurz, D. Letzig, K.U. Kainer, Key Eng. Mater. 473 (2011), p. 335 – 342.
4. C.E. Dreyer, W.V. Chiu, R.H. Wagoner, S.R. Agnew, J. Mat. Process. Techn. 210 (2010), p. 37-47.
5. S. Gall, S. Müller, W. Reimers, Extrusion of Magnesium sheets with increased widths, Steel Research International – Special Edition: 10th ICTP, (2011) p. 286-290.
6. V. Hasek, Blech Rohre Profile Vol. 25 (10) (1978), p. 213-220, p. 285-292, p. 493-499, p. 613-627.
7. W. Volk, P. Hora, Int. J. Mater. Form. 4 (2011) 339 - 346
8. V. Kree, J. Bohlen, D. Letzig and K.U. Kainer: Praktische Metallographie Vol. 41 (2004), p. 233-246.
9. J. Bohlen, P. Dobron, K. Hantzsche, D. Letzig, F. Chemlik, K.U. Kainer, Int. J. Mat. Res. 100 (2009) 790-795.
10. J. Bohlen, M.R. Nürnberg, J.W. Senn, D. Letzig, S.R. Agnew, Acta Mater. 55 (2007) 2101–2112.
11. X. Li, T. Al-Samman, S. Mu, G. Gottstein, Mater. Sci. Eng. A 528 (2011) 7915 – 7925.
12. S.E. Ion, F.J. Humphreys, S.H. White, Acta Metall. 30 (1982) 1909-1919.
13. A. Styczynski, C. Hartig, J. Bohlen, D. Letzig, Scripta Mater. 50 (2004), p. 943-947.
14. J.P. Hadorn, K. Hantzsche, S. Yi, J. Bohlen, D. Letzig, J.A. Wollmershauser, S.R. Agnew, Metall. Mater. Trans. A 43 (2012) 1347 – 1362.
15. S.R. Agnew, in: H.I. Kaplan (ed.) „Magnesium Technology 2002", TMS Annual Meeting Seattle (2002), p. 169 – 174.

Magnesium Technology 2016
Edited by: Alok Singh, Kiran Solanki, Michele V. Manuel, and Neale R. Neelameggham
TMS (The Minerals, Metals & Materials Society), 2016

Prediction of Magnesium Alloy Formability: The Role of Texture

V.M. Miller[1], T.D. Berman[2], I.J. Beyerlein[3], T.M. Pollock[1]

[1]Materials Department, University of California Santa Barbara; Santa Barbara, CA 93106-5050, USA
[2]Materials Science and Engineering Department, University of Michigan, 2300 Hayward Street, Ann Arbor, MI, 48109 USA
[3]Theoretical Division, Los Alamos National Laboratory, Los Alamos, NM, 87545 USA

Keywords: crystallographic texture, formability, viscoplastic self-consistent model (VPSC), magnesium alloy

Abstract

Control of crystallographic texture is critical for the development of wrought magnesium sheet product with sufficient levels of ductility and formability. However, it is not feasible to systematically study mechanical behavior as a function of texture experimentally because texture cannot be varied independent of other parameters important for formability. A protocol for generating synthetic textures in conjunction with a viscoplastic self-consistent (VPSC) polycrystalline plasticity model to predict deformation behavior and formability has been developed. The ability to isolate and systematically vary a single texture characteristic is demonstrated for the strength of the basal peak. The changes in slip system activity, texture stability, and predicted formability are presented.

Introduction

Magnesium alloys are desirable for use in many applications, such as transportation, where reduction in weight is a critical concern due to their high specific strength. However, wrought magnesium product in particular has been rarely adopted because of its poor forming behavior and limited ductility at room temperature.[1] One of the primary reasons for the poor forming behavior in magnesium alloys is the development of a strong crystallographic texture during deformation processing. In rolled sheet material, this manifests as a basal texture, with the crystallographic c-axes of the individual grains aligned with the sheet normal direction. This texture is detrimental to formability because it results in highly anisotropic behavior and the orientations it contains lack available deformation mechanisms to accommodate the forming operations.[2–5] Strong basal textures demonstrate increased propensity to form shear bands, which ultimately lead to fracture.[6]

The experimental investigation of forming behavior in magnesium alloy sheet has generally utilized r-values as a simple quantitative metric for formability. The r-value of a material is defined as $r = \epsilon_w/\epsilon_t$, where ϵ_w and ϵ_t are the strains in the width and thickness directions, respectively, after a uniaxial tensile test performed to a pre-defined strain threshold, usually 10% for Mg alloys.[7] These ratios are calculated for specimens of multiple orientations within the sheet plane, typically with the loading axis 0°, 45°, and 90° from the sheet rolling direction. These three measurements give an indication of the in-plane plastic anisotropy of the sheet.

In hcp metals, including magnesium alloys, low r-values indicate good formability.[8] This is in contrast with the behavior of cubic metals, where high values are desired because they indicate a resistance to sheet thinning.[9;10] In commercial magnesium sheet, r-values greater than three are commonly reported.[5;11]

While experimental investigations of the link between crystallographic texture and formability in magnesium alloys have provided invaluable insights, it is not feasible to systematically explore the effect of different texture characteristics with experiments alone. The development of crystallographic texture during thermomechanical processing is dependent on numerous other factors including grain size,[12;13] the presence of intermetallic particles,[14] and alloy composition.[2;15] Each of these effects has been shown to independently influence the forming behavior as well as the texture.

Polycrystal plasticity simulations used in conjunction with tools that allow for the creation of synthetic datasets offer the opportunity to systematically vary texture characteristics while guaranteeing other parameters (e.g. grain size and critical resolved shear stress values) remain constant. In this work, a framework for generating and testing the forming behavior of synthetic textures is described. Textures are

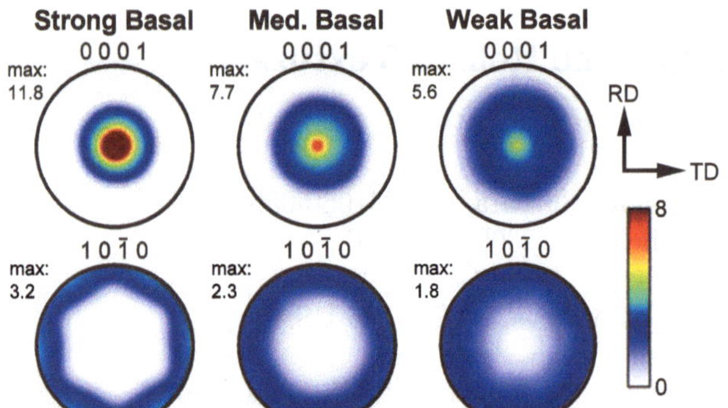

Figure 1: Sythetic starting textures with varied basal peak strength generated using DREAM.3D.

generated using the DREAM.3D software package,[16] then uniaxial tensile tests are simulated using a viscoplastic self-consistent (VPSC) polycrystal plasticity model.[17;18] The case of varied basal peak strength is examined to demonstrate the utility of the framework. Changes in macroscopic stress-strain behavior, slip system activity, texture evolution, and r-value indicators of formability are reported. Future extension of this framework to examine other texture characteristics and the effect of grain size is discussed.

Simulation Procedure

Synthetic Texture Generation

Textures to be used as inputs for the VPSC model were generated with DREAM.3D Version 5.2 software package.[16] First, the desired texture components were specified in Stats Generator, an auxiliary tool in the DREAM.3D software package. Stats Generator then generates a set of goal statistics for the microstructure as a *.dream3d file. With the exception of texture, all microstructural parameters were left at the default values.

This goal statistics file is loaded into DREAM.3D, and a synthetic volume 150 voxels on edge with a voxel spacing of 0.5μm is created. Ellipsoidal grains matching the specified microstructure and texture statistics are populated into the volume. DREAM.3D is then used to output a Los Alamos FFT file, which contains the voxel locations and orientations. A Matlab script generated with the MTEX toolbox[19] was used to import the Los Alamos FFT files as orientation density functions and convert to VPSC texture files containing 10,000 grains. Using this procedure, three textures of varied basal strength were generated, as shown in Figure 1.

Table 1: Voce hardening parameters for AZ61L

		τ_0	τ_1	θ_0	θ_1
basal	$\{0001\}\langle 2\bar{1}\bar{1}0\rangle$	63	20	190	100
prism	$\{10\bar{1}0\}\langle 2\bar{1}\bar{1}0\rangle$	165	20	300	90
pyramidal	$\{11\bar{2}2\}\langle \bar{1}\bar{1}23\rangle$	230	20	230	90
tensile twin	$\{10\bar{1}2\}\langle \bar{1}011\rangle$	110	0	100	30

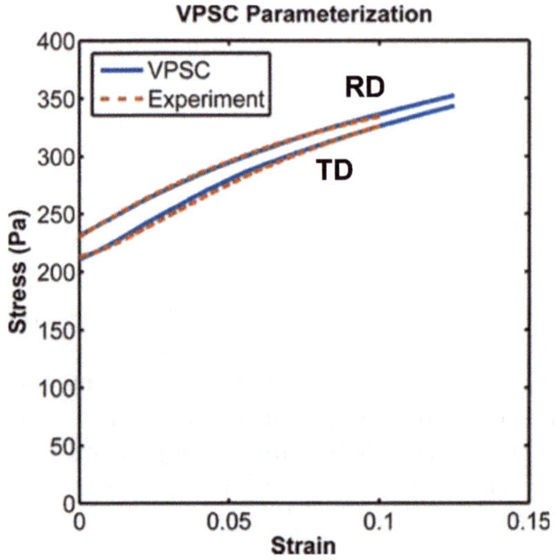

Figure 2: Comparison of experimental AZ61L and VPSC-fit uniaxial tension stress-strain curves for loading along the sheet RD and TD.

Viscoplastic Self-Consistent Modeling

The mechanical response at room temperature of these synthetic textures was calculated using the VPSC polycrystal plasticity framework developed at Los Alamos National Laboratory.[17;18]

An extended Voce hardening law is used to describe the response of each slip system, where the critical resolved shear stress on deformation mode s after shear strain Γ is given by:

$$\tau^s = \tau_0^s + (\tau_1^s + \theta_1^s\Gamma)\left(1 - exp\left(-\frac{\theta_0^s\Gamma}{\tau_1^s}\right)\right) \quad (1)$$

In this equation, τ^s and θ^s represent phenomenological fitting parameters describing the hardening behavior of the sth slip system. It should be noted that τ_0 can be regarded as an effective critical resolved shear stress at yield. Twinning is modeled using a Schmid-type law that accounts for directionality, and the Predominant Twin Reorientation scheme proposed by Tomé *et al.*[20] incorporated into the VPSC model accounts for the effect of twinning on texture.

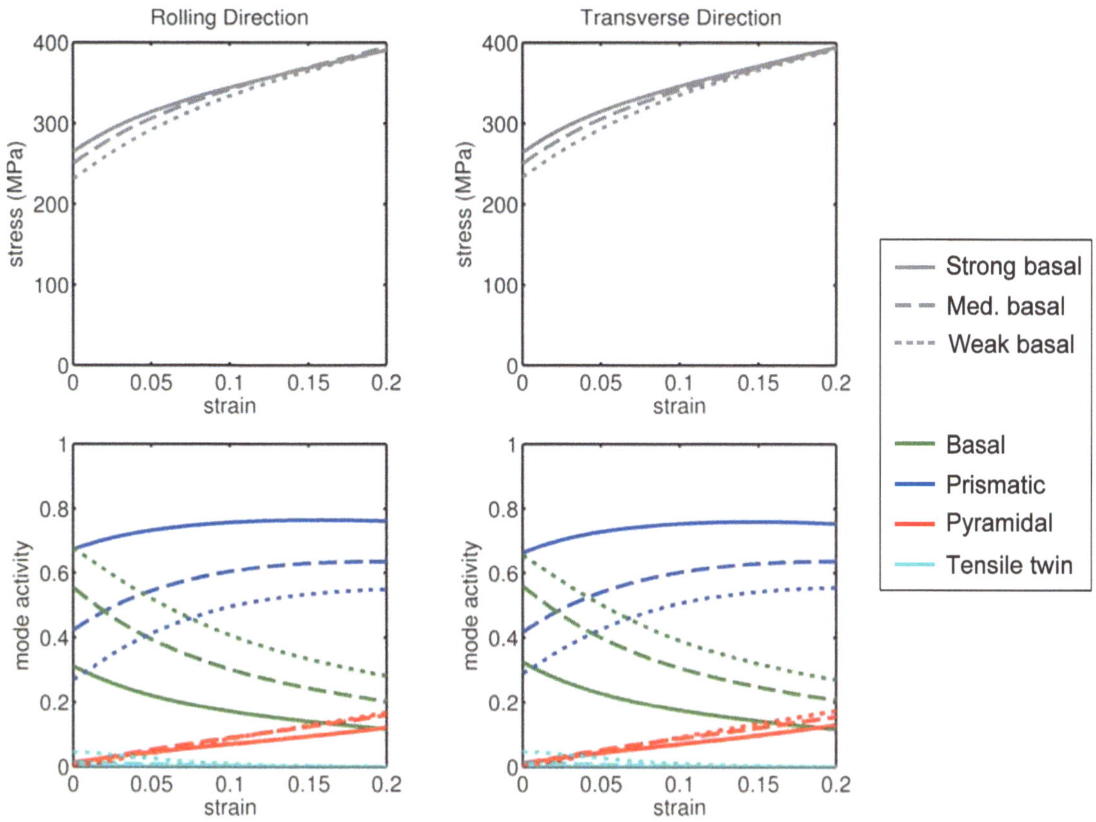

Figure 3: Stress-strain behavior and relative deformation mode activity for uniaxial tension along the rolling direction (left) and transverse direction (right).

The values of these hardening parameters, presented in Table 1, are selected to reproduce the in-plane anisotropy of the stress-strain behavior of fine-grained thixomolded and thermomechanically processed AZ61L after complete recrystallization at 285°C. The full description of the experimental processing pathway is presented elsewhere.[4] A comparison of the mechanical responses along the sheet rolling direction (RD) and transverse direction (TD) are shown in Figure 2. The ratio of basal to prismatic critical resolved shear stress values used in this work, 1:2.6, falls within the range of values (1:1.1 to 1:6) commonly reported for polycrystal studies in the literature.[11;21;22] Twinning is observed to be relatively hard in comparison to basal slip, likely because the experimental data is from material with a fine grain size, which is known to inhibit deformation twinning.[23;24]

Using these hardening parameters, uniaxial tensile tests with the tensile axis inclined in the sheet plane in 15° increments are simulated for each texture to 20% tensile elongation at a strain rate of 10^{-3}/s. The r-value calculations are performed after 10% strain for consistency with experiments reported in the literature.[4;25] The resulting deformation texture, stress-strain response, relative slip activity, and r-values are examined.

Results and Discussion

The effects of varied basal peak strength on deformation behavior in uniaxial tension were simulated for the starting textures shown in Figure 1. Stress-strain curves and relative deformation mode activity for uniaxial tests along the RD and TD are shown in Figure 3. The tests along the RD and TD demonstrated very similar behavior, especially at low strains, because of the symmetry of the peak in the basal pole figure.

Above approximately 8% strain, after easy basal slip is exhausted, slight differences in the RD and TD stress-strain curves are apparent due to the asymmetry of the prism plane distribution in the starting textures. These directional dependencies are small-

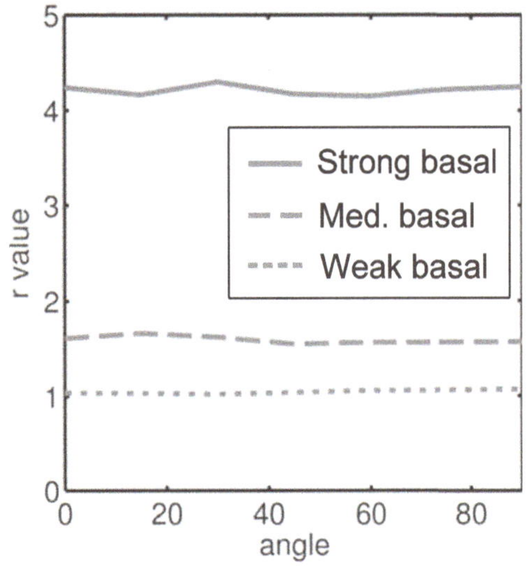

Figure 4: Calculated r-values as a function of angle within the sheet plane for each of the three basal textures.

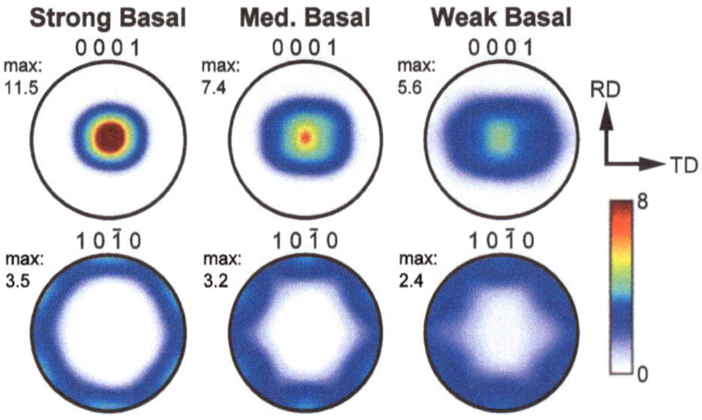

Figure 5: Deformation textures after 20% uniaxial strain along the rolling direction for each of the starting textures.

est for the weakest basal texture, because the prism plane distribution is closer to isotropic.

For all three basal textures, deformation was primarily accommodated by a combination of basal and prismatic slip. Small amounts of tensile twinning were also observed for the weaker basal textures early in the deformation, while minor contributions to deformation were made by pyramidal slip later in deformation. A decrease in basal peak intensity in the starting texture is accompanied by an increase in the activity of easy basal slip and a decrease in the harder prismatic slip mode. This tradeoff also accounts for the differences in yield behavior observed in the stress-strain curves.

The observed differences in slip activity result in dramatically different predicted r-values, shown in Figure 4. The r-values remained approximately constant within the sheet plane, again as a result of the symmetry of the basal peak. Additionally, they increased with increasing basal strength, indicating a reduction in formability. The variation in r-value with basal peak strength results from the differences in slip activity. The sharp basal textures rely heavily on prismatic slip during deformation. In sheet material with a basal texture, prismatic slip primarily causes strain in the width direction of the tensile specimen. Conversely, pyramidal and basal slip result in thickness strains. Consequently, high r-values are a direct result of the dominance of prismatic slip.

These differences in slip activity also result in differences in texture evolution during deformation. The deformation textures after 20% uniaxial strain along the rolling direction are shown in Figure 5. In all cases, asymmetry developed in the basal pole figure during deformation with the basal poles spreading toward the TD, leading to a decrease in the peak intensity of the basal pole figure. The basal peak is also observed to begin to split toward the RD, particularly for the weakest basal texture. This basal peak splitting has been attributed to the activity of pyramidal slip,[26] which correlates well with the slip activity predictions in Figure 3. The substantial prismatic slip activity also leads to a strengthening of the six-fold symmetry in the prismatic pole figure. These results are qualitatively similar to the texture evolution during experimental deformation of the AZ61L material used to parameterize this work.[4]

These results comprise a methodology of generating arbitrary textures and simulating their plastic response to evaluate their forming behavior. In this study, this framework is applied to a set of synthetic textures with varied basal peak strength. The results agree well with the wealth of data in literature[2–6] indicating that increasing basal peak strength is detrimental to forming behavior. These results also complement previous geometric analyses[27] that considered the tradeoff between basal and prismatic slip while simultaneously allowing pyramidal slip and twinning to play a role in deformation.

This framework is a promising approach to consider the effects on deformation behavior of other texture and microstructure characteristics that would not be feasible to isolate experimentally. For instance, less commonly studied texture characteristics, such as the distribution of the prismatic planes or the symmetry of the basal peak, can be isolated and compared at a constant basal peak strength. Additionally, the VPSC model allows for independent control over the hardening parameters for the various slip systems. This enables future comparison of the present results with those for a coarse-grained material where twinning is effectively softened relative to the other deformation modes.

Conclusions

In this study, a framework was established to generate synthetic textures with arbitrary characteristics and evaluate their forming behavior using the a viscoplastic polycrystal plasticity model. The VPSC model was parameterized based on experimental tensile data for thixomolded and thermomechanically processed AZ61L. To demonstrate the utility of this framework, the effect of basal peak strength on macroscopic stress strain behavior, slip system activity, texture evolution, and r-value was examined. Increasing basal peak strength is demonstrated to correlate with increasing r-values and consequently decreasing formability. This poor forming behavior results from the dependence on prismatic slip during deformation. This framework for generating synthetic textures and evaluating their forming behavior is a promising tool to evaluate the effect of other texture characteristics that are difficult to isolate experimentally.

Acknowledgements

The authors would like to express their gratitude to nanoMAG, LLC for providing the material used in this study. V. Miller additionally thanks the University of California Santa Barbara for the support of a Regents' Special Fellowship and T.M. Pollock acknowledges the support of the Institute for Collaborative Biotechnologies through grant W911NF-09-0001 from the Army Research Office.

References

[1] Polmear I. Light Alloys: From Traditional Alloys to Nanocrystals. Burlington, MA: Butterworth-Heinemann, 2006, 4 edition.

[2] Stanford N, Barnett M. Materials Science and Engineering: A 2008;496:399.

[3] Stanford N, Atwell D, Beer A, Davies C, Barnett M. Scripta Materialia 2008;59:772.

[4] Berman TD, Pollock TM, Jones JW. Metallurgical and Materials Transactions A 2015;46:2986.

[5] Agnew S, Horton J, Lillo T, Brown D. Scripta Materialia 2004;50:377.

[6] Gehrmann R, Frommert MM, Gottstein G. Materials Science and Engineering: A 2005;395:338.

[7] Lankford W, Snyder S, Bauscher J. Transactions of the American Society for Metals 1950;42:1197.

[8] Bohlen J, Nürnberg MR, Senn JW, Letzig D, Agnew SR. Acta Materialia 2007;55:2101.

[9] Hosford W. The Mechanics of Crystals and Textured Polycrystals. Number 32 in Oxford Engineering Science Series. New York: Oxford University Press, 1993.

[10] Choi SH, Brem JC, Barlat F, Oh KH. Acta materialia 2000;48:1853.

[11] Agnew SR, Duygulu O. International Journal of Plasticity 2005;21:1161.

[12] Berman TD, Donlon W, Miller VM, Decker R, Huang J, Pollock TM, Jones JW. Magnesium Technology 2012 w/CD 2012;:339.

[13] Berman TD, Donlon W, Hung C, Milligan P, Decker R, Pollock TM, Jones JW. In: Magnesium Technology 2013. TMS (The Minerals, Metals & Materials Society), 2013, 113.

[14] Li X, Jiao F, Al-Samman T, Ghosh Chowdhury S. Scripta Materialia 2012;66:159.

[15] Stanford N. Materials Science and Engineering: A 2010;527:2669.

[16] Groeber MA, Jackson MA. Integrating Materials and Manufacturing Innovation 2014;3:1.

[17] Lebensohn RA, Tomé C. Acta Metallurgica et Materialia 1993;41:2611.

[18] Tomé CN. Modelling and Simulation in Materials Science and Engineering 1999;7:723.

[19] Bachmann F, Hielscher R, Schaeben H. Solid State Phenomena 2010;160:63.

[20] Tomé C, Lebensohn RA, Kocks U. Acta Metallurgica et Materialia 1991;39:2667.

[21] Koike J, Kobayashi T, Mukai T, Watanabe H, Suzuki M, Maruyama K, Higashi K. Acta materialia 2003;51:2055.

[22] Stanford N, Barnett M. International Journal of Plasticity 2013;47:165.

[23] Barnett M, Keshavarz Z, Beer A, Atwell D. Acta Materialia 2004;52:5093.

[24] Koike J, Kobayashi T, Mukai T, Watanabe H, Suzuki M, Maruyama K, Higashi K. Acta Materialia 2003;51:2055.

[25] Agnew SR. In: Magnesium Technology 2002. 2002, 351–356.

[26] Agnew SR, Yoo MH, Tomé CN. Acta Materialia 2001;49:4277.

[27] Koike J, Ohyama R. Acta Materialia 2005; 53:1963.

Magnesium Technology 2016
Edited by: Alok Singh, Kiran Solanki, Michele V. Manuel, and Neale R. Neelameggham
TMS (The Minerals, Metals & Materials Society), 2016

TEXTURE EVOLUTION AND MECHANICAL PROPERTIES OF Mg-Li ALLOY DURING THERMO-MECHANICAL PROCESS

Y. Zou, Y. Zhang, Y. Zhao, S. S. Xu, H. Guo, M. L. Zhang, Z. W. Zhang*
Key Laboratory of Superlight Materials and Surface Technology, Ministry of Education, College of Materials Science and Chemical Engineering, Harbin Engineering University, Harbin 150001, P R China

Keywords: Mg-Li alloy, Recrystallization, Texture, Mechanical properties

Abstract

The effects of rolling reduction and annealing treatment on the microstructure, texture evolution and mechanical behaviors of a duplex Mg-Li base alloy were investigated in this study. The results show that as-cast alloy mainly consists of α-phase and β-phase along with AlLi, Al_2La and Al_4La second phases. For the hcp α phase, the samples exhibit an obvious basal texture, indicating that basal slip should be easily activated during hot-rolling process, the orientation spread of the basal planes is broader towards the transverse direction (TD). For the bcc β phase, the samples exhibit a main $\{001\}<\bar{1}10>$ texture. The average Vickers microhardness (HV) and tensile strength is enhanced with the rolling and further annealing processes due to solution strengthening, intermetallic particles strengthening and Hall-Petch effect.

Introduction

Magnesium-lithium alloys, as the lightest structural alloys, have attracted great interest in the aerospace application and automotive industries [1-3] by virtue of their excellent properties, i.e. low density, high specific strength and elastic modulus, better formability, weak mechanical anisotropy [4, 5]. Based on the Mg-Li binary phase diagram [6], with the Li content between 5 wt% to 11 wt%, Mg-Li alloy have a duplex structure mixed with α and β phases. The existence of soft bcc structured β phase makes Mg-Li alloys possess better ductility than other magnesium alloys, however, lower strength at the same time [7]. Alloying is one of the most effective method to overcome the drawback [8, 9]. Aluminum is a common and useful alloying element in Mg-Li base alloys [10-15] . The tensile strength and hardness of Mg-Li alloys can be improved by means of solid solution and intermetallic compound reinforcements. Moreover, it is well known that addition of rare earth (RE) elements in magnesium alloys shows an effective effect on the mechanical properties. The improvement has mainly been attributed to the formation of highly thermal stable metastable RE-containing phases in grains and along the grain boundaries which significantly increase the creep resistance. On the other hand, the addition of rare earth (RE) elements in Mg alloy can induce more weaker and random texture which promises improved formability [16]. In this study, duplex Mg-9Li-6Al-8La alloy is employed to investigate the microstructure, texture evolution and mechanical properties during thermo-mechanical treatments.

Experimental

Ingots with a nominal composition of Mg-9Li-6Al-8La (wt. %) were prepared under argon atmosphere by induction-melting high purity elemental metals (99.9 Mg, 99.9 Li, 99.9 Al and 30 Mg-La wt. pct.). The induction-melted alloy buttons were drop-cast into plates in a copper mold. To evaluate the effect of rolling and annealing on the texture evolution and mechanical properties, the as-cast ingots were then machined into 100 mm×40mm×10mm rectangular prism for hot rolling. The as-machined specimens were heated at 573K for 30min and then hot-rolled with a total thickness reduction of 40% and 75%, respectively, followed by water quenching. Between passes specimens were replaced in the furnace for 5 min to regain temperature. Some of the 75% thickness reduced specimens were annealed at 573K for 1 h. All the annealing processes were protected by magnesium oxide powder to avoid oxidation. Hereafter, the as-cast specimens are labeled as CS, and as-rolled ones as RS1 and RS2, with thickness reduction of 40% and 75% respectively, and the as-annealed specimens as AS. The microstructure and element composition analysis of the alloys was carried out with optical microscope (OM), scanning electron microscope (SEM, JSM-6480A, Japan Electronics) and energy dispersive X-ray spectroscopy (EDS) (INC250, Japan Electronics). The specimens for observation were polished and etched with a 4 vol.% nital. Phase identification was investigated by X-ray diffraction on Rigaku D/max-TTR-III diffractometer using Cu Kα. The textures were measured using X-ray diffraction (XRD) (XPERT PRO, Cu Kα). As the Mg-9Li-6Al-8La alloy consists of α phase hcp structure and β phase bcc structure, two sets of pole figures were measured, four incomplete pole figures of {0002}, $\{10\bar{1}0\}$, $\{10\bar{1}1\}$ and $\{10\bar{1}2\}$ for α phase, and three incomplete pole figures of {200}, {110} and {211} for β phase. Incomplete pole figures were measured up to a tilt angle of 70° using the reflection method of XRD. We used ResMat-TexTools (Version 3.3) to construct the orientation distribution function (ODF), from which the complete pole figures were derived. Tensile tests were carried out with a strain rate of 1×10^{-3} s^{-1} on an Instron 4505 tensile tester with an extensometer at room temperature. The gauge length, width and thickness of the tensile specimens are 12.5, 4 and 1 mm, respectively.

Results and Discussion

Microstructure

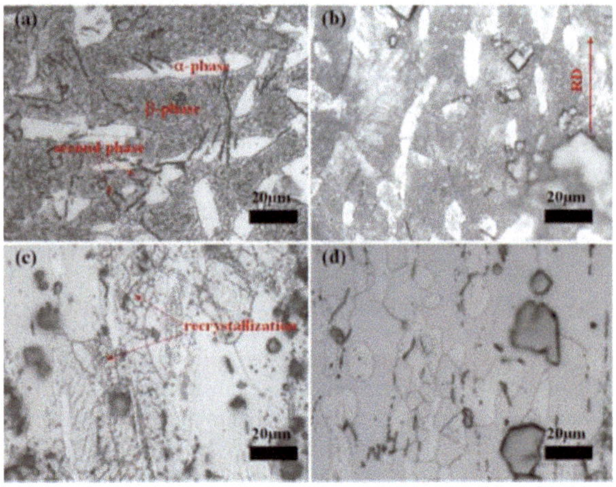

Fig.1. Optical Microstructures of the CS (a), RS1 (b), RS2 (c) and AS (d) alloys.

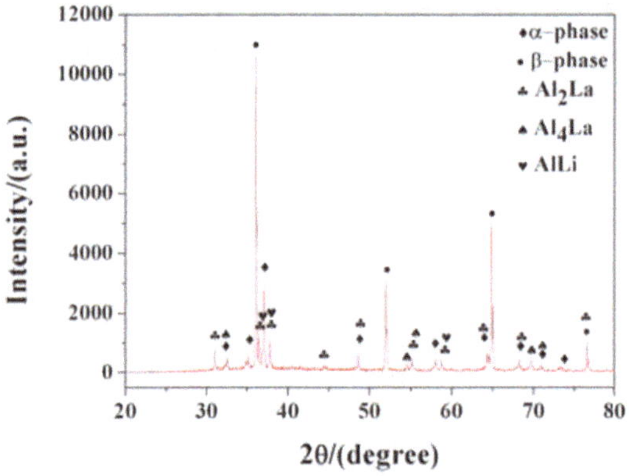

Fig. 2. The XRD patterns of the as-cast alloy (CS).

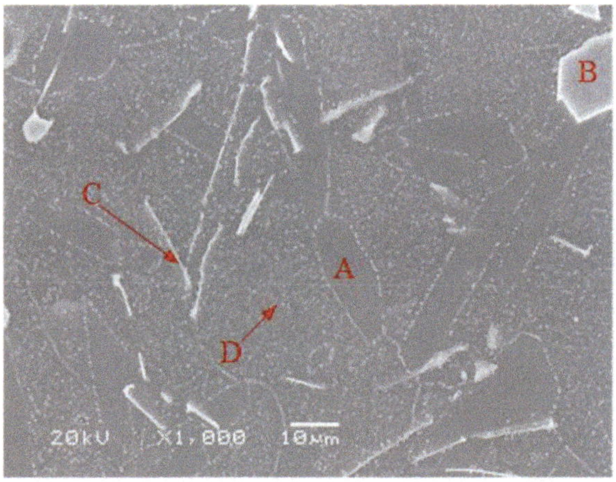

Fig. 3. The SEM morphology and corresponding EDS dots of the as-cast alloy (CS).

The microstructures of CS, RS1, RS2 and AS samples measured by optical microscope are shown in Fig. 1. There are two phases, α-phase (bright color) and β-phase (gray color) in the four samples. It is obvious that the irregular block and long strip α phase as well as second phases uniformly disperse within the β phase matrix (Fig.1a). Hot-rolling makes α phase elongate along the rolling direction (RD), and the percentage of α phase alignment is different for different thickness reduction samples. As shown in Fig. 1b and Fig. 1c, not all α phases are aligned in the 40% thickness reduction RS1 sample, but almost all are aligned in the 75% thickness reduction RS2 sample. Moreover, a lot of small grains with grain size about 3μm can be observed in the elongated β phase (Fig.1c), indicating the occurrence of recrystallization during rolling to a thickness reduction of 75%. The recrystallization can occur during rolling passes or during reheating between passes. In the AS samples, the long strip α phase changes to irregular block-like, and some small grains in β phase grow up to 7~10μm, indicating that static recrystallization occurred in both α phase and β phase after annealing (Fig.1d). Based on the results above, we can conclude that the recrystallization occurs in elongated β phase first during hot-rolling process while in both α phase and β phase after annealing. This may be related to the difference in the recovery and dynamic recrystallization between α and β phases during hot-rolling at 573K. It was reported that stacking fault energy (SFE) plays a key role in controlling the occurrence of dynamic recovery and recrystallization. A high SFE is beneficial to the dynamic recovery and recrystallization during thermo-mechanical processes [17, 18]. Thus, dynamic recovery and recrystallization occurs first in the bcc β phase during hot-rolling process, which has a higher stacking fault energy; while the SFE of α phase is lower, hindering the occurrence of dynamic recovery and recrystallization. After annealing at 573K for 1h, recrystallization can occur in both α phase and β phase.

Table 1. EDS results of the as-cast alloy(CS).

Element / Number	Wt.%			At.%		
	MgK	AlK	LaL	MgK	AlK	LaL
A	100			100		
B	1.8	30.5	67.7	4.3	66.9	28.8
C	62.5	17.3	20.2	76.6	19.1	4.3
D	98.96	1.04		99.06	0.94	

The XRD patterns of the as-cast alloy (CS) were illustrated in Fig. 2. The XRD results confirmed that the Mg-9Li-6Al-8La alloy consisted of α-phase and β-phase along with AlLi, Al_2La and Al_4La second phases. The SEM morphology and corresponding EDS results of the as-cast alloy (CS) were shown in Fig. 3 and Table 1. The Mg-9Li-6Al-8La alloy presented typical duplex structure of α-phase (Dot A) and β-phase. Some block-shaped (Dot B) and needle-like (Dot C) precipitates were evenly distributed in the matrix, which had atomic ratio of Al:La about 2.3 and 4.4, identifying as Al_2La and Al_4La respectively. Moreover, some small spherical particles (Dot D) are evenly distributed in β-phase as well as along the grain boundaries, as Li cannot be detected in EDS, combining results of previous study and XRD analysis, it was AlLi phase.

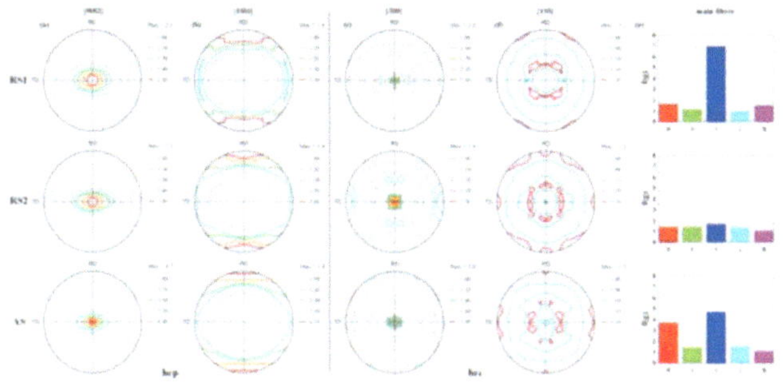

Fig.4. The representative calculated {0002} (a), $\{10\bar{1}0\}$ (b) pole figures for the hcp structure and {200} (c), {110} (d) pole figures and corresponding main fibers (e) for the bcc structure of RS1, RS2 and AS alloy.

Texture Evolution

For analyzing the co-deformation of α and β phases during large plastic deformation, the evolution of texture during rolling and annealing is analyzed. The representative calculated pole figures of {0002}, $\{10\bar{1}0\}$ for the hcp α phase and {200}, {110} for the bcc β phase in RS1, RS2 and AS samples are shown in Fig.4. To characterize the deformation of the bcc β phase during rolling, the main fiber textures of α, γ, ε, ζ and η are also included in Fig.4 since these five fibers generally prevail in the rolled bcc metals [19]. For the hcp α phase, the RS1, RS2 and AS samples exhibit an obvious basal $\{0002\}<1\bar{1}00>$ texture, indicating that basal slip should be easily activated during hot-rolling process. After rolling the samples to a thickness reduction of 75%, the basal texture intensity was enhanced, on the other hand, the splitting of basal poles around rolling direction (RD) was decreased and the pole figures exhibited a flat pattern, which is consistent with the Mg-RE studies [20, 21] that the orientation spread of the basal planes is broader towards the transverse direction (TD) rather than towards the rolling direction (RD). This can be related to the recrystallization during rolling passes or during reheating between passes [21]. After annealing, the basal texture was retained and the intensity was further enhanced. The orientation spread of the basal planes is also broader towards the transverse direction (TD) which also be related to the static recrystallization and grain growth.

For the bcc β phase, the RS1, RS2 and AS samples exhibit a main $\{001\}<\bar{1}\bar{1}0>$ texture. With the further rolling and annealing process, the intensity was decreased. Since the fiber textures can affect the mechanical properties and the evolution of fiber textures in bcc β phase is important. The evolution is calculated and shown in Fig. 4e. For ideal rolling fibers in bcc structure, α, γ, ε, ζ, η are mainly concerned, where α with <110>//RD, γ <111>//ND, ε <110>//TD, ζ <110>//ND and η <001>//RD [19]. The sample RS1 has a strong ε fiber. With the thickness reduction increasing to 75%, the fraction of ε fiber decreases. Annealing induces an increase in α fiber and ε fiber significantly. This can be attributed to static recovery and recrystallization process during annealing treatment.

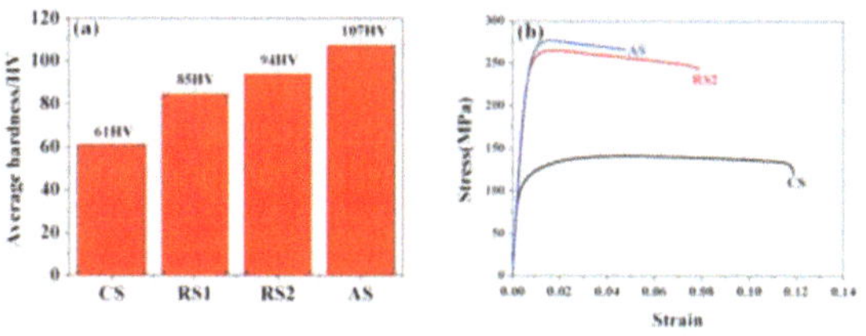

Figure.5. The average hardness of the CS, RS1, RS2 and AS alloys and tensile engineering stress-strain curves of the CS, RS2 and AS alloys at room temperature.

Mechanical Properties

The average Vickers microhardness (HV) test results of the CS, RS1, RS2 and AS alloys is presented in Fig.5a, which is 61HV, 85HV, 94HV and 107HV, respectively. It can be concluded that the average Vickers microhardness (HV) is enhanced with the rolling and further annealing processes. Increase in hardness with thickness reduction and further annealing treatment indicate that Mg-9Li-6Al-8La alloy could be work hardened and thermal strengthened.

The tensile engineering stress-strain curves of the samples CS, RS2 and AS measured at ambient temperature are shown in Fig. 5b. The tensile direction is aligned along the rolling direction. The tensile yield strengths (TYS) of the CS, RS2 and AS are ~108MPa, 238MPa and 242MPa, respectively. The ultimate tensile strengths (UTS) of the CS, RS2 and AS are 142MPa, 265MPa and 277MPa, respectively. As for the ductility, during the tension, fracture elongations of the CS, RS2 and AS are ~10.8%, 7.9% and 4.8%, respectively. It is evident that the strength is enhanced significantly accompanying with the ductility is decreased by the rolling processes and further annealing treatment, which is also consistent with texture evolution in relative harder α phase and softer β phase during the rolling processes and further annealing treatment, that is, the basal texture intensity in relative harder α phase is increased while the $\{001\}<\bar{1}\bar{1}0>$ texture intensity in softer β phase is decreased. Increase in tensile strength for a reduction of 75% alloy (RS2) is approximately 86.6% while the decrease in ductility is only 26.9% as compared with the as-cast alloy (CS). This can be explained as solution strengthening and intermetallic particles strengthening which play an important role to the improvement of strength due to the solute atoms and intermetallic particles can be act as obstacles or barriers for motion of dislocations during deformation. When dislocation segments are held up by the local obstacles and then raise the strength. Moreover, the occurrence of recrystallization during rolling to a thickness reduction of 75% results in lots of small grains with grain size about 3μm in β phase (Fig.1c), according to the classic Hall-Petch relationship [22, 23]:

$$\sigma_y = \sigma_0 + k_y d^{1/2} \quad (1)$$

where σ_y is the yield stress, σ_0 is the friction stress when dislocations move on the slip plane, d is the average grain size, and k_y is the stress concentration factor. The decrease in average grain size during rolling to a thickness reduction of 75% also leads to an improvement in strength. However, precipitates are brittle phase which are gathered at grain boundaries, when the alloy suffers from divorce, the second phase can't deform with the matrix, decreasing the ductility. After annealing treatment, the increase in tensile strength is about 95.1% while the decrease in ductility is 55.6% as compared with the CS, which can be mainly ascribed to the amount of the small AlLi phase dissolved into the matrix during annealing treatment which causes the effect of solution strengthening and the prismatic plate shaped precipitates in large size formed in the alloy which causes more effective intermetallic particles strengthening, while, bad for ductility.

Conclusions

The effects of rolling reduction and annealing treatment on the microstructure, texture evolution and mechanical behaviors of an Mg-9Li-6Al-8La alloy were investigated in this study. The as-cast alloy mainly consists of α-phase and β-phase along with AlLi, Al_2La and Al_4La second phases. For the hcp α phase, the RS1, RS2 and AS samples exhibit an obvious basal texture, indicating that basal slip should be easily activated during hot-rolling process, the orientation spread of the basal planes is broader towards the transverse direction (TD) rather than towards the rolling direction (RD) which can be related to the recrystallization during rolling passes or during reheating between passes. For the bcc β phase, the RS1, RS2 and AS samples exhibit a main $\{001\}<\bar{1}10>$ texture. The average Vickers microhardness (HV) and tensile strength is enhanced accompanying with the ductility is decreased with the rolling and further annealing processes. This can be explained as solution strengthening and intermetallic particles strengthening which play an important role in the improvement of strength due to the solute atoms and intermetallic particles can be act as obstacles or barriers for motion of dislocations during deformation. Dislocation segments are held up by the local obstacles and then raise the strength. Besides, the occurrence of small recrystallization grains during rolling to a thickness reduction of 75% and after annealing treatment are also beneficial to increase strength according to the classic Hall-Petch relationship. However, precipitates are brittle phase which are gathered at grain boundaries, when the alloy suffers from divorce, the second phase can't deform with the matrix, decreasing the ductility.

Acknowledgments

This research was supported by the Fundamental Research Funds for the Central Universities (HEUCFZ1308), the NSFC Funding (51171081, 51371062 and U1460102), NSFHLJ (ZD201411), the Scientific Research Foundation for the Returned Overseas Chinese Scholars, Heilongjiang province and the Project 2013DB04 at NPL, CAEP. Yun Zou is benefited from the International Exchange Program of Harbin Engineering University for Innovation-oriented Talents Cultivation.

References

[1] A. Sanschagrin et al., "Mechanical properties and microstructure of new magnesium-lithium base alloys," *Materials Science & Engineering A*, 220 (1996), 69–77.

[2] Z. Trojanova et al., "Deformation behaviour of Mg–Li alloys at elevated temperatures," *Materials Science & Engineering A*, 410 (2005), 148-151.

[3] S.J. Wang et al., "Microstructures and mechanical properties of 5 wt.% Al2Yp/Mg–Li composite," *Materials Letters*, 60 (2006), 1863–1865.

[4] A.M. Russellet al., "Anomalously high impact fracture toughness in B.C.C. Mg-Li between 4.2 K and 77 K," *Scripta Materialia*, 39 (1998), 1663–1667.

[5] C.W. Yang et al., "Tensile mechanical properties and failure behaviors with the ductile-to-brittle transition of the $\alpha+\beta$-type Mg–Li–Al–Zn alloy," *Scripta Materialia*, 61 (2009), 1141-1144.

[6] T.B. Massalski et al., *Binary alloy phase diagrams*, (ASM International, 1990), 1485.

[7] R. Mahmudi et al., "Effect of Li content on the indentation creep characteristics of cast Mg–Li–Zn alloys," *Materials & Design*, 75 (2015), 184-190.

[8] R. Wu, Y. Deng, and M. Zhang, "Microstructure and mechanical properties of Mg–5Li–3Al–2Zn–xRE alloys," *Journal of materials science*, 44 (2009), 4132-4139.

[9] T. Zhu et al., "Influence of the combined addition of Y and Nd on the microstructure and mechanical properties of Mg–Li alloy," *Materials & Design*, 57 (2014), 245-249.

[10] Y.H. Kim et al., "Microstructure and mechanical properties of Mg–xLi–3Al–1Sn–0.4 Mn alloys (x=5, 8 and 11wt%)," *Journal of Alloys and Compounds*, 583 (2014), 15-20.

[11] B. Jiang et al., "A new approach to grain refinement of a Mg–Li–Al cast alloy," *Journal of Alloys and Compounds*, 492 (2010), 95-98.

[12] T. Liu et al., "Mechanical properties of a two-phase alloy Mg–8% Li–1% Al processed by equal channel angular pressing," *Materials Science & Engineering A*, 360 (2003), 345-349.

[13] Z. Qu et al., "The solution and room temperature aging behavior of Mg–9Li–xAl (x=3, 6) alloys," *Journal of Alloys and Compounds*, 536 (2012), 145-149.

[14] A. Alamo and A. Banchik, "Precipitation phenomena in the Mg-31 at% Li-1 at% Al alloy," *Journal of Materials Science*, 15 (1980), 222-229.

[15] F. Herbstein and B. Averbach, "The structure of lithium-magnesium solid solutions-I: Measurements on the Bragg reflections," *Acta metallurgica*, 4 (1956), 407-413.

[16] S. Yi et al., "Mechanical anisotropy and deep drawing behaviour of AZ31 and ZE10 magnesium alloy sheets," *Acta Materialia*, 58 (2010), 592-605.

[17] A. Jorge, W. Regone, and O. Balancin, "Effect of competing hardening and softening mechanisms on the flow stress curve modeling of ultra-low carbon steel at high temperatures," *Journal of materials processing technology*, 142 (2003), 415-421.

[18] E. Poliak and J. Jonas, "A one-parameter approach to determining the critical conditions for the initiation of dynamic recrystallization," *Acta Materialia*, 44 (1996), 127-136.

[19] M. Hajian and A. Assempour, "Numerical study on the effects of main BCC rolling texture components on the formability of sheet metals," *The International Journal of Advanced Manufacturing Technology*, (2015), 1-9.

[20] J. Bohlen et al., "The texture and anisotropy of magnesium–zinc–rare earth alloy sheets," *Acta Materialia*, 55 (2007), 2101-2112.

[21] L. Mackenzie and M. Pekguleryuz, "The recrystallization and texture of magnesium–zinc–cerium alloys," *Scripta Materialia*, 59 (2008), 665-668.

[22] E.O. Hall, "The deformation and ageing of mild steel: III discussion of results," *Proceedings of the Physical Society Section B*,64 (1951), 747-753.

[23] N.J. Petch, "The cleavage strength of polycrystals," *Journal of the Iron and Steel Institute,* 174 (1953), 25-28.

Magnesium Technology 2016
Edited by: Alok Singh, Kiran Solanki, Michele V. Manuel, and Neale R. Neelameggham
TMS (The Minerals, Metals & Materials Society), 2016

EFFECT OF DYNAMIC RECRYSTALLIZATION ON MICROSTRUCTURE EVOLUTION AND TEXTURE WEAKENING DURING ANNEALING OF HIGH SPEED ROLLED AZ31 MAGNESIUM ALLOY SHEETS

Jing Su[1], Abu Syed H. Kabir[1], Mehdi Sanjari[1], In-Ho Jung[1] and Stephen Yue[1]

[1] Department of Mining and Materials Engineering, McGill University, 3610 Rue University, Montreal, Quebec, Canada H3A 0C5

Keywords: AZ31 Mg alloy, dynamic recrystallization static recrystallization, texture weakening

Abstract

Magnesium AZ31 (Mg-3 wt. % Al-1 wt. % Zn) alloy sheets were rolled at a high speed of 1000 m/min at 100 °C to reductions of 30% and 49%. Annealing was then conducted on the as-rolled specimens at temperatures of 200 °C and 350 °C for different times. The microstructure was characterized by optical microscopy and the macrotexture was analyzed by X-ray diffraction. The as-rolled microstructure of the specimen subjected to the reduction of 30% was heavily twinned and shear banded, while a partially dynamically recrystallized and twinned microstructure was observed at the reduction of 49%. Effect of the initial microstructure on static recrystallization behavior and texture evolution during annealing was studied at different temperatures. Texture weakening was found during annealing at both reductions. However, the weaker texture can be achieved in the specimen subjected to the reduction of 30% than that of 49%.

1. Introduction

It is well known that magnesium alloys have poor ductility and limited formability at low temperatures, which is due to the insufficient number of activated slip systems and the strong basal texture. In order to avoid cracking or fracture, Mg alloy sheets are often rolled at high temperatures [1, 2]. Dynamic recrystallization (DRX) inevitably occurs during hot deformation, which refines the microstructure, but the strong basal texture still forms [3, 4]. This limits the further deformation of Mg alloy sheets. It has been reported that static recrystallization (SRX) of the heavily deformed structure during annealing has potential to weaken the texture [3, 5-11]. It is more effective when the as-rolled structure is heavily twinned or shear band than dynamically recrystallized.

Twins and shear bands are the dominant features during low temperature rolling. However, the challenge is to form Mg alloy sheets at low temperature without fracture. In order to improve the formability of Mg alloys, much research has been conducted to weaken or randomize the texture by alloying additions of rare earth elements, Y or Li, or by using differential speed rolling or equal channelling angular extrusion to introduce high levels of shear. Recently, high speed rolling (HSR) has been proved to improve the formability of Mg sheets [12-14]. Our previous work [15] showed a high reduction of 72% achieved in AZ31 Mg alloy at a high rolling speed of 1000 m/min at a low temperature of 100 °C. The better rollability related to high speed rolling was attributed to initiation of dynamic recrystallization and activation of <c+a> slip [15]. In the present study, the effect of dynamic recrystallization on microstructure and texture evolution during annealing was studied by comparing the fully twinned and shear banded microstructure at the reduction of 30% and the partially recrystallized and twinned microstructure at the reduction of 49%. In addition, effect of annealing temperature on the static recrystallization characteristics of the specimens at both reductions was also investigated.

2. Experimental procedure

A twin roll cast (TRC) AZ31B magnesium alloy (Mg - 3.2 wt. % Al - 0.77 wt. % Zn - 0.34 wt. % Mn) supplied by the Posco Company in South Korea was used in the present research. The TRC plate was warm rolled and homogenized at 300 °C for 1 h so as to break down the as-cast structure. The initial sheets were 3 mm in thickness and 30 mm in width. The homogenized plates were subjected to single-pass rolling at 100 °C at 1000 m/min to reductions of 30% and 49%. Annealing was conducted at temperatures of 200 °C and 350 °C for different times. The specimens were quenched after annealing to freeze in the microstructure.

Microstructural characterization was carried out by optical microscopy. The specimens were mechanically ground with up to 4000 grit SiC paper and polished with a 0.05 μm colloidal silica suspension for the final step. A solution consisting of picric acid (4.2 g), acetic acid (10 ml), water (10 ml) and ethanol (70 ml) was used to etch the specimens to reveal the optical microstructures. X-ray diffraction was utilized to analyze the macrotextures of the specimens, which were measured at the mid-thickness of the sheets. The hardness was measured on the ND-RD cross-section of the specimen by Vickers hardness testing. At least 10 indentations were measured for each sample to calculate the average hardness value.

3. Results and Discussion

3.1 As-rolled microstructure

As-rolled microstructures at reductions of 30% and 49% are shown in Figs. 1 (a) and (b). The specimen at the reduction of 30% presents a heavily twinned and shear banded microstructure. At the reduction of 49%, the small dynamically recrystallized grains can be seen on twins in the shear bands and unrecrystallized twins between shear bands are also visible. It is noted that twins in shear bands recrystallized more preferentially than the twins between the shear bands. Since the orientation of shear bands is favourable for easy glide of basal slip and twin activation [16], therefore, much of the deformation appeared to be concentrated in the shear bands [17]. The higher concentration of strain at shear bands than the region outside the bands results in significant lattice rotation of the grains in the bands [18].

The (0002) pole figures of the specimens subjected to reductions of 30% and 49% are shown in Figs. 2 (a) and (b). Basal pole splitting along rolling direction (RD) can be seen at both reductions. This is due to formation of secondary and contraction twins and/or <c+a> slips [14, 19-21]. The maximum intensity of the basal texture of the specimen at the reduction of 30% is slightly lower than that at the reduction of 49%.

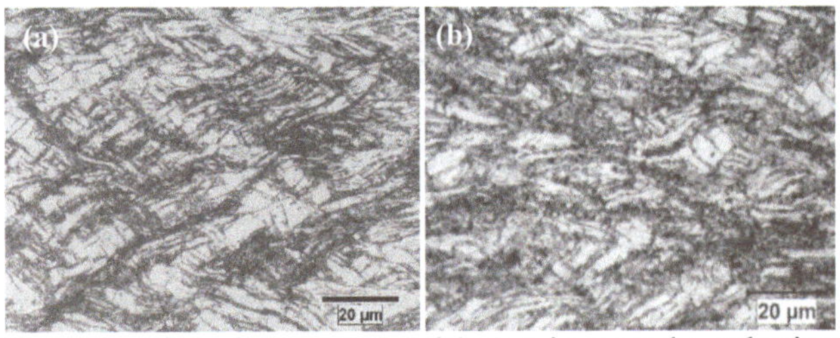

Fig. 1 As-rolled microstructures of the specimens at the reductions of 30% (a) and 49% (b)

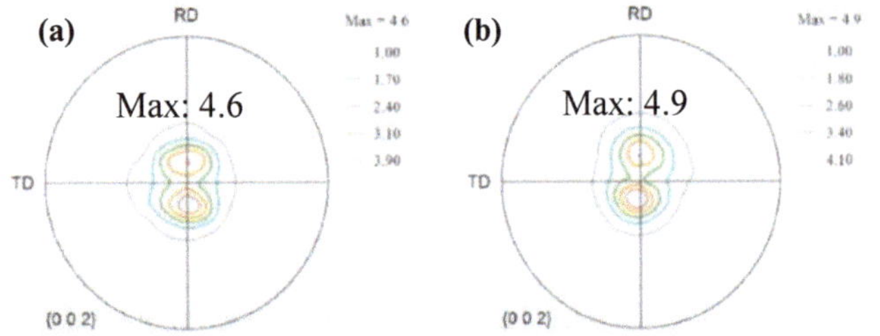

Fig. 2 (0002) pole figures of the specimens subjected to reductions of 30% (a) and 49% (b)

3.2 Microstructure evolution during annealing

3.2.1 Effect of DRX on microstructure evolution during annealing at 200 °C

Optical microstructures of the specimens subjected to the reductions of 30% and 49% during annealing at 200 °C for 30 s and 45 s are compared in Fig. 3. After annealing of the 30% reduction specimen for 30 s, no obvious static recrystallization can be observed at this magnification (Fig. 3 (a)). However, small recrystallized grains became visible in the vicinity of shear bands after 45 s annealing (see the enlarged image of the rectangular region in Fig. 3 (b)). By comparison, the as-rolled microstructure at the reduction of 49% was already dynamically recrystallized on the shear bands. After annealing for both 30 s and 45 s, no significant static recrystallization can be seen in the microstructure (Figs. 3 (c) and (d)).

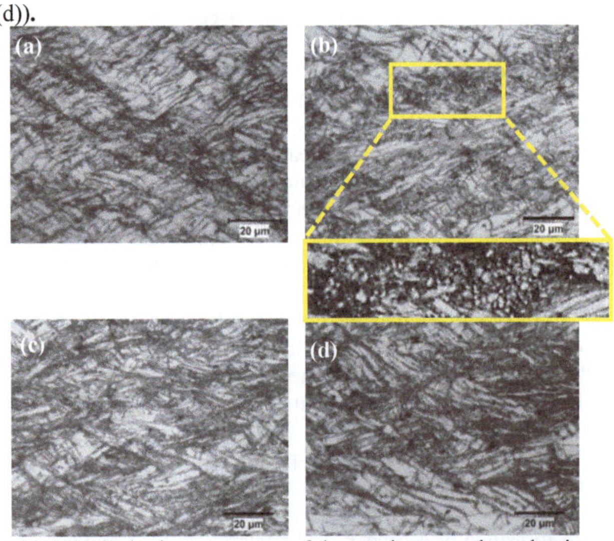

Fig. 3 Optical microstructures of the specimens at the reductions of 30% and 49% during annealing at 200 °C for 30 s ((a) and (c)) and 45 s ((b) and (d))

Optical microstructures of the 30% and 49% reductions specimens annealed at 200 °C for 1 min and 3 min are displayed in Fig. 4. Compared to that after 45s annealing, a higher volume fraction of SRXed grains can be seen in the vicinity of the twins and shear bands, accompany by growth of the recrystallized grains after annealing for 1 min of the specimens at both reductions. After annealing for 3 min, more SRXed grains formed in the specimens at both reductions. It is noticed that the size of the recrystallized grains at the reduction of 30% is larger than that at 49%.

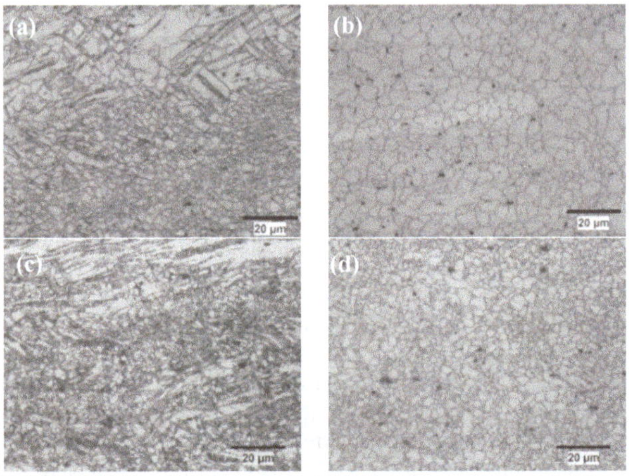

Fig. 4 Optical microstructures of the specimens at the reductions of 30% and 49% during annealing at 200 °C for 1 min (a) (c) and 3 min (b) (d)

With increasing annealing time to 30 min, most of the microstructure was recrystallized in the specimen at the reduction of 49%, but a few large elongated grains were found resistant to recrystallization (Fig. 5 (a)). In the enlarged image of the rectangular region in Fig. 5 (a), small recrystallized grains can be seen on twins in the elongated grain, while the large untwined regions remained unrecrystallized, since there were only small numbers of twins formed in the elongated grains. After annealing for 2 h, these elongated grains still did not recrystallize completely (Fig. 5 (b)).

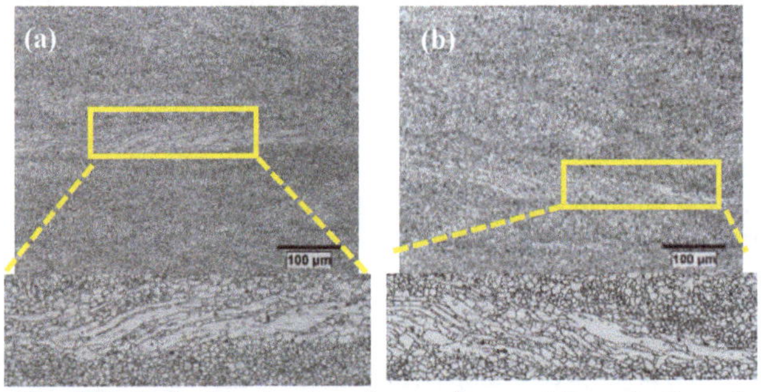

Fig. 5 Optical microstructure the specimens at the reduction of 49% annealed at 200 °C for 30 min (a) and 2 h (b)

3.2.2 Effect of DRX on microstructure evolution during annealing at 350 °C

Optical microstructures of the specimens subjected to the reduction of 49% annealed at 350 °C for 5 s and 10 s are shown in Fig. 6. Small recrystallized grains were observed in the vicinity of the twins and shear bands after annealing for 5 s (the enlarged image of the rectangular regions in Fig. 6 (a)). With increasing annealing time to 10 s, there is no significant change in the microstructure, although more recrystallization could occurred. The shear banding and twinning features still remained.

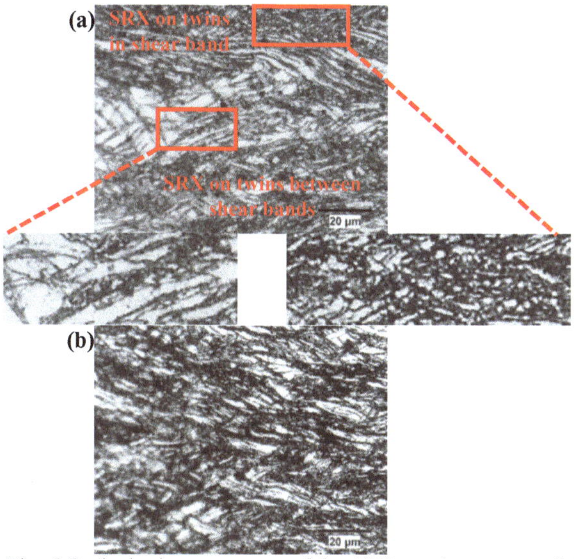

Fig. 6 Optical microstructures of the 49% specimens annealed at 350 °C for 5 s (a) and 10 s (b)

Optical microstructures of the specimens subjected to the reductions of 49% and 30% during annealing at 350 °C for 15 s and 30 s at both high and low magnifications are compared in Fig. 7. After annealing for 15 s on the specimen at the reduction of 49%, massive SRX took place and replaced the twinning structure. It is worth mentioning that the SRXed grains in shear bands show smaller grain size that those between the bands. For the specimen at the reduction of 30%, after annealing for 15 s (Fig. 7 (c)), small recrystallized grains are seen in the vicinity of the twins and shear bands, while the unrecrystallized twins can still be found. Elongated grains can be seen in the low magnification microstructures (right columns). Recrystallization on twins can be observed in the elongated grains in the 49% reduction specimen after annealing for 30 s (the enlarge image of the rectangular region of Fig. 7 (b)).

After annealing for 1 min on the 49% reduction specimen (Fig. 8 (a)), it can be seen that the large elongated grains were recrystallized, which is related to the subgrain growth. However, the grain size of the recrystallized grains inside the elongated grains are much larger than the twin related recrystallized grains. In the specimen subjected to the reduction of 30%, the recrystallization of the elongated grains was also observed after annealing for 5 min (Fig. 8 (c)). Full recrystallization was achieved after annealing for 10 min and 30 min for the specimens at the reductions of 49% (Fig. 8 (b)) and 30% (Fig. 8 (d)), respectively. Thus, the recrystallization rate of the specimen subjected to the reduction of 49% is much higher than that of 30% at the annealing temperature of 350 °C.

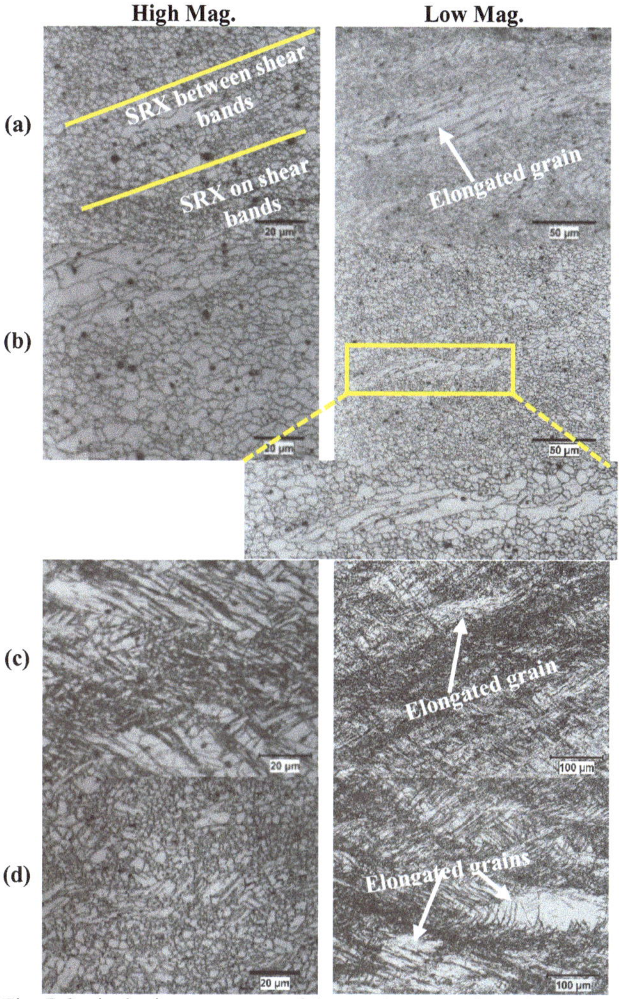

Fig. 7 Optical microstructures of the specimens at the reduction of 49% and 30% annealed at 350 °C for 15 s (a) (c) and 30 s (b) (d) at both high and low magnifications

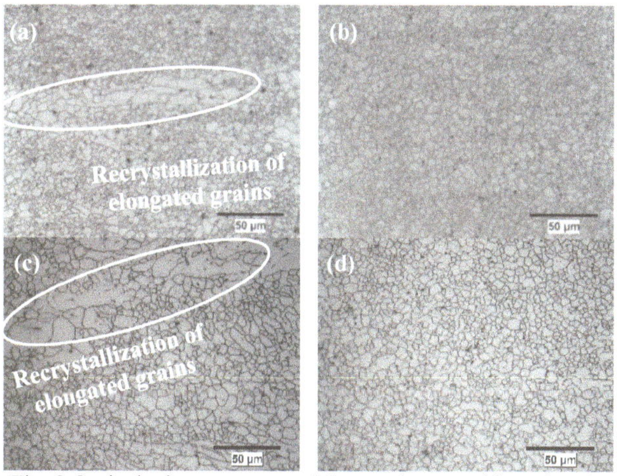

Fig. 8 Optical microstructures of the specimens subjected to the reduction of 49% annealed at 350 °C for 1 min (a) and 10 min (b) and the specimens at the reductoin of 30% annealed at 350 °C for 5 min (c) and 30 min (d)

3.3 Effect of annealing temperature on SRX

The relative hardness ΔH (defined in Equation (1)) indicates the fraction of microstructure softening and thus refers to the volume fraction of recrystallization, X_v.

$$\Delta H = \frac{H_0 - H_i}{H_0 - H_r} \quad (1)$$

where H_0 is the hardness of the as-deformed material, H_i is the measured hardness during isothermal annealing and H_r is the hardness at the annealing time where full recrystallization achieved. Therefore, the plot of the fractional softening with annealing time shows the rate of static recrystallization during annealing.

Figure 9 shows the relationship of fractional softening with annealing time at 200 °C. It can be seen the softening rates of the 30% and 49% specimens are very close. At the annealing time of 45 s, more fractional softening can be seen in the specimen at the reduction of 30% than that of 49%. This is due to the nucleation of the recrystallized grains on shear bands in the specimen at the reduction of 30%, as shear bands possessed much higher stored energy. However, for the specimen at the reduction of 49%, the shear bands have already recrystallized and the annealing temperature was too low to cause obvious grain growth during this short annealing time (45 s). From 1 min to 3 min annealing, the softening rates of the two reductions are almost identical, which is consistent with the microstructure observation. After annealing time longer than 3 min, the softening rates of the 49% specimens are slightly higher than those of the 30% specimens.

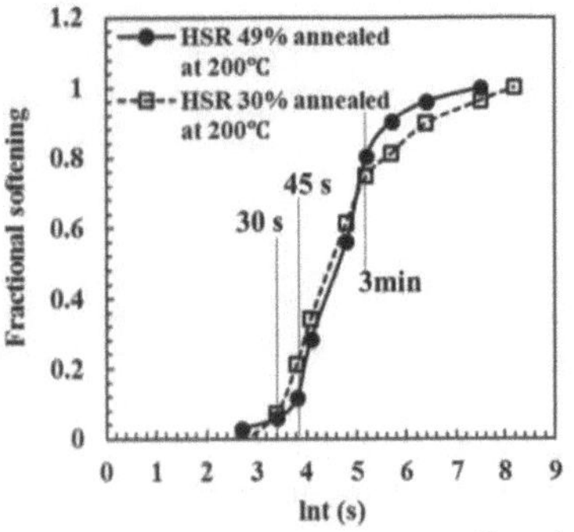

Fig. 9 Fractional softening versus annealing time of the specimens subjected to reductions of 30% and 49% annealed at 200 °C for different times.

The correspondences of the fractional softening with annealing time at 350 °C of the specimens at the reductions of 49% and 30% are compared in Fig. 10. The softening rate of the 49% specimens is much higher than that of the 30% specimens. At the same annealing time of 15 s, only 20% of recrystallization occurred in the specimen at the reduction of 30%, while around 75% recrystallization formed in the specimen at the reduction of 49%. For the specimen at the reduction of 30%, it can be seen that the softening rate increases dramatically during annealing from 15 s to 1 min, and nearly 80% recrystallization was achieved after 1 min. From 1 min to 30 min, the softening rate slows down and the full recrystallization is achieved at 30 min. For the specimens at the reduction of the 49%, the higher softening rate is seen from 5 s to 15 s, followed by a much lower rate from 15 s to 10 min where full recrystallization is reached.

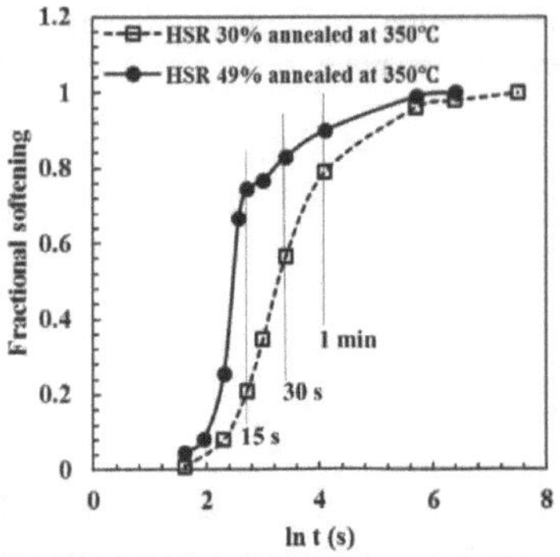

Fig. 10 Fractional softening versus annealing time of the specimens subjected to reductions of 30% and 49% annealed at 350 °C for different times.

In summary, during annealing at 200 °C, there were no big differences of the static recrystallization behavior between the partially DRXed and twinned specimen (49%) and the heavily twinned and shear banded specimen (30%), while the softening rate of the DRXed specimen was much higher than the twinned specimen during annealing at the higher temperature of 350 °C. This is because that the dominant recrystallization mechanism at 200 °C is related to twins and shear bands. But at higher temperatures, the static recrystallization is not only related to twins, but also to the grain/subgrain boundary migration rate, which is greatly affected by temperature. This also explains the recrystallization of the large elongated grains during annealing at 350 °C, but they remained un-recrystallized at 200°C even after 2 h.

During annealing at 200 °C, SRXed grains initially nucleated on twins and shear bands, and then in the non-twinned regions. Grain growth took place at same time, however the growth rate was very low, as the temperature was low. Therefore, at the beginning of the annealing (before 1min), there is no obvious growth of the pre-existing DRXed grains, thus the fractional softening rate of the DRXed specimen is slightly slower than the twinned specimen which underwent nucleation on shear bands. After 3 min when the twins and shear bands were mostly recrystallized, the recrystallization rate of the DRXed specimen was slightly higher than that of the twinned specimen. This may be due to more dislocation accumulated in the 49% specimen. It is evident that the hardness of the as-rolled specimen at the reduction of 49% is higher than that at 30%. Although the 49% specimen has already dynamically recrystallized, the DRXed grains mainly occurred on the shear bands. Thus, with increasing reduction, dislocation mostly accumulated in the non-twinned regions.

However, at the higher annealing temperature of 350 °C, the pre-existing DRXed grains grew immediately during annealing, and the new SRXed grains nucleated and grew at the same time. The recrystallization rate of the specimen at the reduction of 49% was much higher than that at 30%, since the DRXed grains served as the precursor for recrystallization. The recrystallization rate of the non-twinned regions by subgrain growth was also increased, since the mobility of the subgrain boundaries was increased as increasing annealing temperature. In addition, there was more dislocation accumulated in the 49% specimen, which is also contributed to the higher recrystallization rate.

3.2 Texture evolution during annealing

The maximum intensity of basal texture versus annealing time at 350 °C of the specimens subjected to reductions of 30% and 49%

are compared in Fig. 11. As mentioned above, the basal texture of the as-rolled specimens at the reductions of 49% was slightly stronger than at 30%. Texture weakening was seen at both reductions. After annealing for 30 s and 1 min, the values of the maximum intensity of basal texture of the 49% specimen are lower than that of the 30% specimens. This is due to the higher rate of static recrystallization in the 49% specimen than in the 30% specimen at the same annealing time. It was noticed in Fig. 10 that that the volume fraction of recrystallization of the 49% specimen annealed for 30 s was higher than that of the 30% specimen annealed for 1 min.

The weakest basal texture of the 49% and 30% specimens are achieved after 1 min and 5 min, respectively. The weakening rate of the basal texture of the specimens at the reduction of 49% was higher than that at 30%. However, during annealing, the specimen at the reduction of 30% shows the weakest basal texture (3.5), compared to that at the reduction of 49% (3.9). In another word, the texture weakening during annealing of the specimen at the reduction of 30% was more effective than that at 49%. This may be because that the statically recrystallized grains formed on shear bands and twins resulted in more random orientation and thus a weaker texture than the dynamically recrystallized grains on shear bands and twins, which rotated to basal orientations during deformation [5, 6]. The as-rolled microstructure of the specimen subjected to the reduction of 49% was partially recrystallized and twinned, thus texture weakening was also achieved through twinning and shear banding related recrystallization. However, when the initial structure is dynamically recrystallized, the intensities of the basal texture are higher than the heavily twinned structure, which is due to the growth of DRXed grains during subsequent annealing [7-11].

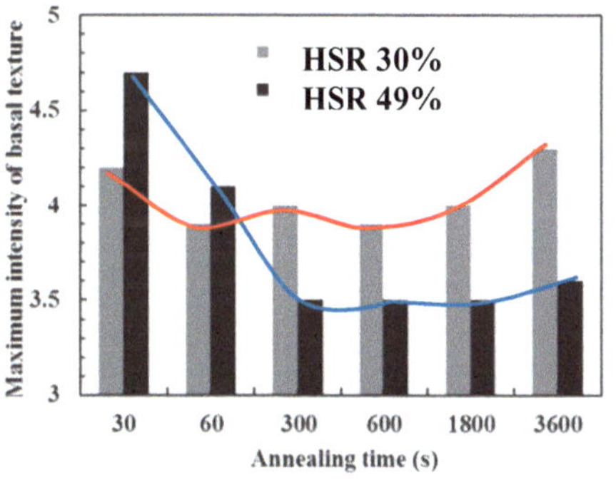

Fig. 11 The maximum intensity of basal texture verse annealing time at 350 °C of the specimens subjected to reductions of 30% and 49%

4. Conclusion

(1) At the low annealing temperature of 200 °C, formation of DRX on shear bands in the as-rolled microstructure did not affect the rate of static recrystallization, although the recrystallized grain size of the partially DRXed and twinned specimen (49%) was smaller than the fully twinned and shear banded specimen (30%).

(2) At the higher annealing temperature of 350 °C, the rate of static recrystallization of the partially DRXed and twinned specimen (49%) was much higher than that of the twinned and shear banded specimen (30%), which is due to the higher mobility of the grain/subgrain boundaries. Thus, the pre-existing DRXed grains serving as the precursor for recrystallization grew immediately during annealing.

(3) The weakening of the maximum intensity of the basal texture was found in the specimens at both reductions during annealing at 350 °C. However, the texture weakening was more effective in the fully twinned and shear banded specimen (30%) than the partially DRXed and twinned specimen (49%), while the weakening rate of the former is slower than the latter, which is related to the higher static recrystallization rate at high temperatures in the latter case.

Acknowledgement

The authors are grateful to the Natural Sciences and Engineering Research Council (NSERC) Magnesium Strategic Research Network (MagNET) in Canada for funding this project and Posco Inc. in South Korea for providing the twin roll cast AZ31B alloy plates. They are also thankful to Prof. Utsunomiya for making available the high speed rolling facility at Osaka University.

Reference

[1] H. Koh, T. Sakai, H. Utsunomiya, S. Minamiguchi, Mater. Trans., 48 (2007) 2023-2027.
[2] F. Zarandi, S. Yue, in: F. Czerwinski (Ed.) Magnesium Alloys - Design, Processing and Properties, InTech, 2011, pp. 297-320.
[3] S. Yi, I. Schestakow, S. Zaefferer, Mater Sci Eng A, 516 (2009) 58–64.
[4] J.A.d. Valle, M.T. Pe´rez-Prado, O.A. Ruano, Mater Sci Eng A, 355 (2003) 68-78.
[5] A. Levinson, R.K. Mishra, R.D. Doherty, S.R. Kalidindi, Acta Mater, 61 (2013) 5966–5978.
[6] X. Li, P. Yang, L.-N.Wang, L. Meng, F. Cui, Mater. Sci. Eng. A, 517 (2009) 160–169.
[7] S. Yi, H.-G. Brokmeier, D. Letzig, J Alloys Compd, 506 (2010) 364–371.
[8] M.T. Perez-Prado, O.A. Ruano, Scr Mater, 46 (2002) 149–155.
[9] M.T. Perez-Prado, O.A. Ruano, Scr Mater, 48 (2003) 59–64.
[10] W.X. Wu, L. Jin, Z.Y. Zhang, W.J. Ding, J. Dong, J Alloys Compd, 585 (2014) 111–119.
[11] X. Huang, K. Suzuki, Y. Chino, Mater. Sci. Eng. A, 560 (2013) 232–240.
[12] H. Koh, T. Sakai, H. Utsunomiya, S. Minamiguchi, Materials Transactions, 48 (2007) 2023-2027.
[13] H. Li, E. Hsu, J. Szpunar, H. Utsunomiya, T. Sakai, J Mater Sci, 43 (2008) 7148–7156.
[14] M. Sanjari, A.S.H. Kabir, A. Farzadfar, H. Utsunomiya, E. Essadiqi, R. Petrov, L. Kestens, S. Yue, J. Mater. Sci., 49 (2014) 1426-1436.
[15] J. Su, M. Sanjari, A.S.H. Kabir, I.-H. Jung, J.J. Jonas, S. Yue, H. Utsunomiy, Mater. Sci. Eng. A, 636 (2015) 582–592.
[16] M.R. Barnett, N. Stanford, Scr. Mater., 57 (2007) 1125–1128.
[17] M.R. Barnett, M.D. Nave, C.J. Bettles, Mater. Sci. Eng. A, 386 (2004) 205–211.
[18] Y.B. Chun, C.H.J. Davies, Mater. Sci. Eng. A, 556 (2012) 253–259.
[19] M. Sanjari, A. Farzadfar, A.S.H. Kabir, H. Utsunomiya, I.-H. Jung, R. Petrov, L. Kestens, S. Yue, J. Mater. Sci., 49 (2014) 1408–1425.
[20] H. Asgari, J.A. Szpunar, A.G. Odeshi, L.J. Zeng, E. Olsson, Mater. Sci. Eng. A, 618 (2014) 310–322.
[21] H. Asgari, A.G. Odeshi, J.A. Szpunar, Mater. Design, 63 (2014) 552–564.

Magnesium Technology 2016

Corrosion

Magnesium Technology 2016
Edited by: Alok Singh, Kiran Solanki, Michele V. Manuel, and Neale R. Neelameggham
TMS (The Minerals, Metals & Materials Society), 2016

NUMERICAL INVESTIGATION OF THE AE44-MILD STEEL GALVANIC STRUCTURAL JOINT

N. Muthegowda, B. Gholami Bazehhour, and K.N. Solanki*
School for Engineering of Matter, Transport, and Energy, Arizona State University, Tempe, AZ 85287
*Corresponding Author: (480) 965-1869; (480) 727-9321 (fax), E-mail: kiran.solanki@asu.edu

Keywords: Galvanic corrosion, Magnesium, Joints

Abstract

Here, we present an experimentally validated numerical model for an AE44 (Magnesium alloy) and mild steel galvanic couple subjected to a mechano-electrochemical process. The model is capable of tracking moving boundaries of the corroding constituent of the couple by employing Arbitrary Langrangian Eulerian (ALE) finite element method. Results show that when an anode is under a purely elastic deformation, there is no significant effect of mechanical loading on the electrochemical galvanic process. However, when the applied tensile load is sufficient to cause a plastic deformation, the local galvanic corrosion activity at the vicinity of the interface is increased remarkably. The effect of other factors, such as electrode area ratios, electrical conductivity of the electrolyte and depth of the electrolyte, are studied. It is observed that the conductivity of the electrolyte significantly influences the surface profile of the anode, especially near the junction.

Introduction

With the increase in use of metals in all fields of technology, it is inevitable to have dissimilar metal joints. The majority of aviation, automobile, electronics industries etc. are in the quest to find lighter metals to increase efficiency and performance of dissimilar metal joints. An example of such an effort was in the early 1980s when Volvo conducted a "Light Component Project", which resulted in a concept car [1]. Galvanic corrosion is one of the major hurdles to the use of magnesium parts in the automobile industry, and the use of magnesium in exterior vehicle components has been acknowledged as a vital issue [2]. In many aircraft and aerospace vehicles, the dissimilar metals are joined in such a way that the electrical resistance is minimum. In such a situation, galvanic corrosion is the primary concern. According to the recent LMI Cost of Corrosion Study [3], the annual cost of corrosion for U.S. Navy Ships and Aviation in 2010 was a combined ~$5.75 billion, and more than 80% of structural failures were due to combined environmental and mechanical loading. Finite Element Modelling (FEM) tools can be used in a complimentary way to describe current densities, distributions and pH evolution at the scale of interest. These models must be first experimentally validated before being used as predictive tools. Several attempts have been made in the literature to train models and experimentally validate their output. Bucaille et.al.[4] found a good agreement between the simulated and the experimental dissolution depth profile of a millimetre-sized bimetallic Al/Al4%Cu electrode after a free corrosion experiment. To date, the traditional approach to understanding the degradation behavior of dissimilar joints has provided a somewhat myopic view because the experimental and modeling behavior analyses are restricted to a specific loading scenario instead of structural life prediction methodologies within a combined loading framework. Hence, understanding the combined environmental and mechanical loading in structural elements and how their combination initiates corrosion, in addition to improving upon predictive capability, would undoubtedly aid in the improvement of availability and mission readiness.

Model development

The corrosion simulations were performed using commercial COMSOL Multiphysics 4.4b software on a geometry modelling a 10 mm wide and 5 mm thick anode (Mg AE44 alloy) and cathode (mild steel) with a 10 mm depth electrolyte of 2.5 Sm^{-1} (1.6 wt% NaCl) conductivity(Dimensions match that of Deshpane's [5] experimental setup). The electrodes are modelled as entire 2 D blocks so as to apply load and see the effect of loading (strain) on galvanic corrosion. The two electrodes (AE44 - mild steel) are assumed to be bonded together. We perform a stationary mechanical elasto-plastic solid stress analysis on the galvanic couple. The left end of the mild steel is fixed and a tensile load of about 160 MPa is applied on the anode end, which is sufficient enough to cause plastic deformation in Magnesium AE44. The boundary conditions and the dimensions are given in Figure 1. We use 2-D triangular mesh elements with a plane stain approximation and large plastic strain plasticity model with isotropic hardening. The adaptive mesh refinement technique was utilized as it eliminated mesh elongation and smearing to provide a smoother and better mesh quality and direct (MUMPS) solver was used. Standard material properties were used for Mg alloy AE44 and mild steel[5][6][7].

The transportation of species i can be represented by the Nernst-Planck equation as given in Equation (1),

$$N_i = D_i \nabla c_i - z_i F u_i c_i \nabla \varphi + c_i V \quad (1)$$

where, N_i is the flux, D is the diffusion coefficient, c_i is the concentration of the species, z_i is the charge, F is the Faraday constant and u_i is the mobility of species i, respectively and φ is the electric potential and V is the solvent/electrolyte. Numerous assumptions are made to simplify the above equation, which takes the form of the Laplace equation for the electric potential and represents the upper bound for the rate of corrosion, as transport by convection and by diffusion are neglected [8]

$$\nabla^2 \varphi = 0 \quad (2)$$

where φ, the electric potential in Equation (2), is solved over the electrolyte domain subject to boundary conditions, as shown schematically in Figure 1. The boundary conditions at the anode and the cathode surfaces are vital in order to predict the correct corrosion rates. The polarization data, obtained experimentally for individual alloys, [5] are used as the boundary condition for the anode and the cathode surfaces.

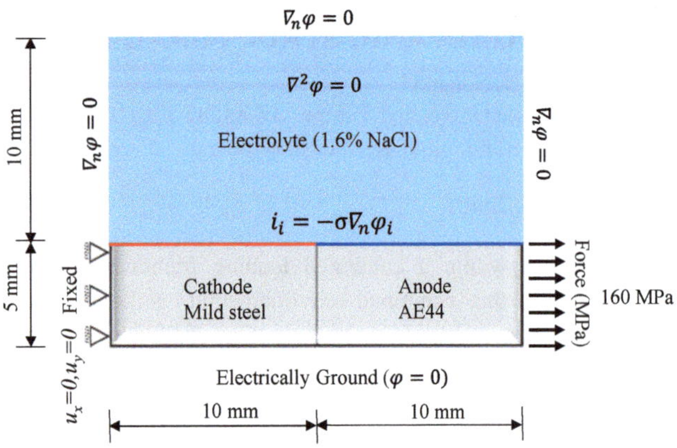

Figure 1: The schematic of a computational domain along with the governing equation and the boundary conditions for galvanic corrosion of an AE44 - mild steel couple under mechanical loading

The effects of elastic and plastic deformations on equilibrium potential of anodic reaction have been derived by Gutman [9], where the overall equilibrium potential of anodic reaction under continuous elasto-plastic deformation is given by,

$$\varphi_{a,eq} = \varphi^0_{a,eq} - \frac{\Delta P V_m}{zF} - \frac{TR}{zF} ln\left(\frac{\upsilon\alpha}{N_0}\varepsilon_p + 1\right) \qquad (3)$$

Where $\varphi_{a,eq}$ is the equilibrium electrical (electrode) potential of the system under loading, V_m is molar volume of the electrode, z is charge number, F is Faraday's constant and ΔP is magnitude of hydrostatic stress, T is absolute temperature, R is ideal gas constant (8.314 J/mol K), υ is the orientation-dependent factor, α is a constant coefficient of 10^9 - 10^{11} cm^{-2} and N_0 is the initial dislocation density [10], [11][12] before plastic deformation. On obtaining the solution for the Laplace equation subjected to the above boundary conditions, we obtain potential and current density distribution in the electrolyte, anode and cathode surfaces. Corrosion rate or interface velocity can then be calculated from current density using Equation (4). The ALE method is employed to incorporate the moving interface during corrosion. The rate of corrosion can be calculated from current density using Faraday's law[13]

$$C_R = \frac{M}{zF\,\rho} i_{cp} \qquad (4)$$

Where M is the molar mass of the corroding species, z is the electron number, F is Faraday's constant, ρ is the density of the corroding species and i_{cp} is the galvanic current density in Am^{-2}.

Results and Discussion

Initially the 2D model is used to study the galvanic corrosion for a simple case under no-load conditions and the results are compared to that of Despande's [5][6] results. Figure 2 shows that the results of AE44 are in good agreement with that of Deshpande's results. The 2-D model of the electrodes previously used to study simple galvanic corrosion is used to study galvanic corrosion under mechanical loading. Initially, the electrolyte part of the model is omitted and only the electrodes are subjected to loading. The two electrodes (AE44 - mild steel) are assumed to be bonded together. We perform a stationary mechanical elasto-plastic solid stress analysis on the galvanic couple. A tensile load of about 160 MPa is applied onto the anode end, sufficient to cause plastic deformation in Magnesium AE44. Since, the plastic strain is a very important factor to consider while studying the mechano-electrochemical effect on galvanic corrosion given by Gutman's equations, Equation (3), the effective plastic strain and Von-Mises stress are the main results to be analyzed. Due to the stress concentration at the electrodes' interface, the anode plastically deforms to a greater extent in the vicinity of the junction, and material away from junction is under elastic strain.

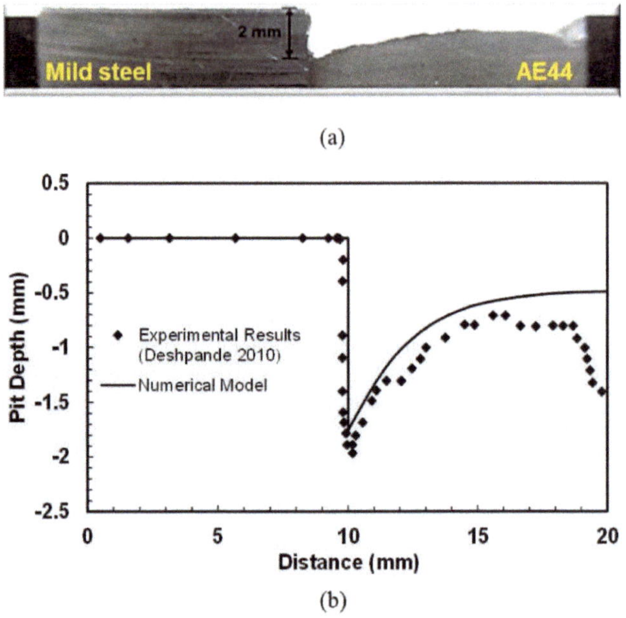

Figure 2: (a) Cross-sectional view of AE44 – mild steel couple after 3 days of immersion in the electrolyte solution (1.6 wt% NaCl solution), (b) The predicted depth profile using the numerical model and the data obtained from the immersion experiment conducted by Deshpande [5], after 3 days of immersion in the electrolyte solution.

The solution obtained for the static study is shown in Figure 3 where the effective plastic strain and von Mises stress on the surface of AE44 is plotted along the length of the anode. It can be observed that a narrow region near the junction has a maximum plastic strain of 1.52% and hydrostatic stress of 184.82 MPa as seen from Figure 3. The plastic strain and hydrostatic stress reduces drastically as we move away from the junction. Now, at the pre-strained condition, the solution for galvanic corrosion of AE44 - mild steel is obtained. The static solution is coupled to that of the corrosion module through the multiphysics field coupling technique in COMSOL Multiphysics®. Initially in static analysis, only the metal electrodes are chosen to participate in the static solid mechanics analysis; whereas, the entire model (electrodes with the electrolyte) is selected for the time dependent corrosion solution.

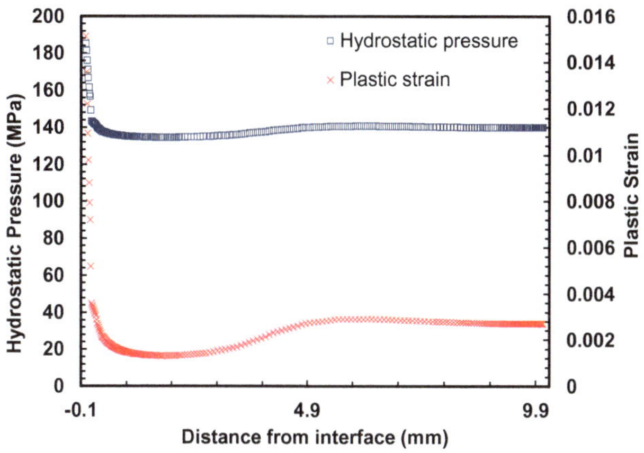

Figure 3: The variation of plastic strain and hydrostatic stress on the surface of the anode (AE44) is plotted along the distance from the interface when subjected to a load of 160 MPa.

According to Gutman's mechano-electrochemical interactions theory, the equilibrium potential of a metal varies under different loading conditions and it was found that when the anode (AE44) is under an elastic deformation, there is no apparent effect on the potential of the electrode [9][14]. However, when the applied tensile strain is sufficient to cause a plastic deformation, the change in the electrode potential is very significant [15][9]. The plastic strain changes the local equilibrium electrode potential, and hence, the corresponding net anodic dissolution current density increases significantly which results in an increased rate of corrosion. The increase in the rate of galvanic corrosion is dependent on the amount of plastic stain. As plastic stain increases, the rate of galvanic corrosion increases. The depth profile obtained for the galvanic couple AE44 - mild steel under a tesile load of 160 MPa is compared with that of a no load condition and immersion experiment result conducted by Deshpande[5] as shown in Figure 4.

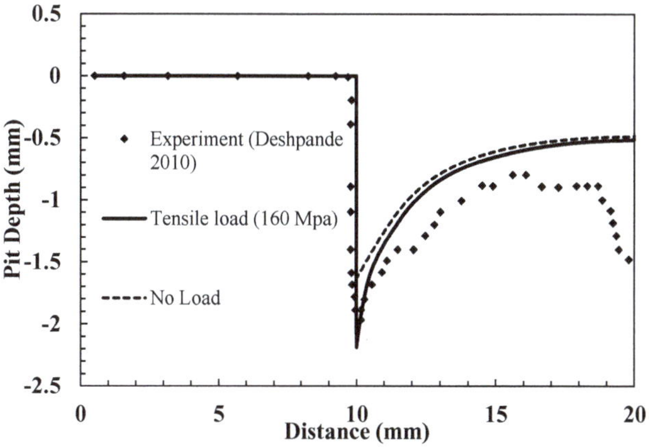

Figure 4. The depth profile predicted for Mg alloy AE44 using the numerical model for both with and without tensile load of 160 MPa. Conditions and results obtained from the immersion experiment conducted by Deshpande, after 3 days of immersion in of 1.6 wt% NaCl electrolyte solution (electrical conductivity of 2.5 Sm-1).

It can be observed form the Figure 4, that the maximum and overall rate of corrosion is more in the case of the plastically deformed anode (AE44). In the case of the plastically deformed anode, the rate of galvanic corrosion was to found to be maximum in the vicinity of the junction due to the localization of plastic strain, as shown in Figures 4 and 3.

Table 5: The corrosion rates with and without load are calculated and compared below.

Galvanic Couples	Corrosion rate (mm y^{-1})				
	Mixed potential theory	ALE method		SVET experiment	Immersion experiment
		No load	160 MPa Load		
AE44–mild steel	231	194	244	197	235

The rate of corrosion for the galvanic couple is calculated and is compared for both with and without load conditions as well as the experimental results conducted by Deshpande [6]. It can be observed that the numerical results closely match with that of the experimental values.

Conclusion

A numerical model for solving the Nernst-Planck equation, which captures the electrochemical process, is employed and this model is capable of tracking the moving boundary for a 2-D electrode during galvanic corrosion and can handle nonlinear boundary conditions. The corrosion rates obtained for the galvanic couple AE44 – mild steel was compared with those calculated by Deshpande [5] from the numerical model and experimental techniques [6]. The validated numerical model is extended to simulate the mechano-electrochemical process through a multiphysics field coupling technique in order to study the effects of mechanical loading on galvanic corrosion of AE44 – mild steel. Results show that, when Magnesium alloy AE44 is under a purely elastic deformation, there is no apparent effect of mechanical loading on the electrochemical galvanic process. However, when the applied tensile load is sufficient to cause a plastic deformation, the local galvanic corrosion activity at the vicinity of the interface is increased remarkably and the rate of galvanic corrosion increased significantly with the increase in plastic strain.

References

[1] R. Mellde, "Volvo LCP 2000 Light Component Project," SAE International, Warrendale, PA, SAE Technical Paper 850570, Feb. 1985.

[2] M. Isacsson, M. Ström, H. Rootzén, and O. Lunder, "Galvanically Induced Atmospheric Corrosion on Magnesium Alloys: A Designed Experiment Evaluated by Extreme Value Statistics and Conventional Techniques," SAE International, Warrendale, PA, SAE Technical Paper 970328, Feb. 1997.

[3] E. F. Herzberg, E. D. Ambrogio, C. L. Barker, E. F. Harleston, W. M. Haver, N. T. O'Meara, R. J. Marafioti, G. L. Stimatze, A. Timko, and J. C. Tran, "The Annual Cost of Corrosion for Army Ground Vehicles and Navy Ships," *LMI Rep. DAC2ITI*, 2012.

[4] P. Bucaille, R. Oltra, and T. Warner, “Theoretical and Experimental Studies of Galvanic Corrosion between Aluminium and Al-4% Cu Alloys,” *Mater. Sci. Forum*, vol. 242, pp. 207–212, 1997.
[5] K. B. Deshpande, “Validated numerical modelling of galvanic corrosion for couples: Magnesium alloy (AE44)–mild steel and AE44–aluminium alloy (AA6063) in brine solution,” *Corros. Sci.*, vol. 52, no. 10, pp. 3514–3522, Oct. 2010.
[6] K. B. Deshpande, “Experimental investigation of galvanic corrosion: Comparison between SVET and immersion techniques,” *Corros. Sci.*, vol. 52, no. 9, pp. 2819–2826, Sep. 2010.
[7] “For Casting Buyers | AFS - American Foundry Society.” [Online]. Available: http://www.afsinc.org/buyer.cfm?navItemNumber=10373. [Accessed: 06-Jul-2015].
[8] M. Verbrugge, “Galvanic corrosion over a semi-infinite, planar surface,” *Corros. Sci.*, vol. 48, no. 11, pp. 3489–3512, Nov. 2006.
[9] E. M. Gutman, *Mechanochemistry of Solid Surfaces*. Singapore ; River Edge, NJ: World Scientific Pub Co Inc, 1994.
[10] T. Shintani, Y. Murata, Y. Terada, and M. Morinaga, “Evaluation of Dislocation Density in a Mg-Al-Mn-Ca Alloy Determined by X-ray Diffractometry and Transmission Electron Microscopy,” *Mater. Trans.*, vol. 51, no. 6, pp. 1067–1071, 2010.
[11] P. Klimanek and A. Pötzsch, “Microstructure evolution under compressive plastic deformation of magnesium at different temperatures and strain rates,” *Mater. Sci. Eng. A*, vol. 324, no. 1–2, pp. 145–150, Feb. 2002.
[12] F. Barlat, M. V. Glazov, J. C. Brem, and D. J. Lege, “A simple model for dislocation behavior, strain and strain rate hardening evolution in deforming aluminum alloys,” *Int. J. Plast.*, vol. 18, no. 7, pp. 919–939, Jul. 2002.
[13] W. S. Tait, *An introduction to electrochemical corrosion testing for practicing engineers and scientists*. Clair, Racine, Wis., 1994.
[14] L. Y. Xu and Y. F. Cheng, “An experimental investigation of corrosion of X100 pipeline steel under uniaxial elastic stress in a near-neutral pH solution,” *Corros. Sci.*, vol. 59, pp. 103–109, Jun. 2012.
[15] L. Y. Xu and Y. F. Cheng, “Corrosion of X100 pipeline steel under plastic strain in a neutral pH bicarbonate solution,” *Corros. Sci.*, vol. 64, pp. 145–152, Nov. 2012.

Magnesium Technology 2016
Edited by: Alok Singh, Kiran Solanki, Michele V. Manuel, and Neale R. Neelameggham
TMS (The Minerals, Metals & Materials Society), 2016

FABRICATION OF A SUPERHYDROPHOBIC FILM WITH SELF-CLEANING PROPERTY ON MAGNESIUM ALLOY AND ITS CORROSION RESISTANCE PROPERTIES

Meng Zhou, Xiaolu Pang, Kewei Gao
Department of Materials Physics and Chemistry,
University of Science and Technology Beijing,
Beijing 100083, China

Keywords: Superhydrophobic, Micro/nano structure, Magnesium alloy, Corrosion resistance, Self-cleaning property

Abstract

A hierarchical superhydrophobic film has been fabricated on a magnesium alloy substrate via a facile chemical substitution method following stearic acid modification. The characteristics of the film were investigated by scanning electronic microscope (SEM), energy dispersive spectroscopy (EDS), Fourier transformed infrared (FT-IR) spectroscopy and X-ray diffraction (XRD). The results showed that the hierarchical film surface with cluster structures composed of micro-sheets and stearic acid had been successfully grafted onto the sample surface. The film wettability was examined by water contact angle (CA) measurement and the obtained CA value was as high as 167.4°. The corrosion resistance of the superhydrophobic film to the substrate was tested with potentiodynamic polarization measurement. Polarizaton measurements revealed that the superhydrophobic film coated magnesium alloy had better corrosion resistance in neutral 3.5 wt. % NaCl solution. Morever, the superhydrophobic film showed good self-cleaning property and chemical stability.

1. Introduction

In nature, the superhydrophobic phenomenon is evident everywhere. Many materials exhibit fascinating characteristics such as lotus leaves, strider legs, rice leaves, butterfly wings and so on. Especially, lotus leaf is a famous example, where water droplets falling onto them bead up and roll off. In the process of water droplets rolling, small particles of dirt is picked up (self-cleaning) and this phenomenon is called "lotus effect" [1, 2]. The surface phenomenon "lotus effect" can be attributed to the hierarchical micro/nano structure in the form of papillose epidermal cells and epicuticular wax with low surface energy. Inspired by the lotus effect, the superhydrophobic surfaces with water contact angle (CA) larger than 150° have been attracting tremendous interests for its potential applications in the fields of military, energy, environment, corrosion protection, self-cleaning, anti-icing and other industrial processes [3-6]. Over the past few years, many researchers have been trying to design and fabricate artificial superhydrophobic surface. The fabrication of artificial superhydrophobic surface is a two step process: first, to achieve a micro/nano structure surface; and second to modify it with low surface free energy materials.

Magnesium (Mg) and its alloys have been applied in the automotive, aerospace and electronic industries because of their high specific strength, low density and electromagnetic compatibility. Especially, it is expected to be an excellent material for reducing vehicle weight and reducing fuel consumption. Furthermore, it is considered to be the most promising green engineering material in the 21st century [7, 8]. However, due to low standard potential, Mg alloys is very active and easily corroded in aqueous solution. Thus, their wide applications is seriously limited [9]. Because the preparation of superhydrophobic surface can inhibit the contact of a surface with environmental humidity, it has been considered as one of the most promising methods to improve magnesium corrosion performance. In addition, the use of superhydrophobic surfaces on the magnesium substrate can confer it the self-cleaning property.

Up to now, various methods have been explored to fabricate superhydrophobic surface, such as micro-arc oxidation, in-situ hydrothermal synthesis, electrochemical deposition, chemical vapor deposition, chemical etching and so on. For example, Liang fabricated a superhydrophobic surface on magnesium alloy via micro-arc oxidation pretreatment and followed by chemical modification [10]. Zhou et al observed that Zn–Al layered double hydroxides superhydrophobic films can be in-situ grown on AZ91D magnesium alloy via a hydrothermal synthesis method followed by modification with stearic acid [11]. A few papers have reported the electrodeposition of superhydrophobic surfaces on magnesium, but they all required the preparation of a trasition-metal layer to fabricate a rough structure and then followed with the low-surface-energy material modification [12-15]. Ishizaki prepared a superhydrophobic surface on AZ31 magnesium alloy via microwave plasma-enhanced chemical vapor deposition [16]. Wang used H_2O_2 and H_2SO_4 to etch pure magnesium substrate and subsequently midified it with stearic acid to gain a superhydrophobic surface [17]. Compared with the other methods, etching method is a more effective and simpler way to obtain a rough surface on magnesium alloy.

On the basis of the above consideration, this work aims to develop a simple, highly effective and low-cost method to fabricate the superhydrophobic magnesium alloy. Firstly, the micro/nanoscale hierarchical surfaces were prepared by immersing the magnesium alloy surface in $CuCl_2$ aqueous solution. Then the as-prepared micro/nanoscale hierarchical surfaces were immersed in ethanol solution of stearic acid After chemical modification, the coating with micro/nanoscale hierarchical structure and high contact angle magnesium alloy was fabricated with desirable corrosion protection and self-cleaning properties.

2. Experimental

10 × 10 × 3 mm and 20 × 30 × 3 mm AZ91D magnesium alloy sheets (composed: Al of 8.5-9.5 wt%, Zn of 0.45-0.9 wt%, Mn of 0.17-0.4 wt%, Si <0.08 wt%, Fe <0.004 wt%, Ni<0.001 wt%, with balance being Mg) were polished mechanically using 2000 grit SiC paper as samples in this study. Prior to the experiments, the samples were rinsed with deionized water, degreased by acetone and dried with cold wind sequentially. In

order to determine the optimum conditions, the cleaned magnesium alloy sheets were immersed in the 0.1 M, 0.3 M and 0.5 M $CuCl_2$ aqueous solution at ambient temperature for 5 seconds. After immersion, the Mg sheets were taken out, washed with deionized water and dried in air. After the reaction, the samples were covered with a layer of green film. For the convenience, the Mg alloy sheets immersed in the 0.1 M, 0.3 M and 0.5 M $CuCl_2$ aqueous solution were marked as sample A, sample B and sample C, respectively. Finally, to achieve surface superhydrophobicity, the as-prepared samples were immersed in an ethanol solution of 0.05 M stearic acid for 3 h at room temperature.

The surface morphology and elemental analysis were performed using a scanning electron microscope (SEM) equipped with energy dispersive X-ray spectrometer (EDS). The films were coated with carbon using a thermal evaporator before imaging to minimize charging problems. The crystalline structure of the samples surface were analyzed by X-ray powder diffraction (XRD) using the Rigaku D/max -rB diffractometer with Cu Ka radiation. The chemical composition of the samples surface were studied using the infrared absorption and transmittance spectra recorded on the Fourier transform infrared spectrometer (FT-IR) at room temperature. The water contact angles were measured at ambient temperature using an optical contact angle meter (Krüss, DSA100, Germany). Ultrapure water droplets (4 μL) were carefully dropped onto the surfaces, and the average value of five measurements obtained at different positions in the samples was used as the final contact angle. To estimate the long-term durability of superhydrophobic properties under atmospheric environment, the samples were exposed in the air and the contact angles were measured.

The corrosion resistance of the film was evaluated using potentiodynamic polarization on the electrochemistry workstation (CS310, Wuhan Corrtest Instrument Co., Ltd., China). A typical three-electrode cell system, with a Ag/AgCl electrode (saturated with KCl) as reference electrode, a superhydrophobic Mg alloy surface (area, 1 cm^2) as the working electrode and platinum electrode as the counter electrode, was used to perform the polarization tests in 3.5 wt % NaCl solution. Before measurement, the working electrode was immersed in the quiescent test solution until the cell open circuit potential (OCP) became stable. The potentiodynamic polarization curves were subsequently measured with respect to the open circuit potential (OCP) at a scanning rate of 1 mV/s. Each test was repeated more than three times to verify reproducibility of the results. All measurements were conducted at room temperature.

3. Results and discussion

3.1 SEM morphologies of the films on magnesium alloy AZ91D

As is known to all, the fabrication of the hierarchical structure is a crucial factor in the fabrication of superhydrophobic surfaces. The morphology of the prepared samples were investigated by SEM, and the images of the films prepared with different concentrations of the $CuCl_2$ aqueous solution are shown in Figure 1. We find that the concentration of the $CuCl_2$ aqueous solution plays a vital role in tailing the morphologies of the resultant films (Fig.1). When the magnesium alloy sheets were immersed in the 0.1 M $CuCl_2$ aqueous solution, the hierarchical structure with micrometer-scale plateaus was observed (Fig.1a). With the increase of the $CuCl_2$ concentration (up to 0.3 M), it was found that the micro/nanoscale protrusions began to appear on micrometer-scale plateaus (Fig.1b). However, the protrusions disappeared from the hierarchical nanometer-scale plateaus when the magnesium alloy sheets were immersed in the 0.5 M $CuCl_2$ aqueous solution (Fig.1c). As shown in Figure 1, our fabricated surface is composed of micro/nanoscale protrusions or hierarchical structure on magnesium alloy AZ91D, which is the basis for the superhydrophobicity.

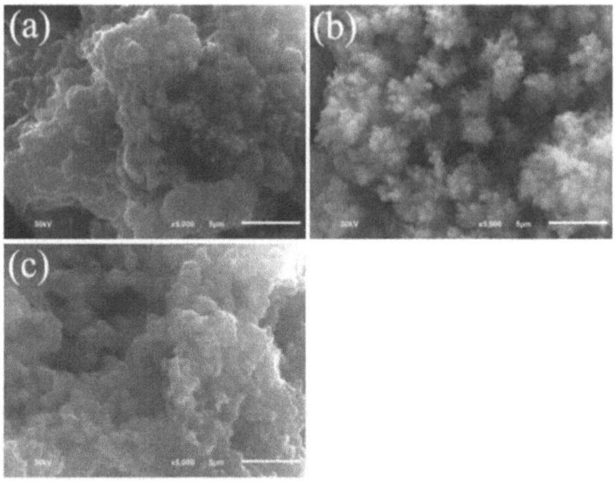

Figure 1. SEM images of the sample surfaces after immersion in $CuCl_2$ aqueous solution for (a) 0.1, (b) 0.3 and (c) 0.5 M.

3.2 Chemical compositions of the films

The chemical composition of the as-prepared surface was studied by XRD and FT-IR. Fig.2 shows the XRD patterns of the sample B surfaces. The XRD pattern shows the characteristic peaks of $AlCu_3$ (JCPDS 28-0005). In addition, some peaks of Mg phase are attributed to the Mg alloy substrate. The EDS spectrum (Fig.3) of the sample B surfaces showed that micro/nanoscale protrusions are composed of C, O, Cl, Mg, Cu and Al. The Al/Cu atomic ratio approximately is 1:3, which confirms the formation of $AlCu_3$ films on the Mg alloy during the chemical substitution reaction. The main reaction mechanisms are exhibited in Scheme 1 and the main chemical reaction on the surfaces is depicted as below:

$$9Mg_{17}Al_{12} + 210Cu^{2+} = 70AlCu_3 + 38Al^{3+} + 153Mg^{2+} \quad (1)$$

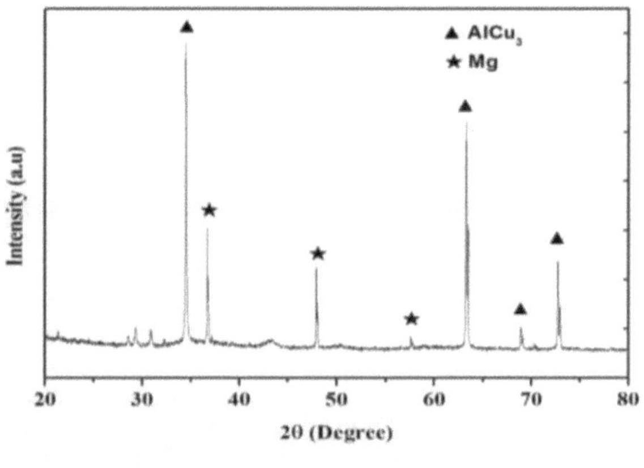

Figure 2. XRD pattern of the sample B surface.

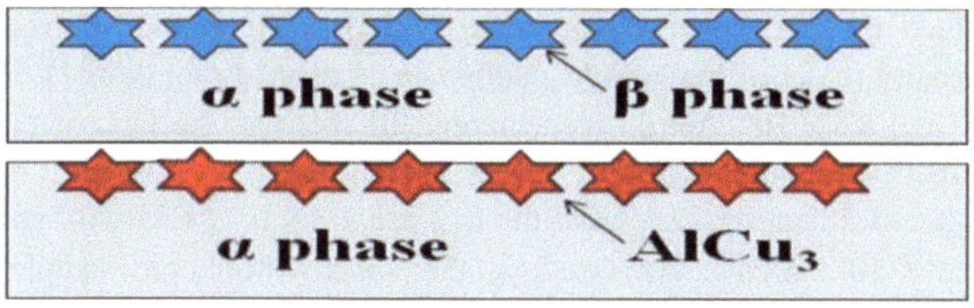

Scheme 1. Schematic illustration of fabricating hierarchical structure on AZ91D magnesium alloy.

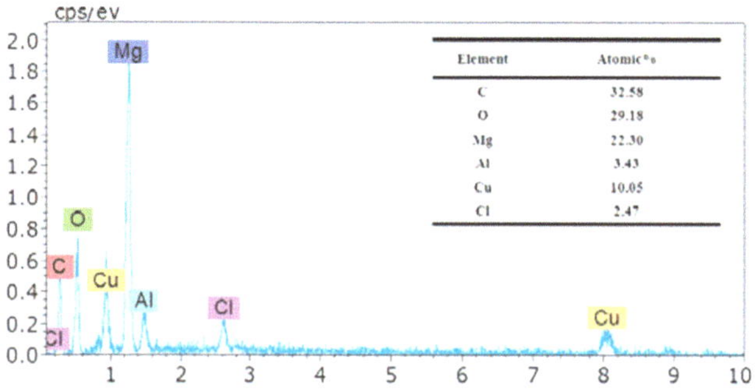

Figure 3. EDS spectra of the sample B surface.

To obtain superhydrophobic surface, roughening and low-surface energy material modification are both important. In order to verify the stearic acid on the surface of coated sample B, the FT-IR spectrum was used. Fig.4 shows the FT-IR spectra of the sample B. In the low frequency region, the superhydrophobic surface exhibits adsorption peaks at 1562 cm^{-1} and 1463 cm^{-1}. Typically, the carboxyl group (-COO) from stearic acid appears at 1702 cm^{-1}, and the two adsorption peaks may stem from asymmetric and symmetric stretches of carboxyl group. In the high frequency region, the adsorption peaks at about 2918 cm^{-1} and 2850 cm^{-1} was corresponding to symmetric and asymmetric stretching vibration of C-H, respectively. The result indicates the existence of the longchain aliphatic groups on the surface. To elucidate the chemical stability of superhydrophobic surface, the FT-IR spectra of the sample surface were retested in 30 days later under atmospheric environment. In addition, the adsorption peaks at 2358 cm^{-1} and 2341 cm^{-1} for samples are the presence of small quantities of CO_2 [17, 18]. As a result, after exposure for 30 days, the adsorption peaks positions were unchanged. This phenomenon demonstrates that the stearic acid had been successfully grafted onto the sample surface and it will be beneficial for superhydrophobic surfaces to obtain good chemical stability under atmospheric environment.

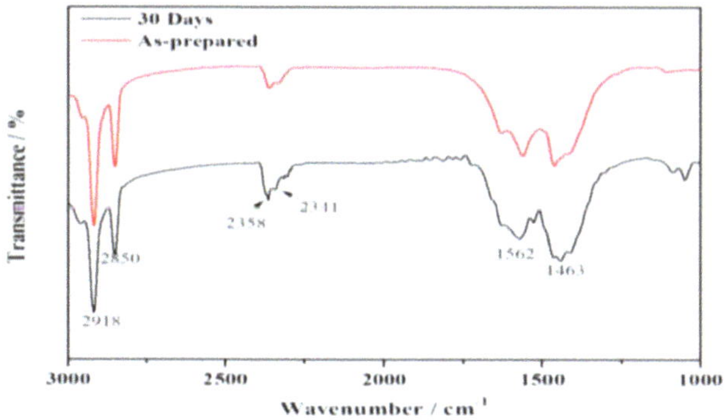

Figure 4. FT-IR spectra of the superhygrophobic surface on sample B.

3.3 Surface wettability and self-cleaning effect of the films

The CA value of Magnesium is about 36°, because it is a type of hydrophilic material with an intrinsic oxidized layer. Figure 5 illustrates the water droplet (4 μl) shapes on different sample surfaces. After superhydrophobic treatment, the CA values of sample A, sample B and sample C were about 144.5°, 167.4° and 152.3° (Fig. 5), respectively. According to the results, sample B has the best wettability and it could be attributed to the micro/nanoscale protrusions structure which is similar to the structure of lotus leaf. Compared with hierarchical structure with micrometer-scale or nanomenter-scale plateaus, the micro/nanoscale protrusions structure of sample B made its surface rougher and caused the increase of CA value. The hydrophobicity mode in our case is governed by Cassie-Baxter equation[19]:

$$\cos\theta = f_s\left(\cos\theta_0 + 1\right) - 1 \tag{2}$$

where θ and θ_0 represent CA of water on a rough surface and on a flat surface. f_s is the fractional interfacial areas of solid/water on the surface. The value of θ_0=36° for the smooth surface is unchangeable, but for the superhydrophobic surface, the trapped air can reduce the f_s. Therefore, the value of θ would increase with the value of the air fraction (1-f_s) increases. For sample B surface, according to the aforementioned Cassie-Baxter formula, the air fraction result is 98.7%. The large fraction of air trapped in the micro/nanoscale protrusions structure would greatly increase the air/liquid interfaces and the water droplet cannot penetrate into the rough structure.

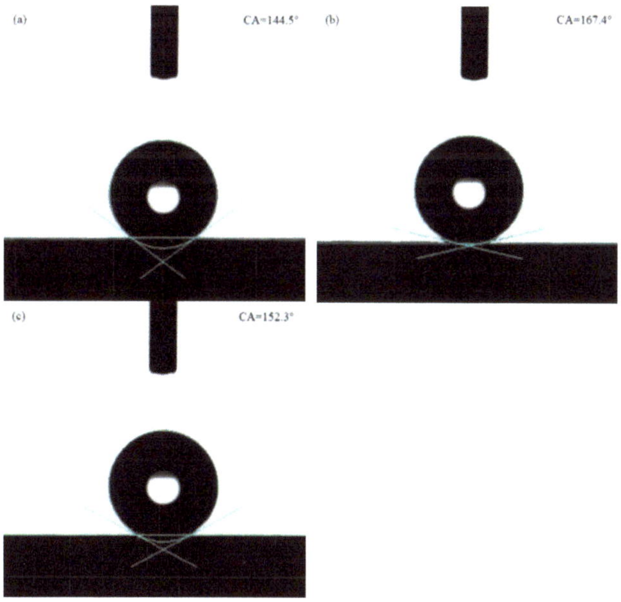

Figure 5. Digital photographs of water droplet (4 μl) shapes: (a) 0.1, (b) 0.3 and (c) 0.5 M.

It is well-known that the self-cleaning effect is important for superhydrophobic surfaces in their applications. In our study, the superhydrophobic surface is carried out by deliberately spreading bentonite powder on the sample B surface before self-cleaning test. The sample was placed in a horizontal table with a title angle of about 4°. Fig. 6 shows the process of a water droplet dripped on the sample B surface and bounced to the desktop. This phenomenon could be attributed to the micro/nanoscale protrusions structure and the extremely low energy of the surface modified by stearic acid. When the water droplet dripped on the surface, the dust can be easily carried off by the water surface tension during the rolling process.

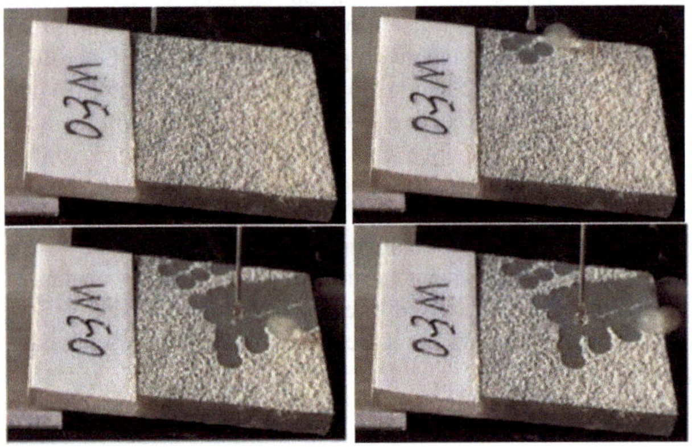

Figure 6. The process of self-cleaning behavior of sample B superhydrophobic surface.

In self-cleaning test, the larger title angles the more easily rolling off. Compared with results of Li (10°) and Liu (15°), our results indicate that sample has a great self-cleaning performance [4, 13]. Moreover, the durability of the superhydrophobic sample also is very important factor, which is a basic property to realize long-team self-cleaning performance. To elucidate the stability of the sample surface, the CA of the sample B was monitored. Figure 7 shows the relationship between the water contact angles and exposure time of the sample B when it was exposed at atmosphere environment. After exposure for more than 90 days, the CA only slightly decreased from 167.4° to 163.5°. Based on the FT-IR spectra data analysis and above results, this phenomenon indicate that the sample B surface have good long-team stability.

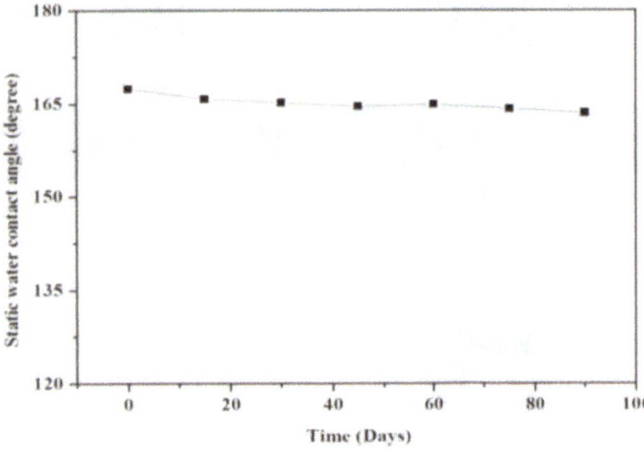

Figure 7. Water contact angles of the sample B surface with the different exposure time .

3.4 Corrosion resistance performance of the films

It is well-known that the polarization curve is a useful method to determine the instantaneous corrosion rate of a substrate. Figure 8 give the electrochemical polarization curves for bare substrate and the various superhydrophobic films on the magnesium alloy in a 3.5 wt.% NaCl aqueous solution at room temperature. In a typical polarization curve, a higher corrosion potential or a lower corrosion current density indicate a lower corrosion rate and a better corrosion resistance. From Table 1, it's noted that the corrosion potential (E_{corr}) and corrosion current density (i_{corr}) of the bare Mg substrate as control sample is -1.51 V and 8.8×10^{-3} A/cm^2, respectively. In contrast, the E_{corr} of the superhydrophobic samples reveal a significant positive shift (-1.33, -1.28 and -1.22 V) and the i_{corr} are 4.3×10^{-5}, 1.3×10^{-5} and 2.1×10^{-5} A/cm^2. As for i_{corr}, the superhydrophobic samples displayed current density of lower magnitude by more than two orders than the bare Mg substrate. These results indicates that the sample with superhydrophobic surface is very effective in the protection of the Mg alloy substrate from corrosion. Based on the data analysis of Table 1, sample B exhibited best corrosion resistance compared with both sample A and sample C. This corrosion resistance behavior could be attributed to the air trapped in the micro/nanoscale protrusions structure of sample B. Therefore, the trapped air could serve as an effective barrier to keep corrosive ions away from the bare substrate and generate the best corrosion protection effect.

Table 1 Corrosion potential (E_{corr}) and corrosion current densities (i_{corr}) derived from the polarization measurements of the samples

Sample	E_{corr}(V vs.Ag/AgCl)	i_{corr}(A/cm^2)
Bare substrate	-1.51	8.8×10^{-3}
Sample A	-1.22	2.1×10^{-5}
Sample B	-1.28	1.3×10^{-5}
Sample C	-1.33	4.3×10^{-5}

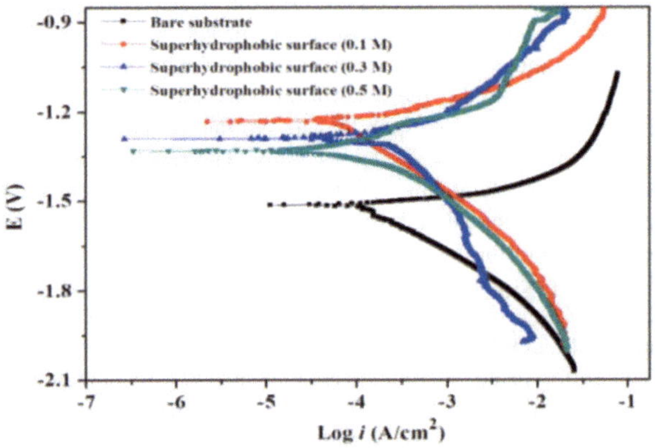

Figure 8. Potentiodynamic polarization curves of the bare Mg alloy substrate, sample A, sample B, sample C.

4. Conclusions

In summary, a superhydrophobic surface was successfully fabricated on Mg alloy via chemical substitution method following stearic acid modification. The obtained surface is composed of micro/nanoscale protrusions or hierarchical structure and has a maximum contact angle of 167.4°. According to the results of corrosion resistance tests, the superhydrophobic film shows an excellent performance of corrosion resistance when immersed in neutral 3.5 wt. % NaCl solution. Based on FT-IR analysis, the as-prepared superhydrophobic surface showed self-cleaning property and good chemical stability as well as long-term durability under atmospheric environment. This method has great potential value for industrial applications to fabricate a stable superhydrophobic surface on magnesium alloy.

Acknowledgements

This project is supported by National Natural Science Foundation of China (No. 51271024) and Beijing Natural Science Foundation Key Project (No. 2131004)

References

[1] X.J. Feng, L. Jiang, "Design and Creation of Superwetting/Antiwetting Surfaces," Advanced Materials, 18 (2006) 3063-3078.

[2] L. Feng et al., Petal Effect: "A Superhydrophobic State with High Adhesive Force," Langmuir, 24 (2008) 4114-4119.

[3] L. Zhao et al., "One-step method for the fabrication of superhydrophobic surface on magnesium alloy and its corrosion protection, antifouling performance," Corrosion Science, 80 (2014) 177-183.

[4] W. Li, Z. Kang, "Fabrication of corrosion resistant superhydrophobic surface with self-cleaning property on magnesium alloy and its mechanical stability," Surface and Coatings Technology, 253 (2014) 205-213.

[5] J. Wang et al., "Fabrication of hydrophobic surface with hierarchical structure on Mg alloy and its corrosion resistance," Electrochimica Acta, 55 (2010) 6897-6906.

[6] W. Li et al., "In situ growth of superhydrophobic and icephobic films with micro/nanoscale hierarchical structures on the aluminum substrate," Journal of Colloid and Interface Science, 410 (2013) 165-171.

[7] C. Berziou et al., "Corrosion behaviour of dc magnetron sputtered Fe1−xMgx alloy films in 3 wt% NaCl solution," Corrosion Science, 49 (2007) 4276-4295.

[8] E. Aghion, B. Bronfin, "Magnesium alloys development towards the 21st century," Materials Science Forum, 350 (2000) 19-30.

[9] J.E. Gray, B. Luan, "Protective coatings on magnesium and its alloys — a critical review," Journal of Alloys and Compdounds, 336 (2002) 88-113.

[10] J. Liang et al., "Fabrication of Superhydrophobic Surface on Magnesium Alloy," Chemical Letters, 36 (2007) 416-417.

[11] M. Zhou et al., "Insitu grown superhydrophobic Zn–Al layered double hydroxides films on magnesium alloy to improve corrosion properties," Applied Surface Science, 337 (2015) 172-177.

[12] Z. She et al., "Researching the fabrication of anticorrosion superhydrophobic surface on magnesium alloy and its mechanical stability and durability," Chemical Engineering Journal, 228 (2013) 415-424.

[13] Y. Liu et al., "A electro-deposition process for fabrication of biomimetic super-hydrophobic surface and its corrosion resistance on magnesium alloy," Electrochimica Acta, 125 (2014) 395-403.

[14] Z. Wang et al., "Facile and fast fabrication of superhydrophobic surface on magnesium alloy," Applied Surface Science, 271 (2013) 182-192.

[15] Z. She et al., "Novel Method for Controllable Fabrication of a Superhydrophobic CuO Surface on AZ91D Magnesium Alloy," ACS Applied Materials & Interfaces, 4 (2012) 4348-4356.

[16] T. Ishizaki et al., "Corrosion resistance and chemical stability of super-hydrophobic film deposited on magnesium alloy AZ31 by microwave plasma-enhanced chemical vapor deposition," Electrochimica Acta, 55 (2010) 7094-7101.

[17] Y. Wang, et al., "Super-hydrophobic surface on pure magnesium substrate by wet chemical method," Applied Surface Science, 256 (2010) 3837-3840.

[18] W.F. Ng, M.H. Wong, F.T. Cheng, "Stearic acid coating on magnesium for enhancing corrosion resistance in Hanks' solution," Surface and Coatings Technology, 204 (2010) 1823-1830.

[19] A.B.D. Cassie, S. Baxter, "Wettability of porous surfaces," Transactions of the Faraday Society, 40 (1944) 546-551.

Magnesium Technology 2016
Edited by: Alok Singh, Kiran Solanki, Michele V. Manuel, and Neale R. Neelameggham
TMS (The Minerals, Metals & Materials Society), 2016

THE SURFACE FILMS AND THEIR POSSIBLE ROLES IN Mg CORROSION

Guang-Ling Song[1*], Paul E. Gannon[2]
[1]College of Materials, Xiamen University, 422 S. Siming Rd, Xiamen, Fujian 361005, China,
*guangling.song@hotmail.com; glsong@xmu.edu.cn
[2]Montana State University, Bozeman, MT 59717, USA

Keywords: Mg, surface film, corrosion

Abstract

The poor corrosion performance of Mg alloys stems from the non–protectiveness of their surface films. It is important to understand how the corrosion resistance of surface films is affected by their composition and microstructure. This paper reports on the Mg surface film microstructure and its dependence on polarization potential, the differences of the film from the anodized coatings formed at high anodic potentials, the influence of passivating elements on the passivity and corrosivity of single phase Mg alloys, and the possibility of forming a sacrificial anode coating to protect Mg from corrosion attack.

Introduction

The chemically active Mg surface is normally covered by an oxide/hydroxide film [1, 2]. Improving the film protectiveness is an obvious but difficult approach to corrosion resistant Mg alloys. It is well known that the stability and protectiveness of a surface film is determined by the film's composition and microstructure, which are eventually dependent on the conditions in which the films are formed [3-7]. A comprehensive understanding of the film's composition, microstructure and performance is critical for successful applications of Mg alloys in practice. This paper briefly summarizes the effect of anodic potential/current density, alloying element and bath composition on the film's formation and dissolution behavior. Based on these, the possible roles of different films in corrosion are also discussed.

Anodic surface film

The polarization potential or current density at which a surface film is formed is the primary parameter that can affect the integrity of the surface film, making the substrate Mg or its alloy more active or passive [8-11]. The effect of potential or current density on Mg surface film performance is of great interest from a practical point of view, as in industrial applications Mg alloys are normally in contact with other engineering metals and thus under an anodic polarization condition by the coupling metals [12]. Although the anodic polarization potentials or current densities galvanically applied by the coupling metals may be relatively low, their influence on surface film should not be overlooked.

The effect of polarization potential on a surface film microstructure was illustrated by pure Mg anodically polarized at different potentials. A discontinuous corrosion product film was formed after the pure Mg was immersed in saturated $Mg(OH)_2$ solution at its open-circuit potential (OCP) for 24 hours (see Figure 1 (a)~(f)). The significantly corroded areas were covered by a film that contained mainly O and Mg (Figure 1 (b)~(c)), while the insignificantly corroded surface zones were basically metallic Mg (Figure 1 (b)~(d)) (there might be a thin oxide/hydroxide film on them, which was too thin to be observed via the SEM). The corrosion product film on the noticeably corroded region was about ~10 μm thick, containing cracks (Figure 1(e)~(f)). At a higher potential, for example, 0.5V/SCE, the film became significantly thicker (a few hundred μm), contained more cracks, and even loosely ruptured in some areas (Figure 1 (g) and (h)).

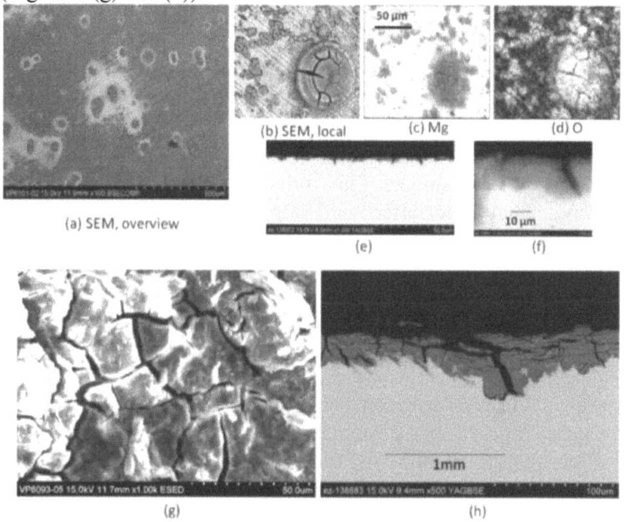

Figure 1. The surfaces and films of Mg at different potentials after 24 hours of immersion in saturated $Mg(OH)_2$ solution at different potentials:

(a) SEM topographic image of a surface region after immersion at OCP,
(b) SEM topographic image of a local corrosion spot on the surface after immersion at OCP,
(c) Element Mg distribution over the local surface area,
(d) Element O distribution over the local surface area,
(e) SEM cross-section image of a surface area after immersion at OCP,
(f) SEM cross-section image of an enlarged zone in the local surface area (e) after immersion at OCP,
(g) SEM topographic image of a surface region after immersion at anodic potential 0.5 V/SCE,
(h) SEM cross-section image of a Mg surface region after immersion at anodic potential 0.5 V/SCE.

The detailed microstructures of the surface films formed on minimally corroded surface areas at different potentials were further revealed by TEM (Fig.2(a) and (b)). The photos clearly show that the films consist of a porous inner layer and a more loosely filamentous outer layer. Electron diffraction analyses indicated that both the less porous inner and more porous outer layers were a mixture of MgO+$Mg(OH)_2$; the inner dominated by MgO and the outer was mainly $Mg(OH)_2$[13].

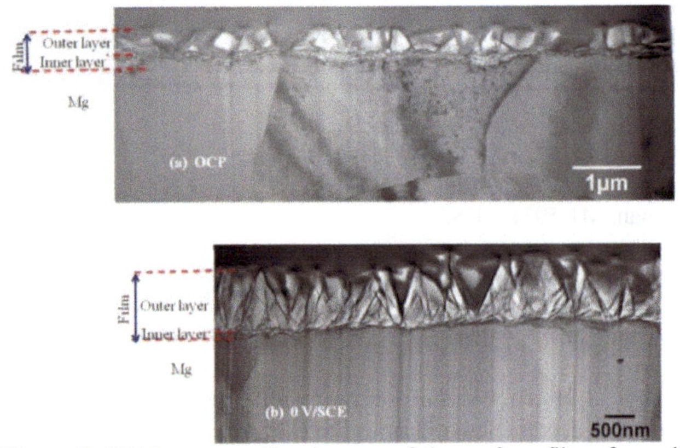

Figure 2. TEM cross-section images of the surface films formed on Mg after 24 hours of immersion in saturated $Mg(OH)_2$ solution at (a) OCP and (b) 0.5 V/SCE.

The anodic potentiostatic polarization curve presented in Fig.3 has been IR-drop corrected for the pure Mg in the saturated $Mg(OH)_2$ solution. It shows that the anodic current density gradually increases with increasing potential. This indicates that the presence of the surface film formed at the OCP and anodic potentials could not effectively stop Mg anodic dissolution. In addition, Fig.3 shows that the hydrogen evolution rate from the film-covered Mg surface increases with increasing potential, suggesting that the porous surface film on Mg could not retard the hydrogen evolution process. These results support that both Mg dissolution and hydrogen evolution can only occur from the substrate Mg surface exposed to the solution in the surface film pores. When polarization potential increases, the surface film becomes thicker, but also more porous and cracked. As the polarization potential across the Mg/film interface increase with increasing polarization potential, the Mg dissolution and hydrogen evolution processes at the Mg/film interface are also accelerated by the increased polarization potential.

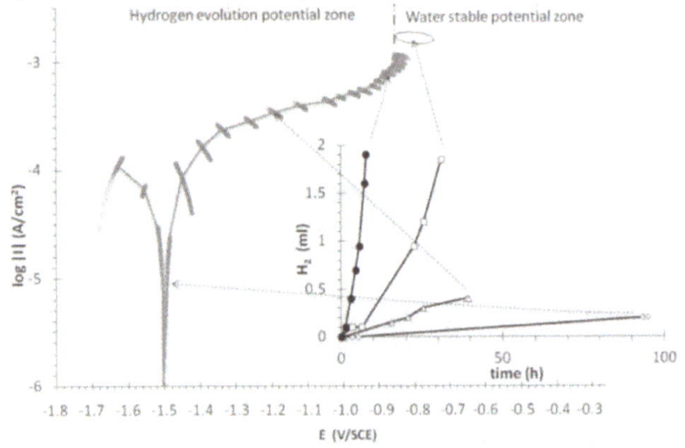

Figure 3 IR-drop corrected polarization curve and hydrogen evolution volume vs. time at different potentials for Mg immersed in saturated $Mg(OH)_2$ solution [13].

The accelerated hydrogen evolution process on Mg at a high anodic potential is unlikely to be a result of a cathodic hydrogen reaction in corroding surface areas, as a cathodic reaction normally becomes slower at a higher potential. It cannot be caused by enlarging corroded surface area with increasing potential either, because the corroded surface area could not further enlarge after the whole Mg surface had been fully corroded. It was observed in experiment [13] that corrosion damage spread over the whole Mg surface in a few hours at ~-0.6V/SCE (IR drop-corrected potential), but the hydrogen evolution rate still increased with increasing potential. Figure 3 clearly shows that the hydrogen evolution rate became dramatically higher when the polarization potential was close to or higher than the hydrogen equilibrium potential (-0.60 V/SHE) in the saturated $Mg(OH)_2$ solution (pH10.5). If the hydrogen evolution from corroding surface areas was a cathodic reaction, it should have significantly slowed down, or stopped at such a high potential. At least, it should not accelerate.

The abnormal hydrogen evolution phenomenon, termed as "anodic hydrogen evolution" (hydrogen evolution rate increases with increasing potential from corroding surface areas), can only be explained by the model of Mg^+ ion involved dissolution in the surface film pores [13, 14]. In this model, the surface film only blocks some of the Mg surface, partially preventing Mg from dissolution and stopping the dissolved Mg^+ ions from reacting with water generating hydrogen. Thus, the anodic Mg dissolution and hydrogen evolution are only partially limited. In other words, the porous surface film cannot effectively stop Mg dissolution and anodic hydrogen evolution because of its porosity or non-protectiveness.

Surface film model

The reactions involved in the surface film during corrosion are schematically illustrated in Figure 4. The anodic dissolution of Mg into Mg^+ occurs on the bare Mg surface exposed to the solution in the film pores. The intermediate ion Mg^+ reacts with water immediately, producing Mg^{2+}, OH^- and also H_2. The dissolved Mg^{2+} and OH^- can form $Mg(OH)_2$ and deposit on the Mg surface to build up the outer layer of the film while the original film is being chemically dissolved into the solution. Meanwhile, on the film covered area, some of the metallic ions are oxidized and incorporated into the film. There is an equilibrium reaction between MgO and $Mg(OH)_2$ in the film. The ratio of these two compounds is determined by the water content in the film.

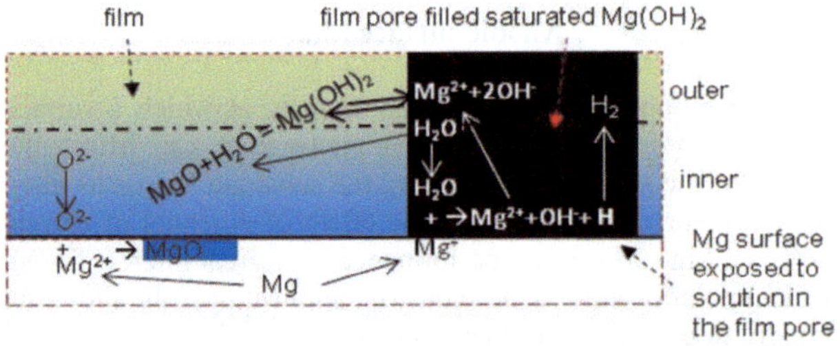

Figure 4. Schematic illustration of the chemical and electrochemical processes within the porous film a Mg electrode [13].

It should be noted that the surface film as described in Figure 4 is a kinetically steady layer. The steady state is achieved when the film dissolution rate is equal to the film formation (or growth) rate. The simultaneous film formation and dissolution processes have recently been experimentally verified through tracing isotopes of elements O and H in the film [15, 16].

The surface film model (Figure 4) can successfully explain most of the corrosion behaviors of Mg and its alloys. For example, the corrosion rate of Mg decreasing first and then increasing with time in a simulated body fluid (Figure 5) can be simply ascribed to the initial deposition of Ca-containing compounds on the Mg surface and the later accelerated dissolution of the surface film by chlorides from solution [17].

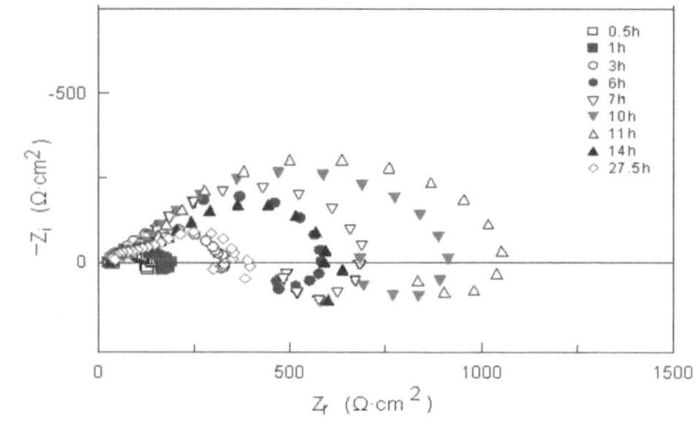

Figure 5. Electrochemical Impedance Spectra for Mg measured at different times in a simulated body fluid [17]. The impedance increases in the first 11 hours and then decreases, indicating that the corrosion slows down initially and then accelerates.

The surface film model (Figure 4) also predicts that the film will be more filamentous and porous with a thinner inner layer and a thicker outer layer in a more aggressive solution due to the rapid dissolution of Mg. This has recently been observed on AZ31 and E717 alloys in a chloride containing solution [18].

Anodized films

The model cannot be used to predict the film formation behaviour if the polarization potential or current density is too positive (e.g. >60V or >10 mA/cm^2), because many more complicated and unclear reactions are involved. In that case, a surface film or coating with a very different morphology will be formed on Mg and its alloys in some bath solutions. This surface film is known as anodized coating, which is much thicker (at least a few micrometers) and appears to be more porous than the above mentioned surface film formed at a relatively low anodic potential or current density. The differences in morphology and porosity of the anodized coating result from its sparking/arcing reaction at the high potential or current density involved in the anodizing process.

There are already many publications on the development of anodized coatings. The reported coatings are formed in various conditions. To better understand the influence of applied potential or current density on the coating porosity and morphology, a series of coatings formed under different anodizing current waveforms but in the same bath solution [19] are selected and analyzed in this paper. Figure 6 shows how the anodizing potential waveform affects the coating porosity. It can be concluded that the coating porosity and roughness increase with increasing anodizing current density and time. Interestingly, most of the pores actually have not penetrated through the coatings reaching the substrate alloy. The coating cross-section image presented in Figure 7 reveals that even in an arcing spot, i.e., the most porous area, most of the pores are completely closed or sealed. Moreover, there is a thin but continuous and less porous inner layer between the porous layer and the substrate. Therefore, corrosive solution cannot easily and directly reach the substrate directly in corrosion.

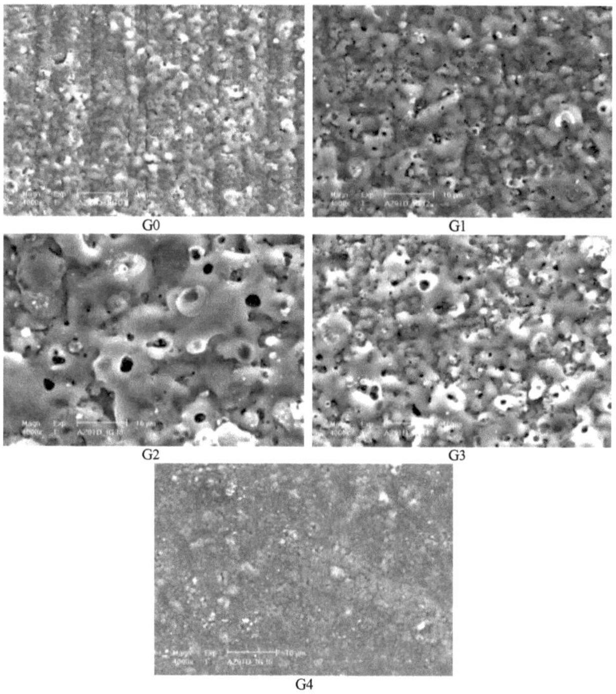

Figure 6. SEM topographic view of anodized coatings formed on AZ91D in bath solution 1%KOH+1.6%K_2SiO_3 using the following anodizing potential waveforms: [19]:
G0: the anodizing current density was 20 mA/cm^2 for 10 minutes,
G1: the anodizing current density was 15mA/cm^2 for 20 minutes.
G2: the anodizing current density was 20 mA/cm^2 for 7 minutes, 15 mA/cm2 for 7 minutes and 10 mA/cm^2 for 7 minutes.
G3: the anodizing current density was 25 mA/cm^2 for 7 minutes, 20 mA/cm2 for 4 minutes and 15 mA/cm^2 for 7 minutes.
G4: the anodizing current density was 20 mA/cm^2 for 5 minutes.

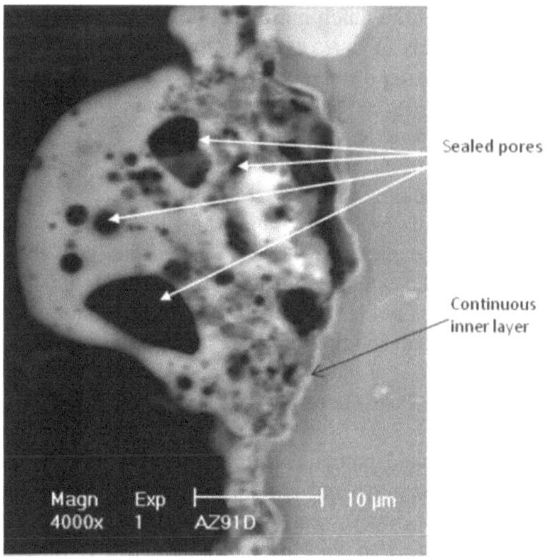

Figure 7. SEM cross-section image of a localized arcing spot on anodized AZ91D alloy under G5 anodizing condition [19].

In spite of the morphology, the anodized coating formed at high anodic potential or current density in effect has a microstructure similar to the anodic surface film formed at low potential or current density. They both are porous and have a very porous outer layer and a relatively less porous inner layer. However, the outer layer of the anodized coating is more porous and thicker, and the inner is denser. Thus, the anodized coating is more corrosion resistant, and to a greater degree inhibits the anodic dissolution of the substrate Mg alloy. For example, the polarization curves of non-anodized and anodized ZE41 coupons in saturated $Mg(OH)_2$ + 0.1 wt.% NaCl are compared in Figure 8. The anodized coating on the anodized alloy significantly suppresses the anodic current densities compared with the surface film on the non-anodized alloy.

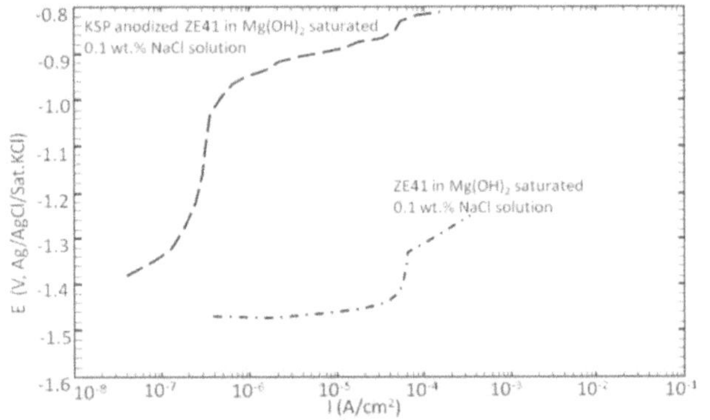

Figure 8. Typical anodic polarization curves for ZE41 and anodized ZE41 in saturated $Mg(OH)_2$ + 0.1 wt.% NaCl solution. The anodization was performed in 1.6 wt.% K_2SiO_3+1wt.% KOH at ambient temperature; the anodizing current density was controlled at 20 mA/cm^2 for 10 minutes, then 10 mA/cm^2 for 10 minutes and finally 5mA/cm^2 for 10 minutes [20].

Although the surface model (Fig. 4) cannot be applied to the coating formation due to complicated reactions involved in the anodization process, it can be used to interpret the coating's corrosion resistance. For example, the roles of the anodized coating in suppressing the anodic current density of the anodized alloy in a chloride containing solution can be explained by this model. Due to the significantly thicker outer and the evidently denser inner, as well as the much more chemically stable coating composition, the ingress of corrosive solution into an anodized coating and the dissolution of the coating and substrate are much slower than those in a surface film according to this film model. More specifically, the beneficial effect of the anodized coating on corrosion performance of an anodized Mg alloy can be ascribed to the "retarding", "blocking" and "passivating" effects [20]. Thereby, an anodized Mg alloy has much lower anodic polarization current densities than an unanodized one. All these beneficial effects of the anodized coating result from the high potential or current density that is applied to form the coating and the carefully selected bath solution in which the coating grows.

Possible passivity

In addition to the applied potential/current that can modify the electrochemical reactions involved in the film formation, both the substrate and solution directly participate in the process, and hence have a critical effect on the film stability and protectiveness. Since the film normally contains some of the elements from the substrate, it would be of great scientific interest to know whether a passive film could be formed on a Mg alloy if the matrix phase of the substrate contained a sufficiently high concentration of strong passivating elements, such as Ti, Cr, Ni, etc.. This possibility was recently explored using magnetron-sputtered single phase Mg-Ti alloys [21]. Detailed deposition conditions and resulting film morphologies and compositions have been reported before [22]. It was found that the surface coatings on the Mg based Mg-Ti alloy were thick and non-protective. The coating eventually became a thin continuous protective passive film mainly containing Mg and Ti oxide on a Ti based alloy (see Figure 9). The transition from a corrosive Mg-Ti alloy to a passive Mg-Ti alloy with increasing Ti content in the alloy is further demonstrated in Figure 10. The polarization curves of these alloys clearly show a significantly enhanced OCP and dramatically decreased anodic dissolution current densities when the Ti content increases. There is a dramatic drop in anodic current density when Ti concentration becomes higher than 50 wt.%. These suggest that the corrosivity of Mg-Ti alloy decreases as Ti content increases, and the alloy becomes passive when the Ti level (atomic %) is higher than the Mg content.

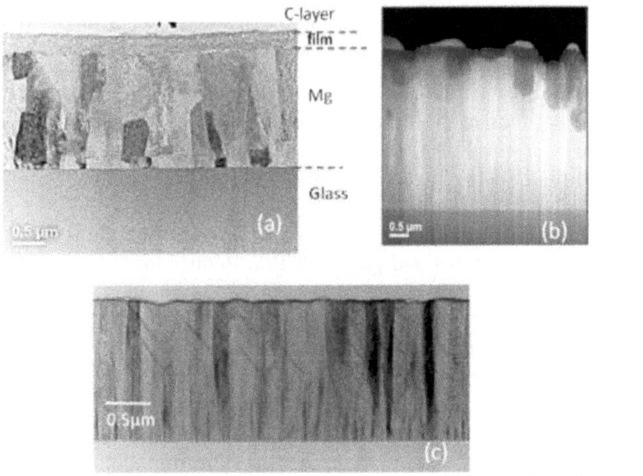

Figure 9. Bright Field STEM image of the cross-section for the magnetron-sputtered Mg-Ti alloys (a) Mg93Ti0, (b) Mg68Ti19 and (c) Mg52Ti38 after 5 h of immersion in saturated $Mg(OH)_2$+0.1 wt.% NaCl. The films on Mg93Ti0 and Mg68Ti19 are non-passive surface film, which are thick and loose. The film on Mg52Ti38 is a passive one that is compact and very thin (barely seen in this photo, but can be clearly detected though element mapping).

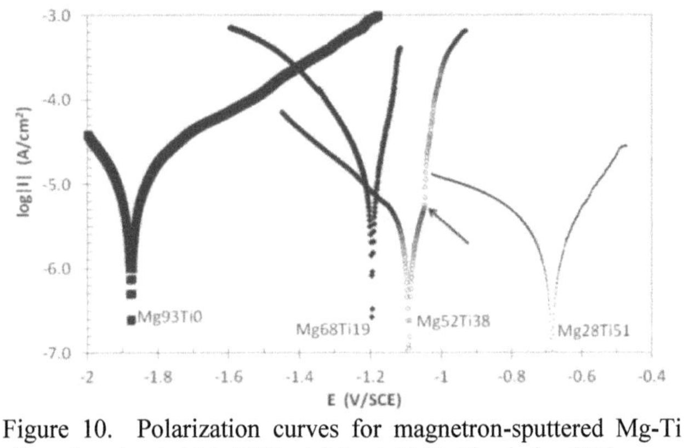

Figure 10. Polarization curves for magnetron-sputtered Mg-Ti alloys in saturated $Mg(OH)_2$ + 0.1 wt. % NaCl solution [21].

Similar beneficial passivating effect of Cr has also been observed on single phase magnetron-sputtered Mg-Cr alloys (see Figure 11). The anodic polarization current densities decrease with increasing Cr content in the alloy. The alloy does not show a passivation tendency until the Cr content is higher than 70 wt.%. The similarity in the influence of passivating element on polarization curve between Mg-Cr and Mg-Ti alloys implies that the Mg-Cr alloy had a similar thin passive film formed when the Cr content was sufficiently high.

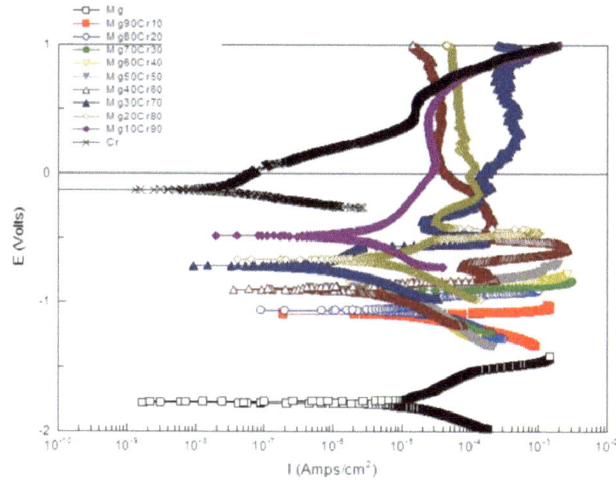

Figure 11. Polarization curves of magnetron-sputtered Mg-Cr alloys in saturated $Mg(OH)_2$ + 0.1 wt. % NaCl solution.

The above polarization curves suggest that passivity can only be achieved when the alloy contains a sufficiently high content of passivating element (higher than the content of Mg). In this case, the alloy has already become a passivating element based alloy. Therefore, it might be unlikely to passivate Mg through alloying with strong passivating elements in practice. Nevertheless, the exploration of passivity of Mg alloy opens up a possible way to corrosion resistant Mg alloys. Surface alloying may make a Mg alloy passive without altering the alloy's bulk composition and properties [23]. It is quite possible for a Mg alloy to have a surface containing large amounts of passivating elements, on which the formation of a thin, continuous, stable and protective passive film will become possible. This kind of surface alloying may be realized through ion-implantation [24, 25] or other methods.

It should be noted that although the composition, stability and protectiveness of the film on such a single phase substrate can be modified to a great degree by the alloying elements from the substrate, the film is not an anodized coating. Before the film becomes passive, it is not as corrosion resistant as an anodized coating, and its formation can still be described by the surface film model (Fig.4). To form a passive film, the concentrations of passivating elements from the substrate should be high enough to stabilize the film and close off the film pores.

Sacrificial anode coating

Equally important to film performance is the solution in which the film is formed. According to the film model (Figure 4), the bath solution can also strongly influence the film composition and corrosion resistance, because the deposition of compounds from solution is an important step in film formation. Hence, modification of the bath composition is a straightforward approach of enhancing the film protectiveness. In fact, the importance of bath solution to the film formation has been demonstrated by anodization. An anodized coating can only be formed in a solution with certain compositions at a very high potential or current density, not in any solutions.

Following this train of thought, a corrosion resistant film could be formed even without applying anodizing potential or current, as long as the bath solution is carefully selected. For example, after Mg was immersed in a metaphosphoric acid solution (200g HPO_3 dissolved in 200g distilled water) for 10 seconds and then in sodium metaphosphate solution (60g $NaPO_3$ dissolved in 500g distilled water) for 24 hours, a metaphosphated coating was formed on the Mg surface. This film could be a magnesium metaphosphate or phosphate layer with Mg, P and O as its main constituents. It was highly defective and discontinuous, and did not appear to be protective for the substrate Mg (see Figure 12). However, surprisingly, such a surface film can cathodically protect the substrate Mg [26]. The open-circuit-potential (OCP) of the film-covered Mg was measured in a saturated $Mg(OH)_2$ solution. It had a more negative OCP than bare Mg (Figure 13), suggesting that the metaphosphated film could act as a sacrificial anode coating on Mg. SEM observation confirmed that the metaphosphated film has two different layers during corrosion, which is similar to a normal surface film as presented earlier, and it can effectively and uniformly protect the substrate Mg from corrosion attack [26].

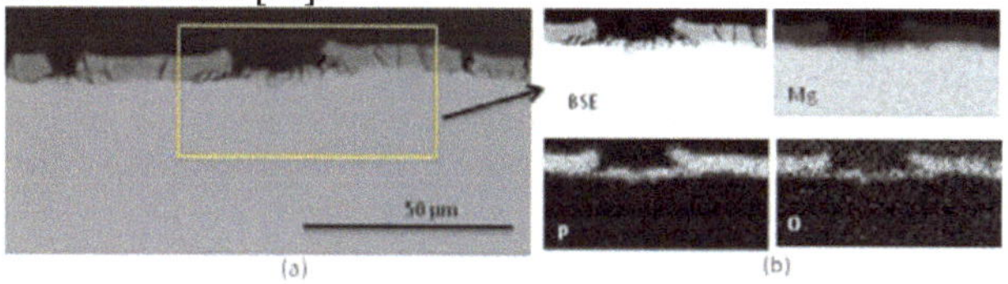

Figure 12. (a) BSE image of the cross-section of metaphosphated Mg, and (b) Mg, P and O distributions in the surface film formed on Mg [26].

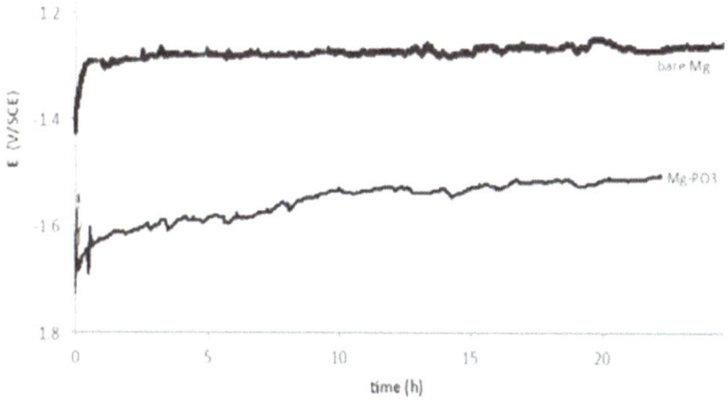

Figure 13. OCPs of bare and metaphosphated Mg in saturated $Mg(OH)_2$ solution [26].

Mg is the most active engineering metal, and it is often used as a sacrificial anode/coating to protect other engineering metals from corrosion attack. So far, no sacrificial anode coating has been developed or considered for Mg. These results demonstrated the possibility of forming a sacrificial coating for Mg. It is also worthwhile to note that the protection mechanism of the sacrificial film is different from that of a surface film or anodized coating. A less porous film may more effectively retard the ingress of corrosive solution. After the corrosive solution reaches the substrate, it may not necessarily cause corrosion damage to the substrate because of the galvanic protection effect offered by the film. The film is always preferentially dissolved before the substrate, and the surface film model (Fig.4) cannot be used to explain the protectiveness of this sacrificial coating, although it describes the film formation.

Summary

The normal surface films on Mg and its alloys are porous and non-protective. The film porosity and morphology as well as the film stability and protectiveness can be affected by applied anodic potential or current density. At a high potential or current density, a much thicker and more protective anodized coating can be formed on Mg in some solutions. When the polarization potential or current density is high enough, Mg anodization will occur in some bath solution. The anodized film is much thicker and more corrosion resistant. Some of the ions from the bath solution can react with Mg and form less soluble compounds deposited on Mg surface. The corrosion performance of a surface film can also be improved if the substrate Mg is alloyed with passivating elements. However, the surface film is still porous or non-protective enough to make the Mg based alloy passive when the alloying element concentrations are lower than the Mg content in the matrix phase. Surface film stability and protectiveness can be significantly improved through modifying the solution composition. A sacrificial anode coating may be formed on Mg surface to protect Mg from corrosion attack. In summary, the surface films formed on Mg and its alloys under different conditions have different compositions and corrosion performance. However, their microstructures appear to be similar, usually consisting of two layers. Their corrosion performance is to some degree associated with their porosity.

Acknowledgements

The authors would like to thank Dr. Zhiming Shi, Dr. Michael P. Brady and Dr. Kinga A. Unocic for their helpful discussions. The support from the National Environmental Corrosion Platform is also acknowledged.

References

1. G.-L. Song, Z. Xu, Crystal Orientation and Electrochemical Corrosion of Polycrystalline Mg, Corrosion Science 63(2012) 100-112.
2. G.-L.Song, Chapter 1: Electrochemistry of Mg and its Alloys, in G.-L. Song (ed.), Corrosion of Magnesium alloys, Woodhead publishing Limited, UK, 2011, pp.3-65
3. P.L.Bonora, M.Andrei, A.Eliezer, E.M.Guatman, Corrosion behaviour of stressed magnesium alloys, Corrosion Science 44(2002)729-749.
4. E.Ghali, Chapter 2: Activity and passivity of Magnesium (Mg) and its Alloys, in G.-L. Song (ed.), Corrosion of Magnesium alloys, Woodhead publishing Limited, UK, 2011, pp.66-114
5. G.-L. Song, Effect of tin modification on corrosion of AM70 magnesium alloy, Corrosion Science 51 (2009) 2063-2070.
6. Y.Choi, S.Salman, K.Kuroda, M.Okido, Improvement in corrosion characteristics of AZ31 Mg alloy by square pulse anodizing between transpassive and active regions, Corrosion Science 63 (2012) 5-11.
7. S.Moon, Y.Nam, Anodic oxidation of Mg–Sn alloys in alkaline solutions, Corrosion Science 65 (2012) 494 -501.
8. P.L.Bonora, M.Andrei, A.Eliezer, E.M.Guatman, Corrosion behaviour of stressed magnesium alloys, Corrosion Science 44(2002)729-749.
9. E.Ghali, Chapter 2: Activity and passivity of Magnesium (Mg) and its Alloys, in G.-L. Song (ed.), Corrosion of Magnesium alloys, Woodhead publishing Limited, UK, 2011, pp.66-114
10. Y.Choi, S.Salman, K.Kuroda, M.Okido, Improvement in corrosion characteristics of AZ31 Mg alloy by square pulse anodizing between transpassive and active regions, Corrosion Science 63 (2012) 5-11.
11. S.Moon, Y.Nam, Anodic oxidation of Mg–Sn alloys in alkaline solutions, Corrosion Science 65 (2012) 494 -501.
12. G.-L. Song, Potential and current distributions of one-dimensional galvanic corrosion systems, Corrosion Science 52 (2010) 455–480.
13. G.-L.Song, K.A.Unocic The anodic surface film and hydrogen evolution onMg, Corrosion Science 98 (2015) 758–765
14. G. Song, Recent progress in corrosion and protection of magnesium alloys, Advanced Engineering Materials 7 (2005)563-586.
15. M.P. Brady, M.Fayek, H.H. Elsentriecy, K.A. Unocic, L.M.Anovitz, J.R.Keiser, G.L. Song, B. Davis, Tracer film growth study of hydrogen and oxygen from the corrosion of magnesium in water, Journal of the Electrochemical Society 161(9) (2014)C395-C404
16. M. P. Brady, M. Fayek, H.M. Meyer, III, D.N. Leonard, H. H. Elsentriecy, K. A. Unocic, L. M. Anovitz, E. Cakmak, J. R. Keiser, G.L. Song, and B. Davis, Tracer Study of Oxygen and Hydrogen Uptake by Mg Alloys in Air with Water Vapor, Scripta Materialia (accepted on 4/8/2015)
17. Z. Li, G.-L. Song, S. Song, Effect of bicarbonate on biodegradation behaviour of pure magnesium in a simulated body fluid, Electrochimica Acta 115 (2014) 56– 65
18. M.Brady, G.Rother, L.Anovitz, K.Littrell, K.Unocic, H.Elsentriecy, G.-L.Song, J.Thomson, N.Gallego, B.Davis, Film breakdown and nano-porous Mg(OH)2 formation from corrosion of magnesium alloys in salt solutions, Journal of the Electrochemical Society 162 (4) (2015) C140-C149
19. Z. Shi, G.-L. Song, A. Atrens, Influence of anodising current on the corrosion resistance of anodised AZ91D magnesium alloy, Corrosion Science 48 (2006) 1939–1959
20. G.-L. Song, Z. Shi, Corrosion mechanism and evaluation of anodized magnesium alloys, Corrosion Science 85 (2014) 126 - 140
21. G.-L. Song, K. A. Unocic, H. Meyer, E. Cakmak, M. P. Brady, P. E. Gannon, P. Himmer, Q. Andrews, The Corrosivity and Passivity of Sputtered Mg-Ti Alloys (submitted to Corrosion Science)
22. G.-L. Song, D. Haddad, The Grains and Topographies of Magnetron Sputter-deposited Mg-Ti Alloy Thin films, Materials Chemistry and Physics 125(2010)548-552
23. L. Zhu, G. Song, Improved corrosion resistance of AZ91D magnesium alloy by an aluminium-alloy coating, Surface and Coatings Technology 200 (2006) 2834-2840
24. C. Liu, Y. Xin, X. Tian, J. Zhao and P. K. Chu, Corrosion resistance of titanium ion implanted AZ91 magnesium alloy, J. Vac. Sci. Technol. A 25(2007)334-339
25. M. Vilarigues, L.C. Alves, I.D. Nogueira, N. Franco, A.D. Sequeira, R.C. da Silva, Characterisation of corrosion products in Cr implanted Mg surfaces, Surface & Coatings Technology 158 (2002) 328-333.
26. G.-L. Song, N. J. Dudney, J. Li, R. L. Sacci, J. K. Thomson, The possibility of forming a sacrificial anode coating for Mg, Corrosion Science 87 (2014) 11–14

Magnesium Technology 2016
Edited by: Alok Singh, Kiran Solanki, Michele V. Manuel, and Neale R. Neelameggham
TMS (The Minerals, Metals & Materials Society), 2016

Micro-arc Oxide Film of Aluminum Coating Pre-sprayed on AZ31 Magnesium Alloy

Suyuan Yang , Lin Zhou, Xingwang Cheng
(School of Materials Science and Engineering, Beijing Institute of Technology, Beijing 100081, China)

Keywords: Magnesium alloy, Micro-arc oxidation, Microstructure, Corrosion resistance

Abstract

Using the micro-arc oxidation (MAO) in silicate electrolyte, ceramic film was formed on pure Al coating previously sprayed on AZ31 magnesium alloy. The surface and cross-sectional morphologies, distribution of elements, phase composition, microhardness, bonding strength and corrosion resistance of the film were systematically investigated. The results show that the MAO film is uniform and compact, mainly comprised of α-Al_2O_3,γ-Al_2O_3 and $3Al_2O_3 \cdot 2SiO_2$. The ceramic film adheres well to the Al coating with the average hardness of 1197 HV. The results of polarization curves in 3.5wt.% NaCl solution show that the present technology can obviously improve the corrosion resistance of AZ31 magnesium alloy.

Introduction

As the lightest metallic structural material, magnesium and its alloys show attractive prospect for application in aero-space, automobile, electronic and military parts, due to their low density, high strength-to-weight ratio and good damping characteristic. However, magnesium has lower standard electrode potential(-2.363V) as well as a loose and porous surface, which makes it easy to corrod in most corrosion environments. Meanwhile the lower hardness and abrasive resistance are also the critical issues limiting their further applications [1-3]. Therefore, it is necessary to provide a coating to protect magnesium and its alloys from corrosion.

There are quite a number of available coating processes for protecting metal and its alloys, such as chemical conversion coatings[4-5], electroplating[6] and electroless plating[7-8], anodizing[9] and micro-arc oxidation[10-12]. Micro-arc oxidation(MAO) is an effective surface treatment technology developed recently[13-14]. The ceramic coating can be directly formed on the surface of metals such as Al, Mg or Ti by electrochemical process. The formed oxide film exhibit a compact structure and good bonding strength which can enhance the surface hardness and improve the corrosion resistance and wear resistances. MAO treatment for improving the corrosion resistance of magnesium alloys has been established[15-18]. However, the MAO technology for Mg alloy is not mature yet for Al and Ti alloys, which results in the worse effect than the others[19]. In this paper, a technology of micro-arc oxidation for Al coating pre-sprayed on magnesium alloy was employed to improve its corrosion resistance, which has not been reported before.

1. Experimental

A commercial AZ31 Mg alloy with composition(wt.%) of Al2.5-3.5, Zn0.7-1.3, Mn0.20-0.35, Cu0.05, Fe0.005, Si0.10, Ni0.005 was selected as the substrate in the present study. Prior to spraying, the 100×50×7mm samples were polished with alumina waterproof abrasive paper from 400 to 2000 grits, degreased with acetone, rinsed with distilled water then dried at room temperature. A pure aluminum coating about 200μm thick was plasma sprayed on AZ31 alloys using pure aluminum powders in -200~400mesh range.

After degreasing mentioned above, the aluminum coated surface was treated by micro-arc oxidation in an electrolyte mainly consisting of silicate, with a constant voltage of 500V at temperature between 15- 35℃. Parts of samples were heated at 100℃ in a furnace, then sealed by microcrystalline wax.

A Hitachi S-4800 scanning electron microscope(SEM) was employed for observing the surface and cross-section morphologies of coatings. The Energy-dispersive X-ray spectroscopy(EDX) was used for qualitative elemental chemical analysis. The phase analysis of the coatings were carried out using a X' Pert PRO MPD equipment(CuK_α radiation).

Microhardness was measured by using a HVS-1000 Digital Micro Hardness Tester. The bond strength was tested by tensile experiment and the bonding agent was E-7-2resin.

The corrosion resistance of the coated and uncoated samples in 3.5wt.%NaCl solution was evaluated by potentiodynamic polarization test through PARSTAT 2273. The electrochemical measurement was conducted using a conventional three-electrode electrochemical cell with a saturated calomel electrode (SCE) as the reference electrode and a Pt foil as the auxiliary electrodes respectively, and the samples with the area of 1 cm^2 as the working electrode. When the open circuit potential(OCP) became steady (about 30 min), potential scanning range was conducted at a rate of 0.33mV/s from −0.15 V to 1.4 V (vs OCP).

2. Results and Discussion

2.1 Morphologies and composition

Fig.1 shows the typical surface morphology of oxide film. It can be seen that the surface is composed of molten particles in dozens of microns in diameter. Beside, it shows lots of residual micro-sized pores, which are not only the reacting channel of solution and aluminum layer but also the overflow channel of molten oxides during oxidation. In the process of micro-arc oxidation, the weak parts of initial oxide film is broken down with arc discharge in micro-area. Meanwhile, we can see numerous swimming arc-points or sparks. With instantaneous high temperature generated by the discharge, the oxides or aluminum layer were melted or even vaporized in this area, molten products would solidify and form a non-metallic ceramic coating when encountering the electrolyte at low temperature while the molten products overflew along the broken down pores.

Figure.1 Surface morphology of AZ31 after microarc-oxidation.

Fig.2 is the cross-sectional micrograph of the samples. The interfaces among MAO film, aluminum layer and substrate were clearly observed. As the ceramic particles are generated directly on the aluminum layer during micro-arc oxidation, the boundary of ceramic film and aluminum layer is chemical bond. Compared to thermal spraying, its interface is compacted without any crack and partial shedding revealing a higher bond strength. Although obviously occlusion phenomenon along the interface formed by thermal spraying are observed, there are evidences of cracks and partial shedding. The oxide film was divided into two layers, i.e. a 40μm outer loose layer (A) and a 100μm inner dense layer (B).There are several bigger pores and cracks distributed non-uniformly in the loose layer, and parts of cracks even run through the whole layer. The cracks might be caused by MAO or the following cutting process. There are some holes in dense layer, but they fail to form any crack because of limited quantity and small volume. The thickness and density of dense layer may have a great impact on the improvement of corrosion resistance.

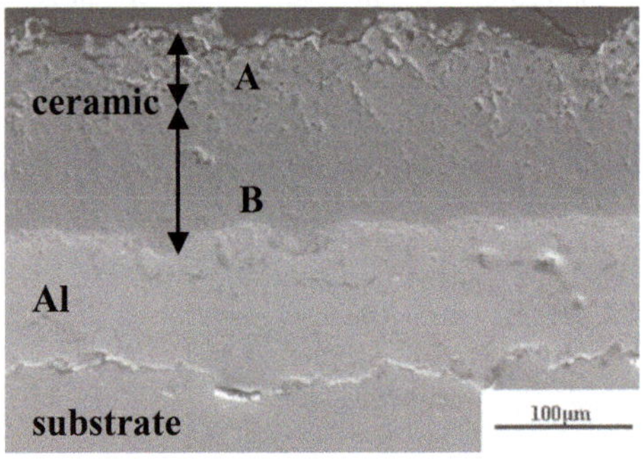

Figure.2 Cross-section microstructure after microarc-oxidation.

The EDX line-scan result shows that the contents of aluminum and oxygen increase gradually from outer to inner layer in micro-arc oxide film (Fig.3).

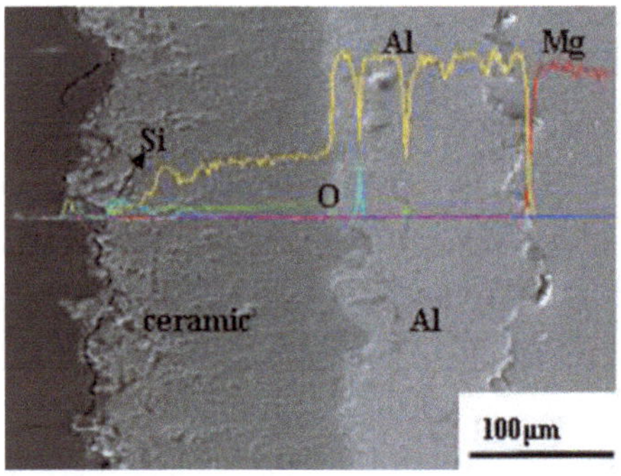

Figure.3 EDX line scan on cross-section.

According to other reports[20], the line-scan result indicates that the B layer mainly composed of aluminum-oxides, and closer to the substrate a higher quantity of aluminum-oxides are generated. Moreover, the quantity of silicon in loose layer is higher, possibly caused by the diffusion of silicon from electrolyte or the reaction between elements in loose layer and electrolyte during plasma reaction.

2.2 Phase structure

Fig. 4 shows the phase analysis of the MAO film. It is indentified that the film is mainly composed of γ-Al_2O_3, α-Al_2O_3, mullite ($3Al_2O_3 \cdot 2SiO_2$). As pointed in Ref.[21], from outer to inner of oxide layer, the amount of γ-Al_2O_3 is getting smaller while that of α-Al_2O_3 increases. In outer loose layer, the main phase is γ-Al_2O_3, and the inner dense layer contains α-Al_2O_3. During the MAO process, the discharge of Al atoms played a main role in forming the oxide film. Besides, the formation of mullite was caused by interaction of alumina and silicon, of which the latter was generated from electrolyte through the processes of transmission, absorption and deposition.

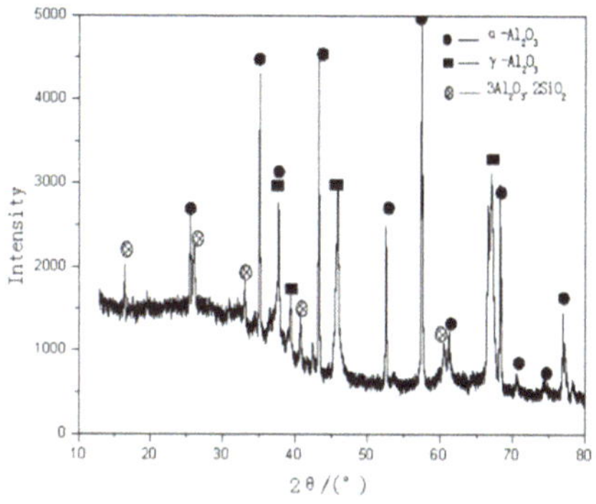

Figure.4 XRD pattern of the surface layer of ceramic film.

2.3 Hardness and bond strength

The microhardness between loose and dense layers in ceramic film is quite different. Their average hardness are 751HV and 1197HV with a highest value 1287HV in the dense layers close to aluminum layer. It is related to the phase composition and density in different layers. The inner layer exhibits higher hardness because of higher content and density of α-Al_2O_3. Meanwhile, higher hardness possibly facilitates wear resistance.

The average bond strength is 10.2MPa for all the samples and thus the weak interface only mechanically combined was always separated when tensile-tested.

2.4 Corrosion resistance

Fig.5 shows the potentiodynamic polarization curves of MAO-treated AZ31 alloy samples with and without sealing. The calculated electrochemical parameters are listed in Table1.

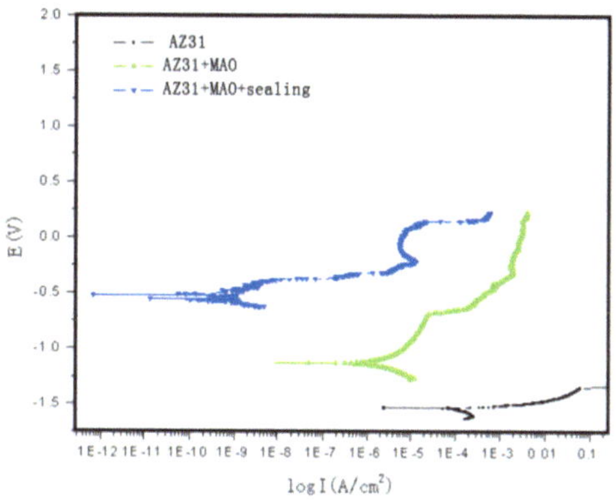

Figure. 5 Polarization curves of AZ31 substrate before and after Al spray and micro-arc oxidation treatment.

It can be seen from Fig.5, that the stable potentials, E_{corr}, of AZ31 alloy and the Al+MAO before sealing are more negative than that of with sealing, and the AZ31 alloy has the lowest stable potential. The higher stable potential, the higher is the anticorrosion tendency, which means the protective oxide film either with or without sealing can provide moderate protection to the substrate alloy. The current density, I_{corr}, of anodic corrosion for AZ31 alloy increases rapidly when the corrosion potential exceeds the self-corrosion potential, leading to the activation and dissolution. However, in such a case, the I_{corr} of Al+MAO sample slowly increases with increasing potential till the potential is high enough, revealing good corrosion resistance. After sealing the anodic part of polarization curve contains a passivation range, in which the current density keeps constant when the potential increases, and the appearance of passivation range has a positive effect on improving corrosion resistance. Fitting the curves in Fig.5, the calculated self-corrosion potential of Al+MAO sample is 404mV higher comparing to AZ31 substrate and the corrosion current density declines about two orders of magnitude as shown in Table 1.

Table1 Electrochemical parameters after curves fitting

Samples	E_{corr}/V	I_{corr}/ ($A \cdot cm^{-2}$)
AZ31	-1.535	4.38×10^{-4}
Al+MAO	-1.131	3×10^{-6}
Al+MAO+sealing	-0.539	1.2×10^{-8}

After sealing, the self-corrosion potential is 539mV higher than the unsealed one, and the corrosion current density also declines by two orders of magnitude. It is clear that micro-arc oxide film on sprayed aluminum layer can greatly improve the corrosion resistance of magnesium alloy in Cl^{-1} containing solutions.

Fig.6 shows the morphology of Al+MAO layer after the electrochemical corrosion. It can be seen from Fig.6 that the interface between Al layer and AZ31 alloy shows reaction with lots of loose products, the thickness of ceramic layer becomes thinner than that before corrosion, and no obvious corrosion is found in MAO layer or interface between MAO and Al layers. The thinning of MAO layer may be caused by peeling of coating when cutting the samples or the corrosion reaction between parts of impurities existed in MAO layer and ionic solutions, which makes the area loosen and dissolve in electrolyte.

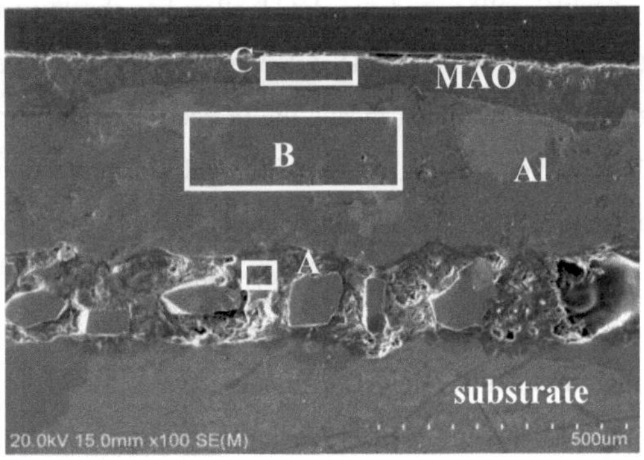

Figure.6 Cross section microstructure of Al+MAO layer after electrochemical corrosion.

Table2 EDS results of MAO layers after electrochemical corrosion

Area	Al	Mg	Cl	O
A wt%	54.92	21.19	1.21	22.68
at%	46.70	20.00	0.78	32.52
B wt%	96.39			3.61
at%	94.07			5.93
C wt%	53.87			46.13
at%	40.91			59.09

The results of EDS analysis on corroded area marked in Fig.6 is showed in Table2. It is found that the composition of area A consists of Al, Mg, O and a small amount of Cl. In area B and C, only Al and O are detected. The results indicate that no obvious corrosion occurs in areas B and C which only serve as the channel for ionic solutions. Though the layer produced by MAO has low porosity, but the thickness of ceramic layer becomes thinner and more loose than that before corrosion, the plasma spray Al layer has high porosity, So the Cl^- still can penetrate to the interface between Al layer and AZ31 substrate. As the lower electrode potential of Mg (Mg-2.363, Al-1.67), corrosion occurs at the substrate and Al layer interface firstly. Owing to the great brittleness of ceramic, the corrosion products with their expansibility when pouring out can make the MAO layer crack or peel in the loose or weak parts. Obviously, the MAO+Al can fulfill greater potential in corrosion resistance of magnesium alloys under the effect of sealing.

3 Conclusions

In the present study ceramic film was created on pure Al coating previously sprayed on AZ31 magnesium alloy by the micro-arc oxidation（MAO）in silicate electrolyte. An oxide film was formed on the surface of Al coating plasma sprayed on AZ31 alloy. It has a two-layer structure with the microhardness of the inner dense layer reaching up to 1197HV. The bonding between two layers is quite good. The microarc oxide film is mainly consisted of γ-Al_2O_3, α-Al_2O_3 and mullite ($3Al_2O_3 \cdot 2SiO_2$). The microarc oxidation significantly improved the corrosion resistance of the AZ31. We believe that the present investigation will open up a new horizon on protecting Mg alloys in industries.

References

[1] Wang F, Kang Z X, Li Y Y. Corrosion protection on the surface of magnesium[J].Titanium Technology Progress, 2007,24(1):20.

[2] Chen Z H. Magnesium alloys[M].Beijing: Chemical Industry Press, 2004.

[3] Clow B B. Magnesium industry overview[J]. Advanced Materials Processes, 1996, 150(4):33-34.

[4] Cheng Y L,Wu H L,Chen Z H,Wang H M, Zhang Z, Wu Y W. Corrosion properties of AZ31 magnesium alloy and protective effects of chemical conversion layers and anodize

coatings[J]. Trans. Nonferrous Met. Soc. China, 2007, (17):502-508.
[5] Zucchi F, Frignani A, Grassi V,Trabanelli G, Monticelli C. Stannate and permanganate conversion coatings on AZ31 magnesium alloy[J].Corrosion Science, 2007,(49):4542-4552.
[6] Gray J E, Luan B. Protective coatings on magnesium and its alloys: A critical review [J]. Journal of Alloys and Compounds, 2002, 336(88):113.
[7] Fairweather W A. Electroless nickel plating of magnesium[J]. Transactions of The Institute of Metal Finishing, 1997, 75(3):113-117
[8] Lawrie B. UK company leads the way in magnesium plating[J]. Finishing, 1994, 18(11): 22-23.
[9] Zhang Y J, Yan C W, Wang F H. Study on the environmentally friendly anodizing of AZ91D magnesium alloy[J]. Surface and Coatings Technology, 2002, (161) :36–43.
[10] Cao F H, Cao J L, Zhang Z, Cao C N. Plasma electrolytic oxidation of AZ91D magnesium alloy with different additives and its corrosion behavior[J]. Materials and Corrosion, 2007,58(9):696-703.
[11] Song Y L, Liu Y H, Yu S R, Zhu X Y, Wang Q. Plasma electrolytic oxidation coating on AZ91 magnesium alloy modified by neodymium and its corrosion resistance[J]. Applied Surface Science,2008,254: 3014–3020.
[12] Guo H F, An M Z. Growth of ceramic coatings on AZ91D magnesium alloys by micro-arc oxidation in aluminate–fluoride solutions and evaluation of corrosion resistance[J]. Applied Surface Science, 2005,246:229–238.
[13] Wirtz G P, Brown S D, Kriven W M. Ceramic coatings by spark deposition[J] . Materials and Manufacturing Processes,1999,6(1):87-115.
[14] Dittrich K H, Leoard L G. Micro-arc oxidation of aluminium alloy Components[J]. Crystal Research and Technology, 1984, 19(1):93-96.
[15] Duan H P, Du K Q, Yan C W. Electrochemical corrosion behavior of composite coatings of sealed MAO film on magnesium alloy AZ91D[J]. Electrochimica Acta, 2006, 51: 2898-2908.
[16] Hao J M , Chen H, Zhang R J. Corrosion resistances of magnesium alloys by micro-arc oxidation and anodic oxidation[J]. Material Protection,2003,36(1):20-21.
[17] Inho H, Jai H C , Bao H Z, Hong K B, Lee I S. Micro-arc oxidation in various concentration of KOH and structural change by different cut off potential[J]. Current Applied Physics, 2007, 7(S1):23-27.
[18] Liu Y P, Duan L H, Ma S X. Solution on microstructure and corrosion protection of ceramic coatings formed on magnesium alloy during micro-arc oxidation[J]. Journal of Chinese Society for Corrosion and Protection, 2007,27(4):202-204.
[19] Hou W A , Song X J. Preparing ceramic coating on the surface of magnesium alloy by hot spray and micro-arc oxidation[J]. Nonferrous Metal, 2006,(s1):113.
[20] Xue W B , Wang C, Chen R Y. Characterization of structure and composition of microarc oxidation coating on Al-Si alloy[J]. Journal of Beijing Normal University, 2003 , 39(5):618-622.
[21] Xue W B, Deng Z W, Wang X F, Chen R Y. Analysis of morphology and phase composition of ceramic films formed by the microarc oxidation of aluminum alloys[J]. Journal of Beijing Normal University, 1996 , 31(1):67–70.

Magnesium Technology 2016

Poster Session

Magnesium Technology 2016
Edited by: Alok Singh, Kiran Solanki, Michele V. Manuel, and Neale R. Neelameggham
TMS (The Minerals, Metals & Materials Society), 2016

Study on Fatigue Mechanism of Mg-0.6at%Y Alloy by Cyclic Tensile Test

Qinghuan Huo*, Daisuke Ando, Junichi Koike, Yuji Sutou
Department of Materials Science, Tohoku University, 6-6-11-1016 Aoba, Aramaki, Aoba-ku, Sendai 980-8579, Japan

Keywords: magnesium alloy, fatigue, twinning, dislocation.

Abstract

In this investigation, the fatigue mechanisms of Mg-0.6at%Y dilute alloy were studied by cyclic tensile tests. After annealing at 723 K for 8 h, average grain size of 180 μm was obtained. Yield stress and ultimate tensile strength were 65 and 140 MPa, respectively. Cyclic tensile tests were carried out at room temperature for 10 cycles. Stress was continuously cycled from nil to peak value and subsequently unloading back to nil. Peak stress was controlled according to yield stress: below yield stress, near yield stress and above yield stress. The results showed that basal slip and prismatic slip are the main fatigue mechanisms of Mg-Y alloy. {10-12} tension twinning only occurs when peak stress is above yield stress. Furthermore, compared with previous work on AZ31 Mg alloy, prismatic dislocation slip promotes cyclic strain hardening more obvious than {10-12} tension twinning does.

Introduction

Recently, researchers have found that rare earth (RE) element addition is an excellent way to weaken anisotropy and activate non-basal slip systems of magnesium (Mg) alloys, and yttrium (Y) is among the attractive RE elements [1, 2]. It has been manifested that basal slip, {10-12} twinning, prismatic slip, and pyramidal slip are the main deformation mechanisms of Mg-Y alloy [1-3].

Besides subjected to stress from static loading, Mg-Y alloy is also expected to serve as structural material subjected to cyclic loading and unloading. Thus, fatigue mechanisms become important and need to be clearly understood. However, up to now, there are few reports about fatigue mechanisms of Mg-Y alloy. As is well known in textured Mg alloys, there always exist obvious tension-compression yield asymmetry and asymmetric hysteresis loops [4-6]. Under low-cycle fatigue condition, {10-12} twinning-detwinning are indicated to be the main impact factor [5, 6]. For high-cycle fatigue, {10-12} twinning-detwinning is the primary factor under low stress amplitude, whereas prismatic dislocation slip dominates cyclic deformation under high stress amplitude [6, 7].

In contrast, Mg-Y alloy has weak texture and less preferential orientation apt to activate {10-12} twinning [1, 2]. So it is suggested that different mechanisms control fatigue process of Mg-Y alloy. In this paper, high-cycle fatigue mechanisms of Mg-Y alloy were investigated by cyclic tensile tests at room temperature. It is expected that the results in the present work can enrich the understanding of Mg-Y alloy fatigue and extend the application of Mg-Y alloy.

Experimental Procedure

The samples used in this study were hot-extruded sheets of Mg-0.6at%Y alloy, which had random texture distribution. Since twinning can easily take place in large grains, homogenising annealing was performed at 723 K for 8 h. Attained average grain size was 180 μm. Samples for cyclic tensile tests were machined with a gauge dimension of 15 mm in length (rolling direction), 4 mm in width (transverse direction) and 2 mm in thickness (normal direction). The surfaces of the samples were polished and etched only once before tensile tests. No further surface treatment was carried out after tensile tests in order to keep as much as possible detailed surface information. All tensile tests were carried out at room temperature with a strain rate of 1×10-3 s-1 for 10 cycles. Stress was continuously cycled from nil to different peak values and subsequent unloading back to nil. Based on triplicate tests, the yield strength and ultimate tensile strength of Mg-Y alloy were 65 and 140 MPa, respectively. Then the peak stress values were selected at 40, 70, and 100 MPa. In other words, stress amplitudes were decided according to the yield stress. The crystallographical orientations were examined by scanning electron microscope (SEM) and electron backscatter diffraction (EBSD). Observation of internal microstructures was realized by focused ion beam microscope (FIB) and transmission electron microscope (TEM).

Results and Discussion

Fig. 1 shows the microstructure and {0001} pole figure after homogeneous annealing. Equiaxed grains with an average grain size of about 180 μm are obtained. The basal texture distribution is random and multi-peak.

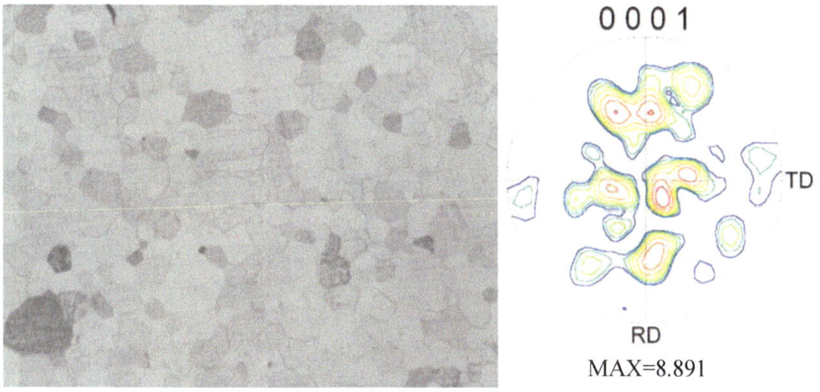

Fig. 1 Microstructure and {0001} pole figure after homogeneous annealing.

Fig. 2 shows the engineering stress-strain curves of different peak stresses. When peak stress is equal to 40 MPa, hysteresis loops can be observed with a little cyclic hardening. When peak stress increases to 70 MPa, evident cyclic hardening happens. After deformation for 10 cycles, strain increment is about 0.26%. Increasing peak stress to 100 MPa, then more notable cyclic hardening occurs. Strain increment is about 0.48%, nearly 2 times of 70 MPa. Compared with our previous work on AZ31 Mg alloy [7], it is manifested that cyclic hardening in Mg-Y alloy is more obvious than that in AZ31 Mg alloy. In the case of AZ31, cyclic hardening does not happen until peak stress increases to 100 MPa (above its yield strength). Strain increment of AZ31 Mg alloy is about 0.07%, only 1/7 of Mg-Y alloy at 100 MPa. That is to say, cyclic strain hardening is more obvious in Mg-Y alloy than in AZ31 Mg alloy.

Fig. 3 and 4 show TEM images of peak stress equal to 70 and 100 MPa. After 10-cycle tensile tests, large number of dislocations are generated. By using Burgers vector equal to [0001] direction, <c> dislocation disappears. However, when Burgers vector is equal to [1-210] direction, some dislocations still remain. Then it can be concluded that <a> dislocation occurs during cyclic tensile tests.

Fig. 5 shows typical SEM and OIM maps rapidly after tensile tests without any surface treatment. When peak stress is controlled at 70 MPa, obvious slip traces can be easily observed. With the assistance of basal plane traces and hexagonal prisms, it can be suggested that both basal slip and prismatic slip occur during cyclic deformation. Through selecting the grain within prismatic slip, high Schmid factor of {1-100}<11-20> can be achieved while Schmid factor of {0001}<11-20> is low. Meanwhile, abrupt misorientation changes of about 86° reveal that {10-12} tension twinning also happens, although twinning always tends to be inhibited by weakened basal texture.

Fig. 6 shows typical OIM maps after 10-cycle tensile test. As {10-12} twinning-detwinning has been confirmed to be one of the main fatigue mechanisms [4, 7], here in this work, residual twins are found when peak stress is equal to 70 and 100 MPa (blue lines corresponding to a misorientation of 86°). However, no residual twins are found when peak stress is equal to 40 MPa. Furthermore, residual twins are remained the most at peak stress of 100 MPa. Hence, {10-12} twinning-detwinning performs the most drastically at 100 MPa, and detwinning can be finished when stress amplitude is below yield stress.

Thus, the obvious strain hardening in Mg-Y alloy is derived from both {10-12} residual twins and prismatic slip. Compared with our previous work on AZ31 Mg alloy [7], the impact factor for strain hardening is only {10-12} residual twins, resulting in little strain hardening. It can be concluded that the extra strain hardening in Mg-Y alloy is induced by prismatic slip.

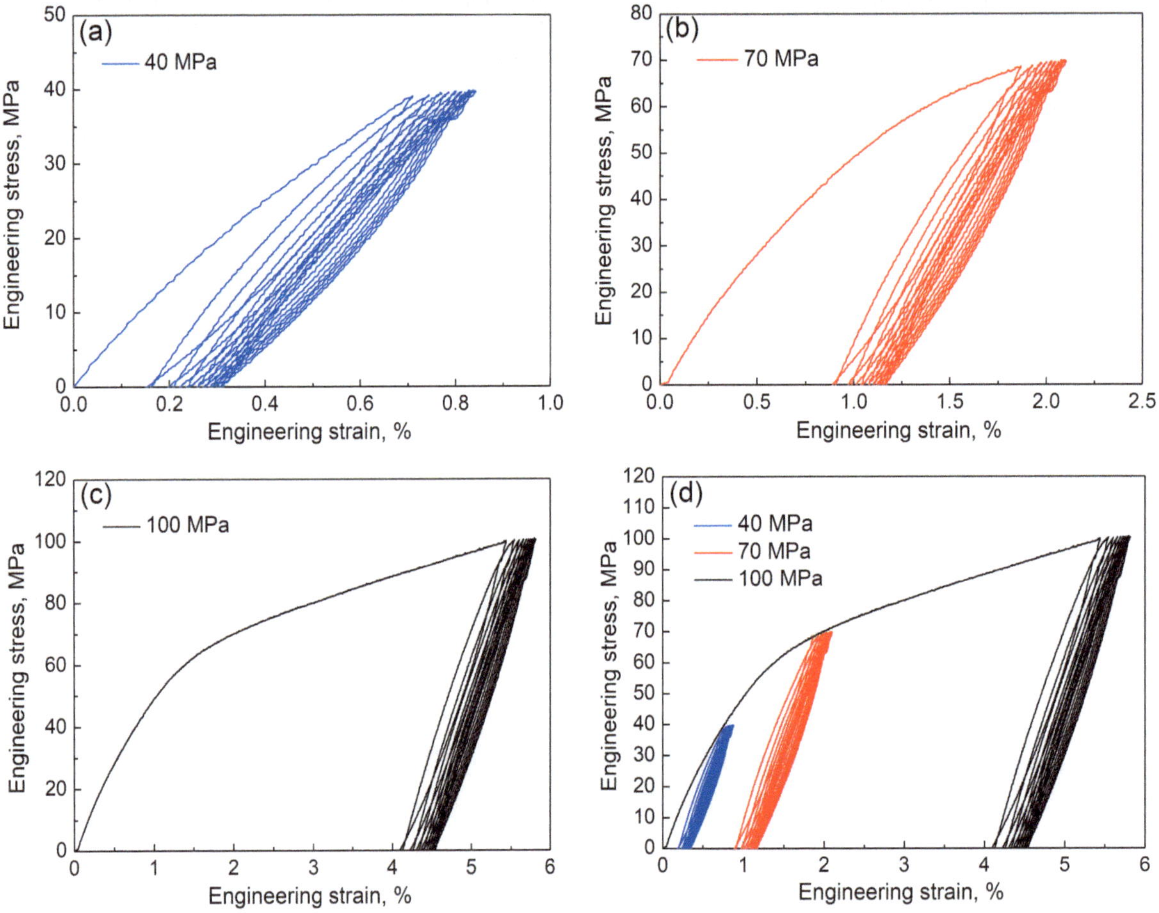

Fig. 2 Engineering stress-strain curves of peak stress of (a) 40 MPa, (b) 70 MPa, (c) 100 MPa and (d) all peak values.

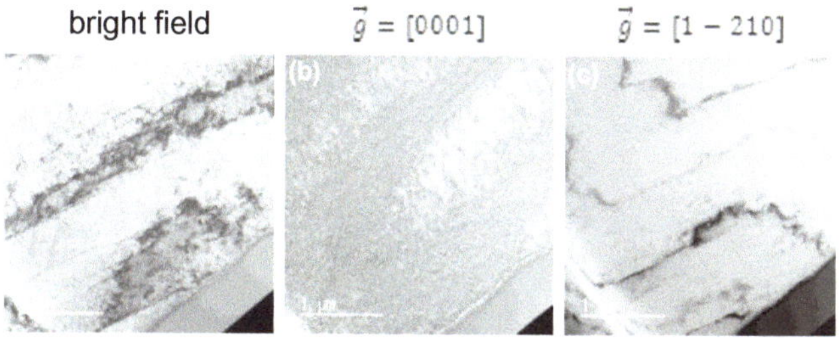

Fig. 3 TEM images after 10-cycle tensile test at 70 MPa.

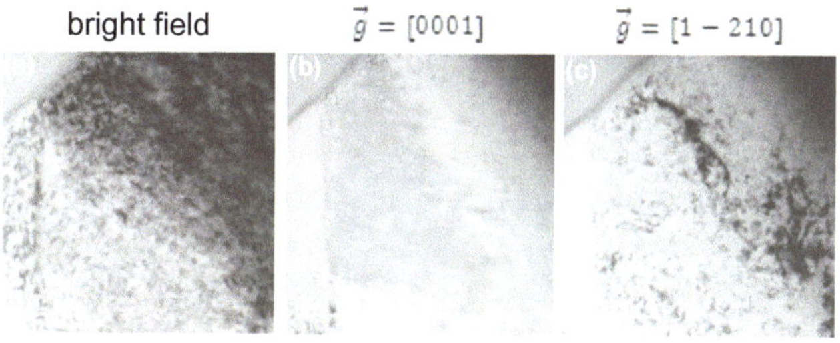

Fig. 4 TEM images after 10-cycle tensile test at 100 MPa.

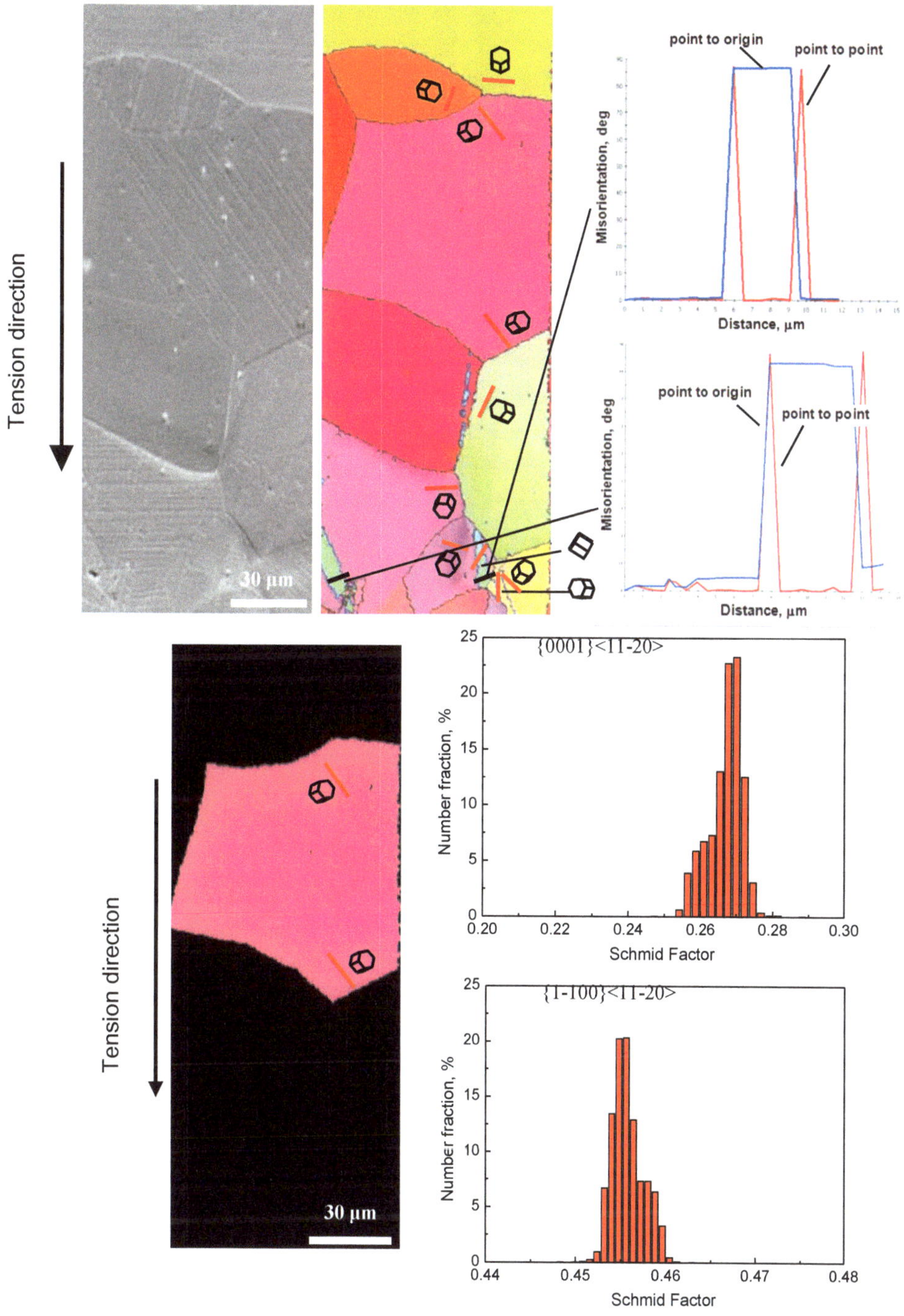

Fig. 5 Typical SEM and OIM images of 10-cycle tensile tests at 70 MPa.

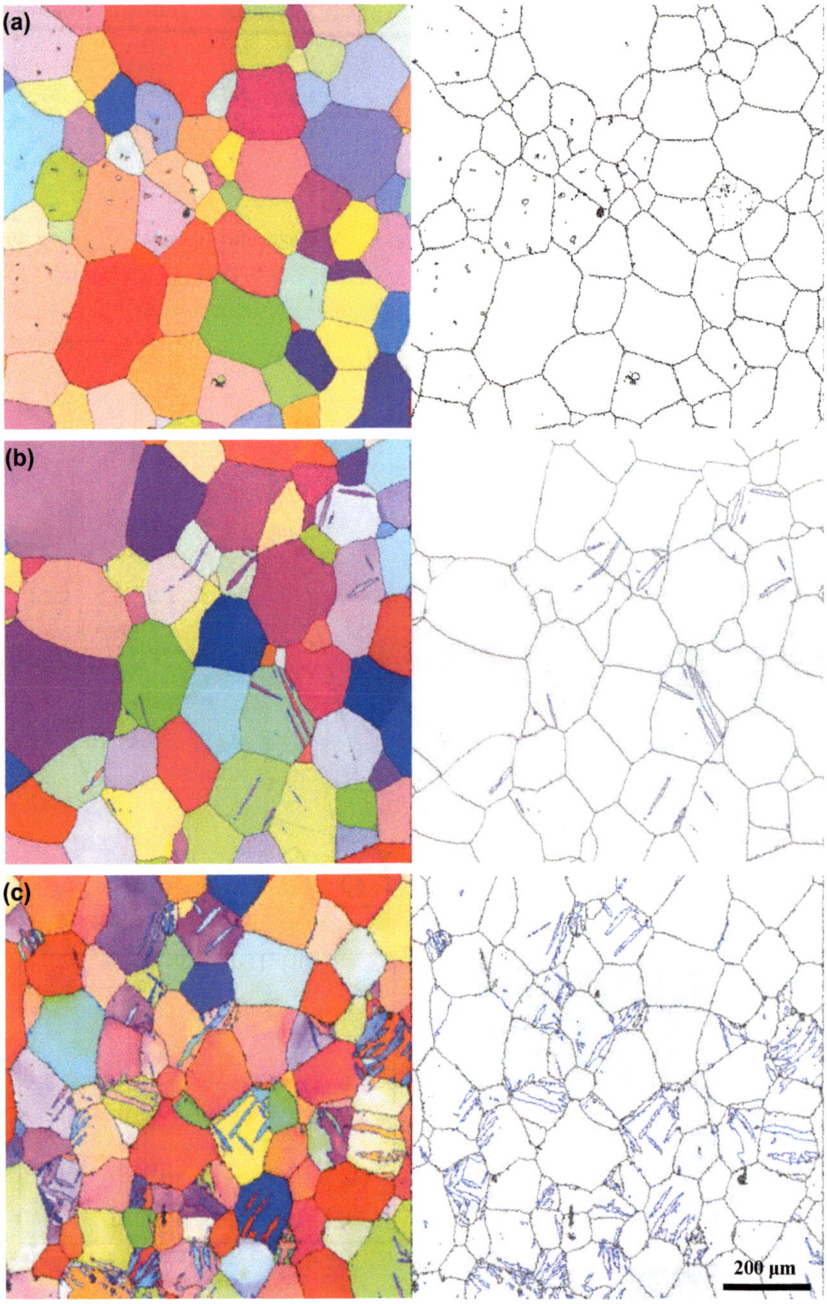

Fig. 6 Typical OIM maps after 10-cycle tension test at (a) 40 MPa, (b) 70 MPa, and (c) 100 MPa.

Conclusions

The fatigue mechanisms of Mg-Y alloy were studied by cyclic tensile tests at room temperature. Basal slip, prismatic slip and {10-12} twinning-detwinning are the main fatigue mechanisms of Mg-Y alloy. {10-12} residual twins and prismatic slip can be found when peak stress is near or above yield stress. Prismatic slip promotes more obvious strain hardening than only {10-12} tension twinning.

References

[1] Z. Zhao, Q. Chen, Y. Wang, D. Shu, "Microstructures and mechanical properties of AZ91D alloys with Y addition," *Materials Science and Engineering A*, 515 (2009), 152-161.

[2] M. Suzuki, T. Kimura, J. Koike, K. Maruyama, "Strengthening effect of Zn in heat resistant Mg-Y-Zn solid solution alloys," *Scripta Materialia*, 48 (2003), 997-1002.

[3] S. Sandlöbes, S. Zaefferer, I. Schestakow, S. Yi, R.

Gonzalez-Martinez, "On the role of non-basal deformation mechanisms for the ductility of Mg and Mg-Y alloys," *Acta Materialia*, 59 (2011), 429-439.

[4] Y.J. Wu, R. Zhu, J.T. Wang, W.Q. Ji, "Role of twinning and slip in cyclic deformation of extruded Mg-3%Al-1%Zn alloys," *Scripta Mater*ialia, 63 (2010), 1077-1080.

[5] Q. Li, Q. Yu, J. Zhang, Y. Jiang, "Effect of strain amplitude on tension-compression fatigue behavior of extruded Mg6Al1ZnA magnesium alloy," *Scripta Materialia*, 62 (2010), 778-781.

[6] S.M. Yin, H.J. Yang, S.X. Li, S.D. Wu, F. Yang, "Cyclic deformation behavior of as-extruded Mg-3%Al-1%Zn," *Scripta Materialia*, 58 (2008), 751-754.

[7] J. Koik, N. Fujiyama, D. Ando, Y. Sutou, "Roles of deformation twinning and dislocation slip in the fatigue failure mechanism of AZ31 Mg alloys," *Scripta Materialia*, 63 (2010), 747-750.

Magnesium Technology 2016
Edited by: Alok Singh, Kiran Solanki, Michele V. Manuel, and Neale R. Neelameggham
TMS (The Minerals, Metals & Materials Society), 2016

Mechanical Response of a Gravity Cast Mg-9Al-1Zn-0.2Sc Alloy at Strain Rates from 10^{-4} to 10^{3} /s

R.B. Blessington[1], A.D. Brown[1], A. Lock[1], J.P. Escobedo[1], P.J. Hazell[1], D. East[2], M.Z. Quadir[3]
[1]School of Engineering and Information Technology, The University of New South Wales, Canberra, ACT, Australia
[2]CSIRO Manufacturing Flagship, Clayton, VIC, Australia
[3] Mark Wainright Analytical Centre, Electron Microscope Unit, UNSW Australia, NSW, Australia

Keywords: magnesium, scandium, Split Hopkinson Pressure Bar, microstructure, strain rate

Abstract

A magnesium alloy of nominal composition Mg-9Al-1Zn-0.2Sc was formed into plates by die casting and underwent annealing and T4 condition heat treatments to investigate the mechanical response of varying microstructures at strain rates from 10^{-4}-10^{3}/s in tension and compression. Full microstructural characterization was performed using optical microscopy, electron backscatter diffraction and energy dispersive x-ray spectroscopy. Quasi-static and dynamic testing was performed using a universal testing machine and a Split Hopkinson Pressure Bar in conjunction with digital image correlation for strain field mapping. Characterization and mechanical testing indicates that the T4 condition has the highest overall strength due to small equiaxed grains, a decrease in the size of β-$Mg_{17}Al_{12}$ phase at the grain boundaries, and an increase in the size of scandium intermetallics. Testing indicates an increase in strain hardening for dynamic compression and strain rate dependence in tension; failing suddenly due to casting defects dominating the fracture mechanics.

Introduction

Due to the need for lightweight metals in the aerospace and automotive industries, the interest in magnesium alloy systems has received renewed attention [1]. However, the degradation of mechanical properties at elevated temperatures, as low as 373K, limits their use in many applications [14].The addition of rare earth metals to magnesium alloys, specifically Scandium, has been shown to improve yield strength at both room and elevated temperatures during quasi-static testing [2-4, 15]. The strengthening effects have been concluded to be a result of limiting the formation of β-$Mg_{17}Al_{12}$ phase due to the formation of the Mg_5Al_4Sc intermetallic compound at grain boundaries, which is effective at inhibiting dislocation motion at room and elevated temperatures [2].

The effects of scandium on the dynamic properties of magnesium alloys has yet to be explored significantly, as well as the effects of heat treatments have on the mechanical properties of scandium containing magnesium alloys [2]. The primary aim of this study is to investigate the effects of scandium on the mechanical properties of magnesium alloys for quasi-static and dynamic loading conditions for multiple thermal processing conditions.

Experimental Methods

A magnesium alloy of nominal percent weight composition Mg-9Al-1Zn-0.2Scwas cast as a plate in a tool steel mold was received from the Commonwealth Scientific and Industrial Research Organization (CSIRO).Two further heat treatments were performed on the received material in an inert CO_2 atmosphere to investigate effects on the mechanical properties due to grain size effects and the growth of second phases and intermetallics at the grain boundaries (GBs). The heat treatments investigated were classified as: fully annealed (723K for 24h then furnace cooled) and the T4 condition (648K for 3h, 688 K for 18h, then quenched in water) [2,5].

The characterization of the material has been undertaken using optical microscopy, electron backscatter diffraction (EBSD) and energy dispersive x-ray spectroscopy (EDS). The EDS was performed in parallel with the EBSD data collection, which required the specimen to be at a tilt angle of 70°, theoretically making the data slightly less quantitatively accurate than standard EDS procedure carried out at 0° tilt. Analysis of the EBSD data of for all three material conditions revealed a distinct lack of crystallographic texture as shown by the inverse pole figures (IPFs) in Fig. 1a through 1c; which was expected due to the cast nature of the material. The average area grain sizes for each microstructure were as follows: 380μm for as-cast, 319μm for annealed, and 127μm for T4. The grain shapes for all material conditions were fairly equiaxed, with the T4 condition having a slightly dendritic grain structure. Data obtained from EDS clearly indicates significant differences in the composition of phases at the GBs within the annealed and T4 material conditions, as indicated by Fig. 1d through 1g. The annealed material condition contains greater amounts of the β-$Mg_{17}Al_{12}$ phase at the GBs than compared to the T4 condition. Additionally, the size of the strengthening Mg_5Al_4Sc intermetallic compound particles greatly increased in size from the annealed to T4 material conditions. This coupled with their respective β-phase amounts and differences in grain sizes, should lead to an increase in the mechanical properties in the T4 material from mechanical testing due to better GB and dislocation pinning. Significant casting defects in the form of voids were found though the thickness of the plates from optical microscopy. Optical microscopy revealed these casting defects to be as large as several millimeters in size on rare occasion. Samples tested in this work were visually inspected for the presence of defects at the surface, but it is impossible to discern if pores exist through the thickness before testing.

Tensile testing of the MgSc alloy was performed at a strain rate of 10^{-4} s^{-1} using a Shimadzu 50kN universal testing machine for all three material conditions at room temperature. The samples were cut from two different cast plates, A and B, and manufactured in accordance with ASTM B557M-14 sub-size specimens: cut with a 25mm gauge length, 6mm gauge width, and a thickness of 5mm. The gauge length was tapered to the grips with 6mm radius fillets. Strain was measured by laser extensometer and by digital image correlation (DIC) recorded using a Canon EOS 70D and analyzed using the commercially available VIC 2D software package and the open source nCorr software package. Black spray paint was used to create an optimized random speckle pattern required for DIC analysis.

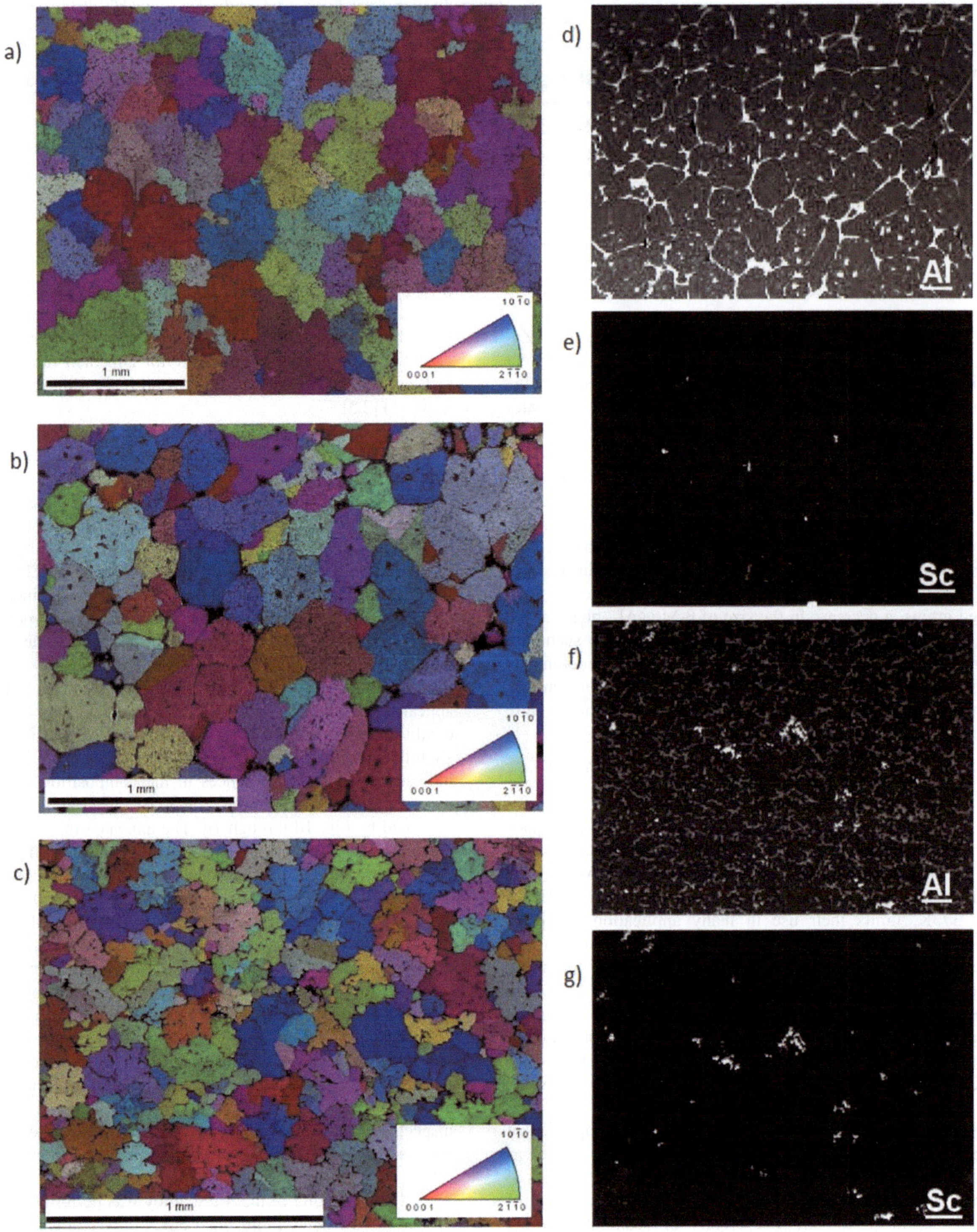

Figure 1. EBSD and EDS characterization of MgSc Microstructure: a) EBSD of the as-cast condition; b) EBSD of the annealed condition; c) EBSD of the T4 condition; d) EDS of aluminum content in the annealed microstructure; e) EDS of scandium content in the annealed microstructure; f) EDS of aluminum content in the T4 microstructure; and g) EDS of scandium content in the T4 microstructure.

Dynamic testing was performed at room temperature in compression and tension on a Split Hopkinson Pressure Bar (SHPB) using 12.7mm and 19.05mm C-350 maraging steel bars, respectively, equipped with strain gauges 914.4mm from the sample-bar interfaces to read the voltage history during testing. Compression samples were constrained by the height of the cast plate and measured 8.7mm in diameter and 5.2mm in length, meeting the design standards of maintaining a diameter to length ratio between 0.5 and 1 as well as maintaining a sample diameter to bar diameter ratio between 0.5 and 1 [12]. The faces of the compression specimens were lubricated to prevent friction from inhibiting strain in the radial direction and ensure that specimens were only subjected to uniaxial stress [10]. Tensile specimens were designed as dogbone specimens to fit into mechanical grips that thread directly onto the 19.05mm incident and transmission bars. The dogbone specimens contained a gauge length of 10mm with a rectangular gauge cross section of 5mm in height and 3.2mm in depth, tapered with 6mm fillets. A Phantom v710 high speed camera operated at 60,000 frames per second was used to ensure the samples underwent uniform dynamic loading and DIC was attempted for dynamic tensile testing. The strain rates tested were: 700 and 1500 s^{-1} in compression and 800 s^{-1}in tension. Stress-strain relationships were calculated using the standard conversion from incident and transmission bar strain gauge voltage-time histories and may be found in Gray III [12].

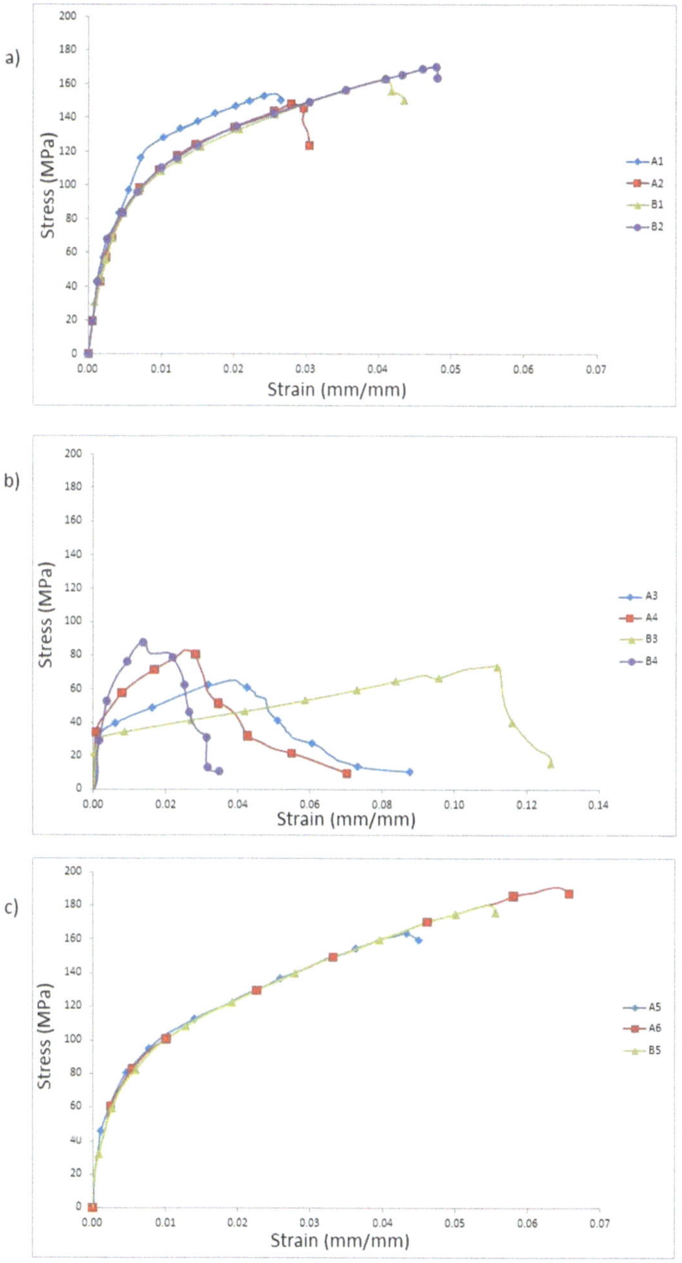

Figure 2. Quasi-static tension results: a) as-cast material; b) annealed material; and c) T4 heat treated material.

Results and Discussion

The stress-strain data obtained from quasi-static tensile tests exhibit a range of responses for the three material conditions, as seen in Fig. 2. The average 0.2% offset yield strength obtained from samples tested from plates A and B was highest in the T4 condition with a value of 61.9 MPa, with average yield strengths of 51.8 MPa and 36.1 MPa for the as-cast and annealed material conditions, respectively. The average ultimate tensile strength (UTS) was also highest in the T4 condition across both plates with a value of 174.8 MPa, with average UTS values of 158.6 MPa and 74.2 MPa for the as-cast and annealed material conditions, respectively. Using the as-cast condition as a reference material, there was a reduction in yield strength and UTS by ~30% and 53%, respectively, for the annealed material, whereas there was an increase in the yield strength and UTS by ~20% and 10%, respectively, for the T4 material.

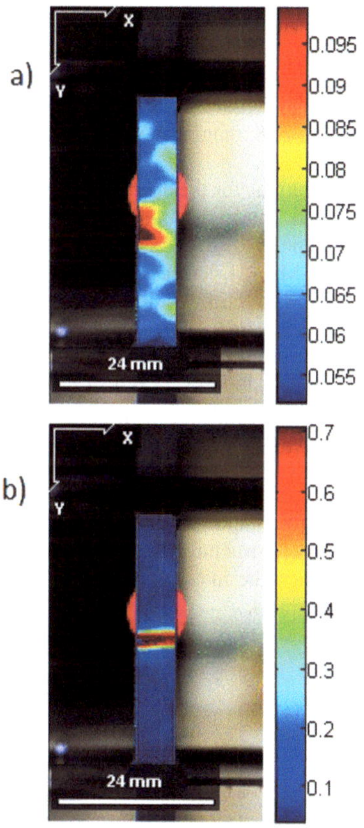

Figure 3. DIC images of quasi-static tension testing: a) immediately before failure; and b) after failure has occurred.

In a general sense the yield strength of the material was found to be closer in comparison to that of pure magnesium (55 MPa) than the yield strength expected from commercial magnesium alloys, such as AZ31 (113 MPa) [6,7]. Additionally, the yield strength and the UTS for similar materials undergoing the same annealing process as samples A3, A4, B3 and B4 have reported to be approximately 216 MPa and 341 MPa, respectively [2]. A significant contributor to the low yield strength values recorded is likely due to the presence of casting defects in the form of voids, which are known contributors to weakening similar materials in tension [8]. In Fig. 2b, the two specimens tested in the annealed condition from plate B exhibit drastically different stress-strain responses likely caused by local damage initiation, growth, and fracture around interior voids. Digital image correlation analysis showed competing strain regions at multiple points along the gauge length for most the samples tested, from which one such localized strain region became the source of failure, as can be seen by strain maps just before and immediately after failure of sample A6 in Fig. 3. The peak localized strain just before failure in sample A6 was around 0.095, about 70% of the total strain to failure captured by the laser extensometer.

As shown in Fig. 2a and 2c, the as-cast and T4 material conditions exhibit an increase in flow stress throughout the deformation process with failure occurring just after the UTS for all specimens, indicative of strain hardening mechanisms followed by a brittle fracture mode. The average strains to failure for the as-cast and T4 material conditions were 0.038 and 0.056, respectively. Although the annealed material condition had much lower yield strength and UTS, the samples experienced similar, but high in variance, strain hardening as the other material conditions up to similar strains. However, the annealed samples did not fracture immediately after reaching the UTS as the other two material conditions and continued to strain to an average strain failure of ~0.08, ranging from 0.036 to 0.127, indicative of increased ductility.

The explanations for the decrease in yield strength and UTS for the annealed material condition lies primarily with the microstructure along the GBs. As previously stated, an increase in the size of the β-$Mg_{17}Al_{12}$ eutectic phase at the GBs is present in the annealed microstructure, which is the mechanical weak link within the material. The T4 microstructure has a smaller overall grain size, increasing the strength of the material from the well-known Hall-Petch relationship and also resulting in a more even dispersion of smaller regions of β-phase along the GBs. The Mg_5Al_4Sc intermetallic strengthening compounds are also larger in size for the T4 condition, resulting in an increase in GB and dislocation pinning capabilities of the material. These scandium-laden intermetallic compounds are smaller and more dispersed in the annealed microstructure, which does not allow them to properly strengthen the material given the large average size of the β-phase and grains.

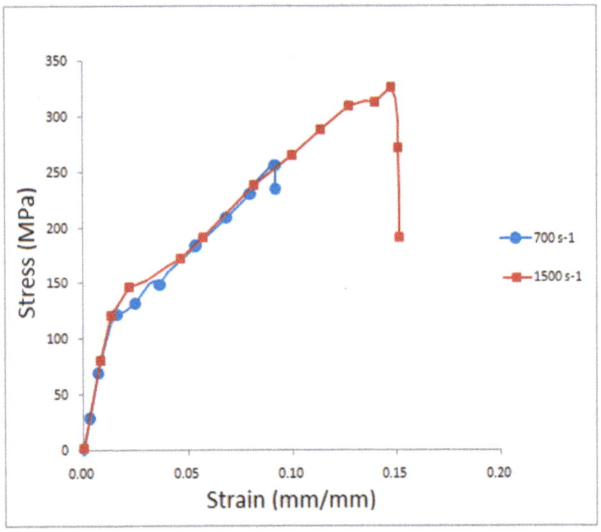

Figure 4. Dynamic compression results for the as-cast material.

Interestingly, there is an increase in ductility in the annealed material, which is hypothesized to be caused by the fracture surface preferring to follow the weak β-phase through the material, resulting in a potentially longer path to fracture indicated by a jagged, inclined fracture surface from visual inspection. A study by Yan et al. [9] showed that a magnesium alloy under low triaxiality constraints of high shear stresses resulted in a ductile fracture surface of increased inclination with the loading direction as compared to high triaxiality constraints leading to brittle fracture perpendicular to the loading direction. Though the apparent embrittlement was due solely to high constraint conditions in [9], there are comparisons to be made with the state of shear stress components and increased ductility from the inclined fracture surfaces in the annealed samples here.

Figure 4 shows the mechanical response of the as-cast material condition in compression at strain rates of 700 s^{-1} and 1500 s^{-1} via SHPB testing. The material exhibits negligible strain rate dependence for the two strain rates tested. A Phantom v710 high speed camera confirmed that the addition of a large mass momentum trap on the incident bar ensured a single deformation pulse imparted on the samples, thus, the drop in stress near the maximum strains shown in Fig. 4 are not material failure, but the samples beginning their free-fall descent away from the pressure bars. The material shows strain hardening up to approximately double that of the estimated yield stress of ~125-150 MPa, which is similar to values known for AZ31 [7] and similar peak stresses obtained from other studies of a similar material [11]. The increase in total strain for the 1500 s^{-1} shot condition is to be expected since there is a larger force imparted on the specimen from an increase in impactor velocity and nearly identical stress pulse duration from the 700 s^{-1} shot. These high strain rate compression tests were able to obtain yield strengths approximately double that of those found from the quasi-static uniaxial tensile testing for the same material condition, providingpromising preliminary evidence for low velocity impact resistance of this alloy.

The mechanical response of the material subject to 700 s^{-1} strain rate tensile SHPB testing indicate poor performance, as the material fractured with negligible gauge length elongation and at a low yield stress. Figure 5a shows the stress-strain response of all three material conditions. Note that it is difficult to impossible to accurately determine the precise Young's modulus and yield stress in SHPB testing, particularly in tension, since the sample does not achieve a uniform state of stress until somewhere after 1-2% plastic strain [12]. It is also important to note that the stress-strain data shown in Figs. 4 and 5 have not undergone any post-processing. The annealed sample is the only one where the yield strength can be somewhat estimated: between 40-70 MPa, which is similar to that material conditions' average quasi-static tensile yield strength of 36.1 MPa.

The samples had an applied speckle pattern for DIC analysis, however, only a few frames exist over the duration of the failure process resulting and was abandoned for the current study. Figures 5b and 5c show the a sample just before and just after fracture, note the lack of elongation in the gauge length. The other specimens exhibited similar fracture processes, but failed more centrally in the gauge section. Two more specimens were tested in tension, but the signal was too weak in the transmission strain gauge to obtain any stress values. It is apparent that the casting defects dominate the fracture mechanics at intermediate strain rate tensile testing as compared to the quasi-static testing. Further studies using casts with limited defects are required to obtain useful high strain rate tensile data.

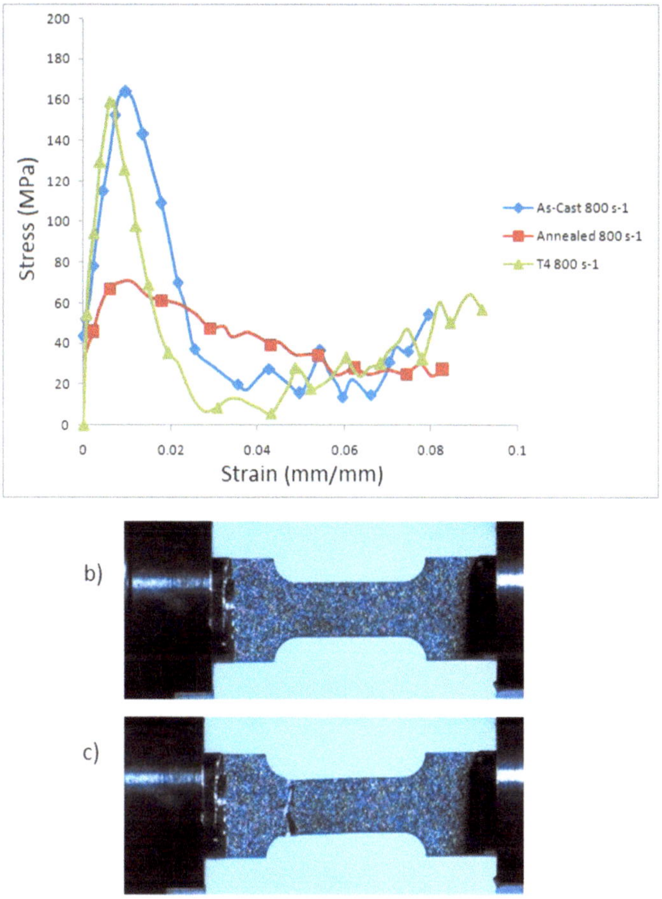

Figure 5. Dynamic SHPB tensile testing: a) stress-strain data for all three material conditions; b) high speed camera footage immediately before failure for the as-cast condition; and c) immediately after failure for the as-cast condition.

Conclusions

Quasi-static and dynamic testing of a cast MgSc alloy of varying microstructures was performed over a range of strain rates from 10^{-4} to 10^{3} in both compression and tension using a conventional universal testing machine and the SHPB technique. Quasi-static tensile tests indicate that the T4 material condition exhibits the most promising mechanical properties due to small β-phase present at the GBs, an increase in the size of scandium intermetallics at the GBs, and a decrease in the average grain size compared to the as-cast and annealed material conditions. Compression testing at two higher strain rates demonstrated an increase in approximate yield stress and more pronounced strain hardening mechanisms over the load duration. Preliminary dynamic tensile testing revealed that the casting defects present in the material dominate the fracture mechanics at intermediate to high strain rates, requiring stock material with fewer defects to properly characterize the material response.Future mechanical testing will be conducted at elevated temperatures for all material conditions and strain rates to gain a broader understanding of the mechanical strengthening potential of scandium in magnesium alloys.

Acknowledgements

This work was partially funded by a UNSW Canberra Rector's Start-Up Grant. The authors would like to acknowledge the technical support and guidance provided by members of the technical service group at UNSW Canberra: Jim Baxter, Patrick Nolan, David Sharp and Thomas Thompson.

References

[1] Mordike, B., & Ebert, T. (2001).Magnesium, Properties – applications – potential. *Materials Science and Engineering.*302 (1), 37-45.

[2] Xiao, D., Song, M., Zhang, F., & He, Y. (2009). Characterization and preparation of Mg-Al-Zn alloys with minor Sc. Journal of Alloys and Compounds. 484 (1), 416-421.

[3] Shalomeev, V., Lysenko, N., Tsivirko, E., Lukinov, V. & Klochikhin, V. (2008). Structure and Properties of Magnesium Alloys with Scandium.Metal Science and Heat Treatment. 50 (1-2), 34-37.

[4] Wang, S., Chou, C., Fann, Y. (2008). Microstructures and mechanical properties of modified AZ31-Zr-Sc alloys. Materials Science and Engineering A. 485 (1), 428-438.

[5] Cizek, L., Greger, M., Pawlica, L., Dobrzanski, L., &Tanski, T. (2004). Study of selected properties of magnesium alloy AZ91 after head treatment and forming. Journal of Materials Processing Technology. 157-158, 466-471.

[6] Diez, M., Kim, H., Serebryany, V., Dobatkin, S., & Estrin, Y. (2014). Improving the mechanical properties of pure magnesium by three-roll planetary milling.*Materials Science & Engineering A.* 612, 287-292.

[7] Dowling, N. (2013). *Mechanical Behavior of Materials.*(4th ed.). (P.65). New Jersey, NJ: Pearson Education.

[8] Weiler, J., Wood, T., Klassen, R., Maire, E., Berkmortel, R. & Wang, G. (2005). Relationship between internal porosity and fracture strength of die-cast magnesium AM60B alloy. Materials Science and Engineering A. 395 (1-2), 315-322.

[9] Yan, C., Ma, W., Burg, V. & Chen, M. (2007). Experiemental and numerical investigation on ductile-brittle fracture transition in a magnesium alloy. Journal of Materials Science. 42 (1), 7702-7707.

[10] Gray, G. (2012). High-Strain-Rate testing of Materials: The Split-Hopkinson Pressure Bar. In Kaufmann, E. (Ed.), *Characterization of Materials.*(2nd ed.). (pp 1-15). New Jersey, NJ: John Wiley & Sons.

[11] Ishikawa, K., Watanabe, H., & Mukai, T. (2005). High strain rate deformation behaviour of an AZ91 magnesium alloy at elevated temperatures.*Materials Letters.*59 (12), 1511-1515.

[12] Gray, G., *ASM HandbooksMechanical Testing and Evaluation*, 2000, vol. 8, pp. 462-476

[13] Ramesh, K., *Springer Handbook of Experimental Solid Mechanics*, 2008

[14] Zhang, J., Guo, Z., Pan, F., Li, Z., & Luo, X. (2007). Effect of composition on the microstructure and mechanical properties of Mg-Zn-Al alloys.*Materials Science and Engineering A.* 456 (1), 43-51.

[15] Von Buch, F., Lietzau, J., Mordike, B., Pisch, A., &Schmid-Fetzer, R. (1999). Development of Mg-Sc-Mn alloys. *Materials Science and Engineering A.* 263 (1), 1-7.

Magnesium Technology 2016

SYMPOSIUM:

Magnesium-based Biodegradable Implants

Magnesium Technology 2016

MAGNESIUM-BASED BIODEGRADABLE IMPLANTS

Materials and Processing / Surface Modification and Corrosion

Magnesium Technology 2016
Edited by: Alok Singh, Kiran Solanki, Michele V. Manuel, and Neale R. Neelameggham
TMS (The Minerals, Metals & Materials Society), 2016

FABRICATION, TESTING AND PERFORMANCE OF RARE EARTH-CONTAINING MAGNESIUM BIODEGRADABLE METALS

Dong Bian[1] and Yufeng Zheng[1]
[1] Department of Materials Science and Engineering, College of Engineering, Peking University, China

Keywords: rare earth, magnesium, biodegradable metals

Abstract

In the present work, various Mg-RE (Sc, Y, La, Ce, Pr, Nd, Sm, Eu, Gd, Tb, Dy, Ho, Er, Tm, Yb, Lu) binary alloys were fabricated by casting. The microstructures and mechanical properties have been studied, and their degradation behaviors in Hank's solution have also been tested. The aim of this work is to screen the appropriate alloying elements for the rare earths elements in biomedical magnesium.

Introduction

For a long tradition, the addition of rare earth elements (REEs) into metallurgy is used to improve magnetic, electrical, mechanical and some other properties of various alloys [1]. In magnesium, the addition of REEs can refine grains, purify melt, improve the corrosion resistance, and improve mechanical property [2]. For magnesium alloys devoted for biomedical purposes, the addition of REEs is supposed to increase mechanical property, enhance corrosion resistance and improve biocompatibility. Elements, such as Y, Nd and Gd, have been widely added into biodegradable magnesium [3]. However, only limited studies were performed on other REEs. In this work, a comparative study of REEs in magnesium was carried out, trying to find out the appropriate REEs in biomedical magnesium.

Materials and methods

Various Mg-RE binary alloys were fabricated by conventional gravity casting using pure magnesium (99.95 wt.%) and commercial pure REEs. Their microstructures were observed under an optical microscope (Olympus BX51M). Phase constituents were characterized via x-ray diffraction (XRD, Rigaku DMAX 2400, Japan). Mechanical properties were characterized by micro-hardness tests (HMV-2T, Shimadzu corporation, Japan). Electrochemical measurements were carried out in Hank's solution using an electro-chemical workstation (PGSTAT 302N, Metrohm Autolab). A three electrodes cell with a platinum counter-electrode and a saturated calomel electrode (SCE) as the reference electrode was utilized for electrochemical test. Immersion corrosion tests were also carried out to study the corrosion behaviors in Hank's solution.

Results

Microstructures and phase constituents

Figure 1 displays the typical microstructures of the as-cast Mg-RE alloys. Microstructures of the remaining alloys are quite similar to Mg-1Sc or Mg-3Y. Except for Mg-3Ce and Mg-3Eu, coarse grains (grain size in millimeter scale) can be found in all the remaining as-cast Mg-RE alloys. Large grains were derived from the low cooling rate during casting because natural cooling in the furnace was adopted. Plenty of second phases can be found in alloys with high RE addition, higher than their solid solubility limits. Phase constituents of the binary alloys are summarized in Table 1.

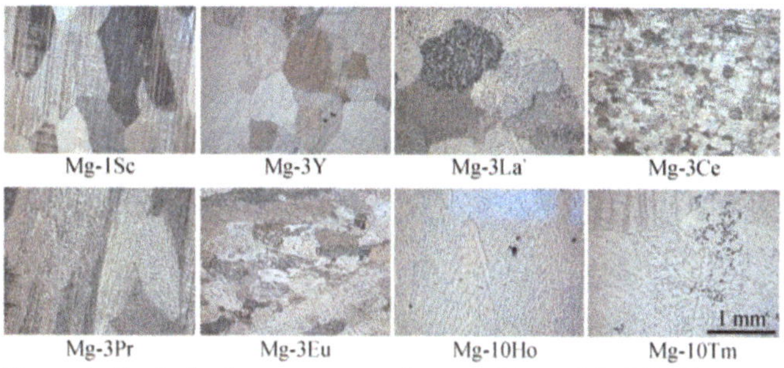

Figure 1. Typical microstructures of as-cast binary Mg-RE alloys.

Table 1. Phase constituents of the as-cast Mg-RE alloys identified by XRD

Alloy	Mg	Mg-1Sc	Mg-3Sc	Mg-1Y	Mg-3Y
Phases	α	α	α	α	α+$Mg_{12}Y$
Alloy	Mg-1La	Mg-3La	Mg-1Ce	Mg-3Ce	Mg-0.5Pr
Phases	α+$Mg_{12}La$	α+$Mg_{12}La$	α+$Mg_{12}Ce$	α+$Mg_{12}Ce$	α
Alloy	Mg-1Pr	Mg-3Pr	Mg-1Nd	Mg-3Nd	Mg-1Sm
Phases	α+$Mg_{12}Pr$	α+$Mg_{12}Pr$	α	α+$Mg_{12}Nd$	α+Mg_2Sm
Alloy	Mg-3Sm	Mg-1Eu	Mg-3Eu	Mg-1Gd	Mg-5Gd
Phases	α+Mg_2Sm	α+$Mg_{17}Eu_2$	α+$Mg_{17}Eu_2$	α	α
Alloy	Mg-1Tb	Mg-5Tb	Mg-1Dy	Mg-10Dy	Mg-1Ho
Phases	α	α	α+$Mg_{12}Dy$	α+$Mg_{12}Dy$	α
Alloy	Mg-10Ho	Mg-1Er	Mg-10Er	Mg-1Tm	Mg-10Tm
Phases	α	α	α	α	α+$Mg_{24}Tm_5$
Alloy	Mg-1Yb	Mg-3Yb	Mg-1Lu	Mg-10Lu	/
Phases	α	α+Mg_2Yb	α	α	/

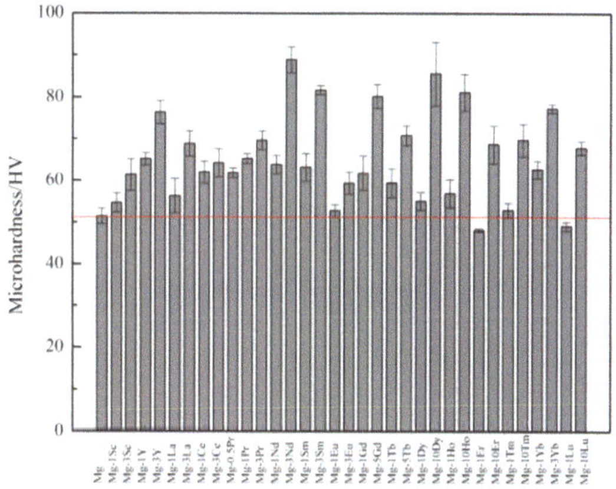

Figure 2. Micro-hardness of the as-cast Mg-RE alloys.

Mechanical properties

Mechanical properties of the binary Mg-RE alloys were evaluated by micro-hardness tests, seeing Figure 2 here. Except for Mg-1Er and Mg-1Lu, the addition of REEs all improves the microhardness of magnesium. For the same RE, the more it was added, the better the micro-hardness is. Mg-3Nd exhibits the highest micro-hardness with the value of 89 HV, significantly higher than that of pure magnesium (51 HV).

Electrochemical corrosion

Figure 3 shows the corrosion rates of various Mg-RE alloys in Hank's solution, calculated from electrochemical corrosion data. As-cast Mg-1Sm exhibits the slowest corrosion rate of 0.19 mm/y. The highest corrosion rate is found in Mg-10Er alloy with the value of 0.94 mm/y and it shows the largest error fluctuation. Large fluctuation may be caused by the high Er content and its inhomogeneous distribution. Within the range of RE addition in this work, Nd, Sm, Eu, Gd, Ho and Yb all improve the corrosion resistance of magnesium (99.95 wt.%).

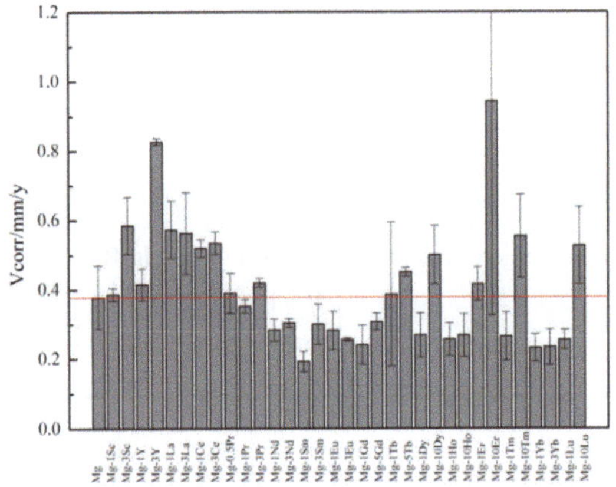

Figure 3. Corrosion rates of various as-cast Mg-RE alloys in Hank's solution calculated from electrochemical corrosion data.

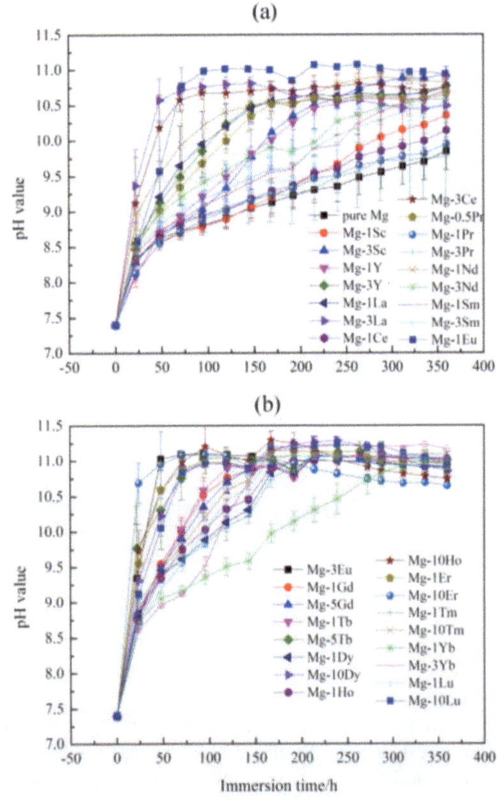

Figure 4. pH value monitor during the immersion corrosion test of as-cast Mg-RE alloys.

Immersion corrosion

During the immersion corrosion tests, pH values of the Hank;s solution were monitored. Figure 4 shows the pH value variation as a function of immersion time. It is clear that pH value variations of Mg-1Sc, Mg-1Pr, Mg-3Pr and Mg-1Ce are slow and similar to pure magnesium during the whole immersion period, showing relatively high corrosion resistance. Mg-1Y, Mg-3Sc, Mg-3Nd, Mg-1Sm, Mg-3Sm and Mg-1Yb show a little bit faster in the pH increasing than the aforementioned ones. The remaining alloys show a fast pH increasing in the first 100-150 h immersion and then pH become gradually steady around the value of 11, implying a fast degradation.

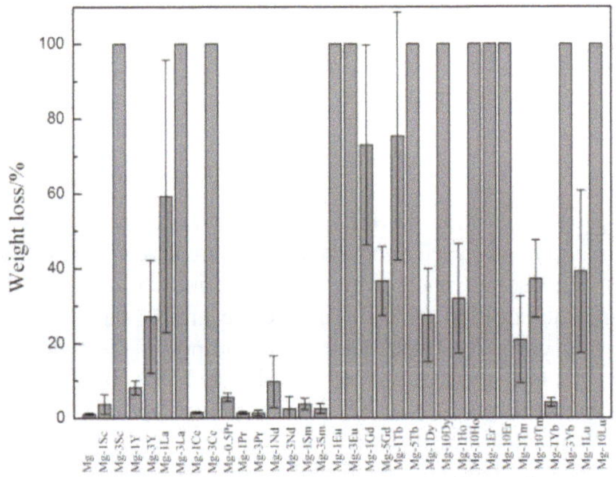

Figure 5. Weight loss of as-cast Mg-RE alloys after immersing in Hank's solution for 20 days.

Weight loss after 20 days immersion shows a similar trend to pH value variations, seeing Figure 5. Several alloys, such as Mg-3Sc, Mg-3La, Mg-3Ce and so on, have been completely dissolved or broken into particles.

Conclusions

(1) Coarse grains were formed in as-cast Mg-RE alloys because of the low cooling rate during natural cooling in the furnace.

(2) Except for 1 wt.% Er and 1wt.% Lu, the addition of REEs all improved the micro-hardness of magnesium.

(3) Some different results were found among the electrochemical corrosion and immersion corrosion tests and that could be related to the test duration because immersion test is a long time corrosion results. Corrosion resistance of Mg-1Sc, Mg-1Pr, Mg-3Pr, Mg-3Nd, Mg-1Sm and Mg-1Yb are superior to other Mg-RE cast alloys based on both electrochemical corrosion and immersion corrosion results.

References

1. R. Ferro, A. Saccone, G. Borzone, "Rare earth metals in light alloys," *Journal of Rare Earths,* 15 (1997), 45.

2. Y. Chen, Z. Xu, C. Smith, J. Sankar, "Recent advances on the development of magnesium alloys for biodegradable implants," *Acta Biomaterialia,* 10 (2014), 4561-4573.

3. N. Hort, Y. Huang, D. Fechner, M. Störmer, C. Blawert, F. Witte, et al., "Magnesium alloys as implant materials - Principles of property design for Mg-RE alloys," *Acta Biomaterialia,* 6 (2010), 1714-1725.

MANUFACTURING OF OSTEOSYNTHESIS SYSTEMS MADE OF MAGNESIUM ALLOY AZ91

Britta Hering, Andi Wippermann, Tobias Mörke, Thilo Grove, Berend Denkena
Leibniz Universität Hannover, Institute of Production Engineering and Machine Tools, An der Universität 2, 30823 Garbsen, Germany

Keywords: Mg implant, corrosion test, machining

Abstract

Resorbable magnesium alloys are promising materials for osteosynthesis systems. The work describes the influence of the manufacturing process on the degradation of Mg implant systems at different zones. To figure out if the degradation process can be influenced by surface properties, deep-rolling and ball-end milling processes are applied to machine a variety of surface integrity conditions. Corrosion behaviour is analysed by conventional immersion test as well as by drip testing. The latter process is developed in order to acquire timely results, and is based on measuring material loss during corrosion by using a profilometer.

Introduction

Magnesium alloys were initially tested as implant materials by the end of the 19th century. Edward C. Huse used magnesium wires as ligature for bleeding vessels [1]. Lambotte fixed an osteosynthesis plate made of magnesium with steel screws. Due to the development of extensive subcutaneous gas cavities, he conducted further investigations on supracondylar fractures of children, inserting Mg nails. Though the volume of the extra articular implant was smaller, he still discovered gas cavities. However, as they did not harm the patient and disappeared after several weeks, his results were promising overall [2, 3]. The most important advantage was the obsolete surgery for removing implants. Besides, Young's modulus as well as the mechanical properties of Mg and its alloys are similar to natural bone. Thus, stress shielding can be avoided. Magnesium ions released during degradation are essential for the human body and the excess above physiological level can be excreted by kidneys and intestine [4, 5].
Although magnesium is in general suitable as implant material, some requirements need to be fulfilled by an appropriate alloy. Biocompatibility of all alloying metals is most important. Ductility is required for ease of fabrication and strength is needed to offer adequate mechanical support to the injured tissue. Degradation rate should be moderate in order to maintain the mechanical integrity of the implant during the healing process [6]. At best, the implant bears the whole load at the beginning of the healing process but degrades gradually to stimulate the recovery of the bone. The degradation can be adjusted by the composition and microstructure of the material and can also be controlled by the surface integrity of the implants introduced into the workpiece during machining [7, 9]. According to Lucas, a large microstructured surface accelerates the corrosion process, whereas compressive residual stresses reduce the degradation rate of MgCa-alloys. Nevertheless, this effect strongly depends on the composition of the alloy [7]. The machining process, tools and setting parameters can determine surface characteristics. Experiments conducted by Salahshoor et. al., Lucas and Denkena et al. confirm this fact [7, 9, 10, 11].
Additional challenges arise for implant systems, as different areas have to be distinguished when tailoring the corrosion behaviour. For example, the surfaces of intramedullary nails remain exposed to bone marrow. Only the heads of screws used to fixate the nail lie extraosseous and are consequently in contact with the blood flow. Osteosynthesis plates are fixated externally onto the bone. Thus, the major part of their surface is in contact with the blood flow and might show different corrosion behaviour to the contact zone between screw head and counterbore.
In order to find appropriate machining processes for different zones of an implant system, the effect of ball-end milling and deep-rolling on the surface properties is determined and resulting corrosion behaviour is investigated in this work. On the one hand, an immersion test is used to evaluate the corrosion behaviour of the contact zone. On the other hand, a new test is developed allowing for a quick feedback on the degradation behaviour of the surface after a corrosion time of three minutes.

Material and Methods

In this work, the Mg alloy AZ91 is used that contains 89.0 wt% magnesium, 9.0 wt% aluminum and 0.5 wt% zinc. The alloy is produced with pure magnesium (Magnesium Electron Limited, UK, 99.94 %) by gravity die casting and subsequent hot extrusion process. Cuboid samples (40 mm x 15 mm x 15 mm) are cut out of the resulting bar stock as this geometry can safely be clamped in a milling machine (MC16, HELLER).

Machining Process

Deep-rolling and milling processes with spherical end mills are investigated in these experiments. The samples are face-milled first (R220.13-0050-12, SECO, d = 50 mm) with four inserts (SEKR1203AFFN). The cutting velocity v_c is fixed at 470 m/min, the feed per tooth f_z at 0.1 mm and the depth of cut a_p is 0.7 mm. Subsequently, the samples are either deep-rolled or processed by a ball-end miller. For deep-rolling a HG3 ball (ECOROLL) is used. The process pressure varies between 50 bar and 400 bar and the overlap of adjacent lines lies between 0 and 75 %. For ball-end milling two different tool diameters, namely 1 mm and 3 mm, are used. The overlap of adjacent lines is fixed at 50 % or 75 %, and the feed per tooth f_z is 0.003 mm or 0.006 mm. In order to represent the contact zone of screw head and counterboring in a simplified form, cuboid workpieces are processed on one face and are put face to face to another prepared sample for further investigations.

Surface Analytic

Surface roughness is investigated by a tactile measurement device of MAHR according to DIN EN ISO 4288. To investigate surfaces of immersed samples before and after corrosion, a digital microscope type VHX-600DSO (KEYENCE) is used. The cross section of the corroded cuboids is also documented. In order to analyse the grain size, samples are embedded in epoxy raisin, the cross section is polished and etched with nitric acid (2 %), before investigating via microscope.
An optical profilometer (NANOFOCUS µScan) is used to analyse the surfaces of samples used for the quick test.

Investigation of Corrosion Behaviour

Quick test In order to determine corrosion resistance of a surface, a drop test is used as described basically by Denkena et al. [8]. It is based on the fact that during corrosion magnesium is oxidized and Mg^{2+}-Ions are released, which results in a change in the surface structure.

In this work the method has been extended. A solution of simulated body fluid (SBF) is prepared according to Kokobu et al. [12]. Firstly, the surface is cleaned with ethanol and an area A of approximately 1 mm^2 (length x: 25 mm, width y: 4 mm) is scanned by µScan (NANOFOCUS) with a resolution of 10 µm. Secondly, three droplets of SBF are applied by syringe. The initial pH value is 7.4 but it is not controlled during the test. After a reaction time of three minutes, the fluid is removed and the same area is investigated again. During the reaction, a degradation of magnesium as well as the forming of solid products like $Mg(OH)_2$, MgO and apatic phases are expected, leading to a change in the surface structure. The change in volume is calculated by Matlab Version R2014b. The routine is developed in-house. Standard Matlab tools such as the fitting toolbox including the fit function and fittype ('poly11') is applied to align the surfaces. If surface measurements show small tilting or translational shift, the data are rotated or shifted by common matrix operators. Following recommendations of the supplier of our measurement system, the data are filtered before processing. Again, common 3 by 3 median filter is used for noise elimination. A 7 by 7 median filter is used to smoothen the data. The procedure is explained in more detail and illustrated in figures 1-3.

As it is not possible to position samples totally parallel into the mounting bracket, the first step consists of aligning the investigated areas. Figure 1 shows on the left-hand side raw data of the scanned surfaces before (top) and after the corrosion test (bottom) and on the right-hand side after alignment.

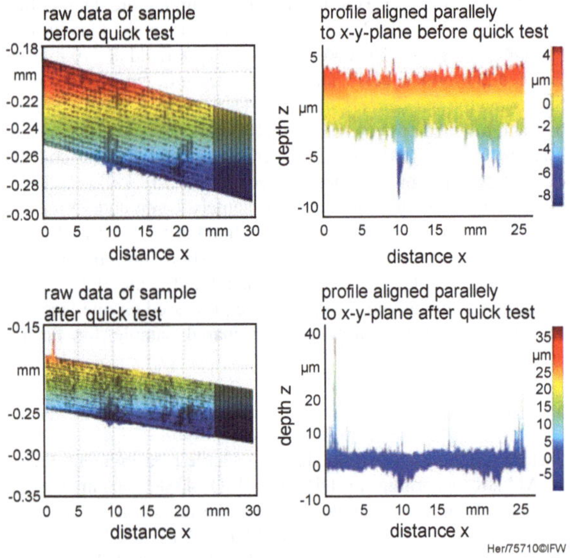

Figure 1. Surface information before and after corrosion. Left: raw data, right: aligned data.

In the following, two significant points that are found in the scans before and after corrosion are marked (Figure 2). Then, both scans are rotated and shifted until they match.

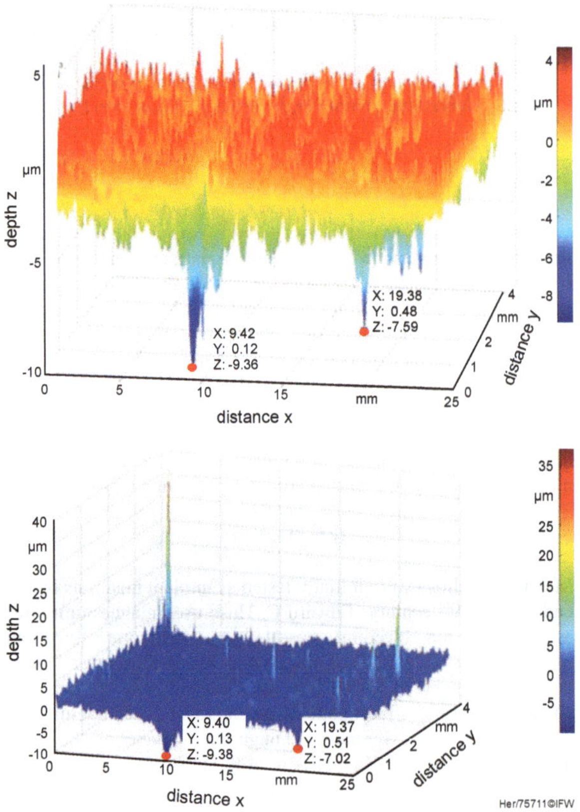

Figure 2. Marking significant points before shifting and turning both scans.

Afterwards the surface data of the second scan is subtracted from the data of the first. Figure 3 shows the resulting image.

The areas of the droplets can easily be distinguished from the remaining surface. Within these uncorroded areas three squares are marked (Figure 3). Their surface information helps to define the initial medium height level. In the last step, the difference in volume before and after corrosion is calculated. Therefore, at least three squares per sample are chosen within the droplet areas (not shown).

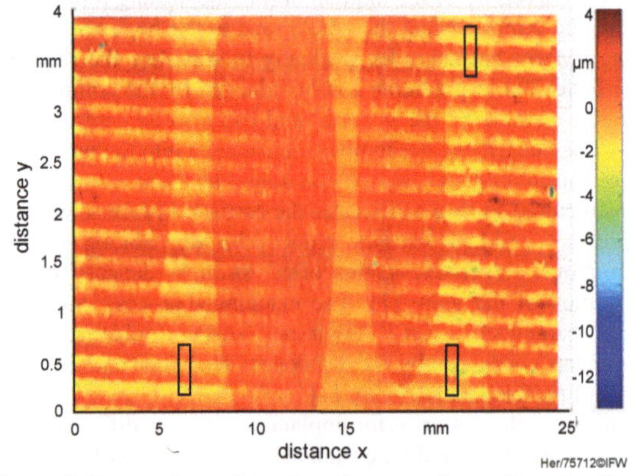

Figure 3. Image after subtracting the second scan from the first.

Immersion Test In addition to the corrosion resistance of the surface, the corrosion process in the contact zone of screw head and counterbore is investigated. Two processed surfaces are pressed against each other by using a holder consisting of plates and screws. A torque of 3 Nm is applied as the resulting pressure corresponds to the pressure between screw and counterbore when fixating an implant onto tibiae of rabbits [13]. To avoid corrosion of unprocessed parts, they are protected by a lacquer. Eight samples are tested at the same time. These are put into 25 l SBF tempered at 37 °C possessing a pH-value of 7.4 (Figure 4).

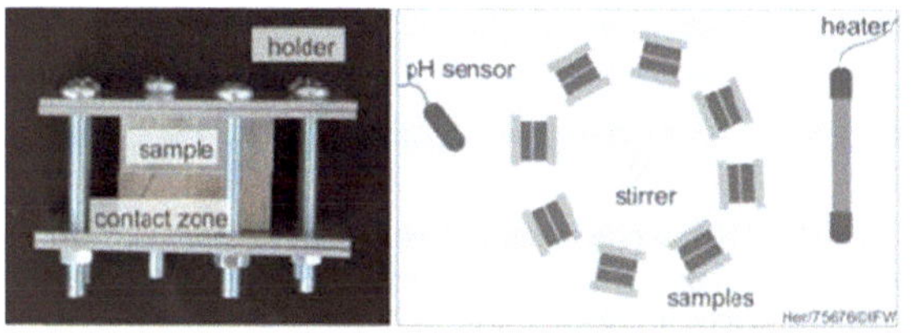

Figure 4. Left: two cuboids arranged face to face in holder; right: whole set-up in box filled with SBF.

Temperature, pH-value and level of the slightly stirred (100 rpm) solution are checked and adjusted daily. Samples are removed after 144 h, washed with distilled water and ethanol and dried at 80 °C for 48 h (oven type: UT6060, HERAEUS). In order to evaluate the corrosion process, contact areas as well as polished cross sections are investigated under the microscope.

Results and Discussion

Influence of Deep-rolling and Ball-end Milling on Surface Properties

During deep-rolling, passive forces depend on rolling pressure p_w. According to Röttger et al. [14], the theoretical force (F_p) basing on the tool radius r_b is calculated by:

$$F_p = p_w \cdot \pi \cdot r_b^2 \qquad (1)$$

In Figure 5 theoretical values are compared to experimental data. As the overlap does not have an effect on the passive force, a medium value is given.

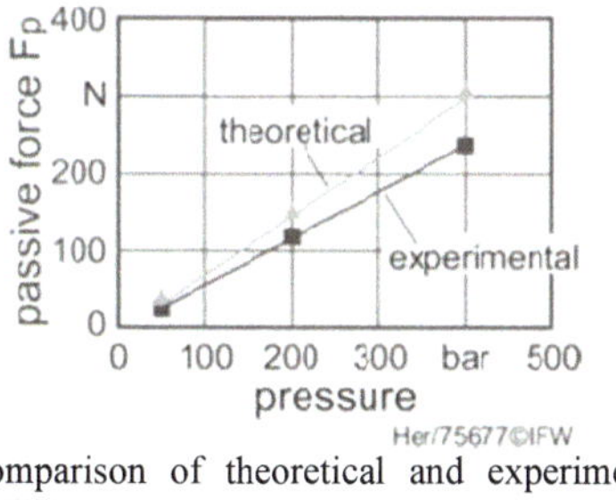

Figure 5. Comparison of theoretical and experimental forces during deep-rolling.

Calculated and experimental forces are directly proportional to pressure. Nevertheless, the theoretical value is always higher.
In the following, the relation between deep-rolling parameters and surface roughness is investigated. Figure 6 shows that with increasing pressure as well as with decreasing overlap the roughness of the samples increases. For deep-rolling with 200 bar and 400 bar the roughness is reduced to 1/5 when augmenting the overlap from 0 % to 75 %. Additionally, a higher pressure leads to a wider range of roughness.

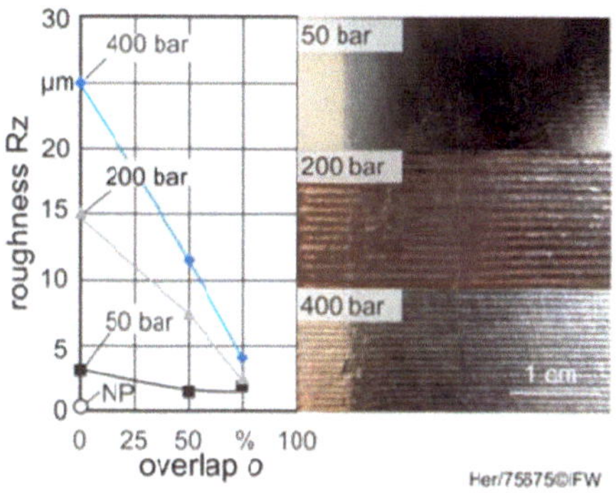

Figure 6. Left: influence of deep-rolling parameters on roughness. For comparison a non-processed (NP), only pre-milled sample (f_z = 0.2 mm) is given. Right: photos of samples before corrosion test.

Figure 7 shows the influence of ball-end milling on the surface properties. In comparison to deep-rolled surfaces, the roughness is much higher. As expected, it depends on the diameter of the tool as well as on the overlap of adjacent lines.

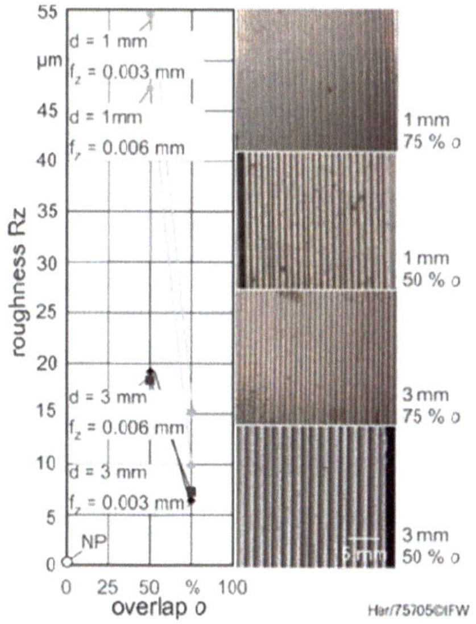

Figure 7. Left: influence of ball-end milling parameters on roughness. For comparison a non-processed (NP), only pre-milled sample (f = 0.2 mm) is given. Right: photos of samples before corrosion test.

In case of the 1 mm tool, the roughness is less than a third when augmenting the overlap from 50 % to 75 %. For the 3 mm ball-end miller the roughness is halved for the higher overlap. Thus, the tool diameter has a vital impact on the roughness.

Effect of Surface on Corrosion in Contact Zone

When fixating osteosynthesis plates, different environments can be distinguished that might lead to different corrosion processes.

In contrast to the main part directly exposed to blood, the exchange of ions is limited in the contact zones between counterboring of the plate and head of the fixating screw. Thus, during corrosion a concentration gradient develops. In order to represent the resulting conditions, a simplified set-up is used (Figure 4). Figure 8 shows the deep-rolled samples after the immersion test (144 h).

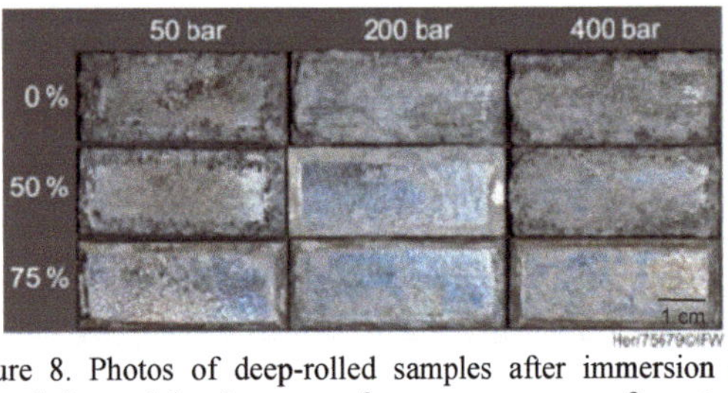

Figure 8. Photos of deep-rolled samples after immersion test. From left to right: increase of process pressure; from top to bottom: increase of overlap.

A smaller overlap generates worst surfaces for corrosion resistance. Besides, samples deep-rolled with a pressure of 50 bar or 400 bar and an overlap of 75 % look spotty and surfaces processed with 200 bar possess largest intact areas. Although machining paths form grooves along the surface, no direction of propagation of corrosion is preferred.

Besides, polished cross sections of all samples are investigated under the microscope (Figure 9). The microscope images support the findings already mentioned. Additionally, Figure 9 reveals how pitting corrosion proceeds. However, in case of sample presented in the center, the results mean a large discrepancy to the findings mentioned above. Conspicuously, an overlap of 75 % is necessary to generate surfaces that can resist the corrosive attack. Combining all observations, it is concluded that promising results are achieved with an overlap of 75 % at 200 bar. Though the surface processed at 50 bar seems to be intact at first glance, a homogenous removal is obvious with a closer look to the attacked surface (Figure 8).

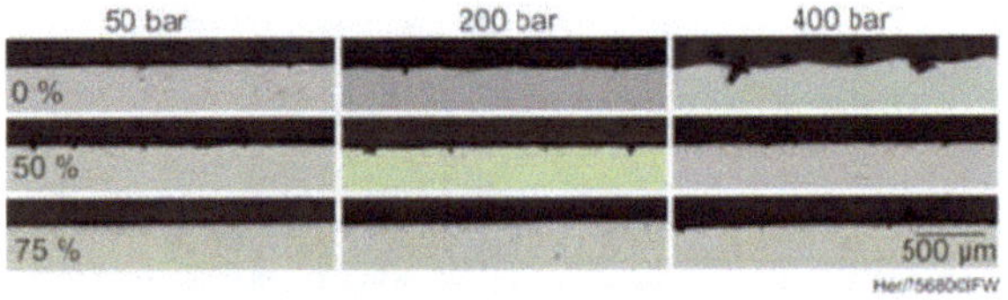

Figure 9. Microscope images of polished cross sections of deep-rolled samples after immersion test. From left to right: increase of process pressure; from top to bottom: increase of overlap.

In comparison to deep-rolled samples, surfaces machined by ball-end milling show lower resistance to corrosion in the contact area. Images of the latter are shown in Figure 10.

Figure 10. Photos of ball-end milled samples after immersion test. Effect of tool diameter, overlap and feed are shown.

After corrosion all surfaces look severely attacked. Within this test series, surfaces prepared with a tool of a diameter of 3 mm remain more intact. But from the cross-section given in Figure 11 pitting corrosion is obvious. Here, surfaces of all samples machined with a lower feed (second row) are less corroded. The sample with wavy profile (second row, second from left) remains nearly intact. Besides, a higher overlap and therefore a smoother surface does not implicit an improved resistivity for ball-end milled samples. This is a contrast to the results observed with deep-rolled samples.

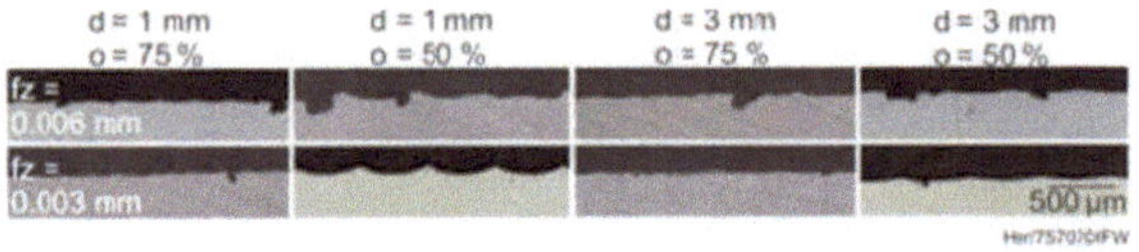

Figure 11. Microscope images of polished cross sections of ball-end milled samples after immersion test. Effect of tool diameter, overlap and feed are shown.

Effect of Surface Properties on Corrosion Behaviour Investigated via Quick Test

All samples are investigated using a quick test in order to get a timely result concerning corrosion resistivity of the surface. As no increase of volume can be detected for any sample, the deposition of solid corrosion products on the surface, as measure for the degradation process, is not useful. Consequently, the loss of volume due to corrosion is more meaningful. It is given in dependence of roughness Rz in Figure 12 and Figure 13. The deep-rolled samples are more corroded than the ball-end milled surfaces, though the latter have a much rougher surface. Not-processed surfaces (NP) show a high volume change.

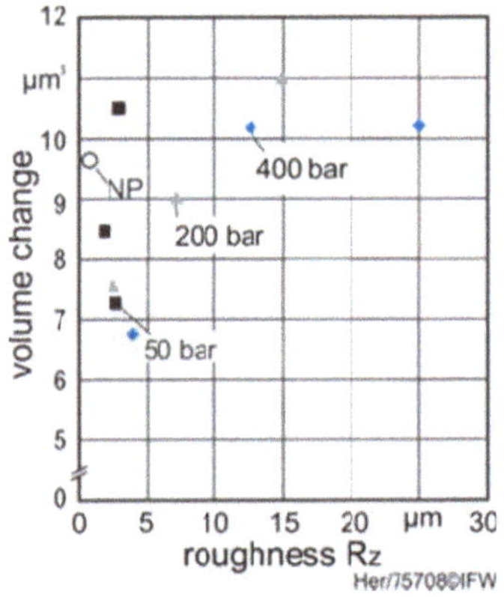

Figure 12. Volume change within droplet area due to corrosion of deep-rolled surfaces. Margin of error 0.1-1.0 μm^3. Overlap is varied (0 %, 50 %, 75 %) within series.

As already mentioned, during ball-end milling axial forces are comparably low. Here again, the overlap determines the roughness. Though samples prepared with the 1 mm tool are rough, the loss of volume due to corrosion is relatively low. Surfaces processed with a lower feed (0.003 mm) have higher resistivity against corrosion. This correlation is also discovered for samples investigated by immersion test (Figure 11). Therefore, the higher contact time of the tool densifies the surface.

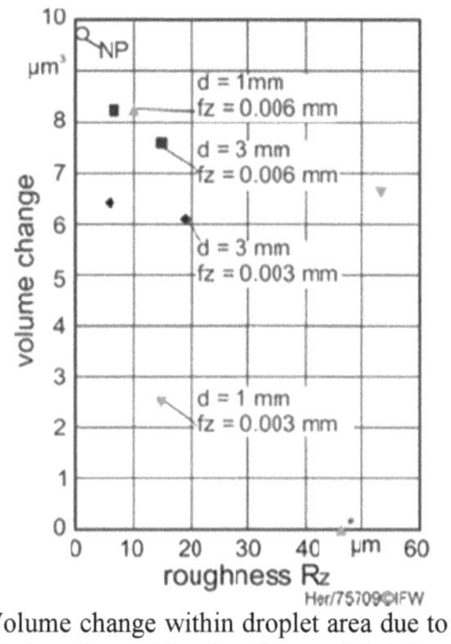

Figure 13. Volume change within droplet area due to corrosion of ball-end milled surfaces. Margin of error 0.1-1.0 μm^3; *Sample too wavy for method.

Thus, deep-rolling has a profounder impact on the surface than ball-end milling. During deep-rolling the axial force is much higher (25 N – 143 N, depends on the selected pressure) than during ball-end milling. The latter process keeps the forces below 5 N. As described in [7] and [15], higher forces during deep-rolling augment micro hardness and residual compressive stresses leading to a higher resistivity against corrosion. However, in this work ball-end milling has the opposite effect. Although microscope images show glassy faces with no indication of scratches or cracks (see also Figure 6), higher process forces might have destroyed the microstructure and consequently reduced the resistivity against corrosion. Because of grain sizes larger than 500 µm in diameter, the investigation of residual stresses is not possible for AZ91 samples. For example, the beam diameter of the X-ray diffractometer is 2 mm. Thus, only few grains are in the measured area and resulting reflections do not give reliable results. Therefore, a statement about the correlation of residual stresses and degradation behavior is impossible. Nevertheless, when comparing samples prepared with the same pressure, it is obvious that higher overlap with lower roughness reduces the volume change due to corrosion (Figure 12).

In contrast to work by Lucas et al. and Denkena et al. [7, 10], results presented here show that a smooth surface does not imply reduced reactivity of the surface. Lucas states that the effect of deep-rolling on the degradation behaviour depends on the alloy and its microstructure, whereas our work in addition suggests the dependency on the considered area. Deep-rolled, smooth surfaces in the contact zone block the entering of medium and prevent crevice corrosion more effectively than in case of the wavy profiles. Nevertheless, a wavy profile protects surfaces in direct contact to the medium. Probably, the rough surface helps to stabilize the protection layer formed during corrosion as described in [8] and [16].

The use of different methods to investigate the corrosion behaviour reveals that the standard immersion test is not sufficient for complete characterization of an implant system. Most of the authors concentrate on surfaces in direct contact to the corrosion medium [7, 8, 11, 17]. But this approach is only recommendable for investigating the corrosion behaviour of the alloy itself.

Though eudiometric investigations are common for evaluating the corrosion behaviour, these tests are also prone to error, as hydrogen molecules easily dissipate due to their size. Small variation in temperature and pH-value influences the solubility of hydrogen in the corrosion medium. Thus, the whole amount of hydrogen cannot be monitored. Although hydrogen is one of the corrosion products, Kuhlmann et al. have shown by in-vivo tests that gas cavities around implants are not filled with high concentrations of hydrogen [17]. Their results disprove the common misbelief and show how quickly hydrogen exchanges with the surrounding tissue. Additionally, they reveal the difficulty to draw conclusions from in-vitro experiments to in-vivo experiments. A quick test based on the change in the surface structure circumvents the hurdle of hydrogen formation and gives the result within a short period of time.

Conclusion

Results support the theory that the effect of the surface structure on the corrosion resistance depends on the corrosion experiment and consequently on the investigated area of an implant. It is shown that deep-rolled samples processed at 200 bar with an overlap of 75 % perform best in contact zones, whereas ball-end milled surfaces prepared with a feed of 0.003 mm show higher resistance in direct contact to the medium. As samples could neither be investigated by X-ray diffraction method nor by hole drilling method due to their grain size, a statement about the influence of residual stresses is not possible, yet. Consequently, future work will concentrate on transferring these findings on Mg alloys with appropriate microstructure, which are also more common as implant materials.

Acknowledgement

We thank the DFG for financial support of this work within the collaborative research centre 599. We thank Johannes Walbaum and for assisting in the development of the MatLab tool.

References

1. E.C. Huse, "A new ligature?," *Chicago Med J Exam*, (1878),171-172.

2. A. Lambotte, "Technique et indications de la prothèse perdue dans la traitement des fractures," *Presse Med Belge*, 17 (1909), 321–323.

3. A. Lambotte, "L'utilisation du magnésium comme matériel perdu dans l'ostéosynthèse," *Bull Mém Soc Nat Cir*, 28 (1932), 1325–1334.

4. M.P. Staiger et al.,"Magnesium and its alloys as orthopedic biomaterials: a review," *Biomaterials*, 27 (2006), 1728–1734.

5. F. Witte et al., "Degradable biomaterials based on magnesium corrosion," *Curr Opin Solid State Mater Sci.*, 12 (2008), 63–72.

6. L. Yang, Y. Huang, Q. Peng, "Mechanical and corrosion properties of binary Mg–Dy alloys for medical applications," *Materials Science and Engineering B*, 76 (2011), 1827–1834.

7. A. Lucas, "Funktionsangepasste mechanische Bearbeitung von Magnesium als resorbierbarer Implantatwerkstoff," (Ph.D. thesis, Leibniz Universität Hannover, 2013).

8. B. Denkena et al., "Influence of stress on the degradation behavior of MgLAE442," *Procedia CIRP*, 5 (2013), 189-195.

9. B. Denkena, A. Lucas, "Biocompatible Magnesium Alloys as Absorbable Implant Materials Adjusted Surface and Subsurface Properties by Machining Processes," *Annals CIRP*, 56 (1) (2007), 113-116.

10. B. Denkena et al., "Biocompatible Magnesium Alloys as Degradable Implant Materials - Machining Induced Surface and Subsurface Properties and Implant Performance," *Special Issues on Magnesium Alloys*, ed. by W.A. Monteiro, (2011), 109-127.

11. M. Salahshoor, Y.B. Guo, "Biodegradation control of magnesium-calcium biomaterial via adjusting surface integrity by synergistic cutting-burnishing" *Procedia CIRP*, 13, (2014), 143-149.

12. T. Kokubo, H. Takadama, "How useful is SBF in predicting in vivo bone bioactivity?," *Biomaterials*, 27 (2006), 2907-2915.

13. L. Wolters et al., "Degradation behaviour of LAE442-based plate-screw-systems in an in vitro bone model" *Materials Science and Engineering C*, 49 (2015), 305–315.

14. K. Röttger, G. Wilcke, S. Mader, "Festwalzen – eine Technologie für effizienten Leichtbau. Materialwissenschaft und Werkstofftechnik," (Weinheim, WILEY-VCH Verlag GmbH & Co. KGaA, 2005).

15. J. Winkler, "Herstellung rotationssymmetrischer Funktionsflächen aus Magnesiumwerkstoffen durch Drehen und Festwalzen," (Ph.D. thesis, Universität Hannover, 2000).

16. B. Hering et al., "The influence of machining on residual stresses and corrosion behaviour of MgLa2 osteosynthesis plates" (Paper presented at Biomedizinische Technik/Biomedical Engineering in Graz, Sept. 2013).

17. Julia Kuhlmann et al., "Fast escape of hydrogen from gas cavities around corroding magnesium implants" *Acta Biomaterialia*, 9(10), (2013), 8714-8721.

Magnesium Technology 2016
Edited by: Alok Singh, Kiran Solanki, Michele V. Manuel, and Neale R. Neelameggham
TMS (The Minerals, Metals & Materials Society), 2016

ABSORBABLE FILAMENT TECHNOLOGIES: WIRE-DRAWING TO ENABLE NEXT-GENERATION MEDICAL DEVICES

Adam J. Griebel[1], Jeremy E. Schaffer[1]

[1]Research & Development, Fort Wayne Metals Research Products Corp., Fort Wayne, IN, 46809, USA

Keywords: wire drawing, magnesium wire, absorbable wire, composite, medical wire

Abstract

This presentation discusses the primary advantages and challenges associated with the production of absorbable medical wire, with a particular emphasis on magnesium-based technology. Purposeful discussion of some alternative materials such as iron as well as hybrid design possibilities are included for comparison. Nutrient metal technology may revolutionize the medical device industry by alleviating long-term side-effects associated with permanent metallic implants. Alloys of nutrient elements such as magnesium, iron, and zinc are being pursued as material solutions to enable absorbable devices. Efforts by the research community have focused on improving properties including corrosion rate, strength, and ductility of materials in each of the aforementioned material classes. Robust evaluation of candidate materials is a challenge that requires material in the correct form and device-accurate process-structure-function state. For medical devices, forms will include wires for orthopedic fixation (~1 mm diameter), surgical staples (~0.30 mm), stents (~150 µm), and aneurysm occlusion (~20 µm). Through a well-engineered wire-drawing process, substantial grain-refinement and strengthening are possible. Results of ongoing work include strength levels exceeding 500 MPa, with greater than 4% uniform elongation and formability on par with conventional implantable stainless steel achieved in cold drawn magnesium alloy wire.

Introduction

Wire, generally defined as an elongated metallic article, has found considerable use in many applications throughout history. The process used to produce wire, wire drawing, involves pulling material through progressively smaller and smaller conical dies. This process, dating back centuries[1], has seen considerable advancements over time. In more recent decades, drawing techniques and melt practices have advanced such that wires suitable for the medical industry, with pristine surface finishes and excellent, repeatable mechanical properties, are regularly produced today in a wide range of alloys, encompassing stainless steels, cobalt-based superalloys, titanium, precious metals, and Nitinol. To date, these alloys have been designed to be extremely resistant to the harsh environment of the human body.

However, a paradigm shift is taking place. Many current devices which remain implanted indefinitely (e.g. stents, staples, etc.) are only needed for a short period of time, after which their continued presence could lead to long-term complications. This reality has led to recent efforts to develop materials that are absorbed by the body after they have served their purpose[2]. Towards this end, nutrient metals, whose elements exist naturally in the body, have been investigated. Of the potential nutrient metals, magnesium has without question received the most attention and research[3,4], in large part due to its substantial presence in and tolerance by the body. Iron, and more recently zinc, have also been considered as potential solutions[5–8]. Efforts by the research community have focused on improving properties including corrosion rate, strength, and ductility of materials in each of these material classes. The microstructural and mechanical properties of a given alloy in the cast state will vary after extrusion, and can vary still further once subjected to cold-drawing. Variations in the parameters of any of these processes will impact the final properties. Therefore, robust evaluation of candidate alloys requires material in the correct form and device-accurate process-structure-function state.

For many medical devices, the correct form could include wire for orthopedic fixation (~1 mm diameter), surgical staples (~0.30 mm), stents (~150 µm), and aneurysm occlusion (~20 µm). Until recently, however, obtaining magnesium alloy wires in these dimensions has been challenging, due in large part to the limited ductility available at room temperature.

The history of absorbable wire is long and discrete. Iron wires were used as early as the 17th and 18th centuries[9,10] and reports of magnesium wires for blood vessel ligatures date as early as 1878[11]. There were numerous reports of magnesium implantation occurring up through the first half of the 20th century[12,13]. A patent from 1937 describes a suture material made through drawing and standing of magnesium wires[14]. However, once stainless steels, with their predictable behavior in the body, were developed, the idea of absorbable metals for temporary implant applications was set aside.

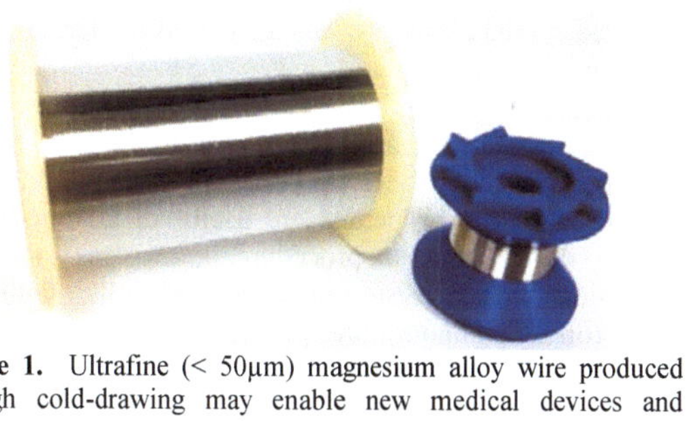

Figure 1. Ultrafine (< 50µm) magnesium alloy wire produced through cold-drawing may enable new medical devices and treatments.

With the renewed push for absorbable materials, several groups have worked to overcome the low ductility of magnesium and produce the wires necessary for medical devices on an academic scale[15–18]. These works highlight the need for medical-grade absorbable wire.

Wire-based Medical Devices

Once wire has been drawn to the correct diameter, there are many secondary operations that take place to add value and produce useful structures. Braiding, weaving, stranding, coiling, and forming of wires are all extensively used in the medical device industry today. With absorbable wire materials, these same techniques can be employed to replace existing medical devices, or even leverage their temporary nature to enable entirely new treatments. A few examples will be discussed below.

Textile approaches, such as braiding, weaving, or knitting, are often used to produce wire-based stents (Fig 2, top). Braided stents are often self-expandable and can offer excellent flexibility and kink-resistance. These features make them ideal for peripheral stenting. Developing stents made of absorbable metals is a promising innovation, and clinical trials with Biotronik's DREAMS Mg-based stent are encouraging[19]. The DREAMS stent is laser-machined from tubing, but it is logical to assume that devices made of wire will also find success. Planar woven meshes made from wire have been investigated in tissue engineering applications[20], and magnesium-based wires may provide the perfect union of temporary strength and long-term tissue healing.

Wire forming broadly encompasses processes that utilize plastic deformation and thermal treatments to form a length of wire into a desired shape. Wire-formed components can be as complex as stents (Fig 2, bottom), or as simple as staples. Staples today made of stainless steel or titanium, are deployed in applications ranging from gastric partitioning[21] to skin closure[22]. Nutrient metal wire may offer an appealing material choice for these instruments, and preliminary investigations have been reported[23].

Wires twisted together form a strand, and strands twisted together form cables, sometimes called wire ropes. Medical cables are extensively used in cardiac rhythm management as biostimulation leads and increasingly in orthopedic fixation. A common cable configuration is a 7 x 7, consisting of 7 strands of 7 wires. Recently, 110 µm Mg alloy wires were used to produce a 7 x 7 cable with 1 mm overall OD (Fig 3). Similar cables made of Titanium or Cobalt-Chrome alloys are regularly used in sternal closure and femoral cerclage.

Helical coils formed from wire generally have high radial stiffness and hoop strength, but low flexural and axial stiffness. These properties are useful in guidewire tips, pacing leads, and a procedure known as aneurysm occlusion. In aneurysm coil occlusion, coils of ultrafine wire (often platinum) are placed inside the aneurysm, initiating a clotting response to seal off the aneurysm and reduce risk of rupture. Ultrafine magnesium-based wire may offer an attractive alternative to the permanent materials used today.

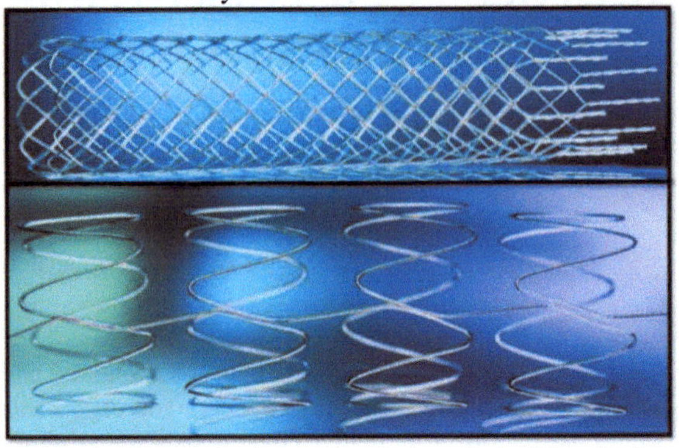

Figure 2. Stents can be produced through braiding (top) or forming (bottom) of wires. Images courtesy Admedes Schuessler.

Figure 3. 1 mm diameter 7x7 cables made from magnesium alloy wires could be used for sternal repair or femoral cerclage.

Nerve repair is an area where high-quality magnesium wire has the potential to expand the limits of medical technology. When nerves have been severed during an injury, there is a critical gap distance beyond which the nerve cannot be repair, even with today's best grafts and conduits[24]. Absorbable polymer fibers have shown promise in extending this gap[25], and recently Mg wire has been proposed as a means to improve healing. Mg wire provides an immediate conductive path across the nerves, and may improve nerve growth[26,27].

Wire Drawing

There are many processing parameters to consider when designing a wire-drawing process. The amount of reduction per pass (generally 10-25%) and the total reduction between anneals are highly dependent on material type and processing history, as are the temperature and time of those anneals. These inter-pass anneals can be batch processes or continuous, spool-to-spool operations. Proper selection of lubricant type (e.g. wet or dry), die material (e.g. diamond or carbide), and die geometry (e.g. angle, bearing length) is vital to wire quality. And while the principle of drawing is the same whether the wire diameter is 12 mm or 12 µm, the technical challenges associated with material handling vary greatly.

Drawing of absorbable ferrous wire is relatively straightforward (being similar to the austenitic stainless steels used today), and a thorough analysis of the influence of wire processing on mechanical, fatigue, and corrosion properties can be found elsewhere[7,28]. To summarize, strengths surpassing 2 GPa can be achieved, fatigue life suffers in a corrosive environment, and annealed materials tend to pit while cold-worked samples tend to suffer from corrosion-assisted cracking.

Zinc-based materials have only recently been considered for absorbable applications[8,29,30], and the potential benefits of zinc are high ductility and a corrosion rate that strikes a balance between that of magnesium and iron. The low strength of pure Zn and the tendency to dynamically and statically recrystallize at room temperature necessitate further alloy development. Trials are underway to explore the influence of wire drawing processes on the performance of several zinc alloys.

Robust production of magnesium alloy wire is challenging. Due to the limited deformation systems available in the hexagonal close packed structure, wire is very prone to cracking. Recently, Fort Wayne Metals has been working to overcome these challenges and produce high-quality cold drawn magnesium alloy wire on a commercial scale with good manufacturing practices to meet the needs of the medical device industry. First working with Mg-Li alloys[31], it was soon determined that the added ductility of lithium was unnecessary, and wires of type AZ, ZM, WE, ZE, and many others have been produced. This work has resulted in unprecedented levels of cold work (98%) and strength (695 MPa)[31,32].

This large amount of cold-work and associated hardening is one of the primary advantages of wire drawing over other production methods. The dislocations introduced through cold-working lead to increases in yield and ultimate tensile strength. For example, extruded, drawn, and annealed 0.51 mm ZM21 wires were progressively cold-drawn to 0.10 mm using oil-based lubrication and diamond dies, giving tensile strength increases from 200 MPa to over 400 MPa (Fig 4). Here, tensile tests used a gauge length of 127 mm and cross-head speed of 1.27 mm/min.

Cold-worked wire often responds very well to thermal treatment, as evidence in Figure 5. Here, 0.127 mm WE43B wires were produced with 90% cold work and then subjected to isochronal (< 10 min, Argon cover gas) heat treatments. At temperatures ranging from 250 to 450°C, precipitation hardening and recrystallization were observed. This indicates that for many magnesium alloys, with the correctly engineered process, finely dispersed precipitates or highly refined grains can be attained.

Diameters as small as 20 µm have been achieved in magnesium alloy wire (Figure 1). At these ultrafine diameters, material cleanliness and surface roughness become extremely important, as even several micron features (be they inclusions, intermetallic particles, or surface defects) become problematic.

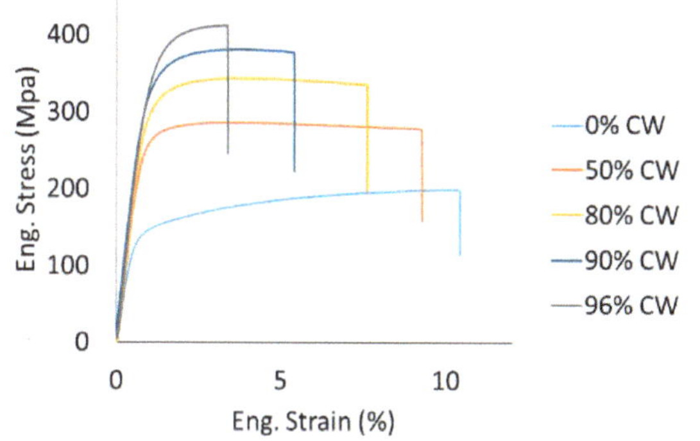

Figure 4. The strength of ZM21 can be increased from 200 MPa to 400 MPa through cold working (CW). Diameters range from 0.510 mm (0%CW) to 0.100 mm (96%CW).

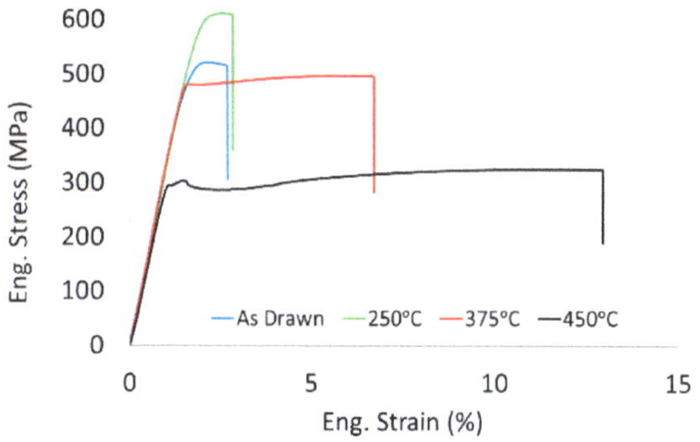

Figure 5. 0.127 mm WE43 wires with 90% cold work subjected to isochronal heat treatments show a wide range of properties.

Degradation Control via Wire Composites

Perhaps the biggest single issue to overcome in absorbable metals is that of non-uniform corrosion, be it pitting or stress-corrosion cracking, which can lead to premature fracture and device failure. One potential avenue to reduce this risk is through the use of wire composites. Drawn-filled-tube (DFT®) composite wire places a core of one material inside a tube or shell of another, to advantageously combine properties. In an absorbable wire, materials with dissimilar degradation rates or galvanic potentials can be combined to control a device's corrosion sequence.

Considerable work by this group has gone into investigating ferrous materials, seeking to take advantage of the higher strength and ductility available as compared to magnesium. Pure Fe and Fe-Mn solid wires as well as DFT wires with magnesium cores showed a range of mechanical, fatigue, and corrosion properties[7,28]. Later work sought to improve device safety through production of DFT wires with Fe35Mn shells and thin, non-absorbable cores of either Nitinol or Tantalum[33]. While a small amount of material was left behind, premature wire fracture was effectively eliminated (Fig 6) in both phosphate buffered saline and bovine serum at body temperature.

In an effort to achieve the risk mitigation of the non-degrading core while still having a material that was completely absorbable, composite wires with magnesium alloy shells and either pure iron, pure zinc, or more noble magnesium alloy cores were produced[34]. After placing coils of these 0.2 mm wires in 37°C bovine serum, it was observed that the nobler core was indeed left behind while the anodic magnesium shell dissolved. (Fig 7). Ongoing tests show promise for fracture prevention, but galvanically accelerated corrosion of the shell must be managed.

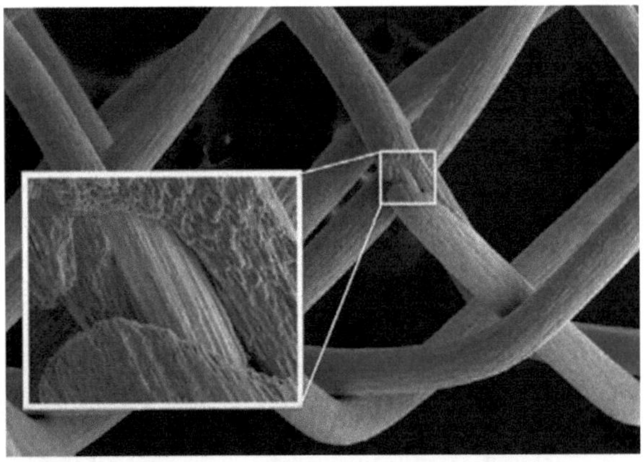

Figure 6. The NiTi core of Fe35Mn-DFT-NiTi composite wires prevented premature wire fracture in phosphate-buffered saline.

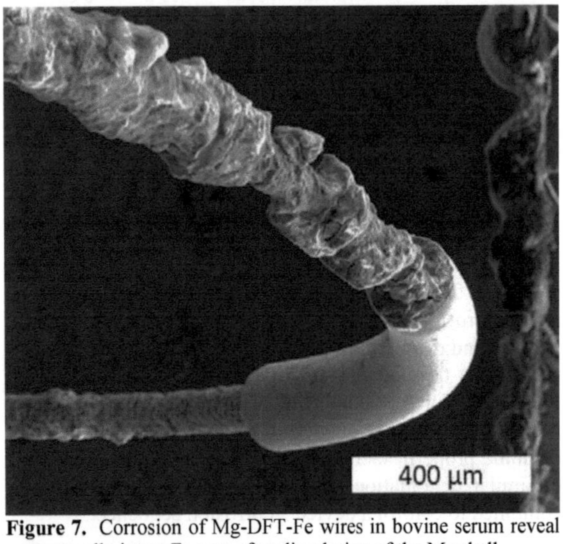

Figure 7. Corrosion of Mg-DFT-Fe wires in bovine serum reveal a structurally intact Fe core after dissolution of the Mg shell.

Conclusion

In summary, high-quality cold-drawing of magnesium and other nutrient metal filaments will be a critical component of successful absorbable device advancement. In particular, strength levels exceeding 500 MPa, with greater than 4% uniform elongation and formability (e.g. stranding, cabling, and coiling) on par with conventional implantable stainless steel are possible in cold drawn magnesium alloy wire. Hybrid absorbable designs that take advantage of galvanic coupling between magnesium and other species such as iron can also be useful to discourage corrosion-assisted fracture. Recent advances in processing and compositing of these absorbable materials may enable next-generation medical devices.

References

1. Theophilus, *On Divers Arts* (New York, NY: Dover Publications, 1979).
2. Y. F. Zheng, X. N. Gu, and F. Witte, "Biodegradable metals," *Materials Science and Engineering: R,* 77 (2014), 1–34.
3. M. P. Staiger et al., "Magnesium and its alloys as orthopedic biomaterials: A review," *Biomaterials,* 27 (2006), 1728–1734.
4. F. Witte et al., "In vitro and in vivo corrosion measurements of magnesium alloys," *Biomaterials,* 27 (2006), 1013–1018.
5. M. Peuster et al., "A novel approach to temporary stenting: degradable cardiovascular stents produced from corrodible metal—results 6–18 months after implantation into New Zealand white rabbits," *Heart,* 86 (2001), 563–569.
6. H. Hermawan et al., "Iron–manganese: new class of metallic degradable biomaterials prepared by powder metallurgy," *Powder Metallurgy,* 51 (2008), 38–45.
7. J. E. Schaffer, E. A. Nauman, and L. A. Stanciu, "Cold drawn bioabsorbable ferrous and ferrous composite wires: An evaluation of in vitro vascular cytocompatibility," *Acta Biomaterialia,* 9 (2013), 8574–8584.
8. P. K. Bowen, J. Drelich, and J. Goldman, "Zinc Exhibits Ideal Physiological Corrosion Behavior for Bioabsorbable Stents," *Advanced Materials,* 25 (2013), 2577–2582.
9. E. Crubzy et al., "False teeth of the Roman world," *Nature,* 391 (1998), 29–29.
10. P. Laing, "Clinical Experience with Prosthetic Materials: Historical Perspectives, Current Problems, and Future Directions" ASTM Standard STP684, (1979).
11. E. Huse, "Magnesium Ligatures," *Chic. Med. J. Exam,* 37 (1878), 171–172.
12. M.G. Seelig, "A study of magnesium wire as an absorbable suture and ligature material," *Archives of Surgery,* 8 (1924), 669–680.
13. F. Witte, "The history of biodegradable magnesium implants: A review. *Acta Biomater.* 6, 1680–1692 (2010).
14. B. Blumenthal, and H. Hadenfeldt, "Material for surgical ligatures and sutures," U.S. Patent 2,094,578 (1937).
15. A. Milenin, and P. Kustra, "Numerical and Experimental Analysis of Wire Drawing for Hardly Deformable Biocompatible Magnesium Alloys," *Arch. Metall. Mater,* 58, (2013).
16. J.-M. Seitz et al., "The Manufacture of Resorbable Suture Material from Magnesium," *Adv. Eng. Mater.* 12 (2010), 1099–1105.
17. J. Bai et al,. "Preparation, microstructure and degradation performance of biomedical magnesium alloy fine wires," *Progress in Natural Science: Materials International,* 24 (2014), 523–530.

18. P. Maier et al., "Mechanical and Corrosive Properties of Two Magnesium wire: Mg4Gd and Mg6Ag," *Magnesium Technology 2015* (2015), 393-398.
19. M. Haude et al., "Safety and performance of the drug-eluting absorbable metal scaffold (DREAMS) in patients with de-novo coronary lesions: 12 month results of the prospective, multicentre, first-in-man BIOSOLVE-I trial," *The Lancet,* 381 (2013), 836–844.
20. S. H. Alavi, and A. Kheradvar, "A Hybrid Tissue-Engineered Heart Valve," *Ann. Thorac. Surg.,* 99 (2015), 2183–2187.
21. W. G. Pace et al., "Gastric partitioning for morbid obesity," *Ann. Surg.,* 190, (1979) 392–400.
22. J.H. Roth, and B.H. Windle, "Staple versus suture closure of skin incisions in a pig model," *Canadian Journal of Surgery,* 31 (1988), 19–20.
23. S. Zhang et al., "In vivo degradation and biocompatibility of linear cutter staples made of high purity magnesium," *Eur. Cell. Mater.* 28 (2014), 76.
24. W. Daly et al., "A biomaterials approach to peripheral nerve regeneration: bridging the peripheral nerve gap and enhancing functional recovery," *J. R. Soc. Interface*, 67 (2013), 202-221.
25. T.-T. B. Ngo et al., "Poly(L-Lactide) microfilaments enhance peripheral nerve regeneration across extended nerve lesions," *J. Neuroscience Research,* 72 (2003), 227–238.
26. J. J. Vennemeyer et al., "Initial observations on using magnesium metal in peripheral nerve repair" *J. Biomater. Appl.* 29 (2015), 1145–1154.
27. J. J. Vennemeyer et al., "Effects of elevated magnesium and substrate on neuronal numbers and neurite outgrowth of neural stem/progenitor cells in vitro," *J. Neuroscience Research* 84 (2014), 72–78.
28. J. E. Schaffer, E. A. Nauman, and L. A. Stanciu, "Cold-Drawn Bioabsorbable Ferrous and Ferrous Composite Wires: An Evaluation of Mechanical Strength and Fatigue Durability," *Metall. Mater. Trans., B* 43 (2012), 984–994.
29. P. K. Bowen et al., "Metallic zinc exhibits optimal biocompatibility for bioabsorbable endovascular stents," *Mater. Sci. Eng. C* 56 (2015), 467–472.
30. X. Liu et al., "Microstructure, mechanical properties, in vitro degradation behavior and hemocompatibility of novel Zn-Mg-sr alloys as biodegradable metals," *Materials Letters* (In Press, Accepted Manuscript), doi:10.1016/j.matlet.2015.07.151
31. A.J. Griebel, and J.E. Schaffer, "Development of high-strength bioabsorbable Mg alloys suitable for conventional cold-working processes," *Eur. Cell. Mater.* 26 (2013), 2.
32. A.J. Griebel, and J.E. Schaffer, "Cold-drawn ZM21 and WE43 wires exhibit exceptional strength and ductility," *Eur. Cell. Mater.* 28 (2014), 2.
33. J.E. Schaffer, and A.J. Griebel, "Embolic risk mitigation in absorbable implants by composite design," *Eur. Cell. Mater.* 28 (2014), 80.
34. A.J. Griebel, & J.E. Schaffer, "Cold-drawn Mg alloy composite wires minimize risk through galvanic coupling," *Eur. Cell. Mater.* 30 (2015), 13.

Magnesium Technology 2016
Edited by: Alok Singh, Kiran Solanki, Michele V. Manuel, and Neale R. Neelameggham
TMS (The Minerals, Metals & Materials Society), 2016

PLASMA SURFACE MODIFICATION OF MAGNESIUM-BASED AND RELATED MATERIALS

Paul K Chu
Department of Physics and Materials Science, City University of Hong Kong, Tat Chee Avenue, Kowloon, Hong Kong, China

Keywords: plasma surface modification; magnesium alloys; corrosion resistance, biocompatibility

Abstract

The interactions between biomaterials and living tissues are crucial to clinical success especially for biodegradable magnesium alloys. Since magnesium-based biomaterials do not have the desirable surface properties and desirable functions, plasma-based surface modification plays an important role in tailoring the surface of biomaterials to allow better adaptation to the physiological surroundings and deliver the required clinical performance. In this distinguished invited talk, recent progress related to research of magnesium-based and related biomaterials in the Plasma Laboratory of City University of Hong Kong is reviewed.

Introduction

Owing to increasing clinical demand, biomaterials are constantly refined both in sophistication and diversity. With regard to magnesium-based alloys which possess the unique properties of natural degradability and good mechanical properties [1], the main focus is control of the degradation rate while the biocompatibility is not compromised [2,3]. The surface of biomedical implant must act as a shield to protect against undesirable degradation and at the same allows the proliferation of cells in orthopedic applications. In this respect, plasma-based surface modification which can modify selective surface properties while preserving the favorable bulk attributes of the materials is very useful [4,5].

Plasma Surface Modification

Magnesium alloys as biodegradable metals are very attractive in load-bearing orthopedic and dental applications [6]. The Young's modulus (E = 41-45 GPa) of Mg alloys is similar to that of bones (E = 3-20 GPa) and stress shielding effects can be mitigated. More importantly, a follow-up operation to remove the implant from the patient can be omitted due to favorable biodegradation. After healing, faster degradation is allowed and the mechanical strength of the implant decreases gradually because the surrounding tissues can now bear the partial load. This strategy requires sufficient corrosion resistance for controlled degradation initially to ensure adequate mechanical strength and structure integrity in the initial healing stage. A stable surface without producing unacceptable hydrogen evolution and other side-effects is thus essential to spur cell/tissue growth. Unfortunately, *in vitro* and *in vivo* tests conducted on most commercial Mg alloys show signs of severe and rapid corrosion immediately after surgical insertion and so the surface corrosion resistance must be improved in a controllable fashion. Herein, recent progress made in our laboratory pertaining to the use of plasma-based technology to improve the surface characteristics of Mg-based alloys is reviewed.

In addition to application of coatings, plasma-based technology such as plasma immersion ion implantation and deposition (PIII&D) is important to the developing biodegradable Mg alloys. Because of energetic ion bombardment during implantation, the modified surface layer does not have an abrupt interface and layer delamination does not pose a serious problem. Chromium is an important element in corrosion resistant alloys such as stainless steels. However, when chromium is incorporated into magnesium by metal ion implantation, strong galvanic corrosion occurs because Cr exists in the metallic state in the implanted layer. Fortunately, surface degradation of pure magnesium can be retarded if oxygen ion implantation is conducted afterwards by PIII to produce a compact surface layer composed of chromium oxide [7]. In contrast, if only oxygen PIII is conducted on Mg-Nd-Zn-Zr alloys, no significant improvement in the corrosion resistance is observed [8-10]. Consequently, we have proposed an alloying principle to modify the surface by ion implantation before O-PIII. In addition to the improved corrosion resistance, the cytocompatibility of the plasma-treated Mg is quite good as shown in Figure 1 [11]. Furthermore, after surface alloying with Cr, Ti, Al, and Zr, the corrosion resistance of Mg-Y-RE (rare earth), Mg-Zn-Zr, and Mg-Nd-Zn-Zr is improved due to the formation of chromium oxide, titanium oxide, aluminum oxide, and zirconium oxide in the near surface, respectively. We have extended this principle to Mg-Ca and Mg-Sr alloys by conducting zirconium and oxygen ion implantation [12]. The process not only improves the corrosion resistance and *in vitro* biocompatibility significantly, but also enhances the bacteria resistance which is important to minimize post-surgery infection. We have also conducted C_2H_2 PIII to produce a thin layer of diamond-like carbon (DLC) film on the Mg-Nd-Zn-Zr alloy to enhance the corrosion resistance in the 0.9 % NaCl solution [13]. The use of a metallic interlayer is a common way to improve adhesion between the diamond-like carbon (DLC) and Mg substrate, but the disadvantage is that the galvanic cells generated in the defects accelerate corrosion. Fortunately, in ion implantation, layer adhesion is not an issue. The defects in the film can provide the possibility of eventual degradation of the bulk alloy but it should be noted that the DLC film is not fully biodegradable and remains in the human body. To avoid this phenomenon or mitigate the severity, we have performed carbon ion implantation using graphite as a cathodic arc source to modify pure magnesium [14]. A composite oxide film containing amorphous carbon formed on the surface enhances the corrosion resistance in the simulated physiological environment.

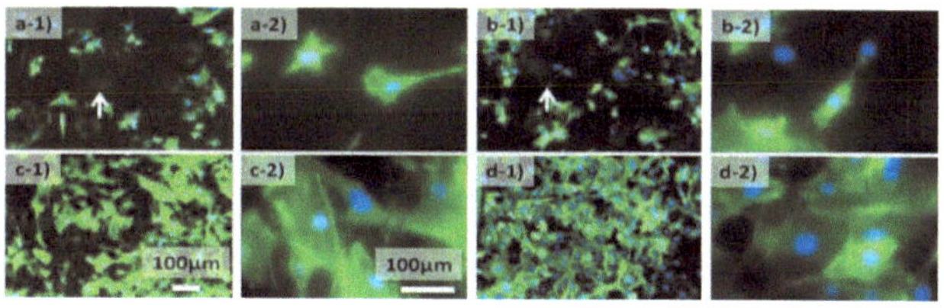

Figure 1. Osteoblast morphology: (a) Untreated Mg after 1 day, (b) Untreated Mg after 3 days, (c) Cr/O PIII Mg after 1 day. (d)

Cr/O PIII Mg after 3 days (1 stands for low magnification and 2 stands for high magnification) [11].

Conclusion

In summary, biodegradable Mg alloys are attractive due to their special properties such as natural biodegradability and mechanical properties. However, many phenomena are still not well understood and more *in vitro* and *in vivo* investigations are needed. Based on current results, plasma-based surface modification encompassing ion implantation and deposition is important to the development of biodegradable Mg alloys. With the combination of modern metallurgy and advanced surface treatment technology, Mg-based alloys have large potential as next-generation materials in biodegradable biomedical implants.

References

1. N. Peppas and R. Langer, "New Challenges in Biomaterials", *Science*, 263 (5154) (1994) 1715-1720.
2. G. S. Wu, P. H. Li, H. Q. Feng, X. M. Zhang, and P. K. Chu, "Engineering and Functionalization of Biomaterials *via* Surface Modification", *J Mat Chem B*, 3(10) (2015) 2024-2042.
3. G. S. Wu and P. K. Chu, "Surface Design of Biodegradable Magnesium Alloys for Biomedical Applications", *Surface Modification of Magnesium and Its Alloys for Biomedical Applications, Vol. 1: Biological Interactions, Mechanical Properties and Testing*, ed. T. S. N. Sankara Narayanan, Il-Song Park, and Min-Ho Lee (Woodhead Publishing, 2015), Chapter 3, 89-119.
4. X. Y. Liu, P. K. Chu, and C. X. Ding, "Surface Modification of Titanium, Titanium Alloys, and Related Materials for Biomedical Applications", *Mat Sci Eng Reports*, 47 (2) (2004) 49-121.
5. X. Y. Liu, P. K. Chu, and C. X. Ding, "Surface Nano-Functionalization of Biomaterials", Mat Sci Eng Reports, 70 (3–6) (2010) 275-302.
6. R. Zeng, W. Dietzel, F. Witte, N. Hort and C. Blawert, "Progress and Challenge for Magnesium Alloys as Biomaterials", *Adv Eng Mat*, 19 (2008) B3-B14.
7. R. Z. Xu, G. S. Wu, X. B. Yang, T. Hu, Q. Y. Lu, and P. K. Chu, "Controllable Degradation of Biomedical Magnesium by Chromium and Oxygen Dual Ion Implantation", *Mat Letts*, 65 (14) (2011) 21710-2173.
8. G. S. Wu, K. Feng, A. Shanaghi, Y. Zhao, R. Z. Xu, G. Y. Yuan, and P. K. Chu, "Effects of Surface Alloying on Electrochemical Corrosion Behavior of Oxygen-Plasma-Modified Biomedical Magnesium Alloy", *Surf Coat Technol*, 206 (14) (2012) 3186-3195.
9. Y. Zhao, G. S. Wu, J. Wu, Q. Y. Lu, J. Wu, R. Z. Xu, K. W. K. Yeung, and P. K. Chu, "Improved Surface Corrosion Resistance of WE43 Magnesium Alloy by Dual Titanium and Oxygen Ion Implantation", *Thin Solid Films*, 529 (2013) 407-411.
10. M. Jamesh, G. S. Wu, Y. Zhao, D. R. McKenzie, M. M. M. Bilek, and P. K. Chu, "Effects of Zirconium and Oxygen Plasma Immersion Ion Implantation on The Corrosion Behavior of ZK60 Mg Alloy in Simulated Body Fluids", *Corrosion Science*, 82 (2014) 7-26.
11. R. Z. Xu, X. B. Yang, J. Jiang, P. H. Li, G. S. Wu, and P. K. Chu, "Effects of Chromium Ion Implantation Voltage on The Corrosion Resistance and Cytocompatibility of Dual Chromium and Oxygen Plasma-Ion-Implanted Biodegradable Magnesium", *Surf Coat Technol*, 235 (2013) 875-880.
12. Y. Zhao, J. Mohammed Ibrahim, W. K. Li, G. S. Wu, Y. F. Zheng, K. W. K. Yeung, and P. K. Chu, "Enhanced Antimicrobial Properties, Biocompatibility, and Corrosion Resistance of Plasma-Modified Biodegradable Magnesium Alloys", *Acta Biomat*, 10 (1) (2014) 544-556.
13. G. S. Wu, X. M. Zhang, Y. Zhao, J. Mohammed Ibrahim, G. Y. Yuan, and P. K. Chu, "Plasma Modified Mg-Nd-Zn-Zr Alloy with Enhanced Surface Corrosion Resistance", *Corrosion Science*, 78 (2014) 121-129.
14. R. Z. Xu, X. B. Yang, P. H. Li, K. W. Suen, G. S. Wu, and P. K. Chu, "Eelectrochemical Properties and Corrosion Resistance of Carbon-Ion-Implanted Magnesium", *Corrosion Science*, 82 (2014) 173-179.

Magnesium Technology 2016
Edited by: Alok Singh, Kiran Solanki, Michele V. Manuel, and Neale R. Neelameggham
TMS (The Minerals, Metals & Materials Society), 2016

DEGRADATION OF MgF_2-COATED AND UNCOATED MgNd2 SPECIMENS IN CONTACT WITH NASAL MUCOSA

Rainer Eifler[1], Martin Durisin[2], Christian Klose[1], Thomas Lenarz[2], Hans Jürgen Maier[1]

[1]Institut für Werkstoffkunde (Materials Science), Leibniz Universität Hannover, An der Universität 2, 30823 Hannover, Germany
[2]Clinic for Laryngology, Rhinology and Otology, Medical School of Hanover, Carl-Neuberg-Str. 1, 30625 Hanover, Germany

Keywords: Magnesium, Coating, Corrosion, In-vivo

Abstract

The magnesium-neodymium alloy MgNd2 offers an elongation of 30%. Therefore, it is well suited as a material for resorbable stents which can be used for treating chronic sinusitis. While the alloy's corrosion properties were already investigated in vitro, the present study focuses on the in vivo degradation behavior of fluoride-coated and uncoated specimens in direct contact with porcine nasal mucosa during a period of 180 days. The study showed a promising and controlled biodegradation, which was investigated by measurements of the MgNd2 specimens' mass and volume losses as well as element analyses on the surface to obtain information about the composition of the degradation layer. The degradation rates of the uncoated samples in direct contact with the physiological tissue showed a maximum during the first 45 days which then decreased for the remaining implantation period. In contrast, the degradation rates of the fluoride-coated samples increased gradually during the implantation duration.

Introduction

Owing to their very good biocompatibility, [1] primary stability [2] and biodegradability magnesium alloys provide the possibility for developing resorbable novel stent sutures and other orthopedic applications [2].
Biodegradable materials offer the possibility to eliminate the need for a second surgical procedure to remove the implant or device. In the last years a lot of alloy development and mechanical analyses as well as in vitro and in vivo investigations have been carried out in the field of biomedical engineering [3]. On this account magnesium alloys possess improved mechanical and degradation properties. A fast degradation of Magnesium implants leads to a loss in structural integrity and the production of a plenty of hydrogen gas. This could induce an increased pH in the implant's environment. All this effects can be reduced by alloying and by coating the implant [4].
Due to their favorable properties magnesium alloys with rare earth elements have become the focus of numerous studies [3]. Especially Neodymium has become increasingly important during the last years. It offers a good solubility in magnesium and improves the deformation behavior [5, 6].
Consequently the utilization of neodymium as an alloying element offers the possibility to produce stents and other implant material which need to be resorbable and deformable. In this regard, Seitz et al. have already shown the suitability of magnesium alloys containing neodymium for this kind of applications [7].
In contrast to the increased ductility, the effect of neodymium on the corrosion behavior of magnesium alloys is differently discussed [8, 9]. Südholz et al. have shown that magnesium alloys with a small amount of neodymium contain the intermetallic phase Mg_3Nd [10]. This phase has just a slight influence on the formation of galvanic cells and doesn't accelerate the corrosion of neodymium containing magnesium alloys. It could also be determined that neodymium demonstrate a good biocompatibility in small quantities [11]. In summary, neodymium increases the ductility, offers a good biocompatibility and shows also the most promising corrosion behavior of all lanthanides. Therefor the chosen MgNd2 alloy reveals new perspectives for bioresorbable stents for chronic sinusitis [5].
Annually millions of people are affected by chronic sinusitis [12]. This widespread illness is mainly based on periodic limitations in the anatomic section between the frontal sinus and the nasal cavity [13, 14]. Headaches, pressure sensation, nasal respiratory obstruction, and nasal secretion are the commonly appearing symptoms [4]. In the last decades a lot of operation methods were developed to solve this problem, but the results failed to meet expectations [15].
Actually used plastic implants may cause infections, dislocations, local scar tissue, prevention of reepithelization, and of course breathing difficulties [12, 16]. In contrast to plastic implants, magnesium based stents can ensure the patency of the frontal sinus and degrade during the phase of wound healing. Accordingly, a controlled degradation can warrant the sufficient opening of the constriction and the postoperative reepithelization process of nasal mucosa can proceed undisturbed [4, 17]. Patients with chronic sinusitis frontalis offers this advantages over the current limitations of non-degradable stenting systems [15]. After an operation due to a chronic sinusitis frontalis the wound healing takes about 3 months [18]. However, the period of implantation of a nasal stent to prevent restenosis is controversy discussed. Information concerning the period range from 6 to 24 weeks [19]. The biocompatibility properties of the magnesium alloy depend not only on the alloying components, it is therefore important to collect also information about the geometry [2], the anatomy and the physiological environment around the implant [4].
Another possibility to influence the corrosion behavior is the application of coatings on the implants. One of the most discussed and investigated coating on degradable magnesium implants is magnesium fluoride (MgF_2). Both elements are naturally occurring in the human body and the degradation products remain harmless [20].
MgF_2 protective layers on magnesium implants have shown enhanced corrosion resistance and improvements in in vivo and in vitro biological testing [21]. MgF_2 disposes of a very brittle and crystalline structure and is therewith normally unsuitable for application on deformable implants [7].
However, this is unproblematic, because the deformed parts are no longer necessary after the stent has been well integrated. The information about the degradation behavior of degradable magnesium implants in this region of the human body are very limited. Within this work the in vivo biocompatibility of the uncoated and coated binary alloy MgNd2 during its direct contact

with nasal mucosa of the frontal sinus was analyzed. This should help to estimate the in vivo corrosion behavior and the period of degradation of coated and uncoated specimens. The influences on the surrounding tissue, the losses in mass and volume as well as the degradation layer on the specimens were also investigated.

Materials and Methods

The magnesium alloy MgNd2 (Mg 90 wt.%, Nd 2 wt.%) was chosen due to its biocompatibility, the mechanical properties and the suitable corrosion behavior. A master Mg-alloy containing 40 wt.% neodymium (Nd40) was initially produced [7]. This was necessary because of the various melting points of magnesium and neodymium (melting points: Nd 1021 °C, Mg 650 °C). During the melting process of the master alloy stirring was carried out to garantee the dissolving of the neodymium (99.5 % purity, Treibacher Industrie AG, Althofen, Austria) in the magnesium (99.94 % purity, Magnesium Elektron, Manchester, UK). The composition of the master alloy Nd40 was chosen near to the eutectic point of the binary Mg-Nd system to lower the temperature of re-melting while producing the main alloy. The required amounts of neodymium and magnesium were filled in a titanium crucible to minimize the impurities. The melting was carried out at a temperature of 750 °C. To ensure the entire dissolution of the neodymium in the magnesium, the melt was stirred for 180 minutes. The main alloy MgNd2 was produced in an open, tilting crucible furnace, which was filled with the required amounts of the Nd40 master alloy and pure magnesium. After the components were melted at a temperature of 750 °C, a mechanical stirrer was inserted into the crucible for 30 minutes. All the casting and melting processes were carried out according to the high reactivity of the liquid magnesium in a shielding, dynamically circulating ARCAL 1 gas atmosphere (Air Liquide, Düsseldorf, Deutschland, volumetric flow rate of 2 l/min). For the main casting process a boron nitride coated steel die (Ø = 130 mm) was heated to 450 °C and purged with shielding gas to reduce the reaction of the melt with the air and to decrease the formation of cavities and pores in the resulting billet.

The composition of the main alloy was determined by means of inductively coupled plasma optical emission spectrometry (ICP-OES). A Nd-content of 1.84 wt.% ± 0.01 for the alloy was analyzed. After the casting process the cast billets were turned down to a diameter of 120 mm. This guaranteed billets without cavities, pores, impurities or other defects. In order to obtain a fine-grained alloy a subsequent hot extrusion process was realized with a continuous ram speed of 1.4 mm/s and a maximum extrusion force of 9.7 MN. Prior to the extrusion, the turned billets, the 10 mm diameter extrusion die and the extruder's container were preheated to a temperature of 350 °C. The extruded rods cooled down at room temperature. For the implantation cylindrical specimens with a diameter of 5 mm and a height of 5 mm were chosen. The specimens were furnished with two intersecting bore holes (1 mm diameter). Uniform turning parameters were used to manufacture specimens with comparable surface roughnesses. The geometry was chosen due to an adequate surface volume ratio and it also offers the possibility to fixate the specimens to the mucous membrane of the pigs' noses with non-resorbable suture material.

In order to verify the possibilities of a corrosion protective layer, half of the specimens (36 specimens) were coated with magnesium fluoride. To produce a sufficient MgF_2 layer, the specimens were boiled in NaOH (200 g·L−1) for 120 min. Afterwards, the specimens were positioned in hydrofluoric acid (40 %) for four days to achieve the MgF_2 coating (chemicals acquired from Carl Roth GmbH & Co. KG, Karlsruhe, Germany). After the hydrofluoric acid treatment the specimens were rinsed with distilled water and ethanol to remove the remains of the chemicals [7]. The 72 uncoated and coated specimens were weighed prior to and after the implantation using a precision balance (Analytical Balance A204). To acquire information about the corrosion process during the experiment, endoscopic examinations were performed every 45 days to visualize the changes on and around the specimens.

After the extraction of the implants from the minipig all specimens were cleaned in ethanol and finally air-dried. The surfaces of three specimens per extraction point were subsequently determined by means of energy dispersive X-ray spectroscopy (EDS) using a SUPRA 40 VP (Carl Zeiss AG, Oberkochen, Germany) to analyze the element composition. Before weighing the specimens and determining the loss of volume after explantation, the corrosion products were removed from the specimens' surfaces by means of an acid treatment. First, the specimens were immersed in an ethanol ultrasonic bath for three minutes and then treated in chromic acid according to ISO 8407:1991 (E) C.5.2 for another three minutes (chemicals acquired from Carl Roth GmbH & Co. KG, Karlsruhe, Germany) to remove all corrosion products from the surface [22]. Lastly, the specimens were rinsed with distilled water and ethanol. The loss in volume of the specimens was determined by means of micro computed tomography (μCT). For this reason, a μCT scanner (μCT80, Scanco Medical AG, Switzerland) was used to volumetrically measure the specimens after the in vivo corrosion. A current of 144 μA, a tube voltage of 55 kV, a layer thickness of 20 μm, and a duration of 1 s per layer were used for the measurements. In order to determine the volume loss per extraction, three samples were analyzed and the average loss was calculated by means of the image-subtraction method.

Results and Discussion

In vitro studies have already shown a suitable bending force of uncoated and coated MgNd2 specimens after immersion in SBF [7]. In addition, a proper in vivo biocompatibility of coated specimens could also be observed [4]. Both studies underline the valuable effect of the coating. Accordingly, the analyses in this study focus on the influence of the MgF_2 coating on the in vivo degradation behavior and the growing corrosion layers in direct contact with porcine nasal mucosa.

The 72 coated and uncoated specimens were evenly distributed implanted into the frontal sinus of 24 minipigs. Each surgery was performed without any anaesthesiological or surgical complication. The healing showed no signs of inflammatory reactions or neurological changes. The endoscopic examinations of the coated specimens showed after 45 days none or in a few cases moderate mucosa hyperplasia. Between the 45th and the 180th day the observed hyperplasia increased to moderate or severe. Altogether the hyperplasia was a bit more severe for the uncoated specimens.

After the explantations of the specimens the element composition on the surfaces was analyzed for each extraction point (45, 90,135 and 180 days). The analyses of the uncoated specimens showed that not only the alloying elements (magnesium and neodymium) were found on the corroded specimens' surfaces after explantation (Figure 1). Further, additional elements located in the physiological environment (phosphor and calcium) were found (Figure 1). The evaluation of the data suggests a corrosion period and alloy depending concentration of elements on the uncoated specimens' surfaces. Thereby, the alloying elements are dominating at early specimen extraction points while physiologically originating elements occur in higher concentrations after half of the period. The corrosion layer of the uncoated specimens showed from the beginning a drop in the magnesium concentration and simultaneously an increase in phosphor and calcium concentration. The neodymium content remains during the 180 days almost constant. The fluctuations in percentage of the elements in the layer between 90 and 180 days can be attributed to an in vivo study, because every minipig possesses a slightly different physiological tissue around the specimens.

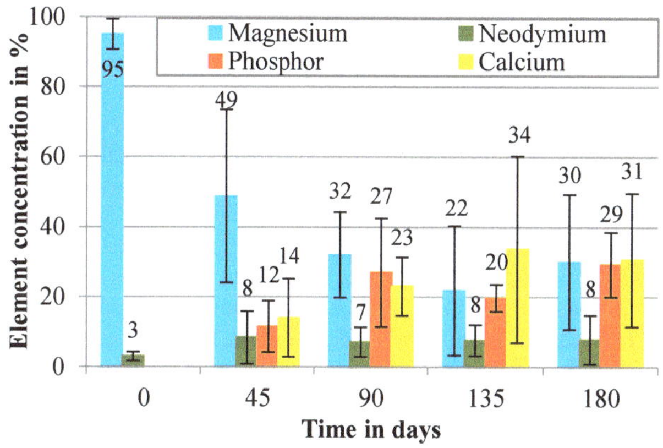

Figure 1: Element concentrations on the uncoated specimens' surface attached to the mucous membrane after 45, 90, 135 and 180 days.

The analyses of the coated specimens' surfaces showed besides the alloying and coating elements (magnesium, neodymium and fluorine) also the physiological elements originating from the physiological environment (phosphor and calcium) (Figure 2).

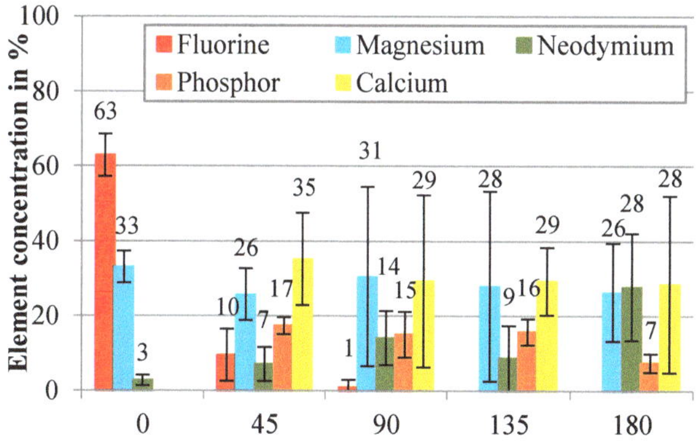

Figure 2: Element concentrations on the coated specimens' surface attached to the mucous membrane after 45, 90, 135 and 180 days.

For the coated specimens the compounds of the corrosion layer depends also strong from the implantation duration. During the first 45 days the content of calcium and phosphor increased, whereby the content of calcium rose almost twice as fast. The content of fluorine disappeared completely during the first 135 days. This can be seen as an indication that the corrosion protection layer of magnesium fluoride is decomposing in this period. Between the 45th and the 180th day the composition of the corrosion layer remained approximately stabile. Over the whole period an increase in neodymium concentration could also be established. The reason for this could be the enrichment of the poorly soluble neodymium in the corrosion layer.

In general, the corrosion layers on the uncoated and coated specimens result mainly from the formation of calcium and phosphor on the surface. Deposition of these corrosion products on the specimens surfaces have already been described as depending on the protection layer and the pH value [23]. Witte et al. assumed also that the growing layers on magnesium implants in such environments compose of calcium phosphate precipitates and other corrosion products such as magnesium oxide or magnesium hydroxide [2]. The analyses of the corrosion layers showed also a difference in the percentage of calcium and phosphor. The coated specimens possess a higher content of calcium and lower content of phosphor compared to the uncoated samples. This could be connected with the accumulation of calcium fluoride on the coated specimens' surface at the beginning of implantation. Therefor the growing corrosion layer clearly depends on the protection layer. The average losses in specimens' mass and volume were also measured in this study. Figure 3 depicts the measured percentage weight losses of the coated and the uncoated specimens relative to the initial weight.

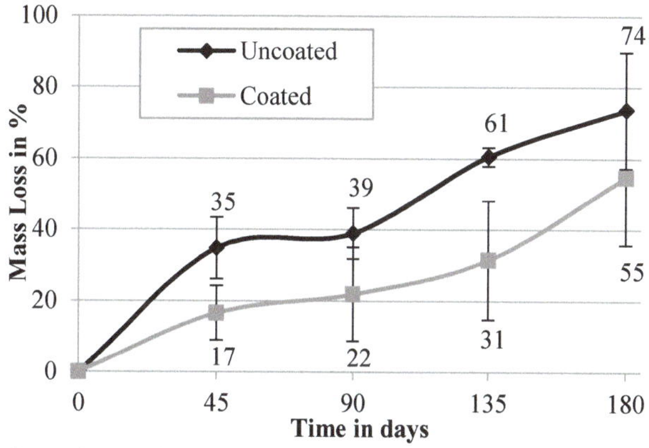

Figure 3: Loss in mass relative to the initial mass of the coated and uncoated specimens after 45, 90,135 and 180 days.

The coated samples show a loss in mass of 17 % within the first 45 days and the uncoated a loss of 35%. In the following 45 days the loss in mass increases initially slighter before it rises again between the 90th and the 180th day. After 45 days of implantation the distance between the losses in mass of the coated and uncoated specimens is nearly the same. The loss in volume shows a comparable behavior to the loss in mass (Figure 4).

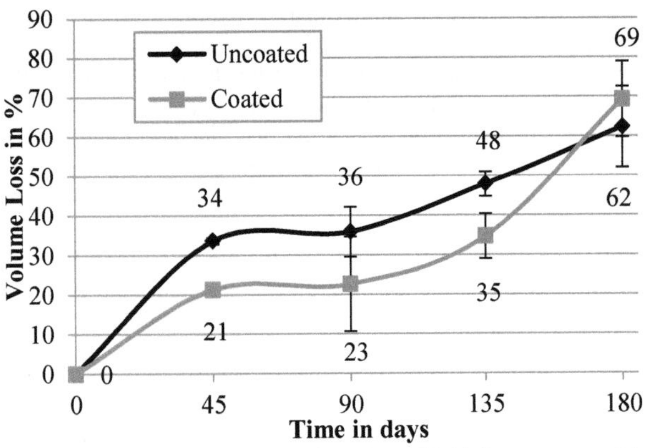

Figure 4: Loss in volume as a function of initial geometry of the coated and uncoated specimens after 45, 90, 135 and 180 days.

After 45 days the loss in volume of the coated specimens is also 13 % lower. Within the following 135 days the distance between the losses of mass remains almost the same, with the exception of the last analyze after 180days. The disparities between the losses in volume and in mass could result from the use of different ways to quantify the losses. The mass loss was analyzed after the corrosion products were removed by means of a chromic acid treatment. This treatment, however, is limited in case of complex surface structures (e.g. pitting, boreholes). In comparison, the loss in volume was determined by means of µCT. This process offers the opportunity to differentiate between areas of bare, none corroded metal and remaining corrosion products for a second time.

Conclusion

The results made within this in vivo study prove that over a period of 180 days the in vivo losses in mass and volume tended to be higher for the uncoated specimens. However, the magnesium fluoride coating has only an initial protecting influence on the specimens. With an increasing time of implantation the effect of the coating reduced and the influence of the layer with physiologically originating elements increased. Furthermore, the corrosion layers on the uncoated specimens provided with increasing time a better protection against corrosion. In addition, it could be shown that the structure of the growing corrosion layer depends on the protection layer. Magnesium fluoride coated specimens showed a much larger percentage of calcium in the corrosion layer.

Acknowledgements

All work carried out for this study was done within the Collaborative Research Centre 599 "Sustainable bioresorbable and permanent implants of metallic and ceramic materials" Subproject R1 (Magnesium Degradation), which was funded by the German Research Foundation (DFG).

References

1. Kraus T, Fischerauer SF, Hänzi AC, Uggowitzer PJ, Löffler JF, Weinberg AM (2012) Magnesium alloys for temporary implants in osteosynthesis: in vivo studies of their degradation and interaction with bone. In: Acta biomaterialia 8 (3), S. 1230–1238.
2. Witte F, Kaese V, Haferkamp H, Switzer E, Meyer-Lindenberg A, Wirth C, Windhagen H (2005) In vivo corrosion of four magnesium alloys and the associated bone response. In: Biomaterials 26 (17), S. 3557–3563.
3. Witte F, Fischer J, Nellesen J, Vogt C, Vogt J, Donath T, Beckmann F (2010) In vivo corrosion and corrosion protection of magnesium alloy LAE442. In: Acta biomaterialia 6 (5), S. 1792–1799.
4. Weber CM, Eifler R, Seitz J, Maier HJ, Reifenrath J, Lenarz T, Durisin M (2015) Biocompatibility of MgF2-coated MgNd2 specimens in contact with mucosa of the nasal sinus – A long term study. In: Acta biomaterialia 18 (0), S. 249–261.
5. Birbilis N, Cavanaugh MK, Sudholz AD, Zhu SM, Easton MA, Gibson MA (2011) A combined neural network and mechanistic approach for the prediction of corrosion rate and yield strength of magnesium-rare earth alloys. In: Corrosion Science 5 (1), S. 168–176.
6. Kopp V, Lefebvre W, Pareige C (2011) Determination of the Mg-Rich Phase Boundary of the Binary Mg-Nd Phase Diagram by Means of Atom Probe Tomography. In: Journal of Phase Equilibria and Diffusion 32 (4), S. 298–301.
7. Seitz J, Eifler R, Stahl J, Kietzmann M, Bach F (2012) Characterization of MgNd2 alloy for potential applications in bioresorbable implantable devices. In: Acta biomaterialia 8 (10), S. 3852–3864.
8. Birbilis N, Easton MA, Sudholz AD, Zhu SM, Gibson MA (2009) On the corrosion of binary magnesium-rare earth alloys. In: Corrosion Science 51 (3), S. 683–689.
9. Krishnamurthy S, Khobaib M, Robertson E, Froes FH (1988) Corrosion behavior of rapidly solidified Mg-Nd and Mg-Y alloys. In: Materials Science and Engineering 99 (1-2), S. 507–511.
10. Südholz AD, Kirkland NT, Buchheit RG, Birbilis N (2011) Electrochemical Properties of Intermetallic Phases and Common Impurity Elements in Magnesium Alloys. In: Electrochemical and Solid-State Letters 14 (2), S. C5-C7.
11. Drynda A, Deinet N, Braun N, Peuster M (2009) Rare earth metals used in biodegradable magnesium-based stents do not interfere with proliferation of smooth muscle cells but do induce the upregulation of inflammatory genes. In: Journal of biomedical materials research. Part A 91 (2), S. 360–369.
12. Chester AC, Antisdel JL, Sindwani R (2009) Symptom-specific outcomes of endoscopic sinus surgery: a systematic review. In: Otolaryngology–head and neck surgery: official journal of American Academy of Otolaryngology-Head and Neck Surgery 140 (5), S. 633–639.
13. Macdonald KI, McNally JD, Massoud E (2009) Quality of life and impact of surgery on patients with chronic rhinosinusitis. In: Journal of Otolaryngology - Head & Neck Surgery 38 (2), S. 286–293.
14. Bhattacharyya N (2003) The economic burden and symptom manifestations of chronic rhinosinusitis. In: American Journal of Rhinology 17 (1), S. 27–32.
15. Freeman SB, Blom ED (2000) Frontal sinus stents. In: The Laryngoscope 110 (7), S. 1179–1182.
16. Hunter B, Silva S, Youngs R, Saeed A, Varadarajan V (2010) Long-term stenting for chronic frontal sinus disease: case series and literature review. In: The Journal of Laryngology & Otology 124 (11), S. 1216–1222.
17. Seitz J, Eifler R, Weber C, Lenarz TH, Maier HJ, Durisin M (2015) In vivo degradation effects of alloy MgNd2 in contact

with mucous tissue. In: Journal of biomedical materials research. Part A 103(7), 2427-2440.
18. Hosemann W (2003) Postoperative Rezidivprophylaxe bei chronischer Pansinusitis und Polyposis nasi. In: HNO 51 (4), S. 279–283.
19. Weber R, Mai R, Hosemann W, Draf W, Toffel P (2000) The success of 6-month stenting in endonasal frontal sinus surgery. In: Ear, Nose, & Throat Journal 79 (12), S. 930-2, 934, 937-8.
20. Zheng Y, Wu J, Ng JC, Wang G, Lian W (2002) The absorption and excretion of fluoride and arsenic in humans. In: Toxicology Letters 133 (1), S. 77–82.
21. Seitz J, Bormann U, Collier K, Wulf E, Eifler R, Bach FW (2012) Application of a Bioactive Coating on Resorbable, Neodymium Containing Magnesium Alloys, and Analyses of their Effects on the In Vitro Degradation Behavior in a Simulated Body Fluid. In: Adv. Eng. Mater. 14 (6), S. B311.
22. DIN EN ISO 8407 Corrosion of metals and alloys – Removal of corrosion products from corrosion test specimens. Wien: Austrian Standards plus GmbH; ISO 8407:2009.
23. Chiu KY, Wong MH, Cheng FT, Man HC (2007) Characterization and corrosion studies of fluoride conversion coating on degradable Mg implants. In: Surface and Coatings Technology 202 (3), S. 590–598.

Magnesium Technology 2016
Edited by: Alok Singh, Kiran Solanki, Michele V. Manuel, and Neale R. Neelameggham
TMS (The Minerals, Metals & Materials Society), 2016

Flow Induced Biodegradation Behavior of Magnesium Metal: From Bioreactors to *In Vivo* Models

Juan Wang[1,2], Nan Huang[1], Yeoheung Yun[2], Jagannathan Sankar[2]

1 Department of Materials Science and Engineering, Southwest Jiaotong University, Chengdu 610064, China
2 NSF-ERC for Revolutionizing Metallic Biomaterials, North Carolina A & T State University, Greensboro, NC 27411, US
Address correspondence to huangnan1956@163.com

Keywords: Magnesium, Biodegradation, Bioreactor, *In vitro, In vivo*, Flow

Abstract

Absorbable metals have been widely tested in various *in vitro* environments to evaluate their suitability as a vascular stent material. However, there exists a gap between *in vivo* and *in vitro* test results. A key step to solve this issue is to identify and test the relevant biochemical and biophysical microenvironments and parameters in test-systems. The purpose of this study was to develop a series of biomimetic flow bioreactors similar to *in vivo*/clinical conditions to reveal biodegradation behavior and mechanism of magnesium (Mg) metals or stents, such as *in vitro* varied fluid flow bioreactor with computational fluid dynamic calculation, *in-situ* or *real-time* electrochemical monitoring bioreactor, porcine aortal bioreactor as well as *in vivo* rat aortal system. Hydrodynamics as a biophysical vascular parameter directly affects on corrosion kinetics, types, rates and products. Establishing an appropriate methodology for accurate determinations of degradation parameters plays a vital part in predicting the fate of Mg-based stent.

Introduction

There is currently a paradigm shift in the medical implant device industry where absorbable metallic materials made from Mg and Mg alloys are being reviewed to replace permanent metal implants [1, 2]. Even with the heavy volume of research that has been conducted, the current standards without modification from American Society for Testing and Materials (ASTM) and International Organization for Standards (ISO) for testing absorbable metals are not appreciate without modification, since the traditional *in vitro* test methods cannot predict the *in vivo* results [3]. A key step of the development of current standards is to identify and test the relevant microenvironments and parameters in test-systems for the specific application [4]. Furthermore, with respect to clinical applications of absorbable Mg-based vascular stents, there needs to be a full understanding of the mechanisms of biodegradation of Mg-based stents.

The vascular stent in blood vessel are exposed to various hemodynamic forces associated with blood flow. These include fluid shear, the tangential force derived from the friction of blood flowing across the luminal cell surface, tensile stress due to deformation of the vessel wall by transvascular flow, and normal stress caused by the hydrodynamic pressure differential across the vessel wall [5]. Those should affect the degradation rate and mode of Mg. The main objective of this research is to better understand the effect of fluid dynamics on the Mg degradation, mimicking the vascular environment. We developed a series of vascular bioreactors to simulate *in vivo*/clinical conditions to reveal degradation mechanisms, such as *in vitro* varied fluid flow system with computational fluid dynamic calculation, in-situ or real-time electrochemical monitoring system, a porcine aortal bioreactor as well as a rat aortal *in vivo* system.

Model 1: *In vitro* varied fluid flow system [6]

The aim of this model was to study corrosion behavior of Mg alloys (MgZnCa plates and AZ31 stents) under varied fluid flow conditions representative of the vascular environment. Experiments revealed that fluid hydrodynamics, fluid flow velocity and shear stress play essential roles in the corrosion behavior of absorbable Mg-based stent devices. Based on the results (Figure 1), it was discovered that: (i) flow-induced shear stress (FISS) accelerates the overall corrosion (including localized, uniform, pitting and erosion corrosions) due to the increased mass transfer and mechanical force; (ii) FISS increased the average uniform corrosion rate, the localized corrosion coverage ratios and depths and the removal rate of corrosion products inside the corrosion pits; (iii) for the strut regions in the Mg stent, areas facing the flow direction have higher shear stress, which caused corrosion product detachment and led to strut fracture. This study demonstrates that flow-induced corrosion needs be understood so that Mg-based stents in vascular environments can be effectively designed.

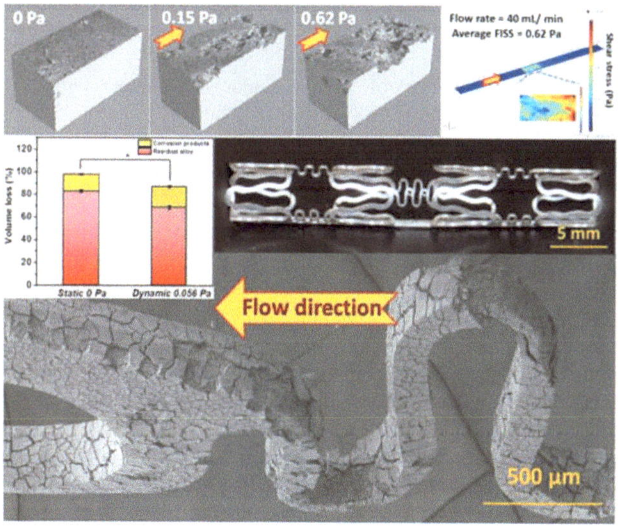

Figure 1. X-ray micro-CT 3-D renderings of full view MgZnCa plates under the FISS values with a single computational fluid dynamics contour and a 2-D slice of test area; SEM image of AZ31 stent under the flow condition for 7 days and volume ratio of

residual AZ31 alloy and corrosion products to initial stent after the static and dynamic degradation tests for 7 days [6].

Model 2: *In-situ* or *real-time* electrochemical monitoring system [7]

An *in-situ* and *real-time* electrochemical study in a vascular bioreactor was designed to analyze corrosion mechanism of Mg alloy (MgZnCa) under mimetic hydrodynamic conditions (Figure 2). Effect of hydrodynamics on corrosion kinetics, types, rates and products was analyzed. FISS accelerated mass and electron transfer, leading to an increase in uniform and localized corrosions. FISS increased the thickness of uniform corrosion layer, but filiform corrosion decreased the resistance of this layer at high FISS conditions. FISS also increased the removal rate of localized corrosion products owing to increased mass transfer effect and mechanical force. Impedance-estimated and linear polarization-measured polarization resistances provided a consistent correlation to corrosion rate calculated by volume loss. The results suggest that the established *in-situ* and *real-time* electrochemical technique can be a useful tool to reveal the flow effect on the magnesium alloy corrosion process and therefore provide a deeper understanding of corrosion product formation and corrosion damage of the alloy.

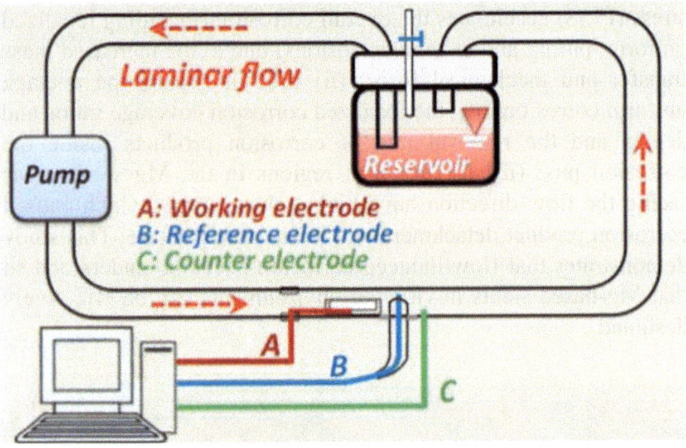

Figure 2. Schematic of the vascular bioreactor with an electrochemical monitoring system.

Model 3: Porcine aortal bioreactor and rat aortal *in vivo* test [7]

Compared with the above-mentioned bioreactors, the porcine aortal bioreactor emphasized to be closer to the physiologically hydrodynamic microenvironments found in the different vascular remodeling stages after stent implantation, i.e. pre- and post-endothelialization stages. A porcine aortal bioreactor and a rat aortal *in vivo* model are developed to study the flow convection and diffusion induced biodegradation behavior of Mg, as well as, it emphasizes to establish a comparison and correlation of biodegradation results from standardized immersion, arterial bioreactor and *in vivo* assessments (Figure 3).

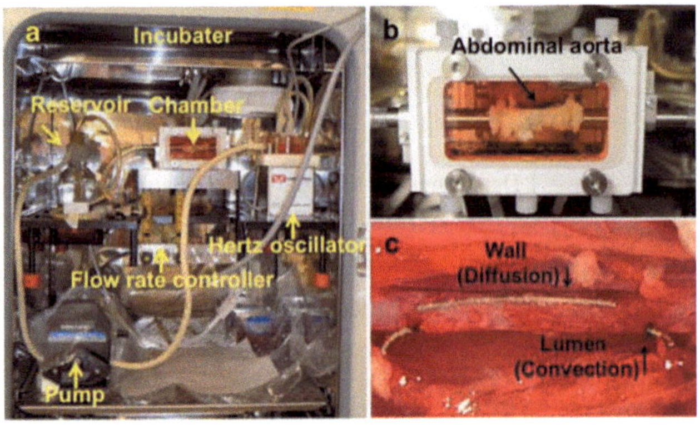

Figure 3. (a) Physical photograph of Lume Gen Bioreactor. The Mg wires were implanted into the wall and lumen of a porcine aorta in the chamber (b) and a rat aorta *in vivo* (c).

As shown in Figure 4, the results revealed the *in vivo* degradation was slower than the *in vitro* degradation both in the standardized static condition and in the flow bioreactor condition. In terms of the aortal bioreactor, the flow convection on the lumen surface severely accelerated Mg degradation due to the increase of mass transfer, fluid shear stress and pulsatile stress on the Mg, compared with the low Mg degradation rate at the flow diffusion environment of the wall. In the aortal in vivo model, the degradation rate of Mg wire in the wall is faster than that in the lumen at in vivo aorta model, and calcification was observed around Mg surrounding tissue in the wall. The established porcine aorta porcine aortal bioreactor and rat aortal *in vivo* model are expected to establish an appropriate methodology for accurate and standardized determinations of stent degradation parameters in the pre- and post- endothelialization stage, which plays a vital part in predicting the fate of Mg-based stent.

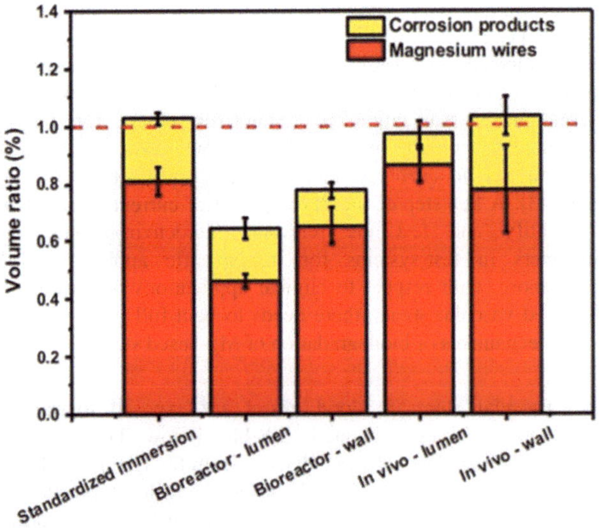

Figure 4. Volume ratios of residual Mg wires (red columns) and corrosion products (yellow columns) in the static no aorta, and the lumen and wall of the bioreactor and *in vivo* conditions for 3 days.

Summary and recommendations for future research

The aim of this work was to provide relevant test methods and data to build standardization strategies on absorbable metals. In particular, our results give an insight into the effect of fluid dynamics on the Mg degradation behavior under mimicking vascular environment. We developed a series of vascular bioreactors to simulate *in vivo*/ physiological conditions to reveal degradation mechanisms, such as *in vitro* varied fluid flow system with computational fluid dynamic calculation, *in-situ* or *real-time* electrochemical monitoring system, porcine aortal bioreactor model as well as rat aortal *in vivo* model system.

Although there are many teams across the globe working to advance biodegradable metals, there are still many great strides to be made in order to develop approved (by the FDA, SFDA, and CE) and commercializable medical devices. The current standards set forth by the ISO and ASTM are simply not suitable for biodegradable metallic biomaterials and new standards must be developed. Thus far, the absence of correlation between *in vitro* and *in vivo* degradation behavior is a major limitation, which proves the need for: i) improved *in vitro* models that mimic as closely as possible the *in vivo* conditions; ii) the identification of factors that influence the corrosion rate *in vitro*, which can help to predict *in vivo* degradation behavior; and iii) the identification of corrosion mechanisms *in vivo*. The latter is especially difficult to determine because the monitoring of degradation as well as physiological parameters *in vivo* and online is currently more or less impossible.

To develop these standards, scientists and researchers will need to develop highly reproducible experimental approaches that can be used by different companies and laboratories in both the industrial and academic world. Although this is a great challenge ahead, when engineers, doctors, scientists and researchers are brought together for one cause, it will lead to the development of approved biomaterial candidates in the near future. Long before any type of approval is possible for biodegradable vascular stents, there needs to be a full understanding of the mechanisms of biodegradation of vascular stents, along with developing reliable test-methods, test-systems and test-strategies.

Acknowledgements

The authors wish to sincerely thank Prof. Deling Kong, Dr. Zhihong Wang and Ms. Yifan Wu from Nankai University for assessments and discussion in animal tests. This study was supported by the Engineering Research Center for Revolutionizing Metallic Biomaterials (NSF-0812348) from the National Science Foundation, USA, and National Natural Science Foundation of China under Grant No. 81330031. X-ray CT imaging was made possible through an USA-NSF Award No. 0959511.

References

[1] Y.F. Zheng, X.N. Gu, F.Witte, "Biodegradable Metals," *Material Science Engineering R,* 77 (2014), 1-34.

[2] F.Witte, "The History of Biodegradable Magnesium Implants: A Review," *Acta Biomaterialia,* 6(2010), 1680-1692.

[3] P.K. Bowen et al, "Rates of *In Vivo* (Arterial) and *In Vitro* Biocorrosion for Pure Magnesium," *Journal of Biomedical Materials Research Part A,* 103(2015), 341-9.

[4] J. Wang et al, "Absorbable Magnesium-based Stent: Physiological Factors to Consider for *In Vitro* Degradation Assessments," *Regenrative Biomaterials,* 2015, no.1-11.

[5] C.F. Buchanan et al, "Flow Shear Stress Regulates Endothelial Barrier Function and Expression of Angiogenic Factors in a 3D Microfluidic Tumor Vascular Model," *Cell adhesion & migration,* 8(2014), 517-24.

[6] J. Wang et al, "Flow-induced Corrosion Behavior of Absorbable Magnesium-based Stents," *Acta Biomaterialia,* 10(2014), 5213-5223.

[7] J. Wang, "Biodegradation Behavior and Biological Responses of Magnesium-based Materials for Vascular Stents in the Hydrodynamic Near-Physiological Ambient Conditions," (Ph.D. thesis, Southwest Jiaotong University, 2015), 68-102.

Magnesium Technology 2016

MAGNESIUM-BASED BIODEGRADABLE IMPLANTS

Corrosion / Market and Clinic

Magnesium Technology 2016
Edited by: Alok Singh, Kiran Solanki, Michele V. Manuel, and Neale R. Neelameggham
TMS (The Minerals, Metals & Materials Society), 2016

Understanding Corrosion-Assisted Cracking of Magnesium Alloys for Bioimplant Applications

R.K. Singh Raman[1,2] and Shervin Eslami Harandi[1]

[1] Department of Mechanical & Aerospace Engineering, Monash University, Melbourne, VIC, 3800, Australia
[2] Department of Chemical Engineering, Monash University, Melbourne, VIC, 3800, Australia

Keywords: Biodegradable implants, Magnesium alloys, Stress corrosion cracking, Corrosion fatigue

Abstract

Magnesium (Mg) alloys possess great potential for their use as temporary implants and devices (such as pins, wires, screws, plates, and stents). Use of Mg alloys as temporary implants will completely avoid the cumbersome procedure of second surgery (which is required when such implants are constructed out of traditional materials such as titanium alloys or stainless steels). However, there are some limitations of Mg as a temporary implant. Firstly, the high corrosion rates of Mg alloys in the physiological environment may lead to loss in mechanical integrity of the implants. Secondly, the simultaneous action of the corrosive human-body-fluid and the mechanical loading can cause sudden and catastrophic fracture due to corrosion-assisted expedited cracking, such as stress corrosion cracking (SCC) and/or corrosion fatigue (CF). SCC and CF of Mg alloy implants are vastly unexplored research areas. This article provides an overview of the experimental results on SCC and CF of different Mg alloys in corrosive environments including simulated body fluid (SBF), and discusses associated fracture mechanisms.

Introduction

Temporary implants support living tissues until regeneration or healing is completed. Identifying a material that degrades in a physiological environment without harmful effects would be highly beneficial for the temporary implant application since an extra surgical procedure for implant removal after it has served its purpose can be averted. In this context, magnesium (Mg) alloys as temporary human implants have attracted forefront research attention [1-4]. Among metallic engineering materials, Mg and its alloys possess excellent biocompatibility and their mechanical properties are closer to those of natural bone compared to traditional metallic implant materials (Table I) [4-8].

However, in such use, the alloys must possess adequate resistance to cracking/fracture under the simultaneous action of the corrosive human body fluid (HBF) and the mechanical loading characteristics of the human body. Human-body fluid-assisted fracture has always been among concerns [9] in use of implants of traditional materials. A recent study on this topic [10] resoundingly establishes such fracture to be among the major concerns, yet it is a vastly underexplored research area in use of Mg alloys as bio-implants. First studies including the one at the authors' group indicated the susceptibility of a Ca-containing Mg-alloy (AZ91) to human-body-fluid-assisted fracture of Mg alloys [11], and subsequent comprehensive studies [12-14] have confirmed such fracture in traditional Mg alloys.

Table I: Comparison of the physical and mechanical properties of Mg with various implant materials and natural bone [5-8]

Material	Density (g/cm^3)	Compressive yield strength (MPa)	Tensile strength (MPa)	Young's modulus (GPa)	Fracture toughness ($MPam^{1/2}$)	Elongation at fracture (%)	Degradation
Natural bone	1.7-2	130-180	30-280	5-20	3-6	3-4	-
Ti-based alloys	4.4-4.5	758-1117	930-1140	110	55-115	8-15	Biostable
Co-Cr-based alloys	8.3-9.2	450-1000	-	220	-	-	Biostable
Stainless steel	7.9-8.1	170-310	480-620	165-200	50-200	30-40	Biostable
Mg-based alloys	1.74-2	65-100	100-400	50	15-40	2-6	Biodegradable
Fe-based alloys	7.87-8	150	200-1400	200	-	40	Biodegradable
Zn-based alloys	6.8- 7	-	100-400	90	-	-	Biodegradable

Stress Corrosion Cracking (SCC)

SCC is caused by the simultaneous action of tensile loading and corrosive environment. There is limited literature on SCC of Mg alloys in body fluid [11-13, 15, 16]. These studies have invariably suggested the tested alloys to be susceptible to SCC under *in-vitro* condition. SCC of Mg alloys in chloride solutions (such as, human body fluid) usually involves hydrogen embrittlement (HE) [16- 18]. Mg and its alloys are known to readily generate hydrogen. As Mg alloys also suffer rapid anodic dissolution (which concerns a separate SCC mechanism), there have been some speculations as to whether hydrogen or anodic dissolution is the predominant mechanism for SCC mechanism of Mg alloys. This aspect was systematically addressed in a recent study on SCC of AZ91D in *m*-SBF [12]. In this study, SSRT tests were conducted at a strain rate of $2.2 \times 10^{-7}\,s^{-1}$ under different conditions: (a) strained in air, (b) strained in modified simulated body fluid (*m*-SBF) solution, (c) immediately strained in air after pre-immersion in *m*-SBF solution, and (d) continuously cathodically charged and simultaneously pulled in *m*-SBF. Besides investigating SCC mechanism (anodic dissolution or hydrogen embrittlement), the tests under the different conditions were also aimed at establishing whether the loss of the mechanical property is indeed a result of the synergistic effect of stress and corrosive environment (i.e. SCC), or it can simply be attributed to the continuously reducing cross sectional area of the specimen due to high corrosion rate of Mg alloys. Time to-failure data for specimens tested under the conditions (a)-(d), are shown in Fig.1. Straining in *m*-SBF resulted in a considerable reduction in time-to-failure as compared to straining in air. As the specimen failed in 28 h when strained in *m*-SBF (Fig. 1), a pre-immersion time of 28 h was selected for straining the pre-immersed AZ91D tensile specimen in air (assuming the stress-independent corrosion damage to be similar to that of specimen strained in *m*-SBF solution). However, time-to-failure for the specimens pulled in air and that pre-immersed in *m*-SBF for 28 h and then strained in air were very similar (Fig. 1). On this basis, the loss of mechanical property in the case of specimen pulled in *m*-SBF was attributed to the simultaneous effect of stress and corrosive *m*-SBF (i.e., SCC) (Fig. 1), and hence, the inference that the stress-independent corrosion contributed only marginally.

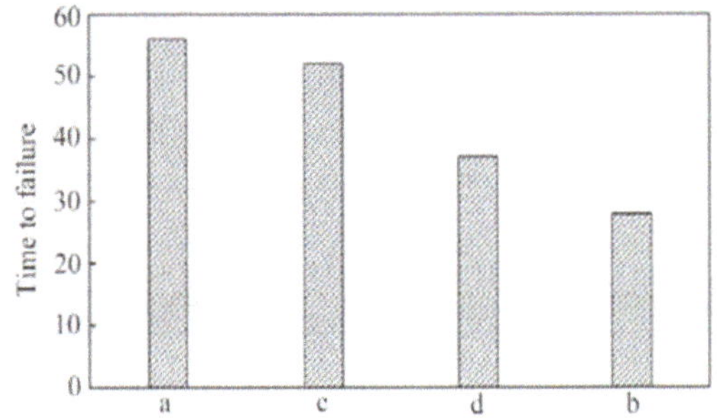

Figure 1. Time to-failure data for specimens strained at $2.2 \times 10^{-7} s^{-1}$ while exposed to different test environments: (a) in air, (b) in *m*-SBF solution, (c) pre-immersed in *m*-SBF for 28h and then immediately strained in air, and (d) continuously charged at 200 mV cathodic to open circuit potential in *m*-SBF during the test. (This figure was developed on the basis of data extracted from Ref 12, where the authors have not reported the statistical variance of the SCC data, however, the variance in SCC susceptibility data in strain rate of $10^{-7} s^{-1}$ is reported in Ref 19).

Also, the continuous cathodic charging condition (i.e., when anodic dissolution is minimal), caused a considerable decrease in the time-to-failure (Fig. 1) that inferred hydrogen-assisted stress corrosion cracking (HASCC) to have contributed to SCC. However, because the maximum loss in mechanical property was observed in the case of the specimen strained in *m*-SBF solution at open circuit conditions (and not under the cathodic charging condition), it was inferred that it was the combined effect of HASCC and anodic dissolution that caused the *m*-SBF-assisted loss in mechanical properties (Fig. 1). Fractography of the specimen tested in air revealed dimples over entire fracture surface whereas the specimen tested in *m*-SBF showed the localized attack at the specimen circumference and one of the distinctive features of SCC in the areas of fracture surface, viz. transgranular cracking and localized cracks.

Corrosion Fatigue

Simultaneous action of cyclic loading and corrosion that leads to sudden failure of metallic materials is called CF. Considering the rapid degradation of Mg alloys in physiological environments and the cyclic nature of the common load spectra in human body, CF should be among the critical design parameters for biodegradable implant application of Mg alloys. Though there are several instances of CF fracture of biomedical implants of traditional materials [20-22], very little is reported on SCC of Mg alloys in body fluid. The authors' group has recently carried out experimental investigation on CF of AZ91D alloy in air and modified simulated body fluid (m-SBF) [14]. The alloy tested in m-SBF suffered a considerable reduction in fatigue resistance as compared to that tested in air (Fig. 2). Microscopy of the fatigue tested specimens showed the intermetallic inclusions to have causes corrosion pits that facilitated easy crack initiation in m-SBF and rapider failure.

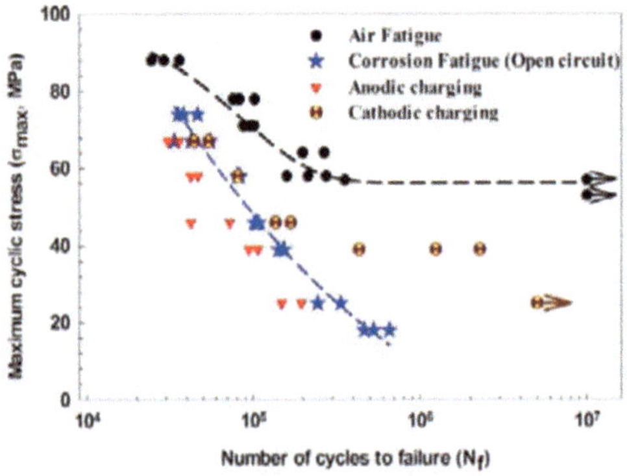

Figure 2. S-N curves for AZ91D Mg alloy under different conditions [14]

This suggestion is consistent with the reported role of surface defects in fatigue cracking of Mg alloys where fatigue cracks have been found to initiate at stress concentration sites and manufacturing defects, such as notches, corners, pores, shrinkage voids and inclusions [23-25] (slip bands and twin boundaries being other initiation sites [26, 27]). The role of pits in fatigue cracking was also duly validated in the authors' group [14] by the tests under an applied cathodic potential that suppressed pitting and thereby, improved fatigue resistance. However, as shown in Fig. 2, the role of the applied cathodic potential in improved fatigue resistance is particularly pronounced in the low stress regime which is explained on the basis that it will take a greater pit depth to initiate/propagate CF cracks at low stresses.

Similar to SCC, the mechanism for CF crack propagation involves hydrogen assisted cracking or/and anodic dissolution at the crack tip. Zeng et al. [28] reported hydrogen embrittlement as the main fatigue crack propagation mechanism of AZ61 Mg alloy in wet air. Uematsu et al. [29] have also suggested hydrogen embrittlement to be the primary and anodic dissolution to be the secondary factor for faster crack propagation in controlled humidity. However, Shipilov [30] reported dissolution mechanism as the main factor of faster fatigue crack propagation of VMD10 and IMV7-1 Mg alloys in different solutions. CF and SCC studies on Mg alloys in the context of bio-implant applications, including those shown in Fig. 1 and 2, have been performed by relatively simplistic approaches to the mechanical loading as well as the test environment. Testing parameters such as frequency and mode of loading, chemistry of the test solutions, and geometry of the specimens in these studies and those under actual *in-vivo* conditions are considerably different. There is a need for more carefully selecting the laboratory testing parameters that will enable more meaningful and realistically comparable *in-vitro* and *in-vivo* results. Another article in this proceeding (by S.E. Harandi and R.K. Sinh Raman) provides an elaboration of a specific approach to address this concern.

Conclusions

Due to their specific properties including biocompatibility, mechanical properties and corrosion resistance, Mg-alloys have generated greater interest among potential biodegradable candidates for temporary implant applications. Surface flaws of the implants when present with corrosive body fluid can lead to fracture under static and cyclic loadings. There has been very little research on corrosion fatigue (CF) and stress corrosion cracking (SCC) of Mg alloy in body fluids. Tests in simulated body fluid have shown a magnesium alloy to be susceptible to both SCC and CF, and pitting to be the common crack initiator.

References

1. B. Zberg, P.J. Uggowitzer, J.F. Löffler, "MgZnCa Glasses Without Clinically Observable Hydrogen Evolution for Biodegradable Implants', *Nature Materials*, 8 (2009), 887-891.
2. E. Ma, J. Xu, "Biodegradable Alloys: The Glass Window of Opportunities, *Nature Materials*, 8 (2009), 855-857.
3. F. Witte, J. Fischera, J. Nellesenb, H. A. Crostackb, V. Kaesec, A. Pischd, F. Beckmanne, H. Windhagen, "In Vitro and in Vivo Corrosion Measurements of Mag Alloys", *Biomaterials,* 27 (2006), 1013-1018.
4. M.P. Staige, A. M. Pietaka, J. Huadmaia, G. Dias, " Mg and its Alloys as Orthopedic Biomaterials: A Review", *Biomaterials,* 27 (2006), 1728-1734.
5. J. Kubásek, I. Pospíšilová, D. Vojtěch, E. Jablonská, T. Ruml, "Structural, Mechanical and Cytotoxicity Characterization of As-Cast Biodegradable Zn-xmg (x = 0.8–8.3 %) Alloys. Materials and Technology, 5(2014), 623-629.
6. D. Vojtech, J. Kubasek, J. Čapek, I. Pospisilova, "Magnesium, Zinc and Iron Alloys for Medical Applications in Biodegradable Implants', *In 23rd International Conference on Metallurgy and Materials*, Thomson Reuters: Brno, Czech Republic, EU 2014.
7. L. N. Zhang, Z. T. Hou, X. Ye, Z. B. Xu, X. L. Bai, P. Shang, 'The Effect of Selected Alloying Element Additions on Properties of Mg-Based Alloy as Bioimplants: A Literature Review", *Frontiers of Materials Science*, 7 (2013), 227-236.
8. S. Jafari, S. E. Harandi, R. K. Singh Raman, "A Review of Stress Corrosion Cracking and Corrosion Fatigue of Magnesium Alloys for Biodegradable Implant Applications", *The Minerals, Metals & Materials Society*, 67 (2015), 1143-1153.
9. M. Sivakumar, S. Rajeswari, "Investigation of Failures in Stainless Steel Orthopaedic Implant Devices: Pit-Iinduced Stress Corrosion Cracking", *Materials Science Letters*, 11 (1992), 1039-1042.
10. R. A. Antunes, M.L.C de Oliveria, "Corrosion Fatigue of Biomedical Metallic Alloys: Mechanisms and Mitigation", *Acta Biomaterialia*, 8 (2012), 937-962.
11. M. B. Kannan, R.K. Singh Raman, "In Vitro Degradation and Mechanical Integrity of Calcium-Containing Magnesium Alloys in Modified-Simulated Body Fluid", *Biomaterials,* 29 (2008), 2306-2314.
12. L. Choudhary, R. K. Singh Raman, "Magnesium Alloys as Body Implants: Fracture Mechanism under Dynamic and Static Loadings in a Physiological Environment, *Acta Biomaterialia,* 8 (2012), 916-923.
13. M.B. Kannan, R.K. Singh Raman, "Evaluation of SCC Behaviour of AZ91 Alloy in Modified-Simulated Body Fluid for Orthopaedic Implant Application", *Scripta Materialia*, 59 (2008), 175-178.
14. S. Jafari, R. K. Singh Raman, C. H. J. Davies, "Corrosion Fatigue Cracking of a Magnesium Alloy in Modified Simulated Body Fluid", *Engineering Fracture Mechanics,* 137 (2015), 2-11.
15. M. B. Kannan, R. K. Singh Raman, F. Witte, C. Blawert, W. Dietzel. "Influence of Circumferential Notch and Fatigue Crack on the Mechanical Integrity of Biodegradable Magnesium-Based Alloy in Simulated Body Fluid. *Biomedical Materials Research Part B: Applied Biomaterials*, 96 (2011), 303-309.
16. L. Choudhary, R. K. Singh Raman, J. Hofstetter, P. J. Uggowitzer, In-vitro Characterization of Stress Corrosion Cracking of Aluminium-Free Biodegradable Magnesium Alloys for Temporary Bio-Implant Applications, *Materials Science and Engineering C,* 42 (2014), 629-636.
17. D. G. Chakrapani, E. N. Pugh, "Hydrogen Embrittlement in a Mg-Al Alloy", *Metallurgical Transactions A*, 7 (1976), 173-178.
18. N. Winzer, A. Atrens, G. Song, E. Ghali, W. Dietzel, K. Kainer, N. Hort, C. Blawert, "A Critical Review of the Stress Corrosion Cracking (SCC) of Magnesium Alloys", *Advanced Engineering Materials*, 7 (2005), 659-693.
19. L. Choudhary, "Stress Corrosion Cracking of Magnesium Alloys for Temporary Bio-implant Applications" (Ph.D. thesis, Monash University, 2013), 89.
20. C. M. Rimnac, T. M. Wright, D. L. Bartel, R. W. Klein, A. A. Petko, "Failure of Orthopedic Implants: Three Case Histories", *Materials Characterization*, 26 (1991), 201-209.
21. G. K. Triantafyllidis, A. V. Kazantzis, K. T. Karageorgiou, "Premature Fracture of a Stainless Steel 316L Orthopaedic Plate Implant by Alternative Episodes of Fatigue and Cleavage Decoherence", *Engineering Failure Analysis,* 14 (2007), 1346-1350.
22. H. Amel-Farzad, M. T. Peivandi, S. M. R. Yusof-Sani, "In-Body Corrosion Fatigue Failure of a Stainless Steel Orthopaedic Implant with a Rare Collection of Different Damage Mechanisms", *Engineering Failure Analysis,* 14 (2007), 1205-1217.

23. C. Potzies, K. U. Kainer, "Fatigue of Magnesium Alloys", *Advanced Engineering Materials*, 6(2004), 281-289.
24. T. S. Shih, W. S. Liu, Y. J. Chen, "Fatigue of As-Extruded AZ61A Magnesium Alloy", *Materials Science and Engineering: A*, 325(2002), 152-162.
25. F. Yang, F. Lv, X. M. Yang, S. X. Li, Z. F. Zhang, Q. D. Wang, "Enhanced Very High Cycle Fatigue Performance of Extruded Mg–12Gd–3Y–0.5Zr Magnesium Alloy", *Materials Science and Engineering A*, 528 (2011), 2231-2238.
26. S. M. Yin, F. Yang, X. M. Yang, S. D. Wu, S. X. Li, G. Y. Li, "The Role of Twinning-Detwinning on Fatigue Fracture Morphology of Mg–3%Al–1%Zn alloy", *Materials Science and Engineering A*, 494(2008), 397-400.
27. M. F. Horstemeyer, N. Yang, K. Gall, D. McDowell, J. Fan, P. Gullett, "High Cycle Fatigue Mechanisms in a Cast AM60B Magnesium Alloy", *Fatigue and Fracture of Engineering Materials and Structures*, 25(2002), 1045-1056.
28. R. C. Zeng, E. H. Han, W. Ke, "Effect of Temperature and Relative Humidity on Fatigue Crack Propagation Behaviof of AZ61 Magnesium Alloy", *Materials Science Forum*, 546-549(2007), 409-412.
29. Y. Uematsu, T. Kakiuchi, M. Nakajima, Y. Nakamura, S. Miyazaki, H. Makino, "Fatigue Crack Propagation of AZ61 Magnesium Alloy under Controlled Humidity and Visualization of Hydrogen Diffusion along the Crack Wake", *International Journal of Fatigue*, 59(2014), 234-243.
30. S. A. Shipilov, "Mechanisms for Corrosion Fatigue Crack Propagation", *Fatigue & Fracture of Engineering Materials & Structures*, 25(2002), 243-259.

Magnesium Technology 2016
Edited by: Alok Singh, Kiran Solanki, Michele V. Manuel, and Neale R. Neelameggham
TMS (The Minerals, Metals & Materials Society), 2016

IN VITRO CORROSION AND CYTOCOMPATIBILITY PROPERTIES OF Mg-2Gd-X(Ag, Ca) ALLOYS

Yiyi Lu, Yuanding Huang, Frank Feyerabend, Regine Willumeit-Römer, Karl Urich Kainer, Norbert Hort
Institute of Materials Research, Helmholtz-Zentrum Geesthacht, Max-Planck-Strasse 1, 21502 Geesthacht, Germany

Keywords: Mg-2Gd-x(Ag, Ca) alloys, solution treatment (T4), in vitro corrosion, cytocompatibility.

Abstract

The effects of solution treatment (T4) on the in vitro corrosion properties were investigated for Mg-2Gd-xAg (x=1, 2 wt. %), Mg-2Gd-xCa (x=0.4, 0.8 wt. %) alloys. The cytocompatibility of pure Mg, Mg-2Gd, Mg-2Gd-2Ag and Mg-2Gd-0.8Ca alloy was studied as well. All alloys were prepared by permanent mould casting. T4 heat treatment was carried out at 510 °C for 48 h under Ar atmosphere and specimens were quenched in water at room temperature. Corrosion experiments were performed in cell culture medium under cell culture conditions. Cytocompatibility was evaluated by direct cell adhesion and live/dead staining tests. After T4 heat treatment, the amount of intermetallic phases is significantly decreased, which leads to significant improvement of corrosion properties by a very high factor (5-30). The results of live/dead staining reveal that the cells are mostly in viable state and homogenously distributed over the surface. The corrosion layer shows no cytotoxicity to primary human osteoblasts.

1. Introduction

As new degradable medical materials, Mg and Mg alloys have many advantages and are attracting more and more attentions in the biological implant applications. Compared with other metal implant materials currently commonly used in clinical, Mg alloys have good mechanical compatibility and load-bearing ability; their density (about 1.75-1.85 g/cm^3) is similar to that of human bone and their Young's modulus (about 44 GPa) is close to that of human bone (20 GPa) as well, which meets the requirements of ideal bone plate [1-3]. However, they still have some disadvantages such as fast corrosion and production of hydrogen gas during corrosion process. Moreover, knowledge on the effects of alloying elements and intermetallic phases in biological environments is necessary to be further understood. Therefore, large number of new biomedical Mg alloys has been developed to improve the property profiles. Mg-REs (rare earth elements) alloys show increased mechanical properties and corrosion resistance [4-9]. Among all RE elements, Gd has a high solid solubility in Mg, with a maximum solubility of 23.49 wt. % at the eutectic temperature [10]. From the biological point of view, REs were used in anticancer drugs due to their anti-carcinogenic properties [11-12]. Moreover, it was reported that Mg-Ag alloys show a good biocompatibility and satisfactory antibacterial properties [13]. As we all know, the major element of human bone is Ca. It plays an important role in physiological processes such as nerve, muscle stress, nerve impulse transmission, maintenance of the heart rhythm, blood coagulation, and cell adhesion, etc. Meanwhile Ca has a small density (1.55 g/cm^3), which makes the density of magnesium alloy closer to that of human bone. Coexistence of magnesium and calcium helps bone healing and can avoid stress shielding [14].

Besides determination of mechanical properties and biocompatibility, corrosion behavior is always a key issue to magnesium alloys and for a preselection only corrosion properties are of interest. However, Mg implants display uncontrolled and high corrosion rates that need to be modified in order to be used as biomedical implant [1, 15]. In this work, the solid solution effect on the in vitro corrosion for Mg-2Gd-xAg (x=1, 2 wt. %) and Mg-2Gd-xCa (x=0.4, 0.8 wt. %) alloys were studied. The corrosion behaviors were investigated in in vitro conditions using immersion tests. The weight loss method was used to calculate the in vitro corrosion rate. Compared to other electrolytes, Dulbecco's modified eagle medium (DMEM) supplemented with fetal bovine serum (FBS) may be a more credible electrolyte for in vitro corrosion tests [16-17]. In addition, the cytocompatibility was examined using primary cells by direct cell adhesion and live/dead staining tests [8, 18].

2. Experimental Procedure

Permanent mould casting [19] was used to prepare all the alloys with real chemical compositions and density listed in Table I. High-purity Mg (MEL, UK, 99.95 wt. %) was molten in a mild steel crucible under a protective atmosphere (Ar + 0.3% SF6). Pure Gd (Grirem, China, 99.5 wt. %) for Mg-2Gd alloy and Mg-33.3 wt% Gd master alloy together with pure silver Ag (99.99%, ESG Edelmetall-Handel GmbH & Co. KG, Germany) or pure Ca (Alfa Asear, Germany, 99.51 wt.%) for ternary alloys were added at a melt temperature of 720 °C. The melt was stirred at 200 rpm for 30 min. After that, the melt was poured into a mild steel mould preheated to 550 °C. A filter (Foseco SIVEX FC, Foseco GmbH, Borken, Germany) was used to assure the cleanliness of the cast ingots. The filled mould was held at room temperature and afterwards the ingot was extracted from the mould after air cooling. The ingot was cuboid block with a size of 30 cm ×21 cm × 3 cm. All samples for each measurement were not taken from the middle of ingot but from the same position near surface of the ingot. The contents of Gd, Ag and Ca were determined by X-ray fluorescence spectrometer (Bruker AXS S4 Explorer, Bruker AXS GmbH., Germany). The contents of Fe, Cu and Ni were determined by spark emission spectrometer (Spectrolab M, Spektro, Germany). The density of the alloys was determined in ethanol by Archimedean principle. For all alloys, a solution treatment (T4) was performed at 510 °C for 48 h, followed by water quenching. The microstructure was investigated using VEGA3 TESCAN scanning electron microscope (SEM) equipped with energy-dispersive X-ray spectroscopy (EDS) at an accelerative voltage of 20 kV. The metallographic specimens for SEM were prepared by grinding with SiC waterproof abrasive paper, polishing with a lubricant containing 1 μm diamond particles and 0.05 μm colloidal silica (OPS), and etching in an etchant containing 30 ml deionized water, 140 ml ethanol, 7 ml glacial acetic acid and 8 g picric acid.

In order to investigate the influence of solution treatment on the corrosion properties, the in vitro corrosion rate was measured. The samples with following geometries (l=10 mm, w=10 mm, h=4 mm) were taken from the plate ingots. The specimens for in vitro corrosion tests were prepared by mounting and grinding with silicon carbide emery paper up to 2500 grit. Five specimens were tested for each alloy. Before sample sterilization the initial weight, the geometries and the density of the samples was recorded. All the samples were sterilized in a 70% ethanol solution using an ultrasonic bath for 20 min. Each sample was immersed in 2 ml o DMEM+10%FBS (cell culture medium, CCM) for 168 h under cell culture conditions (CCC) (37°C, 20% O_2, 5% CO_2, 95% rH). The samples lay on surface during the immersion. The immersion medium was changed every 2-3 days to present a semi-static immersion test. After immersion, the subsequently formed corrosion products were removed via immersing the corroded specimens in chromic acid (180 g/L in distilled water, VWR international, Darmstadt, Germany) for 20 min at room temperature. Then the degradation rate was calculated in mm/year using the equation:

$$\mathrm{CR} = \frac{8.76 \cdot 10^4 \cdot \Delta g}{A \cdot t \cdot \rho}$$

Where Δg is the weight change in grams, A is the surface area of the sample in cm^2, t is the immersion time in hours, and ρ is the density in g/cm^3 [19-20].

The cytocompatibility was studied by direct cell adhesion and live/dead staining tests. Specimens with the geometry (l=10 mm, w=10 mm, h=1.5 mm) were prepared by electrical discharge machining. All surfaces of specimens were ground with 2500 grit emery paper and ultrasonically cleaned in ethanol for 5 min and then dried in air. Pure Mg specimens were also investigated as a reference to evaluate the influence of alloying elements on the cytocompatibility. In vitro qualitative analysis of cell coverage and viability was performed by using LIVE/DEAD (Life technologies, Darmstadt, Germany) assay. In the present work, the primary osteoblasts were used to perform the test for three days. The isolation of cells was approved by the local ethics committee and performed according to the Declaration of Helsinki as described in Fischer et al. [21]. Shortly speaking, the cancellous bone was removed from the hip joint and transferred to a cell culture flask. Then cancellous particles were covered with cell culture medium and incubated until around 70-80% confluent growth then divided. The used osteoblasts were the third passage. Agarose coating was prepared on 12-well-plate (Nunc, Wiesbaden, Germany) to avoid growing of cells on the plates. Before cytotoxicity tests, all specimens were sterilized in 70% ethanol for 20 min and pre-incubated for 24 h under CCC in CCM, then 10^5 primary osteoblasts were seeded on each sample surface in 12-well-plate for 3 days. Thereafter, 3 ml CCM was added to each well. Afterwards, the staining solution was prepared by adding 4 µl Calcein AM (LIVE) and 10 µl Ethidium homodimer-1 (DEAD) to 10 ml of phosphate buffered saline (PBS). The samples were first washed with PBS to eliminate the non-adherent cells, followed by immersing each sample with 1.5 ml of staining solution and incubate it under CCC for 20 minutes. Then the staining solution was replaced by DMEM and samples were visualized by fluorescent microscope (Nikon GmbH, Düsseldorf, Germany), the applied filters were FITC (Ex: 460-500 nm; Em: 510-560 nm; Mirror at 505 nm) and Texas red (Ex: 540-580nm; Em: 600-660; Mirror at 595 nm).

3. Results and discussion

3.1 Solid solution effects on microstructure and in vitro corrosion rate

Fig. 1 shows SEM microstructures of as-cast Mg-2Gd-2Ag alloy (a) and Mg-2Gd-0.8Ca alloy (c), and both alloy after solid solution treatment (b), (d), respectively. The SEM microstructures were observed before and after T4 to investigate the influences of solid solution on in vitro corrosion. It reveals that both the as-cast Mg-2Gd-2Ag and Mg-2Gd-0.8Ca alloys are mainly composed of continuous equiaxed dendrites with the segregation of solute atoms. The segregation mainly locates at dendritic boundaries. In contrast, for both alloy, after solid solution treatment the morphologies of dendrites become indistinguishable. The amount of the intermetallic phases (IMPs) decreases drastically, indicating that the IMPs mostly dissolved inside the matrix. The second phases also disappeared at the grain boundaries after the solution treatment. However, some quite small IMPs were still homogenously distributed (Fig. 1b, 1d). The T4 heat treatment has led to a large change in the microstructures. The grain size is 817.9 ± 677.6 µm for Mg-2Gd-1Ag alloy, 387.1 ± 267.4 µm for Mg-2Gd-2Ag alloy, 349.6 ± 229.6 µm for Mg-2Gd-0.4Ca alloy and 688.1 ± 534.5µm for Mg-2Gd-0.8Ca alloy in as-cast condition.

The in vitro corrosion rates of the as-cast and the T4-treated Mg-2Gd-xAg and Mg-2Gd-xCa ternary alloys are shown in Fig. 2. The estimated corrosion rates in mm/year within one week revealed the lowest degradation rates for Mg-2Gd-1Ag alloy with a value of 0.51 ± 0.11 and for Mg-2Gd-2Ag alloy with a value of 1.30 ± 0.05. Whereas Mg-2Gd-0.4Ca alloy exhibits a rate of 6.09 ± 1.32 mm/year and Mg-2Gd-0.8Ca alloy 4.52 ± 0.38 mm/year in as-cast condition. After solid solution treatment, the in vitro corrosion rate in mm/year is 0.09 ± 0.02 for Mg-2Gd-1Ag alloy, 0.08 ± 0.01 for Mg-2Gd-2Ag alloy, 0.28 ± 0.14 for Mg-2Gd-0.4Ca alloy and 0.15 ± 0.03 for Mg-2Gd-0.8Ca alloy, respectively. For both alloy systems, the smaller grain size leads to higher corrosion rate. In the as-cast condition, the in vitro corrosion rate increases with the increment of Ag content for Mg-2Gd-xAg alloys due to the increased formation of IMPs. However, for Mg-2Gd-xCa alloys with more addition of Ca the in vitro corrosion

Table I Real chemical compositions and the density of the alloys.

Alloys	Real chemical composition (wt. %)							Density (g/cm^3)
	Gd	Ag	Ca	Cu	Fe	Ni	Mg	
Mg-2Gd-1Ag	2.17	1.18	—	0.0020	0.0020	<0.0026	Balance	1.798
Mg-2Gd-2Ag	2.12	2.47	—	0.0022	0.0020	<0.0027	Balance	1.815
Mg-2Gd-0.4Ca	2.33	—	0.43	0.0023	0.0024	<0.0023	Balance	1.780
Mg-2Gd-0.8Ca	2.21	—	0.75	0.0021	0.0018	<0.0027	Balance	1.779

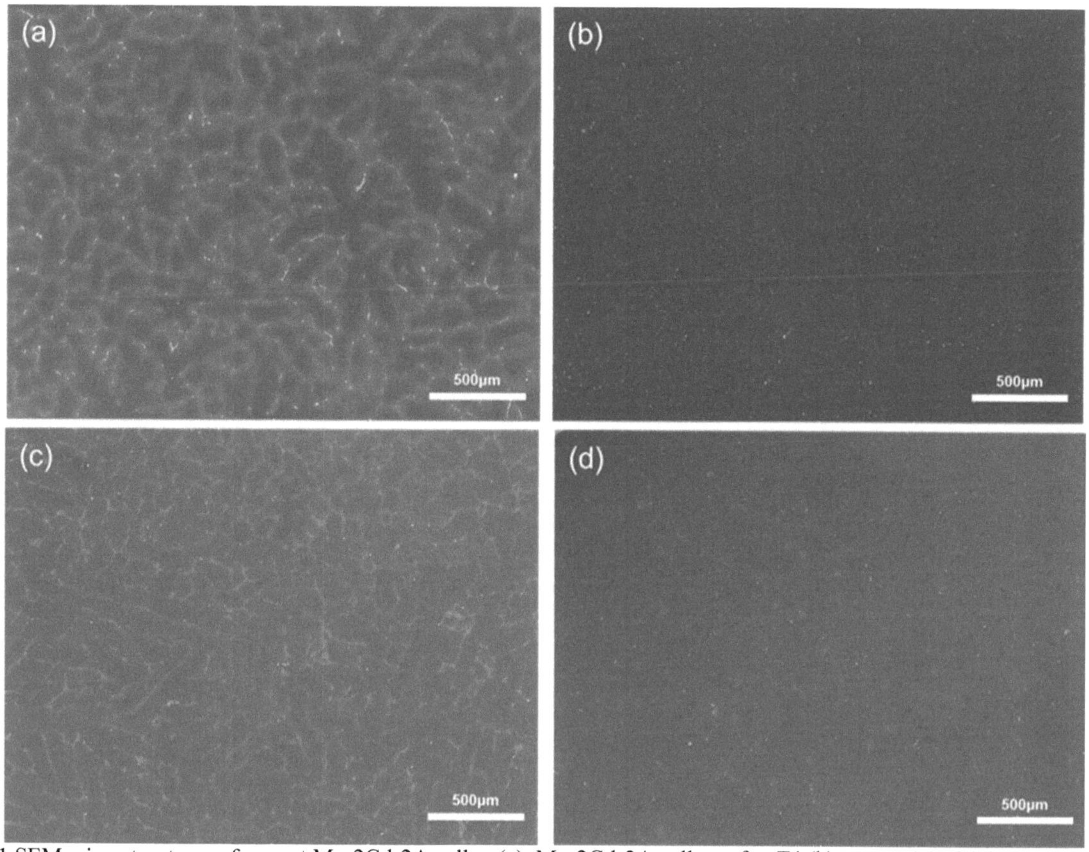

Fig. 1 SEM microstructures of as-cast Mg-2Gd-2Ag alloy (a), Mg-2Gd-2Ag alloy after T4 (b), as-cast Mg-2Gd-0.8Ca alloy (c) and Mg-2Gd-0.8Ca alloy after T4 (d).

rate shows a slight decrease. Fig. 2 reveals the effects of alloying elements (Ag, Ca) and solid solution treatment on the in vitro corrosion rate of the alloys. After T4 heat treatment, the in vitro corrosion rate largely reduces for both Mg-2Gd-xAg and Mg-2Gd-xCa ternary alloys. Their corrosion resistance is obviously improved by a very high factor from 5 to 30 due to the large reduction in the amount of IMPs. Compared with Mg-2Gd-xCa alloys, Mg-2Gd-xAg alloys have a lower corrosion rate in both as-cast and T4 conditions because of their different types and distributions of IMPs.

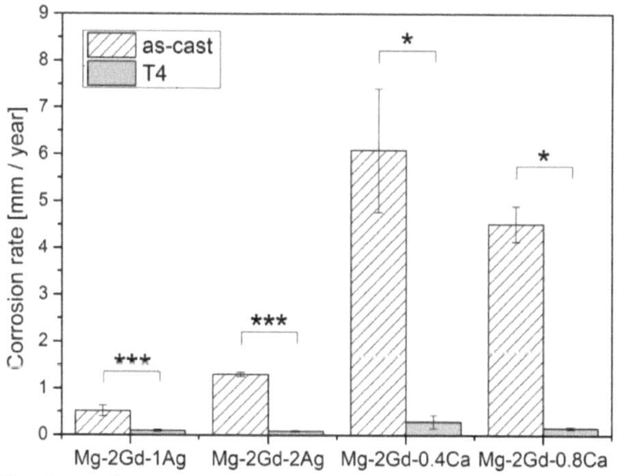

Fig. 2 In vitro corrosion rate of Mg-2Gd-x (Ag, Ca) alloys (mean ± SD, significance level $p<0.001$=*** with t-test, $p<0.05$=* with Mann-Whitney U-test).

3.2 Evaluation of cytocompatibility

Fig.3 shows fluorescence microscope images after LIVE/DEAD staining of primary pre-osteoblasts cultured on as-cast pure Mg, T4 treated Mg-2Gd, Mg-2Gd-2Ag and Mg-2Gd-0.8Ca alloys. The T4 treated alloys were used to study the effect of alloying elements (Gd, Ag and Ca) on cytocompatibility due to their lower corrosion rate compared with as-cast alloys, which makes the live/dead test easier because of less bubbles. The in vitro cell tests were performed to evaluate cell viability and the ability to adhere to the surface of the alloys. Pure Mg, Mg-2Gd, Mg-2Gd-2Ag and Mg-2Gd-0.8Ca alloys were used to culture the primary cells for 3 days. The viability and cell distribution were checked subsequently by LIVE/DEAD staining. The viable cells are marked in green and dead cells in red, respectively.

After exposure to osteoblast cultures for 3 days, a large number of cells was found on the surfaces of pure Mg, Mg-2Gd, Mg-2Gd-2Ag and Mg-2Gd-0.8Ca alloys. In addition, a large number of live cells and only a few dead cells are observed on pure Mg and all the other three alloys after 3 days. The number of dead cells is very few in comparison with that of live cells. The cells are well-spread and adhered to the corrosion layer on both the surfaces of pure Mg and Mg-2Gd alloys. The enrichment of Gd in the corrosion layer shows no negative effect on the adhered cells, demonstrating that no cytotoxicity was observed. In comparison with Mg-2Gd alloy, fewer dead cells are observed for Mg-2Gd-2Ag and Mg-2Gd-0.8Ca alloys. The live cells spread even further. The results of live/dead staining reveal that the cells are predominantly healthy and spread throughout the surface area. The cells are mostly in viable state and homogenously distributed over the surface. Therefore, the corrosion layer shows no cytotoxicity to primary human osteoblasts. The addition of Ag or Ca to Mg-2Gd alloy shows a positive effect on the cell adhesion.

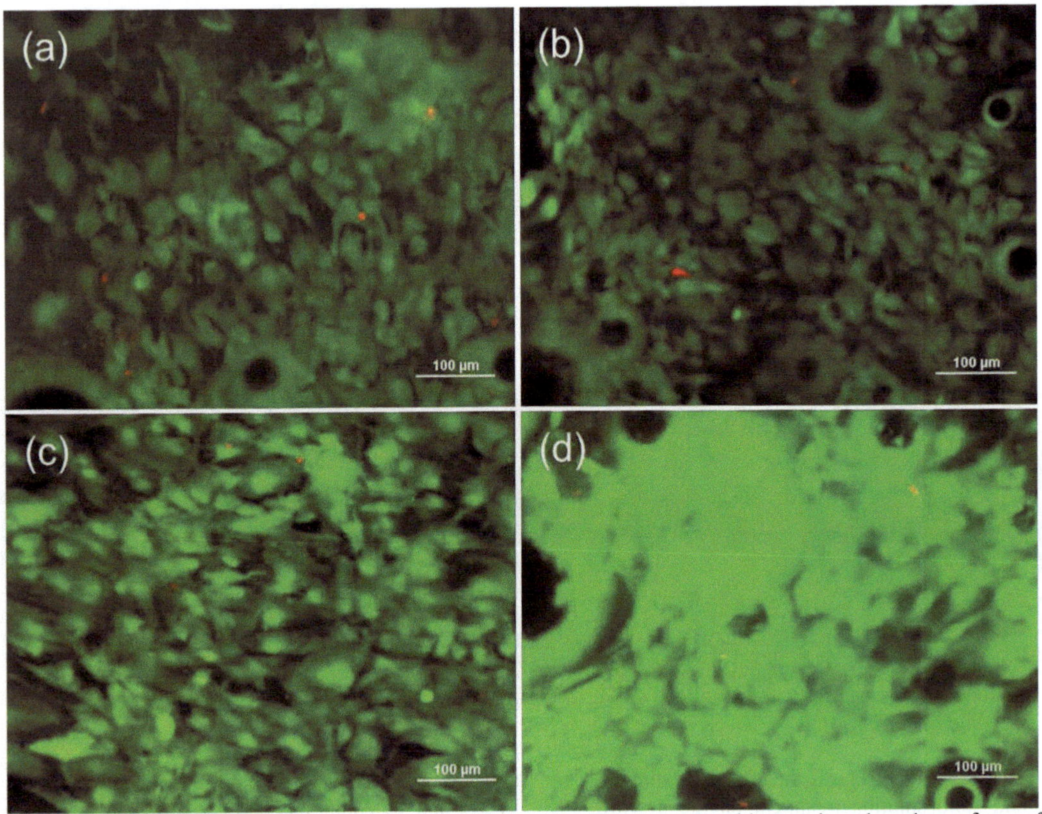

Fig. 3 Fluorescence microscope images after LIVE/DEAD staining of primary pre-osteoblasts cultured on the surfaces of pure Mg (a), Mg-2Gd alloy (b), Mg-2Gd-2Ag alloy (c) and Mg-2Gd-0.8Ca alloy (d) for 3 days, where live cells are labelled in green and dead cells in red.

4. Conclusions

Effects of solid solution treatment on the in vitro corrosion rate were investigated for Mg-2Gd-xAg and Mg-2Gd-xCa alloys. For both alloy systems, most of the IMPs can dissolve into the matrix after T4 heat treatment (510 °C, 48 h), resulting in a large improvement of in vitro corrosion resistance with a very high factor (5-30). The reduction of in vitro corrosion rate is mostly a result of dissolving IMPs, but also due to a change in grain size and in getting rid of the dendritic structure.

Through the evaluation of cytocompatibility we can conclude that a good viability and a homogenous distribution of the cells on the alloy surface could be observed. The corrosion layer enriched with Gd shows no cytotoxicity to primary human osteoblasts. The additions of Ag and Ca elements show a positive effect on the cell adhesion with respect to Mg-2Gd binary alloy.

Acknowledgement

This research was financially supported by the Helmholtz Virtual Institute VH-VI-523 (In vivo studies of biodegradable magnesium based implant materials).

Reference

1. M. P. Staiger, A. M. Pietak, J. Huadmai and G. Dias, "Magnesium and its alloys as orthopedic biomaterials: a review," *Biomaterials*, 27 (2006), 1728-1734.

2. Y. Chen, Z. Xu, C. Smith and J. Sankar, "Recent advances on the development of magnesium alloys for biodegradable implants," *Acta biomaterialia*, 10 (2014), 4561-4573.

3. N. Li and Y. Zheng, "Novel Magnesium Alloys Developed for Biomedical Application: A Review," *Journal of Materials Science & Technology*, 29 (2013), 489-502.

4. L. Yang, Y. Huang, F. Feyerabend, R. Willumeit, C. Mendis, K. U. Kainer and N. Hort, "Microstructure, mechanical and corrosion properties of Mg-Dy-Gd-Zr alloys for medical applications," *Acta biomaterialia*, 9 (2013), 8499-8508.

5. L. Hou, Z. Li, Y. Pan, L. Du, X. Li, Y. Zheng and L. Li, "In vitro and in vivo studies on biodegradable magnesium alloy," *Progress in Natural Science: Materials International*, 24 (2014), 466-471.

6. A. C. Hänzi, P. Gunde, M. Schinhammer and P. J. Uggowitzer, "On the biodegradation performance of an Mg–Y–RE alloy with various surface conditions in simulated body fluid," *Acta biomaterialia*, 5 (2009), 162-171.

7. P. Gunde, A. C. Hänzi, A. S. Sologubenko and P. J. Uggowitzer, "High-strength magnesium alloys for degradable implant applications," *Materials Science and Engineering A*, 528 (2011), 1047-1054.

8. F. Feyerabend, J. Fischer, J. Holtz, F. Witte, R. Willumeit, H. Drucker, C. Vogt and N. Hort, "Evaluation of short-term effects of rare earth and other elements used in magnesium alloys on primary cells and cell lines," *Acta biomaterialia*, 6 (2010), 1834-1842.

9. A. Atrens, M. Liu and N. I. Zainal Abidin, "Corrosion mechanism applicable to biodegradable magnesium implants," *Materials Science and Engineering: B*, 176 (2011), 1609-1636.

10. A. A. Nayeb-Hashemi, J. B. Clark and A. S. M. International: *Phase diagrams of binary magnesium alloys*. (ASM International, Metals Park, Ohio, 1988).

11. X.-N. Gu and Y.-F. Zheng, "A review on magnesium alloys as biodegradable materials," *Frontiers of Materials Science in China*, 4 (2010), 111-115.

12. D. Persaud-Sharma and A. Mcgoron, "Biodegradable Magnesium Alloys: A Review of Material Development and Applications," *Journal of biomimetics, biomaterials, and tissue engineering*, 12 (2012), 25-39.

13. D. Tie, F. Feyerabend, W. D. Muller, R. Schade, K. Liefeith, K. U. Kainer and R. Willumeit, "ANTIBACTERIAL BIODEGRADABLE Mg-Ag ALLOYS," *Eur Cells Mater*, 25 (2013), 284-298.

14. M. Salahshoor and Y. Guo, "Biodegradable Orthopedic Magnesium-Calcium (MgCa) Alloys, Processing, and Corrosion Performance," *Materials*, 5 (2012), 135-155.

15. G. L. Song, "Control of biodegradation of biocompatable magnesium alloys," *Corrosion Science*, 49 (2007), 1696-1701.

16. D. Tie, F. Feyerabend, N. Hort, R. Willumeit and D. Hoeche, "XPS Studies of Magnesium Surfaces after Exposure to Dulbecco's Modified Eagle Medium, Hank's Buffered Salt Solution, and Simulated Body Fluid," *Advanced Engineering Materials*, 12 (2010), B699-B704.

17. R. Willumeit, J. Fischer, F. Feyerabend, N. Hort, U. Bismayer, S. Heidrich and B. Mihailova, "Chemical surface alteration of biodegradable magnesium exposed to corrosion media," *Acta biomaterialia*, 7 (2011), 2704-2715.

18. R. Willumeit, A. Möhring and F. Feyerabend, "Optimization of Cell Adhesion on Mg Based Implant Materials by Pre-Incubation under Cell Culture Conditions," *Int J Mol Sci*, 15 (2014), 7639-7650.

19. N. Hort, Y. Huang, D. Fechner, M. Stoermer, C. Blawert, F. Witte, C. Vogt, H. Druecker, R. Willumeit, K. U. Kainer and F. Feyerabend, "Magnesium alloys as implant materials - Principles of property design for Mg-RE alloys," *Acta biomaterialia*, 6 (2010), 1714-1725.

20. L. Yang, N. Hort, R. Willumeit and F. Feyerabend, "Effects of corrosion environment and proteins on magnesium corrosion," *Corros Eng Sci Techn*, 47 (2012), 335-339.

21. J. Fischer, D. Pröfrock, N. Hort, R. Willumeit and F. Feyerabend, "Improved cytotoxicity testing of magnesium materials," *Materials Science and Engineering: B*, 176 (2011), 830-834.

Magnesium Technology 2016
Edited by: Alok Singh, Kiran Solanki, Michele V. Manuel, and Neale R. Neelameggham
TMS (The Minerals, Metals & Materials Society), 2016

APPROPRIATE CORROSION-FATIGUE TESTING OF MAGNESIUM ALLOYS FOR TEMPORARY BIOIMPLANT APPLICATIONS

Shervin Eslami Harandi[1] and R.K. Singh Raman[1,2]

[1] Department of Mechanical & Aerospace Engineering, Monash University, Melbourne, VIC, 3800, Australia
[2] Department of Chemical Engineering, Monash University, Melbourne, VIC, 3800, Australia

Keywords: Corrosion fatigue, Three-point bending, Magnesium alloys

Abstract

Magnesium (Mg) alloys are attractive for their great potential as temporary bio-implants. However, Mg alloys can suffer sudden cracking/fracture under the simultaneous action of cyclic loading and the corrosive physiological environment, i.e., corrosion fatigue (CF). Though there are reports on CF of Mg alloy, it is necessary that the investigations of such fracture should be performed under conditions that appropriately simulate those in actual human body conditions. This article describes a relatively more accurate testing procedure and preliminary data generated using this procedure.

Introduction

Despite highly advantageous properties of Mg-based alloys as temporary body implants, they have not been used widely for such applications. Mg alloys suffer rapid degradation accompanied by undesirable amounts of hydrogen gas generation even in mildly corrosive medium such as simulated body fluid (SBF). Moreover, due to the presence of the second phase precipitates in the alloy which are highly cathodic to the alloy matrix phase, most Mg alloys are susceptible to pitting both *in-vivo* and *in-vitro*. Pits generally provide the bare Mg surface through which atomic hydrogen can enter and embrittle the matrix to cause premature cracking [1].

Implant devices are regularly subjected to mechanical loading during their service in human physiological environment. For example, a femoral implant experience cyclic bending loading during normal walking [2] and a cardiovascular stent undergoes cyclic tension loading due to heart beat [3]. Simultaneous action of tensile or cyclic loadings along with the presence of aggressive physiological environment can cause sudden cracking/fracture of Mg alloys due to stress corrosion cracking (SCC) and corrosion fatigue (CF), which often occur at stresses considerably below design stresses for the components operating in the non-corrosive environment. Because the brittle SCC/CF fractures can be sudden, catastrophic and premature, they are believed to be the most dangerous forms of corrosion-assisted failures. Therefore, it is essential for any implant material to have adequate resistance to cracking or fracture in actual body environments. However, CF and SCC studies on Mg alloys in the context of bio-implant applications have been performed by relatively simplistic approaches to the mechanical loading as well as the test environment. Substantial differences have been observed between *in-vitro* testing parameters such as frequency and mode of loading, chemistry of the test solutions, and geometry of the specimens in these studies and those under actual *in-vivo* conditions. For *in-vitro* studies to be more meaningful and realistically comparable with the *in-vivo* results, the laboratory testing parameters should be selected carefully. This approach will help in optimizing the *in-vitro* test, as well as in reducing the number of relevant animal experiments. This article describes the development of an experimental approach for appropriate mechanochemical condition for corrosion fatigue testing of Mg alloys for temporary bio-implant applications and some preliminary CF data.

Appropriate Mechano-Chemical Conditions for Corrosion Fatigue Testing

A bone is a matrix of inorganic salts and connective tissues. Healing of a fractured bone is profoundly governed by the biological and mechanical factors. Bones and implants experience acute and complex loadings during service. For example, a cantilever bending in a loaded femur occurs when femoral head is exposed to compressive forces resulting in bending moment in diaphyseal bone shaft along with the axial compressive effect. Hence, for such applications, it is imperative to evaluate the mechanical integrity of the Mg alloys in the femoral midshaft gap by using the bending CF test. The test can be conducted in either three-point or four-point loading configuration. Under three-point bending, the bending moment rises linearly from zero at the supports to a maximum at the central loading points, while the length of specimen receives a constant shear force. Under four-point bending, bending moment increases linearly from zero at the support points to a constant load between these two points. In contrast to three-point bending, where the failure emerges at the middle point of force applications, fractures in four-point bending is observed at the weakest point between the two inner forces but not necessarily at the midpoint. The femoral implant in the midshaft gap experiences loading similar to that in three-point cycling bending.

For *in-vivo* studies, it is appropriate that the dimension of implants be chosen based on the bone and animal size. According to international standard (ISO) [4], a cylindrical implant into tibial and femoral diaphysis bone shaft should not have its length greater than 2 mm and its diameter greater than 50 mm. Therefore, for more meaningful tests to simulate *in-vivo* condition, the dimensions of the specimens for *in-vitro* tests were selected similar to those for *in-vivo* application. Accordingly, samples of a magnesium alloy (AZ91D) were machined from an as-cast billet to a diameter of 2 mm and length of 50 mm (Fig. 1). For three-point bending tests on these samples, a fixture was designed to fit into the testing machine for accommodating the samples of proper size (Fig 2a). Bending tests create both tensile and compression stresses respectively on the convex and concave sides of the specimen. As a result, an area of shear stress can be produced along the midline. In order to ensure that the initial fracture is predominantly produced as a result of tensile or compression stress, the shear stress must be minimized by controlling the span-to-diameter ratio to 16:1. In addition, to prevent the stress concentration, the loading nose and supports should be designed with a circular surface.

For the three-point bending CF test, a corrosion chamber made of acrylic was designed and fabricated (Fig. 2b and 2c). In order to avoid the galvanic corrosion during the test, the supports of the bending fixture were located outside the corrosion chamber and the loading nose was made of stainless steel coated with alumina (Al_2O_3) ceramic (Fig. 2d). The test medium was maintained at the nominal body temperature of 37 °C, using a water bath. Besides, to simulate *in-vitro* flow of body fluid, a submersible pump was used for continuously circulating the Hanks' solution through the corrosion chamber (Fig. 2b and 3).

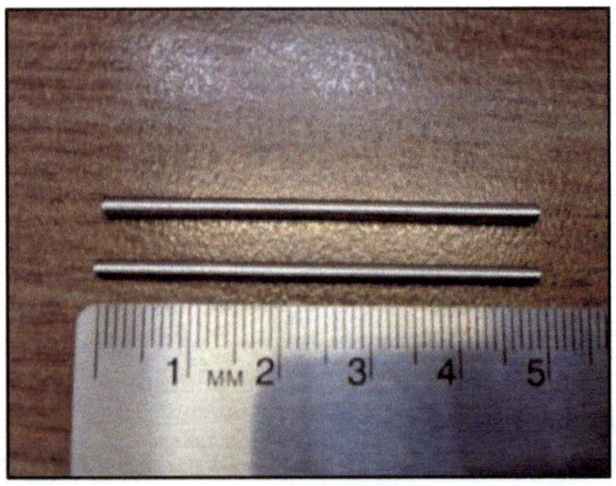

Figure 1. Specimens for three-point bending corrosion fatigue tests

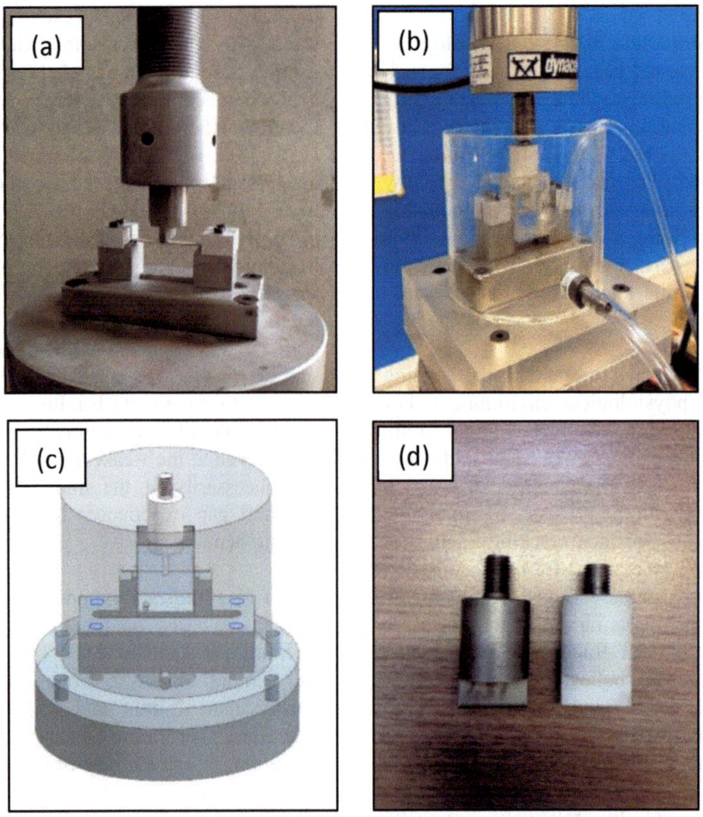

Figure 2. (a) Three-point bending fixture, (b) acrylic corrosion chamber (placed within a jacket that contains the test solution in the event of any leakage), (c) a schematic of set-up for loading, corrosion chamber and jacket, and (d) bare and coated loading nose.

Different pseudo-physiological solutions that mimic the composition of body fluids have been employed for *in-vitro* experiments that includes modified simulated body fluid (m-SBF), Hanks', Dulbecco's Modified Eagle's medium (DMEM), Ringer's solution etc. Compositions and ion concentrations of common solutions are listed in Table I. In the tests reported here, Hanks' balanced salt solution (HBSS) , the recipe of the original Hanks' solution (Table II) [5], which is the most commonly used media for *in-vitro* study. However, in addition to inorganic ions, blood plasma also contains organic compounds such as proteins, amino acids, and glucose that need to be accounted for while conducting *in-vitro* studies on biodegradable Mg alloys. Besides, rapid degradation of Mg alloys during corrosion testing dramatically increase alkalinity of pseudo-physiological solutions resulting in the pH increase. This issue can be addressed by either sufficiently frequent replacement of the used test solution with the fresh one or maintaining the pH of the existing solution by adding common buffering agents such as HEPES or Tris. However, these buffering agents accelerate the degradation rate of Mg alloy because they consume the generated OH^- that would otherwise provide corrosion protection. Bubbling CO_2 through the test solution can increase the level of carbonic acid, and thereby, a controlled purging of the solution with CO_2 can be employed for controlling the pH at the required level. For this purpose, a set of CO_2-pH controller, CO_2 gas cylinder, and a solenoid valve were used for controlling the pH of the test solution. Digital pH controller monitors pH of solution using a pH probe. Once the pH increases even slightly beyond the set level, the controller opens gas solenoid valve allowing CO_2 gas to be bubbled into the solution, thereby lowering the pH back to the required level (Fig. 3).

Table I. Ion concentrations and organic compounds in common solutions [2]

		Blood Plasma	Hanks'	Ringers'	DMEM	Original SBF	m-SBF
Na^+	(mmol L^{-1})	142	142	130	127.3	142	142
K^+	(mmol L^{-1})	5	5.9	4	5.3	5	5
Ca^{2+}	(mmol L^{-1})	2.5	1.3	1.4	1.8	2.5	2.5
Mg^{2+}	(mmol L^{-1})	1.5	0.8	-	0.8	1.5	1.5
HCO_3^-	(mmol L^{-1})	27	4.2	-	44.1	4.2	10
Cl^-	(mmol L^{-1})	103	145	109	90.8	148.8	103
HPO_4^{2-}	(mmol L^{-1})	1	0.8	-	0.9	1	1
SO_4^{2-}	(mmol L^{-1})	0.5	0.8	-	0.8	0	0.5
Buffer	-	-	-	-	HEPES	Tris	HEPES
Amino acids	(g L^{-1})	-	-	-	1.6	-	-
Glucose	(g L^{-1})	0.65-1.1	1	-	4.5	-	-
Albumin	(g L^{-1})	30-55	-	-	-	-	-
α-globulins	(g L^{-1})	5-10	-	-	-	-	-
β-globulins	(g L^{-1})	6-12	-	-	-	-	-
γ-globulins	(g L^{-1})	6.6-15	-	-	-	-	-
α_1-lipoproteins	(g L^{-1})	6-12	-	-	-	-	-
Fibronogen	(g L^{-1})	1.7-4.3	-	-	-	-	-
Total cholestrol	(g L^{-1})	1.2-2.5	-	-	-	-	-
Fatty acids	(g L^{-1})	1.9-4.5	-	-	-	-	-
Lacate	(mmol L^{-1})	0.5-2.2	-	-	-	-	-
Urea	(mmol L^{-1})	3-7	-	-	-	-	-

DMEM = Dulbecco's modified Eagle's medium; SBF = simulated body fluid; m-SBF = modified simulated body fluid; HEPES = hydroxyethyl-piperazine ethanesulafonic acid

Three-point bending fatigue tests were performed using a low-capacity servo-hydraulic fatigue machine and employing a sinusoidal wave form at a frequency of 1 Hz (i.e., similar to the normal walking frequency for an adult). Since biodegradable Mg alloys will be required to maintain their mechanical integrity for a considerable period of time depending on their applications, high cycle fatigue (HCF) is of primary concern. Therefore, the tests were continued until fracture of the specimens, or stopped when the specimens did not fail in 10^6 cycles (which is considered as the average activity in a year for a patient).

Table II. Compositions of the Hanks' balanced salt solution

Component	g/L
Sodium Chloride (NaCl)	8.00
Potassium Chloride (KCl)	0.40
Potassium Phosphate (KH_2PO_4)	0.06
Sodium Phosphate ($Na_2HPO_4.2H_2O$)	0.06
Magnesium Sulfate ($MgSO_4.7H_2O$)	0.20
Calcium Chloride ($CaCl_2$)	0.14
Sodium Bicarbonate ($NaHCO_3$)	0.35

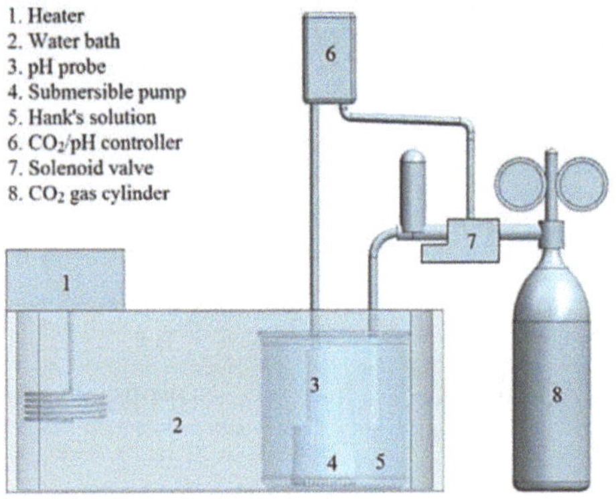

Figure 3. Schematic of the set-up for pumping test solution into corrosion chamber as shown in Fig. 2b and for controlling the pH of the test solution during the test

Test Results

The S-N curves of the as-cast AZ91D alloy tested in air and Hanks' solution are shown in Fig. 4. It can be seen that the fatigue limit of the alloy in air at 10^6 cycles is 142 MPa which is indicative of a non-propagating crack while the corrosion fatigue strength of the alloy in Hanks' solution is much lower than in air (approximately 30×10^3 cycles). The drastic reduction in fatigue strength of the alloy in Hanks' solution has been attributed mainly to the accelerated crack initiation as a result of localized corrosion such as pitting [6]. Pitting of Mg alloys that is profoundly influenced by the intermetallic phase particles accelerates fatigue crack propagation due to dissolution or/and hydrogen embrittlement at the crack tip [6].

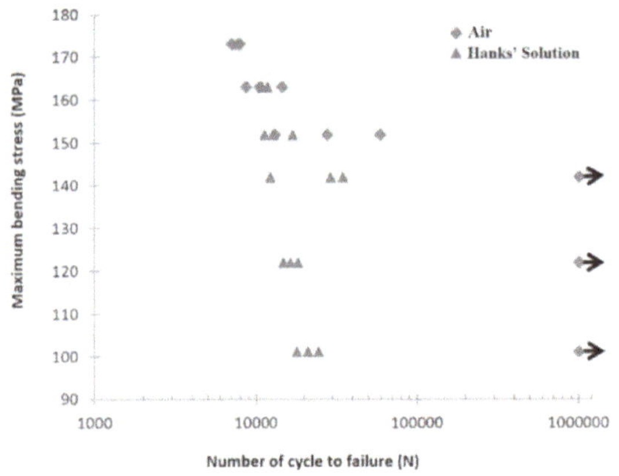

Figure 4. S-N curves for AZ91D magnesium ally in air and Hanks' solution

Conclusion

Mg alloys are reported to be susceptible to CF and SCC in corrosive physiological environments, and pitting is the common cause for crack initiation. Although, there are reports evaluating SCC of Mg alloys in simulated body fluids (SBF), there is very little reported on CF of Mg alloys in body fluid. Moreover, the available data have not necessarily been generated under conditions actually simulating loading conditions experienced in the human body. An experimental set-up has been established for investigation of CF of biodegradable Mg alloys under appropriate simulation of mechanochemical conditions. Preliminary CF data using this set-up has been generated.

References

1. R. K. Singh Raman, S. Jafari, S. E. Harandi, "Corrosion Fatigue Fracture of Magnesium Alloys in Bioimplant Applications: A Review", *Engineering Fracture Mechanics*, 137 (2015), 97-108.
2. S. E. Harandi, R. K. Singh Raman, "Appropriate Mechanochemical Conditions for Corrosion-Fatigue Testing of Magnesium Alloys for Temporary Bioimplant Applications", *The Minerals, Metals & Materials Society*, 67 (2015), 1137-1142.
3. L. Choudhary, R. K. Singh Raman, "Magnesium Alloys as Body Implants: Fracture Mechanism under Dynamic and Static Loadings in a Physiological Environment", *Acta Biomaterialia*, 8 (2012), 916-23.
4. I. S. ISO, "Biological evaluation of medical devices", 10993-6 (1994), 1-11.
5. J. Hanks, "Hanks' Balanced Salt Solution and pH Control", *Tissue Culture Association Manual*, 1 (1975), 3-4.
6. S. Jafari, R. K. Singh Raman, C. H. J. Davies, "Corrosion Fatigue of a Magnesium Alloy in Modified Simulated Body Fluid", *Engineering Fracture Mechanics*, 137 (2015), 2-11.

Magnesium Technology 2016
Edited by: Alok Singh, Kiran Solanki, Michele V. Manuel, and Neale R. Neelameggham
TMS (The Minerals, Metals & Materials Society), 2016

STANDARDIZED GUIDANCE FOR THE PRECLINICAL EVALUATION OF ABSORBABLE METAL IMPLANTS

Byron K. Hayes[1]
[1]W.L. Gore & Associates, Inc.; 3450 West Kiltie Lane; Flagstaff, AZ 86005, USA

Keywords: Standards, ISO, ASTM, absorbable, bioabsorbable, degradable, biodegradable, magnesium, iron, zinc, implant

Abstract

Absorbable metallic implants constitute a technology in its infancy, with new cardiovascular and orthopedic devices (principally magnesium based) only now entering the marketplace. The technology brings a fundamental shift in the paradigm of corrosion, moving from a historical focus on its complete prevention to one of managing the entire corrosion process, including the generated degradation products. Since an implant needs to perform its structural function and also remain compatible with the body's physiology, complete implant corrosion brings a multitude of new questions (and potential understandings). Since any medical implant needs to demonstrate a predictable level of safety, identification of appropriate standards for evaluating an implant that intentionally corrodes is needed. A collaborative effort between ASTM and ISO has been initiated to develop standardized guidance for researchers, regulators, and product developers to appropriately evaluate the composition, degradation, and biocompatibility of these new materials and their relatively complex *in vivo* degradation processes.

Introduction

Absorbable implantable materials can be broadly categorized to include: natural polymers (e.g. collagen), synthetic polymers & copolymers (typically the α-hydroxy-esters PGA,PGA/PLA), and corrodible metals and alloys (typically Mg^{o} or Fe^{o} based). While these absorbable materials are often perceived as being a new medical device technology, a closer look at history indicates otherwise. Sutures are likely oldest & most common medical device after the knife, with records of wound closure dating back to 5000-3000 BC and the origins of surgery[1]. The first use of absorbable sutures dates to the Roman Empire ~120 AD when Galen documented the use of sheep gut guitar strings for wound closure[2]. However, scientific observations were lacking until Philip Syng Physick (1768-1837), the first professor of surgery at the University of Pennsylvania, observed the leather band aids appeared to absorb, with later confirmation of *in vivo* absorption in the horse reported by Lister in 1867[2].

The concept of an absorbable metal implant is also not new, with records of the use of iron sutures dating from the mid-1800s[3]. During that same general time period, magnesium was also being produced as photography flash wires, which led to their availability and reported use in 1878 as ligatures to stop bleeding vessels in human patients[4]. Regardless, due to limitations of earlier metal processing technologies, magnesium implants are only now being commercially introduced to the marketplace.

Absorbable metals are currently in the process of crossing a modern day threshold with European CE Mark approval having recently been granted to both the MAGNEZIX bone compression screw (Syntellix AG, Hanover, Germany) in 2013 and the VELOX CD vascular closure device (Transluminal Technologies, Syracuse, NY US) in 2014. Along with each of these approvals comes the need for determinations by both the respective regulators and the market itself regarding what is truly important in the function of an absorbable metallic device – be it coming from a regulatory perspective of public safety or directly from the doctors, payers, and patients who make up the market and, ultimately, will determine the long term demand for the device and its embedded technologies. Thus, now that these devices are at their commercial infancy, standards can assist in guiding a more efficient and rapid future development – focusing both product developers and regulators toward those properties of absorbable metals that are important and how best to measure them.

Benefits and Challenges of the Technology

Absorbable metals are a promising but complex technology that could provide unique benefits beyond those being currently supplied by absorbable polymers, be they natural or synthetic. Of foremost benefit is the relatively high strength that emanates from the absorbable metal's metallic nature, which – when compared to polymers - at least initially provides increased strength for any delivered volume – effectively minimizing implant size for any particular mechanical support application. However, while this aspect of absorbable metals is highly attractive, implant degradation through a surface-based corrosion process is much more complex and less predictable when compared to the bulk degradation mechanism of ester hydrolysis found in the common absorbable PLA/PGA based polyesters. The rate of the metal's surface corrosion can also be affected by protein deposition *in vivo*, an environment that has been found to be very difficult to model outside of the body. Additionally, each respective metal - be it magnesium, iron, or zinc - possesses its own unique corrosion mechanisms, which can then be affected by both electrolyte and atmospheric composition. Beyond the readily measurable impact an absorbable metal's alloy composition has on an implant's mechanical properties, its microstructural and phase aspects can also affect overall degradation and thereby performance, ranging from the effect inclusions carry on mechanism of corrosion to the level of local and systemic toxicity generated from both soluble and insoluble degradation products.

The aforementioned and relatively unique aspects of absorbable metal degradation can, in general, be considered as additive to the fundamental performance requirements that have already established for nonabsorbable implants. Thus, with multiple new and experimentally complex factors now needing consideration, questions inevitably arise regarding what testing is truly appropriate and meaningful to assure both functional performance and patient safety – a recognition that drives both the need for and the development of standards.

The Value of Standards

The ultimate value of medical research comes to full realization when it becomes an accepted procedure and/or commercialized product that truly helps to address people's afflictions. Such scientific contributions can range from revolutionary practice changing concepts to relatively smaller but highly necessary components of much larger chain of events, such as the development and establishment of a manufacturing process or the successful completion of a clinical trial. While many new drug or device related therapeutic concepts are introduced and evaluated, few actually make it into the medical marketplace – being driven by inadequate performance and/or prohibitive development costs. Even if a concept does successfully enter the marketplace, for long term survival it ultimately needs to demonstrate its practical therapeutic value. Thus, while the monetary rewards of commercial success can be significant, new medical product development carries both significant risk and up-front investment.

To earn entry into the marketplace, both an implant's design and implementation (including manufacturing and surgical placement procedure) must inevitably be evaluated for its safety by each country's respective regulatory authority. Such regulatory evaluation can be especially difficult when a utilized technology, be it of the overall device or a component, is new and both the regulators and the companies themselves struggle to define what aspects are important to protect the public health and how those properties should be evaluated. This quandary is especially pronounced in applications where the function of the device or drug means life or death for the patient or when an adverse outcomes could generate significant and/or long lasting patient morbidity.

When both developing and evaluating a product and its related technologies, the presence of relevant standards helps to speed both the acquisition and use of the product or component. For example, the presence of established and accepted specifications for structural steel I-beams facilitates their sales and use, provides design engineers with assurance that their selected components will perform as expected, and allows inspectors a broadly accepted and efficient means to check for both the quality and reliability of a structure's components. Similarly, standards for medical devices can help to speed product development by better defining what is important for evaluation and how to appropriately test to assure the needed performance. However, since a country's national regulatory authority retains ultimate responsibility for the safety of medical devices within their borders, they also retains full jurisdiction of the final requirements for commercial release – something that is likely more than what is covered by standards alone. Thus, while relevant standards can potentially expedite product development and regulatory approval by providing a generally accepted framework for device evaluation, compliance with a standard does not, in itself, imply completion of an exhaustive evaluation and does not provide any guarantee of regulatory acceptance or device approval.

Who Develops Standards

While standards can broadly include laws and regulations, standards within the current context are consensus standards, where the affected parties confer to achieve a consensus regarding what is important and relevant for evaluation – each contributing from their own perspective and related concerns. Active standards development typically includes those with a working knowledge in the respective field, be they researchers, manufacturers, or regulators. Participation in such standards development is voluntary, with the benefit being future recognition of a more expedited and better understood product development process.

At a global level there are many different standards development organizations, which range from those sponsored directly by governments to others that are independent non-profit organizations that achieve much of their funding through the sale of standards. While there are many organizations that generate standards for medical devices, the following provides a brief description of the two organizations involved with the current effort toward developing standards for absorbable metals.

International Standards Organization (ISO) – ISO was established in 1947 to facilitate the international coordination and unification of industrial standards to thereby enhance potential for free trade. Thus, the ISO process is both international by design and one of national participation, with each country providing its own independent review and input. Organizational funding is also by country, with the cost of ISO member subscriptions indexed on gross national product. In turn, each country controls its own internal membership and deliberative processes, with participating experts in the respective standardization topic nominated as national representatives to the respective working group within the relevant ISO Technical Committee. Regardless of the number of experts contributing to document development, each participating country is allowed a single vote that is formally cast by its national standards body toward ISO established approval thresholds that vary dependent on the type of document or action being balloted. Relevant Technical Committees for the development of standards for absorbable metals are TC 150 – Implants for Surgery and TC 194 – Biological and Clinical Evaluation of Medical Devices, which began actively developing medical device standards in 1971.

ASTM-International – ASTM is a US based non-profit organization that has been developing a broad range of standards since 1898 and medical device standards for over 50 years. Membership in ASTM is open to individuals of all nationalities for only a nominal annual membership fee. ASTM deliberates primarily at a Subcommittee level, but requires the content of a proposed standard to achieve an individual level Committee-wide consensus prior to publication, with all identified technical issues requiring full resolution prior to publication. The relevant entity within ASTM in the development of standards for absorbable metals is Committee F04 – Medical and Surgical Devices.

Standards with Absorbable Metal Provisions

While some already published standards address some of the absorbable metal related concerns, development of specific standards for absorbable metals has only relatively recently commenced. Currently published standards that contain notable absorbable metal specific provisions are listed in Table 1.

Table 1 – Currently Published Standards with Absorbable Metal Relevant Content

Standard	Title/Description
ISO/TS 17137	***Cardiovascular implants and extracorporeal systems – Cardiovascular absorbable implants.***
	Provides absorbable specific guidance when evaluating a surgically or interventionally placed cardiovascular implant.
ISO/TR 37137	***Biological evaluation of medical devices — Guidance for absorbable implants***
	Provides absorbable specific guidance regarding the biological evaluation of an absorbable device (polymeric or metallic) in accordance with ISO 10993.
ASTM F1983	***Standard Practice for Assessment of Selected Tissue Effects of Absorbable Biomaterials for Implant Applications***
	Provides absorbable specific guidance regarding *in vivo* study design and observations.
ISO 10993-6	**Tests for local effects after implantation**
	Covers the *in vivo* assessment of the biological safety of a material, with numerous absorbable provisions.

Absorbable Metal Standards Under Development

Absorbable metal specific standards that are currently under development are listed in Table 2. To clarify the overall objective, these draft standards are not intended to list or address the specific requirements for a particular type of device, which are likely already covered by existing device-specific standards. However, these draft standards are intended to augment the existing device-specific standards by providing general guidance regarding how to appropriately assess the absorbable metal specific aspects of the device.

Table 2 – Absorbable Metal Specific Standards
(Note: All five listed documents are under development and have not yet been published)

Standard	Tentative Title/Description
ISO/TS 20721 (Doc. 1)	***Implants for surgery — Standard guide to assessment of absorbable metallic implants***
	Now an accepted Work Item within ISO TC150. Provides absorbable metal specific guidance. Refers to other relevant standards if needed consideration is not directly addressed.
ASTM WK46455 (Doc. 2)	***Standard Guide for Metallurgical Characterization of Absorbable Metallic Materials for Medical Implants***
	Currently under ASTM ballot for publication. Provides listing of appropriate considerations and characterizations when selecting or evaluating an absorbable metal.
Doc. 3	***Guide to in vitro Degradation Testing of Absorbable Metals***
	Provides a listing of considerations when testing an absorbable metal.
Doc. 4	***Guide to Corrosion Fatigue Evaluation of Absorbable Metals***
	Development activity deferred until completion of previously listed ***in vitro* Degradation Testing** document.
Doc. 5	***Guide for the Biological Evaluation of Absorbable Metals***
	Accepted for consideration as a new Work Item proposal within ISO TC194. Provides more detail regarding evaluation of absorbable metals, with inclusion of procedure for adjustment of pH and osmolality during cytotoxicity evaluations.

While most of the above listed draft standards are in an early draft form, the ASTM Materials standard is already nearing publication. However, regardless of current status, the content of all these standards cannot be considered as final until published. Additionally, eventual publication can also remain uncertain for a variety of reasons, which can include a lack of action or involvement by those with appropriate expertise or due to lack of a consensus regarding critical technical aspects of the document.

Conclusion

In summary, standards are an established means to develop a consensus regarding what is truly important and how best to measure those properties. However, standards development is an active and collaborative process where participation is key to the quality of both the input and the outcome. Thus, if you possess expertise and insights regarding what is truly important when evaluating absorbable metal implant, become personally involved in standards development either through participation in ISO and/or ASTM absorbable metal related activities. You can individually become involved in ISO TC150 by contacting your own country's national standards body, and mentioning the described ISO activities. As an individual, you can also become a member of ASTM Committee F04, with registration details available through the ASTM-International web site. Regardless, it is active involvement that will allow generation of those absorbable metal standards that will help facilitate the transition of this new and promising technology into the marketplace where it can then be realized for its true therapeutic value.

[1] M.W. King. "Overview of opportunities within medical textiles," *Canadian Textile Journal*, 118 (4) (2001 July/Aug):34-36.
[2] David Mackenzie, "The History of Sutures," *Medical History*, 17 (02) (1973, Apr.): 158-168.
[3] Simpson, "Art. 140.— On the use of Iron-thread Sutures and Splints in Yesico-vaginal Fistula," *Medical Times and Gazette*, Dec. 4, 1858.
[4] Frank Witte, "The history of biodegradable magnesium implants: A review," *Acta Biomaterialia* 6 (2010) 1680–1692.

Magnesium Technology 2016
Edited by: Alok Singh, Kiran Solanki, Michele V. Manuel, and Neale R. Neelameggham
TMS (The Minerals, Metals & Materials Society), 2016

The Industrial Challenges of Manufacturing Bioabsorbable Magnesium

Robert Thornton[1], Paul Lyon[1], I Syed[1], M Turski[1]
1.Magnesium Elektron, Rake Lane, Manchester, UK

Keywords: Process, Manufacturing, Casting, Extrusion

Abstract

Research in absorbable magnesium has become hugely popular, with groups around the world looking at bespoke biomedical alloy compositions and processing routes. However designing alloy chemistry and thermomechanical processing is only half the development process to producing a successful commercial alloy. A manufacturing route which answers the demands of the alloy's chemistry and processing design as well as economic models and importantly risk must be created to ensure commercial viability and patient safety.

During scale up operations, certain alloys which show outstanding mechanical properties were actually highly unstable during casting because of cracking issues not seen on small diameter billets. Another alloy series had very tight parameters for extrusion in order to achieve the properties including a very low extrusion ratio. This meant the economics and ability to achieve dimensional specifications was incredibly difficult. This presentation details the steps taken to bring research ideas to a full industrial process.

Introduction

Magnesium Elektron is a manufacturer of high performance magnesium alloys having supplied into a range of markets from aerospace to nuclear, however over the last 10 years has been heavily involved in the research and development of alloys for absorbable implants. During this time over 100 alloy designs were investigated for their suitability as implants looking at biocompatibility, in vitro corrosion, mechanical properties and critically their combination in the form of stress corrosion cracking.

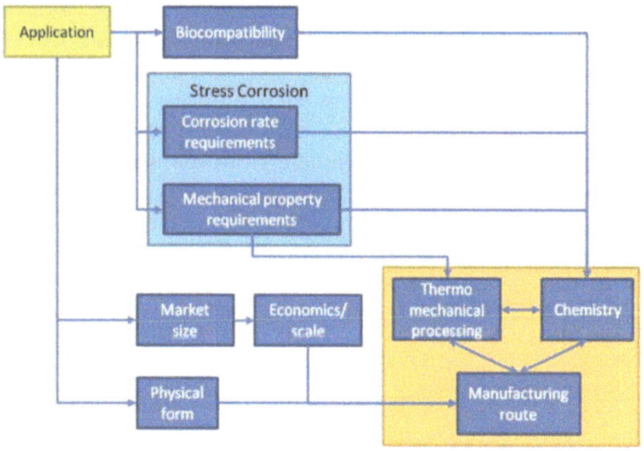

Figure 1: The various factors and complex interaction between them in order to produce commercially successful magnesium alloy

However, designing an alloy system does not end at the discovery of successful chemical, biocompatible, mechanical and stress corrosion properties; but in fact is a complex interaction driven by the application, economics and physical form (see figure 1). For a truly successful commercial alloy, the end goal is a manufacturing route, chemistry and thermo-mechanical process which work together.

Discussion

During scale up investigations, it was found that certain alloys which show outstanding mechanical properties were actually highly unstable for the casting process due to cracking issues not seen on small diameter billets. When developing alloy systems, often very simple casting routes are employed, usually small in diameter. When up scaling to the final manufacturing route, these simple small diameter casting routes, due to economics of scale, are replaced with processes such as direct chill (DC) casting. When diameters are increased, temperature gradients during casting become a critical factor[1]. If these build up too quickly, large internal stresses (figure 2) can establish and lead to catastrophic failures of the casting (figure 3). This failure mode is known as cold cracking[2].

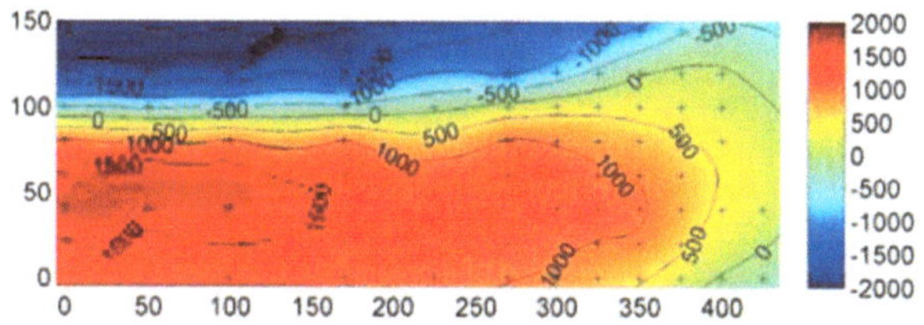

Figure 2: A map of elastic strain (units x10^{-6}) measured by neutron diffraction across a section of DC cast billet[3]. The figure shows the buildup of tension (red) in the middle of the casting and compression (blue) at the fast cooling surface. Cracking occurs if these stresses exceed the yield stress of the material.

Temperature gradients are an inevitable feature of the casting process, where the cooling rate from the surface of the sample exceeds the thermal flow across the billet radius, the latter being affected by chemical composition. Alloys designed with high alloy content can have inherently lower thermal conductivity, this leads to larger temperature gradients and therefore thermal strains and stresses during casting. Table 1 shows a summary of this relationship listing properties from three alloys which have undergone DC casting trials. It shows that casting sensitivity and therefore the amount of process development required increases as thermal conductively goes down.

Thermal strains across the casting can also lead to a failure mode known as hot cracking where alloy systems with lower melting point phases are particularly venerable[4]. Casting sensitivity when using larger scale production routes occurs in a number of alloying systems and shows how important it is to consider the final casting route when designing the alloy chemistry.

Table 1: Alloy content and thermal conductivity of three rare earth alloys. The casting sensitivity indicates how often cracking issues were encountered during casting trials and process development.

	Rare Earth content / wt%	Thermal conductivity / $Wm^{-1}K^{-1}$	Casting sensitivity
Alloy 1	13	35	****
Alloy 2	7	46	**
Alloy 3	4	73	*

Figure 3: Catastrophic failure of a DC cast billet caused by residual stresses during casting.

Alloy systems which have low alloying contents, often for reasons of corrosion[5-7] may have mechanical strength too low for a desired application. This is normally increased by thermo-mechanical processing, specifically to decrease grain size[8]. There should be caution regarding over reliance on such strengthening mechanisms due to the effects of processing variability.

Table 2: Chemical composition of a precipitate free Rare Earth alloy

	Yttrium	Rare Earths	Zirconium	Magnesium
wt%	2.8	1.5	0.6	Bal.

During processing, changing a number of parameters can lower the grain size and in turn increase mechanical properties. One parameter used is ram speed. Figure 4 shows the relationship between ram speed and mechanical properties on a Mg-Y-RE alloy (composition table 2) designed with low neodymium content for a precipitate free microstructure. At 0.05 mms^{-1} and 0.1 mms^{-1} ram speeds the grain size is reduced to 1.2 μm and 2.0 μm respectively.

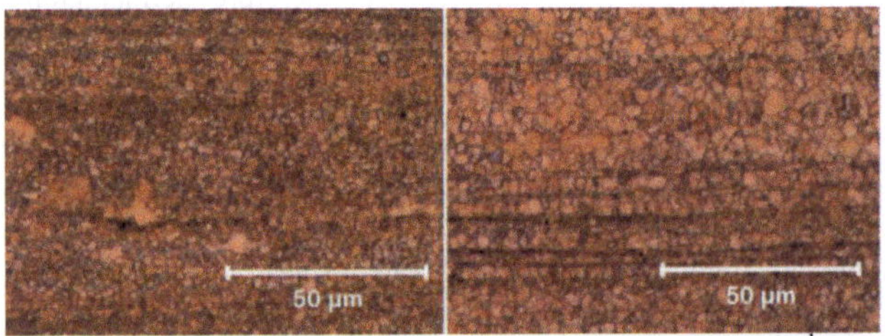

Figure 4: Microstructures of a Mg-RE alloy extruded at 0.05 mms^{-1} and 0.1 mms^{-1} ram speeds.

Figure 5 shows that the relationship between ram speed and mechanical properties is logarithmic. This results in very slight changes in the ram speed leading to significant differences in tensile strength for example. This effect is highly desirable for tailoring properties in development stages, however the slight changes can occur batch to batch or even within the same extrusion meaning large property differentials along the length of the extrusion and between extrusions.

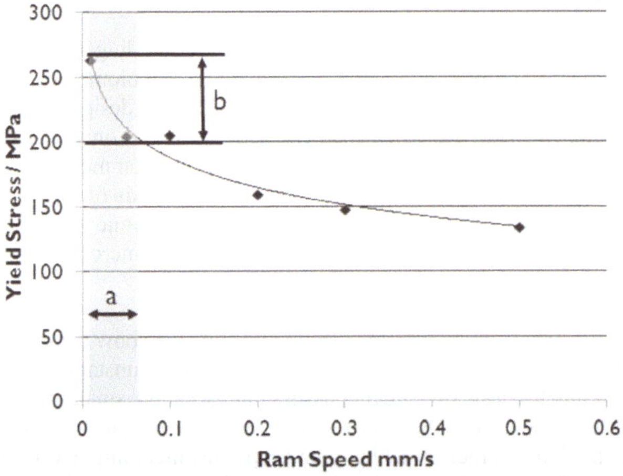

Figure 5: The relationship between ram speed and tensile strength. Grain size increases from left to right. 'a' represents the process variability inherent to manufacture both batch to batch and also during one extrusion. 'b' shows the corresponding differences in tensile strength.

Process sensitively is often overlooked at development stage as only small batch numbers are produced. However manufacturing routes that incorporate sensitivities, such as very small grain size, can lead to an inconsistent product, high testing requirements, high scrap rates and high costs. There are methods to overcome these issues, but all involve high capital cost. Magnesium Elektron installed a bespoke extrusion press to enable such processing control. Designing an alloy with a quantity of strengthening elements to give reasonable solute and precipitate strengthening can lead to a system which avoids the need for sensitive thermo-mechanical processing

Another issue with using fine grain size and work to increase mechanical properties is the influence on stress corrosion cracking[9]. Table 3 shows the two alloys, one which has only 0.5% Zr and the other a complex rare earth alloy with >5% content. The mechanical properties have been tailored on the press to achieve similar yield strengths and both have similar corrosion rates in simulated body fluid (SBF) solution. However even thought they have similar mechanical properties in air, they react differently when stressed in an SBF environment. Figure 6 shows the outcome of a stress corrosion cracking (SCC) susceptibility trial where samples are held in flowing SBF at 37°C under a fixed load. The Mg-0.5Zr alloy failed around 25 times faster than the Mg-RE alloy.

The difference between the two alloys under the combination of stress and load is the result of how both alloys achieve their strength and shows that reliance on grain size and deformation alone could lead to cracking issues in stressed implants.

Table 3: Mechanical and corrosion properties of Mg-RE and Mg-0.5Zr. The corrosion rate is measured in SBF solution at 37°C.

	YS / MPa	UTS/ MPa	Elongation / %	Corrosion rate / mpy
Mg-0.5Zr	219	251	24	283
Mg-RE	183	268	25	247

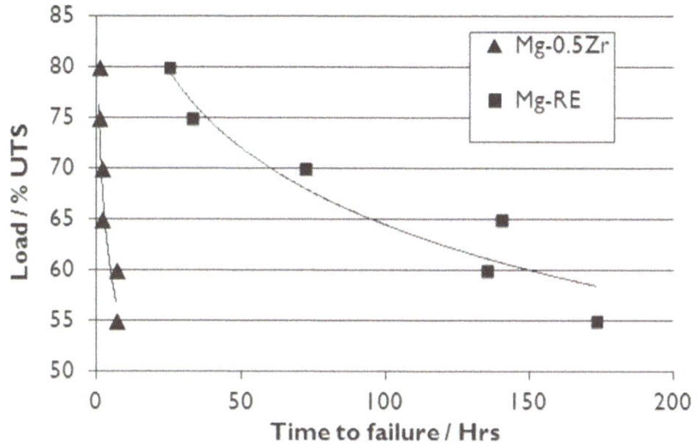

Figure 6: Stress corrosion cracking resistance of alloys Mg-RE and Mg-0.5Zr. Samples are subjected to a fixed load represented as a percentage of their UTS. They are also exposed to SBF solution at 37°C. Both alloys have similar yield strengths. Mg-0.5Zr fails ~25x faster under the combination of load and an SBF environment.

Increasing the alloy content is often the only method to improve the stress corrosion cracking resistance[10] while maintaining acceptable properties for the application. However different elements and combination of elements have vastly different effects on the SCC.

Figure 7 shows three alloys systems (composition listed in table 4) all with similar mechanical strengths; however their resistance to stress in a SBF environment is considerably different. AZ31 and ZX20 are shown due to their prevalence in the literature[11-14] and in some devices[15]. Alloying with aluminum greatly reduces the loads which can be tolerated and the time to failure. The failure mode is by transgranular cracking and occurs almost catastrophically. ZX20 fails by a more predictable pitting type mechanism and is therefore far more suitable for load bearing applications than the use of aluminum. It should be noted however, the grain size for this ZX20 was 10 μm which is far higher than that used in the literature. It would be predicted that as previously mentioned, decreasing the grain size and using work to improve properties will lead to high SCC susceptibility.

Novel compositions have been developed at Magnesium Elektron which use SCC resistant alloying additions to strengthen, meaning that the alloy does not have to rely on deformation and grain size, but simultaneously, the additions resist crack initiation and pit formation.

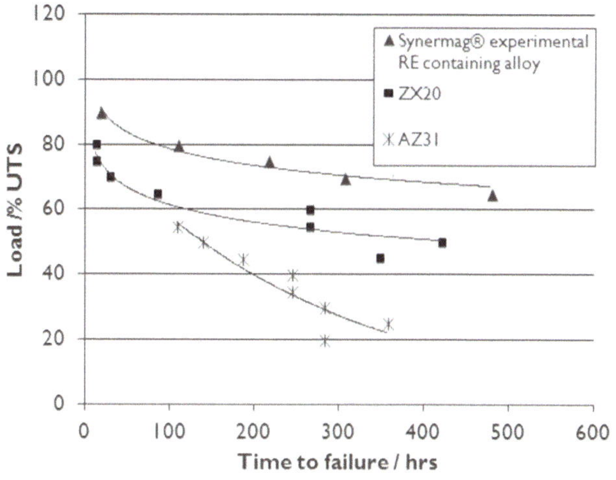

Figure 7: Time to failure plots of three alloys under constant load in flowing SBF solution.

Table 4: Compositions of AZ31 and ZX20 alloys described in figure 7.

	Aluminum	Zinc	Calcium	Magnesium
AZ31	3.0	0.82		Bal.
ZX20		1.9	0.3	Bal.

Process sensitivity can also affect the ability to produce material to the correct shape and form. An alloy series investigated at Magnesium Elektron had high alloy content and strengthening mechanisms that required very cold extrusion. The alloy itself was very promising for load bearing applications with yield strengths in excess of 300MPa and corrosion rates below other good performing alloys; however the extrusion conditions lead to very narrow processing parameters, and the need for a lower extrusion ratio. Outside of these parameters and a satisfactory straight extrudate would not be formed at all (figure 8). As previously mentioned, such process sensitively leads to high costs, but the additional requirements of low extrusion ratios and low temperatures result in small extrusions lengths and makes achieving relevant dimensions incredibly difficult. The outcome is that only simple profiles could be extruded and to lengths which would lead to inefficient manufacture.

Figure 8: Outcomes of extrusion trials using different container temperatures. A 15°C increase in container temperature lead to hot shorting. A 15°C decrease in temperature resulted in no extrusion.

Conclusions

When developing an alloy system is it important to consider the mechanical stresses and corrosion in unison, many alloys look promising when these properties are investigated in isolation, however they can be highly susceptible to the stress corrosion. Alloy developers must also consider how the application and market influence the manufacture route and final form. Designing a system in the lab independent of these considerations can lead to substantial hurdles in its later life. Hurdles such as process sensitivity can result in quality control issues and the inability to meet final form requirements. Both of which will hamper the commercial success of the alloy. Considerations to the full scale process design must be made at the initial stages of development. Often alloy systems which work well at lab scale cannot be practically or economically manufactured on an industrial scale

References

1. J. Grandfield, D. G. Eskin, I. Bainbridge. Direct-Chill Casting of Light Alloys, Wiley publications, (2013)
2. Suyitno, D.G. Eskin, and L. Katgermam. Hot tearing criteria evaluation for direct-chill casting of an Al-4.5 pct Cu alloy. Mater. Sci. Eng. A, (2006), vol. 420, pp. 1–7.
3. M. Turski, A. Paradowska, S. Y. Zhang, D. Mortensen, H. Fjaer, J. Grandfield, B. Davis, R. DeLorme, Validation of Predicted Residual Stresses within Direct Chill Cast Magnesium Alloy Slab. *Metall and Mat Trans A.* (2012), vol 41 A, (6).
4. M. Lalpoor, D.G. Eskin, D. Ruvalcaba, H.G. Fjaer, A. Ten Cate, N. Ontijt, and L. Katgerman. Cold cracking in DC-cast high strength aluminum alloy ingots: An intrinsic problem intensified by casting process parameters. Mater. Sci. Eng. A, (2011), vol. 528 (6), pp. 2831–42.
5. Y Wan, G Xiong, H Luo, F He, Y Huang, X Zhou. Preparation and characterization of a new biomedical magnesium–calcium alloy. *Materials and Design* 29 (2008) 2034–2037
6. N. T. Kirkland, N. Birbilis, J. Walker, T Woodfield, G. J. Dias, M. P. Staiger. In-vitro dissolution of magnesium–calcium binary alloys: Clarifying the unique role of calcium additions in bioresorbable magnesium implant alloys. *J. Bio. Mat. Res: applied biomaterials* (2010) vol 95b, issue 1
7. Z. G. Huan, M. A. Leeflang, J. ZhouL. E. Fratila-Apachitei, J. Duszczyk. In vitro degradation behavior and cytocompatibility of Mg–Zn–Zr alloys. *J Mater Sci: Mater Med.* (2010)
8. B.P. Zhang, L. Geng,* L.J. Huang, X.X. Zhang and C.C. Dong. Enhanced mechanical properties in fine-grained Mg–1.0Zn–0.5Ca alloys prepared by extrusion at different temperatures. *Scripta Materialia* 63 (2010) 1024–1027
9. N Winzer, A Atrens, G Song, E Ghali, W Dietzel, K U Kainer, N Hort, C Blawert. A critical review of the stess corrosion cracking (SCC) of magnesium alloys. *Adv Eng Mat* (2005) 7 No. 8
10. M. B Kannan , W. Dietzel, C. Blawerta, A. Atrens, P Lyon. Stress corrosion cracking of rare-earth containing magnesium alloys ZE41, QE22 and Elektron 21 (EV31A) compared with AZ80. *Mats Sci and Eng: A.* Vol 480, Issues 1–2, 15 May (2008), Pages 529–539
11. J. Hofstetter, M. Becker, E. Martinelli, A.M. Weinberg, B. Mingler, H. Kilian, S. Pogatscher, P.J. Uggowitzer, J.F. Loffler. High-Strength Low-Alloy (HSLA) Mg–Zn–Ca Alloys with Excellent Biodegradation Performance. *JOM*, April (2014), Volume 66, Issue 4, pp 566-572
12. B. Zhang, Y Hou, X Wang,Y Wang, L Geng. Mechanical properties, degradation performance and cytotoxicity of Mg–Zn–Ca biomedical alloys with different compositions. *Mat Sci and Eng: C.* Vol 31, Issue 8, 1 December (2011), Pages 1667–1673
13. D. Zander, N A, Zumdick. Influence of Ca and Zn on the microstructure and corrosion of biodegradable Mg–Ca–Zn alloys. *Corrosion Science* Volume 93, April (2015), Pages 222–233
14. R. Montoya,C. Iglesias, M. L. Escudero, M. C. García-Alonso. Modeling in vivo corrosion of AZ31 as temporary biodegradable implants. Experimental validation in rats. *Mater Sci Eng C Mater Biol Appl.* 2014 Aug 1;41:127-33
15. S M. Green, B Fairman, H F. Brodt, D K. Boger. Closure device, deployment apparatus, and method of deploying a closure device. Patent No. US 8137380 B2. (2008)

Magnesium Technology 2016

SYMPOSIUM:

Strip Casting of Light Metals

Magnesium Technology 2016

STRIP CASTING OF LIGHT METALS

Strip Casting Process

Magnesium Technology 2016
Edited by: Alok Singh, Kiran Solanki, Michele V. Manuel, and Neale R. Neelameggham
TMS (The Minerals, Metals & Materials Society), 2016

MICROSTRUCTURE INVESTIGATIONS OF INVERSE SEGREGATIONS IN TWIN-ROLL CAST AZ31 STRIPS

Christina Krbetschek[1, *], Franz Berge, Matthias Oswald, Madlen Ullmann and Rudolf Kawalla
[1]Faculty of Material Science and Technology, Institute of Metal Forming, Technische Universität Bergakademie Freiberg, Bernhard-v.-Cotta-Str. 4, 09599 Freiberg, Germany
*Christina.Krbetschek@imf.tu-freiberg.de
*corresponding author

Keywords: twin-roll casting (TRC), AZ31, inverse segregations, crack initiator, EDX, X-ray phase analysis

Abstract

During twin-roll casting (TRC) of magnesium alloy AZ31 specific process conditions may promote the formation of inverse segregations. This phenomenon is characterized by pressing remaining melt, which is enriched with alloying elements, into small cavities within the rapidly solidified material at the chill surface. The occurrence of inverse segregations, which remain in the microstructure after an annealing treatment, may have an adverse effect on the mechanical properties. Under mechanical load, the inverse segregations in AZ31 strips in annealed conditions act as a crack initiator and lead to material failure below the maximum deformation limits. In order to clarify the mechanisms, which are responsible for the development of inverse segregations, fundamental investigations regarding composition, influencing process parameters and effect on material properties were carried out. For determining the qualitative chemical composition of the inverse segregations, SEM investigations using the EDX analysis were carried out. In order to specify the individual phases of the segregations, the qualitative X-ray phase analysis was applied. Potential generation mechanisms of inverse segregations during TRC, based on the present results, were discussed. Via tensile tests at room temperature, the influence of the inverse segregations on the elongation to failure A_{80} in heat-treated AZ31 TRC sheets was proved. Magnesium sheets with and without inverse segregations were compared. Interventions for prevention of inverse segregations were discussed.

Introduction

Concerning the requirements of modern lightweight construction, magnesium alloys provide significant advantages over conventional construction materials. Mg alloys have already been used for a long time as casting materials, but wrought magnesium products are rare to find. Low formability and a distinctive crystallographic texture are responsible for a disproportionate price [1]. By TRC, firstly, magnesium strips can be manufactured economically and ecologically efficient [3]. Secondly, a material with a finer microstructure, better homogeneity, reduced segregations and less texture is possible. TRC of Mg alloys has been tested in practice for years, but research activities regarding the microstructure evolution of the TRC sheets, formed during solidification and deformation in the roll gap, is not very advanced yet. Previous studies work usually with pure phenomenological approaches [4], [7]. The relevant characteristics of the TRC strips are thereby explained solely on the variations of the process conditions [5], [8]. The precise processes during solidification and following metal forming in the roll gap cannot be observed directly and are therefore considered only indirectly. Lack of knowledge regarding the process sequence effects, that many of the real phenomena during TRC, as the occurrence of rectilinear inverse segregations parallel to the TRC direction, can be explained only inadequately to date. This ignorance inhibits the advancement of TRC technology significantly. Inverse segregations in TRC strips cause lasting damage to the material and can lead to premature failure of the strips in annealed conditions as well as in the finished rolled strips. For this reason, future developments in the field of TRC technology should be on the analysis and prevention of TRC induced microstructure defects.

Microstructure evolution during TRC

The evolution of microstructure in TRC magnesium depends heavily on the settings of process parameters, such as the setback (Figure 1), the rolling force, the rolling speed plus the pressure and temperature whereby the melt reaches the nozzle.

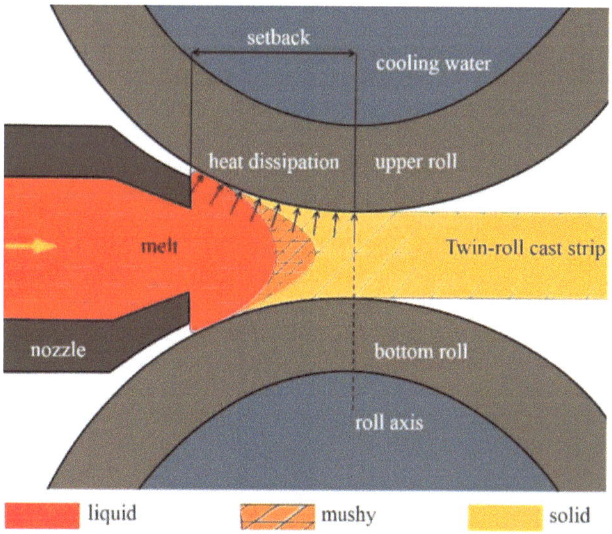

Figure 1. Schematic sketch of the roll gap (similar to [2]) with indicated states of the alloy during twin-roll casting.

During TRC, the used AZ31 magnesium alloy, is being melted in protective gas and is brought into a casting channel. A casting nozzle is located at the end of this channel. Through the nozzle, the melt is fed into the roll gap. Between rotating and cooled rolls, the melt solidifies very quickly and is deformed simultaneously. The area between the nozzle exit, respectively the melt exit and the roll axis can be divided into zones which have three states of matter: liquid, semi-solid (mushy-zone) and solid (Figure 1) [6].

Considered metallurgical, following processes occur: The solidification starts by nucleation on the roll surface. Randomly oriented small grains are formed. Because of the heat dissipation over the rolls, the solidification takes place in the perpendicular direction to the rolling surface. Originating from the crystallites of the quenching zone, columnar crystals or dendrites grow into the strip. Following the transport of the melt and the heat dissipation, the pikes of the dendrites are inclined against the rolling direction (RD). The interdendritic spaces are enriched with high-melting impurities such as Al_8Mn_5 particles and alloying elements. The globular inner zone in the TRC-strips is usually caused by impurities in the melt [6]. Having high melting points, these impurities accumulate in the remaining melt. Heterogeneous nucleation occurs, which also leads to a grain structure with random orientation.

During solidification, the individual fringe areas of the strip grow towards another. After touching of the upper and bottom side of the solidified strip, the material can absorb forces. A forming process with a deterministic degree of deformation follows. A AZ31 strip is formed, with a microstructure similar to an ingot cast structure. However, due to the rapid solidification of the melt between cooled rolls, the formed columnar crystals are very fine and generally more homogeneous than the ingot cast. Because of shear stress during forming, the TRC microstructure is partially destroyed. Simultaneously, the high temperature leads to a partial dynamic recrystallization. Moreover, compared with conventionally cast alloys, due to the high cooling rates, only very small and homogeneously distributed precipitates arise [6], [9], [10]. The evolution of segregations and phases in AZ alloys is considered in [6], [7] and [8]. A thermodynamic description of Mg-Al-Zn system can be found in [11] and [12]. Sometimes, during the sensitive TRC process, inverse segregations partially occur on the bottom side of the TRC strip, which is focused in this work.

Experimental procedure

The TRC process was carried out analogously to the investigations of [13], [14], [15]. TRC strips with the chemical composition Mg-3Al-1Zn (AZ31) were investigated. During this TRC process, the roll gap was set at 4.5 mm, the TRC velocity was 1.45 m/ min and the setback 68 mm. TRC strips of 5.4 mm thickness were manufactured.

For determining the qualitative chemical composition of the inverse segregations in the TRC AZ31 microstructure specimens, SEM investigations using the EDX analysis (energy dispersive X-ray spectroscopy) were carried out. In order to specify the individual phases of the segregations, the qualitative X-ray phase analysis was applied.

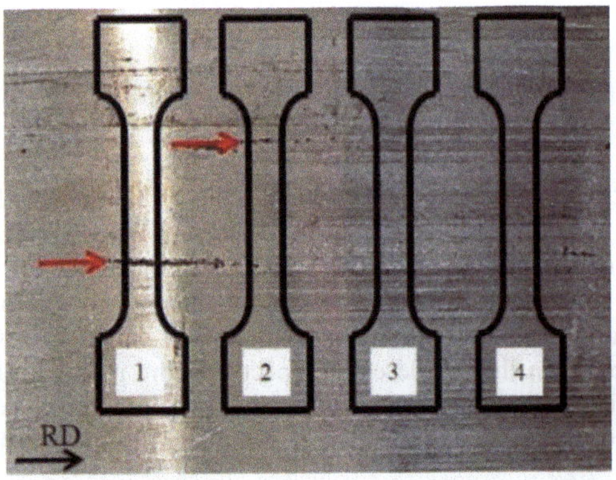

Figure 2. Heat-treated TRC strip. Bottom side. Test specimens were taken from areas with and without inverse segregations.

The heat treatment of the magnesium strips has been conducted in a circulating air oven at a material temperature of 430 °C for 8 h. Tensile-testing specimens with a gauge length of 80 mm were then milled from the strips perpendicular (90°) to the direction of TRC. Quasistatic uniaxial tensile tests were carried out at room temperature. A series of four tensile specimens (90°) was tested. The test specimens were taken from areas with, following entitled as fault strip, and without inverse segregations (Figure 2).

Results

AZ31 strips in TRC conditions

Figure 3 presents scanning electron micrographs of the TRC strip cross-section (BSD contrast). The cross section through the TRC microstructure specimen is showing an inhomogeneous chemical composition. These areas, which show open cracks after a heat treatment, have a strongly increased content of alloying elements with high atomic numbers. Up to 2 mm deep into the strip, deviations in the chemical composition can be recognized. The surrounding material shows a typical TRC microstructure, by way of example in [6]. A distinctive center segregation does not exist. In the faulty areas, the material is characterized by a grain structure (b in Figure 3). Grain boundaries are enriched with smallest particles and alloying elements. Below the surface of the strip, small cavities can be observed. The cavities are not connected to the strip surface (c in Figure 3). In deeper areas, concentrated with alloying elements, micro cracks reveal partly (a in Figure 3).

For determining the qualitative chemical composition of the inverse segregations, EDX analysis (point analysis) were carried out at the marked areas (1-5, Figure 3). Table 1 to Table 4 contain the measured chemical composition of the TRC specimen, in the areas marked with 1-4.

Furthermore, particles (< 2 µm) have been observed, whose chemical composition is indicated to be Al_8Mn_5 intermetallic phases (Point 5, Figure 3). However, due to the small size of the particles the exact determination of the chemical composition is hardly possible because a fraction of the surrounding material is always included in the detection. The investigated microstructure specimens were not homogenized by annealing. For this reason, the matrix material can be divided into two main areas. The range 1 corresponds to the interdendritic space; area 2 is designated as the impoverished dendritic plane. The interdendritic spaces (point 1, Figure 3) are enriched with aluminum and zinc. On the contrary, the proportion of these elements in the matrix is impoverished (dendritic plane, point 2, Figure 3). This structural constitution complies with a rapidly solidified ingot cast structure. During homogenization, the diffusion processes compensate concentration differences of such magnitude. The investigated segregations are very small and interfused by the matrix material. Therefore, it can be assumed, that also an increased amount of magnesium is measured in the analyzed point 3 and 4. An analysis at point 3 (Figure 3) revealed a chemical composition of the γ-phase ($Mg_{17}(Al, Zn)_{12}$, Table 3). Based on the measurement in point 4, a chemical composition could be ascertained, which can be assigned with the so-called φ-phase ($Al_2Mg_5Zn_2$, point 4, Figure 3, Table 4). In order to specify the individual phases of the segregations, the qualitative X-ray phase analysis was applied.

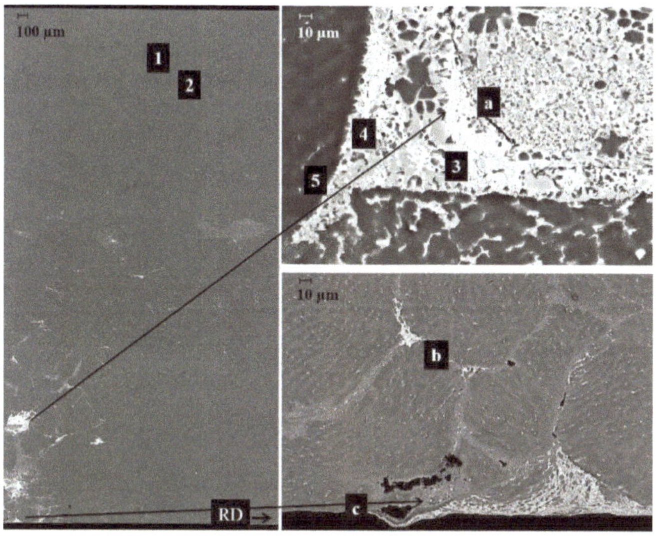

Figure 3. Scanning electron microscope overview image (BSD contrast) of the strip cross-section towards the twin-roll cast direction and detailed view of the inverse segregation.

Table 1. Determined chemical composition by EDX analysis in point 1 in Figure 3 (interdendritic spaces).

Element	Normalized concentration [atom%]	Normalized concentration [weight%]	Relative error [%]
Magnesium	94.9	92.7	5.4
Aluminium	4.0	4.4	0.3
Zinc	1.1	2.9	0.2

Table 2. Determined chemical composition by EDX analysis in point 2 in Figure 3 (plains of the dendrites).

Element	Normalized concentration [atom%]	Normalized concentration [weight%]	Relative error [%]
Magnesium	98.2	97.2	5.5
Aluminium	1.3	1.5	0.1
Zinc	0.5	1.3	0.1

Table 3. Determined chemical composition by EDX analysis in point 3 in Figure 3 (γ-phase ($Mg_{17}(Al, Zn)_{12}$)).

Element	Normalized concentration [atom%]	Normalized concentration [weight%]	Relative error [%]
Magnesium	64.6	56.8	3.1
Aluminium	29.1	28.4	1.4
Zinc	6.3	14.8	0.5

Table 4. Determined chemical composition by EDX analysis in point 4 in Figure 3 (φ-phase ($Al_2Mg_5Zn_2$)).

Element	Normalized concentration [atom%]	Normalized concentration [weight%]	Relative error [%]
Magnesium	56.8	39.8	2.4
Aluminium	19.1	14.8	0.8
Zinc	24.1	45.4	1.5

The qualitative X-ray phase analysis of the TRC microstructure specimens were performed on the expected inverse segregation and on the matrix. However, the local measuring range cannot be precisely adjusted with the X-ray phase analysis. A measurement in the absolute center of the inverse segregation is not guaranteed. On the described areas of the specimens, respectively a diffraction pattern was captured. Using the position of lines in the diffraction image, the individual reflections could assign specific distances between lattice planes and thus the associated phases. In the examined matrix, mainly interferences of the magnesium solid solution phase can be determined. In addition, reflections of the γ-phase ($Mg_{17}(Al, Zn)_{12}$) plus the intermetallic compound Al_8Mn_5 were found. In the area of inverse segregation, not only the mentioned phases could be detected, but also another phase, which equates to a high probability to a φ-phase ($Al_2Mg_5Zn_2$). By means of the measured intensity ratios, it is clearly shown, that the γ-phase (Mg_{17} (Al, Zn)$_{12}$) occurs more frequently in the affected area of the analyzed fault strip (Figure 4).

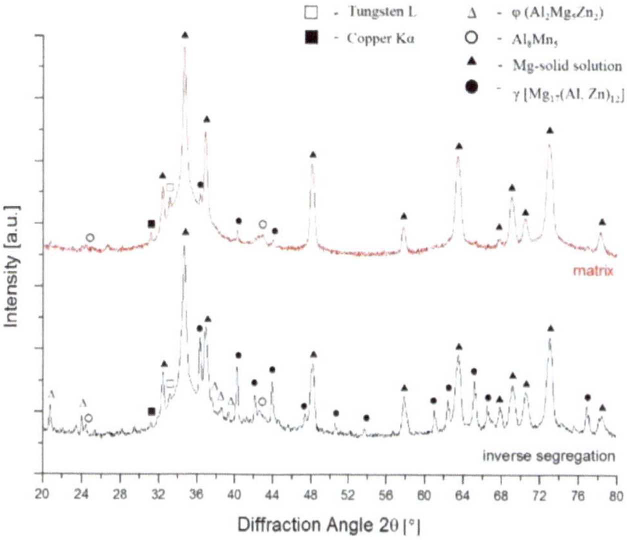

Figure 4. X-ray diffraction pattern of TRC AZ31 specimen, measured in the intact matrix area (red curve) and in the faulty area (black curve). Reflections of each identified phase are entered into individual diffraction pattern (see chart legend).

AZ31 strips in annealed conditions

Sometimes, during the sensitive TRC process, inverse segregations partially occur on the bottom side of the TRC strip. Thereby, remaining melt, enriched with alloying elements, is pressed into small cavities within the rapidly solidified material at the chill surface. In the current twin-roll cast process, statements about the presence and formation of inverse segregation can be taken only conditionally, since the surface of the TRC strip does not have visible impurities initially. The visible formation of material-damaging defect structures will only occur after annealing, when superficial surface defects and impurities appear with rectilinear alignment parallel to the TRC direction (Figure 2, Figure 5a). Often the formation of magnesium oxides is observed in these areas (Figure 2, Figure 5b). The defect structures show different characteristics and variations from skin-deep and hardly visible to deep (from a few µm to several mm) and wide.

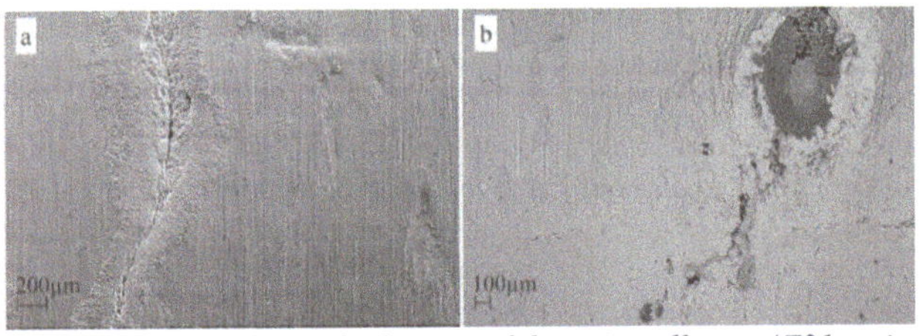

Figure 5. a) Cracks on the bottom of the twin-roll cast AZ31-strip, b) TRC AZ31 strip with MgO impurities/ deposits.

Figure 6 shows a scanning electron micrograph (BSD contrast) of the cross section through a faulted area in the TRC AZ31 specimen in annealed condition. The surface is rough, cavities under the surface are still present (a in Figure 6). It can be seen, there is a disturbed, inhomogeneous composition at the grain boundaries (for instance b in Figure 6). The grain boundaries are enriched with particles (Al_8Mn_5, for example point 1 in Figure 6). An increased concentration of aluminum and zinc can still be measured in the faulty area (for instance point 2 in Figure 6, Table 5). The flawless matrix of the strip has a chemical composition, which complies with the AZ31 alloy design. A chemical composition corresponding to the γ-phase ($Mg_{17}(Al, Zn)_{12}$) and the φ-phase ($Al_2Mg_5Zn_2$), can no longer be observed.

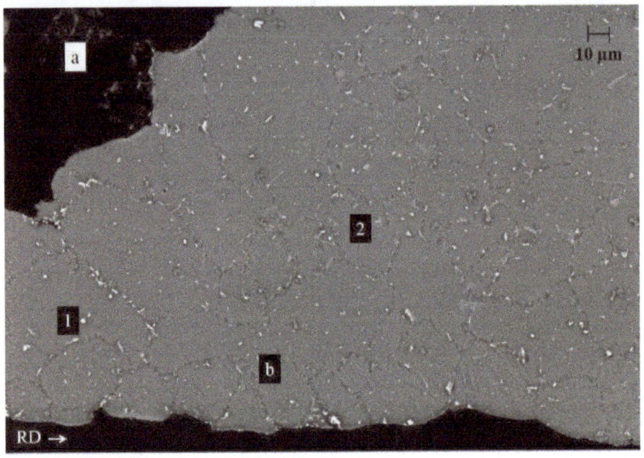

Figure 6. Scanning electron microscope image (BSD contrast) of the TRC AZ31 strip (bottom) in annealed condition. Cross-section towards the twin-roll cast direction. Detailed view of the inverse segregation in annealed condition.

Table 5. Determined chemical composition by EDX analysis in point 2 in Figure 6.

Element	Normalized concentration [atom%]	Normalized concentration [weight%]	Relative error [%]
Magnesium	93.1	90.3	4.9
Aluminium	5.4	5.8	0.3
Zinc	1.5	3.9	0.2

Under mechanical load, the defect structures in AZ31 strips in annealed conditions act as crack initiators and lead to material failure below the maximum deformation limits. In Table 6, the elongation to failure (A_{80}), the yield strength ($R_{p0.2}$) and the tensile strength (R_m) of the tested tensile specimen series are listed. In addition, it is noted in Table 6 if the respective specimen is broken at a defect structure (marking *, Table 6).

Specimen 1 and 2 were taken from areas with inverse segregations. Specimen 3 and 4 originate from faultless strip areas (Figure 2.). It is becoming clear that specimens with inverse segregations having a lower elongation to failure and a decreased tensile strength than test specimens taken from areas without inverse segregations (Figure 7, Table 6).

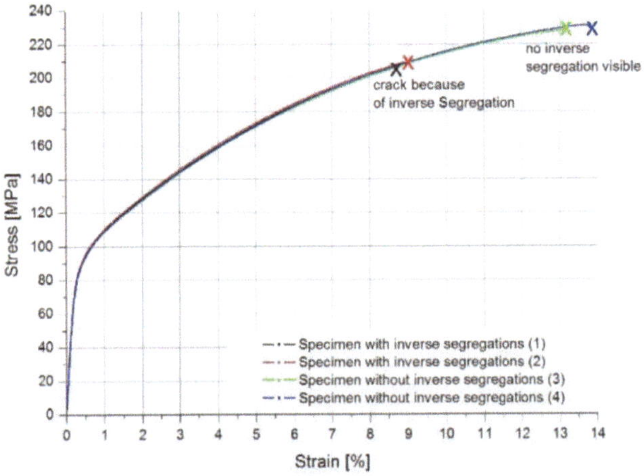

Figure 7. Stress-strain-curves of the heat-treated tensile-testing specimens. Specimen were taken perpendicular (90°) to the TRC direction

Table 6. Mechanical properties of the heat-treated tensile-testing specimens. Specimen were taken perpendicular (90°) to the TRC direction.

TRC strips in annealed condition, thickness 5,4 mm				
	Specimen 1	Specimen 2	Specimen 3	Specimen 4
$R_{p0.2}$	92	92	91	91
R_m	206	209	228	231
A_{80}	8.5*	9*	13	13.5

* specimen is broken directly to a visible defect

Discussion

To prevent the development of inverse segregations, possible formation mechanisms should be clarified. One reason for the generation of inverse segregations could be seen in local nucleation sites on the rolls. It is presumed, that there are tiny local deposits on the rolls, caused by surface brushing or by temperature-induced changes. Smallest amounts of material act as a nucleus, where the solidification in the TRC process begins. Local, a rapid solidification is favored, leading to a rupture of the peripheral shell of the TRC strip before entering the forming zone. The different characteristics of the inverse segregation with respect to the depth and expression support this thesis. A further influential factor on the formation of inverse segregations is the configured setback-value (Figure 1). Due the position of the nozzle in the roll gap, the shape of the forming meniscus is substantially influenced. In a symmetrical position of the nozzle in the roll gap, due to gravity, the contact length between the bottom roll and the melt is enlarged compared to the upper side (Figure 8). At nucleation sites on the bottom roll, rapid solidification of the melt is effected locally. The solidification is not even over the roll body. The shrinkage of the material during solidification leads to the forming of hot cracks at the crystallized peripheral shells. Form and characteristics of the observed inverse segregation, indicate a pressing of the remaining melt, enriched with alloying elements into these cracks within the solidified and shrunken peripheral shell regions of the strip. During progressive peripheral shell crystallization, shrinkage cracks may only partially be filled with remaining melt before entering the forming zone within the roll gap. In this manner, under the strip surface smallest cavities, filled with process gases, can form (c, Figure 3).

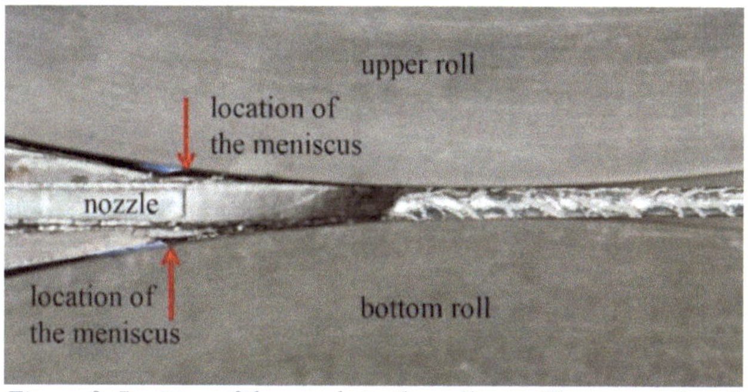

Figure 8. Position of the nozzle in the roll gap. The contact length between bottom roll and melt is enlarged.

The chemical composition and the crystallographic phases obtained by EDX and qualitative X-ray phase analysis demonstrate, that the inverse segregations observed in the TRC strips results from the remaining melt. The γ-phase ($Mg_{17}(Al, Zn)_{12}$) and the φ-phase ($Al_2Mg_5Zn_2$) form during solidification of AZ31 before entry into the roll gap. The development of the phases during the solidification process in TRC strips will be described subsequently. Figure 9 shows the liquidus surface projection of the ternary system Mg-Al-Zn. The chemical composition of the base alloy (AZ31, Mg-3Al-1Zn) is marked with a red dot. Red arrows mark changes in composition of the melt during solidification. When the melt cools down, first primary magnesium solid solution (Mg) crystallizes. The remaining melt is impoverished in magnesium and concentrated with aluminum and zinc until the remaining melt reaches a composition on the mono-variant line e7. In addition to magnesium solid solution (Mg) the γ-phase ($Mg_{17}(Al, Zn)_{12}$) crystallized. On further cooling, the remaining melt is depleting on both, magnesium and aluminum. Currently, the remaining melt is following upon further solidification of the mono-variant line to the transition reaction U2. In addition, during this reaction, the φ-phase ($Al_2Mg_5Zn_2$) is crystallized (Figure 9).

$$L \rightarrow L + (Mg) \rightarrow L + (Mg) + \gamma \rightarrow L + \gamma \leftrightarrow (Mg) + \varphi \qquad (1)$$

The γ-phase ($Mg_{17}(Al, Zn)_{12}$) may be generated over a temperature range of 436 °C until the temperature of the transition reaction. The temperature of the transition reaction U2 is specified in [16] with 368°C.

In strips without inverse segregations, solidification occurs more quickly than in strips, which offer inverse segregations. Consequently, the formation of the φ-phase ($Al_2Mg_5Zn_2$) is inhibited.

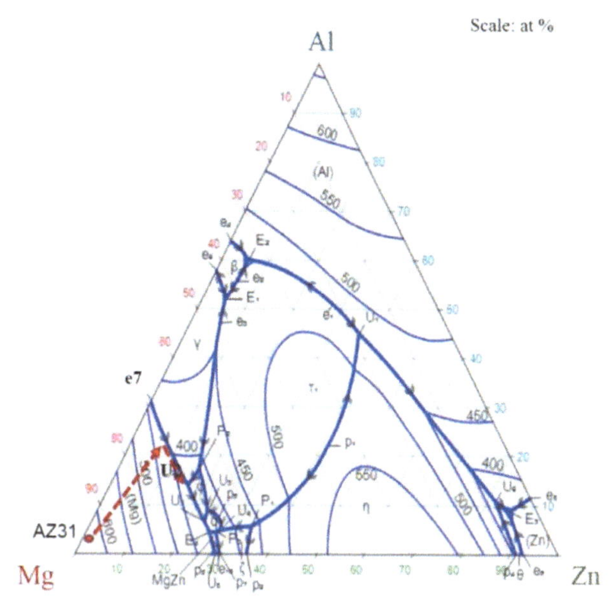

Figure 9. Liquidus surface, calculated using the data of [12] of a ternary chemical composition Mg-Al-Zn. Red circle indicates the starting chemical composition AZ31 (Mg-3Al-1Zn). The dashed arrows characterize the solidification path. [16]

The φ-phase ($Al_2Mg_5Zn_2$) is typically observed in higher zinc-containing magnesium alloys such as in [17] and in higher aluminum-containing alloys such as AZ91 (for example in [18], [19]). Under the test conditions described, the solidification terms during TRC, locally leads to the formation of the φ-phase ($Al_2Mg_5Zn_2$) even when AZ31 is used. If a heat treatment of the strip is implemented above the melting temperature of the φ-phase, the material is locally fused and reacts with the surrounding furnace atmosphere. In this case, the strips suffer irreversible damage. Crack initiation areas arise. In the furnace atmosphere, a percentage of the molten phase reacts to MgO.

The results of the tensile tests show that the accessible elongation to failure (A_{80}) and the tensile strength (R_m) of the heat-treated TRC strips highly depend on the presence and severity of inverse segregation. Specimens without formation of inverse segregation can achieve higher elongation to failure (A_{80}) and increased tensile strength (R_m) compared to specimens, which have serious defects by inverse segregations. Specimens without imperfections have the highest elongation at fracture (A_{80}). The mechanical properties A_{80} and R_m of the specimens without imperfections are improved by ≈ 40 % and 10 % respectively compared to the achievable specific values of the specimens with inverse segregations. This proves, inverse segregations in AZ31 strips in annealed conditions act as a crack initiator and lead to material failure below the maximum deformation limits.

Various strategies are used to prevent inverse segregation in AZ31 TRC strips and thereby in the heat-treated strips. With a more adapted nozzle geometry, decreased setback values can be realized. Decreasing the setback value may lead to a more symmetrical formation of the meniscus shape at the upper and bottom roll. By varying the TRC velocity and the level of melt in the casting launder, the formation of a symmetric meniscus shape can also be influenced decisively (Figure 10). By setting an appropriate TRC velocity, the solidification can be affected. Due to the interaction between the formation of the meniscus and a targeted solidification by adjusting the TRC velocity, inverse segregation could be prevented. Under ideal TRC conditions, strips with high mechanical properties can be produced. For instance, specimens (thickness = 5.3 mm) in annealed conditions, milled from the strip perpendicular (90°) to the direction of TRC offer good mechanical properties with $A_{80} = 17 \pm 1$, $R_{p0,2} = 94 \pm 1$ and $Rm = 239 \pm 0.9$.

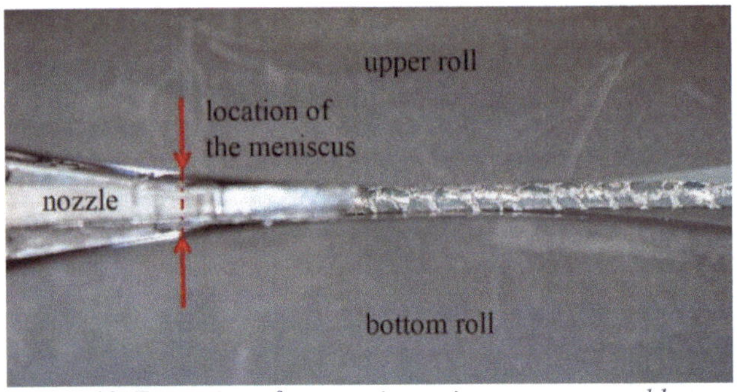

Figure 10. Formation of symmetric meniscus at upper and bottom roll, based on an adjusted position of the casting nozzle in the roll gap, a suitable TRC velocity and a corresponding level of the melt in the launder.

Conclusion

Depending on TRC conditions, AZ31 strips shows barely visible surface defects. When the strip is subjected to subsequent heat treatment, superficial surface defects and impurities appear with rectilinear alignment parallel to the TRC direction. The inverse segregations are principally formed on the bottom side of the TRC strip and remain clearly visible after several steps of strip-rolling and intermediate annealing. Based on tensile testing, it was proven, that the accessible elongation to failure (A_{80}) and the tensile strength (R_m) of the heat-treated TRC strips is highly depend on the presence and severity of inverse segregations. Specimens without inverse segregations exhibit the highest elongation at failure (A_{80}) and tensile strength (R_m). These values are improved by ≈ 40 % and 10 % respectively compared to the achievable specific values of the specimens with strong imperfections. This proves inverse segregations in annealed AZ31 strips act as crack initiator and lead to material failure below the maximum deformation limits.

The individual phases of the inverse segregations were identified. The phases of magnesium solid solution (Mg), the γ-phase ($Mg_{17}(Al, Zn)_{12}$), the intermetallic phase Al_8Mn_5 as well as the φ-phase ($Al_2Mg_5Zn_2$) exist. The γ-phase ($Mg_{17}(Al, Zn)_{12}$) and the φ-phase ($Al_2Mg_5Zn_2$) are formed during the solidification and can be titled as remaining melt. Heat treatment of the AZ31 strip above the melting temperature of the φ-phase, lead to local fusing and reactions with the surrounding furnace atmosphere. The strips suffer irreversible damage. In the furnace atmosphere, a percentage of the molten phase is burned to MgO. In TRC strips in annealed conditions, the γ-phases ($Mg_{17}(Al, Zn)_{12}$) and the φ-phase ($Al_2Mg_5Zn_2$), can no longer be observed.

The generation and formation of the inverse segregations at the bottom of the AZ31 strip indicates nucleation sites on the rolls. Localized, there are smallest amounts of material acting as a nucleus. At this nucleation sites, the solidification begins. At these points, a rapid solidification is supported, leading to a rupture of the peripheral shell of the TRC strip. In interdendritic channels, remaining melt is pressed into small cavities of rapidly solidified material. During progressive solidification of the peripheral shell, shrinkage cracks may be filled partially with remaining melt before the material enters the forming zone. In this manner, under the strip surface smallest cavities, filled with process gases, may occur.

The development of inverse segregations in TRC AZ31 strips can be prevented by improving the nozzle design and the position of the nozzle in the roll gap. A symmetrical formation of the meniscus shape at the upper and bottom roll is necessary. Optimizing the TRC velocity and the level of the melt in the launder help to improve the surface quality of the TRC strips.

Acknowledgment

The authors wish to acknowledge the financial support of the LeiKa project ("Leichtbau-Karosserien", promotional reference: 03WKCA01B) by the Federal Ministry of Education and Research. They also wish to thank Ms. I. Diegel, Mr. H. Winderlich, Mr. G. Schreiber, Mr. R. Kaminke and team for experimental support and Dr. J. Grigoleit and K. Neh for their contribution.

References

[1] K. Neh, R. Kawalla, Key Eng. Mater. 623-623 (2014) 575–580.

[2] B. Frischknecht, Aluminium 77 (2001) 746–751.

[3] R. Kawalla, M. Oswald, C. Schmidt, M. Ullmann, H.-P. Vogt, N. D. Cuong, Metalurgija 195–198.

[4] J.H. Bae, M.S. Shim, B.C. Suh, D.-W. Kim, S.H. Park, N.J. Kim, Materials Letters 132 (2014) 361–364.

[5] B. Forbord, B. Andersson, F. Ingvaldsen, O. Austevik, J.A. Horst, I. Skauvik, Mater. Sci. Eng. A 415 (2006) 12–20.

[6] M. Ullmann, F. Berge, K. Neh, R. Kawalla, Metalurgija 54 (2015) 711–714.

[7] Y. Nakaura, A. Watanabe, K. Ohori, MATERIALS TRANSACTIONS 47 1743–1749.

[8] G. Kurz, J. Bohlen, D. Letzig, K.U. Kainer, Mater. Sci. Forum 765 (2013) 205–209.

[9] I. Bayandorian, Y. Huang, Z. Fan, S. Pawar, X. Zhou, G.E. Thompson, Metall. Mat. Trans. A 43 (2012) 1035–1047.

[10] S.S. Park, W.-J. Park, C.H. Kim, B.S. You, N.J. Kim, JOM 61 (2009) 14–18.

[11] H. Liang, Chen, S.-L., Chang, Y. A., Metall. Mat. Trans. A 28 (1997) 1725–1734.

[12] P. Liang, T. Tarfa, J.A. Robinson, S. Wagner, P. Ochin, M.G. Harmelin, H.J. Seifert, H.L. Lukas, F. Aldinger, Thermochimica Acta 314 (1998) 87–110.

[13] E. Essadiqi, I. Jung, M. Wells, in: Advances in Wrought Magnesium Alloys Fundamentals of Processing, Properties and Applications, 2012, pp. 272–303.

[14] K. Neh, M. Ullmann, M. Oswald, F. Berge, R. Kawalla, Mater. Today Proc. 2 (2015) 45–52.

[15] F. Berge, L. Krüger, H. Ouaziz, C. Ullrich, Trans. Nonferr. Met. Soc. 25 (2015) 1–13.

[16] U. Lukas, in: G. Effenberg (Ed.), MSI, Materials Science International Services GmbH, Stuttgart, 2004.

[17] S.-k. Guan, C.-x. Zhang, L.-g. Wang, L.-h. Wu, P.-l. CHEN, Y.-l. Tang, Trans. Nonferr. Met. Soc. 18 (2008) 593–597.

[18] Ren, Qin, Pei, Li, Guo, Zhao, Transactions of Nonferrous Metals Society of China (English Edition) 241–245.

[19] S.M. Liang, Ma, Y. Q., Chen, R. S., E.H. Han, in: M. Pekguleryuz, N.R. Neelameggham, R.S. Beals (Eds.), Magnesium Technology 2008, 2008, pp. 331–336.

Magnesium Technology 2016

STRIP CASTING OF LIGHT METALS

Strip Casting: Properties

SUBSTITUTION OF RARE EARTH ELEMENTS IN MAGNESIUM ALLOYS FOR THE SHEET PRODUCTION VIA TWIN ROLL CASTING

Kurz, G.[1]; Petersen, T.[1]; Portugal Gonzales, I.[1]; Hoppe, R.[1]; J. Bohlen[1]; Letzig, D.[1]
[1]Magnesium Innovation Centre; Max-Planck-Straße 1, 21502 Geesthacht, GERMANY

Keywords: magnesium, aluminum free, twin roll casting, magnesium sheet, calcium containing alloys

Abstract

Sheet properties of magnesium like ductility, strength and corrosion resistance are improved by alloying with rare earth elements. However, the availability of these elements is limited. So, by using innovative production and processing methods like Twin Roll Casting, high-strength and ductile Mg sheet materials free of rare earth elements should be developed with competitive properties. Various approaches are being used, which include the complete process chain of the alloy development through production of semi-finished products and components to be integrated into the final product.

This paper reports on the substitution of rare earth elements in high-strength and ductile magnesium alloys. This is the aim of a government funded project SubSEEMag of Germany. The paper will also show first results of casting and rolling experiments on aluminum containing and aluminum-free magnesium alloys which contain calcium as a substitute of rare earth. The influence of different alloying elements on the microstructure of the sheets is presented and discussed with respect to arising texture.

Introduction

Presently, the automotive industry begins to implement magnesium sheets in series production. Figure 1 shows possible magnesium sheet applications [1].

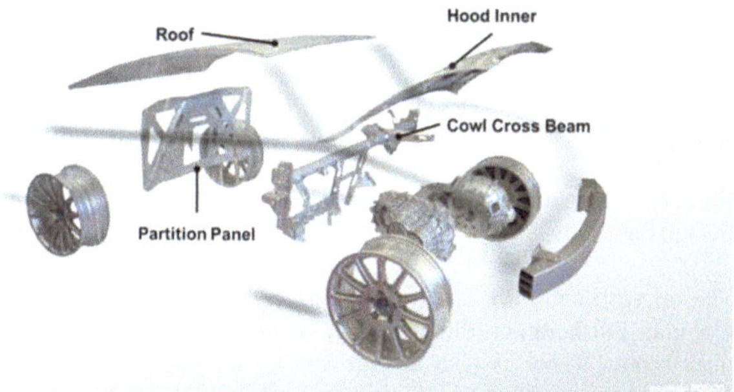

Figure 1: possible magnesium sheet applications in the car

One reason for the limitations in plastic deformation of magnesium alloys is the crystallization of magnesium with a hexagonal close packed lattice structure. It is principally accepted that the comparably low formability of sheets is based on the resulting limitation in the number of active deformation mechanisms, at least in comparison to cubic metals. Such low formability [2, 3] of magnesium alloys also has a significant impact on the process ability during massive deformation. Magnesium alloys containing rare earth elements could partially compensate these disadvantages such as poor formability and corrosion properties [4].

Magnesium sheet materials with the addition of rare earths show a randomized texture and fine-grained microstructure [5-7]. So the addition of rare earths as alloying elements improves the ductility of these alloys. On the other hand the addition of these elements is problematic. Because of their strategic importance for many industrial applications and the tight supply situation rare earth elements are classified as a critical resource group whose supply is associated with high financial costs and leads to economic dependence on imports of these elements. In addition, the production of these substances requires enormous effort and major environmental problems that are to be avoided in terms of sustainable material selection for industrial use. Alloying addition of Ca into Mg-Zn-based alloys shows a similar texture evolution and texture behavior than RE containing alloys [8,9].
This work will report on the effects of alloys with calcium as a substitution of rare earth investigated as part of a funded project by the German Federal Ministry of Education and Research entitled „Substitution of rare earth elements in high strength and ductile Magnesium sheet material - SubSEEMag". The global aim of this project is the development of calcium containing, rare earth element-free Mg sheet materials with advantageous properties, which are currently reachable only by the use of rare earth element-containing alloys by using innovative production and processing methods like twin roll casting. Various approaches are being used, which include the complete process chain of the alloy development through production of semi-finished products and components to be integrated into the final product. This paper presents the first results of this project.

Experimental Procedure

In the first step of the project 5 alloys were cast in billets (Ø 105 mm x 220 mm) by a modified gravity casting process. The cast alloys and their alloy composition are listed in table 1:

Table 1: cast alloys and their alloy composition

Alloy	Composition	Heat treatment
ZMX210	2 m% Zn, 1 m% Mn < 1 m% Ca	350 °C @ 16 h
MX20	2 m% Mn, < 1 m% Ca	450 °C @ 16h
MX21	2 m% Manganese, 1 m% Calcium	450 °C @ 16h
ZA21	2 m% Zn, 1 m% Al	400 °C @ 16h
ZAX210	2 m% Zn, 1 m% Al,< 1 m% Ca	450 °C @ 16h

After casting the billets were heat treated to homogenize the microstructure and to reduce the amount of precipitates. For the rolling trials specimens of the dimensions 100 mm x 150 mm x 20 mm were machined and rolled down on a small cold rolling mill, Metz M205, figure 2.

Figure 2: rolling mill at HZG

The different alloys were rolled in a temperature range between 300 °C and 450 °C and the rolling procedure consisted of 13 passes with different degrees of deformation between $\varphi = 0.1$ and $\varphi = 0.3$, leading to a final gauge of approximately 1.7 mm. φ is given as

$$\varphi = -\ln\left(h_{n+1}/h_n\right)$$

where n is the number of the pass and h_n is the sample thickness after pass n. The rolling schedule is listed in table 2.

Table 2: parameters of the rolling trials

Rolling pass	Deformation φ	Annealing time [min]
		30
1	0.1	15
2	0.1	15
3	0.1	15
4	0.1	15
5	0.2	15
6	0.2	15
7	0.2	15
8	0.2	10
9	0.2	10
10	0.2	10
11	0.2	10
12	0.3	10
13	0.3	10

In order to see how the different alloy compositions influence the mechanical properties of the sheets, tensile tests were performed according to ISO 6892-1. Because of the large amount of unrecrystallized grains in all sheets, a heat treatment of 1h at temperatures between 450 °C and 350 °C was carried out. All samples were prepared in the sheet rolling direction after heat treatment.

After all processing steps, the microstructures of the strips were analyzed using optical microscopy. Standard metallographic sample preparation techniques were employed and an etchant based on picric acid was used to reveal grains and grain boundaries [10]. Texture measurements were performed on the sheet mid-planes using a Panalytical X-ray diffractometer setup. The pole figures were measured up to a tilt angle of 70° which allowed recalculation of full pole figures based on an MTEX software routine [11]. The (1000) and (10-10) pole figures are used in this work to present the texture of the strips at midplane.

Casting Trials

Figures 3 to 7 display the microstructures of the cast billets in the as cast condition and after the heat treatment. Figure 3 shows the microstructures of the alloy ZMX210. The alloy ZMX210 exhibits a fine grained microstructure with a homogeneous grain size distribution. A high number of precipitates could also be observed located at the grain boundaries but also located in the grains. A heat treatment of 16 h at 350 °C reduces the number of precipitates at the grain boundaries, but does not increase the average grain size.

The microstructures of the alloys MX20 and MX21 are displayed in figure 4 and 5. In comparison to the MX20 the microstructure of the MX21 is much finer and more homogeneous because of the higher Ca amount. But the microstructure of the MX21 is dendritic. The precipitates in the MX20 are small and homogeneous distributed in the grains. The precipitates in the MX21 are more located at the grain boundaries. After the homogenization of 16 h at 450 °C the dendritic microstructure disappear und the precipitates at the grain boundaries are reduced. The precipitates in the MX20 are after the heat treatment coarser. The grain size in both alloys is not significantly influenced by the heat treatment.

Figure 3: microstructure of the alloy ZMX210 in as cast and heat treated condition

The microstructure of ZA21 (figure 6) in the as cast condition is globular. Furthermore, the grain size is very large, partially larger than 1 mm. There is also a high amount of precipitates in the microstructure with a homogeneous distribution in the grains. Looking at microstructure after the heat treatment at 400 °C for 16 h it could be seen that the particles were almost disappeared and grains were not significantly grown.

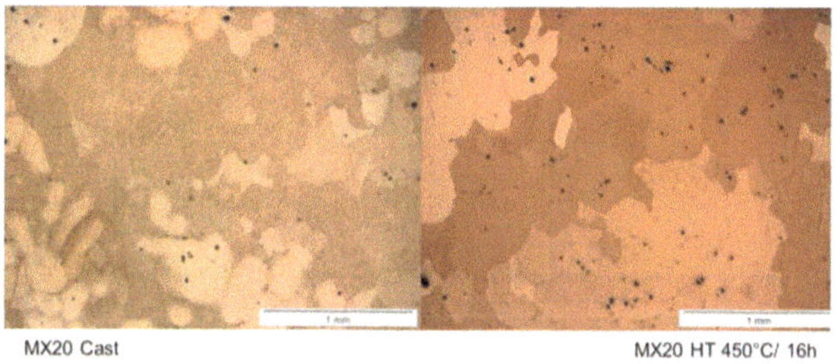

Figure 4: microstructure of the alloy MX20 in as cast and heat treated condition

Figure 5: microstructure of the alloy MX21 in as cast and heat treated condition

Figure 6: microstructure of the alloy ZA21 in as cast and heat treated condition

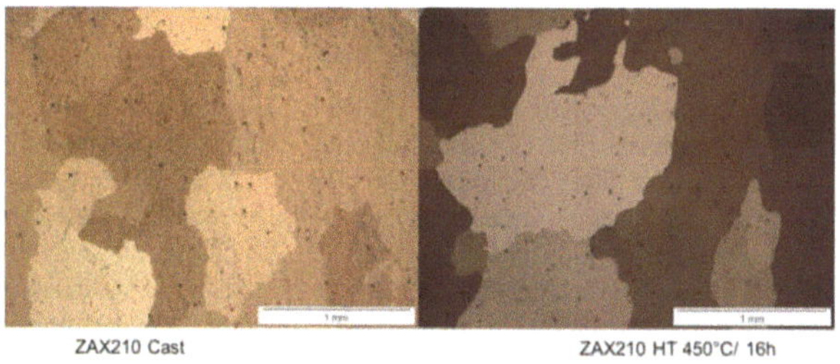

Figure 7: microstructure of the alloy ZAX210 in as cast and heat treated condition

ZAX210 differs from the ZA21 only by a small amount of additional Ca. The alloy ZAX210 has got in the as cast condition, like the alloy ZA21, a very coarse, dendritic microstructure (grain size larger than 1 mm) (figure 7). Also a high amount of precipitates can be observed located between the dendrite arms and the grain boundaries, but the precipitates are not so fine distributed like in the ZA21 alloy. After the heat treatment at 450 °C for 16 h the dendritic microstructure disappeared and the size of the precipitates increased.

Rolling trials

The heat treated slabs were rolled down to final gauge by the above mentioned rolling procedure. In general the first rolling temperature was the annealing temperature of each alloy. After the first test in same cases a review of the results let to an adjustment of the rolling temperature, see Table 3 where the results of the rolling trials are shown.

Table 3: results of the rolling trials

Alloy	450 °C	400 °C	350 °C	300°C
ZMX210		broken		broken
MX20	passed			
MX21	passed			
ZA21		passed	passed	
ZAX210	broken		passed	passed

During rolling the alloy ZMX210 at 400 °C after the first rolling pass hot cracks appeared, see figure 8. So, the rolling temperature was decreased from 400 °C to 300 °C with the same result. The reason for this behavior is the presence of the phase $Mg_6Ca_2Zn_3$ in the material. The melting point of this phase is around 400 °C, so that the additional heat coming from the forming operation leads to hot cracks in the slab.

Figure 8: hot cracks of alloy ZMX210 during rolling at 400 °C and 300 C

The two alloys MX20 and MX21 were rolled at 450°C. Both alloys could be rolled to sheets without any cracks. The microstructures and the (0001) and (10-10) textures of the alloys MX20 and MX21 are presented in figure 9 and 10. A very fine grained and homogeneous microstructure was archived in the alloys MX20 and MX21. But grain size in the MX21 is 26 µm and slightly higher than in the MX20 (15 µm). In both alloys not all grains are fully recrystallized and in the MX21 sheet there are more precipitates, aligned in small lines.

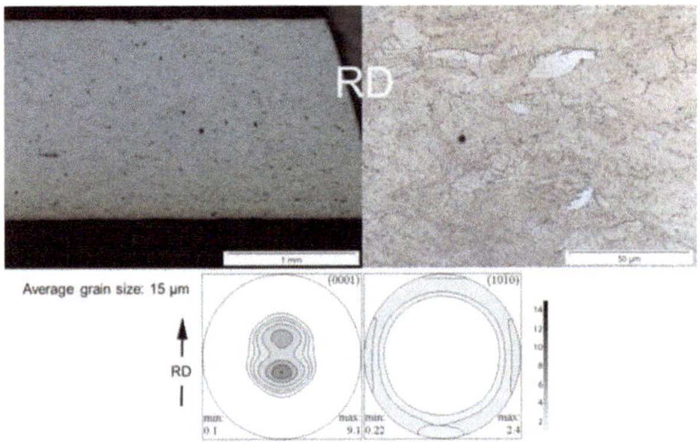

Figure 9: microstructure and texture of the MX20 sheets rolled at 450 C

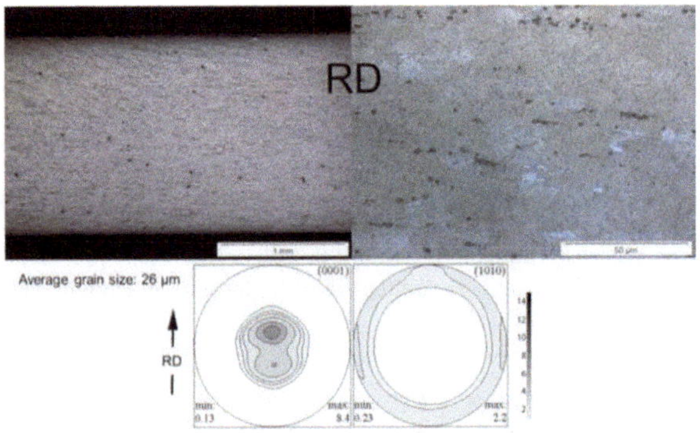

Figure 10: microstructure and texture of the MX21 sheets rolled at 450 °C

A basal texture with two peaks was observed in the (0001) pole figures of both alloys. The texture intensity in the MX20 with $I_{max} = 9.1$ is slightly higher than that of MX21 with $I_{max} = 8.4$ caused by the higher Ca contend in the MX21.

The rolling process of the alloy ZA21 started with a temperature of 350 °C. At this temperature some small cracks at the edges appeared. That is why the temperature was decreased to 400 °C. With these process parameters error-free sheets could be produced. Figure 11 shows the microstructure and texture of the ZA21 sheets rolled at 400 °C. It can clearly be seen that the microstructure is locally deformed, in these areas the grains are very fine. That means the deformation occured very local and in this areas dynamic recrystallization took place. The average grain size is 35 µm, the coarsest microstructure of all four investigated sheets. Figure 11 displays also the (0001) and (10-10) pole figures of the ZA21. The (0001) pole forms a basal texture with a double peak in rolling direction, similar to the texture of the MX20.

The last alloy investigated was ZAX210. The specimen in the first rolling trial at 450 °C was broken after the second rolling pass because of hot cracks. In order to avoid hot cracks the rolling temperature was decreased to 300 °C. At 300 °C the slab could be rolled to final gauge, but a strong tendency to form edge cracks was observed. So the rolling temperature was increased to 350 °C and slab of ZAX210 was rolled to final gauge. The number of edge cracks was reduced.

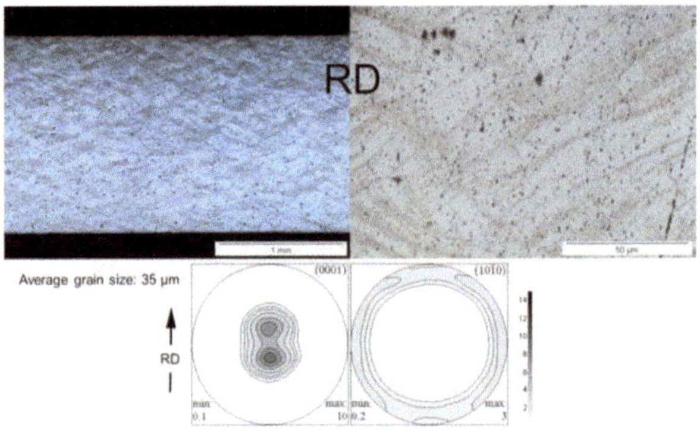

Figure 11: microstructure and texture of the ZA21 sheets rolled at 400 °C

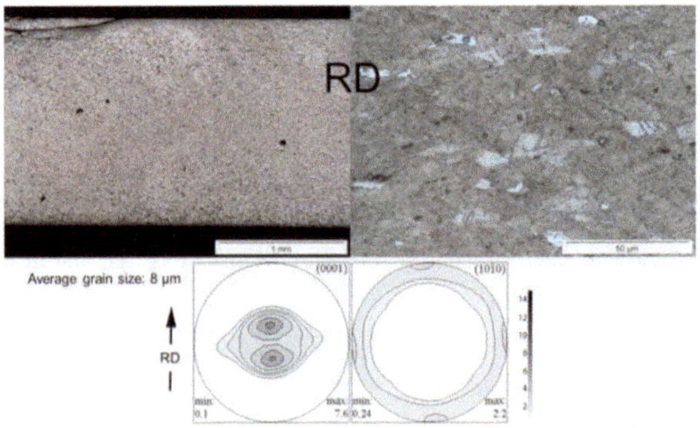

Figure 12: microstructure and texture of the ZAX210 sheets rolled at 350 C

Figure 12 shows the microstructure and texture of this alloy rolled at 350 °C. The microstructure is completely different in comparison to the ZA21, because of the small amount of Ca in the alloy. The alloy ZAX210 forms a very fine grained and homogeneous microstructure without local deformed areas after rolling. The measured grain size of 14 µm is together with the alloy MX20 (15 µm) the finest of all investigated alloys. However the microstructure of the ZAX210 is not fully recrystallized after the rolling procedure.

After the rolling process a weak texture is observed in the (10-10) planes. The basal (0001) plane, however, exhibits a weak texture with two peaks along the rolling direction (RD). The texture also features a characteristic enlargement in the transversal direction (TD). This texture differs a lot from the ZA21 texture with its strong basal texture. The small Ca content in the ZAX210 leads to a stronger distribution in transversal direction and a weaker intensity. This ZAX210 texture is similar to Zn-based RE containing alloys like ZE10, which show a higher formability than the AZ31 alloy.

Tensile Tests

In order to see how the different alloying elements influence the formability of the sheets, tensile tests were performed. Because of the large amount of unrecrystallized grains in the microstructure of all sheets, a heat treatment was carried out. Table 4 shows the parameters of the heat treatments.

Table 4: heat treatment after rolling

Alloy	Heat treatment
MX20	450 °C @ 1h
MX21	450 °C @ 1h
ZA21	400 °C @ 1h
ZAX210	350 °C @ 1h

Table 5 summarizes the results of these tests in stress-strain diagrams. The comparative stress-strain diagrams of all sheets are shown in Figure 13 to 16.

Table 5: results of the tensile tests

Alloy	TYS (MPa)	UTS (MPa)	Uniform elongation (%)	Elongation (%)
MX20	146	209	8.4	8.6
MX21	144	213	7.7	7.8
ZA21	98	214	18.0	19.0
ZAX210	154	245	18.5	27.1

The Zn-Al based alloys sheets exhibit significantly higher elongations compared to those of Mn-Ca based alloys. The differences in the mechanical properties of MX20 and MX21 are marginal, because the textures and the microstructures are quite similar. The difference seen in the as cast microstructure of the alloys MX20 and MX21 disappear after rolling. The finer microstructure of MX21 after casting caused by the higher Ca amount does not result in better mechanical properties after rolling.

The Zn-Al based sheets show a different result. Ca as additional alloying element in the ZAX210 effects a different texture evolution and therefore better mechanical properties. The elongation at fracture of ZA21 is 19 % and the elongation at fracture of ZAX210 is 27.1 %. It is remarkable that the better formability does not go along with lower strength. The UTS and TYS of the ZAX210 are much higher than those of ZA21

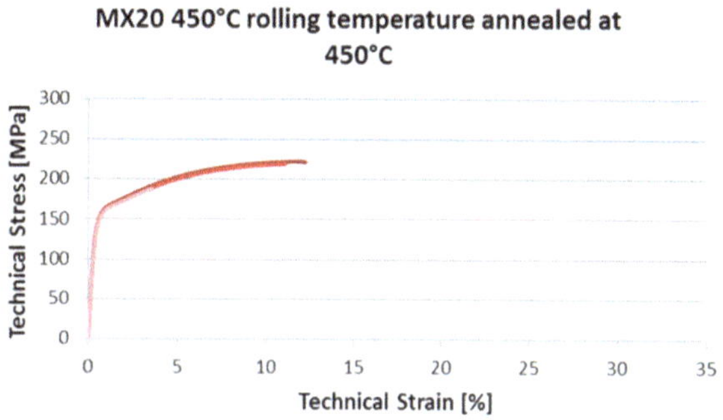

Figure 13: stress-strain curve of MX20 rolled at 450 °C and annealed at 450 °C

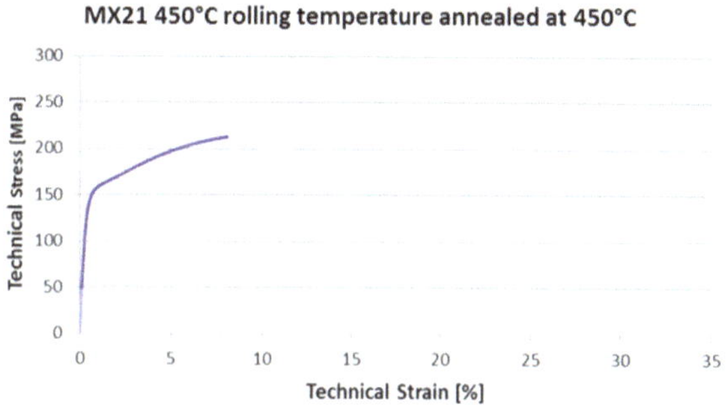

Figure 14: stress-strain curve of MX21 rolled at 450 °C and annealed at 450 °C

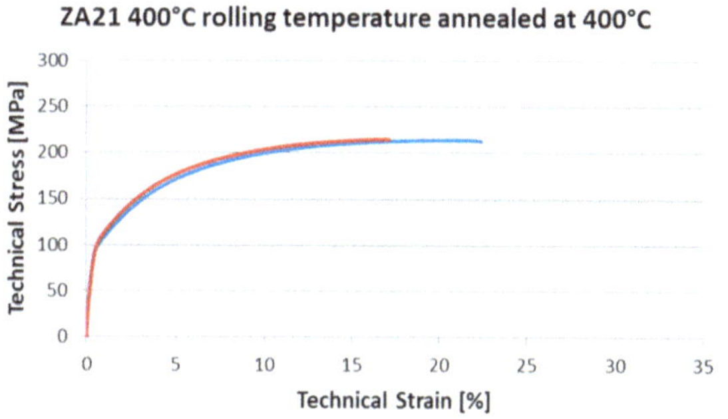

Figure 15: stress-strain curve of ZA21 rolled at 400 °C and annealed at 400 °C

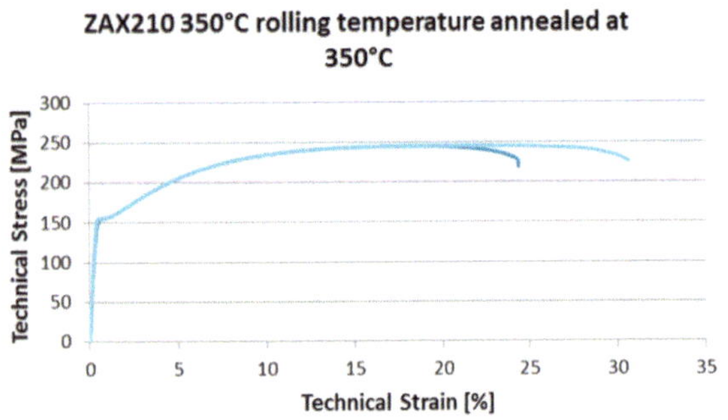

Figure 16: stress-strain curve of ZAX210 rolled at 350 °C and annealed at 350 °C

Conclusion

The results reveal that the alloy composition has got a significant influence on the resulting sheet properties. These results are a first impression of the resulting sheet properties of the investigated alloys. But the results show clearly that in Mn-based alloys a Ca content higher than 1 w% does not result in better mechanical properties in comparison to the MX20 with less than 1 w% Ca. The addition of Zn in this alloy family leads to the formation of a $Mg_6Ca_2Zn_3$ phase that has got a very low melting point. This low melting point makes deformation processes like rolling difficult.

In Al-Zn based alloys the addition of around 1 w% of Ca leads to significant advantages in formability and strength of the sheets. The texture of ZAX210 is similar to Zn-based RE containing alloys like ZE10 and shows also a high formability and a good strength. In the next step the tested alloys have to show how the production process via twin roll casting improves the sheet properties. Furthermore, the alloy composition and the complexity of the rolling process offer chances to tune the product by modifying the process parameters (e.g. temperature or degree of deformation).

Acknowledgements
This research was performed as a joint research project in the research program "r^3-Innovative Technologien für Ressourceneffizienz-Strategische Metalle und Mineralien" within the framework "Forschung für nachhaltige Entwicklung". The project „Substitution of rare earth elements in high strength and ductile Magnesium sheet material - SubSEEMag" was financially supported by the German Federal Ministry of Education and Research

References

[1] D. Klaumünzer, Application Potential of Magnesium within the Volkswagen Group, 2nd German-Korean Workshop, 26.- 30.10.2014, Incheon-Songdo, Korea

[2] L. Stutz, J. Bohlen, G. Kurz, D. Letzig, K.U. Kainer, Key Eng. Mater. 473 (2011), p. 335 – 342.

[3] C.E. Dreyer, W.V. Chiu, R.H. Wagoner, S.R. Agnew, J. Mat. Process. Techn. 210 (2010), p. 37-47.

[4] Stutz, L., Bohlen, J., Letzig, D., Kainer, K.U.: Formability of magnesium sheet ZE10 and AZ31 with respect to initial texture, In Sillekens, W., Agnew, S.R., Mathaudhu, S. N. Neelameggham, N.R., Magnesium Technology 2011, Warrendale, The Minerals, Metals & Materials Society TMS, 2011, 373 – 378

[5] K. Hantzsche, J. Wendt, K. U. Kainer, J. Bohlen und D. Letzig: Effect of process parameters and alloy composition on texture development and mechanical properties of magnesium sheets, JOM, 61, 2009, S. 43-47.

[6] J. Bohlen, M. Nürnberg, J. W. Senn, D. Letzig und S.R. Agnew: The texture and anisotropy of magnesium-zinc-rare earth alloy sheets, Acta Materialia, 55, 2007, P. 2101-2112.

[7] S.B. Yi, D. Letzig, R. Gonzalez-Martinez, K. Hantzsche, J. Bohlen, I. Schestakow und S. Zaefferer: Improvement of magnesium sheet formability by alloying addition of rare earth elements, Thermec 2009, Berlin (2009), P. 1506-1511.

[8] J. Bohlen, J. Wendt, M. Nienaber, K.U. Kainer, L. Stutz, D. Letzig: Calcium and zirconium as texture modifiers during rolling and annealing of magnesium-zinc alloys, Materials Characterization, Volume 101, April 2015, P. 144-152

[9] S. Yi, D. Letzig, K.U. Kainer, O.D. Kwon, J.H. Park and J.J. Kim, in The 10th international conference on Mg alloys and their applications, will be published

[10] V. Kree, J. Bohlen, D. Letzig, K.U. Kainer, "Practical Metallography 5", 2004, 233-246.

[11] F. Bachmann, R. Hielscher, H. Schaeben, "Texture Analysis with MTEX - Free and Open Source Software Toolbox", Solid State Phenomena (2010), 160, 63-68.

Magnesium Technology 2016
Edited by: Alok Singh, Kiran Solanki, Michele V. Manuel, and Neale R. Neelameggham
TMS (The Minerals, Metals & Materials Society), 2016

Microstructure and Mechanical Properties of Ca Containing AZX310 Alloy Sheets Produced via Twin Roll Casting Technology

Sangbong Yi[1], Jun Ho Park[2], Dietmar Letzig[1], Oh Duck Kwon[2], Karl Ulrich Kainer[1], Jae Joong Kim[2]
[1] Helmholtz-Zentrum Geesthacht, Magnesium Innovation Centre, Max-Planck-Str. 1, 21502 Geesthacht, Germany
[2] POSCO, 100 Hondu-ri, Haeryong-myon, Suncheon-si, Jeonnam-do, 540-856, Korea

Keywords: Magnesium sheet, Formability, Corrosion, Twin roll casting

Abstract

AZX310 alloy (Mg-3Al-1Zn-0.3Mn-0.5Ca in wt.%) sheets were produced via twin roll casting and subsequent warm rolling. The eutectic lamellae at the mid-thickness of the strip are changed during the rolling to fine spheroidal particles. The rolled sheets shows a large amount of fine Ca-rich particles which are homogeneously distributed through the whole sheet thickness. The AZX310 sheet shows excellent room temperature formability with the Erichsen index of 7.4, while the strength and ductility are maintained to the similar values of AZ31 sheets, i.e. tensile yield strength 189MPa, UTS 250 MPa and fracture strain 20%. The AZX310 sheet shows a relatively high corrosion resistance with the corrosion rate of 0.25mm / year, evaluated by neutral salt spray test at room temperature. The excellent mechanical performance of the AZX310 alloy sheet is related to the microstructure and texture. The high room temperature formability can be understood as a result of the texture randomization, while the relatively high strength and ductility of the AZX310 sheet is related to the existence of the fine Ca-rich particles.

Introduction

In the last decades magnesium (Mg) alloys have been widely investigated due to their positive characteristics for a lightweight structure, e.g. low density, high machinability and excellent damping capacity. The poor formability of Mg sheets, especially at room temperature, is one of the main reasons which retard industrial application of semi-finished products from this lightest structural metallic material. The traditional Mg-based wrought alloys, e.g. from the Mg-Al-Zn system such as AZ31, have a strong tendency of developing strong basal-type texture during the sheet rolling process. This basal-type texture, in which most grains have their c-axes in the sheet normal direction (ND), causes limited sheet formability because of restricted activities of <a> dislocations slip, especially under loading in ND. Though strain along the ND can be accommodated by pyramidal <c+a> slip with a Burgers vector of <11-23>, this deformation system can be activated at relatively high temperature based on the high critical resolved shear stress (CRSS) at low temperature. To improve the formability of the Mg sheets, it is essential to provide a way of weakening the texture and, more in general, alternating the dominant basal-type texture. In case of a weak basal-type texture, it is expected that <a>-dislocations with relatively low CRSS contribute more to accommodating the subjected deformation. It has been reported that texture weakening can be achieved by alloying Mg with yttrium (Y) and rare earth (RE) elements such as cerium (Ce) or neodymium (Nd) [1-3]. Indeed, an increased ductility and strength have been observed in the sheet having a weaker texture. Even though the alloying addition of RE elements is an effective way of texture weakening and improvement of sheet formability at low temperature, the Mg alloys sheets without RE elements are much beneficial in ecological and economic viewpoint.

Addition of Ca into Mg alloys is known as an effective way to improve the high temperature mechanical properties and the ignition-proof behaviour. Mg alloys containing Ca with their improved non-flammable characteristics have attracted much attention for research activities and industrial applications [4,5]. Recent studies showed that Ca addition leads to a significant weakening of the crystallographic texture in Mg alloys after thermomechanical treatments, such as extrusion [6] or rolling [7,8], and consequently the improvement of the room temperature formability. Moreover, Ca-containing alloys can be further strengthened by an optimised aging hardening scheme [9,10]. It is to be mentioned that the Ca-added alloys investigated for wrought processes handles mostly with Al-free Mg alloys, e.g. Mg-Zn-Ca and Mg-Mn-Ca systems.

Besides the research on the new alloys, the recent developments in processing technologies have opened up new opportunities for wide applications of Mg alloys. Twin roll casting (TRC), by which a thin strip is fabricated directly from molten metal in a single step, enables the production of Mg strips and sheets with improved properties, e.g. microstructure homogeneity and fine grain structure due to fast cooling rate, while it ensures also environmental and economic benefits [11]. Comparing to the commercial hot rolling process that requires a large number of rolling passes at relatively low pass reduction, TRC has been recognized as a sheet processing route with a high cost and time efficiency. Together with the alloy developments towards a highly formable alloy without addition of rare earth elements, this cost-effective sheet production method is promising to obtain the optimised sheet properties and to widen the commercial applications of Mg alloys as wrought products.

In the present study, the mechanical properties and formability of AZX310 sheet produced via TRC technology, which is a modified AZ31 alloy by Ca addition, are examined with giving a focus on analysing the microstructure and property relationship.

Experimental procedures

The TRC strip of the investigated AZX310 (Mg-3Al-1Zn-0.5Ca in wt. %) alloy was produced at POSCO, Korea, with 400 mm in width and 4.6 mm in thickness. The optical micrographs of the TRC strip are shown in Fig. 1. A relatively homogeneous microstructure with well-developed dendritic structure is observed in the TRC strip, without columnar grains along the thickness direction. The optical micrographs taken at the longitudinal plane, Fig. 1 (a), and cross-section, Fig. 1 (b), indicate the stringer-like centreline segregation zone, with some cm in length and 50 ~ 150 μm in thickness, at the mid thickness area of the strip. The centreline segregation zone is mainly composed of eutectic lamellae with matrix grains containing a high amount of alloying elements, i.e. Al (9.7 wt. %), Zn (4.9 wt. %) and Ca (5.9 wt. %) measured by SEM-EDX. The lamellae structure with its thickness

Fig. 1. Optical micrographs of the TRC strip: (a) longitudinal section, (b) cross section (c) magnified images at mid-thickness zone.

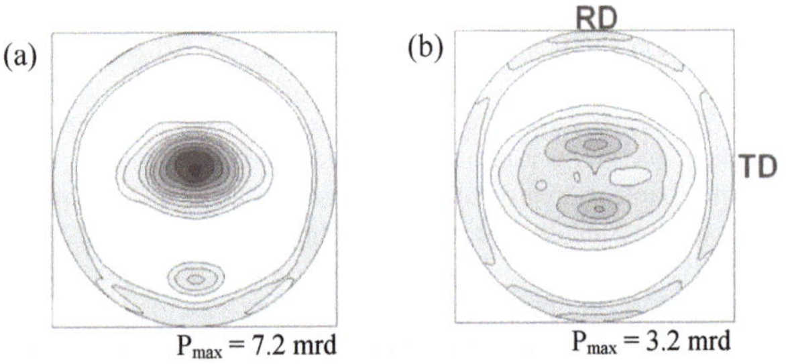

P_{max} = 7.2 mrd P_{max} = 3.2 mrd

Fig. 2. Recalculated (0002) pole figures of the TRC strip measured (a) the strip surface and (b) mid thickness zone (mrd = multiple random degree)

of ~ 80 µm has a high concentration of Al and Ca. Fig. 2 presents the (0002) pole figures recalculated from the XRD measurements at the strip surface and at the mid-thickness area.

A relatively strong basal-type texture is developed at the strip surface with max pole density of P_{max_0002} = 7.2, which indicates a strong rolling deformation during the TRC procedure. A symmetric split of the basal poles in the rolling direction is observed in the mid-thickness zone, P_{max_0002} = 3.2. Microstructural features observed in the TRC strip, e.g. twin bands and fine recrystallised grains along the twin bands, support also that the strips experienced relatively large deformation. It is understood from the different texture sharpness that the deformation degree at the mid-thickness is lower than the strip surface. Slight shift of the basal pole in the rolling direction observed at the strip surface indicates the shear deformation at the near-surface area. The TRC strip was rolled to the final gauge of 1 mm in thickness, and the recrystallization annealing of the rolled sheet was carried out at 400°C for 60 min.

Global texture information on the rolled sheets and after recrystallization annealing was obtained using X-ray diffraction (Panalytical, 40 kV and 40 mA). The microstructure and the texture of the rolled sheets were investigated using electron backscatter diffraction (EBSD) in a field emission gun scanning electron microscope (Zeiss, Ultra 55, working at 15 kV, equipped with an EDAX/TSL EBSD system with a Hikari detector). Tensile samples in the RD, at 45° and in the transverse direction (TD) with a gauge length of 50 mm and a width of 12.5 mm were prepared using discharge machining. Tensile tests were conducted using a universal testing machine, Zwick 050, at room temperature with an initial strain rate of $1x10^{-3}$ s^{-1}. The stretch formability of the sheets was examined by Erichsen tests of the as-rolled and annealed sheets. The tests were performed in accordance to the DIN EN ISO 20482 using the blank diameter of 100mm, punch diameter of 20mm at the punch speed of 5mm/min and blank hold force of 10 kN. The Erichsen index (IE) was determined by the punch stroke corresponding to the max load. The results from the tensile and Erichsen tests are given as the average values from 3 samples, at least, for each condition. The corrosion performance was evaluated by neutral salt spray test with 5 % NaCl solution according to EN ISO 9227. The corrosion rate was determined by weight loss after 48 hrs of immersion in the salt spray chamber.

Results and Discussion

The optical micrographs of the as-rolled and recrystallized sheets are demonstrated in Fig. 3. A strongly deformed microstructure with a large amount of twin bands is observed in the as-rolled sheet. The second phase particles are homogeneously distributed in the rolled and recrystallized sheets, except the mid-thickness area of the sheet where some highly concentrated particles are observed. The EDX analysis on the second phase particles reveal high concentration of Al and Ca, approximately 10 ~ 25 wt.% and 3 ~ 9 wt.% respectively. It should be mentioned that the EDX analysis contains also the information from the matrix volume,

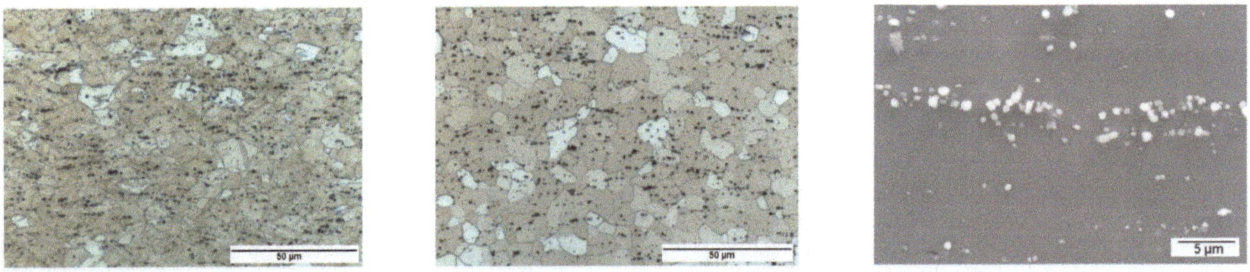

Fig. 3. Optical micrographs of (a) the as-rolled sheet and (b) recrystallized AZX310 sheet. (c) SEM image at the mid-thickness zone of the recrystallized sheet, at the former centreline segregation area in the TRC strip.

regarding the particle size and effective volume of the EDX measurement. From the EDX analysis and phase diagram of the corresponding alloy composition, most particles observed in the recrystallized sheet seem to be $(Mg,Al)_2Ca$ phase. Even though these band-shaped agglomerates of the particles are succeeded from the former centreline segregation of the TRC strips, their morphologies and microstructural features are significantly changed by the rolling and recrystallization annealing. The SEM micrographs of the particles observed at the former centreline segregation are shown in Fig. 3 (c). The eutectic lamellae observed in the TRC strip, Fig. 1 (c), become an agglomerate of fine spheroidal particles with 0.5 ~ 1 µm in diameters. While the eutectic lamellae in the TRC strip act as the crack initiation sites, which leads a premature failure during mechanical loading, the fine spheroid particles in the recrystallized sheet should have a strengthening effect.

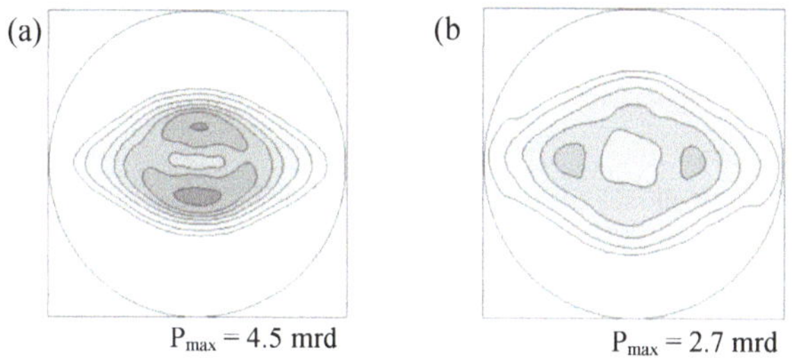

Fig. 4. Recalculated (0002) pole figures of (a) as-rolled sheet and (b) after recrystallization annealing at 400°C for 60min.

Fig. 4 shows the recalculated (0002) pole figures from the XRD measurements of the rolled and annealed AZX310 sheets. The rolled sheet has a relatively weak texture with the P_{max_0002} = 4.5, which is much weaker than those of the TRC strip. The texture of the rolled sheet is represented by the basal pole split in RD together with a wide spread in TD. The recrystallisation annealing results in a further texture weakening to P_{max_0002} = 2.7, and the formation of the basal pole split into TD. The development of recrystallization texture with the TD-split component has been reported in various RE or Ca containing Mg alloys sheets. It should be mentioned that the formation of such weak texture with TD-split texture component has been observed only in Al-free Mg alloys [1,7,8,12], to the authors knowledge. The Al addition into Mg-RE or Mg-Ca alloys causes the formation of coarse Al_XRE_Y or Al_XCa phases, such that the texture weakening effect related to the RE or Ca addition might be diminished in Mg-Al-RE or Mg-Al-Ca alloys. The present alloy produced via TRC should possess fine interdendritic phases and highly dissolved alloying elements in Mg matrix at the as-cast status, due to a rapid solidification and high cooling rate accompanying extended solid solubility of alloying elements [11,13]. The as-cast structure of TRC strip supports the formation of homogeneously distributed fine particles during the following thermomechanical treatments, assisting a formation of finer grain structure in the rolled sheet. Moreover, the texture weakening attributed to the restriction of the grain boundary motion becomes more probable in such TRC materials, as the fine dispersed particles act as a barrier against the boundary motion.

The EBSD inverse pole figure map, measured at the longitudinal sections of the annealed sheet is presented in Fig. 5 together with the distributions of the main alloying elements at the corresponding area. The EBSD measurement was conducted at the area of 300 x 600 μm^2 with 0.4 µm scanning step. Average grain size estimated from the EBSD measurement of the annealed sheet is 9.1 µm. The EBSD results confirm that the weak texture with the TD-split component is uniformly developed through the thickness in the recrystallized sheet. Moreover, the elements distribution maps show clearly that the fine particles with a high Al and Ca concentration are homogeneously distributed within the whole sheet thickness. The agglomerates of fine particles at the former centreline segregation, i.e. at the sheet mid-thickness area, are marked on the elements maps, Fig. 5 (b) and (c). The elements distribution maps indicate again that the observed particles are Al-Ca rich phases, most probably Al_2Ca or $(Mg, Al)_2Ca$ phase [14,15].

The mechanical properties and anisotropic behaviors of the annealed sheet were evaluated by the tensile tests at room temperature. The stress-strain curves measured from different sample directions are presented in Fig. 6. The yield strength in the sheet rolling direction (RD) is higher than the other sample directions, while the ultimate tensile strength values are comparable in all sample directions. The differences in the yield strength between the RD and the transverse direction (TD) amount to 33 MPa. The yield anisotropy can be understood from the texture of the annealed sheet. That is, the basal <a> slip is easier in TD than in RD, based on the tiled basal poles in TD that results in a higher Schmid factor. The high uniform elongation values achieved in all sheet directions, 15 ~ 19 %, indicate that the sheet has a high stretch formability. It should be mentioned that the mechanical properties of the AZX310 sheet, i.e. YS 189 MPa, UTS 250 MPa and fracture strain 20 % at room temperature, are similar to those of AZ31 sheets.

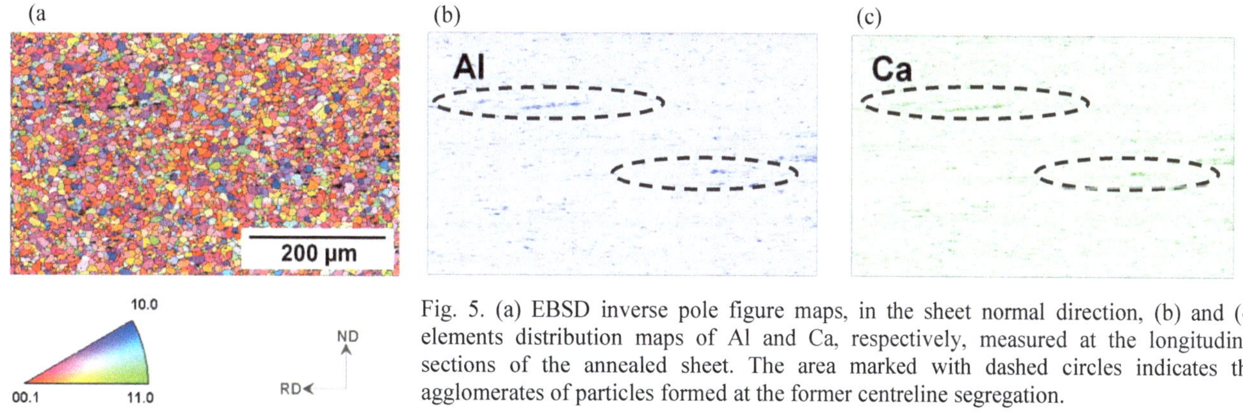

Fig. 5. (a) EBSD inverse pole figure maps, in the sheet normal direction, (b) and (c) elements distribution maps of Al and Ca, respectively, measured at the longitudinal sections of the annealed sheet. The area marked with dashed circles indicates the agglomerates of particles formed at the former centreline segregation.

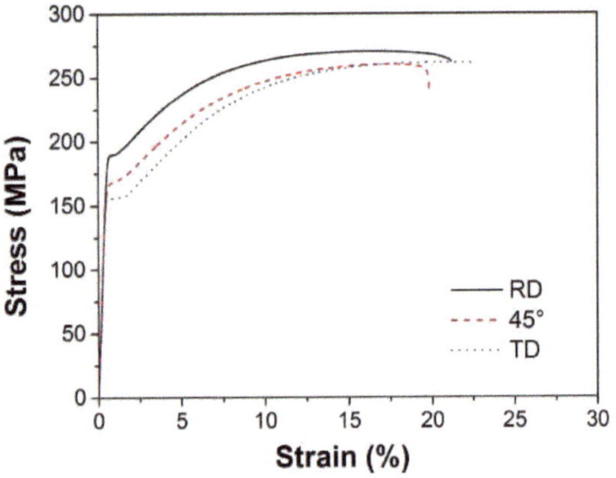

Tensile properties of annealed sheet			
	RD	**45°**	**TD**
YS (MPa)	189	165	156
UTS (MPa)	269	259	262
A_g (%)	15	17	19
E_f(%)	22	20	21
n	0.21	0.27	0.32

Fig. 6. Tensile stress-strain curves measured in three different directions, i.e. RD, 45° to the RD and TD, of the annealed sheet. Mechanical properties are summarized in the table (YS: yield strength, UTS: ultimate tensile strength, A_g: max uniform elongation, E_f: elongation to fracture, n: strain hardening exponent).

The annealed sheet shows a relatively high strain hardening exponent in sheet planar directions, n = 0.21 ~ 0.32. The strain hardening exponent and uniform elongation (A_g) have a proportional relationship, i.e. a large work-hardening exponent leads to a large uniform elongation. Based on the previous study [16] investigating the relation between tensile behavior and formability of Mg alloys sheet, it is expected that the AZX310 sheet has a high room temperature formability.

Fig. 7 presents the samples after Erichsen tests and the Erichsen Index (IE) of as-rolled and annealed sheets. The annealed sheet has the IE of 7.4. This excellent stretch formability of the recrystallized sheet is related again to the weak texture with the TD-split component, which is advantageous for the material flow in the sheet thickness direction due to the high activity of basal <a> slip, consequently, for a sheet formability. Furthermore, the fine grain structure with average grain size of 9.1 μm and the fine particles distributed homogeneously enhance the sheet formability.

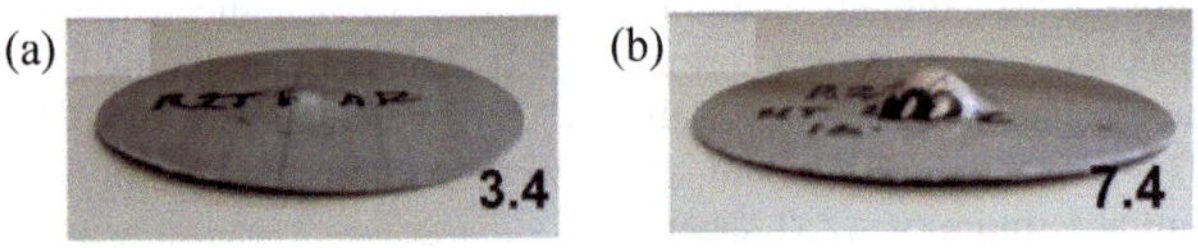

Fig. 7. Samples after Erichsen tests with Erichsen Index, (a) as rolled and (b) recrystallized sheet.

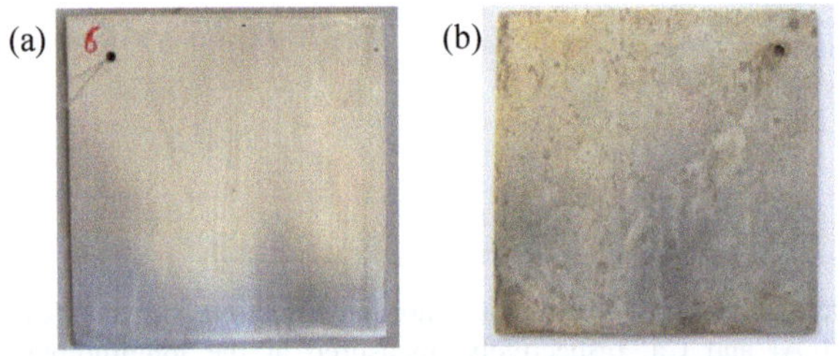

Fig. 8. Surface appearance of the annealed AZX310 sheet (a) before and (b) after the neutral salt spray test for 48 hrs.

The surface appearance of the corrosion sample is shown in Fig. 8. Before the corrosion testing the samples were mechanically cleaned by 1200 grit emery paper. The corrosion rate estimated from the salt spray test (SST) for 48 hrs is 0.25 mm/year. The sample after the SST indicates a mixture of general and mild filiform type of corrosion, while a localized pitting corrosion is almost prevented, Fig. 8 (b).

Summary

The AZX310 alloy (Mg-3Al-1Zn-0.5Ca in wt. %) sheets which were produced via TRC and subsequent warm rolling and annealing have been investigated. The centreline segregation of the TRC strip was changed to fine spheroidal particles with the size of 0.5 ~1 μm by the rolling. The excellent room temperature formability with the IE of 7.4 of the AZX310 alloy sheet is to be understood as a result of a texture randomization. The high strength comparable to that of AZ31 sheet are related to the existence of the fine Al-Ca rich particles, such as Al_2Ca and $(Al, Mg)_2Ca$ phases, where their homogenous distribution is assisted by the TRC process.

It is obvious that the AZX310 sheet produced via TRC technology shows an excellent sheet performance, in terms of formability and corrosion resistance. To identify the fundamental mechanisms leading such high performance, further studies are necessary on the correlations between the microstructures developed by thermomechanical treatments and the resulting properties.

Acknowledgements

The authors appreciate Dr. G. Kurz, Mr. S. Koch and Ms. Y.K. Shin at HZG for technical supports during the warm rolling and microstructure analysis. This work has been financially supported by POSCO under the contract number HZG-Agreement 2012/79.

References

[1] Bohlen J, Nürnberg MR, Senn JW, Letzig D, Agnew SR. Acta Mater., Vol. 55, 2007, 2101.

[2] Stanford N, Barnett M. Scripta Mater., Vol. 58, 2008, 179.

[3] Chino Y, Kado M, Mabuchi M. Mater. Sci. Eng., Vol. A494, 2008, 343.
[4] You B-S, Park W-W, Chung IS, Scripta Mater., Vol. 42, 2000, 1089.
[5] Lee J-K, Kim SK, Tans. Nonferrous Met. Soc. China, Vol. 21, 2011, 23.
[6] Stanford N, Mater. Sci. Eng. Vol. A528, 2010, 314.
[7] Kim D-W, Suh B-C, Shim M-S, Bae JH, Kim DH, Kim NackJ, Metall. Mater. Trans., Vol. A44, 2013, 2950.
[8] Chino Y, Huang X, Suzuki K, Mabuchi M, Mater. Trans., Vol. 51, 2010, 818.
[9] Bhattacharjee T, Suh B-C, Sasaki TT, Ohkubo T, Kim NackJ, Hono K, Mater. Sci. Eng., Vol. A609, 2014, 154.
[10] Hofstetter J, Becker M, Martinelli E, Weinberg AM, Mingler B, Kilian H, Pogatscher S, Uggowitzer PJ, Löffler JF, JOM, Vol. 66, 2014, 566.
[11] Bae GT, Bae JH, Kang DH, Lee H, Kim NackJ, Met. Mater. Int., Vol. 15, 2009, 1.
[12] Yuasa M, Hayashi H, Mabuchi M, Chino Y, Acta Mater., Vol. 65, 2014, 207.
[13] Wang Y, Kang SB, Cho J, J. Alloys and Compounds, Vol. 509, 2011, 704.
[14] Jiang B, Liu W, Qiu D, Zhang M-X, Pan F, Materials Chemistry and Physics, Vol. 133, 2012, 611.
[15] Kondori B, Mahmudi R, Mater. Sci. Eng., Vol. A527, 2010, 2014.
[16] Kang DH, Kim D-W, Kim S, Bae GT, Kim KH, Kim NackJ, Scripta Mater., Vol. 61, 2009, 768.

Magnesium Technology 2016

STRIP CASTING OF LIGHT METALS

Poster Session

Magnesium Technology 2016
Edited by: Alok Singh, Kiran Solanki, Michele V. Manuel, and Neale R. Neelameggham
TMS (The Minerals, Metals & Materials Society), 2016

MICROSTRUCTURE AND PROPERTIES OF SiC_p/Al MATRIX COMPOSITE STRIP FABRICATING BY TWIN-ROLL CASTING PROCESS

Huagui Huang[1], Zengwei Lv[1], Shengpeng Song[1], Fengshan Du[1]
[1] National Engineering Research Center for Equipment and Technology of Cold Strip Rolling, Yanshan University; 438 Hebei Avenue; Qinhuangdao, Hebei 066004, China

Keywords: SiC_p/Al matrix composite strip, twin-roll casting, mechanical stirring, uniformity of particle distribution

Abstract

In the present study, a short flow process based on twin-roll casting (TRC) was presented to fabricate the SiC_p/Al matrix composite strip, and a pilot experiment was carried out to explore the feasibility and potential benefits of this new technique. Firstly, the composite slurries were prepared by using mechanical stirring to incorporate the SiC reinforcing particles within the molten 6061 aluminum matrix, and then, the slurries were fed through a ceramic nozzle into the gap between two counter rotating water-cooled rolls of a TRC mill. The composite strip was hence fabricated with rapid solidification and hot rolling. Microscopic investigations on the samples of different position of the cast rolling zone were carried out to check the particle distribution and the porosity of matrix. Particles cluster and void defects were found at metal pool, but decrease along the rolling direction because of the severe hot deformation. The experimental results show that the SiC_p/Al matrix composite strip with enhanced uniformity of particle distribution, compactness and mechanical properties can be achieved by TRC.

1. Introduction

The SiC_p/Al matrix composite material has many advantages such as the outstanding mechanical, physical, and high wear resistance properties, so that it has been widely used as cheaper and lighter engineering components in aerospace, automotive, high-speed trains, electronic packaging, and other industry fields [1, 2]. In the past years, many fabrication methods have been presented to overcome the weak wettability between SiC particles and aluminum matrix, such as stir casting [3], squeeze casting [4], powder metallurgy [5, 6], pressure infiltration [7], spray-forming [8], and so on. Followed by the subsequent deformation such as extrusion, forging and rolling, etc, the engineering components were produced.

As a high value added engineering construction material, the composite strip is currently drawing interests, efforts have been focused on the uniformity of particles distribution, workability of the as-cast composite and mechanical property of the rolling strip. For example, R. Mitra et al. [9] carried out the study on the rolling workability of as-cast SiC-aluminum metal of commercial purity composite materials prepared by stir-casting, the particles size was from 39 μm to 157 μm, and the volume fraction was 10%. It was reported that the brittle reaction products Al_4C_3 at the Al-SiC interfaces is the main cause of interfacial crack appeared in the subsequent rolling process. Although oxidation of SiC and alloying of Al matrix with Mg was adopted to restrict the formation of the Al_4C_3 at the interfaces in stir casting, particle-cracking has been observed widely in the samples of as-cast subjected to either cold or hot rolling [9]. It has been found that composite containing the small SiC particles exhibited a superior resistance to fatigue crack growth as compared to the composite with larger SiC particles [10]. So, aiming at stir-cast Al 6061 and Al 6082 alloys-SiC fine particulates reinforced composites with SiC particles size of 8 and 15μm and volume fraction of 0-20%, hot rolling combined with intermediate heat treatment was successfully applied in sheet fabrication by A. El-Sabbagh et al. [11, 12]. The experimental results indicated that the hot rolling contributed to the decomposition of SiCp agglomeration and the redistribution of SiCp by aligning in the rolling direction, decreasing porosity, improved particle/matrix interface and matrix straining [12]. In general, the workability and microstructure of as-cast composite in the subsequent forming process mainly depends on the particles size and volume fraction [13-15].

Twin-roll casting is a short flow process which applies direct conversion of liquid metal into strip with the integration the casting and hot rolling process. It has been widely applied in fabrication of aluminum alloy and magnesium alloy strip [16, 17]. If we combine the stir casting with TRC process, it will bring a new opportunity to fabricate the SiCp/Al matrix composite strip in a low cost and high efficiency way. However, there are several factors that need considerable attention, including distribution uniformity of the SiC reinforcing particles, porosity in the as-rolled metal matrix, chemical reactions and possible cracking between the SiC particle and the matrix aluminum alloy.

The aim of this work is to explore this new technique. A pilot experiment was hence carried out on a modified twin-roll caster. And the microscopic investigation was conducted to check the particles distribution, agglomeration and porosity in the product through SEM (Scanning Electron Microscope) observation.

2. Experiment procedures

2.1 Experiment design

As shown in Figure 1, the whole experimental equipment consists of two components from top to bottom, they are SiC particles reinforcement composite slurries preparation unit and twin-roll caster respectively. For the composite slurries preparation unit, the aluminum was put into the graphite crucible, and then it was heat up to melt by a cylindrical resistance-heated furnace. It's well known that it is very difficult to incorporate the SiC reinforcing particles within the molten aluminum matrix due to their weak wetting properties, so the mechanical stirring have to be used, and the stirrer rod can move up and down along its' axis so that the whole molten aluminum in the graphite crucible can be stirred in the process. Thermocouple was setup in the melt to monitor and control the temperature, which ensure the temperature of the molten aluminum in a reasonable range. To avoid oxidization caused by the inhaled air during the stirring process, the whole unit was designed as a sealed space protected by argon shielding.

According to stir-casting process, the molten aluminum was stirred until a vortex is formed. SiC particles were added into the formed vortex through a special tunnel while continuing stirring to ensure the complete insertion of particles.

There was a hole at the bottom of the graphite crucible, which was blocked with a stopper made of refractory material in the stirring process. While the composite slurries were ready, the stopper would be moved away, and the slurries were fed through a ceramic nozzle into the gap between two counter rotating water-cooled rolls of the TRC mill. The SiC_p/Al composite strip was finally fabricated by combination of rapid solidification with hot rolling.

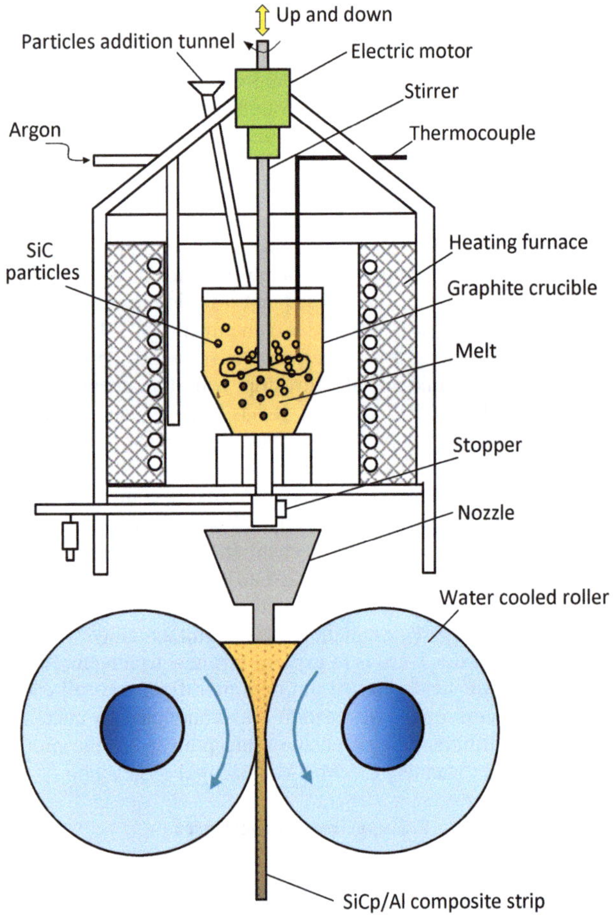

Figure 1. Sketch of the TRC process for SiCp/Al strip fabricating

The experiments were carried out on the vertical Ø160×150 mm twin-roll caster at National Engineering Research Center for Equipment and Technology of Cold Strip Rolling (NCSR), Yanshan University, which is shown in Figure 2. The maximum rolling force and torque of this caster were 50 kN and 1.2 kN·m respectively, and the motor power was 3 kW.

Considering the small volume of the graphite crucible and mechanical stirring, a small amount of aluminum was used in this experiment to prevent the melt splashed out of the crucible. Accordingly, a ceramic nozzle with width of 50 mm but not 150 mm was used. An additional heat tape was wrapped around the intermediate channel to compensate the heat loss, so as to avoid the melt solidified in the nozzle. Since the change of metallostatic head in the outlet of crucible would lead to fluctuation of the liquid level of cast-rolling area. In order to keep the liquid level stable in process, manual control was applied on the outlet flow by adjusting the stopper opening.

Figure 2. The experimental equipment

2.2 Materials

SiC particles with average size of 100 μm and purity greater than 95% were used as reinforcement elements. The 6061 aluminum alloy was used as metal matrix, which is a precipitation hardening aluminum alloy and commonly used for construction of aircraft structures, automotive parts, etc. As shown in Table I, it contains magnesium and silicon as its major alloying elements.

Table I. The chemical composition of 6061 Al alloy (wt%)

Cu	Mn	Mg	Zn	Cr
0.15~0.4	0.15	0.8~1.2	0.25	0.04~0.35
Ti	Si	Fe	Al	/
0.15	0.4~0.8	0.7	Bal.	/

2.3 Experiment procedure

Oxidization pre-treatment was conducted on SiC particles at 900 ℃ to generate a layer of SiO_2 on the particles surface in resistance furnace for 6 hours. Experimental results shown that the surficial SiO_2 layer and addition of Mg are both helpful to improve the wettability between particles and molten aluminum [11]. The raw material of 6061 aluminum alloy was melt at 700 ℃ in the crucible with argon shielding, and then, 1% Mg was added into the molten aluminum. Before adding the SiC particles, the molten aluminum was stirred at 200 rpm for up to 1 min, and then the particles were added into the formed vertex. Continuous stirring for 3 minutes was needed to ensure the particles disperse in the molten alloy uniformly. For adding more SiC-particles, the melt was keep 700 ℃ by the temperature control system from the thermocouple, to prevent the increasing viscosity due to temperature loss in mechanical stirring.

The composite slurry with volume fraction of about 8% was finally prepared and cast into the twin-roll caster with roll speed of 100 rpm. As shown in Figure.3, a composite strip with width of 50mm and thickness of 2 mm was fabricated.

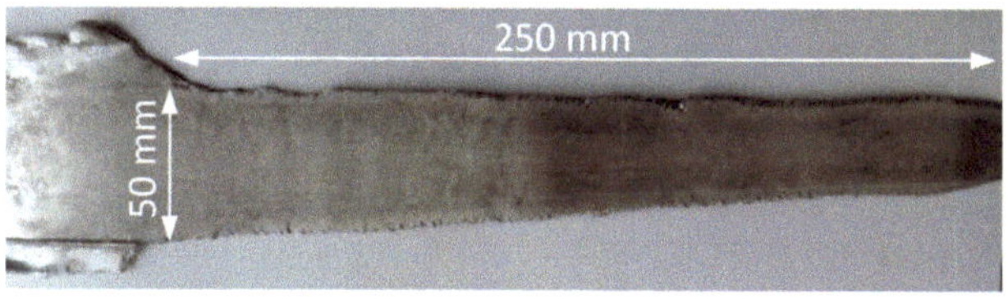

Figure 3. SiCp/Al matrix composite strip fabricated by TRC

3. Results and discussion

3.1 Microstructure evolution in the cast-rolling zone

In order to investigate the microstructure in different position of the cast-rolling zone, the caster was stopped during casting to acquire the samples contain both as-cast and as-rolled composite in the roll bite (see Figure 4(a)). In this experiment, the volume fraction of SiC particles was 8%. The microstructure investigation was subsequently performed with SEM observation. As shown in Figure 4(b), it's well known that the whole cast-rolling zone can be divided into three areas according to the different metal states. They were liquid area, mushy area and solid area respectively.

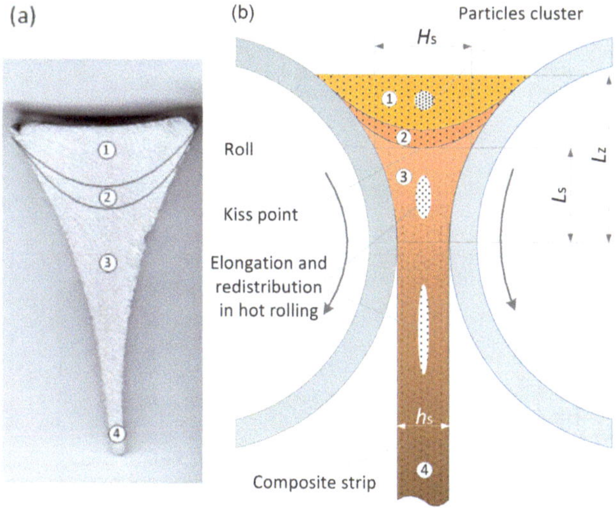

Figure 4. Acquired sample (a) and schematic diagram (b) of cast-rolling zone, including liquid area ①, mushy area ②, solid area ③ and as-rolled area ④

In order to demonstrate the microstructure evolution in the cast-rolling zone clearly, SEM pictures of typical position in liquid area and as-rolled area were presented as Figure 5 and Figure 6. Since no deformation was imposed on the liquid and mushy area, microstructure of these areas has two typical characteristics of as-cast composite, particles aggregation (see Figure 5(a)) and matrix voids (see Figure 5(b)). Both the improper preparation process of composite slurries and undesirable flow field of the melt pool in cast-rolling process would lead to formation of particles aggregation. Many studies on the uniformity of the particles distribution in stir-casting has been carried out in the pass years [3, 11, 18]. However, the effect of turbulent flow and temperature in the melt pool on uniformity of particles distribution was still unknown. The coupled simulation of flow and thermal based on the two-phase fluid would help to understand the particles redistribution in the melt pool with different nozzle structure and process parameters. And, this has been regarded as one of the major tasks in the future.

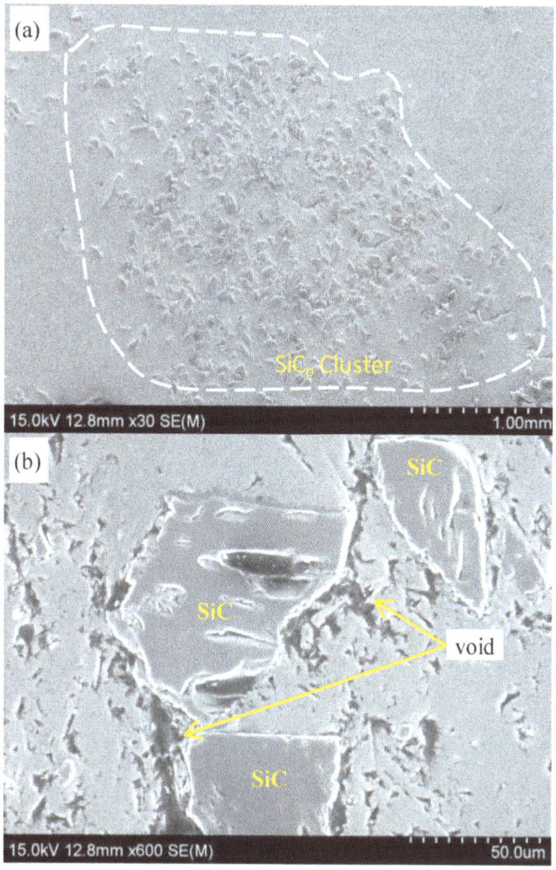

Figure 5. SEM photograph of liquid area for (a) general view and (b) partial view of SiC particle morphology

Besides, the high porosity was harmful to the mechanical properties of composite, so further deformation was needed to increase the density. Fortunately, below the kiss point, the severe deformation was applied on the composite through hot rolling. As shown in Figure 6, the microstructure of as-rolled samples shown that the particles aggregation and matrix porosity were both decreased in comparison of that in as-cast sample. Hereinto, the cross section of the SiC particles cluster in liquid area was near round shape (see Figure 5(a)), and it was elongated along the rolling direction and finally crushed to be elliptical (see Figure 6(a)) due to the hot rolling. It was evidence that the hot rolling deformation leads to decomposition and redistribution of the particles cluster by aligning in the rolling direction. According to rolling principle, the elongation of SiCp cluster was proportional to the rolling reduction. In twin-roll casting process, the deformation stage appeared below the kiss point. As shown in Figure 4(b), the initial inlet thickness was denoted as H_s, the exit thickness was h_s, so the rolling elongation can be formulated as following.

$$\eta = H_s/h_s \tag{1}$$

The rolling elongation depended on many structure parameters such as the roller diameter and roll gap, and process parameters such as melt temperature, heat transfer between composite and water cooled roller, the liquid level of the melt pool denoted as L_z, etc.

On the other hand, the particle cracking has been observed in the as-rolled sample (see Figure 6(b)), which was accordance with that reported in literature [11, 12], but the crack mechanism was still unknown similarly. Due to the temperature drop in strip edge, coupled with the low workability of the composite, edge cracking was found (see Figure 3).

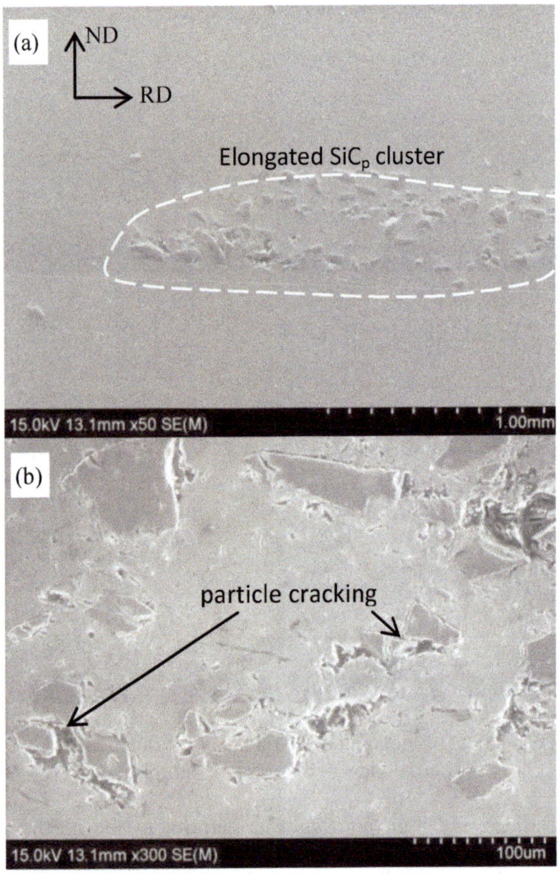

Figure 6. SEM photograph of as-rolled area for (a) general view and (b) partial view of SiC particle morphology

SEM-EDS (Energy Dispersive Spectrometer) study revealed that Al, Si, C, O, Mg were the elemental compositions in the particle/matrix interface. The enrich oxygen and magnesium in the interface was able to create a diffusion barrier to prevent the interfacial reactions and formation of Al_3C_4.

3.2 Fracture morphology analysis for tensile test

In order to prepare the tensile specimen free of defects, the strip edge with cracking was trimmed. The composite strip was held on as-rolled state without temper or any other heat treatments. The tensile test was performed on a material testing machine, four specimens with volume fraction of SiC particles 0%, 4%, 8% and 8% were tested (see Figure 7(a)), and the tensile stress-strain curves were presented in Figure 7(b). It was found that the fracture did not all appear at middle of specimen, and with the increase of SiCp volume fraction, the breaking elongation rate decreased, however, the yield stress had no definite trend which was not in accord with that of AA6061–SiC composites prepared through spark plasma sintering and hot rolling [19]. Actually, the yield stress increased with the increasing of the volume fraction of SiC particles in literature [19]. It can be preliminarily ascertained from the experimental results that the no definite trend may be caused by the inhomogeneity of SiCp distribution in the strip. On the other hand, the addition of Mg element and the mechanical stirring may also lower the cleanliness of matrix. So, works have to be carried out next to improve the uniformity of SiCp distribution and the matrix purity in this technology.

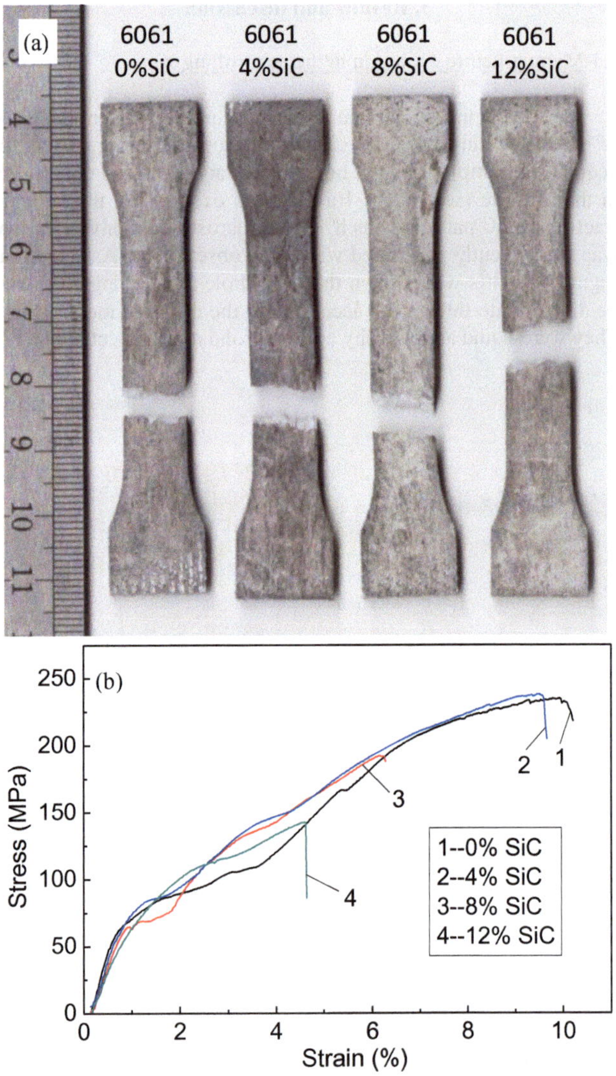

Figure 7. Tensile test with (a) specimens and (b) tensile curve

The SEM picture about the fracture morphology for specimen with 8% SiC particles was presented in Figure 8. The tensile fracture of test composite has mixed-rupture characteristics of quasi-cleavage and dimples. Hereinto, the quasi-cleavage was most appeared around the SiC particle (see Figure 8(b)), and the dimples were mainly appeared in aluminum matrix, which indicated that it was ductile fracture at the matrix and brittle fracture at the particle/matrix interface.

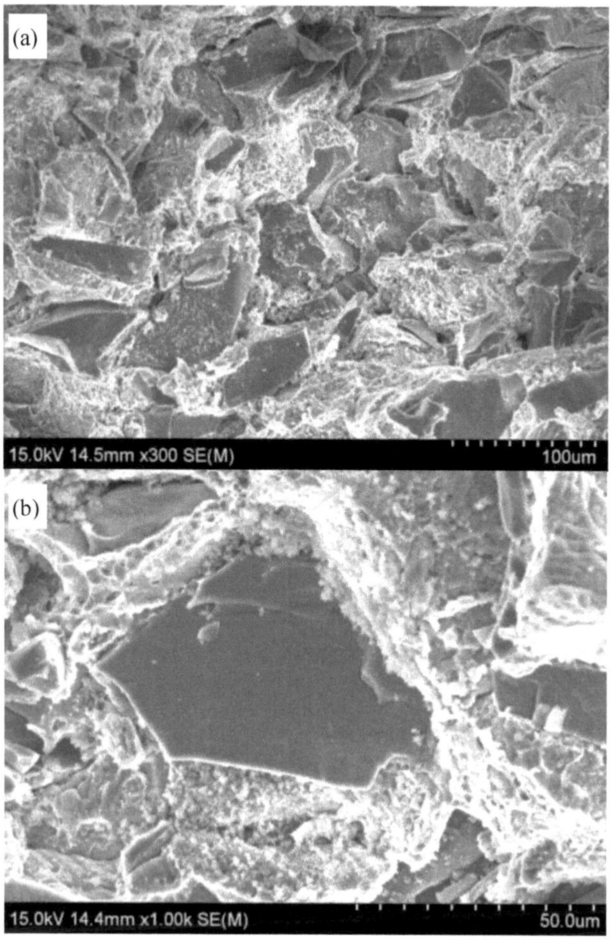

Figure 8. Fracture morphology for (a) full view and (b) partial view around one SiC particle in sample with volume fraction of 8%

4. Conclusion

(1) Combining the preparation of SiCp/Al matrix composite slurries by mechanical stirring with twin-roll casting process, the SiC particulate-reinforced 6061 aluminum matrix composite strip with thickness of 2 mm and width of 50mm was fabricated. The volume fraction of SiC particles was 8% and the average size was 100 μm.

(2) As common defects in as-cast composite, SiC particles cluster and matrix porosity were found in melt pool of the cast-rolling zone, but they can be eliminated partly or entirely by severe deformation in cast-rolling process, which makes it possible to produce strip with satisfactory quality.

(3) As a new attempt in this area, there are many problems need to be resolved, such as the formation mechanism of particles cluster in the liquid area of cast-rolling zone, optimization of structure and process parameters, the uniformity of SiCp distribution and the matrix purity control, and so on, which will become the key points in the future study.

Acknowledgements

The authors gratefully acknowledge the support from the National Natural Science Foundation of China (No.51459255, NO. 51374184) and the Natural Science Foundation of Hebei Province (No. E2013203377, QN2015214).

References

1. D. B. Miracle, "Metal matrix composites – From science to technological significance", *Composites Science and Technology*, 65(2005), 2526-2540.
2. I. A. IBRAHIM, F. A. MOHAMED and E. J. LAVERNIA, "Particulate reinforced metal matrix composites - a review", *Journal of materials science*, 26(1991), 1137-1156.
3. J. Hashim, L. Looney, M. S. J. Hashmi, "Metal matrix composites production by the stir casting method", *Journal of Materials Processing Technology*, 92-93(1999), 1-7.
4. S. M. Seyed Reihani, "Processing of squeeze cast Al6061–30vol% SiC composites and their characterization", *Materials and Design*, 27(2006), 216-222.
5. H. Wang, R. Zhang, X. Hu, C. A. Wang and Y. Huang, "Characterization of a powder metallurgy SiC/Cu-Al composite", *Journal of Materials Processing Technology*, 197(2008):43-48.
6. S. Bathula, R. C. Anandani, A. Dhar, A. K. Srivastava, "Microstructural features and mechanical properties of Al 5083/SiCp metal matrix nanocomposites produced by high energy ball milling and spark plasma sintering", *Materials Science and Engineering: A*, 545(2012),97-102.
7. Y. N. Cui, L. F. Wang, J. Y. Ren, "Multi-functional SiC/Al Composites for Aerospace Applications", *Chinese Journal of Aeronautics*, 21(2008) 578-584.
8. M. M. Sharma, C. W. Ziemian, T. J. Eden, "Fatigue behavior of SiC particulate reinforced spray-formed 7XXX series Al-alloys", *Materials & Design*, 2011,32(8-9), 4304-4309.
9. R. Mitra, V.S. Chalapathi Rao, R. Maiti, M. Chakraborty, "Stability and response to rolling of the interfaces in cast Al-SiC_p and Al-Mg alloy-SiC_p composites", *Materials Science and Engineering A*, 379 (2004), 391-400.
10. W. Li, H. Liang, J. Chen, S. Q. Zhu, Y. L. Chen, "Effect of SiC particles on fatigue crack growth behavior of SiC particulate-reinforced Al-Si alloy composites produced by spray forming", *Procedia Materials Science*, 3(2014), 1694-1699
11. A. El-Sabbagh, M. Soliman, M. A. Taha, H. Palkowski, "Hot rolling behaviour of stir-cast Al 6061 and Al 6082 alloys – SiC fine particulates reinforced composites", *Journal of Materials Processing Technology*, 212(2012),497-508.
12. A. El-Sabbagh, M. Soliman, M. A. Taha, H. Palkowski, "Effect of rolling and heat treatment on tensile behaviour of wrought Al-SiCp composites prepared by stir-casting", *Journal of Materials Processing Technology*, 213 (2013), 1669-1681.
13. B. C. Ko, G. S. Park, Y. C. Yoo, "The effects of SiC particle volume fraction on the microstructure and hot workability of SiCp/AA 2024 composites", *Journal of Materials Processing Technology*, 95(1999), 210-215
14. L. J. Zhang, F. Q, J. G. Wang, Q. C. Jiang, "High strength and good ductility at elevated temperature of nano-SiCp Al2014 composites fabricated by semi-solid stir casting combined with hot extrusion", *Materials Science & Engineering A*, 626(2015), 338-341
15. A. Ahmed, A. J. Neely, K. Shankar, P. Nolan, S. Moricca, T. Eddowes, "Synthesis, Tensile Testing, and Microstructural Characterization of Nanometric SiC Particulate-Reinforced Al

7075 Matrix Composites", *Metallurgical and Materials Transactions A*, 41(2010), 1582-91.
16. T. Haga, K. Tkahashi, M. Ikawaand, H. Watari, "Twin roll casting of aluminum alloy strips", *Journal of Materials Processing Technology*, 153-154(2004), 42-47
17. A. Hadadzadeh, M. A. Wells, I. Jung, "Scale-up modeling of the twin roll casting process for AZ31 magnesium alloy", *Journal of Manufacturing Processes*, 2014,16(4), 468-478
18. S. Balasivanandha Prabu, L. Karunamoorthy, S. Kathiresan, and B. Mohan, "Influence of stirring speed and stirring time on distribution of particles in cast metal matrix composite", *Journal of Materials Processing Technology*, 171(2), 2006, 268-273
19. X. P. Li, C. Y. Liu, M. Z. Ma and R. P. Liu, "Microstructures and mechanical properties of AA6061–SiC composites prepared through spark plasma sintering and hot rolling", *Materials Science and Engineering: A*, In Press, 2015 (doi:10.1016/j.msea.2015.10.015)

AUTHOR INDEX
Magnesium Technology 2016

A

B

C

D

E

F

G

H

I

J

K

W

X

Y

Z

SUBJECT INDEX
Magnesium Technology 2016

N

O

P

R

S

T

U

V

W

X

Z

MIX
Papier aus verantwortungsvollen Quellen
Paper from responsible sources
FSC® C105338

If you have any concerns about our products,
you can contact us on
ProductSafety@springernature.com

In case Publisher is established outside the EU,
the EU authorized representative is:
Springer Nature Customer Service Center GmbH
Europaplatz 3, 69115 Heidelberg, Germany

Printed by Libri Plureos GmbH
in Hamburg, Germany